AF539641

A Collection of Problems in Mathematical Physics

A Collection of Problems in Mathematical Physics

B.M. BUDAK, A.A. SAMARSKII,
and A.N. TIKHONOV

Translated by
A.R.M. Robson

Translation Edited by
D.M. BRINK
Clarendon Laboratory, Oxford

Dover Publications, Inc.
New York

Published in Canada by General Publishing Company, Ltd., 30 Lesmill Road, Don Mills, Toronto, Ontario.

This Dover edition, first published in 1988, is an unabridged republication of the work first published by the Pergamon Press Ltd., Oxford, England, in 1964 under the title *A Collection of Problems on Mathematical Physics* as volume 52 in the *International Series of Monographs in Pure and Applied Mathematics.* It is a translation of the original Russian *Sbornik zadach po matematicheskoi fizike* published by Gostekhizdat, Moscow, in 1956. It is reprinted by special arrangement with Pergamon Books Ltd., Headington Hill Hall, Oxford OX3 OBW, England.

Manufactured in the United States of America
Dover Publications, Inc., 31 East 2nd Street, Mineola, N.Y. 11501

Library of Congress Cataloging-in-Publication Data

Budak, B. M. (Boris Mikhailovich)
A collection of problems in mathematical physics.

Translation of: Sbornik zadach po matematicheskoi fizike.
Reprint. Originally published: A collection of problems on mathematical physics. Oxford, England : Pergamon Press, 1964. Originally published in series: International series of monographs in pure and applied mathematics ; v. 52.
Includes index.
1. Mathematical physics—Problems, exercises, etc. I. Samarskii, A. A. (Aleksandr Andreevich) II. Tikhonov, A. N. (Andrei Nikolaevich), 1906- . III. Title.

QC20.82.B813 1988 530.1'5'076 88-25686
ISBN 0-486-65806-6 (pbk.)

CONTENTS

† The number of the pages referring to the answers and solutions is given in italics

TRANSLATION EDITOR'S NOTE

A NUMBER of the more uninteresting problems which involve the method of images and the use of special functions have been removed from the English translation. The collection is still very large and a student should attempt only a few problems from each section for himself but will have the solutions of the remaining problems for reference.

D. M. BRINK

PREFACE

THE PRESENT book is based on the practical work with equations of mathematical physics done in the Physics Faculty and the external section of Moscow State University. The problems set forth were used in the course "Equations of Mathematical Physics" by A. N. Tikhonov and A. A. Samarskii, and in "A Collection of Problems on Mathematical Physics" by B. M. Budak. However, in compiling the present work the range of problems examined has been considerably enlarged and the number of problems several times increased. Much attention has been given to problems on the derivation of equations and boundary conditions. A considerable number of problems are given with detailed instructions and solutions. Other problems of similar character are given only with the answers. The chapters are divided into paragraphs according to the method of solution. This has been done in order to give students the opportunity, by means of independent work, of gaining elementary technical skill in solving problems in the principal classes of the equations of mathematical physics.

Therefore this book of problems does not claim to include all methods used in mathematical physics. For example, the operational method, variational and differential methods and the application of integral equations are not considered.

It is hoped, however, that this book will be useful not only to students but also to engineers and workers in research institutions.

For convenience a set of references is given at the end of the book. The book "Equations of Mathematical Physics" by A. N. Tikhonov and A. A. Samarskii is most often referred to, as the terminology used, and the order in which the material is set out in this book, most closely corresponds with our own.

In conclusion the authors consider it necessary to point out that although B. M. Budak and A. N. Tikhonov worked on one group of chapters and A. A. Samarskii and A. N. Tikhonov on the other group, the joint working out of the general structure of the book and the joint discussion of the chapters written make each author responsible in equal measure for its contents.

B. M. Budak, A. A. Samarskii, A. N. Tikhonov

A Collection of Problems In Mathematical Physics

CHAPTER I

CLASSIFICATION AND REDUCTION TO CANONICAL FORM OF SECOND ORDER PARTIAL DIFFERENTIAL EQUATIONS

IN THIS chapter problems are set on the determination of the type and on the reduction to canonical form of equations in two and more independent variables.

In the case of two independent variables equations with constant and variable coefficients are considered. In the case of three or more independent variables only equations with constant coefficients are considered, since for three or more independent variables the equation with variable coefficients cannot, generally speaking, be reduced to canonical form by the same transformation, in the entire region, in which the equation belongs to a given type. In § 1 problems are given for an equation in two independent variables, and in § 2 for three or more independent variables.

§ 1. The Equation for a Function of Two Independent Variables

$$a_{11}u_{xx}+2a_{12}u_{xy}+a_{22}u_{yy}+b_1u_x+b_2u_y+cu=f(x, y)$$

1. The Equation with Variable Coefficients

1. Find the regions where the equation

$$(l+x)u_{xx}+2xyu_{xy}-y^2u_{yy}=0$$

is hyperbolic, elliptic and parabolic and investigate their dependence on l, where l is a numerical parameter.

In problems Nos. 2–20 reduce the equation to canonical form in each of the regions.

2. $u_{xx}+xu_{yy}=0.$

3. $u_{xx}+yu_{yy}=0.$

4. $u_{xx}+yu_{yy}+\frac{1}{2}u_y=0.$

5. $yu_{xx}+xu_{yy}=0.$

6. $xu_{xx}+yu_{yy}=0.$

7. $u_{xx}+xyu_{yy}=0.$

8. $u_{xx}\operatorname{sign} y+2u_{xy}+u_{yy}=0.$

9. $u_{xx}+2u_{xy}+(1-\operatorname{sign} y)u_{yy}=0.$

10. $u_{xx}\operatorname{sign} y+2u_{xy}+u_{yy}\operatorname{sign} x=0.$

11. $y^2u_{xx}-x^2u_{yy}=0.$

12. $x^2u_{xx}-y^2u_{yy}=0.$

13. $x^2u_{xx}+y^2u_{yy}=0.$

14. $y^2u_{xx}+x^2u_{yy}=0.$

15. $y^2u_{xx}+2xyu_{xy}+x^2u_{yy}=0.$

16. $x^2u_{xx}+2xyu_{xy}+y^2u_{yy}=0.$

17. $4y^2u_{xx}-e^{2x}u_{yy}-4y^2u_x=0.$

18. $x^2u_{xx}+2xyu_{xy}-3y^2u_{yy}-2xu_x+4yu_y+16x^4u=0.$

19. $(1+x^2)u_{xx}+(1+y^2)u_{yy}+xu_x+yu_y=0.$

20. $u_{xx}\sin^2 x-2yu_{xy}\sin x+y^2u_{yy}=0.$

2. The Equation with Constant Coefficients

By means of a substitution $u(x, y) = e^{\alpha x+\beta y}v(x, y)$ and reduction to canonical form simplify the following equations with constant coefficients.

21. $au_{xx}+4au_{xy}+au_{yy}+bu_x+cu_y+u=0.$

22. $2au_{xx}+2au_{xy}+au_{yy}+2bu_x+2cu_y+u=0.$

23. $au_{xx}+2au_{xy}+au_{yy}+bu_x+cu_y+u=0.$

§ 2. The Equation with Constant Coefficients for a Function of n Independent Variables

$$\sum_{i,k=1}^{n} a_{ik}u_{x_i x_k}+\sum_{i=1}^{n} b_i u_{x_i}+cu=f(x_1,\ldots,x_n)$$

Reduce to canonical form equations 24–28.

24. $u_{xx}+2u_{xy}+2u_{yy}+4u_{yz}+5u_{zz}+u_x+2u_y=0.$

25. $u_{xx}-4u_{xy}+2u_{xz}+4u_{yy}+u_{zz}=0.$

26. $u_{xx}+u_{tt}+u_{yy}+u_{zz}-2u_{tx}+u_{xz}+u_{ty}-2u_{yz}=0.$

27. $u_{xy}+u_{xz}-u_{tx}-u_{yz}+u_{ty}+u_{tz}=0.$

28. (a) $\sum_{i=1}^{n} u_{x_i x_i}+\sum_{i<k}^{n} u_{x_i x_k}=0.$

(b) $\sum_{i<k}^{n} u_{x_i x_k}=0.$

29. Eliminate terms with lowest derivatives in the equation

$$\sum_{i=1}^{n} a_i u_{x_i x_i}+\sum_{i=1}^{n} b_i u_{x_i}+cv=f(x_1,x_2,\ldots,x_n),\ a_i\neq 0,\ i=1,\ldots,n.$$

CHAPTER II

EQUATIONS OF HYPERBOLIC TYPE

PROBLEMS on vibrations of continuous media (string, rod†, membrane, gas, etc.) and problems on electromagnetic oscillations are reducible to equations of hyperbolic type.

In the present chapter the statement and solution of boundary-value problems for equations of hyperbolic type (see footnote †) are considered, in the case where the physical processes under consideration can be described by functions of two independent variables: one spatial coordinate and time.

Chapter VI is devoted to equations of hyperbolic type for functions with a larger number of independent variables.

§ 1. Physical Problems Reducible to Equations of Hyperbolic Type; Statement of Boundary-value Problems

In the first group of problems of this chapter the continuity and homogeneity of the media are assumed, and also the continuity of the distribution of forces.

In the second group of problems a discontinuity in the medium and a discontinuity of both the characteristics of the medium and the density of the distribution of forces are allowed.

The third group of problems is devoted to establishing a similarity between different oscillatory processes.

---

† Transverse vibrations of a flexible rod reduce to a parabolic equation of fourth order, while the longitudinal vibrations reduce to a hyperbolic equation of second order. However boundary-value problems for transverse vibrations of a rod are closely related to boundary-value problems for longitudinal vibrations and therefore are considered in the present chapter.

There are a number of important physical problems, reducible to equations of hyperbolic type for functions, not dependent on time; for example, in the steady flow around a body of a supersonic stream of gas an equation of hyperbolic type is obtained for the velocity potential.

Stating the boundary-value problem, corresponding to a physical problem, means in the first place, choosing a function characteristic of the physical process†, and then

(1) deriving the differential equation for this function,

(2) formulating the boundary conditions for it,

(3) formulating the initial conditions‡.

1. Free Vibrations in a Non-resistant Medium; Equations with Constant Coefficients

In an investigation of small vibrations in homogeneous media§ we arrive at differential equations with constant coefficients.

1. *Longitudinal vibrations of a rod.* A flexible rectilinear rod is disturbed from its equilibrium state by small longitudinal displacements and velocities imparted to its cross-sections at time $t = 0$. Assuming that the cross-sections of the rod always remain plane, state the boundary-value problem for determining the displacements of the cross-sections of the rod for $t > 0$. Consider the case where the ends of the rod

(a) are rigidly fixed,

(a′) move in a longitudinal direction according to a given law,

(b) are free,

(c) are flexibly attached, i.e. each end is subject to a longitudinal force, proportional to its displacement and directed oppositely to the displacement.

2. *Small vibrations of a string*#. A string is stretched by a force T_0 and has its ends rigidly fixed. At time $t = 0$ initial displacements and velocities are given to points of the string.

---

† As a rule, this function will be indicated already in the conditions of the problem.

‡ The presence of initial conditions is characteristic of boundary-value problems of hyperbolic and parabolic type. For a discussion of the concepts and definitions, associated with boundary-value problems for equations of hyperbolic type, see [7], pages 32–43, and pages 125–127.

§ For example, in homogeneous rods and strings of constant cross-section.

The derivation of the equation of small transverse and small longitudinal vibrations of a string is similar to that carried out in [7], pages 11–21.

State the boundary-value problem for determining the small displacements of points of the string for $t > 0$.

3. *Torsional vibrations of a flexible cylinder.* A flexible homogeneous cylinder is displaced from its state of equilibrium by giving its cross-sections small angular displacements in planes at right angles to the cylinder axes.

State the boundary-value problem for determining the angles of deflection of cross-sections of the cylinder for $t > 0$; consider the case of free, rigidly attached and flexibly attached ends.

4. *Longitudinal vibrations of a gas in a tube.* An ideal gas enclosed in a cylindrical tube performs small longitudinal vibrations; plane cross-sections, consisting of particles of the gas, are not deformed, and all the gas particles move parallel to the axis of the cylinder.

Form the boundary-value problems to determine (1) the density ρ, (2) the pressure p, (3) the velocity potential ϕ of the gas particles, (4) the velocity v and (5) the displacement u of the gas particles in cases where the ends of the tube are

(a) closed by rigid impermeable surface,

(b) open,

(c) closed by pistons of negligibly small mass, fixed to a spring with coefficient of rigidity ν and slipping without friction inside the tube.

5. *Zhukovskii's problem on a hydraulic hammer.* The inlet of a straight cylindrical tube of length l is connected to a reservoir with an infinite capacity. A compressible liquid flows from the reservoir through the tube with a constant velocity v_0. At the initial time $t = 0$ an outlet section of the tube $x = l$ is closed.

Form the boundary-value problem to determine the velocity and the pressure of the liquid in the tube.

6. At the end $x = l$ of the tube of the preceding problem there is a pneumatic cap (Fig. 1) and apparatus A, controlling the amount of liquid $Q(t)$, flowing out of the tube. $Q(t)$ is a given function of time.

Let Ω_0 and P_0 be the average volume and pressure of the air in the cap; assuming the liquid to be incompressible, and the walls of the cap rigid, and assuming the process of compression and rarefaction of air in the cap isothermic and the change of volume of air in the cap small in comparison with the average volume Ω_0, derive the boundary condition for the end $x = l$.

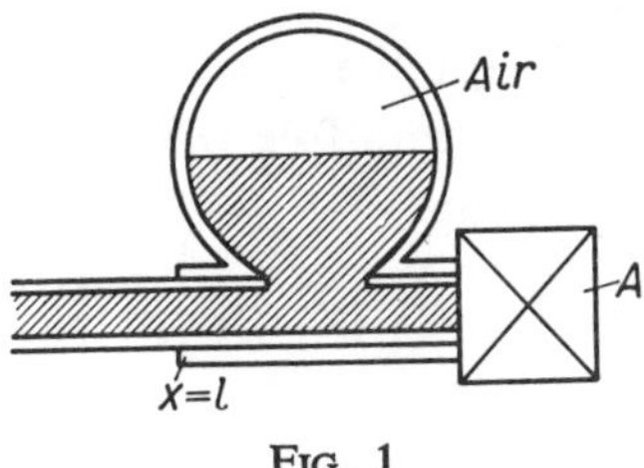

FIG. 1

7. *Gravity in a canal.* Water partially fills a shallow horizontal canal of length l with rectangular cross-section. The depth of the water equals in equilibrium h. The ends of the canal are closed by plane rigid surfaces, perpendicular to its axis.

Let us choose the x-axis along the canal. For small disturbances of the free surface in the canal a wave motion may develop in which the cross-sections, consisting of fluid particles, will be displaced a distance $\xi\,(x, t)$ along the x-axis and there will be a deflection $\eta\,(x, t)$ of the equilibrium free surface of the water.

Let the initial values $\xi(x, t)$ and $\eta(x, t)$ be given at the time $t = 0$.

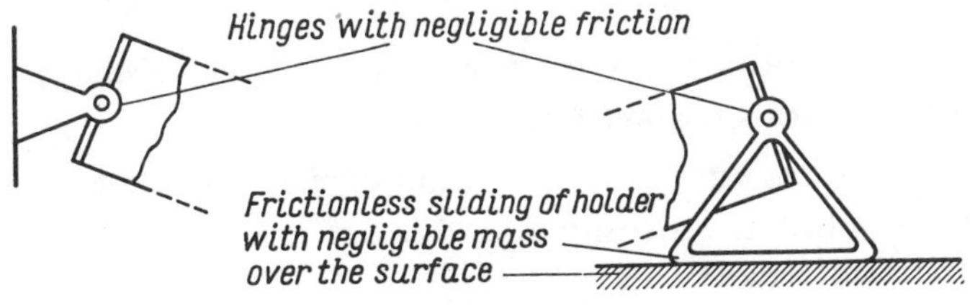

FIG. 2

State the boundary-value problem for determining $\xi(x, t)$ and $\eta(x, t)$ for $t > 0$.

8. *Transverse vibrations of a rod.* Points of a flexible homogeneous rectangular rod freely hinged at the ends (Fig. 2) are given small transverse displacements and velocities in a vertical plane at the initial time $t = 0$.

State the boundary-value problem to determine the transverse displacements of points of the rod for $t > 0$, assuming that the rod performs small transverse vibrations.

9. Consider problem 8 for the case where one end of the rod is rigidly fixed and the other end free (Fig. 3).

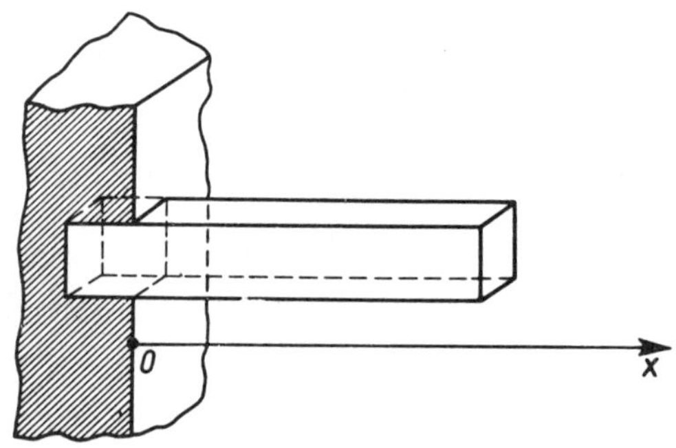

FIG. 3

10. Consider problem 8, assuming that the rod is attached to a flexible surface of negligible mass. The coefficient of elasticity of the surface equals k, i.e. the transverse elastic force per unit length, acting at the point x of the rod, equals $ku(x, t)$ where $u(x, t)$ is the displacement of the point x at time t.

2. Forced Vibrations and Vibrations in a Resistant Medium; Equations with Constant Coefficients

11. Starting at time $t = 0$, a continuously distributed transverse force with linear density $F(x, t)$ is applied to a string, whose ends are rigidly fixed.

State the boundary-value problem which determines the transverse displacements $u(x, t)$ of points of the string for $t > 0$.

12. For $t > 0$ an alternating current of strength $I = I(t)$ passes through a wire $0 \leqslant x \leqslant l$ rigidly fixed at the ends and of negligibly small electrical resistance. The string is placed in a constant magnetic field of intensity H, perpendicular to it. State the boundary-value problem for transverse vibrations of the string produced by the electromagnetic forces acting on the string†.

13. Beginning at time $t = 0$, one end of a linear flexible homogeneous rod performs longitudinal vibrations according to a given law, and a force $\Phi = \Phi(t)$, directed along the axis of the rod is applied to the other end. At time $t = 0$ the rod was at rest in an undeformed state. State the boundary-value problem to determine the small longitudinal displacements $u(x, t)$ of points of the rod for $t > 0$.

14. The upper end of a compressible homogeneous vertical heavy rod is rigidly fixed to the roof of a freely falling lift, which, having reached a velocity v_0, stops instantaneously. State the boundary-value problem for the longitudinal vibrations of this rod.

15. State the boundary-value problem for small transverse vibrations of a string in a medium with a resistance proportional to the velocity, assuming that the ends of the string are fixed.

16. State the boundary-value problem for small transverse vibrations of a linear homogeneous flexible rod in a medium with resistance proportional to velocity, acted on by a continuously distributed transverse force. Assume the ends of the rod rigidly fixed.

17. State the boundary-value problem for small transverse vibrations of a linear homogeneous flexible rod, one end of which is fixed, and the other is acted on by a transverse force, varying with time according to a given law.

18. State the boundary-value problem for small longitudinal vibrations of a homogeneous flexible rod, in a non-resistant medium, if one of its ends is rigidly fixed, and the other is acted on by a resistance proportional to velocity.

19. *Electrical vibrations in conductors.* State the boundary-value problem to determine the current and potential in a thin conductor with a continuously distributed ohmic resistance R, capacitance C, self-inductance L and leakage conductance G‡, if one

† See [17], page 204.

‡ The values R, C, L, G are calculated per unit length; the homogeneity of the conductor indicates that R, C, L and G do not depend on x.

end of the conductor is earthed, and an e.m.f. $E(t)$ is applied to the other end and if the initial current $i(x, 0) = f(x)$ and the initial potential $v(x, 0) = F(x)$ are given.

3. Vibration Problems Leading to Equations with Continuous Variable Coefficients

If the vibrating medium is inhomogeneous, and the functions, describing its properties (volume density, modulus of elasticity, etc.), are continuous functions of position, then the differential equation of the function, describing the oscillations, will have continuous variable coefficients. But other cases can be found leading to equations with continuous variable coefficients.

20. State the boundary-value problem for the longitudinal vibrations of a flexible rod $0 \leqslant x \leqslant l$ of variable cross-section $S(x)$, if the ends of the rod are rigidly fixed, the volume density equals $\rho(x)$, the modulus of elasticity equals $E(x)$, and the vibrations are produced by the initial longitudinal displacements and velocities. Assume the deformation of the cross-sections to be negligibly small.

21. State the boundary-value problem for the longitudinal vibrations of a flexible rod, having the shape of a truncated cone, if the ends of the rod are rigidly fixed and the rod is set in motion by initial longitudinal deflections and velocities at $t = 0$. The length of the rod equals l, the radius of the base $R > r$, the material of the rod is homogeneous.

Neglect the deformation of the cross-sections.

22. Form the boundary-value problem for small transverse vibrations of a homogeneous flexible wedge-shaped rod of rectangular cross-section if its thick end is rigidly fixed, and its thin end is free (Fig. 4). The modulus of elasticity of the rod equals E, the volume density equals ρ.

Neglect the deformation of the cross-sections.

23. State the boundary-value problem for the transverse vibrations of a heavy string displaced from its vertical position of equilibrium, if its upper end is rigidly fixed, and the lower end free.

24. Consider problem 23 assuming that the string rotates with an angular velocity $\omega = \text{const.}$ with respect to the vertical position of equilibrium.

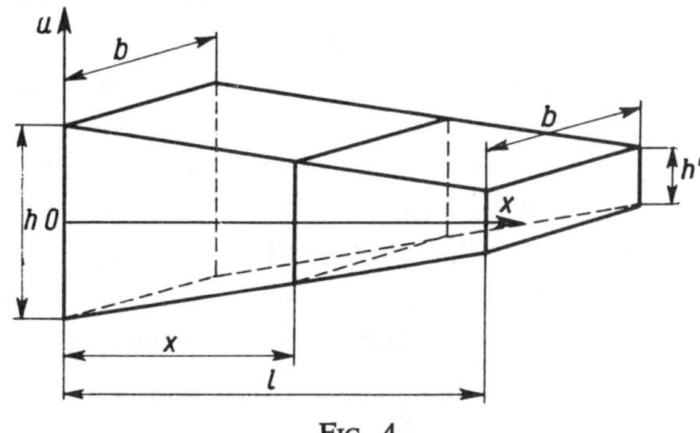

FIG. 4

25. A light string rotating about a vertical axis with constant angular velocity exists in a horizontal plane, one end of the string being attached to some point of the axis, and the other end being free. At the initial time $t = 0$ small deflections and velocities normal to this plane are imparted to points of the string.

State the boundary-value problem for determining the deflections of points of the string from the plane of equilibrium motion.

4. Problems Leading to Equations with Discontinuous Coefficients and Similar Problems (Piecewise Homogeneous Media, etc.)

If the density distribution of a vibrating flexible body or the density distribution of forces applied to it changes abruptly in the neighbourhood of certain points of space, then it is often found useful to assume that at these points a discontinuity of these densities occurs, and, in particular, to introduce concentrated masses or forces, if in the neighbourhood of the certain points the density of the mass or the density of the force is large. Then in the statement of the boundary-value problems differential equations with discontinuous coefficients and with a discontinuous constraint are obtained. If between the points of discontinuity the coefficients of the equation remain constant, then the problem can be reduced to equations with constant coefficients and matching conditions at the points of discontinuity. We are considering only interior points of the medium; if concentrated masses or forces occur at boundary points of the vibrating medium, then these should be included in the boundary conditions†.

26. Two semi-infinite homogeneous flexible rods of identical cross-section are joined at the ends and form one infinite rod‡. Let ρ_1, E_1 be the volume density and modulus of elasticity of one of them, and ρ_2, E_2 of the other.

State the boundary-value problem for determining the deflections of the rod from its equilibrium position, if at the initial moment of time longitudinal displacements and velocities are imparted to cross-sections of the rod.

27. Consider problem 26 for the case of transverse vibrations of a composite infinite rod.

28. Consider the problem, similar to problem 26, for longitudinal vibrations of a gas in an infinite cylindrical tube, if on one side of some cross-section there is a gas with one set of physical characteristics and another gas on the other side.

29. State the boundary-value problem for the wave motion of a liquid in a canal§ of rectangular cross-section, if the dimensions of a cross-section at some point of the canal change abruptly, i.e. the canal "consists" of two semi-infinite canals with different cross-sections.

30. Consider problem 26 assuming that the ends of the constituent rods are joined not directly, but between them there is a heavy weight of negligibly small thickness and mass M.

---

† Problems with a concentrated force at the end of the rod and with a concentrated e.m.f. at the end of the conductor were already considered in the previous section (see problems 13, 19).

‡ If one end of a rod is so far away from the region under investigation, so that in that region and during the time interval being considered it is possible to neglect disturbances, propagating from this end, then the rod may be assumed to be semi-infinite ($x_0 \leqslant x < +\infty$ or $-\infty < x \leqslant x_0$); if both ends of the rod satisfy this condition, then the rod may be assumed infinite ($-\infty < x < +\infty$). Similar conditions hold for a string, tube, saturated gas, etc.

§ See problem 7.

31. Two semi-infinite homogeneous rods of identical rectangular cross-sections are joined at the ends so that they form one infinite rod of constant cross-section, the ends of the semi-infinite rods being joined not directly, but by a weight of negligibly small thickness and mass M.

State the boundary-value problem for the transverse vibrations of such a rod.

32. State the boundary-value problem for the longitudinal vibrations of a homogeneous flexible vertical rod, neglecting the action of gravity on the particles of the rod, if the upper end of the rod is rigidly fixed, and to the lower end is attached a load Q. At the initial time a support is removed from under the load and the load begins to stretch the rod.

33. State the boundary-value problem for the transverse vibrations in a vertical plane of a flexible rectangular homogeneous rod, which is horizontal in an equilibrium state, if one end of the rod is rigidly fixed, and the other end is attached to a load Q, the moment of inertia of which with respect to the mean horizontal line of the adjoining end is negligibly small.

34. State the boundary-value problem for the longitudinal vibrations of a flexible horizontal rod with a load Q at the end, if the other end of the rod is rigidly fixed to a vertical axis, which rotates with an angular velocity, varying with time according to a given law. The bending vibrations are excluded by means of special guides, between which the rod slides.

35. Consider problem 34, assuming that the axis of rotation is horizontal.

36. State the boundary-value problem for the torsional vibrations of a cylinder of length $2l$, consisting of two cylinders of length l, if at the ends of the composite cylinder and between the ends of the connected cylinders there are pulleys (Fig. 5) with given axial moments of inertia.

37. Let an infinite string perform small transverse vibrations under the action of a transverse force, applied, for $t > 0$, at some given point of the string.

State the boundary-value problem to determine the deflections of points of the string from their positions of equilibrium. Consider also the case where the point of application of the force moves along the string in the course of time according to a given law.

38. Consider problem 37 for the transverse vibrations of the rod.

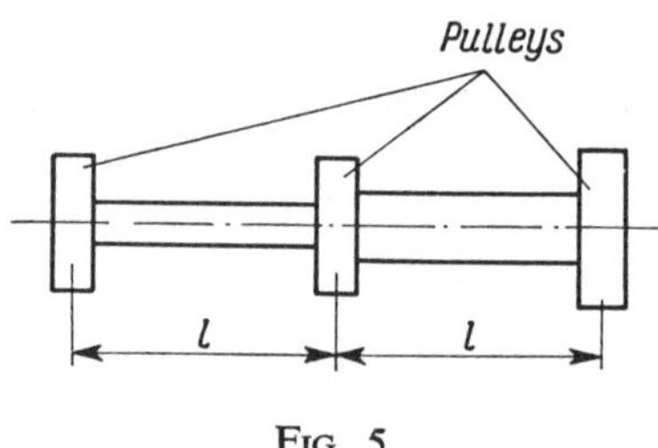

FIG. 5

39. The end of a semi-infinite cylindrical tube, filled with an ideal gas, houses a piston of mass M, which slides in the tube, the frictional resistance being proportional to the speed of the piston with a coefficient of proportionality equal to k^*. Let the piston be mounted on a spring with a coefficient of elasticity k^{**} with its axis directed along the axis of the tube.

State the boundary-value problem for the longitudinal vibrations of the gas in the tube.

40. A bead of mass M is fixed to a point of an infinite string and a spring with coefficient of elasticity k, perpendicular to the equilibrium position of the string (see Fig. 11) attaches it to the axis of the string.

State the boundary-value problem for the transverse vibrations of the string. Consider also the case where the bead is subject to a resistance proportional to the velocity with a coefficient of proportionality k^*.

41. State the boundary-value problem for the electrical vibrations in a conductor of negligibly small resistance and loss, if the ends of the conductor are earthed; one end through a lumped resistance R_0, and the other through a lumped capacity C_0.

42. Consider problem 41, assuming that one end of the conductor is earthed by a lumped self-inductance $L_0^{(1)}$, and an e.m.f. $E(t)$ is applied through a lumped self-inductance $L_0^{(2)}$ at the other end.

43. State the boundary-value problem for the electrical vibrations in a conductor, if the ends of the conductor are earthed through lumped resistances.

44. Form the boundary-value problem for the electrical vibrations in a conductor, if each of its ends is earthed through a lumped resistance and lumped self-inductance connected in series.

Find the relationships which the values of the lumped self-inductances and resistances must satisfy in order that homogeneous boundary conditions of the third kind should hold for $v(x, t)$.

45. State the boundary-value problem for the electrical vibrations in an infinite conductor, obtained by a combination of two semi-infinite conductors through a lumped capacity C_0.

Consider the boundary-value problem for determining the strength of the current in the case where there is no loss.

46. Consider problem 45 for the case where the semi-infinite conductors are joined not by a lumped capacitance, but by a lumped resistance R_0.

47. State the boundary-value problem for the electrical vibrations in a conductor, one end of which is earthed through a lumped resistance R_0 and a lumped self-inductance $L^{(1)}$ connected in parallel, and the other end through a lumped capacitance C_0 and lumped self-inductance $L^{(2)}$ connected in parallel.

48. State the boundary-value problem for the electrical vibrations in a conductor, the ends of which are earthed through

(a) a lumped self-inductance L_0,

(b) a lumped resistance R_0,

(c) a lumped capacitance C_0.

5. Similarity of Boundary-value Problems

Let there be two boundary-value problems (I) and (II), corresponding to physical phenomena of identical or different nature. We denote by $x', t', u'(x', t')$ the spatial coordinate, time and unknown function in the one problem, and by $x'', t'', u''(x'', t'')$

the corresponding values in the other problem. If the equation, initial and boundary conditions of each problem have an identical form, then the problems are said to be similar.

Let us denote by D_{I} the domain of variation $(x,' t')$ in problem (I), and by D_{II} the domain of variation (x'', t'') in problem (II). If there exist constants k_x, k_t, k_u, "coefficients of similarity", such that

$$u'(x', t') = k_u u''(x'', t'') \quad \text{if} \quad x' = k_x x'', \; t' = k_t t'', \tag{1}$$

as (x', t') passes through D_{I}, and (x'', t'') passes through D_{II}, then problem (I) is said to be similar to problem (II) with coefficients of similarity k_x, k_t, k_u†.

It is readily shown that if problem (I) is similar to problem (II), then it is possible to choose new units $x_0', t_0', u_0', x_0'', t_0'', u_0''$ in problems (I) and (II) so that the transition to the dimensionless quantities

$$\xi = \frac{x'}{x_0'}; \qquad \tau = \frac{t'}{t_0'}; \qquad U = \frac{u'}{u_0'}$$

and

$$\xi = \frac{x''}{x_0''}, \qquad \tau = \frac{t''}{x_0''}, \qquad U = \frac{u''}{u_0''}$$

leads to a complete correspondence of both boundary-value problems, viz. the region of variation of the dimensionless coordinates (ξ, τ) in both problems is the same, the coefficients in the equations and the boundary conditions are dimensionless and numerically equal, and the initial values are identically equal. Obviously, a valid and reciprocal statement: if there exists a transformation of dimensions, changing problems (I) and (II) into identically corresponding dimensionless problems, then problems (I) and (II) are similar.

49. Formulate the problem on the electrical vibrations in a conductor, similar to the problem on the longitudinal vibrations of a homogeneous flexible rod, one end of which is rigidly fixed, and the other end free.

† It is possible to consider a more extensive class of transformations including, in addition to extensions and compressions, even parallel displacements, i.e. transformations of the origins of the coordinates of x, t, u.

Establish the necessary and sufficient conditions that the first problem should be similar to the second with given coefficients of similarity.

50. Formulate the problem on the electrical vibrations in a conductor, similar to the problem on the longitudinal vibrations of a homogeneous flexible rod, in the following cases:

(a) one end of the rod is rigidly fixed, and the other end elastically attached;

(b) one end of the rod is free, and the other experiences a resistance proportional to the velocity;

(c) one end of the rod is fixed elastically, and the other end moves according to a given law.

Establish the necessary and sufficient conditions that the first problem be similar to the second.

51. Formulate the problem on the torsion vibrations of a cylinder, similar to problem 41 on the electrical vibrations in a conductor, taking the function characterizing the electrical vibrations, first as the voltage and then as the strength of the current.

Establish the necessary and sufficient conditions that the first problem is similar to the second.

§ 2. Method of Travelling Waves (D'Alembert's Method)

The general solution $u = u(x, t)$ of the wave equation

$$u_{tt} = a^2 u_{xx} \tag{1}$$

may be represented in the form†

$$u(x, t) = \phi_1(x-at)+\phi_2(x+at), \tag{2}$$

where $\phi_1(z)$ and $\phi_2(z)$ are arbitrary functions, $\phi_1(x-at)$ is a forward wave, propagating in the positive direction with respect to

† It is sometimes more convenient to make use of other equivalent forms of representing the solution in the form of a travelling wave, for example,

$$u(x, t) = \phi_1(at-x)+\phi_2(at+x)$$

or

$$u(x, t) = \phi_1\left(t-\frac{x}{a}\right)+\phi_2\left(t+\frac{x}{a}\right).$$

the x-axis with a velocity a, and $\phi_2(x+at)$ is a backward wave, propagating with the same velocity in the negative direction†.

To solve the boundary-value problem for equation (1) by the method of travelling waves means to determine the function $\phi_1(z)$ and $\phi_2(z)$ from the initial and boundary conditions.

In the first part of this section problems for the infinite straight line $-\infty < x < +\infty$ are considered, in the second part, for the semi-infinite straight line with homogeneous and inhomogeneous boundary conditions, in the third part, for an infinite straight line, consisting of two semi-infinite regions, distinguishable by physical characteristics, in the fourth, problems for a finite segment with homogeneous and inhomogeneous boundary conditions.

1. Problems for an Infinite String

52. An infinite string is excited by a localized initial deflection, shown in Fig. 6. Plot (trace) the position of the string for the times‡ $t_k = kc/4a$, where $k = 0, 1, 2, 3, 5$.

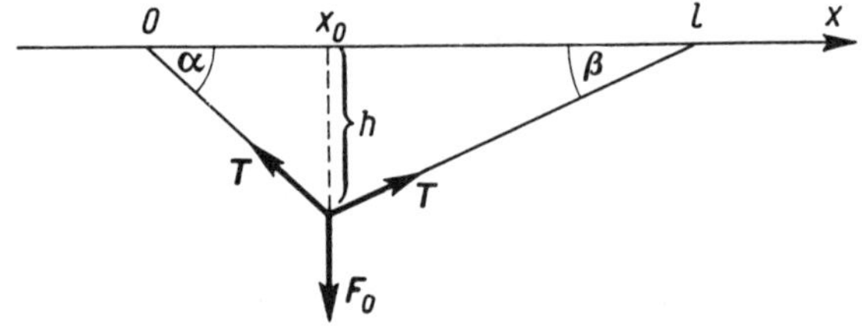

Fig. 6

53. An infinite string is excited by a localized initial deflection having the form of a quadratic parabola (Fig. 7). Find: (a) formulae, describing the profile of the string for $t > 0$, and (b) formulae, representing the law of motion of an arbitrary point x of the string for $t > 0$.

---

† See [7], pages 39–54 and 57–68. Use of solutions in the form (2) for steady-state problems, where t is a geometric coordinate, will be given in chapter V.

‡ Here and in later problems a means the wave velocity appearing in equation (1) $u_{tt} = a^2 u_{xx}$.

54. At time $t = 0$ an infinite string is excited by an initial deflection, having the form described in Fig. 8. At what point x and at what time $t > 0$ will the deflection of the string be a maximum? What is the value of this deflection?

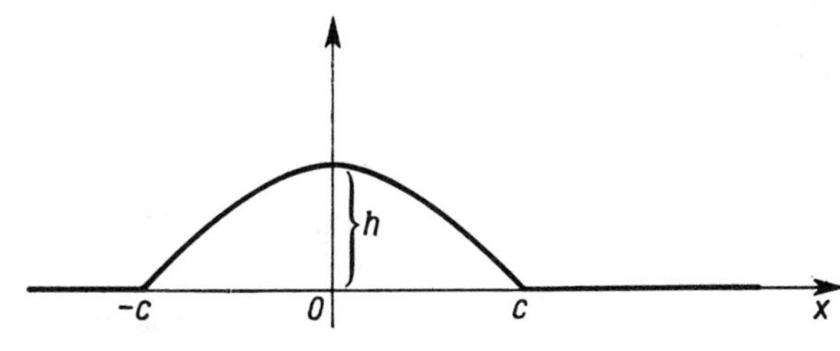

FIG. 7

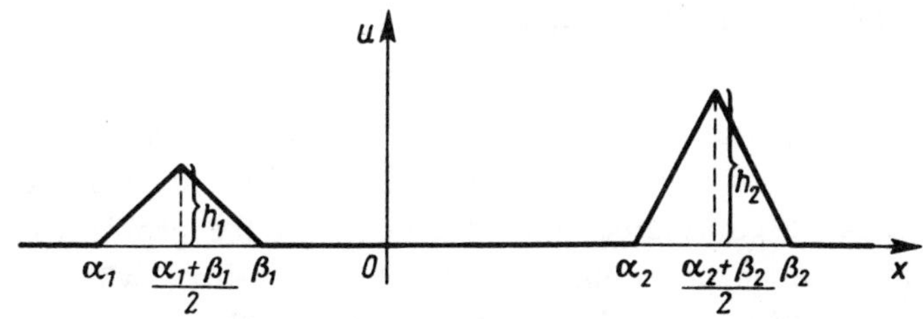

FIG. 8

55. A transverse initial velocity $v_0 = \text{const.}$ is imparted to an infinite string over a section $-c \leqslant x \leqslant c$; outside this section the initial velocity equals zero. Find formulae, describing the law of motion of points of the string with different abscissae for $t > 0$, and plot the positions of the string for the times

$$t_k = \frac{kc}{4a},$$

where $k = 0, 2, 4, 6$.

56. At the initial time $t = 0$ an infinite string receives a transverse blow at the point $x = x_0$, transmitting an impulse I to the string.

Find the deflection $u(x, t)$ of points of the string from positions of equilibrium for $t > 0$ assuming that the initial displacements of other points of the string and their initial velocities equal zero.

57. A wave $\phi(x-at)$ propagates along an infinite string. Knowing the form of the wave at time $t = 0$, find the state of the string for $t > 0$. Compare with results obtained in the solution of problem 52.

58. Solve the problem of propagation of electrical vibrations in an infinite conductor for the condition that

$$GL = CR, \tag{1}$$

where G, L, C, R are the leakage conductance, self-inductance, capacity and resistance per unit length of the conductor†. The voltage and the current in the conductor at the initial time are given.

2. Problems for a Semi-infinite Region

If only one end of a string‡ is far enough from the part of it under investigation so that a reflection from that end is not important in the oscillations of this part, at least during the time interval being considered, then we arrive at the problem of the vibrations of a semi-infinite string $0 < x < +\infty$, where $x = 0$ corresponds to the "near" end of the string. In this case the boundary-value problem consists of the equation, boundary condition and initial conditions§:

$$u_{tt} = a^2 u_{xx}, \quad 0 < x < +\infty, \quad 0 < t < +\infty, \tag{1}$$

$$\alpha_1 u_{tt}(0, t) + \alpha_2 u_t(0, t) + \alpha_3 u_x(0, t) + \alpha_4 u(0, t) = \Phi(t),$$
$$0 < t < +\infty, \tag{2}$$

$$u(x, 0) = \phi(x), \quad u_t(x, 0) = \psi(x), \quad 0 < x < +\infty, \tag{3}$$

† This condition ensures the possibility of passage through the line without deformation of shape. (See for more detail [7], pages 70–71 and the preceding.) Later if this condition for a guide is fulfilled, then we shall call it briefly: a distortionless transmission line.

‡ Or a rod or a conductor.

§ The assignment of two boundary conditions is also possible, if only one initial condition is given. (See for more detail [7], page 77.)

where at least one of the constants α_1, α_2, α_3, α_4, appearing in the boundary condition, must differ from zero†; if $\Phi(t) = 0$, then the boundary condition is homogeneous.

59. A semi-infinite string, fixed at an end, is excited by an initial deflection, described in Fig. 9.

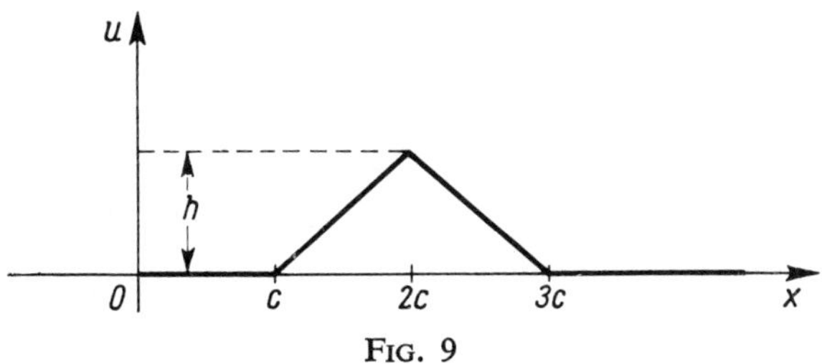

FIG. 9

Plot the shape of the string for the times

$$t = \frac{c}{a}; \quad t = \frac{3c}{2a}; \quad t = \frac{2c}{a}; \quad t = \frac{7c}{2a}.$$

60. An initial longitudinal velocity is imparted to a semi-infinite flexible rod $0 \leqslant x < +\infty$ with a free end $x = 0$, equal to v_0 over the segment $[c, 2c]$ and equal to zero outside this segment.

It is possible to plot the value of the longitudinal displacement $u(x, t)$ of cross-sections of the rod graphically in a direction, perpendicular to the x-axis, i.e. to treat this in the same way as was done in the case of the string. Utilizing this method of representation, trace the curve $u = u(x, t)$ for the times

$$t = 0; \quad \frac{c}{a}; \quad \frac{2c}{a}; \quad \frac{4c}{a}.$$

61. A semi-infinite string $0 \leqslant x < +\infty$ with a fixed end $x = 0$ receives at time $t = 0$ a transverse blow, transmitting an impulse

† If the boundary condition (3) takes the form $u_t(0, t) + \alpha u(0, t) = \Phi(t)$ the value of $u(0, 0)$ being known, then $u(0, t)$ may be calculated and we arrive at the boundary condition of the form $u(0, t) = \Phi(t)$. A similar expression is valid for a boundary condition of the form

$$u_{tt}(0, t) + \alpha u_t(0, t) + \beta u(0, t) = \Phi(t).$$

I to the string over the section $0 \leqslant x \leqslant 2l$, the profile of the distribution of velocity, obtained by the blow, having at time $t = 0$ the form of a half-wave sinusoidal with base $0 \leqslant x \leqslant 2l$. Find the formulae, describing the law of motion of points of the string with different abscissae x for $t > 0$.

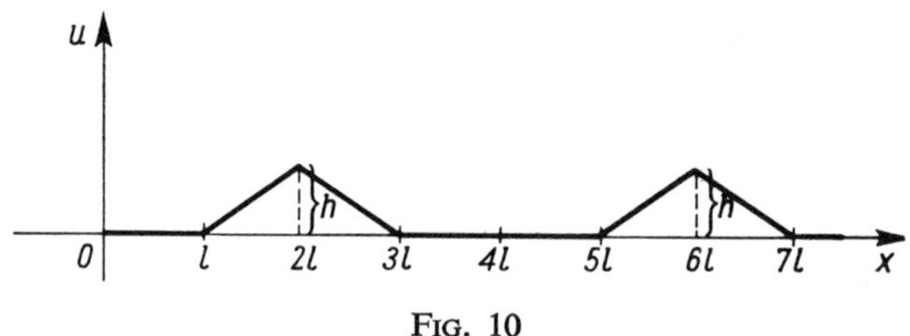

FIG. 10

62. A semi-infinite flexible rod $0 \leqslant x < +\infty$ with a free end at $x = 0$ is perturbed at time $t = 0$ by longitudinal displacements, the profile of which† is depicted in Fig. 10. Find at what points and when for $t > 0$ the displacement reaches its greatest value. What is the value of this greatest displacement?

63. An impulse I is transmitted at a point $x = x_0$ to a semi-infinite string with a fixed end at the initial time $t = 0$ by means of a transverse blow. Find the deflections $u(x, t)$ of points of the string from positions of equilibrium for $t > 0$ if the initial deflections $u(x, 0) = 0$, and the initial velocities at points $x \neq x_0$ also equal zero.

64. Solve problem 63 assuming that the initial impulse I is transmitted to the points $x_n > x_{n-1} > \dots > x_2 > x_1 > 0$.

65. An impulse I is transmitted to a semi-infinite rod with a free end at the initial time $t = 0$ by means of a longitudinal blow at the end.

Find the displacements $u(x, t)$ of points of the rod from the positions of equilibrium $u(x, t)$ for $t > 0$ if the initial deflections $u(x, 0) = 0$ and the initial velocities at points $x > 0$ also equal zero.

66. A load $Q = Mg$, moving with constant speed v_0 parallel to the x-axis, adheres at time $t = 0$ to the free end of a semi-

† See problem 60.

infinite rod $0 \leqslant x < +\infty$ and sticks to it. Find the displacements $u(x, t)$ of cross-sections of the rod from their positions of equilibrium for $t > 0$ if the initial deflections $u(x, 0) = 0$ and the initial velocities equal zero everywhere except at the end $x = 0$ where it equals v_0.

67. Initial longitudinal displacements are imparted to the sections of a semi-infinite flexible rod with its end fixed elastically to a support

$$u(x, 0) = \begin{cases} \sin \dfrac{\pi x}{l} & \text{if} \quad 0 \leqslant x \leqslant l, \\ 0 & \text{if} \quad l \leqslant x < +\infty \end{cases}$$

and initial velocities $u_t(x, 0) = 0$. Find the longitudinal deflections $u(x, t)$ of cross-sections of the rod for $t > 0$.

68. A semi-infinite vertical circular axle $0 \leqslant x < +\infty$ for $t < 0$ rotates with angular velocity $\omega = \text{const}$. At time $t = 0$ its end $x = 0$ touches a horizontal supporting surface and is acted on by a twisting moment of a frictional force, proportional to the angular velocity of the end. Find the deflection angles $\theta(x, t)$ of cross-sections of the axle for $t > 0$, assuming that $\theta(x, 0) = 0$.

69. A wave $u(x, t) = f(x+at)$ travels along a semi-infinite string $0 \leqslant x < +\infty$ for $t < 0$. Find the vibrations of the string for $0 < t < +\infty$ for cases where the end of the string

(a) is rigidly fixed,

(b) is free,

(c) is fixed elastically,

(d) is acted on by a frictional resistance, proportional to the velocity.

70. A wave $u(x, t) = f(x+at)$ travels along a semi-infinite cylindrical tube $0 < x < +\infty$ filled with an ideal gas for $t < 0, f(0) = 0$. At the end of the tube there is a piston of mass M_0, mounted on a spring with a coefficient of rigidity H_0 and of negligibly small internal mass. The piston tightly closes the tube and in motion in the tube experiences a resistance proportional to the velocity. Find $u(x, t)$ for $0 < t < +\infty$.

71. Find the electrical vibrations in a semi-infinite distortionless transmission line for $t > 0$, if for $t < 0$ a wave

$$v(x, t) = e^{-\frac{R}{L}t} f(x+at),$$

$$i(x, t) = -e^{-\frac{R}{L}t} \sqrt{\frac{C}{L}} f(x+at).$$

travels along the line. Consider the case where the end of the transmission line is earthed

(a) through a lumped resistance R_0,

(b) through a lumped capacity C_0,

(c) through a lumped inductance L_0.

Establish under what conditions in case (a) the reflected wave is absent ("complete absorption") and under what conditions the amplitude of the reflected wave is half the amplitude of the incident wave.

72. A constant e.m.f. E is applied to the end $x = 0$ of a semi-infinite distortionless transmission line over a sufficiently long interval of time, so that a steady distribution of voltage and current intensity is established in the line. Then at time $t = 0$ the end of the line is earthed through a lumped resistance R_0.

Find the voltage and current in the line for $t > 0$.

73. The end of a semi-infinite string $0 < x < +\infty$, starting at time $t = 0$, moves according to the law

$$u(0, t) = \mu(t).$$

Find the deflection $u(x, t)$ of points of the string for $0 < t < +\infty$, if the initial velocities and deflections equal zero.

74. A longitudinal force $F(t)$ is applied to the end of a semi-infinite rod at time $t = 0$. Find the longitudinal vibrations of the rod for $t > 0$, if the initial velocities and initial deflections of its points equal zero.

75. A semi-infinite horizontal tube of constant cross-section is filled at $t < 0$ with a fluid at rest. Beginning at time $t = 0$ a pressure pump with a compensating air cap is fitted to its end†. Find the pressure and velocity of the fluid in the tube for $t > 0$.

---

† See problems 5 and 6.

76. Find the longitudinal vibrations of a semi-infinite rod with zero initial conditions, if at the times

$$t_k = kT, \qquad k = 0, \qquad 1, 2, \ldots, n, \ldots,$$

longitudinal impulses are given to the end of the rod

$$I_n = I = \text{const.}$$

and a concentrated mass M is attached to the end.

77. An e.m.f. is applied to the end of a semi-infinite distortionless transmission line $0 < x < +\infty$

$$E(t) = E_0 \sin \omega t; \qquad 0 < t < +\infty.$$

At time $t = 0$ the voltage and current in the line are equal to zero. Find the voltage and current in the transmission line for $t > 0$, separating the steady process of propagation of waves with frequency ω from the transients. Determine the time, for which the amplitude of the transient waves will constitute not more than 10 per cent of the amplitude of the steady state vibrations at a point x of the line.

3. Problems for an Infinite Line, Consisting of Two Homogeneous Semi-infinite Lines

78. An infinite flexible rod is obtained by joining at the point $x = 0$ two semi-infinite homogeneous rods. For $x < 0$ the volume density, modulus of elasticity of the rod and the velocity of propagation of small longitudinal disturbances equal ρ_1, E_1, a_1 and for $x = 0$ they equal ρ_2, E_2, a_2. Let a wave $u_1(x, t) = f[t-(x/a_1)]$, $t \leqslant 0$ from the region $x < 0$ travel along the rod. Find the reflected and transmitted waves. Investigate the solution for $E_2 \to 0$ and for $E_2 \to +\infty$.

79. At the point $x = 0$ of an infinite homogeneous string a concentrated mass M is attached, supported by a spring of rigidity k with negligibly small internal mass (Fig. 11). Find the deflection of the string $u(x, t)$ for $t > 0$, if the string is excited at time $t = 0$ by a transverse impulse $I = Mv_0$, transmitted to the mass M and directed along the axis of the spring.

80. The mass M of the preceding problem, in oscillating, experiences a frictional resistance proportional to the velocity.

Find the reflected and transmitted waves, taking the wave $u_1(x, t) = f(x-at)$ travelling from the region $x < 0$ as the initial condition.

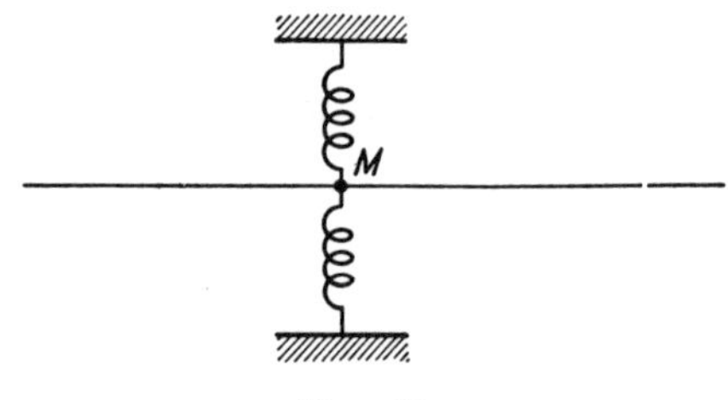

FIG. 11

81. A plane source of small disturbances moves uniformly with speed less than the speed of sound along a cylindrical infinite tube of gas. Assuming that the variation of the pressure at the source for time $t > 0$ is a known function of time, find the vibrations of the gas to the left and right of the source, if initially the gas was in an unperturbed state, and the source was at the point $x = 0$.

82. Solve the problem of the vibrations of an infinite string under the action of a concentrated transverse force $F(t)$ for $t > 0$ if the point of application of the force slides along the string with constant velocity v_0, from the position $x = 0$ where $v_0 < a$ and the initial conditions are zero.

4. Problems for a Finite Segment

83. The ends of a string $x = 0$ and $x = l$ are rigidly fixed; the initial deflection is given by the equation

$$u(x, 0) = A \sin \frac{\pi x}{l} \quad \text{if} \quad 0 \leqslant x \leqslant l,$$

the initial velocities equal zero. Find the deflections $u(x, t)$ for time $t > 0$.

84. Solve the problem of the longitudinal vibrations of a rod, one end of which ($x = 0$) is rigidly fixed, and the other end ($x = l$) is free, if the rod has an initial extension

$$u(x, 0) = Ax, \quad 0 \leqslant x \leqslant l,$$

and initial velocities are zero

$$u_t(x, 0) = 0, \quad 0 \leqslant x \leqslant l.$$

85. Solve problem 84, if the end $x = l$ of the rod is fixed elastically.

86. One end of a rod ($x = 0$) is rigidly fixed, and the other end ($x = l$) is free. At the initial moment of time a longitudinal impulse I is imparted to the free end. Find the vibrations of the rod.

87. One end of a horizontal rod is rigidly fixed and the other end is free. At the initial time $t = 0$ a mass $Q = Mg$ strikes the free end of the rod with a velocity v_0, directed along the axis of the rod, and remains in contact with it until $t = t_0$. Find the longitudinal vibrations of the rod for $t > 0$.

88. Solve the preceding problem for a rod, both ends of which are free.

89. Solve problem 87, assuming that the rod has the form of a truncated cone.

90. Solve problem 88 for a rod having the shape of a truncated cone.

91. Find the longitudinal vibrations of a rod with zero initial conditions, if one of its ends is fixed or free and the other moves according to a given law; consider the case where

(a) the right-hand end is fixed,

(b) the left-hand end is fixed,

(c) the right-hand end is free.

92. Find the pressure vibrations at the end $x = 0$ of a tube for $t > 0$, if it is equal to zero at the end $x = l$, and the input of liquid at the end $x = 0$ is a known function of time. The resistance of the tube is negligibly small, and the pressure disturbance and velocity of the liquid for $t = 0$ equal zero.

93. Solve the problem of an elastic longitudinal impact between two identical rods, moving in the same direction along the same straight line with velocities v_1 and v_2; $v_1 > v_2 > 0$ (Fig. 12). Find the distribution of velocities and tensions in the rods during the impact.

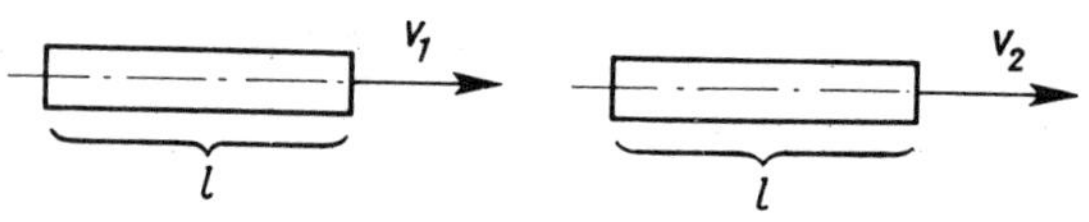

FIG. 12

94. A constant e.m.f. E is applied to the end $x = 0$ of a distortionless transmission line†, starting at time $t = 0$; the end $x = l$ is earthed. The initial voltage and initial current in the line equal zero. Find the electrical vibrations in the line for $t > 0$ and find at what time the current in the line will differ by less than 10 per cent from the limiting value (for $t \to +\infty$).

95. Solve the preceding problem for the condition that the end $x = l$ is insulated.

96. One end ($x = l$) of a conductor of negligibly small resistance and loss is earthed through

(a) a lumped resistance R_0,

(b) a lumped capacity C_0,

(c) a lumped inductance L_0, and an e.m.f. $E =$ const. is applied to the other end ($x = 0$) at time $t = 0$.

Find the voltage $v(x, t)$ at the end $x = l$ for $t > 0$ for all cases.

§ 3. Method of Separation of Variables

In this section problems on vibrations of a finite section of a string with various boundary conditions are considered, and also analogous problems on vibrations from other fields of physics and engineering.

† See the footnote to problem 58.

1. Free Vibrations in a Non-resistant Medium†

97. Investigate the vibrations of a string with fixed ends $x = 0$ and $x = l$, excited by an initial deflection, depicted in Fig. 13 and evaluate the energy of the various harmonics. The initial velocities equal zero.

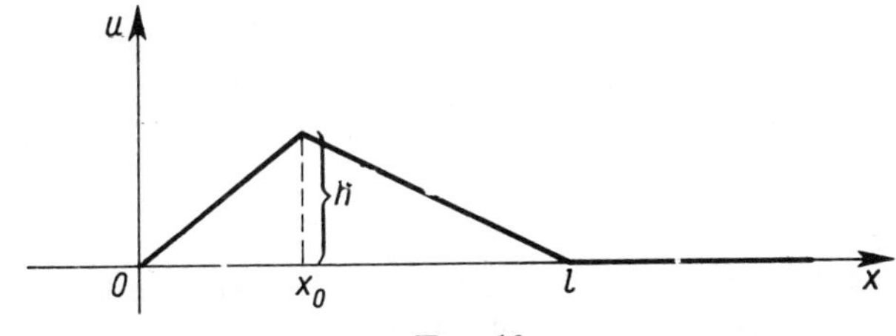

FIG. 13

98. A string $0 \leqslant x \leqslant l$ with fixed ends, up to the time $t = 0$, was in a state of equilibrium under the action of a transverse force $F_0 = \text{const.}$, applied at the point x_0 of the string, perpendicularly to the undisturbed position of the string. At the initial time $t = 0$ the action of the force F_0 ceases instantaneously. Find the vibrations of the string for $t > 0$.

99. The ends of a string are fixed, and the initial deflection has the form of a quadratic parabola, symmetrical with respect to the perpendicular to the mid-point of the string. Find the vibrations of the string, if the initial velocities equal zero.

100. A string‡ with fixed ends is excited by the impact of a rigid plane hammer, which gives it the following initial distribution of velocities:

$$u_t(x, 0) = \psi(x) = \begin{cases} 0, & 0 \leqslant x \leqslant x_0 - \delta, \\ v_0, & x_0 - \delta \leqslant x \leqslant x_0 + \delta, \\ 0, & x_0 + \delta \leqslant x \leqslant l. \end{cases}$$

Find the vibrations of the string, if the initial deflection equals zero. Evaluate the energy of the individual harmonics.

---

† In this and the following two sections the media are assumed to be homogeneous.

‡ See [7], pages 147–150.

101. A string† fixed at the ends is excited by the impact of a sharp hammer, imparting an impulse I at the point x_0. Find the vibrations of the string if the initial deflection equals zero. Evaluate the energy of the individual harmonics.

102. A string‡ fixed at the ends is excited by the impact of a rigid sharp hammer‡, imparting to it an initial distribution of velocities

$$u_t(x, 0) = \begin{cases} 0, & 0 \leqslant x \leqslant x_0 - \delta, \\ v_0, \cos\left(\frac{\pi}{2} \cdot \frac{x - x_0}{\delta}\right), & x_0 - \delta \leqslant x \leqslant x_0 + \delta, \\ 0, & x_0 + \delta \leqslant x \leqslant l. \end{cases}$$

Find the vibrations of the string if the initial deflection equals zero. Evaluate the energy of the individual harmonics.

103. Find the longitudinal vibrations of a rod, one end of which $(x = 0)$ is fixed, and the other $(x = l)$ is free, for the initial conditions

$$u(x, 0) = kx, \quad u_t(x, 0) = 0 \quad \text{for} \quad 0 \leqslant x \leqslant l.$$

104. A rod with a fixed end $x = 0$ exists in a state of equilibrium under the action of a longitudinal force $F_0 = \text{const.}$, applied to the end $x = l$. At time $t = 0$ the action of the force F_0 ceases instantaneously. Find the vibrations of the rod if the initial velocities are zero.

105. Find the longitudinal vibrations of a flexible rod with free ends, if the initial velocities and initial displacements in a longitudinal direction are arbitrary. Consider the possibility of uniform linear motion of the rod.

106. Find the vibrations of a flexible rod with free ends, which has received a longitudinal impulse I at one end at $t = 0$.

107. Solve the preceding problem for the case where the end to which the impulse is not applied, is fixed.

108. One end of a rod is fixed elastically, and the other end is free. Find the longitudinal vibrations of the rod for arbitrary initial conditions.

---

† See [7], pages 147–150.

‡ For the excitation of a string by a supple convex hammer see problem 152.

109. One end of a rod ($x = l$) is fixed elastically, and a longitudinal force $F_0 =$ const. is applied to the other end ($x = 0$). The rod is in a state of equilibrium under the action of this force. Find the vibrations of the rod when the force F_0 instantaneously disappears at the initial time, if the initial velocities equal zero.

110. One end of a rod ($x = l$) is fixed elastically, and the other end ($x = 0$) receives a longitudinal impulse I at the initial time. Find the longitudinal vibrations of the rod if the initial displacement of the rod is zero.

111. Find the longitudinal vibrations of a rod with elastically fixed ends with identical coefficients of rigidity, if the initial conditions are arbitrary.

112. Solve the preceding problem, if the coefficients of rigidity of the connections at the ends of the rod are different.

113. Find the vibrations of the liquid level in a circular canal, the breadth and depth of which are small in comparison with its radius, if the initial displacement of the surface from an equilibrium position and the initial rate of change of this surface are given.

114. Prove the additive nature of the energy of the individual harmonics for the free vibrations of a string in a non-resistant medium with homogeneous boundary conditions of first, second and third kind.

115. Investigate the transverse vibration of a rod $0 \leqslant x \leqslant l$ for arbitrary initial conditions, if the ends of the rod

(a) are fixed by hinges,

(b) are rigidly fixed,

(c) are free.

116. Solve the preceding problem, assuming that the vibrations are produced by a transverse blow at the point $x = x_0$, transmitting an impulse I to the rod.

2. Free Vibrations in a Resistant Medium

117–122. In problems 97, 101, 103, 105, 108, 111, vibrations of strings and rods in a non-resistant medium were considered. We assume now that in these problems the medium produces a resistance proportional to the velocity, we then obtain problems 117, 118, 119, 120, 121 and 122 respectively. Solve problems 117–122, not evaluating the energy of the individual harmonics.

123. An insulated homogeneous electrical conductor $0 \leqslant x \leqslant l$ is charged to some potential $v_0 = \text{const}$. At the initial moment the end $x = 0$ is earthed, and the end $x = l$ continues to be insulated.

Find the distribution of voltage in the conductor if the inductance, resistance and capacity per unit length of the conductor are known†.

124. Find the electrical vibrations in a homogeneous conductor $0 \leqslant x \leqslant l$, if the end $x = 0$ is earthed, the end $x = l$ is insulated, the initial current equals zero and the initial potential equals

$$v(x, 0) = \begin{cases} 0, & 0 < x < a, \\ \dfrac{Q}{C(b-a)}, & a < x < b, \\ 0, & b < x < l. \end{cases}$$

Consider only the case when

$$\frac{\pi}{l\sqrt{CL}} > \left| \frac{R}{L} - \frac{G}{C} \right|$$

and find an expression for the voltage.

125. Find the voltage in a conductor with an initial current and an initial voltage, equal to zero, if at the initial time a concentrated charge Q is liberated at the point $x = x_0$. The other conditions are the same as in the preceding problem.

3. Forced Vibrations under the Action of Distributed and Concentrated Forces in a Non-resistant Medium and in a Resistant Medium

In this section problems with constant constraining forces are considered first, then problems with constraining forces varying harmonically with time and, finally, problems with constraining forces, varying with time according to an arbitrary law.

---

† The leakage conductance $G = 0$ in accordance with the assumption that the conductor is insulated.

126. Solve problem 97 for the condition that the vibrations arise from a gravity field in a medium with a resistance proportional to the velocity, and the ends of the string are fixed at the same height.

127. A vertical flexible rod $0 \leqslant x \leqslant l$ has its upper end $(x = 0)$ rigidly attached to a freely falling lift, which, having attained a velocity v_0, stops instantaneously. Find the longitudinal vibrations of the rod, if its lower end $(x = l)$ is free.

128. Find the longitudinal vibrations of a rod $0 \leqslant x \leqslant l$, if one of its ends is rigidly fixed, and a force $F_0 =$ const. is applied to the other end at time $t = 0$.

129. The input of liquid into the end $x = l$ of a tube $0 \leqslant x \leqslant l$ drops at time $t = 0$ by an amount $A =$ const.; the end $x = 0$ is connected to a large tank in which the liquid pressure remains invariant.

Assuming that until the change of input at the end $x = l$ the pressure and input to the tube were constant, find the change of input into the tube for $t > 0$ and the change of pressure in the section $x = l$ for $t > 0$.

130. Find the voltage in a homogeneous electrical conductor, the resistance, inductance, leakage conductance and capacity per unit length of which respectively equal R, L, G, and C, if the initial current and voltage equal zero, the end $x = l$ is insulated and a constant e.m.f. E is applied to the end $x = 0$ beginning at time $t = 0$.

131. Solve the preceding problem assuming that the end $x = l$ of the conductor is earthed.

132. A constant transverse force F_0 is applied at the point x_0 of a string $0 \leqslant x \leqslant l$ at time $t = 0$. Investigate the vibrations of the string, if its ends are fixed.

133. A continuously distributed force of linear density

$$\Phi(x, t) = \Phi(x) \sin \omega t$$

is applied for $t > 0$ to a string $0 \leqslant x \leqslant l$ with fixed ends. Find the vibrations of the string in a non-resistant medium; investigate the possibility of resonance and find the solution in the case of resonance.

134. Solve the preceding problem when the linear density of the force equals $\Phi(x, t) = \Phi_0 \sin \omega t$, $0 < x < l$, $0 < t < +\infty$, where $\Phi_0 = \text{const.}$

135. Find the longitudinal vibrations of a rod $0 \leqslant x \leqslant l$, the end $x = 0$ of which is rigidly fixed, and the end $x = l$, starting at time $t = 0$, moves according to the law

$$u(l, t) = A \sin \omega t, \quad 0 < t < +\infty.$$

The medium does not produce a resistance to the vibrations.

136. Find the longitudinal vibrations of a rod $0 \leqslant x \leqslant l$ in a non-resistant medium, if the end $x = 0$ of the rod is rigidly fixed, and a force is applied to the end $x = l$, starting at time $t = 0$

$$F(t) = A \sin \omega t, \quad 0 < t < +\infty.$$

137. Solve problem 35 assuming that at the initial time $t = 0$ the rod was in a horizontal position and that $Q = 0$ and $\omega = \text{const.}$ Consider the non-resonant case.

138. Find the vibrations of a string $0 \leqslant x \leqslant l$ rigidly fixed at the ends, if a transverse force is applied at a point $x = x_0$ of this string at time $t = 0$

$$F(t) = A \sin \omega t, \quad 0 < t < +\infty.$$

Consider only the case where the frequency of the constraining force does not coincide with any of the eigenfrequencies.

139. Solve the preceding problem if

$$F(t) = A \cos \omega t, \quad 0 < t < +\infty.$$

140. Solve problem 138 if $F(t)$ is an arbitrary periodic force of period ω, i.e.

$$F(t) = \frac{\alpha_0}{2} + \sum_{n=1}^{+\infty} (\alpha_n \cos n\omega t + \beta_n \sin n\omega t), \quad 0 < t < +\infty.$$

141. A continuously distributed force with linear density $\Phi(x, t) = \Phi_0(x) \sin \omega t$ is applied to a string $0 \leqslant x \leqslant l$ with rigidly fixed ends at time $t = 0$. Find the vibrations of the string for zero initial conditions assuming that the medium produces a resistance proportional to the velocity. Find the steady-state vibrations, representing the principal part of the solution for $t \to +\infty$. (Compare with problem 133.)

Note. The steady-state oscillations have the frequency of the constraining force; oscillations with other frequencies are damped out.

142. Solve problem 136 assuming that the vibrations occur in a medium with a resistance proportional to the velocity. Find the steady-state vibrations representing the main part of the solution for $t \to +\infty$.

143. Solve problem 130 assuming that an e.m.f. $E(t) = E_0 \sin \omega t$, $0 < t < +\infty$ is applied at the end $x = l$ of the conductor ($E_0 = \text{const.}$), and the end $x = 0$ is insulated. Find the steady-state vibrations representing the main part of the solution for $t \to +\infty$,

144. Solve problem 131 assuming that an e.m.f. $E(t) = E_0 \sin \omega t$, $0 < t < +\infty$, $E_0 = \text{const.}$, is applied at the end $x = l$ of the conductor at time $t = 0$, and the end $x = 0$ is earthed. Find the steady-state vibrations.

145. Find the steady-state vibrations of pressure at the end $x = l$ of a tube $0 \leqslant x \leqslant l$, if a damping cap exists at this end (cf. problems 6, 75) and the input of liquid varies harmonically with time. The pressure remains constant at the other end of the tube.

146. Find the vibrations of a string $0 \leqslant x \leqslant l$ with rigidly fixed ends under the action of a force, applied at time $t = 0$ and having a density

$$F(x, t) = \Phi(x)t, \quad 0 \leqslant x \leqslant l, \quad 0 < t < +\infty,$$

assuming that the medium does not produce a resistance to the vibrations.

147. Find the longitudinal vibrations of a rod $0 \leqslant x \leqslant l$, the left end of which is fixed, and a force

$$F(t) = At, \quad 0 < t < +\infty, \quad A = \text{const.},$$

is applied to the right hand end at time $t = 0$, assuming that the medium does not produce a resistance to the vibrations.

148. Find the vibrations of a string $0 \leqslant x \leqslant l$ with fixed ends under the action of a distributed force, applied at time $t = 0$ and having a density

$$F(x, t) = \Phi(x) t^m, \quad 0 \leqslant x \leqslant l, \quad 0 < t < +\infty, \quad m > -1,$$

assuming that the medium produces no resistance to the vibrations.

149. Find the longitudinal vibrations of a rod $0 \leqslant x \leqslant l$ in a non-resistant medium under the action of a force

$$F(t) = At^m, \quad 0 < t < +\infty, \quad A = \text{const.}, \quad m > -1,$$

applied at time $t = 0$ to the end $x = l$, if the end $x = 0$ is rigidly fixed.

150. Solve problem 133 by the method indicated for problem 148.

151. Solve problem 141 by the method indicated for problem 148.

152. Find the vibrations of a string† $0 \leqslant x \leqslant l$ with fixed ends, produced by the impact of a smooth convex hammer assuming that the medium does not exert a resistance to the vibrations. The hammer acts on the string with a force, the linear density of which equals

$$F(x, t) = \begin{cases} A \cos\left(\dfrac{\pi}{2} \dfrac{x - x_0}{\delta}\right) \sin \dfrac{\pi t}{\tau}, & |x - x_0| < \delta, \quad 0 \leqslant t \leqslant \tau, \\ 0, \quad |x - x_0| < \delta, \quad t > \tau, \\ 0, \quad 0 \leqslant x \leqslant x_0 - \delta, \quad x + \delta \leqslant x \leqslant l, \quad 0 < t < \infty. \end{cases}$$

153. Find the vibrations of a string $0 \leqslant x \leqslant l$ with rigidly fixed ends in a non-resistant medium, produced by a transverse blow at the point x_0, $0 < x_0 < l$, at time $t = 0$, transmitting an impulse I to the string‡.

154. Solve problem 146 assuming that the medium produces a resistance proportional to the velocity.

155. Solve problem 153 assuming that the medium produces a resistance proportional to the velocity.

---

† See [7], pages 147–150.

‡ Compare with the solution of problem 101.

156. Find the transverse vibrations of a rod with hinged ends under the action of a constant transverse force P, the point of application of which moves along the rod, starting at time $t = 0$ from the end $x = 0$ to the end $x = l$ with constant velocity v_0, assuming that the vibrations occur in a non-resistant medium.

157. Solve the preceding problem if $P = P_0 \sin \omega t$, $P_0 = \text{const.}$

158. Find the transverse vibrations of a rod under the action of a transverse central force $P = P_0 \sin \omega t$, applied at time $t = 0$ at the point x_0 of the rod, if the ends of the rod are hinged and the medium offers no resistance to the vibrations.

159. Solve the preceding problem, assuming that the vibrations occur in a medium with a resistance proportional to the velocity.

160. The end $x = 0$ of a rod is rigidly fixed, and a constant transverse force $F = F_0 = \text{const.}$ is applied to the free end $x = l$ at time $t = 0$. Find the transverse vibrations of the rod, produced by the force F_0.

161. Solve the preceding problem in the case where the action of the force $F = F_0$ continues only up to a time $t = T > 0$.

162. Solve problem 160 in the case where $F = F_0 \sin \omega t$.

163. The end $x = l$ of a rod is rigidly fixed, and the end $x = 0$ is hinged. Find the transverse vibrations of the rod, produced by a uniformly distributed transverse force with linear density $f_0 \sin \omega t$, applied to the rod at time $t = 0$.

4. Vibrations with Inhomogeneous Media and Other Conditions Leading to Equations with Variable Coefficients; Calculations with Concentrated Forces and Masses

164. Investigate the longitudinal vibrations of an inhomogeneous rod $0 \leqslant x \leqslant l$ of constant cross-section, obtained by joining two homogeneous rods at $x = x_0$, if

(a) the volume density and coefficient of elasticity are respectively equal to

$$\rho(x) = \begin{cases} \bar{\rho}, & 0 < x < x_0, \\ \bar{\bar{\rho}}, & x_0 < x < l, \end{cases} \qquad E(x) = \begin{cases} \bar{E}, & 0 < x < x_0, \\ \bar{\bar{E}}, & x_0 < x < l, \end{cases}$$

where $\bar{\rho}$, $\bar{\bar{\rho}}$, $\bar{E}$, $\bar{\bar{E}}$ are constants;

(b) the initial longitudinal displacements equal

$$u(x, 0) = \phi(x) = \begin{cases} \dfrac{h}{x_0} x, & 0 < x < x_0, \\ \dfrac{h(l-x)}{l-x_0}, & x_0 < x < l; \end{cases}$$

(c) the initial velocities equal zero:

$$u_t(x, 0) = \psi(x) = 0, \quad 0 < x < l;$$

(d) the ends of the rod are fixed:

$$u(0, t) = u(l, t) = 0, \quad 0 < t < +\infty.$$

165. Investigate the steady-state longitudinal vibrations of the composite rod, described in the preceding problem, if its end $x = 0$ is fixed, and a force

$$F(t) = F_0 \sin \omega t, \quad 0 < t < +\infty,$$

is applied to the end $x = l$ at time $t = 0$.

166. Investigate the longitudinal vibrations of the rod, described in problem 164, if one of its ends ($x = 0$) is fixed rigidly, the other end ($x = l$) fixed elastically, and the initial conditions are arbitrary.

167. Investigate the vibrations of a homogeneous string $0 \leqslant x \leqslant l$ with fixed ends and with a concentrated mass M, attached at a point $x = x_0$ of the string, produced by initial deflections

$$u(x, 0) = \phi(x) = \begin{cases} h\dfrac{x}{x_0} & \text{for} \quad 0 < x < x_0, \\ h\dfrac{l-x}{l-x_0} & \text{for} \quad x_0 < x < l. \end{cases}$$

168. The cross-section of the composite rod, described in problem 164, equals $\bar{S}$ in the section $0 \leqslant x \leqslant x_0$ and equals $\bar{\bar{S}}$ in the section $x_0 \leqslant x \leqslant l$; at the join x_0 a mass M is attached; the end $x = 0$ is rigidly fixed and the end $x = l$ is free. Find the longitudinal vibrations of the rod with arbitrary initial conditions.

169. One end of a flexible homogeneous drum is rigidly fixed, and a pulley with an mechanical moment of inertia M is placed on the other. Find the torsion vibrations of the drum for arbitrary initial conditions, if the shear modulus equals G, the geometrical moment of inertia of a cross-section of the drum equals K, and the mechanical moment of inertia per unit length of the drum equals J.

170. Find the steady-state longitudinal vibrations of a conical flexible rod $0 \leqslant x \leqslant l$, produced by a harmonic longitudinal force $F = F_0 \sin \omega t$, applied to the end $x = l$, if the end $x = 0$ is rigidly fixed (see problems 21 and 89).

171. Solve problem 23 assuming that the vibrations of the string are produced by initial deflections, and the initial velocities equal zero.

172. Solve problem 24 for arbitrary initial conditions.

173. Solve problem 25 for arbitrary initial conditions, taking the origin of coordinates at the fixed end of the string.

§ 4. Method of Integral Representations

In the present section problems on vibrations of an infinite, semi-infinite and finite string are considered, and also analogous problems from other fields of physics. The following methods are employed for their solution: the Fourier integral method, transition to a finite interval by the method of images, Riemann's method.

1. The Method of the Fourier Integral

174. Solve the boundary-value problem

$$u_{tt} = a^2 u_{xx} + f(x, t), \quad -\infty < x < +\infty, \quad 0 < t < +\infty, \tag{1}$$

$$u(x, 0) = 0, \quad u_t(x, 0) = 0, \quad -\infty < x < +\infty. \tag{2}$$

175. Solve the boundary-value problem

$$u_{tt} = a^2 u_{xx} + c^2 u^{\dagger}, \quad -\infty < x < +\infty, \quad 0 < t < \infty,$$

$$u(x, 0) = \phi(x), \quad u_t(x, 0) = \psi(x), \quad -\infty < x < +\infty.$$

176. Solve the boundary-value problem

$$u_{tt} = a^2 u_{xx} + c^2 u + f(x, t), \quad -\infty < x < +\infty, \quad 0 < t < +\infty,$$

$$u(x, 0) = u_t(x, 0) = 0, \quad -\infty < x < +\infty.$$

† We recall that the telegraphic equation is reduced to this form by means of the substitution $v(x, t) = e^{-\mu t} u(x, t)$.

177. Solve the boundary-value problem

$$u_{tt} = a^2 u_{xx}, \qquad 0 < x, t < +\infty,$$
$$u(0, t) = 0, \qquad 0 < t < +\infty,$$
$$u(x, 0) = \phi(x), \qquad u_t(x, 0) = \psi(x), \qquad 0 < x < +\infty.$$

178. Solve the boundary-value problem

$$u_{tt} = a^2 u_{xx}, \qquad 0 < x, t < +\infty,$$
$$u_x(0, t) = 0, \qquad 0 < t < +\infty,$$
$$u(x, 0) = \phi(x), \qquad u_t(x, 0) = \psi(x), \qquad 0 < x < +\infty.$$

179. Solve the boundary-value problem

$$u_{tt} = a^2 u_{xx}, \qquad 0 < x, t < +\infty,$$
$$u(0, t) = \mu(t), \qquad 0 < t < +\infty,$$
$$u(x, 0) = u_t(x, 0) = 0, \qquad 0 < x < +\infty.$$

180. Solve the boundary-value problem

$$u_{tt} = a^2 u_{xx}, \qquad 0 < x, t < +\infty,$$
$$u_x(0, t) = \nu(t), \qquad 0 < t < +\infty,$$
$$u(x, 0) = u_t(x, 0) = 0, \qquad 0 < x < +\infty.$$

181. Solve the boundary-value problem for the equation

$$u_{tt} = a^2 u_{xx} + f(x, t)$$

with zero initial conditions and boundary conditions

(a) $u(0, t) = 0$,

(b) $u_x(0, t) = 0$.

182. Solve the boundary-value problem

$$u_{tt} = u_{xx} + c^2 u, \qquad 0 < x, t < +\infty,$$
$$u_x(0, t) = \nu(t), \qquad 0 < t < +\infty,$$
$$u(x, 0) = u_t(x, 0) = 0, \qquad 0 < x < +\infty.$$

183. Solve the boundary-value problem

$$v_{tt} = v_{xx} + c^2 v, \qquad 0 < x, t < +\infty,$$
$$v(0, t) = \mu(t), \qquad 0 < t < +\infty,$$
$$v(x, 0) = 0, \qquad v_t(x, 0) = 0, \qquad 0 < x < +\infty.$$

184. Solve the boundary-value problem

$$u_{tt} = u_{xx} + c^2 u, \quad 0 < x, t < +\infty,$$
$$u_x(0, t) - hu(0, t) = \varkappa(t), \quad 0 < t < +\infty,$$
$$u(x, 0) = 0, \quad u_t(x, 0) = 0, \quad 0 < x < +\infty.$$

185. Prove† that

$$\int_{-\infty}^{+\infty} \bar{f}(\lambda)\bar{g}(\lambda) e^{-i\lambda x}\, d\lambda = \int_{-\infty}^{+\infty} g(s) f(x-s)\, ds,$$

where $\bar{f}(\lambda)$ and $\bar{g}(\lambda)$ are Fourier transforms of the functions $f(x)$ and $g(x)$ with kernel $e^{i\lambda\xi}$.

186. Prove that

$$\int_0^{+\infty} \bar{f}^{(c)}(\lambda)\bar{g}^{(c)}(\lambda) \cos \lambda x\, d\lambda = \frac{1}{2}\int_0^{+\infty} g(s)[f(|x-s|) + f(x+s)]\, ds,$$

where $\bar{f}^{(c)}(\lambda)$ and $\bar{g}^{(c)}(\lambda)$ are Fourier cosine transforms of the functions $f(x)$ and $g(x)$.

187. Prove that

$$\int_0^{+\infty} \bar{f}^{(s)}(\lambda)\bar{g}^{(c)}(\lambda) \sin \lambda x\, d\lambda = \frac{1}{2}\int_0^{+\infty} f(s)[g(|x-s|) - g(x+s)]\, ds,$$

where $\bar{f}^{(s)}(\lambda)$ and $\bar{g}^{(c)}(\lambda)$ are respectively the Fourier sine transform‡ and the Fourier cosine transform of the functions $f(x)$ and $g(x)$.

188. Solve the boundary-value problem

$$u_{tt} + a^2 u_{xxxx} = 0, \quad -\infty < x < +\infty, \quad 0 < t < +\infty,$$
$$u(x, 0) = \phi(x), \quad u_t(x, 0) = a\psi''(x), \quad -\infty < x < +\infty.$$

Consider also the special case where

$$\phi(x) = A e^{-\frac{x^2}{4k^2}}, \quad \psi(x) \equiv 0, \quad -\infty < x < +\infty,$$

---

† This relation is often called Parceval's theorem.

‡ See § 4 of the answers and hints to the present chapter, introducing part of section 1.

189. Solve the boundary-value problem

$$u_{tt}+a^2u_{xxxx}=0, \quad 0<x, t<+\infty,$$

$$u(0, t)=\mu(t), \quad u_{xx}(0, t)=0, \quad 0<t<+\infty,$$

$$u(x, 0)=u_t(x, 0)=0, \quad 0<x<+\infty.$$

190. Prove that a representation of the solution of the boundary-value problem

$$u_{tt}=a^2u_{xx}, \quad 0<x<+\infty, \quad 0<t<+\infty,$$

$$\sum_{k=0}^{N} A_k\frac{\partial^k u}{\partial x^k}=0, \quad 0<t<+\infty, \quad x=0,$$

$$u(x, 0)=\phi(x), \quad u_t(x, 0)=\psi(x), \quad 0<x<l,$$

in the form

$$u(x, t)=\frac{\phi(x-at)+\phi(x+at)}{2}+\frac{1}{2a}\int_{x-at}^{x+at}\psi(z)\,\mathrm{d}z$$

is valid if $\phi(x)$ and $\psi(x)$ are extended over the negative semi-axis of x so that the functions

$$\Phi(x)=\sum_{k=0}^{N} A_k\frac{\mathrm{d}^k\phi(x)}{\mathrm{d}x^k} \quad \text{and} \quad \Psi(x)=\sum_{k=0}^{N} A_k\frac{\mathrm{d}^k\psi(x)}{\mathrm{d}x^k}$$

are odd†.

191. Prove that a representation of the solution of the boundary-value problem

$$u_{tt}=a^2u_{xx}+f(x, t), \quad 0<x, t<+\infty,$$

$$\sum_{k=0}^{N} A_k\frac{\partial^k u}{\partial x^k}=0, \quad 0<t<+\infty, \quad x=0,$$

$$u(x, 0)=0, \quad u_t(x, 0)=0, \quad 0<x<+\infty$$

in the form

$$u(x, t)=\frac{1}{2a}\int_0^t \mathrm{d}\tau\int_{x-a(t-\tau)}^{x+a(t-\tau)} f(\xi, \tau)\,\mathrm{d}\tau$$

† Here and below we do not consider the question of continuity and differentiability.

is valid if the function $f(x, t)$ is extended over the negative semi-axis of x so that the function

$$F(x, t) = \sum_{k=0}^{N} A_k \frac{\partial^k f(x, t)}{\partial x^k}$$

is odd with respect to x.

192. Prove that a representation of the solution of the boundary-value problem

$$u_{tt} = a^2 u_{xx} + c^2 u, \qquad 0 < x, \qquad t < +\infty,$$

$$\sum_{k=0}^{N} A_k \frac{\partial^k u}{\partial x^k} = 0, \qquad 0 < t < +\infty, \qquad x = 0,$$

$$u(x, 0) = \phi(x), \qquad u_t(x, 0) = \psi(x), \qquad 0 < x < +\infty,$$

in the form

$$u(x, t) = \frac{\phi(x-at)+\phi(x+at)}{2} +$$

$$+\frac{ct}{2} \int_{x-at}^{x+at} \frac{I_1\left(c\sqrt{t^2-\frac{(x-\xi)^2}{a^2}}\right)}{\sqrt{t^2-\frac{(x-\xi)^2}{a^2}}} \phi(\xi)\, d\xi +$$

$$+\frac{1}{2a} \int_{x-at}^{x+at} I_0\left(c\sqrt{t^2-\frac{(x-\xi)^2}{a^2}}\right) \Psi(\xi)\, d\xi$$

is valid if the functions $\phi(x)$ and $\psi(x)$ are extended over the negative semi-axis of x, so that the functions

$$\Phi(x) = \sum_{k=0}^{N} A_k \frac{d^k \phi(x)}{dx^k} \quad \text{and} \quad \Psi(x) = \sum_{k=0}^{N} \frac{d^k \psi(x)}{dx^k}$$

are odd.

193. Prove that a representation of the solution of the boundary-value problem

$$u_{tt} = a^2 u_{xx} + c^2 u + f(x, t), \quad 0 < x, t < +\infty,$$

$$\sum_{k=0}^{N} A_k \frac{\partial^k u}{\partial x^k} = 0, \quad 0 < t < +\infty, \quad x = 0,$$

$$u(x, 0) = 0, \quad u_t(x, 0) = 0, \quad 0 < x < +\infty$$

in the form

$$u(x, t) = \frac{1}{2a} \int_0^t d\tau \int_{x-a(t-\tau)}^{x+a(t-\tau)} I_0\left(c\sqrt{(t-\tau)^2 - \frac{(x-\xi)^2}{a^2}}\right) f(\xi, \tau)\, d\xi$$

is valid if $f(x, t)$ is extended along the negative semi-axis of x in such a way that the function

$$F(x, t) = \sum_{k=0}^{N} A_k \frac{\partial^k f(x, t)}{\partial x^k}$$

is odd with respect to x.

1*. Transition to a Finite Interval by the Method of Images

194. Solve the boundary-value problem

$$u_{tt} = a^2 u_{xx} + c^2 u, \quad 0 < x < l, \quad 0 < t < +\infty,$$

$$u(0, t) = 0, \quad u(l, t) = 0, \quad 0 < t < +\infty$$

$$u(x, 0) = \phi(x), \quad u_t(x, 0) = \psi(x), \quad 0 < x < l.$$

195. Solve the boundary-value problem

$$u_{tt} = a^2 u_{xx} + c^2 u, \quad 0 < x < l, \quad 0 < t < +\infty,$$

$$u(0, t) = 0, \quad u_x(l, t) = 0, \quad 0 < t < +\infty,$$

$$u(x, 0) = \phi(x), \quad u_t(x, 0) = \psi(x), \quad 0 < x < l.$$

196. Solve the boundary-value problem

$$u_{tt} = a^2 u_{xx} + c^2 u, \quad 0 < x < l, \quad 0 < t < +\infty,$$

$$u_x(0, t) = 0, \quad u_x(l, t) = 0, \quad 0 < t < +\infty,$$

$$u(x, 0) = \phi(x), \quad u_t(x, 0) = \psi(x), \quad 0 < x < l.$$

197. Solve the boundary-value problem

$$u_{tt} = u_{xx}+c^2u, \quad 0<x<l, \quad 0<t<+\infty,$$
$$u(0, t) = \mu_1(t), \quad u(l, t) = \mu_2(t), \quad 0<t<+\infty,$$
$$u(x, 0) = 0, \quad u_t(x, 0) = 0, \quad 0<x<l.$$

2. Riemann's Method

198. Find the Riemann function for the operator

$$L(u) = \frac{\partial^2 u}{\partial t^2} - a^2\frac{\partial^2 u}{\partial x^2}, \quad a = \text{const.},$$

and using it solve the boundary-value problem

$$\frac{\partial^2 u}{\partial t^2} = a^2\frac{\partial^2 u}{\partial x^2}+f(x, t), \quad -\infty<x<+\infty, \quad 0<t<+\infty,$$
$$u(x,0) = \phi(x), \quad u_t(x, 0) = \psi(x), \quad 0<x<+\infty.$$

199. Find the Riemann function for the operator

$$L(u) = \frac{\partial^2 u}{\partial t^2} - a^2\frac{\partial^2 u}{\partial x^2} \pm c^2u, \quad a = \text{const.},$$

and using it solve the boundary-value problem

$$\frac{\partial^2 u}{\partial t^2} = a^2\frac{\partial^2 u}{\partial x^2} \pm c^2u+f(x, t), \quad -\infty<+\infty, \quad 0<t<+\infty,$$
$$u(x, 0) = \phi(x), \quad u_t(x, 0) = \psi(x), \quad -\infty<x<+\infty.$$

200. Solve the boundary-value problem

$$x^2\frac{\partial^2 u}{\partial x^2} - y^2\frac{\partial^2 u}{\partial y^2} = 0, \quad -\infty<x<+\infty, \quad 1<y<+\infty,$$
$$u\Big|_{y=1} = \phi(x), \quad \frac{\partial u}{\partial y}\Big|_{y=1} = \psi(x), \quad -\infty<x<+\infty.$$

201. Solve the boundary-value problem

$$(l-x)\frac{\partial^2 u}{\partial x^2} - \frac{\partial u}{\partial x} = \frac{1}{a^2}\frac{\partial^2 u}{\partial t^2}, \quad -\infty<x<l,$$
$$0<t<+\infty, \quad l>0,$$
$$u\Big|_{t=0} = \phi(x), \quad \frac{\partial u}{\partial t}\Big|_{t=0} = \psi(x), \quad -\infty<x<l.$$

202. Solve the boundary-value problem

$$(l^2-x^2)\frac{\partial^2 u}{\partial x^2}-2x\frac{\partial u}{\partial x}-\frac{\partial^2 u}{\partial y^2}-\frac{1}{4}u=0, \qquad -l<x<l,$$
$$0<y<+\infty,$$

$$u\bigg|_{y=0}=\phi(x), \quad \frac{\partial u}{\partial y}\bigg|_{y=0}=\psi(x), \qquad -l<x<l.$$

CHAPTER III

EQUATIONS OF PARABOLIC TYPE

EQUATIONS of parabolic type are obtained in a study of such physical phenomena as heat conduction, diffusion, the propagation of electromagnetic waves in conducting media, and the motion of a viscous fluid.

In the present chapter the statement and solution of boundary-value problems for equations of parabolic type are considered in cases where the physical processes being investigated can be described by functions of two independent variables: one spatial coordinate and time. In particular, throughout the whole of the present chapter initial values of the unknown function will be assumed to depend only on one spatial coordinate.

Chapter V, which is a continuation and development of this chapter, is devoted to equations of parabolic type for functions with a larger number of independent variables.

§ 1. Physical Problems Leading to Equations of Parabolic Type; Statement of Boundary-value Problems

In the first group of problems of this section the homogeneity of the medium is assumed, and in the second group deviation from homogeneity of the medium and the presence of sources are considered. The third group is devoted to establishing a similarity between different physical phenomena, leading to equations of parabolic type.

To state the boundary-value problem, corresponding to a given physical problem, means to choose a function describing the physical process, and then

(1) derive the differential equation of this function,

(2) establish boundary conditions for it,

(3) formulate the initial conditions.

A brief summary of the basic principles of heat conduction and diffusion, from which the differential equations and boundary conditions are derived, is given in chapter III, § 1, of the answers and hints.

1. Homogeneous Media; Equations with Constant Coefficients

Throughout all the problems of this section the media are assumed homogeneous and isotropic, and their properties independent of the unknown function and time. Rods, conductors, tubes, etc., here and elsewhere, unless otherwise indicated are assumed to have constant cross-section.

1. State the boundary-value problem for determining the temperature of a rod $0 \leqslant x \leqslant l$ thermally insulated along the sides, if its initial temperature is an arbitrary function of x; consider the case where

(a) the ends of the rod are maintained at a given temperature,

(b) a given heat flow is supplied to the ends of the rod from outside,

(c) at the ends of the rod a convectional heat exchange obeying Newton's law takes place with a medium, whose temperature is given.

2. On the sides of a rod a heat exchange obeying Newton's law takes place with a medium, the temperature of which is a given function of time. Neglecting temperature variation across the rod state the boundary-value problem for the temperature in the rod for the initial and boundary conditions of the preceding problem.

3. State the boundary-value problem for the cooling of a thin ring, at the surface of which a heat exchange obeying Newton's law takes place with the surrounding medium, which has a given temperature. Neglect the non-uniformity of the temperature distribution across the wire.

4. State the boundary-value problem for the heating of a semi-infinite rod, if the end of the rod burns, the combustion front being propagated with constant velocity v_0 and having a known temperature $\phi(t)$.

10. State the boundary-value problem for the motion of a layer of viscous liquid between two parallel planes, if one of them at time $t = 0$ begins to move parallel to the other with a given velocity, having a constant direction. Neglect the effect of gravity.

11. Derive the equation for the propagation of a plane electromagnetic field in a conducting medium. (A medium is said to be conducting if it is possible to neglect displacement currents in comparison with conduction currents.)

2. Inhomogeneous Media; Equations with Variable Coefficients and Matching Conditions

In this section piecewise homogeneous media are considered, leading to equations with piecewise continuous coefficients and to matching conditions. Then problems are considered leading to equations with continuously varying coefficients.

12. An infinite rod of constant cross-section is obtained by joining two semi-infinite homogeneous rods with different coefficients of heat conduction and thermal conductivity.

State the boundary-value problem for determining the temperature in this rod, considering the cases where

(a) the ends of the constituent rods are joined directly (joined end to end),

(b) the ends of the rods are joined by a solid joint with a heat capacity C_0 the material of the joint possessing a very large heat conductivity.

Assume the surface of the rod and the external surface of the joint (not adjoining the rod) to be thermally insulated.

13. A closed cylindrical vessel with impermeable walls is obtained by joining two cylindrical vessels at the initial time. Each of these is filled by a homogeneous medium with a uniformly distributed substance, the concentration of this substance in both constituent vessels being different and the properties of the medium in each vessel being different.

State the boundary-value problem for the diffusion of the substance in the composite cylinder, considering the case where

5. Derive the equation for the temperature of a thin w heated by a constant electric current, if at its surface a convec heat exchange obeying Newton's law takes place with an e ronment having a known temperature. State the boundary-va problem for determining the temperature in this wire, if its en are fixed in solid clamps with a given heat capacity and a ve large thermal conductivity.

6. Derive the equation of diffusion in a stationary mediun assuming that surfaces perpendicular to the x-axis are surface of equal concentration at every moment of time t. Write dow the boundary conditions, assuming that diffusion occurs in th plane layer $0 \leqslant x \leqslant l$; consider the case where

(a) at the boundary surfaces the concentration of diffusing substance is maintained equal to zero,

(b) the boundary surfaces are impermeable,

(c) the boundary surfaces are semi-permeable, diffusion through these surfaces taking place according to a law similar to Newton's law for a convective heat exchange.

7. Derive the diffusion equation in a medium, moving with constant velocity in the direction of the x-axis, if the surfaces perpendicular to the x-axis are surfaces of equal concentration at every moment of time t.

8. Derive the equation of the diffusion of suspended particles by calculating the precipitation, assuming that the velocity of the particles, produced by gravity, is constant, and that the concentration of the particles depends only on one geometric coordinate z (height) and time t. Write down the boundary condition, corresponding to an impermeable membrane.

9. Derive the equations of problem 6 for a substance, the molecules of which

(a) disintegrate (e.g. an unstable gas), the rate of disintegration of the diffusing substance at every point of space being proportional to the concentration;

(b) multiply (for example, diffusion of neutrons), the rate of multiplication of the diffusing substance at every point of space being proportional to the concentration.

(a) the cylinders are joined directly,

(b) the cylinders are joined by a semi-permeable membrane.

14. State the boundary-value problem for the heating of a thin rod, over which a closely fitting electric furnace of constant output slides with constant velocity. The outside surface of the furnace, not adjoining the rod, is thermally insulated, and the heat capacity of the furnace is negligibly small.

15. A molten metal fills a vertical cylindrical vessel, the walls and bottom of which are insulated. At time $t = 0$ the free surface of the metal is maintained at a temperature $v_1 \equiv$ const., which is below the temperature of fusion. State the boundary-value problem for the cooling and solidification of the metal, if its initial temperature equals $v_0 =$ const.

16. State the boundary-value problem for the motion under the action of gravity of a thin vertical infinite plane lamina in a layer of viscous fluid between two rigid planes parallel to it. Neglect the action of the field of gravity on the fluid.

17. State the boundary-value problem for the cooling of a uniformly heated rod, having the shape of a truncated cone, assuming the temperature is constant over a cross-section, if the ends of the rod are thermally insulated, and if there is a lateral heat exchange between the surface and a medium whose temperature equals zero.

3. Similarity of Boundary-value Problems†

18. Formulate the problem of heat conduction, analogous to problem 10 on the motion of a viscous fluid. Determine the necessary and sufficient conditions that the first problem is similar to the second with given coefficients of similarity k_x, k_t, k_u.

19. Formulate the problem on heat conduction 2 (problem (I)), similar to problem 9 (problem (II)) on the diffusion of an unstable gas. Determine the necessary and sufficient conditions for the

† Concerning the idea of the similarity of boundary-value problems see chapter II, § 1 pages 15 and 201.

similarity of problem (I) to problem (II) with given coefficients of similarity.

20. Formulate the problem to determine the electric current in conductor, analogous to the following problem on the determination of the temperature in a rod: "find the temperature of a rod if at one end and at the lateral surface a heat exchange takes place with a medium whose temperature equals zero, and the temperature of the other end varies according to a given law; the initial temperature of the rod equals zero" (problem (II)). Determine the necessary and sufficient conditions for the similarity of problem (I) to problem (II) with given coefficients of similarity.

21. Formulate the problem of heat conduction (problem (I)), analogous to the problem of the propagation of a plane electromagnetic field in a conducting layer $0 \leqslant x \leqslant l''$ (problem (II) for zero initial conditions, assuming that everywhere to the left of the layer a constant homogeneous magnetic field, parallel to the layer, was instantaneously established, the plane $x = l''$ being ideally conducting.

§ 2. Method of Separation of Variables

In the first part† of the present section problems on homogeneous isotropic media are collected together; they reduce to a linear differential equation with constant coefficients. In the second part problems on inhomogeneous media are considered, and also some problems with concentrated factors‡.

1. Homogeneous Isotropic Media. Equations with Constant Coefficients

(a) *Problems of heat conduction with constant boundary conditions*

22. (a) Find the distribution of temperature in a rod $0 \leqslant x \leqslant l$ thermally insulated along the surface, if the temperature of its

† Rods, conductors, cylinders, encountered in this section, are assumed to have constant cross-section.

‡ See the first footnote on page 30.

ends is maintained equal to zero, and the initial temperature equals an arbitrary function $f(x)$.

(b) Consider, in particular, the case where $f(x) \equiv U_0 = \text{const.}$, and give an estimate of the error, made in replacing the sum of the series, representing the solution at the point $x = \frac{1}{2}$, by its partial sum, and determine the time for which the ratio of the sum of all its terms, starting with the second, to the first term will be less than a given $\varepsilon > 0$.

Note. We say that at the point under consideration a steady-state† is set up with relative accuracy ε.

23. The initial temperature of a rod $0 \leqslant x \leqslant l$ thermally insulated along the surface equals

$$U_0 = \text{const.}, \tag{1}$$

and a constant temperature is maintained at its ends

$$u(0, t) = U_1 = \text{const.}, \quad u(l, t) = U_2 = \text{const.}, \quad 0 < t < +\infty. \tag{2}$$

Find the temperature $u(x, t)$ of the rod for $t > 0$; determine also the steady-state temperature

$$\bar{u}(x) = \lim_{t \to +\infty} u(x, t).$$

24. The initial temperature of a rod $0 < x < l$ is an arbitrary function of $f(x)$. The temperatures of the ends are constant:

$$u(0, t) = U_1 = \text{const.}, \quad u(l, t) = U_2 = \text{const.}, \quad 0 < t < +\infty.$$

A heat exchange obeying Newton's law takes place at the surface with a medium whose temperature equals $u_0 = \text{const.}$ Find the temperature of the rod. Consider, in particular, the case where $U_1 = U_2 = 0$, $f(x) = 0$.

25. Determine the temperature of a rod $0 \leqslant x \leqslant l$ thermally insulated along the surface and thermally insulated at the ends, if its initial temperature is an arbitrary function of x. Proceed next to the case in which a convectional heat exchange (obeying Newton's law) takes place between the surface and a medium whose temperature equals zero.

† For more detail concerning the steady-state, see [25].

26. Find the temperature of a rod, at the surface of which a convective heat exchange takes place with a medium of zero temperature, if a constant heat flow is maintained at the ends of the rod, and the initial temperature is an arbitrary function.

27. Determine the temperature of a rod $0 \leqslant x \leqslant l$ thermally insulated along its surface, if one of its ends ($x = 0$) is kept at a given fixed temperature, and a given constant heat flow is maintained at the other end ($x = l$), the initial temperature being arbitrary. Consider, in particular, the case where the initial temperature equals zero, and the end $x = l$ is thermally insulated, and estimate the error made in replacing the sum of the series, representing the solution at the point $x = l$, by its partial sum. Find the time at which a steady-state† is achieved at the end $x = l$ correct to ε.

28. Find the temperature of a rod $0 \leqslant x \leqslant l$ thermally insulated along its surface and thermally insulated at the end $x = 0$, if the initial temperature of the rod equals zero and a constant heat flow is fed into the rod through the end $x = l$. Give an estimate of the error made in replacing the sum of the series representing the solution at the point $x = 0$ by its partial sum.

29. Find the temperature of a rod $0 \leqslant x \leqslant l$ thermally insulated along its surface, one end of which ($x = 0$) is thermally insulated and at the other end ($x = l$) a convective heat exchange takes place with a medium whose temperature equals $U_0 = \text{const}$. The initial temperature of the rod equals zero. Estimate the error made in replacing the sum of the series, representing the solution at the point $x = 0$, by its partial sum; determine the time at which a steady-state‡ will occur at the end $x = 0$ to a degree of accuracy ε.

30. (a) Find the temperature of a rod $0 \leqslant x \leqslant l$ thermally insulated along its surface, if at each end a convective heat exchange

† Concerning the steady-state see the conditions of problem 22 and the corresponding note. Here the ratio of the sum of all terms, dependent exponentially on time, beginning with the second, to the first term, dependent exponentially on time, should be considered.

‡ See problem 22 and the footnote above.

takes place with an external medium having a constant temperature, and the initial temperature is arbitrary.

(b) Consider, in particular, the case where the temperature of the external medium is the same at both ends, and the initial temperature of the rod equals zero, and establish the connection with the solution of problem 29.

31. Solve problem 30 (a) assuming that at the surface of the rod a convective heat exchange takes place with a medium whose temperature equals zero.

32. Determine the temperature distribution in a thin homogeneous ring of unit radius, at the surface of which a convective heat exchange takes place with an environment of constant temperature; the initial temperature of the ring is arbitrary†. Consider, in particular, the case where the ring was uniformly heated at the initial moment of time.

(b) *Problems of heat conduction with variable boundary conditions and free terms, dependent on x and t*

33. Find the temperature distribution in a rod $0 < x < l$ thermally insulated along the surface, if a temperature, equal to zero, is maintained at its end $x = 0$ and at the end $x = l$ the temperature varies according to the law

$$u(l, t) = At, \qquad A = \text{const.}, \qquad 0 < t < +\infty .$$

The initial temperature of the rod equals zero.

34. Find the temperature of a rod $0 \leqslant x \leqslant l$ thermally insulated along the surface, if heat sources of density equal to $\Phi(t) \sin(\pi x/l)$ are continuously distributed over the rod, and the initial temperature of the rod is an arbitrary function $f(x)$ and the temperature of the ends is maintained equal to zero.

35. (a) Find the temperature of a rod $0 \leqslant x \leqslant l$ thermally insulated along the surface, if its initial temperature is an arbitrary function $f(x)$, the temperature of the ends is maintained equal to zero and heat sources of density equal to $F(x, t)$ are continuously distributed over the rod.

---

† See problem 3.

(b) Consider, in particular, the limiting case in which a concentrated source of constant magnitude Q exists at the point x_0, $0 < x_0 < l$, and the initial temperature of the rod equals zero.

36. A heater moves with constant velocity v_0 along a rod $0 \leqslant x \leqslant l$, at the surface of which a convective heat exchange takes place with a medium (the temperature of the medium equals zero). The flow of heat from the heater to the rod equals $q(t) = Ae^{-ht}$, where h is the coefficient of heat exchange appearing in the equation of heat conduction for a rod $u_t = a^2 u_{xx} - hu$. Find the temperature of the rod, if its initial temperature equals zero and the temperature of the ends is at all times maintained equal to zero.

37. Solve problem 35 (a) for a rod $0 \leqslant x \leqslant l$ thermally insulated along its surface, if a convective heat exchange takes place at its ends with a medium whose temperature varies according to a given law.

38. Find the temperature $u(x, t)$ of a rod, solving the boundary-value problem

$$u_t = a^2 u_{xx} - Hu + f(x, t), \quad 0 < x < l, \quad 0 < t < +\infty, \quad (1)$$

$$u_x(0, t) - hu(0, t) = \psi_1(t), \quad u_x(l, t) + hu(l, t) = \psi_2(t),$$
$$0 < t < +\infty, \quad (2)$$

$$u(x, 0) = \phi(x), \quad 0 < x < l \quad (3)$$

by reducing it to a homogeneous boundary-value problem.

39. Find the asymptotic expression as $t \to +\infty$ for the temperature $u(x, t)$ in a rod thermally insulated along the surface, if one of the following boundary conditions is fulfilled at its ends:

(a) $u(0, t) = 0, \quad u(l, t) = A \cos \omega t, \quad 0 < t < +\infty,$

(b) $u(0, t) = 0, \quad u_x(l, t) = Ae^{i\omega t}, \quad 0 < t < +\infty,$

(c) $u(0, t) = 0, \quad u_x(l, t) = hu(l, t) = Ae^{i\omega t}, \quad 0 < t < +\infty.$

40. At the surface of a thin ring of unit radius a convective heat exchange takes place with a medium whose temperature equals zero; the initial temperature of the ring equals zero†.

† See problem 32 and also problem 3.

Q units of heat were liberated at the initial moment of time at some fixed point of the ring. Find the temperature of the ring. Consider a point of the ring, diametrically opposite the point from which the heat was liberated, and estimate the error made in replacing the sum of the series, representing the solution at this point, by its partial sum.

(c) *Problems of diffusion*

41. The pressure and temperature of the air in a cylinder $0 \leqslant x \leqslant l$ is atmospheric; one end of the cylinder at time $t = 0$ is open, and the other end remains closed all the time. The concentration of some gas in the surrounding atmosphere equals $U_0 = \text{const}$. At time $t = 0$ the gas diffuses into the cylinder through the open end. Find the amount of gas, in the cylinder, if its initial concentration in the cylinder equals zero.

42. Solve the preceding problem assuming that both ends of the cylinder are closed by a semi-permeable membrane, across which diffusion takes place.

43. Solve problem 41, assuming that the diffusing gas dissociates, the rate of dissociation at each point being proportional to the concentration of the gas at that point.

44. There exists in a cylinder $0 \leqslant x \leqslant l$ a diffusing substance, the molecules of which multiply, the rate of multiplication at each point being proportional to the concentration of the substance at the same point. Find the critical length of the cylinder† for cases where

(a) a concentration, equal to zero, is maintained at both ends of the cylinder;

(b) a concentration, equal to zero, is maintained at one end, and the other end is tightly closed;

(c) both ends of the cylinder are tightly closed.

(d) *Problems of electrodynamics*

45. Find the electric current in a conductor $0 \leqslant x \leqslant l$, if one end is insulated, and a constant e.m.f. is applied to the other end.

† Concerning the concept of critical dimensions see [7], pages 519–521.

The distributed inductance and leakage conductance in the conductor are negligibly small, the initial potential equals $v_0 = \text{const.}$, and the initial current equals zero.

46. The distributed inductance and leakage conductance of a conductor $0 < x < l$ are zero; the initial potential and initial current also equal zero. Find the voltage in the conductor, if one of its ends $(x = l)$ is earthed through a lumped capacity C_0, and if a constant e.m.f. E_0 is applied to the other end $(x = 0)$.

47. Find the electric current in a conductor $0 \leqslant x \leqslant l$ of negligibly small inductance and leakage conductance, if its end $x = l$ is earthed, the initial current and initial potential equal zero, and if a constant e.m.f. E_0 is applied to the end $x = 0$ through a lumped resistance R_0.

48. A conducting layer $0 \leqslant x \leqslant l$ was free from electromagnetic fields for $t < 0$. At time $t = 0$ a constant homogeneous magnetic field H_0, parallel to the layer, is developed everywhere outside the layer. Find the magnetic field in the layer for $t > 0$. Find the time to reach a steady-state in the middle of the layer with relative accuracy ε.

2. Inhomogeneous Media. Equations with Variable Coefficients and Matching Conditions

49. A rod $0 \leqslant x \leqslant l$ thermally insulated along the surface and of constant cross-section is composed of two homogeneous rods $0 \leqslant x \leqslant x_0$, $x_0 \leqslant x \leqslant l$ with different physical properties. Find the temperature in the rod, if its ends are maintained at zero temperature and if the initial temperature is arbitrary.

50. Find the temperature of a homogeneous rod thermally insulated along the surface, at the point x_0 of which $(0 < x_0 < l)$ there is a concentrated thermal capacity C_0. The initial temperature of the rod is arbitrary, and the ends are maintained at a temperature equal to zero.

51. Find the temperature of a rod $0 \leqslant x \leqslant l$ thermally insulated along the surface, having the shape of a truncated cone (see problem 17) if the temperature of the ends of the rod is maintained equal to zero, and the initial temperature of the rod is arbitrary.

52. Solve the preceding problem for a rod, the surface of which is obtained by rotation of the curve $y = Ae^{-mx}$ about the x-axis.

53. A heavy vertical plane stands in a layer of viscous liquid, enclosed between two fixed vertical planes. At time $t = 0$ the plane starts to fall. Find its velocity and the velocity of particles of the viscous liquid, if the initial velocities equal zero and if the falling plane is equidistant from the boundary planes. Neglect the action of gravity on the liquid.

§ 3. Method of Integral Representations and Source Functions

In the present section the application of integral representations to the solution of boundary-value problems for the equation $u_t = a^2u_{xx}+bu+f(x, t)$ (where b and f may be identically equal to zero) is considered in the case of an infinite straight line, semi-infinite line and finite segment. At first problems on the application of the Fourier integral transform are given. Then problems are given on the formation of source functions (Green's functions) and their application to the solution of boundary-value problems.

1. Homogeneous Isotropic Media. Application of the Fourier Integral Transform to Problems on the Infinite Line and the Semi-infinite Line

Applying the Fourier integral transform†, solve the following boundary-value problems.

54. $u_t = a^2u_{xx}, \quad -\infty < x < +\infty, \quad 0 < t < +\infty,$

$$u(x, 0) = f(x), \quad -\infty < x < +\infty.$$

55. $u_t = a^2u_{xx}+f(x, t), \quad -\infty < x < +\infty,$

$$0 < t < +\infty,$$

$$u(x, 0) = 0, \quad -\infty < x < +\infty.$$

† See answers and hints, chapter II, § 4 page 296.

56. $u_t = a^2 u_{xx}, \quad 0 < x, t < +\infty,$

$$u(0, t) = 0, \quad 0 < t < +\infty,$$
$$u(x, 0) = f(x), \quad 0 < x < +\infty.$$

57. $u_t = a^2 u_{xx}, \quad 0 < x, t < +\infty,$

$$u_x(0, t) = 0, \quad 0 < t < +\infty,$$
$$u(x, 0) = f(x), \quad 0 < x < +\infty.$$

58. $u_t = a^2 u_{xx}, \quad 0 < x, t < +\infty,$

$$u(0, t) = \phi(t), \quad 0 < t < +\infty,$$
$$u(x, 0) = 0, \quad 0 < x < +\infty.$$

59. $u_t = a^2 u_{xx}, \quad 0 < x, t < +\infty,$

$$u_x(0, t) = \phi(t), \quad 0 < t < +\infty,$$
$$u(x, 0) = 0, \quad 0 < x < +\infty.$$

60. $u_t = a^2 u_{xx} + f(x, t), \quad 0 < x, t < +\infty,$

$$u(0, t) = 0, \quad 0 < t < +\infty,$$
$$u(x, 0) = 0, \quad 0 < x < +\infty.$$

61. $u_t = a^2 u_{xx} + f(x, t), \quad 0 < x, t < +\infty,$

$$u_x(0, t) = 0, \quad 0 < t < +\infty,$$
$$u(x, 0) = 0, \quad 0 < x < +\infty.$$

62. Using the equation from problem 186, chapter II, prove that

$$\int_0^{+\infty} \frac{e^{-\alpha\lambda^2} \cos \lambda x}{\lambda^2 + h^2} d\lambda = \frac{\sqrt{\pi}}{4h\sqrt{\alpha}} \int_0^{+\infty} e^{-h\xi} \left(e^{-\frac{(x-\xi)^2}{4\alpha}} + e^{-\frac{(x+\xi)^2}{4\alpha}} \right) d\xi.$$

63. Using the equation from problem 187, chapter II, prove that

$$\int_0^{+\infty} \frac{e^{-\alpha\lambda^2} \lambda \sin \lambda x}{\lambda^2 + h^2} d\lambda = \frac{\sqrt{\pi}}{4\sqrt{\alpha}} \int_0^{+\infty} e^{-h\xi} \left(e^{-\frac{(x-\xi)^2}{4\alpha}} - e^{-\frac{(x+\xi)^2}{4\alpha}} \right) d\xi.$$

64. Applying the Fourier transform with the kernel

$$K(x, \lambda) = \sqrt{\frac{2}{\pi}} \frac{\lambda \cos \lambda x + h \sin \lambda x}{\lambda^2 + h^2},$$

solve the boundary-value problem

$$u_t = a^2 u_{xx}, \quad 0 < x, t < +\infty,$$
$$u_x(0, t) - hu(0, t) = -h\phi(t),$$
$$u(x, 0) = 0, \quad 0 < x < +\infty.$$

65. Applying the Fourier transform with the same kernel as in the preceding problem, solve the boundary-value problem

$$u_t = a^2 u_{xx}, \quad 0 < x, t < +\infty,$$
$$u_x(0, t) - hu(0, t) = 0, \quad 0 < t < +\infty,$$
$$u(x, 0) = f(x), \quad 0 < x < +\infty.$$

2. Homogeneous Isotropic Media. Calculation of Green's Functions

In the present section problems on the calculation and application of Green's functions are gathered together. At first, problems are given for an infinite line, then for a semi-infinite line, the medium being assumed isotropic and homogeneous. Then problems for an infinite line, composed of two homogeneous semi-infinite lines, are considered, with certain other problems with inhomogeneous media for an infinite line and semi-infinite line. Finally problems are given for a finite segment, two different descriptions of the Green's functions being considered: one is obtained by the method of separation of variables (Fourier method), the other by the method of images, and a comparison of them is made.

(a) *Infinite straight line*

66. The surface of an infinite rod $-\infty < x < +\infty$ is thermally insulated and the initial temperature equals zero. At the initial time, Q units of heat are liberated instantaneously at a point $x = \xi$ of the rod. Find the temperature of the rod. (Calculation of the Green's function for the equation $u_t = a^2 u_{xx}$ on the straight line $-\infty < x < +\infty$.)

67. Solve the preceding problem for a rod, on whose surface a convective heat exchange takes place with a medium whose temperature equals zero. (Formation of the Green's function for the equation $u_t = a^2 u_{xx} - hu$ on the straight line $-\infty < x < +\infty$.)

68. Using the source function, obtained in the solution of problem 66, solve the boundary-value problem

$$u_t = a^2 u_{xx} + f(x, t), \quad -\infty < x < +\infty, \quad 0 < t < +\infty,$$
$$u(x, 0) = \phi(x), \quad -\infty < x < +\infty.$$

69. Using the source function, obtained in the solution of problem 67, solve the boundary-value problem

$$u_t = a^2 u_{xx} - hu + f(x, t) \quad -\infty < x < +\infty, \quad 0 < t < +\infty,$$
$$u(x, 0) = \phi(x), \quad -\infty < x < +\infty.$$

70. In problem 66 find the time at which the temperature at point x reaches a maximum, and find this maximum value of the temperature (problem on the propagation of a heat impulse).

71. At the surface of a rod $-\infty < x < +\infty$ a convective heat exchange takes place with a medium whose temperature equals zero; the initial temperature of the rod equals zero; at the point $x = 0$ a heat source of magnitude Q acts continuously. Find the temperature $u(x, t)$ of the rod. Find also the steady temperature

$$\bar{u}(x) = \lim_{t \to +\infty} u(x, t).$$

What would be the steady temperature if the surface of the rod were thermally insulated?

72. By means of the relation obtained in solving problem 68, solve the problem

$$u_t = a^2 u_{xx}, \quad -\infty < x < +\infty, \quad 0 < t < +\infty,$$
$$u(x, 0) = \begin{cases} 0 & \text{for} \quad -\infty < x < -l, \\ U_0 = \text{const.} \neq 0 & \text{for} \quad -l < x < l, \\ 0 & \text{for} \quad l < x < +\infty. \end{cases}$$

73. By means of the relation obtained in solving problem 68, solve the problem

$$u_t = a^2 u_{xx}, \quad -\infty < x < +\infty, \quad 0 < t < +\infty,$$
$$u(x, 0) = \begin{cases} 0 & \text{for} \quad -\infty < x < 0, \\ A e^{-\alpha x} & \text{for } 0 < x < +\infty, \ A = \text{const.}, \ \alpha = \text{const.} > 0. \end{cases}$$

74. Solve the boundary-value problem

$$u_t = a^2 u_{xx} - hu, \quad -\infty < x < +\infty, \quad 0 < t < +\infty,$$

$$u(x, 0) = \begin{cases} 0 & \text{for} \quad -\infty < x < -l, \\ U_0 = \text{const.} & \text{for} \quad -l < x < l, \\ 0 & \text{for} \quad -l < x < +\infty. \end{cases}$$

(Compare with problem 72.)

75. Solve the boundary-value problem 14 on the heating of a rod by a moving heater, for a zero initial condition.

(b) *Semi-infinite straight line*

76. Calculate the source function for the equation $u_t = a^2 u_{xx}$ on the semi-infinite line $0 < x < +\infty$, at the end of which a boundary condition of the first kind is given. Proceed then to the case of the equation $u_t = a^2 u_{xx} - hu$.

77. Solve the preceding problem, if a boundary condition of the second kind is given at the end of the semi-infinite line $0 < x < +\infty$.

78. Solve problem 76, if a boundary condition of the third kind is given at the end of the semi-infinite line $0 < x < +\infty$.

79. Using the source function, solve the boundary-value problem

$$u_t = a^2 u_{xx} + f(x, t), \quad 0 < x, t < +\infty,$$
$$u(0, t) = \phi(t), \quad 0 < t < +\infty,$$
$$u(x, 0) = \psi(x), \quad 0 < x < +\infty.$$

80. Using the source function, solve the boundary-value problem

$$u_t = a^2 u_{xx} + f(x, t), \quad 0 < x, t < +\infty,$$
$$u_x(0, t) = \phi(t), \quad 0 < t < +\infty,$$
$$u(x, 0) = \psi(x), \quad 0 < x < +\infty.$$

81. Using the source function, solve the boundary-value problem

$$u_t = a^2 u_{xx} + f(x, t), \quad 0 < x, t < +\infty,$$
$$u_x(0, t) - hu(0, t) = -h\phi(t), \quad 0 < t < +\infty,$$
$$u(x, 0) = \psi(x), \quad 0 < x < +\infty.$$

82. Prove the validity of the following statement. In order that the solution of the boundary-value problem

$$u_t = a^2 u_{xx}, \quad 0 < x, t < +\infty,$$

$$\sum_{k=0}^{N} A_k \frac{\partial^k u}{\partial x^k} = 0, \quad x = 0, \quad 0 < t < +\infty,$$

$$u(x, 0) = f(x), \quad 0 < x < +\infty,$$

may be represented in the form

$$u(x, t) = \frac{1}{2a\sqrt{\pi t}} \int_{-\infty}^{+\infty} f(\xi) e^{-\frac{(x-\xi)^2}{4a^2 t}} d\xi,$$

it is sufficient to extend the function $f(x)$ for negative x, so that the function

$$\phi(x) = \sum_{k=0}^{N} A_k f^{(k)}(x)$$

is odd.

83. Prove the validity of the following statement. In order that the solution of the boundary-value problem

$$ut = a^2 u_{xx} + f(x, t), \quad 0 < x, t < +\infty,$$

$$\sum_{k=0}^{N} A_k \frac{\partial^k u}{\partial x^k} = 0, \quad 0 < t < +\infty, \quad x = 0, \quad u(x, 0) = 0$$

may be represented in the form

$$u(x, t) = \frac{1}{2a\sqrt{\pi}} \int_{-\infty}^{+\infty} d\xi \int_{0}^{t} \frac{f(\xi, \tau)}{\sqrt{t-\tau}} e^{-\frac{(x-\xi)^2}{4a^2(t-\tau)}} d\tau,$$

it is sufficient to extend $f(x, t)$ for negative x, so that the function

$$F(x, t) = \sum_{k=0}^{N} A_k \frac{\partial^k f(x, t)}{\partial x^k}$$

is odd with respect to x.

84. Solve the boundary-value problem

$$u_t = a^2 u_{xx}, \quad 0 < x, t < +\infty,$$

$$u(0, t) = 0, \quad 0 < t < +\infty,$$

$$u(x, 0) = U_0, \quad 0 < x < +\infty.$$

Draw the graph of the distribution of temperature for the times $t = 1/8a^2$, $t = 1/4a^2$, $t = 1/2a^2$ along the segment $0 \leqslant x \leqslant 4$, and also the changes of temperature at the points $x = 1/4$, $x = 1/2$, $x = 1$ along the time segment $0 \leqslant t \leqslant 1/a^2$.

Find also the velocity of the temperature front αU_0, where $0 < \alpha < 1$, $\alpha = \text{const.}$

85. Solve the boundary-value problem

$$\begin{aligned} u_t &= a^2 u_{xx}, && 0 < x, t < +\infty, \\ u(0, t) &= U_0, && 0 < t < +\infty, \\ u(x, 0) &= 0, && 0 < x < +\infty. \end{aligned}$$

At what time t does the temperature at a point reach the value αU_0, $0 < \alpha < 1$?

86. Solve the boundary-value problem

$$\begin{aligned} u_t &= a^2 u_{xx}, && 0 < x, t < +\infty, \\ u_x(0, t) &= 0, && 0 < t < +\infty, \\ u(x, 0) &= \psi(x) = \begin{cases} U_0, & 0 < x < 1, \\ 0, & 1 < x < +\infty. \end{cases} \end{aligned}$$

87. Solve the boundary-value problem

$$u_t = a^2 u_{xx}, \quad 0 < x, t < +\infty, \tag{1}$$

$$u_x(0, t) - hu(0, t) = 0, \quad 0 < t < +\infty, \tag{2}$$

$$u(x, 0) = U_0 = \text{const.}, \quad 0 < x < +\infty. \tag{3}$$

Derive an asymptotic expression for the temperature of the end of the rod for large values of time

$$u(0, t) \approx \frac{U_0}{\sqrt{\pi}} \left\{ \frac{1}{z} - \frac{1}{2z^3} + \frac{1.3}{2^2 z^5} - \cdots + (-1)^{n-1} \frac{1.3 \ldots (2n-3)}{2^{n-1} z^{2n-1}} \right\}_{z = ah\sqrt{t}}. \tag{4}$$

Give an expression for estimating the error in using formula (4) and find at what time the estimate $u(0, t)$ according to the relation

$$u(0, t) \approx \frac{U_0}{ah\sqrt{\pi t}} \tag{5}$$

gives an error, not exceeding the absolute value of a given $\varepsilon > 0$.

88. Solve the boundary-value problem

$$u_t = a^2 u_{xx} - b^2 e^{-kx}, \quad k > 0, \quad 0 < x, t < +\infty,$$
$$u(0, t) = U_0 = \text{const.}, \quad 0 < t < +\infty,$$
$$u(x, 0) = 0, \quad 0 < x < +\infty.$$

89. Solve the boundary-value problem

$$u_t = a^2 u_{xx}, \quad 0 < x, t < +\infty,$$
$$-u_x(0, t) = q, \quad 0 < t < +\infty,$$
$$u(x, 0) = 0, \quad 0 < x < +\infty.$$

90. Solve the boundary-value problem

$$u_t = a^2 u_{xx} - h(u - U_2), \quad U_2 = \text{const.}, \quad 0 < x, t < +\infty,$$
$$u(0, t) = U_1, \quad 0 < t < +\infty, \quad U_1 = \text{const.},$$
$$u(x, 0) = U_0, \quad 0 < x < +\infty, \quad U_0 = \text{const.}$$

91. The initial current and initial voltage in a semi-infinite homogeneous conductor $0 \leqslant x < +\infty$ equals zero. The inductance per unit length is negligibly small. Starting at time $t = 0$, a constant e.m.f. E_0 is applied to an end of the conductor. Find the voltage in the conductor.

92. Solve the boundary-value problem

$$u_t = a^2 u_{xx}, \quad 0 < x, t < +\infty,$$
$$u_x(0, t) - hu(0, t) = -Ah \cos \omega t, \quad 0 < t < +\infty,$$
$$u(x, 0) = 0, \quad 0 < x < +\infty.$$

93. Find the steady temperature waves in a semi-infinite rod $0 < x < +\infty$ thermally insulated along the surface, if the temperature of the end of the rod varies according to the law

$$u(0, t) = A \cos \omega t.$$

Find the velocity of propagation of a temperature wave with a given frequency ω (dispersion of temperature waves).

94. The initial current and initial voltage in a homogeneous conductor $0 \leqslant x < +\infty$ equal zero. Beginning at time $t = 0$, an e.m.f. $E(t) = E_0 \cos \omega t$ is applied at the point $x = 0$. Find the voltage in the conductor, if the inductance and leakage conductance per unit length are negligibly small.

95. The initial temperature of a semi-infinite rod thermally insulated along the surface is given

$$u(x, 0) = f(x), \quad 0 < x < +\infty.$$

What heat flow must be fed into the rod through its end in order that the temperature of the end should vary according to the given law

$$u(0, t) = \mu(t), \quad 0 < t < +\infty, \quad \mu(0) = f(0)\,?$$

Consider the particular case where $f(x) \equiv 0$.

96. The initial temperature of a semi-infinite rod thermally insulated along the surface is given

$$u(x, 0) = f(x), \quad 0 < x < +\infty,$$

and at the end $x = 0$ a convective heat exchange takes place with the external medium. How should the temperature of the external medium vary in order that the temperature of the end of the rod varies according to the given law

$$u(0, t) = \mu(t), \quad \mu(0) = f(0), \quad 0 < t < +\infty\,?$$

Consider the special case where $f(x) \equiv 0$.

97. Solve problem 95 on condition that a convective heat exchange takes place at the surface of the rod with a medium of zero temperature.

98. Solve problem 96 on condition that a convective heat exchange takes place at the surface of the rod with a medium whose temperature equals zero.

99. Solve the boundary-value problem

$$u_t = a^2 u_{xx} + f(x, t), \quad 0 < t < +\infty, \quad v_0 t < x < +\infty,$$
$$u(x, 0) = 0, \quad 0 < x < +\infty,$$
$$u(v_0 t, t) = 0, \quad 0 < t < +\infty.$$

100. Solve the boundary-value problem

$$u_t = a^2 u_{xx}, \quad 0 < t < +\infty, \quad v_0 t < x < +\infty,$$
$$u(x, 0) = f(x), \quad 0 < x < +\infty,$$
$$u(v_0 t, t) = 0, \quad 0 < t < +\infty.$$

101. Solve the boundary-value problem

$$u_t = a^2 u_{xx}, \quad 0 < t < +\infty, \quad v_0 t < x < +\infty,$$
$$u(x, 0) = 0, \quad 0 < x < +\infty,$$
$$u(v_0 t, t) = \mu(t), \quad 0 < t < +\infty.$$

102. Solve the boundary-value problem

$$u_t = a^2 u_{xx} + f(x, t), \quad 0 < t < +\infty, \quad v_0 t < x < +\infty,$$
$$u(x, 0) = f(x), \quad 0 < x < +\infty,$$
$$u_x(v_0 t, t) = \mu(t), \quad 0 < t < +\infty.$$

(b) *Finite segment*

All problems 103–105 on finding the source function may be solved in two ways: by the method of images and by the method of separation of variables; one of these gives a better description of the source function for small values of time t, and the other for large values of time.

103. Find the source function for a finite rod thermally insulated along the surface, if its ends are maintained at zero temperature. Estimate the remainder of the series representing the solution.

104. Find the source function for a finite rod thermally insulated along the surface, if its ends are also thermally insulated. Estimate the remainder of the series representing the solution.

105. Find the source function for a finite rod thermally insulated along the surface, if one of its ends ($x = 0$) is thermally insulated, and the other end ($x = l$) is maintained at zero temperature. Estimate the remainder of the series representing the solution.

106. (a) Find N, for which the remainder of series (2) of the solution of problem 103 satisfies the inequality

$$|R_N(x, \xi, t)| < \varepsilon \tag{1}$$

for $0 \leqslant x, \ \xi \leqslant l, \ 0 \leqslant t \leqslant t^*$.

(b) Find N, for which the remainder of series (12) of the solution of problem 103 satisfies the inequality (1) for $0 \leqslant x$, $\xi \leqslant l$, $0 \leqslant t \leqslant t^*$.

107. Solve the preceding problem for series (1) and (6) of the answer to problem 104.

108. Solve problems 103, 104, 105 in the case where a convective heat exchange takes place at the surface of the rod with a medium whose temperature equals zero.

109. Using the source function, found in the solution of problem 103, solve the boundary-value problem

$$u_t = a^2 u_{xx} + f(x, t), \quad 0 < x < l, \quad 0 < t < +\infty,$$
$$u(0, t) = \phi(t), \quad u(l, t) = 0, \quad 0 < t < +\infty,$$
$$u(x, 0) = f(x), \quad 0 < x < l.$$

110. Using the source function, found in the solution of problem 104, solve the boundary-value problem

$$u_t = a^2 u_{xx} + f(x, t), \quad 0 < x < l, \quad 0 < t < +\infty, \qquad (1)$$
$$u_x(0, t) = \phi(t), \quad u(l, t) = 0, \quad 0 < t < +\infty, \qquad (2)$$
$$u(x, 0) = f(x), \quad 0 < x < l. \qquad (3)$$

111. The temperature of one end of a rod ($x = 0$) is maintained constant and different from zero, $u(0, t) = U_0 \neq 0$, and the temperature of the other end ($x = l$) is at all times equal to zero, $u(l, t) = 0$. Find the temperature of the rod, if its surface is thermally insulated, and the initial temperature equals zero; give an expression for the temperature of the rod in terms of the error integral.

112. One end of a rod ($x = l$) is thermally insulated, and a constant heat flow $[-\lambda u_x(0, t) = -\lambda q_0]$ is maintained at the other end ($x = 0$). Find the temperature of the rod, if its initial temperature equals zero, and the surface is thermally insulated; give an expression for the temperature of the rod in terms of the error integral.

3. Inhomogeneous Media; Equations with Piecewise Continuous Coefficients and Matching Conditions

113. An infinite rod $-\infty < x < +\infty$ thermally insulated along the surface and of constant cross-section is obtained by joining two homogeneous semi-infinite rods $-\infty < x < 0$ and $0 < x < +\infty$ at the point $x = 0$; the ends of the rods are tightly fixed to one another. The initial temperature, coefficient of thermal conductivity and heat conduction of the left-hand and right-hand rods are respectively equal to $U_1 =$ const., a_1, k_1, $U_2 =$ const., a_2, k_2. Find the temperature of the composite rod.

114. Solve the preceding problem if the initial temperature equals

$$u(x, 0) = \begin{cases} f_1(x), & -\infty < x < 0, \\ f_2(x), & 0 < x < +\infty. \end{cases}$$

115. An infinite rod consists of two semi-infinite rods, as specified in problem 113. Find the temperature of the rod for $t > 0$, if at time $t = 0$, $Q = c_2\rho_2$ units of heat were instantaneously liberated at a point $\zeta > 0$, and the initial temperature of the rod was zero.

116. A ball of heat capacity C_0 and very large heat conductivity is placed at the end of a semi-infinite rod thermally insulated along the surface, so that at each moment of time the ball may be assumed to be at a uniform temperature equal to the temperature of the end of the rod. Let the surface of the ball be thermally insulated.

Find the temperature of the rod if its initial temperature equals

$$u(x,0) = f(x), \quad 0 < x < +\infty,$$

where $f(+0)$ and $f'(+0)$ exist.

117. Let the semispace $x > 0$ be filled with a liquid with coefficients of thermal conductivity and heat conduction k_2, a_2 and with an initial temperature $U_2 =$ const., and let the plane $x = 0$ be maintained at a constant temperature $U_1 < U_2$, U_1 being lower than the temperature of solidification of the liquid.

Find the law of propagation of the freezing front of the liquid, and also the temperature of the liquid and of the solid substance into which the liquid changes on freezing.

CHAPTER IV

EQUATIONS OF ELLIPTIC TYPE

A STUDY of steady-state processes, i.e. processes independent of time, often leads to equations of elliptic type. Steady-state electric and magnetic fields (electrostatic, magnetostatic, and constant electric current fields), the potential flow of an incompressible liquid, steady-state temperature fields and others are treated here.

The simplest equation of elliptic type is Laplace's equation $\Delta u = 0$, to which the present chapter is devoted. Later, in chapter VII, problems on other equations of elliptic type are considered.

§ 1. Physical Problems Leading to Equations of Elliptic Type, and the Statement of Boundary-value Problems

1. Boundary-value Problems for Laplace's and Poisson's Equation in a Homogeneous Medium

In contrast to equations of hyperbolic and parabolic type boundary-value problems for the elliptic equation are characterized by the absence of initial conditions. There are three main types of boundary-value problem for Laplace's equation: the first boundary-value problem (Dirichlet's problem), if $u|_\Sigma = f_1$, the second boundary-value problem (Neumann's problem), if $\left.\dfrac{\partial u}{\partial n}\right|_\Sigma = f_2$, the third boundary-value problem, if $\left(\dfrac{\partial u}{\partial n} + hu\right)_\Sigma = f_3$, where f_1, f_2, f_3 are functions, given at the boundary Σ of the region in which the solution of Laplace's equation is sought.

1. *Steady-state temperature field.* Derive the equation, satisfied by the temperature of a steady-state temperature field in a homogeneous medium; in deriving the equation consider the presence of distributed heat sources, which are invariant with time. Give a physical interpretation of boundary-value problems of first, second and third kinds. Establish the necessary condition for the existence of a steady-state temperature for the second boundary-value problem.

2. *The equation of steady-state diffusion.* Derive the equation of a steady-state diffusion process: (a) in a quiescent homogeneous isotropic medium, (b) in a homogeneous isotropic medium, moving with a given velocity, for example, along the x-axis.

3. *The equation of electrostatics.* Starting from Maxwell's equations, show that the potential of an electrostatic field satisfies Poisson's equation with the right-hand side proportional to the volume density of the charges $\rho(x, y, z)$. Give a physical interpretation of boundary conditions of the first and second kind.

4. *The equation of magnetostatics.* Show that the potential of a steady-state magnetic field satisfies Laplace's equation.

5. *The field of a constant electric current.* Prove that the potential of an electric field of a constant electric current satisfies Laplace's equation. Formulate the boundary conditions (1) at an earthed ideally conducting surface, (2) at the boundary of a dielectric.

6. *Potential flow of an incompressible liquid.* Show that the velocity potential of a steady-state flow of incompressible liquid satisfies Laplace's equation. Write down the boundary condition at the surface of a solid, at rest or moving with some given velocity.

7. *Main problems of electrostatics.* The electrostatic field, produced by a charged conductor carrying finite charge, may be determined (1) by giving the value of the potential of the conductor, (2) by giving the value of the charge on the conductor. These problems are called the first and second fundamental problems of electrostatics. Give a mathematical formulation of these problems.

2. Boundary-value Problems for Laplace's Equation in Inhomogeneous Media

In an inhomogeneous, but isotropic medium the fundamental equation of the steady-state field has the form

$$\operatorname{div}(k \operatorname{grad} u) = 0$$

or

$$\frac{\partial}{\partial x}\left(k\frac{\partial u}{\partial x}\right)+\frac{\partial}{\partial y}\left(k\frac{\partial u}{\partial y}\right)+\frac{\partial}{\partial z}\left(k\frac{\partial u}{\partial z}\right)=0,$$

where the characteristic of the medium $k = k(x, y, z)$ is a variable. If the coefficient k has a discontinuity on some surface, then at this surface the matching conditions

$$u_1 = u_2, \tag{1}$$

$$k_1\left(\frac{\partial u}{\partial n}\right)_1 = k_2\left(\frac{\partial u}{\partial n}\right)_2, \tag{2}$$

are satisfied, where the suffixes 1 and 2 denote respectively the left and right-hand limiting values at the surface of discontinuity.

8. Solve problem 1 assuming that the coefficient of heat conduction is a variable $k = k(x, y, z)$. State the boundary-value problem of heat conduction for the case of a piecewise-homogeneous medium (for the case of a piecewise-constant k), first eliminating the matching conditions (1) and (2). Give a physical interpretation of these conditions.

9. Write down the equation for the potential of an electric field in an inhomogeneous dielectric of dielectric constant

$$\varepsilon = \varepsilon(x, y, z).$$

Assuming $\varepsilon(x, y, z)$ to be piecewise-continuous, derive the matching conditions at the surfaces of discontinuity of the function $\varepsilon(x, y, z)$ and formulate the corresponding boundary-value problem.

10. Solve the problem, similar to problems 8 and 9, for a steady-state magnetic field.

11. Solve the problem, similar to problems 8 and 9, for the electric field of a constant current.

12. *Similarity of different steady-state fields.* Establish the similarity between the field of a constant electric current on the one hand, and temperature, electrostatic, magnetostatic fields, the concentration field of a steady-state diffusion process and the velocity field of the potential flow of an incompressible liquid on the other hand.

Compare the matching conditions at a boundary of discontinuity.

§ 2. Simplest Problems for Laplace's and Poisson's Equations

In this section boundary-value problems are given for Laplace's and Poisson's equations, the solutions of which can be determined directly, by simple matching of boundary conditions, without the application of general methods.

1. Boundary-value Problems for Laplace's Equation

13. Consider a circle of radius a with centre at the origin of coordinates. Let (ρ, ϕ) be the polar, and (x, y) the cartesian coordinates. Find the solution of the first interior boundary-value problem for Laplace's equation, if the following boundary conditions are given:

(a) $u|_{\rho=a} = A$;

(b) $u|_{\rho=a} = A\cos\phi$;

(c) $u|_{\rho=a} = A+By$;

(d) $u|_{\rho=a} = Axy$;

(e) $u|_{\rho=a} = A+B\sin\phi$;

(f) $u|_{\rho=a} = A\sin^2\phi+B\cos^2\phi$,

where A and B are constants.

14. Solve the second interior boundary-value problem

$$\Delta u = 0, \qquad \left.\frac{\partial u}{\partial n}\right|_C = f(\phi)$$

for circle C of radius a with centre at the point $\rho = 0$ for the following special cases:

(a) $f = A$;

(b) $f = Ax$;

(c) $f = A(x^2-y^2)$;

(d) $f = A\cos\phi+B$;

(e) $f = A\sin\phi+B\sin^3\phi$.

15. Find the functions $u(\rho, \phi)$, harmonic outside a circle of radius $\rho = a$ and satisfying the boundary conditions (a)–(e) of problem 13 (the first exterior boundary-value problem for a circle).

16. Find the functions $u = u(\rho, \phi)$, harmonic outside a circle $\rho = a$ and satisfying the boundary conditions of problem 14 (the second exterior boundary-value problem for a circle).

17. Find the function $u = u(\rho, \phi)$, harmonic inside the ring $a < \rho < b$ and satisfying the boundary conditions

$$u\big|_{\rho=a} = u_1, \qquad u\big|_{\rho=b} = u_2.$$

Using the solution of the problem, find the capacity of a cylindrical condenser per unit length.

18. Find the function, harmonic inside a circular sector $0 < < \rho < a$, $0 < \phi < \alpha$, if

$$u\big|_{\rho=a} = \frac{u_0}{\alpha}\phi, \qquad u\big|_{\phi=0} = 0, \qquad u\big|_{\phi=\alpha} = u_0.$$

19. Find the solution of Laplace's equation in the semi-plane $y > 0$, taking boundary-values at $y = 0$; $u = \phi_1$ for $x < 0$; $u = \phi_2$ for $x > 0$, and compare this with the solution of problem 18.

20. Determine the function u, harmonic

(a) inside a sphere of radius $r = a$,

(b) outside a sphere $r = a$

and taking the value u_0 on the sphere.

21. Determine the steady-state distribution of temperature inside a spherical layer $a < r < b$, if the sphere $r = a$ is maintained at a temperature u_1, the sphere $r = b$ at a temperature u_2.

22. Using the solution of problem 21, find the capacity of a spherical condenser, filled with a dielectric of dielectric constant $\varepsilon =$ const. and bounded by the spheres $r = a$ and $r = b$.

23. Find the capacity of a spherical condenser, filled with an inhomogeneous dielectric of dielectric constant

$$\varepsilon = \begin{cases} \varepsilon_1 & \text{for} \quad a < r < c, \\ \varepsilon_2 & \text{for} \quad c < r < b. \end{cases}$$

24. Solve the problem, analogous to the preceding problem, for a cylindrical condenser.

25. Find the potential of the electrostatic field of a sphere of radius a, charged to a potential u_0 and situated in an infinite medium, with the following distribution of dielectric constant:

$$\varepsilon = \begin{cases} \varepsilon_1 & \text{for} \quad a < r < c, \\ \varepsilon_2 & \text{for} \quad r > c. \end{cases}$$

Consider the special cases:

(a) $c = \infty$, (b) $\varepsilon_2 = \infty$, (c) $\varepsilon_1 = \varepsilon_2 = \varepsilon$.

26. Find the electrostatic field of an infinite conducting cylinder of radius $\rho = a$, charged to a potential u_0 and surrounded by a dielectric covering, bounded by a cylindrical surface of radius $\rho = b$, which is maintained at zero potential.

27. Find the function u, harmonic inside a layer, bounded by the planes $z = 0$ and $z = h$, if

$$u|_{z=0} = u_1, \qquad u|_{z=h} = u_2.$$

28. Find the capacity of a plane condenser, per unit area of the plates, if between the plates there is a dielectric of dielectric constant ε. Consider the two cases:

(a) $\varepsilon = \text{const. for } 0 < z < h$,

(b) $\varepsilon = \begin{cases} \varepsilon_1 & \text{for } 0 < z < h_1, \\ \varepsilon_2 & \text{for } h_1 < z < h. \end{cases}$

29. Determine the function $u = u(x, y)$, harmonic inside the rectangle $0 < x < a$, $0 < y < b$, and satisfying the conditions

$$u(x, 0) = u_1, \qquad u(x, b) = u_2, \qquad \left.\frac{\partial u}{\partial x}\right|_{\substack{x=0 \\ x=a}} = 0.$$

2. Boundary-value Problems for Poisson's Equation

30. Find the solution of Poisson's equation $\Delta u = 1$ inside a circle of radius $\rho = a$, if $u|_{\rho=a} = 0$.

31. Solve the equation $\Delta_2 u = A$ inside a circle of radius $\rho = a$ for the boundary condition $\left.\dfrac{\partial u}{\partial n}\right|_{\rho=a} = B$, choosing the constant B so that the problem has a solution.

32. It is required to solve the equation $\Delta u = A$ inside a ring $a < \rho < b$, for the following boundary conditions:

(a) $u|_{\rho=a} = u_1, \qquad u|_{\rho=b} = u_2;$

(b) $u|_{\rho=a} = u_1, \qquad \left.\frac{\partial u}{\partial n}\right|_{\rho=b} = C;$

(c) $\left.\frac{\partial u}{\partial n}\right|_{\rho=a} = B, \qquad \left.\frac{\partial u}{\partial n}\right|_{\rho=b} = C.$

Determine the constants, for which the problems have solutions.

33. Find the solutions:

(a) of the equation $\Delta u = 1$, (b) of the equation $\Delta u = Ar+B$ inside a sphere $r < a$, if the boundary condition

$$u|_{r=a} = 0$$

is satisfied on the sphere.

34. Inside the spherical layer $a < r < b$ find solutions of the equations

(a) $\Delta u = 1$, (b) $\Delta u = A+B/r$

for the boundary conditions

$$u|_{r=a} = 0, \qquad u|_{r=b} = 0.$$

§ 3. The Source Function

The source function (Green's function) provides a very powerful means of solving boundary-value problems for Laplace's and Poisson's equations.

The present section includes problems on calculation of the source function for a number of regions, by the method of images. The source function in infinite space is equal to $(e/4\pi)(1/r)$, where $e/4\pi$ is the magnitude of the source (the charge).

Different physical interpretations of the source function are possible (electrostatic, thermal, etc.). In a formulation of problems we usually use the electrostatic interpretation of the source function, assuming the boundaries of the regions ideally conducting and earthed.

Problems on the calculation of the source function by the method of separation of variables are given in § 4.

1. The Source Function for Regions with Plane Boundaries

35. Find the potential of the field of a point electric charge, situated above an ideally conducting earthed plane $z = 0$, and calculate the density of the induced surface charges. Write down the solution of the first boundary-value problem for Laplace's equation in the semispace $z \geqslant 0$.

36. Find the potential of a point charge inside a layer, bounded by two ideally conducting planes $z = 0$ and $z = l$, which are maintained at a potential equal to zero. Investigate the convergence of the series formed by the method of images, and demonstrate the possibility of a two-fold termwise differentiation of this series.

37. Consider the problem on a point source of current in a conducting layer $0 < z < l$, insulated along the planes $z = 0$ and $z = l$.

Find the components of the electric field and verify that a direct application of the method of images for finding the potential gives a divergent series.

38. Consider problem 37 assuming that one wall is insulated, and on the other the potential of the field equals zero. Investigate the convergence of the series for the potential.

39. Find the source function for the equation $\Delta u = 0$ in the semispace $z > 0$ for a boundary condition of the third kind.

$$\frac{\partial u}{\partial z} + hu = 0 \quad \text{for} \quad z = 0.$$

40. Inside the bihedral angle of value $\alpha = \pi/n$ (n an integer), bounded by ideally conducting walls, a point electric charge is situated.

Find the electric field produced by this charge.

41. Using the source function solve the first boundary-value problem inside a bihedral angle of value $\alpha = \pi/n$, where n is a natural number, if the boundary conditions

$$u|_{\phi=\alpha} = 0, \quad u|_{\phi=0} = V$$

are given along its sides.

42. Find the potential of the electrostatic field produced by a point charge inside an infinite cylindrical cavity, assuming that the boundary of the region is ideally conducting and has zero potential, and a perpendicular section of the cavity has a rectangular shape with sides a and b.

2. The Source Function for Regions with Spherical (Circular) and Plane Boundaries

43. Find the potential of an electrostatic field, produced by a point charge e inside an earthed sphere.

44. Using the solution of problem 43, find the density of the surface charges induced on the sphere and write down the solution of the first interior boundary-value problem for Laplace's equation inside a sphere; hence derive Poisson's formula, giving the solution of the first boundary-value problem for Laplace's equation (see [7], chapter IV, page 358).

45. Find the potential of an electrostatic field, produced by a point charge located outside an earthed sphere.

46. Using the solution of problem 45, calculate the density of the surface charges on the sphere and write down the solution of the first exterior boundary-value problem for a sphere.

47. (a) Inside an infinite cylindrical cavity of circular section an electrostatic field is produced by a charged filament, parallel to the cylinder axis. Find the potential of this field.

(b) Solve the same problem if the charged filament is located outside the cylinder.

(c) Use the solutions of problems (a) and (b) to form the solution of Dirichlet's problem inside and outside a circle.

48. Find the source function for $\Delta u = 0$ inside an earthed hemisphere, and also a quarter sphere.

49. Find the potential of a field, produced by a point charge e inside a spherical layer, bounded by two concentric conducting earthed spheres of radii a and b. Investigate the convergence of the series formed, and also of the series obtained by a two-fold termwise differentiation of the original series.

Consider the limiting cases $a \to 0$ and $b \to \infty$ and compare with the solutions of problems 43 and 44.

50. Find the field of a point charge e in infinite space in the presence of a conducting sphere over which a charge e_1 is distributed. Calculate the density of the surface charges induced on the sphere.

3. The Source Function in Inhomogeneous Media

51. Find the field of a point charge in infinite space, filled with an inhomogeneous dielectric of dielectric constant

$$\varepsilon = \begin{cases} \varepsilon_1 & \text{for} \quad z > 0, \\ \varepsilon_2 & \text{for} \quad z < 0. \end{cases}$$

Calculate the surface density, and also the value of the charge induced on the boundary of separation $z = 0$.

52. The semispace $z > 0$ is filled with an inhomogeneous conducting medium, the conductivity of which equals

$$\sigma = \begin{cases} \sigma_1 & \text{for} \quad z > h, \\ \sigma_2 & \text{for} \quad 0 < z < h. \end{cases}$$

A point source of current is placed at the point $M(0, 0, \zeta)$.

Determine the electric field on the surface of the conductor (at $z = 0$). Consider the case $\zeta = 0$ (the source on the surface).

53. An earthed conducting sheet, lying in the plane y, z has a hemisphere of radius a with centre at the origin of coordinates pressed into it; the entire semispace $y < 0$, lying below the plane x, z, is filled with a dielectric of dielectric constant ε_2; the medium, filling the semispace $y > 0$ above the plane $y = 0$ has a dielectric constant ε_1. Find the potential of a point charge, situated above the plane $y = 0$ at the point $M_0(x_0, y_0, z_0)$, where

$$r_0 = \sqrt{x_0^2 + y_0^2 + z_0^2} > a.$$

54. The semispace $z > 0$ is filled with an inhomogeneous conducting medium, the conductivity of which equals

$$\sigma = \begin{cases} \sigma_1 & \text{for} \quad y < 0, \\ \sigma_2 & \text{for} \quad y > 0. \end{cases}$$

A point source of current of magnitude I_0 is situated at the point $M_0(0, -h, \zeta)$. Find the potential of the electric field, and also the current density for $y = 0$, $\zeta = 0$.

55. Two earthed conducting spheres of radii a and b are placed at the points $(0, \theta_0, \phi_0)$ and (c, θ_0, ϕ_0) in infinite space. A charge e is placed at a point $\rho = \rho_0$ on the line joining the centres of the spheres. Find the potential of the field outside the spheres.

§ 4. The Method of Separation of Variables

In the present section we give boundary-value problems for Laplace's equation, which are solvable by the method of separation of variables.

1. Boundary-value Problems for a Circle, Ring and Sector

56. Write down the solution of the first boundary-value problem for Laplace's equation inside a circle.

57. Write down the solution of the first boundary-value problem for Laplace's equation outside a circle.

58. Write down the solution for the second boundary-value problem for the equation $\Delta u = 0$: (a) inside, and (b) outside a circle.

59. (a) Write down the solution of the third interior boundary-value problem for Laplace's equation in a circle, if the boundary condition is written in the form

$$\frac{\partial u}{\partial \rho} - hu = -f \quad \text{at} \quad \rho = a.$$

(b) Find also the solution of the third exterior boundary-value problem for a circle.

60. An infinite conducting cylinder (cylindrical conductor) is charged to a potential

$$V = \begin{cases} V_1 & \text{for} \quad 0 < \phi < \pi, \\ V_2 & \text{for} \quad \pi < \phi < 2\pi, \end{cases}$$

where V_1 and V_2 are constants.

Find the field inside and outside the cylinder and also the density of the surface charges and the total charge.

61. Find the solution (a) of the interior, and (b) of the exterior boundary-value problem for Laplace's equation, if u takes the following values at the boundary of the circle,

(1) $u|_{\rho=a} = A\sin\phi$,

(2) $u|_{\rho=a} = A\sin^3\phi + B$,

(3) $u|_{\rho=a} \begin{cases} A\sin\phi & \text{for } 0 < \phi < \pi, \\ \frac{1}{3}A\sin^3\phi & \text{for } \pi < \phi < 2\pi. \end{cases}$

62. Find the distribution of temperature in an infinitely long circular cylinder, if there is a heat flow $Q = q\cos\phi$ over its surface per unit length.

63. Solve problem 60 assuming that the cylinder is filled with an inhomogeneous dielectric of dielectric constant

$$\varepsilon = \begin{cases} \varepsilon_1 & \text{for} \quad \rho < a, \\ \varepsilon_2 & \text{for} \quad a < \rho < b, \end{cases}$$

where b is the radius of the cylinder.

64. An infinitely long circular cylinder of radius a moves with constant velocity v_0 perpendicular to its axis in an infinite incompressible fluid, which is at rest at infinity. Find the velocity potential of the fluid.

65. Solve the problem of the flow around a stationary infinite cylinder, if the velocity of the liquid is v_0 at infinity.

66. (a) A solid sphere moves with constant velocity v_0 in an infinite viscous liquid, at rest at infinity.

Find the velocity potential.

(b) Solve the problem of the flow around a stationary solid sphere of liquid, having a velocity v_0 at infinity.

67. A dielectric sphere of dielectric constant ε_1, situated in an infinite homogeneous dielectric of dielectric constant ε_2 ($\varepsilon_2 \neq \varepsilon_1$), is placed in a homogeneous parallel external field of strength E_0. Find the value of the polarization of the sphere and its dipole moment.

68. Solve problem 67 for an infinite dielectric cylinder of circular section ($\rho \leqslant a$), assuming that the external field E_0 is directed perpendicularly to the axis of the cylinder.

69. A conducting sphere is placed in an external electrostatic field E_0. Find the amount of the distortion of the external field.

70. An infinite conducting cylinder is placed in a homogeneous external electrostatic field E_0, directed along the x-axis; the axis of the cylinder is parallel to the z-axis. Find the density of the surface charge on the cylinder.

71. Solve the interior Dirichlet problem for the ring $a \leqslant \rho \leqslant b$.

72. Find the temperature distribution in a solid, bounded by infinite cylindrical surfaces of radii a and b ($a < b$), if a constant temperature u_0 is maintained at the surface of the cylinder $\rho = a$, and a temperature u_0 is maintained at the surface $\rho = b$ for $0 < \phi < \pi$, and a temperature equal to zero for $\pi < \phi < 2\pi$.

73. A constant current of strength I flows through an infinite coaxial cylindrical cable $a < \rho < b$. Find the temperature distribution inside the conductor, if the surface $\rho = a$ is maintained at a temperature equal to zero, and on the external boundary there is a heat flow equal to $A \cos^2 \phi$, where ϕ is the polar angle.

74. At the boundary of a thin lamina having the shape of a circular sector $\rho \leqslant a, 0 \leqslant \phi \leqslant \alpha$ a temperature

$$u = \begin{cases} f(\phi) & \text{for} \quad \rho = a, \\ 0 & \text{for} \quad \phi = 0 \quad \text{and} \quad \phi = \alpha \end{cases}$$

is given. Find the steady-state temperature field in the lamina. Consider the special case

$$f(\phi) = \begin{cases} u_1 & \text{for} \quad 0 < \phi < \dfrac{\alpha}{2}, \\ u_2 & \text{for} \quad \dfrac{\alpha}{2} < \phi < \alpha. \end{cases}$$

75. Find the steady-state distribution of temperature in a thin lamina, having the shape of a circular sector, whose radii are maintained at a temperature u_1, and the arc at a temperature u_2.

76. Find the electrostatic field inside an infinite cylinder, a perpendicular section of which has the form of a semicircle; the surface of the cylinder, corresponding to the diameter of the semicircle, is charged to a potential V_1, and the remaining surface to a potential V_2.

77. Solve Laplace's equation inside a circular sector, bounded by the arcs of the circles $\rho = a$, $\rho = b$, and by the radii $\phi = 0$, $\phi = \alpha$, if the following conditions are given at the boundaries:

$$u = 0 \quad \text{for} \quad \phi = 0, \quad \phi = \alpha,$$

$$u = \begin{cases} f(\phi) & \text{for} \quad \rho = a, \\ F(\phi) & \text{for} \quad \rho = b. \end{cases}$$

Consider the limiting cases $a \to 0$, $b \to \infty$, $\alpha = \pi$.

78. Determine the magnetic field of currents, one of which flows in a long straight conductor in one direction, while the other flows in the opposite direction in a parallel conductor, at a distance a from the first.

79. Let a current flow in an infinite circular cylindrical film parallel to the z-axis, with current density i. Find the vector-potential of the magnetic field produced by this current.

80. A cylinder or conductor of circular section with magnetic permeability μ_1 is situated in a medium of magnetic permeability μ_2. A current I flows through the conductor. An external magnetic field is directed perpendicular to the axis of the conductor and is everywhere parallel and homogeneous. Determine the total magnetic field at points inside and outside the cylinder, assuming the cylinder is infinitely long.

81. Calculate the value of the magnetic induction on the outside of a cylindrical screen of inner and outer radii a and b, having a magnetic permeability μ_2 and surrounding two parallel straight conductors, situated symmetrically with respect to the axis of the cylinder and carrying oppositely directed currents (magnetic screening of a transmission line); the cylinder should be assumed infinitely long; the coordinates of the conductors are $\rho = c_0$, $\theta_0 = 0$, and $\theta_0 = \pi$.

82. A hollow sphere $a < r < b$ is situated in a homogeneous parallel magnetic field. Let μ be the magnetic permeability of the sphere. The magnetic permeability of the external medium is assumed equal to unity.

Find the distorted magnetic field over all space. Compare the field inside the sphere with the external field for the case $\mu > 1$ and for the case $\mu < 1$.

2. Boundary-value Problems for Strips, Rectangles, Plane Layers and Parallelepipeds

83. Find the solution of the general first boundary-value problem for Laplace's equation inside a rectangle.

84. Solve the compound boundary-value problem for Laplace's equation inside a rectangle, if

(a) boundary conditions of the first kind are given along two adjacent sides and conditions of the second kind are given along the two other sides;

(b) conditions of the first kind are given along two opposite sides, and conditions of the second kind are given along the two remaining sides.

85. Find the electrostatic field inside a region, bounded by conducting plates $y = 0$, $y = b$ and $x = 0$, if the plate $x = 0$ is charged to a potential V, the plates $y = 0$, $y = b$ are earthed, and if there are no charges inside the region.

86. Solve problem 85 assuming that the boundary $y = b$ is maintained at a potential V_0. Consider the limiting case $b \to \infty$.

87. Solve the equation $\Delta u = 0$ inside the rectangle $0 \leqslant x \leqslant a$, $0 \leqslant y \leqslant b$ for the following boundary conditions: $u = V$ at $x = 0$, $u = 0$ at $x = a$ and $y = 0$, $u = V_0$ at $y = b$.

Going to the limit as $a \to \infty$, derive the solution of problem 86.

88. The semi-infinite layer of problem 85 is filled with an inhomogeneous dielectric of dielectric constant

$$\varepsilon = \begin{cases} \varepsilon_1 & \text{for} \quad 0 < y < h, \\ \varepsilon_2 & \text{for} \quad h < y < b. \end{cases}$$

Find the electrostatic field in the dielectric.

89. Find the electrostatic field inside an infinite cylindrical tube of rectangular cross-section with sides a and b, filled with an inhomogeneous dielectric with dielectric constant

$$\varepsilon = \begin{cases} \varepsilon_1 & \text{for} \quad 0 < y < h, \\ \varepsilon_2 & \text{for} \quad h < y < b, \end{cases}$$

if the wall $x = 0$ is charged to a potential V, and the remaining walls are earthed. Consider the case where the wall $x = a$ is removed to infinity.

90. Solve problem 89 for the condition that the wall $y = b$ is charged, and the remaining walls are earthed.

91. An amount of heat Q flows in through the side $y = 0$ of an infinite cylinder of rectangular cross-section $0 \leqslant x \leqslant a$, $0 \leqslant y \leqslant$ b and an amount of heat Q flows out through the side $x = 0$.

Find the temperature distribution inside the cylinder, assuming that the heat flow is uniformly distributed over the surface of the side $y = 0$ and over the side $x = 0$ respectively, and the remaining two sides of the body are thermally insulated.

92. Find the temperature distribution inside a rectangular thin lamina, if a constant heat flux q_0 is fed into one of its sides, and the remaining three sides are maintained at a constant temperature u_1.

93. Find the solution of the general first boundary-value problem for Laplace's equation inside a rectangular parallelepiped.

94. Find the electrostatic field inside a rectangular parallelepiped with conducting walls, if its lateral faces and upper end are earthed, and the lower end is charged to a potential V. By means of a limiting transition derive the solutions of problems 85 and 86.

95. Solve problem 94, if the lateral faces are charged to a potential V, and both ends are earthed.

3. Problems Requiring the Application of Cylindrical Functions†

96. Solve the first boundary-value problem for Laplace's equation inside a finite cylinder $\rho \leqslant a$, $0 \leqslant z \leqslant l$, if $u|_{\rho=a} = 0$, $u|_{z=0} = f(\rho, \phi)$, $u|_{z=l} = F(\rho, \phi)$.

97. Solve problem 96 if

$$u|_{z=0} = f(\rho), \quad u|_{z=l} = F(\rho),$$

where f and F are functions depending only on ρ.

98. Find the function $u(\rho, \phi, z)$, harmonic inside the finite cylinder, reducing to zero at its ends and taking the given values on the surface $\rho = a$:

$$u|_{\rho=a} = f(z).$$

Consider the special cases

(a) $f(z) = f_0 = \text{const.}$;

(b) $f(z) = Az(1-z/l)$.

99. Find the solution of the general first boundary-value problem for Laplace's equation inside a finite cylinder.

100. Find an expression for the potential of the electrostatic field inside a cylindrical tank of circular section $\rho \leqslant a$, $0 \leqslant z \leqslant l$, both ends of which are earthed, and the surface is charged to a potential V_0.

Determine the field strength along the axis.

Consider the limiting case $l \to \infty$.

101. Solve problem 100 if the surface and upper end of the tank are earthed, and the lower end is maintained at a constant potential V_0.

By means of a limiting transition derive the solution of the problem for a semi-infinite cylinder.

102. Solve problems 100, 101 for a semi-infinite cylinder, and compare with the results of the corresponding limiting cases of the solutions of problems 100 and 101.

103. Determine the steady-state distribution of temperature inside a solid, having the shape of a finite cylinder, if a constant

---

† In sections 3 and 4 problems are given solvable by the method of separation of variables, but requiring the application of cylindrical and spherical functions. A number of these problems were solved in § 2 by the method of images.

heat flow q is supplied to the lower end $z = 0$, the surface $\rho = a$ and the upper end $z = l$ are maintained at a temperature equal to zero.

104. Solve the preceding problem, assuming that a heat exchange takes place at the surface with a medium whose temperature equals zero.

105. Solve problems 103 and 104 for a semi-infinite cylinder ($l = \infty$) and compare the results obtained with the limit of the solutions of problems 103 and 104 for $l \to \infty$.

106. Find the strength of the electrostatic field inside the toroid $a < \rho < b$, $0 < z < l$, if the outside surface $\rho = b$ is charged to a potential V_0, and the remaining boundary is earthed.

Consider the limiting cases

(1) $l \to \infty$,

(2) $a \to 0$ (compare with the solution of problem 100).

107. The ends of the toroid ($a < \rho < b$, $0 < z < l$) are maintained at the constant temperature u_0, and the surface at a temperature u_1. Find the steady-state distribution of temperature inside the toroid.

108. Find the steady-state distribution of temperature inside a toroid of rectangular cross-section ($a < \rho < b$, $0 < z < l$), if

(1) the surface is thermally insulated, and the ends are maintained at a constant temperature u_0;

(2) the surface is thermally insulated, the temperature of the lower end $z = 0$ equals zero, and the upper end is maintained at a temperature u_1.

109. Solve problem 107, if a constant temperature u_0 is given at the lower end, and the remaining surface of the toroid is maintained at zero temperature.

110. By means of the method of separation of variables derive an expression for the potential of a point charge, situated on the axis of a finite cylinder $\rho \leqslant a$, $0 < z < h$ with conducting walls.

Show that expressions may be obtained from the solution by means of limiting transitions for the potential of a point charge in a layer $0 \leqslant z \leqslant h$, in semispace and in infinite space.

111. Solve the preceding problem for a semi-infinite cylinder $z > 0$; compare the result obtained with the corresponding limit of the solution of problem 110.

112. Solve problem 110 for an infinite cylinder by the method of separation of variables; compare with the limit of the solution of problem 110.

4. Problems Requiring the Application of Spherical and Cylindrical Functions.

113. Solve the first boundary-value problem for Laplace's equation inside a sphere of radius a.

114. Solve the first boundary-value problem for Laplace's equation outside a sphere of radius a.

115. Find the solution of the second boundary-value problem for Laplace's equation

(a) inside a sphere,

(b) outside a sphere.

Consider the case of the simplest boundary condition:

$$\left.\frac{\partial u}{\partial n}\right|_{\Sigma} = A \cos \theta .$$

116. Find the intensity of the electrostatic field inside and outside a sphere, the upper half of which is charged to a potential V_1, and the lower half to a potential V_2.

117. Find the expansion in spherical functions of the surface charges, induced in an earthed ideally conducting charged sphere by a point charge, placed

(a) inside the sphere,

(b) outside the sphere.

118. Solve the preceding problem for an insulated charged sphere, placed in the field of a point charge.

119. (a). A solid sphere moves with constant velocity in an infinite viscous liquid, at rest at infinity. Find the velocity potential.

(b). Solve the problem on the flow around a stationary solid sphere of a liquid, having a velocity v_0 at infinity.

120. A dielectric sphere of dielectric constant ε_1 is placed in an external homogeneous field E_0, parallel to the axis z.

Determine the distortion of the external field produced by the sphere, if the medium surrounding it is a homogeneous dielectric with $\varepsilon = \varepsilon_2$.

121. Solve the problem on the polarization of a dielectric sphere of radius a in the field of a point charge, if the dielectric constant

$$\varepsilon = \begin{cases} \varepsilon_1 & \text{for} \quad r < a, \\ \varepsilon_2 & \text{for} \quad r > a. \end{cases}$$

Consider the two cases:

(a) the charge is placed outside the sphere,

(b) the charge is situated inside the sphere.

122. A conducting sphere of conductivity σ_1 is placed in a medium of conductivity σ_2.

Determine the currents produced by a point source of current of strength I, situated

(a) inside the sphere,

(b) outside the sphere.

123. Solve the preceding problem assuming the sphere ideally conducting. Compare with problem 122.

124. A point source of heat Q is placed outside a non-conducting sphere. Find the steady-state distribution of temperature outside the sphere.

125. A point source of magnitude Q is situated inside a sphere, at the surface of which a heat exchange takes place with a medium of zero temperature. Find the steady-state distribution of temperature inside the sphere.

126. Find the potential of a point charge, situated between earthed conducting concentric spheres $r = a$ and $r = b$. Determine also the density of the surface charges.

127. An inhomogeneous dielectric sphere of radius b with dielectric constant

$$\varepsilon = \begin{cases} \varepsilon_1 & \text{for} \quad r < a, \\ \varepsilon_2 & \text{for} \quad a < r < b \end{cases}$$

is placed in a medium of dielectric constant ε_3.

Determine the field of a point charge, situated
(1) outside the sphere $r > b$,
(2) inside the sphere $r < a$,
(3) in the region $a < r < b$.
Consider the limiting cases.

128. Find the field inside a dielectric sheath, bounded by concentric spheres of radii a and b $(b > a)$, and situated in a homogeneous parallel electrostatic field of strength E_0; the dielectric constant of the sheath is ε_1, the dielectric constant of the medium ε_2.

129. Calculate approximately the distribution of charge on the inner face of a spherical condenser, assuming that the distance between the centres of the inner and outer layer is small but not zero.

130. Find the potential of a charged thin ring, the total charge of which equals e.

131. The spherical coordinates of a circular ring equal $r_0 = a$, $\theta_0 = \alpha$. A sphere of radius b of dielectric with dielectric constant ε_1 is situated so that its centre is located at the origin of coordinates. Find an expression for the potential between the ring and the sphere, if the linear charge density of the ring equals $\varkappa$. The dielectric constant of the medium equals ε_2.

132. Calculate the potential of the electrostatic field of a charged thin ring, situated inside a conducting sphere, if a potential, equal to zero, is maintained on the sphere. Calculate the normal component of the electric field on the sphere.

133. Calculate the potential at all points of a conducting sphere of conductivity σ in the case where the current I enters at one of its poles $\theta = 0$ and flows out from the pole $\theta = \pi$.

134. Find the potential of the field, produced on one side of an infinite dielectric lamina of thickness l by a point charge e, situated on the opposite side of the lamina.

135. A current I is supplied at the surface of the earth $z = 0$ by means of a point electrode. Determine the potential on the earth's surface, assuming that the specific conductivity of the earth at a depth $z = h$ equals σ_1 and at a greater depth it equals σ_2.

Apply the solution obtained to the case of two electrodes, existing at the points $x = a$ and $x = -a$.

136. A spherical electrode of radius a is half buried in the earth, the conductivity of which σ_Γ in the horizontal direction is greater than in the vertical direction σ_B (anisotropy). Find the distribution of potential on the earth's surface assuming that at the surface of the electrode the potential $V = V_0$.

Hint. One should introduce in place of z a new variable

$$t = \alpha z, \qquad \alpha^2 = \frac{\sigma_\Gamma}{\sigma_B}.$$

Then the equation $\sigma_\Gamma(V_{xx}+V_{yy})+\sigma_B V_{zz} = 0$ transforms to the equation $V_{xx}+V_{yy}+V_{tt} = 0$.

§ 5. Potentials and Their Application

In the present section problems are set on the calculation of volume and surface potentials for some of the simplest cases, and also boundary-value problems are given which may be solved by the method of potential theory.

137. Find the volume potential V of a sphere of constant density $\rho = \rho_0$, by stating the boundary-value problem for V and solving it.

138. Solve problem 137 by direct calculation of the volume integral.

139. Find the volume potential

(a) of masses, distributed with constant density in a spherical layer $a \leqslant r \leqslant b$;

(b) of masses, distributed inside a sphere of radius a with constant density ρ_1 and in a spherical layer $a < b < r < c$ with constant density ρ_2;

(c) of masses, distributed inside a sphere of radius $r = c$ with variable density $\rho = \rho(r)$.

Hence derive the solutions of problems 139 (a) and 139 (b).

140. Find the potential of a single layer, distributed with constant density $\nu = \nu_0$ over a sphere.

141. Find the electrostatic field of volume charges, uniformly distributed inside a sphere, situated above an ideally conducting plane $z = 0$.

142. Find the potential in two dimensions of a circle of constant charge density.

143. Find the potential in two dimensions of a single layer of a segment with constant charge density.

144. Find the potential in two dimensions of a double layer of a segment of constant moment.

145. Determine the potential of a single layer, uniformly distributed over a circular disc.

146. Find the vector-potential of a loop of current.

147. Using the potential of a double layer solve Dirichlet's problem

(a) inside a circle,

(b) outside a circle.

148. Find the solution of Neumann's problem for a circle, utilizing the potential of a single layer.

149. Solve the first and second boundary-value problems for Laplace's equation in semispace, utilizing surface potentials.

150. Find the solution of Dirichlet's problem in a semiplane, making use of the potential of a single layer.

151. Let us consider the surfaces Σ of second order, defined by the equation

$$\frac{x^2}{a^2+s}+\frac{y^2}{b^2+s}+\frac{z^2}{c^2+s}=1,$$

where $a > b > c$. If $-c^2 < s < \infty$, then the surfaces are essentially ellipsoids, for $-b^2 < s < -c^2$ they are single-surface hyperboloids, for $-a^2 < s < -b^2$ they are two-surface hyperboloids. For $s = \infty$ we have a sphere of infinite radius, and for $s = -a^2$ the ellipsoid is compressed into an elliptic disc lying in the plane yz.

Show that the surfaces of the family under consideration may be equipotentials, and their potential is given by the relation

$$V = A\int_s \frac{ds}{R(s)}+B, \qquad R(s)=\sqrt{(a^2+s)(b^2+s)(c^2+s)},$$

where A and B are constants, defined by the conditions at infinity and on the surface Σ.

152. Using the solution of the preceding problem, find an expression for the potential of a charged conducting ellipsoid $x^2/a^2+y^2/b^2+z^2/c^2 = 1$, over which a charge e is distributed. (The dielectric constant of the medium is ε.)

Determine the capacity of the ellipsoid, and also the surface charge density over the ellipsoid. Consider an ellipsoid of revolution.

153. Using the solution of problem 152, calculate the surface charge density for an elliptic disc. Determine the potential, capacity and charge density for a circular disc.

154. Show that the gravitational potential of a homogeneous ellipsoid

$$\frac{x^2}{a^2}+\frac{y^2}{b^2}+\frac{z^2}{c^2}=1$$

is given by the integrals

$$V(x, y, z) = \rho_0\int\limits_0^\infty \frac{1-f(x, y, z; s)}{R(s)}\,ds \quad \text{inside the ellipsoid,}$$

$$V(x, y, z) = \rho_0\int\limits_\lambda^\infty \frac{1-f(x, y, z; s)}{R(s)}\,ds \quad \text{outside the ellipsoid,}$$

where

$$f(x, y, z; s) = \frac{x^2}{a^2+s}+\frac{y^2}{b^2+s}+\frac{z^2}{c^2+s},$$

$$R(s) = \sqrt{(s+a^2)\,(s+b^2)\,(s+c^2)},$$

ρ_0 is the volume density of the ellipsoid, λ is an ellipsoidal coordinate, the positive root $s = \lambda$ of the equation $f(x, y, z; s) = 0$.

155. Calculate the gravitational potential

(a) of a prolate ellipsoid of revolution,

(b) of an oblate ellipsoid of revolution (see problem 152). Consider the limiting case of a homogeneous sphere.

156. Find the logarithmic potential of an elliptical region of constant density by evaluating the integrals.

157. A conducting ellipse, defined by the equation

$$\frac{x^2}{a^2}+\frac{y^2}{b^2}=1 \qquad (a\geqslant b>0),$$

is charged to a potential V_0. Determine the potential outside the ellipse, and also the density of charges, distributed over the ellipse.

158. Calculate the force of interaction of two coaxial wire loops C_a and C_b of radii a and b, through which currents I and I' flow. The contours are situated in parallel planes $z=0$ and $z=d$, their centres lie at the points $x=y=z=0$ and $x=y=0$, $z=d$.

159. Calculate the coefficient of mutual inductance of two coaxial wire circuits 1 and 2, utilizing the relation

$$M_{12}=\oint\limits_{1} \boldsymbol{A}_2\,\mathrm{d}\boldsymbol{s}_1=\mu\oint\limits_{1}\oint\limits_{2}\frac{\mathrm{d}\boldsymbol{s}_1\,\mathrm{d}\boldsymbol{s}_2}{r}=M_{21},$$

where $\boldsymbol{A}_2$ is the vector-potential of the field produced by a unit current, flowing through contour 2; μ is the magnetic permeability of the medium.

160. Show that the expression for the potential, produced by a charged ring of radius a has the form

$$V=\begin{cases}\dfrac{2e}{\pi\varepsilon}\displaystyle\int_0^\infty K_0(\lambda a)\,I_0(\lambda\rho)\cos\lambda z\,\mathrm{d}\lambda & \text{for}\quad \rho<a,\\[2ex] \dfrac{2e}{\pi\varepsilon}\displaystyle\int_0^\infty I_0(\lambda a)\,K_0(\lambda\rho)\cos\lambda z\,\mathrm{d}\lambda & \text{for}\quad \rho>a,\end{cases}$$

where e is the charge of the ring.

161. Show that the potential produced in surrounding space by a disc of radius a, carrying a charge e, equals

$$V(\rho, Z)=\frac{e}{\varepsilon a}\int_0^\infty \mathrm{e}^{-\lambda|z|}\,J_0(\lambda\rho)\,\frac{\sin\lambda a}{\lambda}\,\mathrm{d}\lambda.$$

CHAPTER V

EQUATIONS OF PARABOLIC TYPE

EQUATIONS of parabolic type are obtained in an investigation of such phenomena as heat conduction, diffusion, the propagation of an electromagnetic field in conducting media, the motion of a viscous liquid, the motion of subsurface waters and others.

In the present chapter the statement and solution of boundary-value problems for equations of parabolic type are considered in the case where the physical processes under investigation are characterized by functions of two, three or four independent variables; it is a continuation of the third chapter, in which equations of parabolic type for functions of two independent variables are considered.

§ 1. Physical Problems Leading to Equations of Parabolic Type; Statement of Boundary-value Problems

1. The semi-space $z > 0$ is filled with a liquid with a coefficient of thermal conductivity λ, volume density ρ and specific heat c.

State the boundary-value problem for the heating of the liquid, if it moves with a velocity $v_0 =$ const. in the direction of the x-axis, a heat exchange obeying Newton's law takes place between it and the plane $z = 0$, and the temperature of the boundary plane $z = 0$ equals u_0. Consider, in particular, the case of a steady distribution of temperature when the heat transfer in the direction of the x-axis due to heat conduction may be neglected in comparison with the heat transfer due to motion of the liquid.

2. Formulate the diffusion problem, similar to problem 1, assuming the plane $z = 0$ is impermeable to particles of a diffusing substance; state the corresponding boundary-value problems in the non-steady and steady cases.

3. Derive the diffusion equation for a substance whose molecules

(a) disintegrate (for example, the unstable gas, radon) the rate of disintegration at each point of space being proportional to the concentration;

(b) multiply (for example, the diffusion of neutrons in the presence of fission of nuclei), the rate of multiplication at each point of space being proportional to the concentration.

4. State the boundary-value problem on the propagation of an electromagnetic field in infinite space, filled with a conducting medium of conductivity $\sigma =$ const., magnetic permeability $\mu =$ const., and dielectric constant $\varepsilon =$ const.

5. State the boundary-value problem on the cooling of an infinite flat plate, if a convective heat exchange takes place at its surface with the surrounding medium at zero temperature.

Consider, in particular, the case where the change in temperature with depth in the plate is negligibly small.

6. A circular cylindrical tube is filled with a liquid of very high thermal conductivity†; outside the tube there is air at a temperature $U_0 =$ const. State the boundary-value problem to determine the temperature of the tube assuming that it does not vary with position along the tube.

7. An infinite circular cylinder of radius r_0 with moment of inertia K per unit length is placed in a viscous liquid; at $t > 0$ it is set into motion by the action of a moment M per unit length.

Using the expression in cylindrical coordinates for the equation of motion of the viscous liquid and for the components of the stress tensor‡, state the boundary-value problem for the motion of the viscous liquid and cylinder.

8. A layer of earth lies on a waterproof horizontal base and contains subsurface waters. The vector $\boldsymbol{U}$ of the volume rate of

---

† By conductivity is meant the total conductivity, including the transfer of heat by convective currents of liquid.

‡ See answers and hints.

flow of the water is related to the vector $\boldsymbol{V}$ of the velocity of motion of the water by the relation

$$\boldsymbol{U} = m\boldsymbol{V},$$

where the coefficient m is called the porosity of the earth.

The force of resistance acting on the water is

$$\boldsymbol{f} = -\frac{1}{k}\boldsymbol{U},$$

where k is the so-called permeability†.

We say that the excess pressure relative to the specific gravity of the water is the difference between the real and hydrostatic pressure in the subsurface waters.

State the boundary-value problem for the motion of the free surface of the water for the following assumptions:

(1) the horizontal component of the gradient of the excess pressure is negligibly small,

(2) the forces of inertia, acting on the water, are negligibly small.

§ 2. The Method of Separation of Variables

1. Boundary-value Problems not Requiring the Application of Special Functions

In this section boundary-value problems for regions with plane or spherical boundaries are considered. The solutions may be expressed as a series in the simplest (elementary) eigenfunctions of the Laplacian operator for these regions.

(a) *Homogeneous media*

9. Find the temperature of the parallelepiped $0 \leqslant x \leqslant l_1$, $0 \leqslant y \leqslant l_2$, $0 \leqslant z \leqslant l_3$, if its initial temperature is an arbitrary function of x, y, z and the temperature of the surface is maintained equal to zero.

† In connection with the terminology see [23].

10. Solve the preceding problem for a cube of side l, if at the initial time it was uniformly heated. Find the time at which a steady-state will occur at the centre of the cube with relative accuracy $\varepsilon > 0$†.

11. Find the temperature of the parallelepiped $0 \leqslant x \leqslant l_1$, $0 \leqslant y \leqslant l_2$, $0 \leqslant z \leqslant l_3$, at the surface of which a convective heat exchange takes place with a medium of zero temperature, if its initial temperature equals $f(x, y, z)$; consider, in particular, the case where $f(x, y, z) = U_0 \equiv \text{const.}$

12. At the surface of a cube, uniformly heated at the initial moment of time, a convective heat exchange takes place with a medium whose temperature equals zero. Find an expression for the temperature at the centre of the cube and determine the time at which a steady-state will occur at the centre of the cube with a relative accuracy $\varepsilon > 0$.

13. The walls of a semi-infinite rectangular pipe $0 \leqslant x < +\infty$, $0 \leqslant y \leqslant l_1$, $0 \leqslant z \leqslant l_2$, are maintained at zero temperature. A fluid flows through the pipe with a constant velocity v_0 in the direction of the x-axis. Find the temperature of the fluid, neglecting the transfer of heat in the direction of the x-axis by thermal conduction‡ for the following conditions: (1) the process is steady, (2) a heat exchange obeying Newton's law occurs between the fluid and the walls of the pipe, (3) the temperature of the fluid in the section $x = 0$ equals $U_0 \equiv \text{const.}$

14. In the cube $0 \leqslant x, y, z \leqslant l$, a substance is diffusing whose molecules multiply at a rate proportional to the concentration (see problem 3). Find the critical dimensions of the cube, i.e. find the maximum length of a side l, for which the process of multiplication does not acquire an avalanche nature§. Consider the case where

(a) the concentration is maintained equal to zero on all sides,

(b) all the sides are impermeable,

(c) all the sides are semi-permeable.

---

† See chapter III, § 2, problem 22.

‡ See problem 1.

§ For more detail on the concept of critical dimensions see [7], page 519.

15. Find the temperature of a sphere of radius r_0, the surface of which is maintained at a temperature equal to zero. At the initial time the temperature of the sphere was equal to

$$u|_{t=0}=f(r), \quad 0\leqslant r<r_0.$$

16. The initial temperature of a sphere $0\leqslant r\leqslant r_0$ equals

$$u|_{t=0}=U_0\equiv\text{const.},$$

and a temperature $U_1\equiv$ const. is maintained on the surface of the sphere. Find the temperature of the sphere for $t>0$. Determine the time at which a steady-state is reached at the centre of the sphere with a relative accuracy $\varepsilon>0$.

17. The initial temperature of a sphere $0\leqslant r\leqslant r_0$ equals

$$u|_{t=0}=U_0\equiv\text{const.},$$

and a constant heat flow of magnitude q is supplied through its surface. Find the temperature of the sphere for $t>0$.

18. Find the temperature of a sphere of radius r_0, at the surface of which a convective heat exchange takes place with a medium with zero temperature. The initial temperature of the sphere is

$$u|_{t=0}=f(r), \quad 0\leqslant r<r_0.$$

19. The initial temperature of a sphere $0\leqslant r\leqslant r_0$ is

$$u|_{t=0}=U_0\equiv\text{const.},$$

and a convective heat exchange takes place at its surface with a medium of constant temperature $U_1\equiv$ const. Find the temperature of the sphere for $t>0$.

Determine the time to reach a steady-state at the centre of the sphere with relative accuracy $\varepsilon>0$.

20. The initial temperature of a sphere $0\leqslant r<r_0$ is

$$u|_{t=0}=U_0\equiv\text{const.},$$

and at its surface at time $t=0$ a convective heat exchange takes place with a medium whose temperature equals

$$U_0+at, \quad 0<t<+\infty, \quad U_0=\text{const.}, \quad a=\text{const.}$$

Find the temperature of the sphere for $t>0$.

21. Solve the problem of the cooling of a spherical shell $r_1 \leqslant r \leqslant r_2$, at the inner and outer surfaces of which a convective heat exchange takes place with a medium, having zero temperature. The initial temperature of the shell is

$$u|_{t=0} = f(r), \qquad r_1 < r < r_2.$$

22. In a closed spherical vessel $0 \leqslant r \leqslant R$ a substance diffuses. The molecules of this substance multiply at a rate proportional to the concentration (see problem 14). Find the critical dimensions of the vessel. Consider the case where

(a) a concentration, equal to zero, is maintained on the surface of the vessel,

(b) the vessel wall is impermeable,

(c) the vessel wall is semi-permeable.

(b) *Inhomogeneous media*

23. Find the temperature of a beam of rectangular cross-section $0 \leqslant x \leqslant l_1$, $0 \leqslant y \leqslant l_2$, consisting of two homogeneous beams (of different physical properties) with cross-sections $0 \leqslant x \leqslant x_0$, $0 \leqslant y \leqslant l_2$ and $x_0 \leqslant x \leqslant l_1$, $0 \leqslant y \leqslant l_2$; the ends of the beam are thermally insulated, and the surface is maintained at zero temperature. The initial temperature of the beam is

$$u|_{t=0} = f(x, y), \qquad 0 \leqslant x \leqslant l_1, \qquad 0 \leqslant y \leqslant l_2.$$

24. Find the temperature of a rectangular parallelepiped, consisting of two homogeneous rectangular parallelepipeds $[0 \leqslant x \leqslant x_0,\ 0 \leqslant y \leqslant l_2,\ 0 \leqslant z \leqslant l_3]$ and $[x_0 \leqslant x \leqslant l_1,\ 0 \leqslant y \leqslant l_2,\ 0 \leqslant z \leqslant l_3]$ made of different materials. The surface of the composite parallelepiped is maintained at zero temperature, and its initial temperature is

$$u|_{t=0} = f(x, y, z), \quad 0 \leqslant x \leqslant l_1, \quad 0 \leqslant y \leqslant l_2, \quad 0 \leqslant z \leqslant l_3.$$

25. A sphere $0 \leqslant r \leqslant r_1$ consists of a homogeneous sphere $0 \leqslant r \leqslant r_0$ and a homogeneous spherical shell $r_0 \leqslant r \leqslant r_1$, made of different materials. Find the temperature of the sphere, if its surface is maintained at zero temperature, and the initial temperature of the sphere is

$$u|_{t=0} = f(r), \qquad 0 \leqslant r < r_0.$$

26. A liquid of very high thermal conductivity is contained in a thick spherical shell $r_1 \leqslant r \leqslant r_2$, i.e. its temperature at all times equals the temperature of the inner surface of the shell. Find the temperature of the shell, if its outer surface is kept at zero temperature, and the initial temperature is

$$u|_{t=0} = f(r), \quad r_1 \leqslant r \leqslant r_2.$$

2. Boundary-value Problems Requiring the Application of Special Functions

(a) *Homogeneous media*

27. Solve the problem of the heating of an infinite circular cylinder $0 \leqslant r \leqslant r_0$, with zero initial temperature, if a temperature $U_0 =$ const. is maintained on its surface. Find also an approximate expression for the temperature, averaged over a cross-section of the cylinder, in the steady-state.

28. Find the temperature of an infinite circular cylinder when the initial temperature is

$$u|_{t=0} = U_0\left(1 - \frac{r^2}{r_0^2}\right),$$

and its surface is kept at zero temperature. Find an approximate expression for the temperature, averaged over a cross-section of the cylinder, in the steady-state.

29. Find the temperature of an infinite circular cylinder $0 \leqslant r \leqslant r_0$, if its initial temperature is

$$u|_{t=0} = U_0 \equiv \text{const.},$$

and a constant heat flow of magnitude q is supplied from outside to its surface at time $t = 0$.

30. Find the temperature of an infinite circular cylinder of radius r_0, if the initial temperature is

$$u|_{t=0} = f(r), \quad 0 \leqslant r \leqslant r_0,$$

and a convective heat exchange takes place at the surface of the cylinder with a medium, whose temperature equals zero. Consider, in particular, the case where $f(r) = U_0 \equiv$ const., and write down an approximate expression for the temperature in the steady-state.

31. The initial temperature of an infinite circular cylinder $0 \leqslant r \leqslant r_0$ is

$$u|_{t=0} = U_0 \equiv \text{const.},$$

and a convective heat exchange takes place at the surface of the cylinder with a medium, whose temperature equals $U_1 = \text{const.}$

Find the temperature of the cylinder for $t > 0$.

32. Solve the preceding problem, if the temperature of the medium is $U_1 + at$, where U_1 and a are constants.

33. Outside an infinite circular conducting cylinder $0 \leqslant r \leqslant r_0$ a constant magnetic field H_0, parallel to the cylinder axis, was instantaneously established at time $t = 0$.

Find the intensity of the magnetic field inside the cylinder for zero initial conditions; find next the flux of magnetic induction through a cross-section of the cylinder.

34. Solve the preceding problem, if the intensity of the external magnetic field is

$$H = H_0 \cos \omega t, \qquad H_0 = \text{const}, \qquad 0 < t < +\infty.$$

35. The initial temperature of an infinite circular cylindrical tube $r_1 \leqslant r \leqslant r_2$ is

$$u|_{t=0} = f(r), \qquad r_1 \leqslant r \leqslant r_2.$$

Find the temperature of the tube for $t > 0$, if a temperature $U_1 \equiv \text{const.}$ is maintained on its inner surface, and a temperature $U_2 \equiv \text{const.}$ on the external surface.

36. The temperature of an infinite circular cylindrical tube is zero for $t < 0$. At time $t = 0$ a constant heat flow of magnitude q is conveyed through its external surface, and the inner surface of the tube is maintained at a temperature equal to zero.

Find the temperature of the tube for $t > 0$.

37. Solve the problem on the cooling of an infinite circular cylindrical tube, at the outer and inner surfaces of which a convective heat exchange takes place with a medium of zero temperature. At the initial moment of time the tube was uniformly heated.

38. Between two concentric cylinders of infinite length there is a viscous liquid. At time $t = 0$ the outer cylinder begins to rotate with angular velocity $\omega \equiv$ const.

Determine the velocity of motion of the liquid.

39. Find the temperature of an infinite circular cylinder $0 \leqslant r \leqslant r_0$, if its initial temperature equals

$$u|_{t=0} = f(r, \phi), \quad 0 \leqslant r \leqslant r_0, \quad 0 \leqslant \phi \leqslant 2\pi,$$

and the surface is kept at zero temperature.

40. Find the temperature of an infinite circular cylinder $0 \leqslant r \leqslant r_0$, if its initial temperature equals

$$u|_{t=0} = f(r, \phi), \quad 0 \leqslant r \leqslant r_0, \quad 0 \leqslant \phi \leqslant 2\pi,$$

and a convective heat exchange takes place at the surface with a medium, whose temperature is zero.

41. Find the temperature of an infinite circular cylindrical tube $r_1 \leqslant r \leqslant r_2$, if its initial temperature equals

$$u|_{t=0} = f(r, \phi), \quad r_1 < r < r_2, \quad 0 \leqslant \phi \leqslant 2\pi,$$

and a temperature, equal to zero, is maintained on the outer and inner surfaces.

42. Find the temperature of an infinite circular cylindrical tube $r_1 \leqslant r \leqslant r_2$, if its initial temperature equals

$$u|_{t=0} = f(r, \phi), \quad r_1 < r < r_0, \quad 0 \leqslant \phi \leqslant 2\pi,$$

and a convective heat exchange takes place at the outer and inner surfaces with a medium whose temperature equals zero.

43. Find the temperature of an infinite cylindrical sector $0 \leqslant r \leqslant r_0$, $0 \leqslant \phi \leqslant \phi_0$, if a temperature, equal to zero, is maintained on the surface $r = r_0$ and on the sides $\phi = 0$ and $\phi = \phi_0$, and if the initial temperature equals

$$u|_{t=0} = f(r, \phi), \quad 0 < r < r_0, \quad 0 < \phi < \phi_0.$$

44. Solve the problem of the cooling of a sphere of radius r_0, at the surface of which a temperature, equal to zero, is maintained. The initial temperature of the sphere equals

$$u|_{t=0} = f(r, \theta, \phi), \quad 0 \leqslant r < r_0, \quad 0 \leqslant \theta \leqslant \pi, \quad 0 \leqslant \phi \leqslant 2\pi.$$

45. Solve the problem of the cooling of a sphere of radius r_0, if a convective heat exchange takes place at its surface with a medium, whose temperature equals zero. The initial temperature of the sphere equals

$$u|_{t=0} = f(r, \theta, \phi), \quad 0 \leqslant r < r_0, \quad 0 \leqslant \theta \leqslant \pi, \quad 0 \leqslant \phi \leqslant 2\pi.$$

46. Solve the problem on the cooling of a thick spherical shell $r_1 \leqslant r \leqslant r_0$, on the outer and inner surfaces of which a temperature equal to zero is maintained. The initial temperature of the shell equals

$$u|_{t=0} = f(r, \phi, \theta), \quad r_1 < r < r_2, \quad 0 \leqslant \theta \leqslant \pi, \quad 0 \leqslant \phi \leqslant 2\pi.$$

47. Solve the problem on the cooling of a thick spherical shell $r_1 \leqslant r \leqslant r_2$, on the outer and inner surfaces of which a convective heat exchange takes place with a medium whose temperature equals zero. The initial temperature of the shell equals

$$u|_{t=0} = f(r, \theta, \phi), \quad r_1 < r < r_2, \quad 0 \leqslant \theta \leqslant \pi, \quad 0 \leqslant \phi \leqslant 2\pi.$$

(b) *Inhomogeneous media: central factors*

48. An inhomogeneous circular cylinder $0 \leqslant r \leqslant r_1$, $0 \leqslant \phi \leqslant 2\pi$, $0 \leqslant z \leqslant l$ consists of a homogeneous cylinder $0 \leqslant r \leqslant r_0$, $0 \leqslant \phi \leqslant 2\pi$, $0 \leqslant z \leqslant l$, and a homogeneous cylindrical tube $r_0 \leqslant r \leqslant r_1$, $0 \leqslant \phi \leqslant 2\pi$, $0 \leqslant z \leqslant l$, made of different materials.

Find the temperature of the composite cylinder, if its surface is maintained at a temperature, equal to zero, and the initial temperature equals

$$u|_{t=0} = f(r, \phi, z), \quad 0 \leqslant r < r_1, \quad 0 \leqslant \phi \leqslant 2\pi, \quad 0 < z < l.$$

49. Solve the problem on the cooling of an infinite cylindrical tube $r_1 \leqslant r \leqslant r_2$, filled with a coolant, if the temperature of the coolant is at all times equal to the temperature of the inner surface of the tube, and the external surface is thermally insulated. The initial temperature of the tube equals

$$u|_{t=0} = f(r), \quad r_1 < r < r_2.$$

50. Solve the preceding problem assuming that a convective heat exchange takes place at the external surface with a medium whose temperature equals zero.

51. A cylinder of radius r_1 with moment of inertia K per unit length is immersed in a liquid and is set in rotation by a moment $M = \text{const.}$ per unit length.

Determine the motion of the liquid and the cylinder, if the liquid fills the space between the cylinder and the stationary coaxial tube with inner radius $r_2 > r_1$. The cylinder and tube are assumed infinitely long. At the initial time the cylinder and liquid were at rest.

52. Outside a hollow cylindrical conductor $r_1 \leqslant r \leqslant r_2$ of infinite length a constant magnetic field H_0, parallel to the axis of the conductor, was instantaneously established at time $t = 0$.

Find the magnetic field in the conductor for zero initial conditions, assuming that in the inner cavity it is homogeneous, and also that outside and inside the tube there is a vacuum.

53. An inhomogeneous sphere $0 \leqslant r \leqslant r_1$ consists of a homogeneous sphere $0 \leqslant r \leqslant r_0$ and a homogeneous spherical sheath $r_0 \leqslant r \leqslant r_1$, made of different materials.

Find the temperature of the sphere, if its surface is maintained at a temperature, equal to zero, and the initial temperature equals

$$u|_{t=0} = f(r, \theta, \phi), \quad 0 \leqslant r < r_1, \quad 0 \leqslant \theta \leqslant \pi, \quad 0 \leqslant \phi \leqslant 2\pi.$$

§ 3. The Method of Integral Representations

In the present section the application of integral representations to the solution of boundary-value problems on the theory of heat conduction is considered. Firstly there are problems on the application of the Fourier integral, then on the formation and application of Green's functions.

1. The Application of the Fourier Integral

54. Find the temperature distribution in infinite space, if the initial temperature equals

$$u|_{t=0} = f(x, y, z), \quad -\infty < x, y, z < +\infty.$$

Consider also the special case where $f(x, y, z)$ does not depend on z.

55. Find the temperature of an infinite region, produced by continuously acting sources of density $g(x, y, z, t)$; the initial temperature of the region equals zero. Consider also the special case where $g(x, y, z, t)$ does not depend on t and where $g(x, y, z, t)$ does not depend on z.

56. Solve the boundary-value problem

$$u_t = a^2 \Delta u, \qquad -\infty < x, y < +\infty, \qquad 0 < z < +\infty, \quad 0 < t < +\infty,$$

$$u|_{z=0} = 0, \qquad -\infty < x, y < +\infty, \qquad 0 < t < +\infty,$$

$$u|_{t=0} = f(x, y, z), \qquad -\infty < x, y < +\infty, \qquad 0 < z < +\infty.$$

Consider also the particular case where f is independent of y.

57. Solve the boundary-value problem

$$u_t = a^2 \Delta u, \qquad -\infty < x, y < +\infty, \qquad 0 < z < +\infty, \quad 0 < t < +\infty,$$

$$u|_{z=0} = f(x, y, t), \qquad -\infty < x, y < +\infty, \qquad 0 < z < +\infty, \quad 0 < t < +\infty,$$

$$u|_{t=0} = 0, \qquad -\infty < x, y < +\infty, \qquad 0 < z < +\infty.$$

Consider also the special case where f is independent of y.

58. Solve the boundary-value problem

$$u_t = a^2 \Delta u, \qquad -\infty < x, y < +\infty, \qquad 0 < z, t < +\infty,$$

$$u_z|_{z=0} = 0, \qquad -\infty < x, y < +\infty, \qquad 0 < t < +\infty,$$

$$u|_{t=0} = f(x, y, z), \qquad -\infty < x, y < +\infty, \qquad 0 < z < +\infty.$$

59. Solve the boundary-value problem

$$u_t = a^2 \Delta u, \qquad -\infty < x, y < +\infty, \qquad 0 < z, t < +\infty,$$

$$u_z|_{z=0} = f(x, y, t), \qquad -\infty < x, y < +\infty, \qquad 0 < t < +\infty,$$

$$u|_{t=0} = 0, \qquad -\infty < x, y < +\infty, \qquad 0 < z < +\infty.$$

60. Solve the boundary-value problem

$$u_t = a^2 \Delta u, \qquad -\infty < x, y < +\infty, \qquad 0 < z, t < +\infty,$$

$$u_z - hu = 0, \qquad -\infty < x, y < +\infty, \quad z = 0, \quad 0 < t < +\infty,$$

$$u|_{t=0} = f(x, y, z), \qquad -\infty < x, y < +\infty, \qquad 0 < z < +\infty.$$

Consider also the special case in which f is not dependent on y.

61. Solve the boundary-value problem

$$u_t = a^2\Delta u, \quad -\infty < x, y < +\infty, \quad 0 < z, t < +\infty,$$
$$u_z = h[u - f(x, y, t)], \quad -\infty < x, y < +\infty, \quad z = 0, \quad 0 < t < +\infty,$$
$$u|_{t=0} = 0 \quad -\infty < x, y < +\infty, \quad 0 < z < +\infty.$$

Consider also the special case in which f is not dependent on y.

62. Solve the boundary-value problem

$$u_t = a^2\Delta u + f(x, y, z, t), \quad -\infty < x, y < +\infty, \quad 0 < z, t < +\infty,$$
$$u|_{z=0} = 0, \quad -\infty < x, y < +\infty, \quad 0 < t < +\infty,$$
$$u|_{t=0} = 0, \quad -\infty < x, y < +\infty, \quad 0 < z < +\infty.$$

63. Find the temperature of an infinite beam of rectangular cross-section $0 \leqslant x \leqslant l_1$, $0 \leqslant y \leqslant l_2$, $-\infty < z < +\infty$, if its initial temperature equals

$$u|_{t=0} = f(x, y, z), \quad 0 \leqslant x \leqslant l_1, \quad 0 \leqslant y \leqslant l_2, \quad -\infty \leqslant z \leqslant +\infty,$$

and at the surface

(a) a temperature, equal to zero, is maintained,

(b) heat insulation occurs,

(c) a convective heat exchange takes place with a medium of zero temperature.

64. Solve the preceding problem for a semi-infinite beam of rectangular cross-section: $0 \leqslant x \leqslant l_1$, $0 \leqslant y \leqslant l_2$, $0 < z < +\infty$; consider the case, corresponding to the boundary conditions (a) and (b).

65. Find the temperature of an infinite circular cylinder $0 \leqslant r \leqslant r_0$, $0 \leqslant \phi \leqslant 2\pi$, $-\infty < z < +\infty$, if its initial temperature equals

$$u|_{t=0} = f(r, \phi, z), \quad 0 \leqslant r < r_0, \quad 0 \leqslant \phi \leqslant 2\pi, \quad -\infty < z < +\infty,$$

and at the surface one of the following boundary conditions is fulfilled:

(a) the surface temperature is maintained equal to zero,

(b) the surface is thermally insulated,

(c) at the surface a convective heat exchange takes place with the surrounding medium, the temperature of which equals zero.

66. Find the temperature of a semi-infinite circular cylinder $0 \leqslant r \leqslant r_0$, $0 \leqslant \phi \leqslant 2\pi$, $0 \leqslant z \leqslant +\infty$, if its initial temperature equals

$$u|_{t=0} = f(r, \phi, z), \quad 0 \leqslant \phi \leqslant 2\pi, \quad 0 \leqslant r \leqslant r_0, \quad 0 < z < +\infty,$$

and at the surface, one of the following boundary conditions is fulfilled:

(a) the surface temperature is maintained equal to zero,

(b) the surface is thermally insulated.

67. Find the temperature of a lamina, having the shape of an infinite sector $0 \leqslant r < +\infty$, $0 \leqslant \phi \leqslant \phi_0$, if its initial temperature equals

$$u|_{t=0} = f(r, \phi), \quad 0 < r < +\infty, \quad 0 < \phi < \phi_0,$$

and at the sides of the lamina

(a) a temperature, equal to zero, is maintained,

(b) heat insulation occurs.

68. Solve the preceding problem, assuming that one side of the lamina is thermally insulated, and the temperature of the other side is maintained equal to zero.

69. Find the temperature of an infinite wedge with an angle of inclination ϕ_0, if along its sides

(a) a zero temperature is maintained,

(b) heat insulation occurs.

70. Find the temperature of an infinite region with an infinite circular cylindrical cavity, if the initial temperature equals zero, and the temperature at the surface of the cavity is maintained equal to U_0.

2. The Formation and Application of Green's Functions

71. Prove that a solution of the boundary-value problem

$$\frac{\partial u}{\partial t} = a^2\left\{\frac{\partial^2 u}{\partial x^2} + \frac{\partial^2 u}{\partial y^2} + \frac{\partial^2 u}{\partial z^2}\right\}, \quad -\infty < x, y, z < +\infty,$$
$$0 < t < +\infty, \tag{1}$$

$$u|_{t=0} = f_1(x) f_2(y) f_3(z), \quad -\infty < x, y, z < +\infty \tag{2}$$

is the product of the solutions $u_1(x, t)$, $u_2(x, t)$, $u_3(x, t)$ of the boundary-value problems

$$\frac{\partial u_1}{\partial t} = a^2 \frac{\partial^2 u_1}{\partial x^2}, \qquad -\infty < x < +\infty, \qquad 0 < t < +\infty, \tag{1'}$$

$$u_1|_{t=0} = f_1(x), \qquad -\infty < x < +\infty, \tag{2'}$$

$$\frac{\partial u_2}{\partial t} = a^2 \frac{\partial^2 u_2}{\partial y^2}, \qquad -\infty < y < +\infty, \qquad 0 < t < +\infty, \tag{1''}$$

$$u_2|_{t=0} = f_2(y), \qquad -\infty < y < +\infty, \tag{2''}$$

$$\frac{\partial u_3}{\partial t} = a^2 \frac{\partial^2 u_3}{\partial z^2}, \qquad -\infty < z < +\infty, \qquad 0 < t < +\infty, \tag{1'''}$$

$$u_3|_{t=0} = f_3(z), \qquad -\infty < z < +\infty. \tag{2'''}$$

72. Using the expressions for the Green's functions for the straight lines $-\infty < x < +\infty, -\infty < y < +\infty, -\infty < z < +\infty$ and the assumption, formulated in problem 71, write down an expression for the Green's function for the region

$$-\infty < x, y, z < +\infty.$$

73. Using the source function, found in the preceding problem, solve the boundary-value problem

$$u_t = a^2 \Delta u + F(x, y, z, t), \qquad -\infty < x, y, z < +\infty,$$
$$0 < t < +\infty,$$

$$u|_{t=0} = f(x, y, z), \qquad -\infty < x, y, z < +\infty.$$

74. Express the source function of an instantaneous point source of heat for the semispace $-\infty < x,\ y < +\infty$, $0 < z < < +\infty$, obeying the boundary conditions

(a) $u|_{z=0} = 0$,

(b) $u_z|_{z=0} = 0$,

(c) $(u_z - hu)|_{z=0} = 0$,

in terms of the corresponding one-dimensional source function as in the solution of problem 72.

75. Using the Green's functions found in the preceding problem, solve the boundary-value problems

(a) $u_t = a^2\Delta u + F(x, y, z, t), \quad -\infty < x, y < +\infty, \quad 0 < z, t < +\infty,$

$u|_{z=0} = \Phi(x, y, t), \quad -\infty < x, y < +\infty, \quad 0 < t < +\infty,$

$u|_{t=0} = f(x, y, z), \quad -\infty < x, y < +\infty, \quad 0 < z < +\infty;$

(b) $u_t = a^2\Delta u + F(x, y, z, t), \quad -\infty < x, y < +\infty, \quad 0 < z, t < +\infty,$

$u_z|_{z=0} = \Phi(x, y, t) \quad -\infty < x, y < +\infty, \quad 0 < t < +\infty,$

$u|_{t=0} = f(x, y, z), \quad -\infty < x, y < +\infty, \quad 0 < z < +\infty;$

(c) $u_t = a^2\Delta u + F(x, y, z, t) \quad -\infty < x, y < +\infty, \quad 0 < z, t < +\infty,$

$(u_z - hu)|_{z=0} = h\Phi(x, y, t), \quad -\infty < x, y < +\infty, \quad 0 < t < +\infty,$

$u|_{t=0} = f(x, y, z), \quad -\infty < x, y < +\infty, \quad 0 < z < +\infty.$

76. Let D be a finite, semi-infinite or infinite cylindrical region, parallel to the z-axis, and let its intersection with the plane xy be the region D_{xy}. Let boundary conditions of first, second or third kind be given at the surface of the region D.

Prove that the Green's function of an instantaneous point source of heat for the region D is respectively the product of the Green's functions for a finite segment, semi-axis or complete z-axis and the Green's function of an instantaneous point source of heat for the plane region D_{xy}.

77. Using the result of the preceding problem, write down an expression for the Green's function for the plane layer

$-\infty < x, y < +\infty$, $0 < z < l$. Consider the case where the boundary planes $z = 0$ and $z = l$

(a) are maintained at zero temperature,

(b) are insulated,

(c) one of the boundary planes ($z = 0$) is thermally insulated, and on the other ($z = l$) a zero temperature is maintained,

(d) at both boundary planes a convective heat exchange takes place with a medium of zero temperature.

78. Find the temperature distribution in an infinite region produced by the liberation of Q uniformly distributed units of heat at a spherical surface of radius r' at $t = 0$. (The source function of an instantaneous spherical source of heat.)

79. Using the source function, found in the preceding problem, solve the boundary-value problem

$$\frac{\partial u}{\partial t} = a^2\left(\frac{\partial^2 u}{\partial r^2} + \frac{2}{r}\frac{\partial u}{\partial r}\right) + f(r, t), \qquad 0 < r, t < +\infty,$$

$$u(r, 0) = F(r), \qquad 0 < r < +\infty,$$

where $r = \sqrt{x^2+y^2+z^2}$.

80. Find the temperature distribution in an infinite region produced by the liberation at the initial moment of time of Q uniformly distributed units of heat per unit of length of an infinite cylindrical surface of radius r'. (Formation of the source function of an instantaneous cylindrical source of heat.)

81. Using the source function, found in the preceding problem, solve the boundary-value problem

$$\frac{\partial u}{\partial t} = a^2\left(\frac{\partial^2 u}{\partial r^2} + \frac{1}{r}\frac{\partial u}{\partial r}\right) + f(r, t), \qquad 0 < r, t < +\infty, \tag{1}$$

$$u(r, 0) = F(r), \qquad 0 < r < +\infty, \tag{2}$$

where $r = \sqrt{x^2+y^2}$.

82. Find the source function of an instantaneous point source for the diffusion equation, if the medium, in which diffusion takes place, moves with constant velocity $\boldsymbol{v}$ relative to the system of coordinates being considered.

83. Find the source function of a steady point source of constant magnitude for the diffusion equation in a medium, moving with constant velocity $\boldsymbol{v}$ in the direction of the x-axis, if the diffusion process is steady and if one may neglect the transport of the substance in the direction of the x-axis in comparison with the transport due to the motion of the medium (see problem 2).

84. Solve the preceding problem for the semispace $0 < z < +\infty$, considering the cases where

(a) the plane $z = 0$ is impermeable,

(b) a concentration, equal to zero, is maintained on the plane $z = 0$,

(c) the plane $z = 0$ is semi-permeable, a concentration, equal to zero, being maintained under it (i.e. at $z < 0$).

85. Find the concentration of a diffusing substance in infinite space, liberated by a point source of magnitude $f(t)$ with coordinates $x = \phi(t)$, $y = \psi(t)$, $z = \varkappa(t)$, if the initial concentration of this substance in space equals zero.

86. Find the concentration of diffusing substance in infinite space, if the initial concentration equals

$$u|_{t=0} = \begin{cases} U_0 \equiv \text{const.} & \text{for} \quad 0 \leqslant r < r_0, \\ 0 & \text{for} \quad r_0 < r < +\infty, \end{cases}$$

where r is the radius-vector of the spherical system of coordinates.

87. Solve the preceding problem for the semispace $z > 0$, assuming that $z_0 < r_0$; $(0, 0, z_0)$ are the coordinates of the centre of a sphere, in which the initial concentration equals U_0. Consider the cases where

(a) the plane $z = 0$ is impermeable to the diffusing substance,

(b) at the plane $z = 0$ a concentration, equal to zero, is maintained.

88. Find the concentration of a diffusing substance in infinite space, if its initial concentration equals

$$u|_{t=0} = \begin{cases} U_0 \equiv \text{const.} & \text{for} \quad 0 \leqslant r < r_0, \\ 0 & \text{for} \quad r_0 < r < +\infty, \end{cases}$$

where r is the radius-vector of the cylindrical system of coordinates.

89. Solve the preceding problem for the semispace $x \geqslant 0$, assuming that the cylinder is parallel to the z-axis and its axis intersects the plane $z = 0$ at the point $(x_0, 0)$ where $x_0 > r_0$. Consider the case where

(a) the plane $x = 0$ is impermeable to the diffusing substance,

(b) at the plane $x = 0$ a concentration, equal to zero, is maintained.

90. A canal with vertical walls and with an impermeable base is suddenly filled with water so that in one part of it, for $x < 0$, a water level $H_1 = \text{const.}$ is obtained, and in the other, for $x > 0$, a water level $H_2 = \text{const.}$ is obtained, and further these levels are maintained invariant (see the figure in the answer to the problem, the vertical axis of H is perpendicular to the plane of the diagram).

At the initial moment of time the level of the subsurface waters in the ground layer $y > 0$ equals $H_0 = \text{const.}$

Assuming that the layer lies on the same impermeable base, which forms the bottom of the canal, find the level of the subsurface waters $H(x, y, t)$ for $t > 0$ $(y > 0)$.

91. At the surface of a spherical cavity $0 \leqslant r \leqslant r_0$ in infinite space the temperature should vary according to the law $u|_{r=r_0} = \phi(t)$, where $\phi(t)$ is a given function of time; the initial temperature of the space equals zero.

What heat flow must be fed from the spherical cavity into the space in order to guarantee such a law of temperature variation at the surface of the cavity?

CHAPTER VI

EQUATIONS OF HYPERBOLIC TYPE

DYNAMIC problems on the mechanics of continuous media (acoustics, hydrodynamics, aerodynamics, theory of elasticity) and problems on electrodynamics† lead to equations of hyperbolic type. In the present chapter the statement and solution of boundary-value problems of hyperbolic type are considered for functions of two or more independent variables, so that this chapter is a continuation of chapter II, in which problems for functions of two independent variables were considered. As in chapter II, the vibrations of continuous media are assumed small in the accepted sense of the word.

§ 1. Physical Problems Leading to Equations of Hyperbolic Type; Statement of Boundary-value Problems

In this section the statement of boundary-value problems on the mechanics of continuous media is considered. The statement of the boundary-value problems on electrodynamics is considered in chapter IV‡.

1. State the boundary-value problem on the propagation of small disturbances in a homogeneous ideal gas, filling infinite space, taking as the function describing the process, one of the quantities: the gas density ρ, the gas pressure p, the velocity potential U of the gas particles, the velocity vector of the gas particles $\boldsymbol{v} = \boldsymbol{i}v^{(1)}+\boldsymbol{j}v^{(2)}+\boldsymbol{k}v^{(3)}$, the displacement potential of the gas particles Φ, or the displacement vector of the gas particles $\boldsymbol{u} = \boldsymbol{i}u^{(1)}+$

† The equation of the relativistic theory of attraction with known omissions also belongs to the hyperbolic type.

‡ See also [7], pages 489–502.

$+ju^{(2)}+ku^{(3)}$. Show that each of these quantities may be expressed in terms of any other of the quantities.

2. Derive the boundary conditions for the velocity potential U of the gas particles†, the displacement potential Φ, the density ρ and pressure p at a plane, bounding the semispace, filled with this gas. Consider the case where this plane:

(a) is stationary,

(b) moves with approximately the speed of sound in a direction normal to itself according to a given law.

3. A region is filled with two different ideal gases, with the boundary of separation at the surface Σ‡. Assuming that the unperturbed pressures in both gases are the same, state the boundary-value problem for the propagation of small disturbances in the gas.

4. State the boundary-value problem for the transverse vibrations of a membrane with a rigidly fixed end, if in the undisturbed state the membrane is horizontal, and the surrounding medium does not exert a resistance to the vibrations.

Note. The problem on the vibrations of the membrane is the two-dimensional analogue of the problem on the vibrations of a string§.

5. State the boundary-value problem on the vibrations of a membrane, stretched over the mouth of a closed vessel, taking into account the variation of pressure in the vessel, produced by the vibrations of the membrane, and assuming the velocity of propagation of small disturbances in the gas to be considerably greater than the velocity of propagation of waves in the membrane.

6. Derive the equation of propagation of small disturbances in a gas, moving with constant velocity relative to the chosen system of coordinates.

---

† In connection with the symbols see the answer to problem 1.

‡ Geometric surface. It is assumed that, for the time under consideration, the boundary of separation of the gases Σ may be assumed to be an infinitely thin surface.

§ See chapter II, § 1, and also [7], pages 23–27.

7. State the boundary-value problem on the supersonic steady flow around a rigid wedge by the symmetrical plane-parallel flow of an ideal gas.

8. State the boundary-value problem on the supersonic steady flow around a circular cone by an ideal gas in the direction of the axis of the cone, assuming the undisturbed flow to be homogeneous, and the disturbances, caused by the cone, to be small.

9. The surface of an ideal fluid in a tank with a horizontal base and vertical walls equals $h = \text{const.}$ in the unperturbed state. For small vibrations of the free surface a wave motion may develop, in which the liquid particles, lying on a vertical line, move equally in the horizontal direction. Let $\zeta(x, y, t)$ denote the rise of the disturbed surface above the level of the quiescent liquid. Assuming the pressure p in the disturbed liquid to be equal to the hydrostatic pressure, state the boundary-value problem for the propagation of small disturbances, taking as the function describing the process: (1) $\zeta(x, y, t)$, (2) the potential of the (horizontal) velocity of the liquid. The pressure p_0 at the surface of the liquid remains constant (see problem 7, chapter II, § 1).

10. State the boundary-value problem 9 for the case in which p_0 is a given function of x, y, t, taking the horizontal velocity potential as the function, describing the process.

11. Derive the equation of motion of the centre of mass of an element of an elastic medium, taking the element as a rectangular parallelepiped with sides parallel to the axes of coordinates.

12. Using Hooke's law for a homogeneous isotropic elastic medium, write the equation of motion, found in the preceding problem, in a form containing only the components of volume forces and the displacement vector

$$\boldsymbol{U} = \boldsymbol{i}u(x, y, z, t)+\boldsymbol{j}v(x, y, z, t)+\boldsymbol{k}w(x, y, z, t),$$

and prove that the dilation $\Theta = \operatorname{div} \boldsymbol{U}$ and the shear $\boldsymbol{B} = \operatorname{curl} \boldsymbol{U}$ both satisfy D'Alembert's wave equation. $\partial^2\phi/\partial^2 t = a^2\Delta\phi$, where for Θ, $a^2 = (\lambda+2\mu)/\rho$, and for $\boldsymbol{B}$, $a^2 = \mu/\rho$†.

† "Longitudinal" elastic waves are propagated more rapidly than "transverse".

Note. 1. Any vector $\boldsymbol{U}$ is identically defined by its divergence div $\boldsymbol{U}$ and vortex rot $\boldsymbol{U}$ (see [14], page 209).

2. The shape of an element of an elastic medium, having the undeformed shape described in problem 11, is determined by the quantities

$$\varepsilon_x = \frac{\partial u}{\partial x}, \qquad \varepsilon_y = \frac{\partial u}{\partial y}, \qquad \varepsilon_z = \frac{\partial u}{\partial z},$$

$$\gamma_{xy} = \gamma_{yx} = \frac{\partial v}{\partial y} + \frac{\partial u}{\partial y}, \qquad \gamma_{yz} = \gamma_{zy} = \frac{\partial v}{\partial z} + \frac{\partial w}{\partial y},$$

$$\gamma_{zx} = \gamma_{xz} = \frac{\partial w}{\partial x} + \frac{\partial u}{\partial z},$$

forming the strain tensor

$$(D) = \left\| \begin{matrix} \varepsilon_x & \gamma_{xy} & \gamma_{xz} \\ \gamma_{yx} & \varepsilon_y & \gamma_{yz} \\ \gamma_{zx} & \gamma_{zy} & \varepsilon_z \end{matrix} \right\|.$$

In the case where the medium is homogeneous and isotropic, the components of the stress tensor (see the answer to the preceding problem)

$$(H) = \left\| \begin{matrix} \sigma_x & \tau_{xy} & \tau_{xz} \\ \tau_{yx} & \sigma_y & \tau_{yz} \\ \tau_{zx} & \tau_{zy} & \sigma_z \end{matrix} \right\|$$

are connected by the following relations to the components of the strain tensor

$$\sigma_x = \lambda\Theta + 2\mu\frac{\partial u}{\partial x}, \qquad \sigma_y = \lambda\Theta + 2\mu\frac{\partial v}{\partial y}, \qquad \sigma_z = \lambda\Theta + 2\mu\frac{\partial w}{\partial z},$$

$$\tau_{yz} = \tau_{zy} = \mu\left(\frac{\partial w}{\partial y} + \frac{\partial v}{\partial z}\right), \qquad \tau_{zx} = \tau_{xz} = \mu\left(\frac{\partial u}{\partial z} + \frac{\partial w}{\partial x}\right),$$

$$\tau_{xy} = \tau_{yx} = \mu\left(\frac{\partial v}{\partial x} + \frac{\partial u}{\partial y}\right),$$

where $\Theta = \operatorname{div} \boldsymbol{U}$, and λ and μ are Lamé constants, connected in the following way to Young's modulus E and Poisson's coefficient m:

$$E = \frac{\mu(3\lambda+2\mu)}{\lambda+\mu}, \qquad m = \frac{\lambda}{2(\lambda+\mu)}.$$

Poisson's coefficient m describes the ratio of the longitudinal extension to the corresponding transverse compression. The shear modulus

$$G = \mu.$$

13. Representing the vector of the volume forces in the form $\boldsymbol{F} = \operatorname{grad} \Phi + \operatorname{curl} \boldsymbol{B}$ (for the possibility of representing an arbitrary vector in such a form see [14], page 209), prove that, if $\rho \partial^2 \phi / \partial t^2 = (\lambda + 2\mu)\Delta\phi + \Phi$, $\rho \partial^2 \boldsymbol{A} / \partial t^2 = \mu \Delta \boldsymbol{A} + \boldsymbol{B}$, then the vector $\boldsymbol{U} = \operatorname{grad} \phi + \operatorname{curl} \boldsymbol{A}$ satisfies the equation of motion, obtained in problem 12.

14. A problem on the propagation of disturbances in an elastic medium reduces to a plane problem, if the component w of the displacement vector $\boldsymbol{U}$ and the component Z of the vector of the density of volume forces $\boldsymbol{F} = \boldsymbol{i}X + \boldsymbol{j}Y + \boldsymbol{k}Z$ equal zero, and the remaining quantities are independent of z. For example, the problem on the propagation of strains in a thin lamina, produced by forces acting in its plane, is a plane problem†.

Prove that in the case of a plane problem the displacement vector $\boldsymbol{U}$ is expressed in terms of two scalar potentials, each of which satisfies a corresponding wave equation.

15. Express the boundary conditions for the propagation of elastic disturbances in a homogeneous isotropic semispace in terms of the components of the vector $\boldsymbol{U}$ and the tensor (H) (see problem 12), if the boundary plane

(a) is free,

(b) is rigidly fixed.

Express these boundary conditions for the plane problem in terms of scalar potentials (see problem 14).

16. State the boundary-value problem on the radial vibrations of a circular cylindrical tube under the action of a radial force

† For more detail see [26], page 92.

$F(r, t)$, where $F(r, t)$ is the force acting on unit mass, at a distance r from the axis of the tube.

17. Form the boundary-value problem on the radial vibrations of an elastic spherical shell $r_1 \leqslant r \leqslant r_2$ under the action of a variable pressure $p(t)$ in the inner cavity.

18. Derive the differential equation for the deflection from the undisturbed state of points of a thin isotropic homogeneous lamina, performing small transverse vibrations. Consider, in particular, the case where the lamina lies on (and is attached to) a flexible base.

Note. The problem on the transverse vibrations of the lamina is the two-dimensional analogue of the problem on the transverse vibrations of a rod (see § 1, chapter II).

19. Transforming to polar coordinates, state the boundary-value problem on the transverse vibrations of a circular lamina, if the periphery of the lamina is rigidly fixed.

20. At the origin of coordinates of an infinite region there exists an electric dipole, parallel to the z-axis. The dipole moment varies according to the law

$$M = \begin{cases} M_0 = \text{const.}, & -\infty < t \leqslant 0, \\ M_1 = M_0 \cos \omega t, & 0 < t < +\infty. \end{cases}$$

State the boundary-value problem to determine the electromagnetic field, produced by the field for $t > 0$.

§ 2. The Simplest Problems; Different Methods of Solution

21. (a) Solve the boundary-value problem

$$u_{tt} = a^2 \Delta u, \quad -\infty < x, y, z < +\infty, \quad 0 < t < +\infty, \tag{1}$$

$$u|_{t=0} = \phi(r), \quad u_t|_{t=0} = \psi(r), \quad r^2 = x^2 + z^2, \quad 0 \leqslant r < +\infty. \tag{2}$$

(b) Find

$$\lim_{x, y, z \to 0} u(x, y, z, t).$$

22. Solve the boundary-value problem

$$u_{tt} = a^2 \Delta u + f(r, t), \quad r^2 = x^2 + y^2 + z^2, \quad 0 \leqslant r < +\infty, \quad 0 < t < +\infty, \tag{1}$$

$$u|_{t=0} = 0, \quad u_t|_{t=0} = 0. \tag{2}$$

23. Solve the boundary-value problem

$$u_{tt} = a^2 \Delta u, \qquad -\infty < x, y, z < +\infty, \qquad 0 < t < +\infty \tag{1}$$

for the initial conditions

$$\text{(a)}\ u|_{t=0} = \begin{cases} U_0 = \text{const. inside a circle of radius } r_0 \\ 0 \qquad \text{outside the circle,} \end{cases}$$

$$u_t|_{t=0} = 0 \text{ everywhere;}$$

$$\text{(b)}\ u_t|_{t=0} = \begin{cases} U_0 = \text{const. inside a circle of radius } r_0 \\ 0 \qquad \text{outside the circle,} \end{cases}$$

$$u|_{t=0} = 0 \text{ everywhere.}$$

24. At the initial time $t = 0$ a gas is compressed inside a spherical volume of radius r_0, so that there is a perturbation of the density $\tilde{\rho} = \rho_1$, and outside the volume $\tilde{\rho} \equiv 0$. The initial velocity of the gas molecules equals zero over all space. Find the motion of the gas for $t > 0$.

25. Solve problem 23(b) for the semispace $z \geqslant 0$, if the centre of the sphere is at the point $(0, 0, z_0)$ $z_0 > r_0$; consider the special case where

(a) $u|_{z=0} = 0$,

(b) $u_z|_{z=0} = 0$.

26. Solve problem 23(b) for the region $y \geqslant 0$, $z \geqslant 0$, if the centre of the sphere is at the point $(0, y_0, z_0)$, $y_0 > r_0$, $z_0 > r_0$; consider the case where

(a) $u|_{y=0} = 0, \quad u_z|_{z=0} = 0$,

(b) $u_y|_{y=0} = 0, \quad u|_{z=0} = 0$.

27. An infinite region is filled with a quiescent ideal gas. At time $t = 0$ at some fixed point of this region a spherically symmetrical source of gas of magnitude $q(t)$ begins to act continuously. Find the velocity potential of the gas molecules for $t > 0$, assuming the disturbances, produced by the source, to be small.

28. Solve the preceding problem, if the source is placed

(a) inside the angular region π/n, where n is an integer, greater than zero;

(b) inside the plane layer $0 < z \leqslant l$, the boundary plane being rigid.

29. From the solution of the boundary-value problem

$$u_{tt} = a^2 \Delta_3 u + f(x, y, z, t), \quad -\infty < x, y, z < +\infty, \quad 0 < t < +\infty,$$

$$u|_{t=0} = \phi(x, y, z), \quad u_t|_{t=0} = \psi(x, y, z), \quad -\infty < x, y, z < +\infty,$$

by the method of "descent"†, derive the solution of the boundary-value problem

$$u_{tt}^* = a^2 \Delta_2 u^* + f^*(x, y, t), \quad -\infty < x, y < +\infty, \quad 0 < t < +\infty,$$

$$u^*|_{t=0} = \phi^*(x, y), \quad u_t^*|_{t=0} = \psi^*(x, y), \quad -\infty < x, y < +\infty.$$

30. From the solution of the boundary-value problem

$$u_{tt} = a^2 \Delta_3 u \pm cu + f(x, y, z, t), \quad -\infty < x, y, z < +\infty, \quad 0 < t < +\infty,$$

$$u|_{t=0} = \phi(x, y, z), \quad u_t|_{t=0} = \psi(x, y, z), \quad -\infty < x, y, z < +\infty,$$

by the method of "descent"†, derive the solution of the boundary-value problem

$$u_{tt}^* = a^2 \Delta_2 u^* \pm c^2 u^* + f^*(x, y, t), \quad -\infty < x, y < +\infty, \quad 0 < t < +\infty,$$

$$u^*|_{t=0} = \phi^*(x, y), \quad u_t^*|_{t=0} = \psi^*(x, y), \quad -\infty < x, y < +\infty.$$

31. Along a fixed straight line in infinite space, filled with a quiescent ideal gas, sources of gas are continuously distributed, and begin to act at time $t = 0$. The magnitude of the sources per unit length of this straight line are equal to $q(t)$. Find the velocity potential of the gas molecules for $t > 0$, assuming that disturbances,

---

† See [7], pages 455–457; [2], vol II, pages 553–555.

produced by the sources in the surrounding gas, are small (outside a small neighbourhood of the straight line, carrying the sources).

32. Solve the preceding problem for the region $x \geqslant 0$, $y \geqslant 0$, bounded by the absolutely rigid planes $x = 0$, $y = 0$, if the straight line, on which the sources are situated, is parallel to the z-axis and is determined by the coordinates x_0, y_0, $x_0 > 0$, $y_0 > 0$.

33. In infinite space, filled with an ideal quiescent gas, there exists a spherical shell of radius r_0 with centre at a fixed point. Beginning at time $t = 0$, the radius of the spherical surface varies continuously according to a given law, the radial velocity of points of the surface being equal to $\mu(t)$. Find the motion in the case where $\mu(t) = A \sin \omega t$.

34. Solve the preceding problem, if the sphere is placed in semispace, bounded by a rigid plane.

35. A sphere of fixed radius r_0 is placed in an infinite region filled with an ideal quiescent gas. At time $t = 0$ the centre of the sphere begins to move with a velocity $V(t)$, where $|V(t)| \ll a$, where a is the velocity of sound. Find the velocity potential of the gas particles.

36. Solve the problem on the steady symmetrical supersonic flow around a wedge by an ideal gas; find the velocity potential in the perturbed region and the pressure perturbation on the wedge†.

37. Solve the problem on the steady symmetrical supersonic flow around a circular cone with a small vertex angle‡.

38. A propagating plane wave satisfying the equation

$$u_{tt} = a^2 \Delta u + cu, \tag{1}$$

where $\Delta u = \dfrac{\partial^2 u}{\partial x_1^2} + \cdots + \dfrac{\partial^2 u}{\partial x_n^2}$, is a solution of the form

$$u = f\Big(\sum_{i=1}^{n} a_i x_i - bt\Big). \tag{2}$$

† See problem 7.

‡ See problem 8.

The plane wave $u = f\left(\sum_{i=1}^{n} a_i x_i - bt\right)$ has the same constant value on each plane of the family

$$\sum_{i=1}^{n} a_i x_i - bt = \text{const.} \tag{3}$$

The distance from a plane of (3) to the origin of the coordinates $x_1 = 0, x_2 = 0, \ldots, x_n = 0$ equals

$$\frac{bt + \text{const.}}{\left(\sum_{i=1}^{n} a_i^2\right)^{\frac{1}{2}}}. \tag{4}$$

With change of t plane (3) moves with a velocity

$$\frac{b}{\left(\sum_{i=1}^{n} a_i^2\right)^{\frac{1}{2}}}, \tag{5}$$

remaining parallel to its initial position (for $t = 0$)

$$\sum_{i=1}^{n} a_i x_i = \text{const.} \tag{6}$$

In order to simplify the calculation we assume that $\sum_{i=1}^{n} a_i^2 = 1$, i.e. that the a_i are direction-cosines of the normal to plane (3); $Q = \sum_{i=1}^{n} a_i x_i - bt$ is called the phase of the wave (2) and f the wave-form.

Prove that

(1) for the existence of plane waves of arbitrary shape in equation (1), propagating with velocity a in any direction, it is necessary and sufficient that $c = 0$;

(2) for $c \neq 0$ in equation (1) there exist plane wave solutions with any direction of propagation and with any velocity, except the velocity a, but their form is not arbitrary, and is a solution of the differential equation

$$f''(Q)(a^2 - b^2) + f(Q)\, c = 0. \tag{7}$$

39. Solve the problem of the steady flow along an oscillating wall $y = \varepsilon \sin \omega x$, where ε is small, $-\infty < x < +\infty$, by an ideal compressible liquid, whose unperturbed velocity has the direction of the x-axis and equals $U =$ const. Consider the cases:

(a) of a flow velocity less than the speed of sound,

(b) of a supersonic flow velocity.

40. By means of a superposition of plane waves with a front $f(at-\alpha x-\beta y)$, parallel to the z-axis, where α and β are the direction-cosines of the normal to the wave front, derive the cylindrical waves

$$\psi(r, t) = \int_{at-r}^{at+r} \frac{f(\xi)\,d\xi}{\sqrt{r^2-(at-\xi)^2}}, \tag{1}$$

where $r = \sqrt{x^2+y^2}$.

Find an explicit relation for $\psi(r, t)$ when

$$f(\xi) = \begin{cases} 0 & \text{for} \quad -\infty < \xi < -r_0, \\ U_0 = \text{const.} & \text{for} \quad -r_0 < \xi < r_0, \\ 0 & \text{for} \quad r_0 < \xi < +\infty. \end{cases}$$

41. By means of a superposition of spherically symmetrical waves $\dfrac{f_1(at-r)}{r}$ and $\dfrac{f_2(at+r)}{r}$, where $f_1(\xi)$ and $f_2(\xi)$ are arbitrary functions, obtain the cylindrical waves

$$\psi_1(\rho, t) = \int_{-\infty}^{at-\rho} \frac{2f_1(\xi)\,d\xi}{\sqrt{(at-\xi)^2-\rho^2}}, \quad \psi_2(\rho, t) = \int_{at+\rho}^{+\infty} \frac{2f_2(\xi)\,d\xi}{\sqrt{(at-\xi)^2-\rho^2}},$$

$$\rho^2 = x^2+y^2,$$

assuming the integrals converge.

42. Find the cylindrically symmetrical monochromatic waves in infinite space, solving the equation $u_{tt} = a^2\Delta u$, and then obtain these waves by means of a superposition of plane monochromatic waves.

43. By means of a superposition of plane waves derive the spherical wave of the form

$$\frac{\Phi\left(t+\dfrac{r}{a}\right)-\Phi\left(t-\dfrac{r}{a}\right)}{r},$$

44. Solve the problem of the reflection and refraction of a plane monochromatic wave at the plane boundary of separation of two different ideal gases; find the relation between the angles of incidence, reflection and refraction, and also between the amplitudes of the incident, reflected and refracted waves. The unperturbed pressures in both gases are assumed to be the same.

45. Find the relation between the angles of incidence, reflection and refraction of a plane monochromatic electromagnetic wave at the plane boundary of two homogeneous isotropic dielectrics.

46. Considering the case of the normal incidence of a plane monochromatic electromagnetic linearly-polarized wave at the plane of separation of two homogeneous isotropic dielectrics, find the relation between the amplitudes of the incident, reflected and refracted waves, and give an expression for these waves.

§ 3. The Method of Separation of Variables

1. Boundary-value Problems not Requiring the Application of Special Functions

In this section boundary-value problems for regions with plane and spherical boundaries are considered, the solutions of which are expressed as series in the simplest (elementary) eigenfunctions of the Laplacian operator for these regions.

At first the media are assumed isotropic and homogeneous, then a few problems for inhomogeneous media are considered.

(a) *Homogeneous media*

47. Find the transverse vibrations of a rectangular membrane $0 \leqslant x \leqslant l_1$, $0 \leqslant y \leqslant l_2$ fixed round the edge, produced by an initial deflection

$$u(x, y, 0) = Axy(l_1-x)(l_2-y).$$

48. Find the transverse vibrations of a rectangular membrane $0 \leqslant x \leqslant l_1$, $0 \leqslant y \leqslant l_2$, fixed round the edge, produced by an initial distribution of velocities

$$u_t(x, y, 0) = Axy(l_1-x)(l_2-y).$$

49. Find the transverse vibrations of a rectangular membrane $0 \leqslant x \leqslant l_1$, $0 \leqslant y \leqslant l_2$, fixed round the edge, produced by a transverse concentrated impulse K, applied to the membrane at the point (x_0, y_0), $0 < x_0 < l_1$, $0 < y_0 < l_2$.

50. Find the transverse vibrations of a rectangular membrane $0 \leqslant x \leqslant l_1$, $0 \leqslant y \leqslant l_2$, fixed round the edge, produced by a force continuously distributed over the membrane and perpendicular to its surface, with a density

$$F(x, y, t) = A(x, y) \sin \omega t, \qquad 0 < t \leqslant +\infty.$$

51. Find the transverse vibrations of a rectangular membrane $0 \leqslant x \leqslant l_1$, $0 \leqslant y \leqslant l_2$ fixed round the edge, produced by a central transverse force

$$F(t) = A \sin \omega t, \qquad A = \text{const.}, \qquad 0 < t < +\infty,$$

applied at the point (x_0, y_0), $0 < x_0 < l_1$, $0 < y_0 < l_2$.

52. Find the water waves in a rectangular reservoir $0 \leqslant x \leqslant l_1$, $0 \leqslant y \leqslant l_2$ under the action of a variable external pressure on the free surface

$$p_0(x, y, t) = A \cos \frac{\pi x}{l_1} \cos \frac{\pi y}{l_2} f(t), \qquad 0 < t < +\infty, \qquad f(0) = 0,$$

if the depth of the water in the undisturbed state equals h. The function $f(t)$ is assumed to have a continuous derivative†.

53. Solve problem 49, assuming that the surrounding medium exerts a resistance proportional to the velocity.

54. Find the steady-state vibrations of a rectangular membrane $0 \leqslant x \leqslant l_1$, $0 \leqslant y \leqslant l_2$ in a medium with a resistance, proportional to the velocity, under the action of a uniformly distributed transverse force of density

$$F = A \sin \omega t, \qquad 0 < t < +\infty, \qquad A = \text{const.}$$

The contour of the membrane is rigidly fixed.

55. An ideal gas is enclosed between two concentric spheres S_{r_1} and S_{r_2}. The radius of the inner sphere S_{r_1} varies according to the law

$$r(t) = r_1 + \varepsilon \sin \omega t, \qquad -\infty < t < +\infty, \qquad 0 < \varepsilon < r_1,$$

† See problems 9 and 10.

and the outer sphere remains invariant. Find the steady-state vibrations of the gas between the spheres.

56. An ideal gas is enclosed between two concentric spheres S_{r_1} and S_{r_2} of fixed radii r_1 and r_2. Find the vibrations of the gas between the spheres, produced by an initial radial perturbation of density

$$\rho(r, 0) = f(r), \qquad r_1 < r < r_2.$$

(b) *Inhomogeneous media*

57. Find the transverse vibrations of a rectangular membrane $0 \leqslant x \leqslant l_1$, $0 \leqslant y \leqslant l_2$, consisting of two homogeneous rectangular parts $0 \leqslant x \leqslant x_0$, $0 \leqslant y \leqslant l_2$, and $x_0 \leqslant x \leqslant l_1$, $0 \leqslant y \leqslant l_2$, produced by initial transverse perturbations.

58. A spherical cavity of fixed radius r_2 is filled with two different ideal gases, the surface of separation of which is the sphere $S_{r_1}(0 < r_1 < r_2)$, concentric with the surface of the cavity.

Find the vibrations of the gases for the following initial conditions of velocity potential $u(r, t)$ and pressure $p(r, t)$:

$$u(r, 0) = f(r), p(r, 0) = p_0, \ 0 \leqslant r < r_2.$$

2. Boundary-value Problems Requiring the Application of Special Functions

As in the preceding section, first there are problems for homogeneous media, then for inhomogeneous media.

(a) *Homogeneous media*

59. Find the transverse vibrations of a circular membrane fixed round the edge, produced by a radially symmetrical initial distribution of deflections and velocities, assuming no damping.

60. Solve the preceding problem, assuming that the initial deflection has the form of a paraboloid of revolution, and the initial velocities are zero.

61. Find the water waves in a circular vertical cylindrical vessel with a horizontal base, if the initial conditions possess radial symmetry, and the pressure at the free surface of the water remains constant.

62. Find the vibrations of a circular membrane fixed round the edge in a non-resistant medium, produced by a uniformly distributed constant pressure, acting along one side of the membrane at time $t = 0$, assuming that the surrounding medium does not exert any resistance to the vibrations of the membrane.

63. Find the vibrations of a circular membrane $0 \leqslant r < r_0$ fixed round the edge in a non-resistant medium, produced by a variable pressure

$$p = f(r, t), \quad 0 \leqslant r \leqslant r_0, \quad 0 < t < +\infty,$$

applied to one side of the membrane.

64. Find the vibrations of a circular membrane $0 \leqslant r \leqslant r_0$ fixed round the edge in a non-resistant medium, produced by a uniformly distributed pressure

$$p = p_0 \sin \omega t, \quad 0 < t < +\infty,$$

applied to one side of the membrane.

65. For zero initial conditions, find the vibrations of a circular membrane $0 \leqslant r < r_0$ in a non-resistant medium, produced by the motion of its edge according to the law

$$u(r_0, t) = A \sin \omega t, \quad 0 < t < +\infty.$$

66. Solve problem 59 for the case where the surrounding medium exerts a resistance, proportional to the velocity.

67. Find the steady-state vibrations of a circular membrane fixed round the edge in a medium with a resistance proportional to the velocity, under the action of a uniformly distributed (applied to one side of the membrane) pressure

(a) $p = p_0 \sin \omega t, \quad 0 < t < +\infty, \quad p_0 = \text{const.},$

(b) $p = p_0 \cos \omega t, \quad 0 < t < +\infty, \quad p_0 = \text{const.}$

68. Find the steady-state vibrations of a circular membrane $0 \leqslant r \leqslant r_0$ in a medium with a resistance, proportional to the velocity, produced by the motion of its edge according to the law $u(r_0, t) = A \sin \omega t$ (compare with problem 65).

69. Find the vibrations of the circular membrane of a drum†, produced by radially symmetrical initial perturbations.

70. Find the vibrations of the circular membrane of a drum, produced by a uniformly distributed pressure

$$p = \Pi_0 \sin \omega t, \quad 0 < t < +\infty, \quad \Pi_0 = \text{const.},$$

applied to the outer side of the membrane.

71. Find the transverse vibrations of a circular lamina, rigidly fixed at the edges in a non-resistant medium, produced by radially symmetrical initial perturbations.

72. Find the transverse vibrations of the lamina of the preceding problem, produced by a transverse concentrated blow at the centre, which transmits an impulse I to it.

73. Find the transverse vibrations of the lamina of problem 71, produced by a uniformly distributed transverse force of density $p = p_0 \sin \omega t$, applied at time $t = 0$.

74. Find the transverse vibrations of the lamina of problem 71, produced by a central transverse force $P = P_0 \sin \omega t$, applied at the centre of the lamina at time $t = 0$ (vibrations of the membrane of a loud-speaker).

75. Find the transverse vibrations of a circular ring-shaped membrane with fixed edges, produced by radially symmetrical initial disturbances.

76. Find the transverse vibrations of the membrane described in the preceding problem, produced by a uniformly distributed pressure

$$p = p_0 \sin \omega t, \quad 0 < t < +\infty, \quad p_0 = \text{const.},$$

applied to one side of the membrane.

77. Find the vibrations of a liquid in a vessel with a horizontal base, the walls of which are two coaxial circular cylinders, if the depth of the liquid in the unperturbed state is $h = \text{const.}$, and the initial perturbations are radially symmetrical‡.

---

† See problem 5.

‡ See problem 9.

78. Find the vibrations of a gas (velocity potential) in a circular closed cylindrical vessel, produced by radial vibrations of the lateral wall, begun at time $t = 0$, if the velocity of the wall equals

$$f(z)\cos\omega t, \quad 0 \leqslant z \leqslant l \ (l \text{ length of cylinder}), \quad 0 < t < +\infty.$$

The upper and lower ends are rigid.

79. Find the vibrations of a gas in a circular closed cylinder, produced by the transverse vibrations of one of its ends, begun at time $t = 0$, if the velocity of this end equals

$$f(r)\cos\omega t, \quad 0 \leqslant r \leqslant r_0 \ (r_0 \text{ radius of cylinder}), \quad 0 < t < +\infty.$$

The second end and the lateral wall of the vessel are rigid.

80. Find the vibrations of a gas in a closed vessel, formed by two coaxial circular cylinders and by two transverse plane ends, produced by radial vibrations of the outer cylinder, begun at time $t = 0$, if the velocity of particles of this cylinder equals $f(z)\cos\omega t, 0 \leqslant z \leqslant l$; l is the length of the cylinder. The ends and inner cylinder are rigid.

81. Find the vibrations of a gas in the vessel described in the preceding problem, produced by transverse vibrations of one end, beginning at time $t = 0$, if the velocity of this end equals $f(r)\cos\omega t$. $r^* \leqslant r \leqslant r^{**}$, r^* and r^{**} are the radii of the inner and outer cylinders. The second end and cylinders are rigid.

82. Find the transverse vibrations of a circular membrane $0 \leqslant r \leqslant r_0$ fixed at the edges, produced by a concentrated blow, normal to the surface of the membrane, transmitting an impulse K to the membrane at the point (r_1, ϕ_1), $0 < r_1 < r_0$.

Consider the case where the surrounding medium does not produce a resistance to the motion of the membrane.

83. A water vessel, in the form of a vertical circular cylinder with a horizontal base, moves with a velocity $v_0 = \text{const.}$ in a direction perpendicular to the axis of the vessel for $t < 0$.

Find the water waves in the vessel for $t > 0$, if at time $t = 0$ the vessel is stopped instantaneously and if at $t < 0$ the water was stationary relative to the vessel. The pressure at the free surface of the water is assumed to be constant.

84. Find the vibrations of a circular membrane $0 \leqslant r \leqslant r_0$ with fixed edges, produced by a continuously distributed variable pressure

$$p = f(r)\cos(\phi - \omega t), \quad f(r_0) = 0, \quad 0 < t < +\infty,$$

applied to one side of the membrane.

85. Find the steady-state vibrations of the membrane, described in the preceding problem, in a medium with a resistance, proportional to the velocity.

86. Find the vibrations of a circular membrane $0 \leqslant r \leqslant r_0$, produced by vibrations of its edge according to the law

$$u(r_0, \phi, t) = f(t)\cos n\phi, \quad f(0) = f'(0) = 0, \quad n \text{ an integer} > 0,$$
$$0 < t < +\infty.$$

87. Find the vibrations of a circular membrane $0 \leqslant r \leqslant r_0$, produced by vibrations of its edge according to the law $u(r_0, \phi, t) = F(\phi) \sin \omega t$, $F(\phi)$ is an even function of period 2π.

88. Find the vibrations of a gas in a circular closed cylinder $0 \leqslant r \leqslant r_0$, $0 \leqslant z \leqslant l$, produced by radial oscillations of its lateral wall with a velocity, varying according to the law

$$f(z)\cos n\phi \cos \omega t, \quad n \text{ an integer} > 0, \quad 0 < t < +\infty.$$

The ends of the vessel are rigid.

89. Find the vibrations of a gas in a circular closed cylinder $0 \leqslant r \leqslant r_0$, $0 \leqslant z \leqslant l$, produced by transverse oscillations of one end with a velocity, varying according to the law $f(r) \cos n\phi \cos \omega t$, n is an integer > 0, $0 < t < +\infty$.

90. Find the transverse vibrations of a membrane fixed round the edge, produced by an initial concentrated transverse impulse K, imparted to the membrane at some internal point of it, if the membrane has the shape of a circular sector, and the surrounding medium does not exert a resistance to the vibrations.

91. Solve the preceding problem for a membrane, having the shape of a sector of a circular ring.

92. Find the vibrations of a gas in a region, bounded by two coaxial rigid circular cylinders, two planes perpendicular to the axis of the cylinders, and two planes passing through their axis, if these vibrations are produced by initial perturbations, not dependent on z.

93. A spherical vessel containing gas moves uniformly for a long time with a velocity v, and then at time $t = 0$ it is stopped instantaneously and remains stationary. Find the vibrations of the gas in the vessel.

94. A spherical vessel, filled with gas, starting at time $t = 0$, performs small harmonic oscillations in the direction of one of its diameters; a displacement of the vessel in the direction of this diameter equals $A \sin \omega t$, $0 < t < +\infty$. Find the vibrations of the gas in the vessel assuming that at $t < 0$ the gas was at rest.

95. Find the vibrations of a gas in a spherical vessel $0 \leqslant r \leqslant r_0$, $0 \leqslant \theta \leqslant \pi$, $0 \leqslant \phi \leqslant 2\pi$, produced by small deformations of the wall, begining at time $t = 0$, if the velocities of the wall are radial, and equal

$$AP_n(\cos \theta) \cos \omega t\,\dagger.$$

96. Find the vibrations of a gas in a spherical vessel, produced by small oscillations of its wall, begun at time $t = 0$, if the velocities of the wall are radial, and equal

$$P_n(\cos \theta) f(t),$$

where $f(0) = f'(0) = 0$.

97. Find the vibrations of a gas in a spherical vessel, produced by small oscillations of its wall, begun at time $t = 0$, if the velocity of the wall is radial, and equals

$$f(\theta) \cos \omega t, \qquad 0 < t < +\infty.$$

98. Solve the preceding problem when the velocity of the wall equals

$$AP_{nm}(\cos \theta) \cos m\phi \cos \omega t\,\ddagger.$$

99. Solve problem 98, if the velocity of the wall equals

$$f(\theta) \cos m\phi \cos \omega t.$$

---

† $P_n(\xi)$ is a Legendre polynomial.

‡ $P_{nm}(\xi)$ is an associated Legendre function, $m \leqslant n$.

100. Solve problem 98 if the velocity of the wall equals

$$f(t)\, P_n(\cos\theta) \cos m\phi, \qquad f(0) = f'(0) = 0.$$

101. Solve problem 93 for a gas, contained between two concentric spheres S_{r_1} and S_{r_2}, $r_1 < r_2$.

102. Solve problem 94 for a gas, contained between two concentric spheres S_{r_1} and S_{r_2}, $r_1 < r_2$.

(b) *Inhomogeneous media*

103. Find the transverse vibrations of an inhomogeneous circular membrane $0 \leqslant r \leqslant r_2$ fixed round the edge, obtained by joining a homogeneous circular membrane $0 \leqslant r \leqslant r_1$ and a homogeneous ring membrane $r_1 \leqslant r \leqslant r_2$, if the initial transverse perturbations are given.

§ 4. The Method of Integral Representations

In the first part of this section problems on the application of the Fourier integral are given, in the second part on the formation and application of source functions of instantaneous concentrated sources.

1. The Application of the Fourier Integral

(a) *The Fourier transform*

104. Solve the boundary-value problem

$$u_{tt} = a^2 \Delta_2 u, \qquad -\infty < x, y < +\infty, \qquad 0 < t < +\infty^{\dagger}, \qquad (1)$$

$$u|_{t=0} = \Phi(x, y), \quad u_t|_{t=0} = \Psi(x, y), \qquad -\infty < x, y < +\infty. \qquad (2)$$

105. Solve the boundary-value problem

$$u_{tt} = a^2 \Delta_3 u, \qquad -\infty < x, y, z < +\infty, \qquad 0 < t < +\infty^{\ddagger}, \qquad (1)$$

$$u|_{t=0} = \Phi(x, y, z), \quad u_t|_{t=0} = \Psi(x, y, z),$$
$$-\infty < x, y, z < +\infty. \qquad (2)$$

---

† $\Delta_2 = \operatorname{div}\operatorname{grad}$ is the Laplacian operator for the plane; in cartesian coordinates $\Delta_2 = \partial^2/\partial x^2 + \partial^2/\partial y^2$.

‡ $\Delta_3 = \operatorname{div}\operatorname{grad}$ is the Laplacian operator for space; in cartesian coordinates $\Delta_3 = \partial^2/\partial x^2 + \partial^2/\partial y^2 + \partial^2/\partial z^2$.

106. Solve the boundary-value problem

$$u_{tt} = a^2\Delta_2 u + f(x, y, t), \quad -\infty < x, y < +\infty, \quad 0 < t < +\infty, \tag{1}$$

$$u|_{t=0} = 0, \quad u_t|_{t=0} = 0, \quad -\infty < x, y < +\infty. \tag{2}$$

107. Solve the boundary-value problem

$$u_{tt}| = a^2\Delta_3 u + f(x, y, z, t), \quad -\infty < x, y, z < +\infty, \quad 0 < t < +\infty, \tag{1}$$

$$u|_{t=0} = 0, \quad u_t|_{t=0} = 0, \quad -\infty < x, y, z < +\infty. \tag{2}$$

108. Solve the boundary-value problem

$$u_{tt} + b^2\Delta_2\Delta_2 u = 0, \quad -\infty < x, y < +\infty, \quad 0 < t < +\infty^{\dagger}, \tag{1}$$

$$u|_{t=0} = \Phi(x, y), \quad u_t|_{t=0} = \Psi(x, y), \quad -\infty < x, y < +\infty. \tag{2}$$

(b) *The Fourier–Bessel (Hankel) transform*

109. Applying the Fourier–Bessel transform, solve the boundary-value problem

$$\frac{\partial^2 u}{\partial t^2} = a^2\left(\frac{\partial^2 u}{\partial r^2} + \frac{1}{r}\frac{\partial u}{\partial r}\right), \quad 0 \leqslant r < +\infty, \quad 0 < t < +\infty, \tag{1}$$

$$u(r, 0) = \frac{A}{\sqrt{1 + \frac{r^2}{b^2}}}, \quad u_t(r, 0) = 0, \quad 0 \leqslant r < +\infty. \tag{2}$$

110. Find the radially symmetrical transverse vibrations of an infinite plate, by solving the boundary-value problem

$$\frac{\partial^2 u}{\partial t^2} + b^2\left(\frac{\partial^2}{\partial r^2} + \frac{1}{r}\frac{\partial}{\partial r}\right)^2 u = 0, \quad 0 \leqslant r < +\infty, \quad 0 < t < +\infty, \tag{1}$$

$$u(r, 0) = f(r), \quad u_t(r, 0) = 0, \quad 0 \leqslant r < +\infty. \tag{2}$$

Consider, in particular, the case where

$$f(r) = Ae^{-\frac{r^2}{a^2}}, \quad 0 \leqslant r < +\infty. \tag{2'}$$

---

† The biharmonic operator $\Delta_2\Delta_2$, denoting a twofold application of the Laplacian operator Δ_2.

111. Find the radially symmetrical transverse deflections of points of an infinite plate $0 \leqslant r < +\infty$, if after time $t = 0$ the point $r = 0$ of this plate moves according to a given law. Consider, in particular, the case where

$$u(0, t) = \begin{cases} A(t_0 - t), & 0 \leqslant t \leqslant t_0, \\ 0, & t_0 \leqslant t < +\infty. \end{cases} \tag{1}$$

112. Find the purely radially symmetrically transverse deflections of points of an infinite lamina $0 \leqslant r < +\infty$ under the action of distributed transverse forces of density

$$p(r, t) = 16\rho h b f(r)\psi'(t), \qquad -\infty < t < +\infty,$$

where $2h$ is the thickness of the lamina, ρ is the mass density of the lamina, b has the same meaning as in preceding problems†, $\psi'(t) = d\psi(t)/dt$, $\psi(t)$ depends only on t, and $f(r)$ depends only on r.

Consider, in particular, the case where

(a) the lamina is acted on by a central transverse force

$$16\rho h b\psi'(t), \qquad -\infty < t < +\infty,$$

applied at the point $r = 0$;

(b) the lamina is acted on by a transverse force

$$16\rho h b\psi'(t), \qquad -\infty < t < +\infty,$$

uniformly distributed over the circle $0 \leqslant r \leqslant a$;

(c) the force, described in (b), acts for a time t_0, i.e.

$$\psi'(t) = \begin{cases} 0 & \text{for} \quad -\infty < t \leqslant 0, \\ \psi_0 = \text{const.} & \text{for} \quad 0 < t < t_0, \\ 0 & \text{for} \quad t_0 < t < +\infty. \end{cases}$$

Give asymptotic relations for the solution for small and large values of r;

(d) $p(r, t) = \dfrac{4A\rho h}{c^2} e^{-\frac{r^2}{c^2}} f'(t), \qquad -\infty < t < +\infty;$

† For more detail, see problem 18.

(e) find the transverse velocities of points of the lamina for

$$p(r, t) = \frac{4A\rho h}{c^2} \mathrm{e}^{-\frac{r^2}{c^2}} \delta(t), \qquad -\infty < t < +\infty;$$

$\delta(t)$ is the delta-function (i.e. at time $t = 0$ the lamina receives a transverse blow from a continuously distributed impulse $\frac{4A\rho h}{c^2} \mathrm{e}^{-\frac{r^2}{c^2}}$).

2. Formation and Application of the Functions of the Effect of Concentrated Sources

(a) *Green's functions for an impulse*

113. Find the source function of an instantaneous concentrated impulse of unit magnitude for the equation

$$u_{tt} = a^2 \Delta_3 u$$

in infinite space x, y, z, assuming first that the impulse occurs at the origin of coordinates at time $t = 0$. Find the source function, solving the boundary-value problem

$$u_{tt} = a^2 \Delta_3 u, \qquad -\infty < x,\ y,\ z, < +\infty, \qquad 0 < t < +\infty, \quad (1)$$

$$u|_{t=0} = 0, \qquad u_t|_{t=0} = \delta(x)\delta(y)\delta(z), \qquad -\infty < x, y, z < \infty, \quad (2)$$

and then proceed to the case where the impulse occurs at the point (ξ, η, ζ) at time $t = \tau$.

114. Solve the preceding problem for the equation

$$u_{tt} = a^2 \Delta_3 u \pm c^2 u.$$

115. Solve the two-dimensional analogue of problem 113.

116. Solve the two-dimensional analogue of problem 114.

117. By separating the variables find the source function of an instantaneous concentrated impulse for the first, second and third boundary-value problems of the equation $u_{tt} = a^2 \Delta_2 u$

(a) for a rectangular membrane $0 \leqslant x \leqslant l_1$, $0 \leqslant y \leqslant l_2$,

(b) for a circular membrane $0 \leqslant r \leqslant r_0$, $0 \leqslant \phi \leqslant 2\pi$.

118. By the method of images find the source function of an instantaneous concentrated impulse for the equation $u_{tt} = a^2 \Delta_2 u \pm$ $\pm c^2 u$ for the angle $0 \leqslant \phi \leqslant \pi/n$, where n is a whole number, greater than zero, if at the boundary rays $\phi = 0$ and $\phi = \pi/n$, a boundary condition of the second kind is fulfilled.

119. Let the plane region G be bounded by a piecewise smooth contour Γ. Assuming the application of the Green–Ostrogradskii relation, connecting the curvilinear integral to the double integral, to be possible, find the solutions (a) of the first, (b) of the second and (c) of the third boundary-value problems for the equation $u_{tt} = a^2\Delta_2 u \pm c^2 u + f(x, y, t)$ for inhomogeneous initial and boundary conditions, if the source function of an instantaneous concentrated impulse is known for each of these cases.

120. By means of the source function of the instantaneous concentrated impulse, found in the solution of problem 113, derive Kirchhoff's relation† for the equation

$$u_{tt} = a^2\Delta_3 u + f(x, y, z, t).$$

(b) *Source functions of continuously acting concentrated sources*

121. Find the source function of a continuously acting concentrated source of variable magnitude $f(t)$ ($f(t) = 0$ for $t < 0$), located at a fixed point of space, for the equation $u_{tt} = a^2\Delta_3 u$, i.e. solve the boundary-value problem

$$u_{tt} = a^2\Delta_3 u + \delta(x - x_0)\delta(y - y_0)\delta(z - z_0) f(t),$$

$$-\infty < x, y, z < +\infty, \quad 0 < t < +\infty, \tag{1}$$

$$u|_{t=0} = u_t|_{t=0} = 0. \tag{2}$$

122. Find the source function of a continuously acting concentrated source of variable magnitude $f(t)$ ($f(t) = 0$ for $t < 0$), existing at a fixed point of space, for the equation $u_{tt} = a^2\Delta_2 u$, i.e. solve the boundary-value problem

$$u_{tt} = a^2\Delta_2 u + \delta(x - x_0)\delta(y - y_0) f(t),$$

$$-\infty < x, y < +\infty, \quad 0 < t < +\infty, \tag{1}$$

$$u|_{t=0} = 0, \quad u_t|_{t=0} = 0. \tag{2}$$

123. Find the source function of a continuously acting concentrated source of variable magnitude $f(t)$ ($f(t) = 0$ for $t < 0$),

† See [7], pages 461–465.

moving according to an arbitrary law, for the equation $u_{tt} = a^2\Delta_3 u$, i.e. solve the boundary-value problem

$$u_{tt} = a^2\Delta_3 u + \delta\big(x - X(t)\big)\delta\big(y - Y(t)\big)\delta\big(z - Z(t)\big)f(t),$$

$$-\infty < x, y, z < +\infty, \quad 0 < t < +\infty, \tag{1}$$

$$u|_{t=0} = u_t|_{t=0} = 0, \tag{2}$$

where $X(t)$, $Y(t)$, $Z(t)$ are the coordinates of the source; $X(0) = Y(0) = Z(0) = 0$. In particular, find the source function of a concentrated source, moving linearly with constant velocity v; consider the case where (a) $v < a$, (b) $v > a$.

124. If the source possesses constant magnitude q and moves linearly with constant velocity v, then in a system of coordinates, moving along with the source, the process will be stationary. Find the source function of such a source (a) for $v < a$, (b) for $v > a$, omitting terms with products containing time in the wave equation, transformed to this moving system of coordinates.

125. Find the electromagnetic field, produced by an electron, moving in a dielectric linearly with a constant velocity, exceeding the velocity of light in this dielectric (Cherenkov-radiation).

126. Solve boundary-value problem 20.

127. Find the vibrations of an elastic isotropic homogeneous medium, filling the whole of infinite space, produced by a continuously acting force $F(t)$ ($F(t) = 0$ for $t < 0$), applied to a fixed point of the medium and parallel to a fixed direction.

CHAPTER VII

EQUATIONS OF ELLIPTIC TYPE

$\Delta u + cu = -f$

§ 1. Problems for the Equation $\Delta u - \varkappa^2 u = -f$

In the present section we consider some problems on the equation of elliptic type

$$\Delta u - \varkappa^2 u = 0 \quad (\varkappa^2 > 0), \tag{1}$$

which is obtained in problems on the diffusion of an unstable gas, disintegrating in the process of diffusion,

Equation (1) has fundamental solutions:

(a) $u_0(M) = \dfrac{e^{-\varkappa r}}{r}$ in three-dimensional space,

(b) $u_0(M) = K_0(\varkappa r)$ on a plane;

(r is the distance of the point M from the origin of coordinates). The function $K_0(x)$ has at $x = 0$ a logarithmic singularity and decreases exponentially at infinity.

The method of separation of variables in the solution of equation (1) often leads to Bessel's equation for imaginary argument

$$y'' + \frac{1}{x} y' - \left(1 + \frac{\nu^2}{x^2}\right) y = 0,$$

the general solution of which has the form

$$y = AI_\nu(x) + BK_\nu(x),$$

where $I_\nu(x)$ and $K_\nu(x)$ are cylindrical functions of imaginary argument of first and second kind. The function $I_\nu(x)$ is bounded at $x = 0$ and increases exponentially as $x \to \infty$.

1. Determine the steady-state distribution of concentration of an unstable gas in infinite space, produced by a point source of gas of magnitude Q_0.

2. A point source of an unstable gas is situated at a height ξ above a gasproof plane $z = 0$. Find the steady-state distribution of concentration.

3. Find the source function for the equation $\Delta u - \varkappa^2 u = 0$ in a plane and give a physical interpretation of it.

4. Solve problem 3, assuming that the plane $y = 0$ is impermeable to gas.

5. Find the source function for the diffusion equation of an unstable gas, if the source exists inside the layer $(0 \leqslant z \leqslant l)$, bounded by the gasproof planes $z = 0$ and $z = l$.

6. Solve the analogue of problem 5 for the two-dimensional case.

7. A point source of an unstable gas is placed inside an infinite cylindrical tube with gasproof walls. Determine the steady-state distribution of concentration of the gas, assuming that the tube section may have arbitrary shape.

8. Find the source function for the equation $\Delta u - \varkappa^2 u = 0$ inside a sphere for a boundary condition of the second kind.

9. A point source of gas moves in an infinite medium with constant velocity v_0. Find the steady-state distribution of concentration in the gas.

10. Find the steady-state distribution of concentration of an unstable gas inside an infinite cylinder of circular section, if a constant concentration $u|_\Sigma = u_0$ is maintained on the surface of the cylinder.

11. Solve problem 10 for a region, external to the cylinder.

12. Solve problem 10 inside a sphere of radius a, if

(a) $u|_{r=a} = u_0$,

(b) $u|_{r=a} = u_0 \cos \theta$.

13. Solve problem 12 for a region, external to the sphere of radius a.

14. At a depth h under the earth's surface a medium exists, in which a radioactive substance is distributed with constant density. This substance decays giving off a radioactive gas which diffuses through the earth.

Find:

(a) the distribution of radioactive gas in the earth,

(b) the amount of radioactive gas flowing through the earth's surface, assuming that its concentration on the earth's surface is zero.

15. At a depth h under the earth's surface a radioactive substance is concentrated in a certain volume and gives off an amount of radioactive gas, equal to Q_0 per unit time. Find:

(a) the distribution of concentration of the gas in the earth,

(b) the amount of gas flowing through the earth's surface, assuming that the source of the gas is a point source, and its concentration on the earth's surface is zero.

16. Solve the inverse of problem 15. The distribution of flow across the earth's surface $q = q(\rho)$ is known; it is required to find:

(a) the magnitude of the source Q_0,

(b) the position of the source, i.e. the depth h of the radioactive substance.

§ 2. Some Problems on Natural Vibrations

Problems on wave motion in a finite region reduce to the homogeneous equation $L(v)+\lambda\rho v=0, L(v) = \operatorname{div}(k \operatorname{grad} v)$ ($k(x) > 0$, $\rho(x) > 0$) inside some region T with homogeneous conditions at its boundary. In chapter II, and then in chapter V some problems on natural vibrations of strings and membranes were considered. In the present section a fuller list of eigenvalue problems will be given, solvable by the method of separation of variables.

The expression "find the natural vibrations" will mean that it is required to find the eigenvalues and normalized eigenfunctions for the region under consideration.

1. Natural Vibrations of Strings and Rods

17. Solve the problem of the natural transverse vibrations of a homogeneous string $0 \leqslant x \leqslant l$, if

(a) the ends of the string are rigidly fixed,

(b) the ends of the string are free†,

† This means that $\partial u/\partial x$ equals zero at the ends of the string. This occurs for instance, in fixing the ends of the string to links, (of negligibly small mass), slipping without friction along parallel rods.

(c) one end of the string is free, and the other end fixed,

(d) the ends of the string are fixed elastically,

(e) one end of the string is rigidly fixed, and the second end is elastically fixed,

(f) one end of the string is fixed elastically, and the other end is free.

18. Find the natural longitudinal vibrations of a rod of length l, consisting of two rods of lengths x_0 and $l-x_0$, having different densities (ρ_1 and ρ_2) and moduli of elasticity (E_1 and E_2), assuming that the ends of the rod

(a) are rigidly fixed,

(b) are free,

(c) are fixed elastically.

19. A load of mass M is attached to one end of a rod.

Find the natural longitudinal elastic vibrations of the rod, assuming that the second end of the rod

(a) is rigidly fixed,

(b) is free,

(c) is fixed elastically.

Find the orthogonality condition of the eigenfunctions.

For (a) consider the case of small and large loads, finding the corresponding corrections for the unperturbed eigenvalues.

20. Solve the problem of the natural vibrations of a string loaded with a concentrated mass M, suspended from some interior point of the string, assuming that the ends of the string

(a) are rigidly fixed,

(b) are free,

(c) are fixed elastically.

Calculate the corrections to the eigenvalues for problem (a).

21. Find the transverse natural vibrations of a homogeneous rod, if

(a) both ends of the rod are rigidly fixed,

(b) both ends of the rod are free,

(c) one end of the rod is free, and the other rigidly fixed.

Find the first term of the asymptotic expansion of the eigenfrequencies.

2. Natural Vibrations of Volumes

22. Find the natural vibrations of a rectangular membrane

(a) with rigidly fixed boundary,

(b) with a free boundary,

(c) if two opposite sides are fixed, and the two other sides are free,

(d) if two adjacent sides are fixed and the two other sides are free,

(e) with an elastically attached boundary.

23. Solve problem 22 for a circular membrane (cases (a), (b), (e)).

24. Determine the eigenvalues and normalized eigenfunctions for a rectangular parallelepiped with

(a) boundary conditions of the first kind,

(b) boundary conditions of the second kind,

(c) boundary conditions of the third kind.

25. Find the natural vibrations of a sphere with

(a) boundary conditions of the first kind,

(b) boundary conditions of the second kind,

(c) boundary conditions of the third kind.

26. Solve the problem of the natural vibrations of a circular cylinder of finite length with boundary conditions

(a) of the first kind,

(b) of the second kind,

(c) of the third kind.

27. Determine the natural vibrations of a membrane, having the shape of a circular ring $a \leqslant \rho \leqslant b$, if its boundary

(a) is rigidly fixed,

(b) is free,

(c) is fixed elastically.

28. Determine the natural vibrations of a membrane, having the shape of a circular sector $(\rho \leqslant a, 0 \leqslant \phi \leqslant \phi_0)$ if its boundary

(a) is rigidly fixed,

(b) is free,

(c) is fixed elastically.

29. Find the natural vibrations of a membrane, having the shape of a ring-shaped sector ($a \leqslant \rho \leqslant b$, $0 \leqslant \phi \leqslant \phi_0$)

(a) with a rigidly fixed boundary,

(b) with a free boundary,

(c) with an elastically attached boundary.

30. Determine the eigenvalues and eigenfunctions of a toroid of rectangular section ($a \leqslant \rho \leqslant b$, $0 \leqslant z \leqslant l$) for boundary conditions

(a) of the first kind,

(b) of the second kind,

(c) of the third kind.

31. A plane membrane has the shape of a ring of outer radius a and inner radius ε; the boundary of the membrane is rigidly fixed.

Compare the first eigenvalue λ_1 of such a membrane with the first eigenvalue λ_1^0 of a circular membrane of radius a,

(a) show that $\lim\limits_{\varepsilon=0} \lambda_1 = \lambda_1^0$,

(b) calculate the correction $\Delta\lambda = \lambda_1 - \lambda_1^0$ for small ε.

32. A circular membrane of radius a is loaded with a mass M, uniformly distributed over an absolutely rigid circle of radius ε ($r \leqslant \varepsilon$).

Compare the eigenvalues λ_n of this membrane with the eigenvalues λ_n^0 of the unloaded membrane.

Consider the two cases: M small and M large. If $M \to 0$, then $\lambda_n \to \lambda_n^0$. If $M \to \infty$, then $\lambda_n \to \lambda_{n-1}^0$, whereupon $\lambda_1 \to 0$.

33. Solve problem 31 assuming the outer boundary free.

34. Formulate the problem on the natural vibrations of a drum, similar to the problem on the vibrations of a circular membrane with an additional air volume. How does the fundamental frequency depend on the dimensions of the additional volume (see problem 5, chapter VI)?

35. The circular membrane of a large drum has radius

$$r_0 = 50\,\text{cm}, \qquad \rho = 0{\cdot}1\,\text{g/cm}^2, \qquad T = 10^8\,\text{dynes/cm}^2.$$

What will be the fundamental frequency if the membrane vibrates in free space? The addition of some volume of air to the membrane increases the fundamental frequency 1·45 times. Determine the size of the additional volume.

§ 3. Propagation and Radiation of Sound

In this section some problems on the propagation, radiation and scattering of sound in solids will be considered, leading to the wave equation

$$\Delta u + k^2 u = 0 \ (k^2 > 0). \tag{1}$$

In the solution of the wave equation for a cylinder and sphere, spherical functions $Y_n^{(k)} = P_n^{(k)} \cos(\vartheta) \frac{\cos}{\sin} k\phi$, $Y_n(\vartheta) = P_n(\cos\vartheta)$ and different cylindrical functions appear.

In solving exterior problems for a cylinder Hankel functions

$$H_n^{(2)}(\rho) = \begin{cases} \sqrt{\dfrac{2}{\pi\rho}}\, e^{-i\left(\rho - \frac{\pi}{2}n - \frac{\pi}{4}\right)} + \cdots & \text{for large } \rho, \\ \left.\begin{aligned} &\dfrac{i}{\pi}\left(\dfrac{2}{\rho}\right)^n (n-1)!, && n > 0, \\ &\dfrac{2i}{\pi} \ln \dfrac{1}{\rho}, && n = 0 \end{aligned}\right\} & \text{for small } \rho, \end{cases}$$

are utilized; $\rho = kr$. The function $H_n^{(2)}(kr)$ satisfies the radiation condition

$$\lim_{r\to\infty} \sqrt{r} \left(\frac{\partial u}{\partial r} + iku \right) = 0,$$

which corresponds to a time dependence of the form $e^{i\omega t}$. In solving problems on the radiation of sound by a sphere and the scattering of sound by a sphere the functions

$$\psi_n(\rho) = \sqrt{\frac{\pi}{2\rho}} J_{n+\frac{1}{2}}(\rho) = \begin{cases} \dfrac{\sin\left(\rho - \dfrac{\pi}{2} n\right)}{\rho} + \cdots & \text{for large } \rho, \\ \dfrac{\rho^n}{1.3.5 \ldots (2n+1)} + \ldots & \text{for small } \rho, \end{cases}$$

$$\zeta_n^{(2)}(\rho) = \sqrt{\frac{\pi}{2\rho}}\, H_{n+\frac{1}{2}}^{(2)}(\rho) =$$

$$= \begin{cases} \dfrac{e^{-i\left(\rho - \frac{\pi}{2}(n+1)\right)}}{\rho} + \ldots & \text{for large } \rho, \\ i\dfrac{1.3.5 \ldots (2n-1)}{\rho^{n+1}} + \dfrac{\rho^n}{1.3 \ldots (2n+1)} + \ldots & \text{for small } \rho \end{cases}$$

are employed. Later we shall use for small ρ the following notation for the function $\zeta_m^{(2)}(\rho)$:

$$\zeta_m^{(2)}(\rho) = \alpha_m \, e^{i\gamma_m} + \ldots,$$

where

$$\alpha_m = \frac{1.3 \ldots (2m-1)}{\rho^{m+1}}, \qquad \gamma_m = \frac{\rho^{2m+1}}{1^2.3^2 \ldots (2m-1)^2 (2m+1)^2}.$$

The function $\zeta_n^{(2)}(kr)$ satisfies the radiation condition

$$\lim_{r\to\infty} r\left(\frac{\partial u}{\partial r} + iku\right) = 0,$$

corresponding to a time dependence of the form $e^{i\omega t}$.

All the necessary theoretical information on the substance of paragraph 3 may be found in chapter VII, and also in supplements I and II of [7].

1. Point Source

36. Find the source function for the semispace $z > 0$, if in the plane $z = 0$ the solution of the equation $\Delta v + k^2 v = 0$

(a) satisfies a boundary condition of the first kind $v|_{z=0} = f$,

(b) satisfies a boundary condition of the second kind $\left.\frac{\partial v}{\partial z}\right|_{z=0} = f$.

37. Find the source function for the wave equation in the semispace $y > 0$

(a) for the first boundary-value problem,

(b) for the second boundary-value problem.

38. Calculate the energy which is radiated into free space by an isolated point source of sound, vibrating according to a harmonic law. Find also the value of the specific acoustic impedance.

39. A point source of sound is located in the semispace $z < 0$ at a distance a from the absolutely rigid wall $z = 0$.

Find the radiation of the source, and its intensity in the wave zone and compare with the solution of problem 38.

40. Solve problem 39 assuming that the semispace is filled with a liquid, bounded by the free surface $z = 0$, at which the pressure equals zero. Compare with the solutions of problems 38 and 39.

41. Prove the reciprocity principle in acoustics: "If in an air space, partly bounded by stationary bodies extending a finite distance, partly unbounded, sound waves are excited at any point M, then at any other specified point P the velocity potential agrees both in value and in phase with that which would occur at M, if the source of sound existed at P" (see [36]).

42. Show that in an infinite cylindrical tube of arbitrary section with absolutely rigid walls, travelling sound waves may exist under certain conditions. Find the minimum permissible frequency of the travelling wave, the phase velocity of the travelling waves and calculate the energy flow across a section of the pipe (waveguide). Consider the case of a rectangular and circular section.

43. Find the source function for a point source, situated inside a cylindrical pipe of arbitrary section, for the wave equation with boundary conditions

(a) of the first kind,

(b) of the second kind.

Consider the special case of a circular section.

44. Solve problem 43 for a semi-infinite pipe $z > 0$.

45. Find the source function of a point source for a cylindrical resonator $0 \leqslant z \leqslant l$ of arbitrary cross-section. The walls of the resonator are assumed absolutely rigid.

2. Radiation of Membranes, Cylinders and Spheres

46. A membrane is placed at the section $z = 0$ of a tube of circular section, considered in problem 42, and vibrates with velocity $v = v_0 e^{i\omega t}$. Determine the reaction of the pressure of the sound waves on the membrane.

47. Solve problem 46, assuming that the velocity of the exciting membrane varies according to the law

$$v = v_0(r)\, e^{i\omega t},$$

where $v_0(r)$ is a given function.

Consider the special case

$$v_0(r) = AJ_0\left(\frac{\mu_m}{a} r\right),$$

where A is a constant, μ_m is a root of the equation $J_0(\mu) = 0$.

Find the value of the Poynting vector and the value of the acoustic impedance in the piston.

48. Let a cylinder of radius a pulsate, i.e. contract and expand uniformly according to a harmonic law; its velocity at the surface for $r = a$ equals

$$v_0 e^{i\omega t}.$$

Find the pressure, and radial velocity of the gas at large distances from the cylinder axis, and also the energy flow.

49. Solve problem 48, assuming that the radius of the cylinder is small in comparison with the wavelength $\lambda = 2\pi c/\omega$, i.e.

$$ka \ll 1.$$

50. A cylinder of radius a vibrates as a whole perpendicularly to its axis (along the x-axis) with a velocity $v_0 e^{i\omega t}$. Find the pressure and velocity of air; for the case $ka \ll 1$ calculate the specific acoustic impedance and the total output of radiation per unit length.

51. A cylinder of radius a vibrates according to a harmonic law so that the velocity at its surface equals

$$f(\phi)\, e^{i\omega t},$$

where $f(\phi)$ is a given function. Find the pressure and velocity of the air, the energy flow (for small ka, where $k = \omega/c$).

Derive solutions of problems 48–50 from the formulae found.

52. The centre of a sphere of radius a vibrates along the polar axis with a velocity $v_0 e^{i\omega t}$. If $a \ll \lambda$ ($ka \ll 1$), λ is the wavelength, then such an acoustic radiator in the form of a small vibrating sphere is called an acoustic dipole. Find the energy flow and total energy, radiated by the acoustic dipole.

53. The surface of a sphere of finite size vibrates according to the harmonic law $f(\theta)\, e^{i\omega t}$. Find the total reaction of the medium on the sphere for $ka \ll 1$, where $k = \lambda/2\pi$. Consider the particular case

$$f(\theta) = v_0.$$

54. Investigate the sound field of a piston, mounted flush with the surface of the sphere and capable of vibrating without friction. The distribution of velocities over the sphere in the presence of such a piston can be described thus:

$$v(\theta)=\begin{cases} v_0 & \text{for} \quad 0\leqslant\theta\leqslant\theta_0, \\ 0 & \text{for} \quad \theta_0<\theta\leqslant\pi. \end{cases}$$

Consider the case of small θ_0. Give an expression for the pressure at low frequencies.

55. The surface of a sphere vibrates so that the radial component of velocity on the surface equals

$$v_a=\frac{v_0}{4}(1+3\cos 2\theta)\,e^{i\omega t}.$$

Such a source of sound is called a radiator of second order, or a quadrupole source. Calculate the intensity and magnitude of its radiation. Trace the polar diagram of the intensity of the radiation. Consider the case of long waves.

56. A solid circular lamina vibrates according to a simple harmonic law in a circular aperture, equal in area to it, cut in a solid plane lamina extending to infinity. Find the pressure and velocity of particles of air and the magnitude of the energy radiated.

57. Find the reaction of the sound field on the lamina, considered in problem 56. Consider the particular case where the radius of the piston is small in comparison with the wavelength ($ka \ll 1$).

58. Solve problem 56 if at the surface of the piston (lamina) the velocity is variable:

$$v=v(r)$$

(the piston is "flexible"). Consider only a representation of the solution in the wave zone.

3. Diffraction by a Cylinder and Sphere

59. A plane sound wave propagates in a direction, perpendicular to the axis of an infinite rigid cylinder of radius a. Find the scattered wave. Consider the case of large and small distances from the cylinder.

60. Proceeding from the solution of problem 59, calculate the intensity of the scattered wave.

61. Calculate the total energy scattered per unit length of a cylinder, for the limiting cases of short and long waves (see problem 59). Find the force acting on the cylinder.

62. Find the solution of the problem on the scattering of a plane sound wave by a spherical obstacle.

63. Utilizing the solution of problem 62, calculate the intensity of the scattered wave and the total scattering for the case

$$ka \ll 1,$$

where $k = 2\pi/\lambda = \omega/c$, λ is the wavelength, a the radius of the sphere.

Calculate the force acting on the sphere.

64. Solve the problem of the scattering of a plane wave by a sphere of radius $\rho = a$, if the sphere is completely free and moves under the action of air.

65. Solve the problem of the motion of a sphere of radius a under the action of an incident plane wave, if the sphere is fixed elastically, i.e. the restoring force equals

$$X = -M\omega_0^2\xi,$$

where ξ is the coordinate of the centre of the sphere and M is the mass of the sphere.

Neglect air friction.

§ 4. Steady-state Electromagnetic Vibrations

1. Maxwell's Equations. Potentials. Green–Ostrogradskii's Vector Formulae

66. Write down Maxwell's equations in an orthogonal curvilinear system of coordinates (x_1, x_2, x_3), in which the square of an element of length is given by the relation

$$ds^2 = h_1^2 dx_1^2 + h_2^2 dx_2^2 + h_3^2 dx_3^2,$$

where h_1, h_2, h_3 are metric coefficients.

67. Show that the solution of Maxwell's equations

$$\operatorname{curl} \boldsymbol{H} = \frac{1}{c}\frac{\partial \boldsymbol{D}}{\partial t} + \frac{4\pi}{c}\boldsymbol{j}, \qquad \operatorname{div} \boldsymbol{B} = 0, \quad \boldsymbol{B} = \mu \boldsymbol{H} \ (\mu = \text{const.}),$$

$$\operatorname{curl} \boldsymbol{E} = -\frac{1}{c}\frac{\partial \boldsymbol{B}}{\partial t}, \qquad \operatorname{div} \boldsymbol{D} = 4\pi\rho, \ \boldsymbol{D} = \varepsilon \boldsymbol{E} \ (\varepsilon = \text{const.})$$

may be represented in the form

$$\boldsymbol{B} = \operatorname{curl} \boldsymbol{A}, \quad \boldsymbol{E} = -\operatorname{grad} \phi - \frac{1}{c}\frac{\partial \boldsymbol{A}}{\partial t},$$

where $\boldsymbol{A}$ is the vector-potential, ϕ is the scalar potential related by the Lorentz condition

$$\operatorname{div} \boldsymbol{A} + \frac{\varepsilon\mu}{c}\frac{\partial \phi}{\partial t} = 0$$

and satisfying the equations

$$\Delta\phi - \frac{1}{a^2}\frac{\partial^2 \phi}{\partial t^2} = -\frac{4\pi}{\varepsilon}\rho, \qquad a^2 = \frac{c^2}{\varepsilon\mu},$$

$$\Delta \boldsymbol{A} - \frac{1}{a^2}\frac{\partial^2 \boldsymbol{A}}{\partial t^2} = -\frac{4\pi}{c}\mu \boldsymbol{j}.$$

Here $\Delta \boldsymbol{A}$ is the Laplacian operator, acting on the curvilinear components of vector $\boldsymbol{A}$.

Find an expression for $\Delta \boldsymbol{A}$ in curvilinear orthogonal coordinates. Show that for $\rho = 0$, $\boldsymbol{j} = 0$ Maxwell's equations give a solution of the form

$$\boldsymbol{D} = -\operatorname{curl} \boldsymbol{A}', \quad \boldsymbol{H} = -\operatorname{grad} \phi' - \frac{1}{c}\frac{\partial \boldsymbol{A}'}{\partial t},$$

where $\boldsymbol{A}'$ and ϕ' are the so-called antipotentials.

Consider the case where the time dependence has the form $e^{-i\omega t}$.

68. Find equations for the scalar and vector potentials for Maxwell's equations in a homogeneous conducting medium.

69. Find equations for the polarization potential $\boldsymbol{\Pi}$ (Hertz electric vector) for Maxwell's equations in a vacuum, using the relations

$$\boldsymbol{A} = \frac{1}{c}\frac{\partial \boldsymbol{\Pi}}{\partial t}, \qquad \phi = -\operatorname{div} \boldsymbol{\Pi},$$

where $\boldsymbol{A}$ is the vector potential, ϕ is the scalar potential.

Consider the case where the time dependence has the form $e^{-i\omega t}$. By analogy introduce the magnetic Hertz vector $\boldsymbol{\Pi}'$. Determine the Hertz vectors in a conducting medium.

70. If the metric coefficient $h_1 = 1$ and the electromagnetic fields in a vacuum depend on time as $e^{-i\omega t}$, then they may be represented by means of two scalar functions U and U' (Borgnis' functions):

(a) for a field of electric type ($H_1 = 0$) we have:

$$E_1 = k^2U + \frac{\partial^2 U}{\partial x_1^2}, \quad E_2 = \frac{1}{h_2}\frac{\partial^2 U}{\partial x_1 \partial x_2}, \quad E_3 = \frac{1}{h_3}\frac{\partial^2 U}{\partial x_1 \partial x_3} \quad \left(k = \frac{\omega}{c}\right),$$

$$H_1 = 0, \qquad H_2 = \frac{ik}{h_3}\frac{\partial U}{\partial x_3}, \qquad H_3 = -\frac{ik}{h_2}\frac{\partial U}{\partial x_2};$$

(b) for a field of magnetic type ($E_1 = 0$) we have:

$$E_1' = 0, \qquad E_2' = -\frac{ik}{h_3}\frac{\partial U'}{\partial x_3}, \qquad E_3' = \frac{ik}{h_2}\frac{\partial U'}{\partial x_2},$$

$$H_1' = k^2U' + \frac{\partial^2 U'}{\partial x_1^2}; \qquad H_2' = \frac{1}{h_2}\frac{\partial^2 U'}{\partial x_1 \partial x_2}, \qquad H_3' = \frac{1}{h_3}\frac{\partial^2 U'}{\partial x_1 \partial x_3},$$

where U and U' are functions satisfying the equation

$$\frac{\partial^2 U}{\partial x_1^2} + \frac{1}{h_2 h_3}\left[\frac{\partial}{\partial x_2}\left(\frac{h_3}{h_2}\frac{\partial U}{\partial x_2}\right) + \frac{\partial}{\partial x_3}\left(\frac{h_2}{h_3}\frac{\partial U}{\partial x_3}\right)\right] + k^2 U = 0.$$

Prove this statement.

Consider next a spherical and cylindrical system of coordinates. Show that in a cylindrical system of coordinates $\boldsymbol{\Pi}$ U coincides with the z-component of the Hertz vector $\boldsymbol{\Pi} = (0, 0, U)$.

71. Introduce the functions U and U' for the electromagnetic field in a conducting medium, the parameters of which are ε, μ, σ (conductivity).

72. A sphere of radius a with conductivity σ_1, and dielectric constant ε_1 is placed in an infinite medium of conductivity σ_2 and dielectric constant ε_2. Introducing the functions U and U', formulate boundary conditions for them at the surface of the sphere.

73. Prove the validity of the vector analogue of Green's second formula

$$\int_T (\boldsymbol{W}\,\text{curl curl}\,\boldsymbol{U}-\boldsymbol{U}\,\text{curl curl}\,\boldsymbol{W})\,d\tau$$
$$=\int_\Sigma \{[\boldsymbol{U}\,\text{curl}\,\boldsymbol{W}]-[\boldsymbol{W}\,\text{curl}\,\boldsymbol{U}]\}\,\boldsymbol{n}\,d\sigma,$$

where $\boldsymbol{U}=\boldsymbol{U}(x,y,z)$, $\boldsymbol{W}=\boldsymbol{W}(x,y,z)$ are arbitrary, sufficiently smooth vector functions, T is some volume, bounded by the surface Σ, $\boldsymbol{n}$ is the unit vector, normal to the surface Σ.

74. Prove the validity of the vector analogue of Green's fundamental formula

$$\boldsymbol{U}(M_0)=\boldsymbol{U}(x,y,z)$$
$$=\frac{1}{4\pi}\int_T \{\phi(\text{curl curl}\,\boldsymbol{U}-k^2\boldsymbol{U})+\text{grad}\,\phi\,\text{div}\,\boldsymbol{U}\}\,d\tau p-$$
$$-\frac{1}{4\pi}\int_\Sigma \{[\boldsymbol{n}\,\text{curl}\,\boldsymbol{U}]\,\phi+[[\boldsymbol{n}\boldsymbol{U}]\,\text{grad}\,\phi]+\boldsymbol{n}\boldsymbol{U}\,\text{grad}\,\phi\}\,d\sigma p,$$

where $\boldsymbol{U}$ is an arbitrary vector

$$\phi=\frac{e^{ikt}}{r},\qquad r=\sqrt{(x-\xi)^2+(y-\eta)^2+(z-\zeta)^2}$$

is the distance between the points $M_0(x,y,z)$ and $P(\xi,\eta,\zeta)$.

75. Utilizing Green's fundamental vector relation, obtained in the preceding problem, directly, not introducing potentials, write down an expression for $\boldsymbol{E}$ and $\boldsymbol{H}$, the solutions of Maxwell's equations

$$\text{curl}\,\boldsymbol{H}=-ik_0\varepsilon\boldsymbol{E}+\frac{4\pi}{c}\boldsymbol{j},\qquad k_0=\frac{\omega}{c},\qquad k=k_0\sqrt{\varepsilon\mu},$$
$$\text{curl}\,\boldsymbol{E}=ik_0\mu\boldsymbol{H}_0,$$
$$\text{div}\,\boldsymbol{H}=0,$$
$$\text{div}\,\boldsymbol{E}=\frac{4\pi\rho}{\varepsilon}$$

at interior points of some region T in terms of their values at the surface Σ, bounding the volume T.

2. Propagation of Electromagnetic Waves and Vibrations in Resonators

76. Examine the possibility of propagation of electromagnetic waves along the outside of an infinitely long circular cylinder, the conductivity of which is infinitely high. The conductivity of the surrounding medium σ is finite.

77. Solve the preceding problem, assuming that the conductivity of the cylinder is finite and equal to σ_1.

78. Show that inside an infinite hollow cylindrical tube (waveguide) of arbitrary section with ideally conducting walls there can exist a finite number of travelling electromagnetic waves. Find an expression for the phase velocity and energy flow of a travelling wave in the waveguide.

79. Prove the existence of travelling electromagnetic waves inside a coaxial cable bounded by two coaxial cylindrical surfaces $\rho = a$ and $\rho = b$. The walls of the cable are assumed ideally conducting.

Calculate the energy flow and write down an expression for the components of the field for the fundamental wave, corresponding to maximum wavelength.

80. Find the natural frequencies and the corresponding electromagnetic fields of a spherical resonator with ideally conducting walls. Calculate the average energy over a period in the standing wave.

81. Find the natural electromagnetic vibrations of a cylindrical resonator, which is "a section" of a cylindrical waveguide of arbitrary section with ideally conducting walls. Calculate the average energy over a period in the standing wave. Consider the particular cases of resonators

(a) of rectangular section,

(b) of circular section.

82. Determine the natural frequencies of electromagnetic vibrations inside a toroidal resonator of rectangular section, assuming the walls of the resonator ideally conducting.

83. *Diffraction by a cylinder.* A plane electromagnetic wave is incident on an infinite circular cylindrical conductor, the axis of which is perpendicular to the direction of propagation of the wave. Find the diffracted electromagnetic field, assuming the cylinder to be conducting. The cylinder is surrounded by a dielectric of dielectric constant ε_1 and conductivity equal to zero. Consider the case of an ideally conducting cylinder. Assuming that the radius a of the cylinder is small in comparison with the length of the incident wave ($ka \ll 1$, $k = \lambda/2\pi$) calculate the total scattering.

84. *Diffraction by an ideally conducting sphere.* Consider the problem of the scattering of a plane electromagnetic wave by an ideally conducting sphere. Find the electromagnetic field.

85. *Diffraction by a conducting sphere.* A plane electromagnetic wave, propagating in a medium with parameters $\varepsilon = \varepsilon_1$, $\sigma = 0$, $\mu = 1$, encounters in its path a conducting sphere of radius a with parameters $\varepsilon = \varepsilon_2$, $\mu = \mu_2$, $\sigma = \sigma_2 \neq 0$.

Find the electromagnetic field inside and outside the sphere (problem of Mee on the diffraction by a sphere).

3. Radiation of Electromagnetic Waves

86. Find the radiation field of a small electric dipole, existing in an infinite non-conducting region. Calculate the average output of radiation over a period.

87. Solve the preceding problem, utilizing the representation of the components of the electromagnetic field by means of Borgnis' function in a spherical system of coordinates

$$E_r = \frac{d^2U}{dr^2} + k^2U, \quad E_\vartheta = \frac{1}{r}\frac{d^2U}{dr\,d\phi}, \quad E_\phi = \frac{1}{r\sin\theta}\frac{d^2U}{dr\,d\phi},$$

$$H_r = 0, \quad H_\theta = -\frac{ik}{r}\frac{dU}{d\phi}, \quad H_\phi = \frac{ik}{r}\frac{dU}{d\theta},$$

where U satisfies the equation

$$\frac{d^2U}{dr^2} + \frac{1}{r^2\sin\theta}\frac{d}{d\theta}\left(\sin\theta\frac{dU}{d\theta}\right) + \frac{1}{r^2\sin^2\theta}\frac{d^2U}{d\phi^2} + k^2U = 0.$$

so that the function $u = U/r$ satisfies the equation

$$\Delta u + k^2u = 0.$$

88. At the centre of a spherical resonator with ideally conducting walls there is a small electric dipole, directed along the radius.

Determine the electromagnetic field, produced by the dipole inside the resonator.

89. A homogeneous conducting sphere of radius a with constants $\varepsilon_2, \mu_2, \sigma_2$ is situated in a medium with physical constants $\varepsilon_1, \mu_1, \sigma_1$. At the centre of the sphere there exists an electric dipole, oscillating according to a harmonic law $e^{-i\omega t}$. Calculate the field inside the sphere, the average output of radiation over a period and consider the limiting case $\sigma_1 \to \infty$. Consider the special case where $a \to \infty$.

90. The walls of a spherical resonator are made of a homogeneous conducting material of conductivity σ. Let $r = a$ and $r = b$ be the radii of the resonator walls. At the centre of the resonator an electric dipole is situated, vibrating according to the harmonic law $e^{-i\omega t}$. Find the forced electromagnetic vibrations of the resonator, assuming that the region $r > b$ is non-conducting (air).

Consider the limiting cases $b \to \infty$ and $b \to a$.

91. Inside a sphere of radius a an electric dipole is situated, oriented along the radius and at a distance $r = r'$ from the centre of the sphere. Determine the electromagnetic field of radiation inside the sphere, assuming that the sphere is surrounded by a homogeneous medium having a finite conductivity σ. Consider the limiting case $\sigma \to \infty$.

Consider the special cases: the radius of the sphere is small in comparison with the wavelength and $a \to \infty$.

92. *Vertical electric antenna above the spherical earth.* Find the electromagnetic field, produced by an electric antenna, existing above the earth's surface, which is considered as a sphere of radius a, having a finite conductivity σ and dielectric permeability ε. The antenna is assumed to be an elementary dipole, performing harmonic oscillations along the direction of the earth's diameter. The atmosphere is assumed homogeneous and non-conducting ($\varepsilon = \mu = 1$, $\sigma = 0$).

93. *Vertical antenna on the spherical earth.* Solve the preceding problem (92) assuming that the antenna exists on the earth's surface and is directed along the normal to it.

4. Antenna on the Plane Earth

In problems 94–101 the propagation of waves, emitted by antennae, located on the earth's surface is considered. In addition we shall assume the earth to be plane, homogeneous and conducting (sometimes ideally conducting, sometimes having a finite conductivity); we treat the antenna as a dipole, whose moment periodically varies with time with frequency ω: $\boldsymbol{p} = \boldsymbol{p}_0 e^{-i\omega t}$. For simplicity we shall assume $|\boldsymbol{p}_0| = 1$.

In problems 94–97 it is required to introduce the Hertz vector and state the boundary-value problem for its components.

For the solutions of problems 98–101 it is necessary to carry out a calculation of the electromagnetic field, radiated by the antenna, and also the amount of radiation averaged over a period.

Here an expansion in the Fourier–Bessel integral with the use of the Sommerfeld integral

$$\frac{e^{ikR}}{R} = \int_0^\infty J_0(\lambda r)\, e^{-\sqrt{\lambda^2-k^2}\,|z|} \frac{\lambda\, d\lambda}{\sqrt{\lambda^2 - k_1^2}}, \qquad R = \sqrt{r^2+z^2}$$

is essential for a method of solution.

94. *Vertical electric antenna.* On the plane surface of the earth, filling the semispace $z < 0$, a vertical electric antenna is situated, directed along the z-axis.

Introduce the Hertz vector and formulate boundary conditions for it on the earth's surface, and also eliminate the singularity at the origin. For a solution assume $\mu = 1$.

95. *Vertical magnetic antenna.* On the earth's surface $z = 0$ there exists a vertical magnetic antenna (horizontal coil).

State the boundary-value problem for the corresponding Hertz vector, if the earth has finite conductivity.

96. *Horizontal electric antenna.* State the boundary-value problem for a horizontal antenna lying on the earth's surface, the conductivity of which is finite.

97. *Horizontal magnetic antenna.* An elementary magnetic dipole, situated on the earth's surface $z = 0$, is oriented along the y-axis, i.e. the current coil lies in the vertical plane xz. Formulate the corresponding boundary-value problem for the Hertz vector, assuming the earth to be conducting.

98. Find the electromagnetic radiation field of a vertical electric antenna at the surface of the plane earth (see problem 94). Calculate the energy flow of radiation, assuming $\mu = 1$. Consider the case where the earth is ideally conducting and when the earth is replaced by air.

99. Determine the field, emitted by a vertical magnetic antenna, at the surface of the earth (see problem 95).

100. Solve the problem of the propagation of waves, emitted by a horizontal electric antenna at the earth's surface (see problem 96).

101. Find the electromagnetic field, produced by a horizontal magnetic antenna lying on the surface of the earth (see problem 97).

102. A vertical electric dipole is situated in a medium *1*, whose propagation constant equals k_1, at a point $z = z_0$, $r = 0$. Medium *2* has the form of a plane-parallel plate with propagation constant k_2 and boundaries $z = a < z_0$ and $z = 0$. The semispace $z < 0$ is ideally conducting.

Find the polarization potential of the secondary field $\boldsymbol{\Pi}_{\text{sec}}$.

103. Find the electromagnetic field, produced by a finite line current in infinite space, and calculate the field in the wave zone. Determine the resistance to radiation.

104. Determine the radiation resistance of a half-wave dipole in infinite space, and also the reactive part of the input impedance (reactance) of the half-wave dipole.

105. Inside the cylindrical waveguide, considered in problem 78, a point dipole is placed, parallel to the guide axis and oscillating harmonically as $e^{-i\omega t}$.

Find the energy flow averaged over a period, emitted by the dipole. Calculate the radiation resistance. Look for the solution for a waveguide of arbitrary section and then consider a guide of circular section, assuming that the dipole lies along the axis.

106. Find an expression for the electromagnetic field inside a waveguide, produced by a line current of length $2l$, parallel to the guide axis, and calculate the energy flow across a cross-section of the guide for the particular case of a half-wave dipole, lying on the axis of a waveguide of circular section. Find the active and reactive components of the input impedance. The problem may be solved by an approximation, neglecting the effect of the secondary field on the current distribution in the dipole.

107. Use the solution of problem 106 to find the radiation resistance and reactance of a half-wave dipole, lying on the axis of a waveguide of circular section and directed along this axis.

108. Calculate the field, produced inside an infinite rectangular waveguide with ideally conducting walls by an electric dipole, perpendicular to the guide axis and parallel to one side of a perpendicular section, and find the radiation resistance for

(a) an infinitely small dipole,

(b) a half-wave dipole.

HINTS, ANSWERS AND SOLUTIONS

CHAPTER I

CLASSIFICATION AND REDUCTION TO CANONICAL FORM OF SECOND ORDER PARTIAL DIFFERENTIAL EQUATIONS

§ 1. The Equation for a Function of Two Independent Variables $a_{11}u_{xx}+2a_{12}u_{xy}+a_{22}u_{yy}+b_1u_x+b_2u_y+cu = f(x, y)$

1. The Equation with Variable Coefficients

1. The discriminant of the equation $(l+x)u_{xx}+2xyu_{xy}-y^2u_{yy} = 0$ is equal to $a_{12}^2-a_{11}a_{22} = y^2[x^2+x+l] = y^2(x-x_1)(x-x_2)$, where

$$x_1 = -\frac{1-\sqrt{1-4l}}{2}, \quad x_2 = -\frac{1+\sqrt{1-4l}}{2}.$$

Let $l < 1/4$, then x_1 and x_2 are real, and for $x < x_1$ and also for $x > x_2$ the equation is hyperbolic, and for $x_1 < x < x_2$ it is elliptic; the straight lines $x = x_1$ and $x = x_2$ are boundaries of these regions. For $l = 1/4$ the region of ellipticity vanishes, since $x_1 = x_2 = -1/2$; the straight line $x = -1/2$ forms the boundary. For $l > 1/4$ the equation is hyperbolic everywhere.

2. The equation $u_{xx}+xu_{yy} = 0$ for $x < 0$ belongs to the hyperbolic type and by the substitution $\xi = \frac{3}{2}y+(\sqrt{-x})^3$, $\eta = \frac{3}{2}y-(\sqrt{-x})^3$ reduces to the canonical form

$$u_{\xi\eta}-\frac{1}{6(\xi-\eta)}(u_\xi-u_\eta) = 0, \quad \xi > \eta.$$

For $x > 0$ the equation $u_{xx}+xu_{yy} = 0$ belongs to the elliptic type and by the substitution $\xi' = \frac{3}{2}y$, $\eta' = -\sqrt{x^3}$ reduces to the canonical form

$$u_{\xi'\xi'}+u_{\eta'\eta'}+\frac{1}{3\eta'}u_{\eta'} = 0, \quad \eta' < 0.$$

The characteristics of the equation are the curves (Fig. 14)

$$y-c = \pm\frac{2}{3}(\sqrt{-x})^3,$$

where the branches, directed downwards, are given by the equations $\xi = \text{const.}$, and the branches, directed upwards, are given by the equations $\eta = \text{const.}$

3. The equation $u_{xx}+yu_{yy} = 0$ for $y < 0$ is hyperbolic and by the substitution $\xi = x+2\sqrt{-y}$, $\eta = x-2\sqrt{-y}$ reduces to the canonical form

$$u_{\xi\eta}+\frac{1}{2(\xi-\eta)}(u_\xi+u_\eta) = 0, \quad \xi > \eta.$$

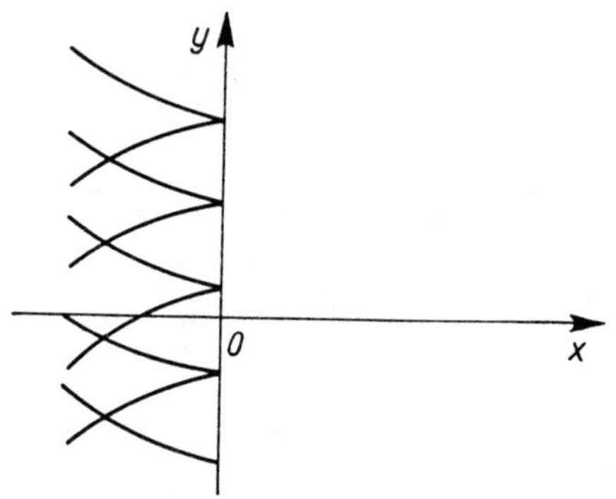

Fig. 14

For $y > 0$ the equation is elliptic and by substituting $\xi' = x$, $\eta' = 2\sqrt{y}$ reduces to the canonical form

$$u_{\xi'\xi'}+u_{\eta'\eta'}-\frac{1}{\eta'}u_{\eta'} = 0, \quad \eta' > 0.$$

The characteristics of the equation are the parabolae (Fig. 15)

$$y = -\frac{1}{4}(x-c)^2.$$

The branches, to the left of the x-axis, are given by the equation $\xi =$ const. and to the right by $\eta =$ const.

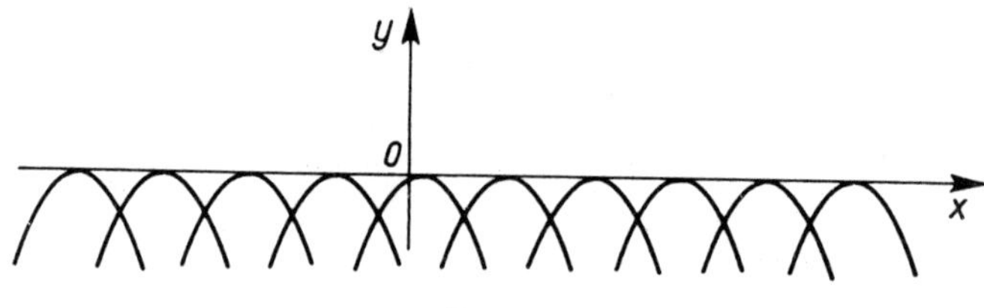

Fig. 15

4. The equation $u_{xx}+yu_{yy}+\frac{1}{2}u_y = 0$ is of a similar type to the equation $u_{xx}+yu_{yy} = 0$, considered in the preceding problem. By the same substitutions as in the equation $u_{xx}+yu_{yy} = 0$, it reduces to the canonical form $\partial^2 u/\partial\xi\partial\eta = 0$ in the region where it is hyperbolic ($y < 0$) and to the canonical form $\partial^2 u/\partial\xi^2 + \partial^2 u/\partial\eta^2 = 0$ in the region where it is elliptic ($y > 0$). The characteristics of the equations $u_{xx}+yu_{yy}+\frac{1}{2}u_y = 0$ and $u_{xx}+yu_{yy} = 0$ coincide.

Note. Comparison of the equations $u_{xx}+yu_{yy}=0$ and $u_{xx}+yu_{yy}+\frac{1}{2}u_y=0$ shows that the presence of terms with lower derivatives modifies the equation essentially since in the one case the coefficients of the equation after reduction to canonical form have a singularity, and in the other case do not.

5. The equation $yu_{xx}+xu_{yy}=0$ is hyperbolic in the second and fourth quadrants and reduces to the canonical form

$$\frac{\partial^2 u}{\partial \xi^2}-\frac{\partial^2 u}{\partial \eta^2}+\frac{1}{3\xi}\frac{\partial u}{\partial \xi}-\frac{1}{3\eta}\frac{\partial u}{\partial \eta}=0$$

by means of the substitution $\xi=(-x)^{3/2}$, $\eta=(y)^{3/2}$ in the second quadrant, $\xi=x^{3/2}$, $\eta=(-y)^{3/2}$ in the fourth quadrant. In the first and third quadrants the equation is elliptic and reduces to the canonical form

$$\frac{\partial^2 u}{\partial \xi'^2}+\frac{\partial^2 u}{\partial \eta'^2}+\frac{1}{3\xi'}\frac{\partial u}{\partial \xi'}+\frac{1}{3\eta'}\frac{\partial u}{\partial \eta'}=0,$$

by means of the substitution $\xi=x^{3/2}$, $\eta=y^{3/2}$ in the first quadrant, $\xi=(-x)^{3/2}$, $\eta=(-y)^{3/2}$ in the third quadrant. The x and y axes are boundaries of the regions. As is well known†, the transition from one canonical form of the hyperbolic equation

$$\frac{\partial^2 u}{\partial \xi \partial \eta}=f\left(\xi,\eta,u,\frac{\partial u}{\partial \xi},\frac{\partial u}{\partial \eta}\right)$$

to the other

$$\frac{\partial^2 u}{\partial \bar{\xi}^2}-\frac{\partial^2 u}{\partial \bar{\eta}^2}=\bar{f}\left(\bar{\xi},\bar{\eta},u,\frac{\partial u}{\partial \bar{\xi}},\frac{\partial u}{\partial \bar{\eta}}\right)$$

is made by the substitution

$$\bar{\xi}=\frac{\xi+\eta}{2},\qquad \bar{\eta}=\frac{\xi-\eta}{2}.$$

6. The equation $xu_{xx}+yu_{yy}=0$ is elliptic in the first and third quadrants and reduces to the canonical form

$$\frac{\partial^2 u}{\partial \xi^2}+\frac{\partial^2 u}{\partial \eta^2}-\frac{1}{\xi}\frac{\partial u}{\partial \xi}-\frac{1}{\eta}\frac{\partial u}{\partial \eta}=0$$

by the substitution $\xi=x^{1/2}$, $\eta=y^{1/2}$ in the first quadrant, $\xi=(-x)^{1/2}$, $\eta=(-y)^{1/2}$ in the third quadrant.

The equation is hyperbolic in the second and fourth quadrants and is reduced to the canonical form

$$\frac{\partial^2 u}{\partial \xi^2}-\frac{\partial^2 u}{\partial \eta^2}-\frac{1}{\xi}\frac{\partial u}{\partial \xi}-\frac{1}{\eta}\frac{\partial u}{\partial \eta}=0$$

by the substitution $\xi=(-x)^{1/2}$, $\eta=(y)^{1/2}$ in the second quadrant, $\xi=(x)^{1/2}$, $\eta=(-y)^{1/2}$ in the fourth quadrant. The x and y axes are boundaries.

† See [7], page 7.

7. The equation $u_{xx}+xyu_{yy}=0$ is elliptic in the first and third quadrants and is reduced to the canonical form

$$u_{\xi\xi}+u_{\eta\eta}+\frac{1}{2\xi}u_{\xi}-\frac{1}{2\eta}u_{\eta}=0$$

by the substitution $\xi=\frac{2}{3}x^{3/2}$, $\eta=2y^{1/2}$ in the first quadrant, and $\xi=\frac{2}{3}(-x)^{3/2}$, $\eta=2(-y)^{1/2}$ in the third quadrant.

The equation is hyperbolic in the second and fourth quadrants and is reduced to the canonical form

$$u_{\xi\xi}-u_{\eta\eta}+\frac{1}{2\xi}u_{\xi}+\frac{1}{2\eta}u_{\eta}=0,$$

by means of the substitution $\xi=\frac{2}{3}(-x)^{3/2}$, $\eta=2y^{1/2}$ in the second quadrant and $\xi=\frac{2}{3}x^{3/2}$, $\eta=2(-y)^{1/2}$ in the fourth quadrant. The x and y axes are boundaries.

8. The equation $u_{xx}\operatorname{sign}y+2u_{xy}+u_{yy}=0$ is parabolic in the first and second quadrants and by the substitution

$$\xi=x+y, \quad \eta=x-y$$

is reduced to the canonical form

$$\frac{\partial^2 u}{\partial\xi^2}=0.$$

It is hyperbolic in the third and fourth quadrants and by the substitution

$$\xi=(1+\sqrt{2})\,x+y, \quad \eta=(1-\sqrt{2})\,x+y$$

is reduced to the canonical form

$$\frac{\partial^2 u}{\partial\xi\,\partial\eta}=0.$$

9. The equation $u_{xx}+2u_{xy}+(1-\operatorname{sign}y)u_{yy}=0$ is hyperbolic in the first and second quadrants and by the substitution $\xi=x-2y$, $\eta=y$ reduces to the canonical form $\partial^2u/\partial\xi\partial\eta$, and it is elliptic in the third and fourth quadrants and by the substitution

$$\xi=x-y, \quad \eta=x$$

reduces to the canonical form $\partial^2u/\partial\xi^2+\partial^2u/\partial\eta^2=0$.

10. The equation $u_{xx}\operatorname{sign}y+2u_{xy}+u_{yy}\operatorname{sign}x=0$ is parabolic in the first and third quadrants and by the substitution $\xi=x+y$, $\eta=x-y$ is reduced to the canonical form $\partial^2u/\partial\xi^2=0$ in the first quadrant and to $\partial^2u/\partial\eta^2=0$ in the third quadrant. The equation is hyperbolic in the second and fourth quadrants and is reduced to the canonical form

$$\frac{\partial^2 u}{\partial\xi\,\partial\eta}=0$$

by the substitution $\xi=-(1+\sqrt{2})x+y$, $\eta=-(1-\sqrt{2})x+y$ in the second quadrant, $\xi=(1+\sqrt{2})x+y$, $\eta=(1-\sqrt{2})x+y$ in the fourth quadrant.

11. The equation $y^2u_{xx}-x^2u_{yy} = 0$ is hyperbolic everywhere, except the coordinate axes, which are boundaries. It is reduced to the canonical form

$$\frac{\partial^2 u}{\partial \xi\, \partial \eta}+\frac{\eta}{4(\eta^2-\xi^2)}\frac{\partial u}{\partial \xi}+\frac{\xi}{4(\eta^2-\xi^2)}\frac{\partial u}{\partial \eta} = 0$$

by the substitution $\xi = y^2-x^2$, $\eta = y^2+x^2$.

12. The equation $x^2u_{xx}-y^2x^2u_{xx}-y^2u_{yy} = 0$ is hyperbolic everywhere, except the coordinate axes, which are boundaries. It is reduced to the canonical form

$$\frac{\partial^2 u}{\partial \xi\, \partial \eta}-\frac{1}{2\xi}\frac{\partial u}{\partial \xi} = 0$$

by the substitution $\xi = xy$, $\eta = y/x$.

13. The equation $x^2u_{xx}+y^2u_{yy} = 0$ is elliptic everywhere except the coordinate axes, which are boundaries. It is reduced to the canonical form

$$\frac{\partial^2 u}{\partial \xi^2}+\frac{\partial^2 u}{\partial \eta^2}-\frac{\partial u}{\partial \xi}-\frac{\partial u}{\partial \eta} = 0$$

by the substitution $\xi = \ln x$, $\eta = \ln y$.

14. The equation $y^2u_{xx}+x^2u_{yy}= 0$ is elliptic everywhere except the coordinate axes, which are boundaries. It is reduced by the substitution

$$\xi = y^2, \quad \eta = x^2$$

to the canonical form

$$\frac{\partial^2 u}{\partial \xi^2}+\frac{\partial^2 u}{\partial \eta^2}+\frac{1}{2\xi}\frac{\partial u}{\partial \xi}+\frac{1}{2\eta}\frac{\partial u}{\partial \eta} = 0.$$

15. The equation $y^2u_{xx}+2xyu_{xy}+x^2u_{yy} = 0$ is parabolic everywhere; by the substitution $\xi = (x^2+y^2)/2$; $\eta = (x^2-y^2)/2$ it is reduced to the canonical form

$$\frac{\partial^2 u}{\partial \xi^2}+\frac{\xi}{2(\xi^2-\eta^2)}\frac{\partial u}{\partial \xi}-\frac{\eta}{2(\xi^2-\eta^2)}\frac{\partial u}{\partial \eta} = 0.$$

16. The equation $x^2u_{xx}+2xyu_{xy}+y^2u_{yy} = 0$ is parabolic everywhere. By the substitution $\xi = y/x$, $\eta = y$ it is reduced to the canonical form

$$\frac{\partial^2 u}{\partial \eta^2} = 0.$$

17. The equation $4y^2u_{xx}-e^{2x}u_{yy}-4y^2u_x = 0$ is hyperbolic. By the substitution $\xi = e^x+y^2$, $\eta = -e^x+y^2$ it is reduced to the canonical form

$$u_{\xi\eta} = -\frac{1}{2(\xi+\eta)^2}(u_\xi+u_\eta).$$

18. The equation $x^2u_{xx}+2xyu_{xy}-3y^2u_{yy}-2xu_x+4yu_y+16x^4u = 0$ is hyperbolic everywhere except the x and y axes, which are boundaries. By the substitution $\xi = xy$, $\eta = x^3/y$ it is reduced to the canonical form

$$\frac{\partial^2 u}{\partial \xi\, \partial \eta}+\frac{1}{4\eta}\frac{\partial u}{\partial \xi}-\frac{1}{\xi}\frac{\partial u}{\partial \eta}+u = 0.$$

19. The equation $(1+x^2)u_{xx}+(1+y^2)u_{yy}+xu_x+yu_y = 0$ is elliptic everywhere. By the substitution $\xi = \ln(x+\sqrt{1+x^2})$, $\eta = \ln(y+\sqrt{1+y^2})$ it is reduced to the canonical form

$$\frac{\partial^2 u}{\partial \xi^2}+\frac{\partial^2 u}{\partial \eta^2} = 0.$$

20. The equation $u_{xx}\sin^2 x-2yu_{xy}\sin x+y^2u_{yy} = 0$ is parabolic everywhere. By the substitution $\xi = y\tan x/2$, $\eta = y$ it is reduced to the canonical form

$$\frac{\partial^2 u}{\partial \eta^2}-\frac{2\xi}{\xi^2+\eta^2}\frac{\partial u}{\partial \xi} = 0.$$

2. The Equation with Constant Coefficients

21. $$\frac{\partial^2 v}{\partial \xi\, \partial \eta}+\frac{4bc-b^2-c^2-12a}{144a^2}v = 0,$$

$$\xi = y+(\sqrt{3}-2)\,x, \quad \eta = y-(\sqrt{3}+2)\,x, \quad u(\xi,\eta) = e^{\alpha\xi+\beta\eta}v(\xi,\eta).$$

$$\alpha = \frac{c-(\sqrt{3}+2)\,b}{12a}, \quad \beta = \frac{c+(\sqrt{3}-2)\,b}{12a}.$$

22. $$\frac{\partial^2 v}{\partial \xi^2}+\frac{\partial^2 v}{\partial \eta^2}+\frac{2}{a}\left(\frac{2bc-b^2-2c^2}{a}+1\right)v = 0,$$

$$\xi = y-\frac{1}{2}x, \quad \eta = \frac{x}{2}, \quad u(\xi,\eta) = e^{\alpha\xi+\beta\eta}v(\xi,\eta),$$

$$\alpha = \frac{b-2c}{a}, \quad \beta = -\frac{b}{a}.$$

23. $$\frac{\partial^2 v}{\partial \eta^2}+\frac{c-b}{a}\frac{\partial v}{\partial \xi} = 0,$$

$$\xi = y-x, \quad \eta = x, \quad u(\xi,\eta) = e^{\alpha\xi+\beta\eta}v(\xi,\eta),$$

$$\alpha = \frac{b^2-4a}{4a(c-b)}, \quad \beta = -\frac{b}{2a}.$$

§ 2. The Equation with Constant Coefficients for a Function of n Independent Variables

$$\sum_{i,k=1}^{n} a_{ik}u_{x_ix_k}+\sum_{i=1}^{n} b_i u_{x_i}+cu = f(x_1,\ldots,x_n)$$

The type of equation

$$\sum_{i,k=1}^{n} a_{ik}u_{x_ix_k}+\sum_{i=1}^{n} b_i u_{x_i}+cu = f(x_1, x_2, \ldots x_n) \tag{1}$$

is determined by the matrix of coefficients of the second derivatives

$$\| a_{ik} \| \tag{2}$$

or by the quadratic form

$$\sum_{i,k=1}^{n} a_{ik} z_i z_k . \tag{3}$$

If in equation (1) one transforms to new independent variables

$$\xi_k = \sum_{i=1}^{n} \alpha_{ki} x_i, \quad k = 1, 2, \ldots, n, \tag{4}$$

then the matrix $\|\bar{a}_{ik}\|$ of the coefficients of the second derivatives in the transformed equation

$$\sum_{i,k=1}^{n} \bar{a}_{ik} u_{\xi_i \xi_k} + \sum_{i=1}^{n} \bar{b}_i u_{\xi_i} + \bar{c} u = 0 \tag{5}$$

will be connected to the matrix $\|a_{ik}\|$ by the relation

$$\| a_{ik} \| = \| \alpha_{ik} \| \cdot \| a_{ik} \| \cdot \| \alpha_{ik} \|^* . \tag{6}$$

The matrix $\|a_{ik}\|$ transforms like the matrix of the quadratic form (3) if in this quadratic form one changes to new variables by the relation

$$z_i = \sum_{k=1}^{n} \alpha_{in}^* s_k , \tag{7}$$

where $\alpha_i^* = \alpha_{ki}$. The matrix of the transformation from the new variables $s_1, \ldots, s_n$ to the old variables $z_1, \ldots, z_n$ in the quadratic form (3) is the transpose of the matrix of the transformation from the old independent variables $x_1, \ldots x_n$ to new independent variables $\xi_1, \ldots, \xi_n$ in equation (1). Thus, in order to find the transformation (4), reducing equation (1) to the canonical form, it is necessary to find the transformation (7), reducing the quadratic form (3) to the canonical form containing only the squares of the variables $s_1, \ldots, s_n$ with coefficients $+1, -1$ or 0. The matrix of the transformation (4) is the transpose of the matrix of the transformation (7).

24. $u_{\xi_1\xi_1} + u_{\xi_2\xi_2} + u_{\xi_3\xi_3} + u_{\xi_1} = 0,$

$$\xi_1 = x, \quad \xi_2 = -x+y, \quad \xi_3 = 2x-2y+z.$$

25. $u_{\xi_1\xi_1} = u_{\xi_2\xi_2} + u_{\xi_3\xi_3},$

$$\xi_1 = x + \frac{1}{2}y - z, \quad \xi_2 = -\frac{1}{2}y, \quad \xi_3 = z.$$

26. $u_{t't'} = u_{x'x'} + u_{y'y'} + u_{z'z'},$

$$t' = \frac{1}{2}t + \frac{1}{2}x - \frac{1}{2}y - \frac{1}{2}z,$$

$$x' = \frac{1}{2}t + \frac{1}{2}x + \frac{1}{2}y + \frac{1}{2}z,$$

$$y' = -\frac{1}{2\sqrt{3}}t + \frac{1}{2\sqrt{3}}x + \frac{1}{2\sqrt{3}}y - \frac{1}{2\sqrt{3}}z,$$

$$z' = \frac{-1}{2\sqrt{5}}t + \frac{1}{2\sqrt{5}}x - \frac{1}{2\sqrt{5}}y + \frac{1}{2\sqrt{5}}z.$$

27. $u_{t't'} = u_{x'x'}+u_{y'y'}+u_{z'z'}$,

$$x' = \frac{1}{\sqrt{2}}x+\frac{1}{\sqrt{2}}y,$$

$$y' = \frac{1}{\sqrt{2}}z+\frac{1}{\sqrt{2}}t,$$

$$z' = \frac{1}{2}x-\frac{1}{2}y-\frac{1}{2}z-\frac{1}{2}t,$$

$$t' = \frac{1}{2\sqrt{3}}x-\frac{1}{2\sqrt{3}}y-\frac{1}{2\sqrt{3}}z+\frac{1}{2\sqrt{3}}t.$$

28. (*a*) $u_{x_1'x_1'}+\sum_{i=2}^{n} u_{x_i'x_i'} = 0,$

$$x_1' = \frac{1}{\sqrt{n(n+1)}}(x_1+\ldots+x_n),$$

(*b*) $u_{x_1'x_1'}-\sum_{i=2}^{n} u_{x_i'x_i'} = 0,$

$$x_1' = \frac{1}{\sqrt{n(n-1)}}(x_1+\ldots+x_n),$$

$$\left.\right\} \quad x_i' = \alpha_{i1}x_1+\ldots+\alpha_{in}x_n, \quad i = 2, 3, \ldots, n,$$

where $(\alpha_{i1}, \ldots, \alpha_{in})$, $i = 1, 2, \ldots, n$ is any orthogonal normalized system of solutions of the equation

$$\alpha_1+\alpha_2+\ldots+\alpha_n = 0.$$

29. $$\sum_{i=1}^{n} a_i v_{x_i x_i}+\left(c-\frac{1}{4}\sum_{i=1}^{n}\frac{b_i^2}{a_i}\right)v = e^{\frac{1}{2}\sum_{i=1}^{n}\frac{b_i}{a_i}x_i} f(x_1, \ldots, x_n),$$

$$u(x_1, \ldots, x_n) = e^{-\frac{1}{2}\sum_{i=1}^{n}\frac{b_i}{a_i}x_i} v(x_1, \ldots, x_n).$$

CHAPTER II

EQUATIONS OF HYPERBOLIC TYPE

§ 1. Physical Problems Reducible to Equations of Hyperbolic Type; Statement of Boundary-value Problems

In the majority of the problems of the present section (as, for instance, in problems on vibrations of strings, rods, gas) only small vibrations are considered. Small vibrations are those for which it is possible to neglect squares, products and higher powers of the functions, describing the vibrations, and of their derivatives.

1. Free Vibrations in a Non-resistant Medium; Equations with Constant Coefficients

In problems of this group the effect of the force of gravity on the vibrations of particles is assumed to be negligibly small in comparison with the effect of other forces, therefore it is possible to neglect the action of gravity†.

1. The axis Ox is directed along a rod; the characteristic function is assumed to be the displacement $u(x, t)$ along the x-axis of a cross-section, whose abscissa equals x in the equilibrium state; in other words, at time t the abscissa of this section equals $\bar{x} = x+u(x, t)$. In order to determine the function $u(x, t)$ we obtain the boundary-value problems:

(a) when the ends of the rod are rigidly fixed, then

$$u_{tt} = a^2 u_{xx} \quad \text{it} \quad 0< x < l, \quad 0 < t <+\infty, \tag{1}$$

$$u(0, t) = u(l, t) = 0 \quad \text{if} \quad 0 < t <+\infty, \tag{2}$$

$$u(x, 0) = f(x), \quad u_t(x, 0) = F(x) \quad \text{if} \quad 0 < x < l, \tag{3}$$

$$a^2 = \frac{E}{\rho_0},$$

where E is the modulus of elasticity, and ρ_0 is the mass density of the rod in the undisturbed state;

---

† We note that in the more general case the force of gravity is not included in the differential wave equation, if the position of equilibrium is assumed to be the static strained state under the action of gravity (compare with [7] page 105).

(a′) when the ends of the rod move according to a given law, then the boundary conditions have the form

$$u(0, t) = \psi(t), \quad u(l, t) = \phi(t) \quad \text{when} \quad 0 < t < +\infty, \tag{2′}$$

where $\psi(t)$ and $\phi(t)$ are given functions of t;

(b) when the ends of the rod are free, then the boundary conditions have the form

$$u_x(0, t) = u_x(l, t) = 0 \quad \text{when} \quad 0 < t < +\infty; \tag{4}$$

(c) when the ends of the rod are fixed elastically then the boundary conditions have the form

$$u_x(0, t) - hu(0, t) = u_x(l, t) + hu(l, t) = 0 \quad \text{when} \quad 0 < t < +\infty, \tag{5}$$

$$h = \frac{k}{ES},$$

where k s the coefficient of elasticity of the connection (it is assumed that it is the same for both ends, otherwise the values of the constant h for the right-hand and left-hand ends would be different), and S is the cross-sectional area.

Method†. We take the x-axis along the rod. Each cross-section of the rod can be described by the coordinate x which it would have in a position of equilibrium‡. Then the section denoted by the coordinate x has a coordinate $\bar{x} = x + u(x, t)$ at time t. Here $u(x, t)$ is the value of the longitudinal displacement of that cross-section which has coordinate x in equilibrium. Thus, the function $u(x, t)$ is expressed in terms of Lagrangian coordinates§.

The differential equation (1) can be obtained by transition to a limit as $\Delta x \to 0$ in the equation of motion expressing Newton's second law for an element $(x, x + \Delta x)$ of the rod, i.e. for an element, whose ends have coordinates x and $x + \Delta x$ in equilibrium. The elastic forces acting on this element are determined from Hooke's law, which is expressed by the equation

$$X = ES_x(x, t),$$

where X is the projection on the x-axis of the force $\boldsymbol{F}$, with which the part of the rod lying to the right of the section under consideration, acts on the part lying to the left of this section, S is the area of this cross-section # and $u_x(x, t)$ is the specific elongation of the rod for that cross-section with coordinate x in equilibrium††.

† Compare with the derivation of the equation in [7], pages 19–21.

‡ The equilibrium state can be the static strained state.

§ See [7], page 19.

The force $\boldsymbol{F}$ is perpendicular to the cross-section and therefore, its direction either coincides with the direction of the axis Ox or is opposite to the direction of the Ox axis.

†† See [7], page 19.

If the ends of the rod are rigidly fixed, then the boundary conditions are obvious. If the ends of the rod are free or fixed elastically, then the boundary conditions may be derived from Newton's second law for boundary elements.

Let us consider, for example, the case where the end $x = l$ is fixed elastically. From the left, the remaining part of the rod acts on the boundary element $(l-\Delta x, l)$, adjoining this end, with a force

$$-ESu_x(l-\Delta x, t),$$

from the right, a flexible support with a force†

$$-ku(l, t).$$

Therefore Newton's second law for this element is expressed by the equation

$$S\rho_0 \Delta x \frac{\partial^2 u}{\partial t^2} = -ESu_x(l-\Delta x, t) - ku(l, t),$$

from which, passing to a limit as $\Delta x \to 0$, we obtain the condition for the end $x = l$

$$ESu_x(l, t) + ku(l, t) = 0$$

or

$$u_x(l, t) + hu(l, t) = 0,$$

where

$$h = \frac{k}{ES}.$$

For the end $x = 0$ the sign of h in the boundary condition will be opposite. In fact, let us consider the element $(0, \Delta x)$. There is a force applied to its left-hand end

$$-ku(0, t),$$

and to the right-hand end

$$ESu_x(\Delta x, t),$$

therefore the equation expressing Newton's second law for this element has the form

$$S\rho_0 \Delta x \frac{\partial^2 u}{\partial t^2} = ESu_x(\Delta x, t) - ku(0, t).$$

Passing to a limit as $\Delta x \to 0$, we obtain:

$$u_x(0, t) - hu(0, t) = 0,$$

where h has the same value as before, if the rod is homogeneous, and the coefficient of elasticity of the connection is the same for both ends.

Note. It is sometimes useful in statements of boundary-value problems to make use of a system of two first order partial differential equations, instead of one second order partial differential equation.

---

† See item (b) of the condition of the problem.

Let us denote by $\hat{p}(x, t)$ the tension on a cross-section with Lagrangian coordinate x, defining it by the relation

$$\hat{p}(x, t) = \frac{X(x, t)}{S},$$

where S is the cross-sectional area of the rod, and $X(x, t)$ is the force, with which the right-hand part of the rod acts on the part of the rod adjoining the section from the left. Let us denote by $u(x,t)$ the displacement from the equilibrium position of the cross-section with Lagrangian coordinate x. For the functions, describing the vibration process, we take

$$\hat{p}(x, t) \quad \text{and} \quad w(x, t) = u_t(x, t).$$

Consider, for example, the case where the left-hand end of the rod is rigidly fixed, and the right-hand end free. We obtain the boundary-value problem

$$\left.\begin{aligned} w_x - \frac{1}{E}\hat{p}_t = 0, \\ -\hat{p}_x + \rho_0 w_t = 0, \end{aligned}\right\} \quad 0 < x < l, \quad 0 < t < +\infty, \tag{1}$$

$$w(0, t) = 0, \qquad \hat{p}(l, t) = 0, \qquad 0 < t < +\infty, \tag{2}$$

$$w(x, 0) = \phi(x), \quad \hat{p}(x, 0) = \hat{\psi}(x), \quad 0 < x < +\infty. \tag{3}$$

2. The axis Ox in a Cartesian system of coordinates is directed along the line of equilibrium of the string. In the position of equilibrium let the point have coordinates $[x; 0; 0]$, and in the deflected position $[x+u_1(x, t); u_2(x, t); u_3(x, t)]$.

To determine the functions $u_1(x,t)$, $u_2(x,t)$, $u_3(x, t)$, describing the vibrations, we must solve the boundary-value problem

$$\left.\begin{aligned} &\frac{\partial^2 u_k}{\partial t^2} = a_k^2 \frac{\partial^2 u_k}{\partial x^2} \quad \text{if} \quad 0 < x < l, \quad 0 < t < +\infty, \\ &u_k(0, t) = u_k(l, t) = 0 \quad \text{if} \quad 0 \leqslant t < +\infty, \\ &u_k(x, 0) = f_k(x), \quad \frac{\partial u_x(x, 0)}{\partial t} = F_k(x) \quad \text{if} \quad 0 \leqslant x \leqslant l, \end{aligned}\right\} \quad k = 1, 2, 3,$$

where $a_1^2 = ES/\rho$, $a_2^2 = a_3^2 = T_0/\rho$, E is the modulus of elasticity, S the cross-sectional area, ρ the linear mass density.

Method. The total tension in the string consists of the initial tensile force T_0 and another force, arising from a relative extension of the elements of the string. For small vibrations of the string it is possible to assume it absolutely flexible, i.e. the force of tension at any point of the string is assumed to be tangential to the string. The differential equations for the functions $u_k(x, t)$ can be derived by a transition to a limit as $\Delta x \to 0$ in the equations of motion for an element $(x, x+\Delta x)$ projected on the coordinate axis. For a discussion of this problem see supplement [7], pagés 15, and also the method of the preceding problem.

3. The axis Ox is directed along the longitudinal axis of the cylinder, and the angle of deflection of a cross-section with abscissa x is denoted by $\theta(x, t)$, the ends of the cylinder being defined by the abscissae $x = 0$ and $x = l$. We obtain the following boundary-value problems for the function $\theta(x, t)$:

(a) in the case of rigidly fixed ends

$$\frac{\partial^2\theta}{\partial t^2} = a^2 \frac{\partial^2\theta}{\partial x^2} \quad \text{if} \quad 0 < x < l, \quad 0 < t < +\infty, \tag{1}$$

$$\theta(0, t) = \theta(l, t) = 0 \quad \text{if} \quad 0 < t < +\infty, \tag{2}$$

$$\theta(x, 0) = f(x), \quad \theta_t(x, 0) = F(x) \quad \text{if} \quad 0 < x < l, \tag{3}$$

$a^2 = GJ/K$, where G is the shear modulus, J the geometric moment of inertia of a cross-section of the cylinder about its axis, K is the mechanical moment of inertia per unit length of the rod (with respect to the same axis);

(b) where the ends of the cylinder are free, then the boundary conditions have the form

$$\theta_x(0, t) = \theta_x(l, t) = 0; \tag{4}$$

(c) when the ends of the cylinder are fixed elastically, then the boundary conditions have the form

$$\theta_x(0, t) - h\theta(0, t) = 0, \quad \theta_x(l, t) + h\theta(l, t) = 0. \tag{5}$$

Method. Show that the moment M of the elastic forces, applied to the cross-section x of the cylinder, can be found from the relation

$$M = GJ\frac{\partial\theta}{\partial x}. \tag{6}$$

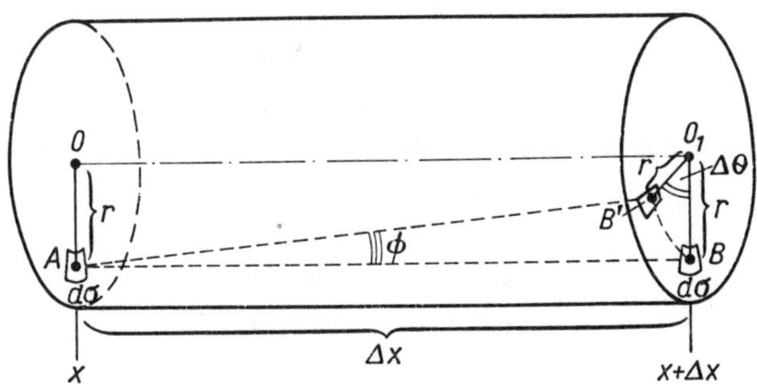

FIG. 16

In order to do this, consider (Fig. 16) the shearing of element AB of the cylinder with bases $d\sigma$ on sections at x and $x+\Delta x$, produced by rotating the section $x+\Delta x$ about the cylinder axis, through an angle $\Delta\theta = (\partial\theta/\partial x)\Delta x$ with respect to the section x, and determine the relation between the angle of shear Φ and $\partial\theta/\partial x$.

The shear stress τ on the base $d\sigma$ of such an element, lying in section x, may be found from Hooke's law

$$\tau = G_{\phi}. \tag{7}$$

The differential equation (1) may be obtained by transition $\Delta x \to 0$ in the equation of rotation† for an element $(x, x+\Delta x)$ of the cylinder.

The boundary conditions are obtained in the same way as in the case of longitudinal vibrations of a rod.

4. The density ρ, pressure p, velocity potential Φ, velocity v of gas particles and the longitudinal displacement u of gas particles satisfy the same differential equation

$$\frac{\partial^2 w}{\partial t^2} = a^2 \frac{\partial^2 w}{\partial x^2}$$

with the same constant

$$a^2 = k \frac{p_0}{\rho_0},$$

where $k = c_p/c_v$ is the exponent of the adiabatic curve

$$\frac{p}{p_0} = \left(\frac{\rho}{\rho_0}\right)^k,$$

equal to the ratio of the specific heat at constant pressure to the specific heat at constant volume, and p_0 and ρ_0 are the pressure and density in the undisturbed gas.

If the ends of the tube are closed, then the boundary conditions for each of the functions u, v, ϕ, p, ρ have the forms

$$\begin{aligned} u(0,t) &= u(l,t) = 0,\\ v(0,t) &= v(l,t), = 0,\\ \phi_x(0,t) &= \phi_x(l,t) = 0,\\ p_x(0,t) &= p_x(l,t) = 0,\\ \rho_x(0,t) &= \rho_x(l,t) = 0. \end{aligned}$$

If the ends of the tube are open, then

$$\begin{aligned} u_x(0,t) &= u_x(l,t) = 0,\\ v_x(0,t) &= v_x(l,t) = 0,\\ \phi(0,t) &= \phi(l,t) = 0,\\ \tilde{p}(0,t) &= \tilde{p}(l,t) = 0,\\ \tilde{\rho}(0,t) &= \tilde{\rho}(l,t) = 0, \end{aligned}$$

where $\tilde{p}(x,t) = p(x,t) - p_0$ is the "pressure deviation" and $\tilde{\rho}(x,t) = \rho(x,t) - \rho_0$ is the "density deviation".

---

† For rotation of a solid about a fixed axis we have: the product of the moment of inertia of the body and the angular acceleration equals the sum of the moments of the forces, applied to the body, with respect to this axis.

If the end $x = l$ of the tube is closed by a gasproof piston of negligible mass mounted on a spring with coefficient of rigidity ν† and slipping inside the tube without friction, we obtain for $u(x, t)$ the boundary condition

$$u_x(l, t)+hu(l, t) = 0,$$

where $h = \nu/Skp_0$, and k is the exponent of the adiabatic curve. Similarly in the presence of such a piston at the end $x = 0$ of the tube we obtain:

$$u_x(0, t)-hu(0, t) = 0.$$

For $v(x, t)$ under the same conditions we have:

$$v_x(0, t)-hv(0, t) = 0, \quad v_x(l, t)+hv(l, t) = 0.$$

For $\tilde{p}(x, t)$ and $\tilde{\rho}(x, t)$ we obtain:

$$\begin{aligned}\tilde{p}_{tt}(0, t)-h^*\tilde{p}_x(0, t) &= \tilde{p}_{tt}(l, t)+h^*\tilde{p}_x(l, t)\\ &= \tilde{\rho}_{tt}(0, t)-h^*\tilde{\rho}_x(0, t) = \tilde{\rho}_{tt}(l, t)+h^*\tilde{\rho}_x(l, t) = 0,\end{aligned}$$

where $h^* = \nu/S\rho_0$. These conditions are also satisfied by

$$p(x, t), \quad \rho(x, t), \quad \phi(x, t).$$

Method. In Lagrangian coordinates ‡ the equation of continuity (1) and the equation of motion (2) ("the fundamental equations of hydrodynamics") have the form

$$\rho_0 = \rho(x, t)\,[1+u_x(x, t)], \tag{1}$$

$$\rho_0 u_{tt}(x, t) = -p_x(x, t). \tag{2}$$

Together with the adiabatic equation

$$p = f(\rho), \quad \text{where} \quad f(\rho) = \frac{p_0}{\rho_0^k}\rho^k, \quad k = \frac{c_p}{c_v}, \tag{3}$$

they form a complete non-linear system of equations for determining the functions $\rho(x, t)$; $u(x, t)$; $p(x, t)$.

Equation (1) expresses the law of conservation of mass of an element of gas, contained between two cross-sections, and equation (2) expresses Newton's second law for this element of gas.

Omitting squares, products and higher powers of the quantities

$$u(x, t), \quad \tilde{p}(x, t) = p(x, t)-p_0, \quad \tilde{\rho}(x, t) = \rho(x, t)-\rho_0$$

---

† The spring will act on the piston with an additional elastic force, equal to $\nu u(l, t)$ for a displacement of the piston equal to u. We speak of an additional elastic force, since in the position of equilibrium an elastic force balancing the unperturbed pressure p_0 already acts on the piston.

‡ See [7], page 19.

and their derivatives, it is easy to derive from equations (1), (2), (3) the linear equations

$$\tilde{\rho}(x,t)+\rho_0 u_x(x,t) = 0, \tag{1'}$$

$$\rho_0 u_{tt}(x,t) = -\tilde{p}_x(x,t), \tag{2'}$$

$$\tilde{p}(x,t) = a^2\tilde{\rho}(x,t), \qquad a^2 = k\,\frac{p_0}{\rho_0}, \tag{3'}$$

a is the velocity of propagation of small disturbances in the gas, i.e. the "velocity of sound"†.

This transition from a non-linear system (1), (2), (3) to a linear system (1′), (2′), (3′) is called "linearization". The equations $\tilde{\rho}_{tt} = a^2\tilde{\rho}_{xx}$ and $\tilde{p}_{tt} = a^2\tilde{p}_{xx}$ are derived from (1′), (2′), (3′) by differentiation and elimination. The potential $\phi(x,t)$ is defined by the relation $\phi_x(x,t) = v(x,t)$ except for an arbitrary function of time. Since $u_t(x,t) = v(x,t)$, from equation (2′) we obtain:

$$\rho_0\phi_{xt}+\tilde{p}_x = 0,$$

i.e.

$$\frac{\partial}{\partial x}(\rho_0\phi_t+\tilde{p}) = 0,$$

therefore, because the velocity potential $\phi(x,t)$ is given except for an arbitrary function of time, it is possible to write:

$$\rho_0\phi_t+\tilde{p} = 0. \tag{4}$$

Relation (4) provides an opportunity of finding the pressure perturbation $\tilde{p}$ if the velocity potential is known. From (1′), differentiating with respect to t, we obtain:

$$\tilde{\rho}_t+\rho_0\phi_{xx} = 0. \tag{5}$$

From (4) and (5) we derive the equation

$$\phi_{tt} = a^2\phi_{xx},$$

differentiation of which with respect to x leads to the equation

$$v_{tt} = a^2 v_{xx}.$$

Similarly equations may be found for the displacement potential and for the displacements.

Considering the motion of a boundary element of gas and using equations (1′), (2′), (3′), (4) and (5), the boundary conditions deduced in the answer are readily obtained.

5. The axis Ox is directed along the pipe, the origin of coordinates O lying in the plane of the inlet section, p_0 is the water pressure in the reservoir. In

† See the introduction to § 2 (conditions) of the present chapter.

order to determine the velocity and pressure $p(x, t)$ we obtain the boundary-value problem

$$\left.\begin{aligned} -\frac{\partial p}{\partial x} &= \frac{\partial(\rho_0 v)}{\partial t} + 2a(\rho_0 v), \quad (1)\\ -\frac{\partial p}{\partial t} &= \lambda^2 \frac{\partial(\rho_0 v)}{\partial x}, \quad (2) \end{aligned}\right\} 0 < x < l; \quad 0 < t < +\infty,$$

$$p(0, t) \cong p_0, \quad v(l, t) = 0; \quad 0 < t < +\infty, \tag{3}$$

$$p(x, 0) \cong p_0, \quad v(x, 0) = v_0; \quad 0 \leqslant x < l, \tag{4}$$

where ρ_0 is the density of the water in the reservoir,

$$\lambda = \frac{1}{\sqrt{\dfrac{\rho_0}{k} + \dfrac{2R_0\rho_0}{E\delta}}}, \tag{5}$$

k is the modulus of elasticity of the water, appearing in Hooke's law

$$\tilde{p} = k\frac{\tilde{\rho}}{\rho_0} \tag{6}$$

$$(\tilde{p} = p - p_0, \quad \tilde{\rho} = \rho - \rho_0),$$

R_0 is the inner radius of the pipe in the unperturbed state, E the modulus of elasticity of the material of the pipe, δ the thickness of the pipe, a the experimentally determined coefficient of frictional resistance per unit length of the pipe†.

In deriving the equations of motion, it is possible, as was shown by N. E. Zhukovskii, to neglect the radial motion of particles‡ for thin tubes with not too large pressure disturbances. At the same time it is necessary to take into account the radially symmetrical tension in the pipe. The force of frictional resistance, acting on an element of water, contained between the cross-sections x and $x+\Delta x$, may be determined by the relation deduced in the footnote.

In deriving the equations of the boundary-value problem we assume the quantities $\tilde{p}$, $\tilde{\rho}$, v to be small and the quantity $p_0/R_0 E\delta$ to be considerably less than unity.

---

† Let S be an inner cross-section of the pipe, then the force of resistance, applied to an element of liquid, contained between the sections x and $x+\Delta x$ equals

$$2a \int_x^{x+\Delta x} S\rho v \mathrm{p} x\,.$$

For a more exact statement of the problem, in which the force of resistance is analysed in greater detail, see, for instance, [44].

‡ In this case the product of the mass of a ring element of the pipe and the radial acceleration is negligibly small.

Let us establish the relation between the inner radius R of the pipe and the pressure p. In order to do this we consider the distance of a semicircular element, cut out from the pipe by the neighbouring cross-sections x and $x+\Delta x$, as shown in Fig. 17. The elastic forces, developed at the ends I and II of this semiring,

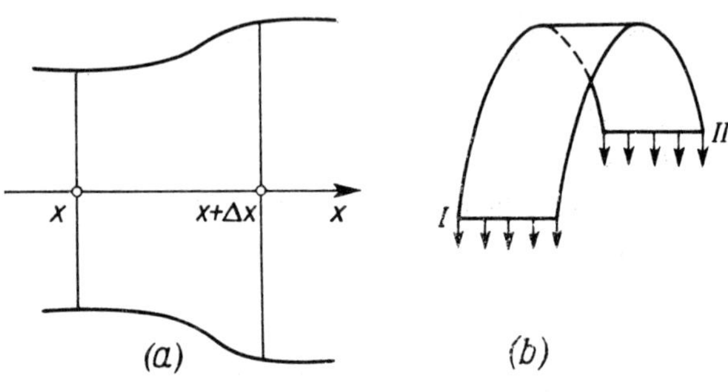

FIG. 17

equal the sum of the projections of the forces of compression of the liquid on the inside surface of the semiring, i.e.

$$2\delta\,\Delta x E\,\frac{R-R_0}{R_0} = 2R\Delta x(p-p_0)$$

or

$$\tilde{p} = \frac{E\delta}{R_0^2}\tilde{R}. \tag{7}$$

Therefore, the quantity $\tilde{R} = R-R_0$ will also be small, and the change of area of the cross-section has the value

$$\tilde{S} = S-S_0 \approx 2\pi\tilde{R}R_0 = S_0\frac{2R_0}{E\delta}\tilde{p}. \tag{8}$$

Let us investigate the equation of continuity, expressing the law of conservation of mass of a substance for a volume, contained between the planes x and $x+\Delta x$ (Fig. 17b):

$$\frac{\partial}{\partial t}\int\limits_x^{x+\Delta x} S\rho\,\mathrm{d}x = (S\rho v)_x-(S\rho v)_{x+\Delta x},$$

i.e.

$$\int\limits_x^{x+\Delta x}\frac{\partial}{\partial t}(S\rho)\,\mathrm{d}x = (S\rho v)_x-(S\rho v)_{x+\Delta x},$$

from which

$$\frac{\partial(S\rho)}{\partial t} = -\frac{\partial(S\rho v)}{\partial x}.$$

Because the quantities $\tilde{S}$, $\tilde{p}$, v are small this equation reduces to equation (2).

Similarly (Fig. 17, a) let us write down Newton's second law for an element of water, contained between the sections x and $x+\Delta x$:

$$\frac{\mathrm{d}}{\mathrm{d}t}\int_x^{x+\Delta x}(S\rho v)\,\mathrm{d}x = (Sp)\,|_x-(Sp)|_{x+\Delta x}-2a\int_x^{x+\Delta x}(S\rho v)\,\mathrm{d}x,\ a=\text{const.},\ a>0,$$

$$\int_x^{x+\Delta x}\left\{\frac{\mathrm{d}}{\mathrm{d}t}(S\rho v)\right\}\mathrm{d}x = (Sp)\,|_x-(Sp)|_{x+\Delta x}-2a\int_x^{x+\Delta x}(S\rho v)\,\mathrm{d}x,$$

from which

$$\frac{\mathrm{d}(S\rho v)}{\mathrm{d}t} = \frac{\partial(Sp)}{\partial x} - 2aS\rho v\,.$$

Because of the smallness of $\tilde{S}$, $\tilde{\rho}$, $\tilde{p}$ and because of the smallness of $2p_0R_0/E\delta$ in comparison with unity this equation (by means of relations (6) and (8)) reduces to equation (1). The initial conditions (4) and the boundary conditions (3), deduced in the answer, are obvious.

In place of the system of differential equations of first order it is possible to obtain one hyperbolic equation, of second order both for $v(x, t)$ and $\rho(x, t)$. Differentiating (1) with respect to x and (2) with respect to t and eliminating v, we obtain:

$$\frac{\partial^2 p}{\partial t^2} = \lambda^2\frac{\partial^2 p}{\partial x^2} - a\frac{\partial p}{\partial t} \quad \text{if} \quad 0<x<l, \quad 0<t<+\infty,$$

$$p(0, t) = p_0 \quad \text{if} \quad 0<t<+\infty.$$

We derive the second boundary condition for $p(x, t)$ from the boundary condition $v(l, t) = 0$ for $0 < t < +\infty$ using equation (1):

$$p_x(l, t) = 0 \quad \text{if} \quad 0<t<+\infty.$$

Further,

$$p(x, 0) = p_0 \quad \text{if} \quad 0<x<l.$$

We derive the second initial condition for $p(x, t)$ from $v(x, 0) = v_0$ for $0 \leqslant x < l$ and equation (2)

$$p_x(x, 0) = 0 \quad \text{if} \quad 0\leqslant x<l.$$

Similarly, we may derive the boundary-value problem for $v(x, t)$.

6. *Solution.* Let ω denote the increase in volume of liquid in the cap, S the area of the inner cross-section of the tube at the end $x = l$. We have:

$$\frac{\mathrm{d}\omega}{\mathrm{d}t} = Sv\Big|_{x=l} - Q(t), \tag{1}$$

$$P_0\Omega_0 = P(\Omega_0-\omega), \tag{2}$$

where v is the flow velocity of the liquid, and P the air pressure in the cap, from which

$$P = \frac{P_0 \Omega_0}{\Omega_0 - \omega} \cong P_0\left(1 + \frac{\omega}{\Omega_0}\right), \tag{3}$$

therefore

$$\frac{d\omega}{dt} = \frac{\Omega_0}{P_0} \frac{\partial p}{\partial t}, \tag{4}$$

where $p = P$ is the pressure of the liquid in the cap. Equation (2) on page 179 gives

$$\frac{d\omega}{dt} = -\lambda^2 \rho_0 \frac{\Omega_0}{P_0} \frac{\partial v}{\partial x}. \tag{5}$$

Substituting (5) in (1), we obtain the boundary condition

$$\frac{\partial v}{\partial x} + \frac{P_0 S}{\Omega_0 \lambda^2 \rho_0} v = \frac{P_0}{\Omega_0 \lambda^2 \rho_0} Q(t) \bigg|_{x=l}. \tag{6}$$

7.

$$\left.\begin{aligned} \frac{\partial^2 \xi}{\partial t^2} &= c^2 \frac{\partial^2 \xi}{\partial x^2}, \quad (1) \\ \frac{\partial^2 \eta}{\partial t^2} &= c^2 \frac{\partial^2 \eta}{\partial x^2} \quad (2) \end{aligned}\right\} \quad \text{if} \quad 0 < x < l, \quad 0 < t < +\infty,$$

where $c^2 = gh$, g the acceleration of gravity,

$$\left.\begin{aligned} \xi(0, t) &= \xi(l, t) = 0, \quad (3) \\ \eta_x(0, t) &= \eta_x(l, t) = 0 \quad (4) \end{aligned}\right\} \quad \text{if} \quad 0 < t < +\infty,$$

$$\left.\begin{aligned} \xi(x, 0) &= f(x), & \xi_t(x, 0) &= \phi(x), \quad (5) \\ \eta(x, 0) &= -hf'(x), & \eta_t(x, 0) &= -h\phi'(x) \quad (6) \end{aligned}\right\} \quad \text{if} \quad 0 < x < l.$$

Method. The change of pressure in the water is negligible for the wave motions considered, i.e. the pressure p at a depth y from the free surface of the water can be assumed to be approximately hydrostatic†. The component v of the velocity of water molecules in the direction of the x-axis may be assumed small, i.e. it is possible to neglect squares, products and higher powers of this function and its derivatives. The liquid may be assumed incompressible.

Solution. At a depth y from the equilibrium free surface of the water, the pressure will be equal to

$$p = p_0 + g\rho(h + \eta - y). \tag{7}$$

Hence we find:

$$\frac{\partial p}{\partial x} = g\rho \frac{\partial \eta}{\partial x}. \tag{8}$$

---

† But it is not possible to neglect its derivative with respect to x.

Let us find the equation of motion for the element $\Delta x\, \Delta y\, \Delta z$ of a section $(x, x+\Delta x)$ (Fig. 18):

$$\Delta x \Delta y \Delta z \rho \frac{\mathrm{d}v}{\mathrm{d}t} = -\Delta x \Delta y \Delta z \frac{\mathrm{d}p}{\mathrm{d}x}\bigg|_{x+\theta\Delta x},$$

where $0 < \theta < 1$, from which after dividing by $\Delta x \Delta y \Delta z$ as $\Delta x \to 0$ we obtain:

$$\rho \frac{\mathrm{d}v}{\mathrm{d}t} = -\frac{\partial p}{\partial x}, \tag{9}$$

or, using (8),

$$\rho \frac{\mathrm{d}v}{\mathrm{d}t} = -g\rho \frac{\partial \eta}{\partial x}, \quad \text{i.e.} \quad \frac{\mathrm{d}v}{\mathrm{d}t} = -g \frac{\partial \eta}{\partial x}. \tag{10}$$

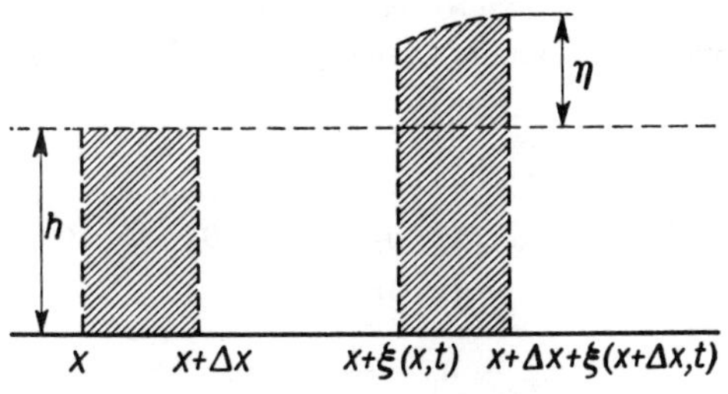

FIG. 18

Further, using the incompressibility of the liquid, we obtain "the equation of continuity" (the equation of conservation of mass)

$$\rho \eta m = -\frac{\partial(\rho m h \xi)}{\partial x}, \tag{11}$$

where m is the width of the canal, from which

$$\eta = -h \frac{\partial \xi}{\partial x}. \tag{12}$$

Linearization of equation (10) gives:

$$\frac{\partial v}{\partial t} = -g \frac{\partial \eta}{\partial x}. \tag{13}$$

But $\partial v/\partial t = \partial^2 \xi/\partial t^2$, therefore

$$\frac{\partial^2 \xi}{\partial t^2} = -g \frac{\partial \eta}{\partial x}. \tag{14}$$

On the other hand, from (12) we obtain:

$$-g \frac{\partial \eta}{\partial x} = gh \frac{\partial^2 \xi}{\partial x^2}. \tag{15}$$

Comparing (14) and (15), we obtain a differential equation for the function $\xi(x, t)$:

$$\frac{\partial^2 \xi}{\partial t^2} = gh \frac{\partial^2 \xi}{\partial x^2} \quad \text{if} \quad 0 < x < l, \quad 0 < t < +\infty. \tag{16}$$

The boundary conditions are readily derived. We note only that the boundary conditions (4) can be obtained from the boundary conditions (3) by differentiating with respect to t and applying (14).

In conclusion we note that in order to determine $\xi(x, t)$ and $\eta(x, t)$ it is sufficient to solve the boundary-value problem (1), (3), (5) and then on finding $\xi(x, t)$ to determine $\eta(x, t)$ by the relation (12).

8. The axis Ox is directed along the longitudinal axis of symmetry of the rod in its equilibrium position, and the transverse deflection $u(x, t)$ of points of the rod from their positions of equilibrium is taken as the characteristic function. We obtain the boundary-value problem for $u(x, t)$

$$\frac{\partial^2 u}{\partial t^2} + a^2 \frac{\partial^4 u}{\partial x^4} = 0\dagger \quad \text{if} \quad 0 < x < l, \quad 0 < t < +\infty, \tag{1}$$

$$u(0, t) = u_{xx}(0, t) = u(l, t) = u_{xx}(l, t) = 0 \quad \text{if} \quad 0 < t < +\infty, \tag{2}$$

$$u(x, t) = f(x), \quad u_t(x, 0) = F(x) \quad \text{if} \quad 0 < x < l, \tag{3}$$

$$a^2 = \frac{EJ}{\rho S}, \tag{4}$$

where E is the modulus of elasticity of the rod, ρ is the mass density, S the cross-sectional area of the rod, J the geometric moment of inertia of a cross-section with respect to a diameter, perpendicular to the plane of vibration.

Method. The derivation of equation (1) is given in [7], pages 150–152.

Considering the motion of the ends of the rod, it is possible to derive boundary conditions.

Let us introduce the boundary conditions in the case of a hinged attachment at the end.

We consider the boundary element $(l-\Delta x, l)$ of a hinged fixed end and write down the equation of rotation with respect to the axis of the hinge (Fig. 19, see also Fig. 2 on page 7)

$$J\rho \Delta x \frac{\partial^2 \phi}{\partial t^2} = F|_{l-\Delta x} \Delta x + M|_{l-\Delta x}. \tag{5}$$

Passing to a limit as $\Delta x \to 0$ on the assumption that there are no infinite angular accelerations $\partial^2\phi/\partial t^2$ and no infinite shear forces F, we obtain $M|_l = 0$, i.e.

$$u_{xx}(l, t) = 0. \tag{6}$$

† This equation is derived on the assumption that the angular accelerations of cross-sections of the rod are absent, i.e. the rod must be sufficiently thin. For the derivation of a more exact equation see[24].

Similarly we may derive the corresponding boundary condition for the left-hand end $u_{xx}(0, t) = 0$. The second boundary condition for the end $x = l$ is obviously: $u(l, t) = 0$.

9. The axis Ox is situated as in the preceding problem. For the determination of $u(x, t)$ we obtain the boundary-value problem

$$\frac{\partial^2 u}{\partial t^2} + a^2 \frac{\partial^4 u}{\partial x^4} = 0 \quad \text{if} \quad 0 < x < l, \quad 0 < t < +\infty, \tag{1}$$

$$u(0, t) = u_x(0, t) = u_{xx}(l, t) = u_{xxx}(l, t) = 0 \quad \text{if} \quad 0 < t < +\infty, \tag{2}$$

$$u(x, 0) = f(x), \quad u_x(x, 0) = F(x) \quad \text{if} \quad 0 < x < l, \tag{3}$$

$$a^2 = \frac{EJ}{\rho S}.$$

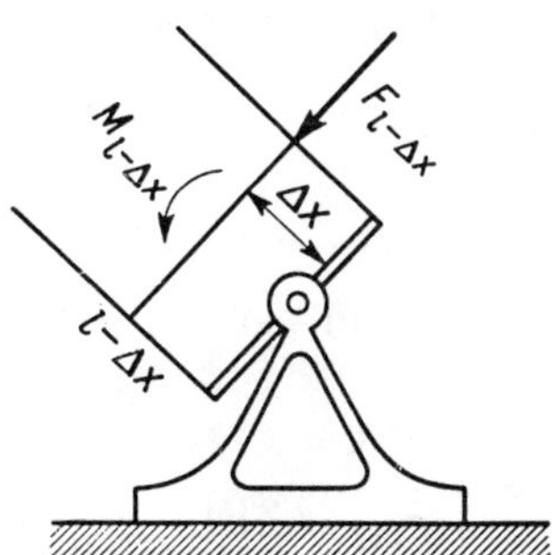

FIG. 19

Method. The boundary conditions for the rigidly fixed end $x = 0$ are the rigidity of the end and the horizontal nature of the tangent. At the free end, as is proved in the usual way, the bending moment and shearing force must be equal to zero.

10. The axis Ox is chosen as in the answer to problem 9. To determine the deflection $u(x, t)$ we obtain the boundary-value problem

$$\frac{\partial^2 u}{\partial t^2} + a^2 \frac{\partial^4 u}{\partial x^4} + \frac{k}{\rho S} u = 0 \quad \text{if} \quad 0 < x < l, \quad 0 < t < +\infty, \tag{1}$$

$$u(0, t) = u_{xx}(0, t) = u(l, t) = u_{xx}(l, t) = 0 \quad \text{if} \quad 0 < t < +\infty, \tag{2}$$

$$u(x, 0) = f(x), \quad u_t(x, 0) = F(x) \quad \text{if} \quad 0 < x < l. \tag{3}$$

Method. Equation (1) is obtained by transition to a limit as $\Delta x \to 0$ in the equation, expressing Newton's second law for an element $(x, x+\Delta x)$ of the rod projected on the axis Ou. In connection with the boundary conditions, see the solution of problem 8.

2. Forced Vibrations and Vibrations in a Resistant Medium; Equations with Constant Coefficients

11.

$$u_{tt} = a^2 u_{xx} + f(x, t) \quad \text{if} \quad 0 < x < l, \quad 0 < t < +\infty, \tag{1}$$

$$u(0, t) = u(l, t) = 0 \quad \text{if} \quad 0 < t < +\infty, \tag{2}$$

$$u(x, 0) = f(x), \quad u_t(x, 0) = F(x) \quad \text{if} \quad 0 < x < l, \tag{3}$$

$f(x, t) = F(x, t)/\rho$, where ρ is the linear mass density of the string.

Method. The differential equation (1) is derived from the equation of motion of an element $(x, x+\Delta x)$ of the string for $\Delta x \to 0$.

12. For the deflections $u(x, t)$ of points of the string from the position of equilibrium we obtain the boundary-value problem

$$u_{tt} = a^2 u_{xx} + \frac{H}{c\rho} I(t), \quad 0 < x < l, \quad 0 < t < +\infty, \tag{1}$$

where $a^2 = T/\rho$, T is the tension of the string, ρ the linear mass density, c the speed of light,

$$u(0, t) = u(l, t) = 0, \quad 0 < t < +\infty, \tag{2}$$

$$u(x, 0) = u_t(x, 0) = 0, \quad 0 < x < l. \tag{3}$$

13.

$$u_{tt} = a^2 u_{xx} \quad \text{if} \quad 0 < x < l, \quad 0 < t < +\infty, \tag{1}$$

$$u(0, t) = \phi(t), \quad u_x(l, t) = \frac{\Phi(t)}{ES} \quad \text{if} \quad 0 < t < +\infty, \tag{2}$$

$$u(x, 0) = 0, \quad u_t(x, 0) = 0 \quad \text{if} \quad 0 < x < l. \tag{3}$$

Method. The boundary condition for the end $x = l$ is obtained by a transition to the limit for $\Delta x \to 0$ in the equation of motion for an element $(l-\Delta x, l)$ of the rod.

14. We obtain the following boundary-value problem for the limiting displacements $u(x, t)$ of cross-sections of the rod

$$u_{tt} = a^2 u_{xx} + g \quad \text{if} \quad 0 < x < l, \quad 0 < t < +\infty, \tag{1}$$

$$u(0, t) = u_x(l, t) = 0 \quad \text{if} \quad 0 < t < +\infty, \tag{2}$$

$$u(x, 0) = 0, \quad u_t(x, 0) = v_0 \quad \text{if} \quad 0 < x < l, \tag{3}$$

where g is the gravitational acceleration, and v_0 the velocity attained by the lift at the moment of stopping.

15. We obtain the following boundary-value problem for the transverse deflections $u(x, t)$ of points of the string from their positions of equilibrium

$$u_{tt} = a^2 u_{xx} - 2\nu^2 u_t \quad \text{if} \quad 0 < t < +\infty, \quad 0 < x < l, \tag{1}$$

$$u(0, t) = u(l, t) = 0 \quad \text{if} \quad 0 < t < +\infty, \tag{2}$$

$$u(x, 0) = f(x), \quad u_t(x, 0) = F(x) \quad \text{if} \quad 0 < x < l, \tag{3}$$

where $2\nu^2 = k/\rho$, ρ is the linear mass density of the string, k "the coefficient of friction", i.e. the coefficient of proportionality in the relation

$$\Phi = -ku_t,$$

giving the frictional force acting on unit length of the string.

Method. Equation (1) is obtained by a transition to the limit as $\Delta x \to 0$ in the equation of motion of an element $(x, x+\Delta x)$ of the string.

16. For the determination of the transverse deflection of points of the rod from their positions of equilibrium let us obtain the boundary-value problem

$$u_{tt}+a^2u_{xxxx}+2\nu^2u_t = f(x,t) \quad \text{if } 0<x<l,\ 0<t<+\infty, \quad (1)$$

$$u(0,t) = u_x(0,t) = u(l,t) = u_x(l,t) = 0 \quad \text{if } 0<t<+\infty, \quad (2)$$

$$u(x,0) = f(x), \quad u_t(x,0) = F(x) \quad \text{if } 0<x<l, \quad (3)$$

here ν has the same meaning as in the preceding problem.

17. We obtain the following boundary-value problem for the transverse deflection $u(x,t)$ of points of the rod from their positions of equilibrium

$$u_{tt}+a^2u_{xxxx} = 0 \quad \text{if} \quad 0<x<l, \quad 0<t<+\infty, \quad (1)$$

$$u(0,t) = u_x(0,t) = 0, \quad u_{xx}(l,t) = 0, \quad EJu_{xxx}(l,t) = -\Phi(t) \quad (2)$$

for $0<t<+\infty$, where $\Phi(t)$ is the transverse force applied to the end $x = l$

$$u(x,0) = f(x), \quad u_t(x,0) = F(x) \quad \text{if} \quad 0<x<l. \quad (3)$$

Method. The last of the boundary conditions (2) may be derived by a transition to a limit as $\Delta x \to 0$ in the equation expressing Newton's second law for an element $(l-\Delta x, l)$ of the rod projected on the axis Ou. For the condition $u_{xx}(l,t) = 0$ see the solution of problem 9.

18. To determine the longitudinal displacements $u(x,t)$ of points of the rod from their position of equilibrium we have the boundary-value problem†

$$u_{tt} = a^2u_{xx} \quad \text{if} \quad 0<x<l, \quad 0<t<+\infty, \quad (1)$$

$$u(0,t) = 0, \quad ESu_x(l,t) = ku_t(l,t) \quad \text{if} \quad 0<t<+\infty, \quad (2)$$

$$u(x,0) = f(x), \quad u_t(x,0) = F(x) \quad \text{if} \quad 0<x<l, \quad (3)$$

where k is the coefficient of friction for the end $x = l$ of the rod.

19. $$v_{xx} = CLv_{tt}+(CR+GL)v_t+GRv \quad \text{if} \quad 0<x<l, 0<t<+\infty, \quad (1)$$

$$v(0,t) = 0, \quad v(l,t) = E(t) \quad \text{if} \quad 0<t<+\infty, \quad (2)$$

$$v(x,0) = F(x), \quad v_t(x,0) = \frac{GF(x)-f'(x)}{C} \quad \text{if} \quad 0<x<l, \quad (3)$$

where $E(t)$ is the given electromotive force applied to the end $x = l$ of the conductor, and L, C, G, R are respectively the coefficient of self inductance,

† For the derivation of the boundary conditions see the method for problem 1.

capacitance, leakage conductance and resistance, calculated per unit of the conductor.

Method. The initial conditions are written in the form (3), if one utilizes the second of the telegraphic equations

$$\begin{aligned} v_x+Li_t+Ri &= 0, \\ i_x+Cv_t+Gv &= 0 \end{aligned} \tag{4}$$

with $t = 0$. The system (4) is derived in [7] on pages 23–24.

3. Vibration Problems Leading to Equations with Continuous Variable Coefficients

20. $$\rho(x)S(x)\frac{\partial^2 u}{\partial t^2} = \frac{\partial}{\partial x}\left[E(x)S(x)\frac{\partial u}{\partial x}\right] \quad \text{if} \quad 0<x<l, \quad 0<t<+\infty, \tag{1}$$

$$u(0,t) = u(l,t) = 0 \quad \text{if} \quad 0<t<+\infty, \tag{2}$$

$$u(x,0) = f(x), \quad u_t(x,0) = F(x) \quad \text{if} \quad 0<x<l. \tag{3}$$

21. The axis Ox is directed along the axis of the cone. For the longitudinal deflections $u(x, t)$ of points of the rod from their positions of equilibrium we obtain the boundary-value problem

$$\left(1-\frac{x}{H}\right)^2\frac{\partial^2 u}{\partial t^2} = a^2\frac{\partial}{\partial x}\left[\left(1-\frac{x}{H}\right)^2\frac{\partial u}{\partial x}\right] \quad \text{if} \quad 0<x<l, \ 0<t<+\infty, \tag{1}$$

$$u(0,t) = u(l,t) = 0 \quad \text{if} \quad 0<t<+\infty, \tag{2}$$

$$u(x,0) = f(x), \quad u_t(x,0) = F(x) \quad \text{if} \quad 0<x<l. \tag{3}$$

Here $a^2 = E/\rho$, E is the modulus of elasticity, ρ the mass density, $H = lR/(R-r)$ the height of the complete cone.

22. We obtain the boundary-value problem for the transverse deflections $u(x, t)$ of points of the rod from their positions of equilibrium

$$\left(1-\frac{x}{H}\right)\frac{\partial^2 u}{\partial t^2} + a^2\frac{\partial^2}{\partial x^2}\left[\left(1-\frac{x}{H}\right)^3\frac{\partial^2 u}{\partial x^2}\right] = 0 \text{ if } 0<x<l, \ 0<t<+\infty, \tag{1}$$

where $a^2 = Eh^2/12\rho$, $H = lh/(h-h')$ is the height of the complete wedge, part of which forms the rod

$$u(0,t) = u_x(0,t) = 0, \quad u_{xx}(l,t) = u_{xxx}(l,t) = 0 \quad \text{if} \quad 0<t<+\infty, \tag{2}$$

$$u(x,0) = f(x), \quad u_t(x,0) = F(x) \quad \text{if} \quad 0<x<l. \tag{3}$$

If for a cross-section with abscissa x the area and moment of inertia (with respect to the horizontal median line of the cross-section) equal $S(x)$ and $J(x)$ respectively, then the equation of the transverse vibrations of the rod will have the form

$$S(x)\rho\frac{\partial^2 u}{\partial t^2} + \frac{\partial^2}{\partial x^2}\left[EJ(x)\frac{\partial^2 u}{\partial x^2}\right] = 0. \tag{4}$$

First it is necessary to derive equation (4) in the same way as in the solution of problem 8 of the present section, and then, substituting the values of $S(x)$ and $J(x)$ for the wedge-shaped rod under consideration, derive equation (1) from equation (4). In connection with the derivation of the boundary conditions (2) see also the solution of problem 8.

23. The axis Ox is directed along the string in the position of equilibrium, and its origin coincides with the free end of the string. In order to determine the transverse deflections $u(x, t)$ of points of the string from their positions of equilibrium we have to solve the boundary-value problem

$$\frac{\partial^2 u}{\partial t^2} = g\frac{\partial}{\partial x}\left(x\frac{\partial u}{\partial x}\right) \quad \text{if} \quad 0 < x < l, \quad 0 < t < +\infty, \tag{1}$$

$$u(0, t) \quad \text{is bounded†} \quad u(l, t) = 0 \quad \text{if} \quad 0 < t < +\infty, \tag{2}$$

$$u(x, 0) = f(x), \quad u_t(x, 0) = F(x) \quad \text{if} \quad 0 < x < l, \tag{3}$$

where g is the gravitational acceleration.

24. In order to determine the transverse deflections $u(x, t)$ of points of the string from positions of equilibrium in a system of coordinates similar to that in the preceding problem, we have to solve the boundary-value problem

$$\frac{\partial^2 u}{\partial t^2} = g\frac{\partial}{\partial x}\left(x\frac{\partial u}{\partial x}\right) + \omega^2 u \quad \text{if} \quad 0 < x < l, \quad 0 < t < +\infty, \tag{1}$$

$$u(0, t) \quad \text{is bounded} \quad u(l, t) = 0 \quad \text{if} \quad 0 < t < +\infty, \tag{2}$$

$$u(x, 0) = f(x), \quad u_t(x, 0) = F(x) \quad \text{if} \quad 0 < x < l, \tag{3}$$

where g is the gravitational acceleration.

25. Let us use the rectangular system of coordinates xOu, whose axis Ox is directed along the string in its equilibrium motion, and the axis Ou is perpendicular to the plane of the equilibrium motion, the origin of coordinates coinciding with the free end of the string. To determine the deflections $u(x, t)$ of points of the string from the plane of equilibrium motion we get the boundary-value problem

$$\frac{\partial^2 u}{\partial t^2} = \frac{\omega^2}{2}\frac{\partial}{\partial x}\left(x^2\frac{\partial u}{\partial x}\right) \quad \text{if} \quad 0 < x < l, \quad 0 < t < +\infty, \tag{1}$$

---

† The requirement of the bounded nature of $u(0, t)$ the deflection of the free end is obvious. This requirement is sufficient from a mathematical point of view to determine a unique solution; a result which depends on the structure of equation (1). By calculating the energy of the vibrating string, it is possible, as in the simplest case of the transverse vibrations of a string, to prove the uniqueness of the solution of the boundary-value problem (1), (2), (3).

$$u(0, t) \quad \text{is bounded} \quad u(l, t) = 0 \quad \text{if} \quad 0 < t < +\infty, \tag{2}$$

$$u(x, 0) = f(x), \quad u_t(x, 0) = F(x) \quad \text{if} \quad 0 < x < l. \tag{3}$$

In connection with the boundary condition for the end $x = 0$ see the footnote to the answer to problem 23 of the present section.

4. Problems Leading to Equations with Discontinuous Coefficients and Similar Problems (Piecewise Homogeneous Media, etc.)

26. The axis Ox is directed along the rod. In the state of equilibrium the plane of the junction of the ends of the semi-infinite rods passes through the origin of coordinates. $u_1(x, t)$ are the longitudinal displacements of points of the first semi-infinite rod, $u_2(x, t)$ of the second. To determine $u_1(x, t)$ and $u_2(x, t)$ we obtain the boundary-value problem

$$\left.\begin{aligned} \frac{\partial^2 u_1}{\partial t^2} &= a_1^2 \frac{\partial^2 u_1}{\partial x^2} \quad \text{if} \quad -\infty < x < 0, \\ \frac{\partial^2 u_2}{\partial t^2} &= a_2^2 \frac{\partial^2 u_2}{\partial x^2} \quad \text{if} \quad 0 < x < \infty, \end{aligned}\right\} \quad \text{if} \quad 0 < t < +\infty, \tag{1}$$

$$u_1(0, t) = u_2(0, t), \quad E_1 \frac{\partial u_1(0, t)}{\partial x} = E_2 \frac{\partial u_2(0, t)}{\partial x} \quad \text{if} \quad 0 < t < +\infty, \tag{2}$$

$$u_1(x, 0) = f(x), \quad \frac{\partial u_1(x, 0)}{\partial t} = F(x) \quad \text{if} \ -\infty < x < 0, \tag{3}$$

$$u_2(x, 0) = f(x), \quad \frac{\partial u_2(x, 0)}{\partial t} = F(x) \quad \text{if} \quad 0 < x < +\infty,$$

$$a_1^2 = \frac{E_1}{\rho_1}, \quad a_2^2 = \frac{E_2}{\rho_2}.$$

Method. The first of the matching conditions (2) indicates that the ends of the semi-infinite rods remain at all times joined together, and the second may be derived for $\Delta x \to 0$ from the equation of motion for an element $(-\Delta x, \Delta x)$ of the composite rod.

27. The axis Ox is chosen as in the preceding problem. To determine the transverse dsiplacements of points of the rod we derive the boundary-value problem

$$\left.\begin{aligned} \frac{\partial^2 u_1}{\partial t^2} + a_1^2 \frac{\partial^4 u_1}{\partial x^4} &= 0 \quad \text{if} \quad -\infty < x < 0, \\ \frac{\partial^2 u_2}{\partial t^2} + a_2^2 \frac{\partial^4 u_2}{\partial x^4} &= 0 \quad \text{if} \quad 0 < x < +\infty, \end{aligned}\right\} \quad \text{if} \quad 0 < t < +\infty, \tag{1}$$

$$\left.\begin{array}{ll} u_1(0,t)=u_2(0,t), & u_{1x}(0,t)=u_{2x}(0,t), \\ E_1u_{1xx}(0,t)=E_2u_{2xx}(0,t), & E_1u_{1xxx}(0,t)=E_2u_{2xxx}(0,t), \end{array}\right\} \text{ if } 0<t<+\infty, \quad (2)$$

$$\left.\begin{array}{llll} u_1(x,0)=f(x), & u_{1t}(x,0)=F(x) & \text{if} & -\infty<x<0, \\ u_2(x,0)=f(x), & u_{2t}(x,0)=F(x) & \text{if} & 0<x<+\infty, \end{array}\right\} \quad (3)$$

$$a_1^2=\frac{E_1J}{\rho_1S}, \quad a_2^2=\frac{E_2J}{\rho_2S}. \quad (4)$$

28. The axis Ox and the functions $u_1(x,t)$ and $u_2(x,t)$ are chosen as in the preceding problem. To determine $u_1(x,t)$ and $u_2(x,t)$ we obtain the boundary-value problem.

$$\left.\begin{array}{lll} \dfrac{\partial^2u_1}{\partial t^2}=a_1^2\dfrac{\partial^2u_1}{\partial x^2} & \text{if} & -\infty<x<0, \\ \dfrac{\partial^2u_2}{\partial t^2}=a_2^2\dfrac{\partial^2u_2}{\partial x^2} & \text{if} & 0<x<+\infty, \end{array}\right\} \text{ if } \quad 0<t<+\infty, \quad (1)$$

$$u_1(0,t)=u_2(0,t), \quad k_1\frac{\partial u_1(0,t)}{\partial x}=k_2\frac{\partial u_2(0,t)}{\partial x} \quad \text{if} \quad 0<t<+\infty, \quad (2)$$

$$\left.\begin{array}{llll} u_1(x,0)=f(x), & \dfrac{\partial u_1(x,0)}{dt}=F(x) & \text{if} & -\infty<x<0, \\ u_2(x,0)=f(x), & \dfrac{\partial u_2(x,0)}{\partial t}=F(x) & \text{if} & 0<x<+\infty, \end{array}\right\} \quad (3)$$

$$a_1^2=k_1\frac{p_0^{(1)}}{\rho_0^{(1)}}, \quad a_2^2=k_2\frac{p_0^{(2)}}{\rho_0^{(2)}},$$

k_1 and k_2 are the adiabatic exponents of the first and second gases, $p_0^{(1)}=p_0^{(2)}$ and $\rho_0^{(1)}, \rho_0^{(2)}$ are the pressures and densities of the first and second gases in the undisturbed state.

Method. The second of the boundary conditions is derived by means of the relations (2′) and (3′) of the solution of problem 4 from the equality of the pressure disturbances

$$\tilde{p}^{(1)}(0,t)=\tilde{p}^{(2)}(0,t),$$

which in its turn is obtained by passing to a limit in the equation of motion for an element $(-\Delta x, \Delta x)$ of the gas.

29. The axis Ox is directed along the canal, the origin of coordinates 0 being located in a plane where the cross-section of the canal varies discontinuously. Let the width and depth† of the left-hand semi-infinite canal equal m_1 and h_1 and of the right-hand equal m_2 and h_2. Then to determine the longitudinal displacements of particles of the liquid and vertical deflections of the free

† The depth read from the free unperturbed surface of the liquid.

surface of the liquid from the equilibrium state we derive the boundary-value problem

$$\left.\begin{aligned} \frac{\partial^2 \xi_1(x,t)}{\partial t^2} &= gh_1 \frac{\partial^2 \xi_1(x,t)}{\partial x^2}, \\ \frac{\partial^2 \eta_1(x,t)}{\partial t^2} &= gh_1 \frac{\partial^2 \eta_1(x,t)}{\partial x^2}, \end{aligned}\right\} \quad \text{if} \quad -\infty < x < 0, \quad 0 < t < +\infty, \tag{1}$$

$$\left.\begin{aligned} \frac{\partial^2 \xi_2(x,t)}{\partial t^2} &= gh_2 \frac{\partial^2 \xi_2(x,t)}{\partial x^2}, \\ \frac{\partial^2 \eta_2(x,t)}{\partial t^2} &= gh_2 \frac{\partial^2 \eta_2(x,t)}{\partial x^2}, \end{aligned}\right\} \quad \text{if} \quad 0 < x < +\infty, \quad 0 < t < +\infty, \tag{1'}$$

$$\eta_1(0,t) = \eta_2(0,t), \quad m_1 h_1 \xi_{1t}(0,t) = m_2 h_2 \xi_{2t}(0,t) \quad \text{if} \quad 0 < t < +\infty, \tag{2}$$

$$\left.\begin{aligned} \xi_1(x,0) &= f(x), & \xi_{1t}(x,0) &= F(x), \\ \eta_1(x,0) &= -h_1 f'(x) & \eta_{1t}(x,0) &= -h_1 F'(x), \end{aligned}\right\} \quad \text{if} \quad -\infty < x < 0, \tag{3}$$

$$\left.\begin{aligned} \xi_2(x,0) &= f(x), & \xi_{2t}(x,0) &= F(x), \\ \eta_2(x,0) &= -h_2 f'(x), & \eta_{2t}(x,0) &= -h_2 F'(x), \end{aligned}\right\} \quad \text{if} \quad 0 < x < +\infty. \tag{3'}$$

Method. The first of the matching conditions (2) follows from the assumption of the continuity of pressure in the liquid at the cross-section $x = 0$, and the second expresses the law of conservation of mass. The first of the conditions (2) may be replaced by the condition

$$h_1 \xi_{1x}(0,t) = h_2 \xi_{2x}(0,t) \tag{4}$$

by using the relations

$$\eta_1(x,t) = -h_1 \xi_{1x}(x,t), \qquad \eta_2(x,t) = -h_2 \xi_{2x}(x,t). \tag{5}$$

Then the boundary-value problem for determining $\xi_1(x, t)$ and $\xi_2(x, t)$ becomes independent of the boundary-value problem for determining $\eta_1(x, t)$ and $\eta_2(x, t)$. We note, finally, that the matching conditions (2) [or the second of the conditions (2) and condition (4)] only approximately describe the phenomenon in the neighbourhood of the cross-section $x = 0$, since both are based on the assumption that the surfaces of the cross-sections $-\Delta x$ and Δx for small Δx are not very different.

30. The axes of coordinates and the characteristic functions are chosen as in problem 26. To determine $u_1(x, t)$ and $u_2(x, t)$ we derive the boundary-value problem

$$\frac{\partial^2 u_1}{\partial t^2} = a_1^2 \frac{\partial^2 u_1}{\partial x^2} \quad \text{if} \quad -\infty < x < 0, \quad 0 < t < +\infty, \tag{1}$$

$$\frac{\partial^2 u_2}{\partial t^2} = a_2^2 \frac{\partial^2 u_2}{\partial x^2} \quad \text{if} \quad 0 < x < +\infty, \quad 0 < t < +\infty, \tag{1'}$$

$$u_1(0,t) = u_2(0,t), \quad M\frac{\partial^2 u_1(0,t)}{\partial t^2} = M\frac{\partial^2 u_2(0,t)}{\partial t^2}$$

$$= E_2\frac{\partial u_2(0,t)}{\partial x} - E_1\frac{\partial u_1(0,t)}{\partial x}, \qquad (2)$$

$$u_1(x,0) = f(x), \quad u_{1t}(x,0) = F(x) \quad \text{if} \quad -\infty < x < 0, \qquad (3)$$

$$u_2(x,0) = f(x), \quad u_{2t}(x,0) = F(x) \quad \text{if} \quad 0 < x < +\infty. \qquad (3')$$

Method. The second of the matching conditions (2) expresses Newton's second law for a layer of mass M. See also the method for problem 26 and the solution of problem 32.

31. The x-axis is directed along the rectilinear position of equilibrium of the rod, the origin of coordinates being situated in the plane of junction of the end of the semi-infinite rods†. To determine the transverse displacements of points of the rod from the position of equilibrium we derive the boundary-value problem

$$\left.\begin{aligned} &\frac{\partial^2 u_1}{\partial t^2} + a_1^2\frac{\partial^4 u_1}{\partial x^4} = 0 \quad \text{if} \quad -\infty < x < 0, \\ &\frac{\partial^2 u_2}{\partial t^2} + a_2^2\frac{\partial^4 u_2}{\partial x^4} = 0 \quad \text{if} \quad 0 < x < +\infty, \end{aligned}\right\} \text{if} \quad 0 < t < +\infty, \qquad (1)$$

$$\left.\begin{aligned} &u_1(0,t) = u_2(0,t); \quad u_{1x}(0,t) = u_{2x}(0,t), \\ &E_1 u_{1xx}(0,t) = E_2 u_{2xx}(0,t)\ddagger, \\ &M u_{1tt}(0,t) = M u_{2tt}(0,t) = E_1 J u_{1xxx}(0,t) - E_2 J u_{2xxx}(0,t), \end{aligned}\right\} \text{if } 0 < t < +\infty. \qquad (2)$$

$$\left.\begin{aligned} &u_1(x,0) = f(x), \quad u_{1t}(x,0) = F(x) \quad \text{if} \quad -\infty < x < 0, \\ &u_2(x,0) = f(x), \quad u_{2t}(x,0) = F(x) \quad \text{if} \quad 0 < x < \infty, \end{aligned}\right\} \qquad (3)$$

$$a_1^2 = \frac{E_1 J}{\rho_1 S}, \quad a_2^2 = \frac{E_2 J}{\rho_2 S}. \qquad (4)$$

32. The axis Ox is directed along the rod, so that its upper end has abscissa $x = 0$. In order to determine the longitudinal deflection $u(x,t)$ of points of the rod we obtain the boundary-value problem

$$u_{tt} = a^2 u_{xx} \quad \text{if} \quad 0 < x < l, \quad 0 < t < +\infty, \qquad (1)$$

$$u(0,t) = 0, \quad \frac{Q}{g} u_{tt}(l,t) = -ESu_x(l,t) + Q \quad \text{if} \quad 0 < t < +\infty, \qquad (2)$$

$$u(x,0) = 0, \quad u_t(x,0) = 0 \quad \text{if} \quad 0 < x < l. \qquad (3)$$

---

† We recall that by the conditions of the problem the thickness of the rigid layer between the ends of the rods is negligibly small.

‡ This matching condition expresses the equality of the bending moments, resulting from the assumption that the cross-sections of the rod are not rotating. For more details on this, see [24].

Solution. We consider the derivation of the second of the boundary conditions (2). We form the equation expressing Newton's second law for a body consisting of a load Q and an element $(l-\Delta x, l)$ of the rod. We obtain (Fig. 20):

$$\left(\frac{Q}{g}+\rho S\Delta x\right)u_{tt}(x_c, t) = -ESu_x(l-\Delta x, t)+Q,$$

where x_c is the coordinate of the centre of mass of the body under consideration. Passing to a limit as $\Delta x \to 0$ we obtain the boundary condition

$$\frac{Q}{g}u_{tt}(x_c, t) = -ESu_x(l, t)+Q.$$

But since the load Q is assumed rigid (non-deformable), then all points of it receive the same longitudinal accelerations for longitudinal vibrations of the rod, therefore in the last relation it is possible to replace x_c by l, then the second of the boundary conditions (2) is obtained.

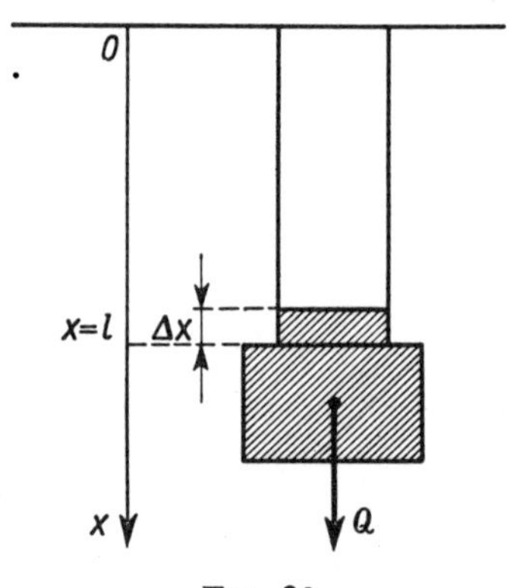

FIG. 20

33. The axis Ox is directed horizontally and, therefore, parallel to the relaxed position of the rod, which we take as the position of equilibrium. In order to determine the transverse deflections $u(x, t)$ of points of the rod from their positions of equilibrium we obtain the boundary-value problem

$$\frac{\partial^2 u}{\partial t^2}+a^2\frac{\partial^4 u}{\partial x^4} = 0 \quad \text{if} \quad 0 < x < l, \quad 0 < t < +\infty, \tag{1}$$

$$u(0, t) = u_x(0, t) = 0, \quad u_{xx}(l, t) = 0;$$

$$\frac{Q}{g}u_{tt}(l, t) = EJu_{xxx}(l, t) \quad \text{if} \quad 0 < t < +\infty, \tag{2}$$

$$u(x, 0) = f(x), \quad u_t(x, 0) = F(x) \quad \text{if} \quad 0 < x < l.$$

Method. The equality $u_{xx}(l, t) = 0$ means that the bending moment is zero which follows from the assumption that cross-sections of the rod do not rotate† (see the derivation of the boundary conditions in the solution of problem 9). The latter of the boundary conditions (2) expresses Newton's second law for the load Q attached to the end of the rod.

34. The axis Ox is directed along the rod, its origin lies on the axis of rotation and coincides with the end of the rod. To determine the longitudinal deflections $u(x, t)$ of points of the rod we obtain the boundary-value problem

$$\frac{\partial^2 u}{\partial t^2} = a^2 \frac{\partial^2 u}{\partial x^2} + \omega^2(x+u) \quad \text{if} \quad 0 < x < l, \quad 0 < t < +\infty, \tag{1}$$

where $a^2 = E/\rho$, E is the modulus of elasticity, and ρ is the mass density of the rod,

$$u(0, t) = 0, \quad \frac{Q}{g} u_{tt}(l, t) = \frac{Q}{g} \omega^2[l+u(l, t)] - ESu_x(l, t) \quad \text{if} \quad 0 < t < +\infty, \tag{2}$$

$$u(x, 0) = f(x), \quad u_t(x, 0) = F(x) \quad \text{if} \quad 0 < x < l. \tag{3}$$

35. The axis Ox is chosen as in the preceding problem. To determine the longitudinal deflections of points of the rod we derive the boundary-value problem

$$\frac{\partial^2 u}{\partial t^2} = a^2 \frac{\partial^2 u}{\partial x^2} + \omega^2(x+u) + g \cos\left(\int_0^t \omega \, dt + \phi_0\right)$$

$$\text{if} \quad 0 < x < l, \quad 0 < t < +\infty, \tag{1}$$

where a^2 has the same meaning as in the preceding problem, and ϕ_0 is the angle between the rod and the vertical direction downwards at time $t = 0$,

$$u(0, t) = 0, \quad \frac{Q}{g} u_{tt}(l, t) = \frac{Q}{g} \omega^2[l+u(l, t)] + Q \cos\left(\int_0^t \omega \, dt + \phi_0\right) -$$

$$- ESu_x(l, t) \quad \text{if} \quad 0 < t < +\infty, \tag{2}$$

$$u(x, 0) = f(x), \quad u_t(x, 0) = F(x) \quad \text{if} \quad 0 < x < l. \tag{3}$$

36. The axis Ox is directed along the rod; the pulley with moment of inertia k_3 has abscissa $x = 0$, the pulley with moment of inertia k_4, mounted between two cylinders, has abscissa $x = l$, finally, the pulley with moment k_5 has abscissa $x = 2l$. In order to determine the deflection angles of cross-sections of the cylinders we have the boundary-value problem

$$\frac{\partial^2 \theta_1}{\partial t^2} = a_1^2 \frac{\partial^2 \theta_1}{\partial x^2} \quad \text{if} \quad 0 < x < l, \quad 0 < t < +\infty, \tag{1}$$

$$\frac{\partial^2 \theta_2}{\partial t^2} = a_2^2 \frac{\partial^2 \theta_2}{\partial x^2} \quad \text{if} \quad l < x < 2l, \quad 0 < t < +\infty, \tag{1'}$$

---

† See also the footnote to problem 8.

$$\left.\begin{aligned} &k_3\frac{\partial^2\theta_1(0,t)}{\partial t^2}=G_1J_1\frac{\partial\theta_1(0,t)}{\partial x},\\ &k_4\frac{\partial^2\theta_1(l,t)}{\partial t^2}=k_4\frac{\partial^2\theta_2(l,t)}{\partial t^2}\\ &\quad=G_2J_2\frac{\partial\theta_2(l,t)}{\partial x}-G_1J_1\frac{\partial\theta_1(l,t)}{\partial x},\\ &k_5\frac{\partial^2\theta_2(2l,t)}{\partial t^2}=G_2J_2\frac{\partial\theta_2(2l,t)}{\partial x}\end{aligned}\right\}\quad \text{if}\quad 0<t<+\infty, \tag{2}$$

$$\left.\begin{aligned}\theta_1(x,0)=f(x),\quad \theta_{1t}(x,0)=F(x)\quad\text{if}\quad 0<x<l,\\ \theta_2(x,0)=f(x),\quad \theta_{2t}(x,0)=F(x)\quad\text{if}\quad 0<x<2l,\end{aligned}\right\} \tag{3}$$

$$a_1^2=\frac{G_1J_1}{k_1},\quad a_2^2=\frac{G_2J_2}{k_2}, \tag{3'}$$

where G_1, J_1, k_1 and G_2, J_2, k_2 are the shear modulus, "geometric" moment of inertia of a cross-section and moment of inertia per unit length for the first and second cylinder† respectively.

37. The axis Ox coincides with the position of equilibrium of the string. To determine the transverse deflections $u(x, t)$ of points of the string we derive the boundary-value problem

$$\left.\begin{aligned}\frac{\partial^2u_1}{\partial t^2}=a^2\frac{\partial^2u_1}{\partial x^2}\quad\text{if}\quad -\infty<x<0,\\ \frac{\partial^2u_2}{\partial t^2}=a^2\frac{\partial^2u_2}{\partial x^2}\quad\text{if}\quad 0<x<+\infty,\end{aligned}\right\}\quad\text{if}\quad 0<t<+\infty, \tag{1}$$

$$u_1(0,t)=u_2(0,t),\quad T_0u_{2x}(0,t)-T_0u_{1x}(0,t)+F(t)=0\ \text{if}\ 0<t<+\infty, \tag{2}$$

$$u_1(x,0)=f(x),\quad u_{1t}(x,0)=F(x)\quad\text{if}\quad -\infty<x<0, \tag{3}$$

$$u_2(x,0)=f(x),\quad u_{2t}(x,0)=F(x)\quad\text{if}\quad 0<x<+\infty. \tag{3'}$$

If the point of application of the force moves along the string according to the law

$$x=\phi(t)\quad\text{if}\quad 0<t<+\infty,\quad \phi(0)=0, \tag{4}$$

then the matching conditions take the form

$$u_1[\phi(t),t]=u_2[\phi(t),t],\quad T_0u_{2x}[\phi(t),t]-T_0u_{1x}[\phi(t),t]+F(t)=0$$
$$\text{if}\quad 0<t<+\infty.$$

38. The axis Ox is chosen as in the preceding problem:

$$\frac{\partial^2u_1}{\partial t^2}+a^2\frac{\partial^4u_1}{\partial x^4}=0\quad\text{if}\quad -\infty<x<0,\quad 0<t<+\infty, \tag{1}$$

† Compare with the answer to problem 3.

$$\frac{\partial^2 u_2}{\partial t^2}+a^2\frac{\partial^4 u_2}{\partial x^4}=0 \quad \text{if} \quad 0<x<+\infty, \quad 0<t<+\infty, \tag{1'}$$

$$\left.\begin{aligned} &u_1(0,t)=u_2(0,t) \quad u_{1x}(0,t)=u_{2x}(0,t) \quad u_{1xx}(0,t)=u_{2xx}(0,t),\\ &EJu_{1xxx}(0,t)-EJu_{2xxx}(0,t)+F(t)=0 \quad \text{if} \quad 0<t<+\infty, \end{aligned}\right\} \tag{2}$$

$$u_1(x,0)=f(x), \quad u_{1t}(x,0)=F(x) \quad \text{if} \quad -\infty<x<0, \tag{3}$$

$$u_2(x,0)=f(x), \quad u_{2t}(x,0)=F(x) \quad \text{if} \quad 0<x<+\infty. \tag{3'}$$

If the point of application of the force moves along the rod according to the law

$$x=\phi(t) \quad \text{if} \quad 0<t<+\infty, \quad \phi(0)=0, \tag{4}$$

then the matching conditions (2) take the form

$$u_1[\phi(t),t]=u_2[\phi(t),t], \quad u_{1x}[\phi(t),t]=u_{2x}[\phi(t),t] \quad \text{if} \quad 0<t<+\infty,$$
$$EJ_{1xxx}[\phi(t),t]-EJu_{2xxx}[\phi(t),t]+F(t)=0 \quad \text{if} \quad 0<t<+\infty.$$

39. The axis Ox is directed along the tube, its origin being at the end of the tube at the equilibrium position of the piston. To determine the longitudinal deflections $u(x,t)$ of particles of the gas we have the boundary-value problem

$$u_{tt}=a^2u_{xx} \quad \text{if} \quad 0<x<+\infty, \quad 0<t<+\infty, \tag{1}$$

$$Mu_{tt}(0,t)=-Skp_0u_x(0,t)-k^*u_t(0,t)-k^{**}u(0,t), \tag{2}$$

$$u(x,0)=f(x), \quad u_t(x,0)=F(x) \quad \text{if} \quad 0<x<+\infty. \tag{3}$$

Method. The boundary condition (2) expresses Newton's second law for the piston†.

40. The axis Ox coincides with the position of equilibrium of the string, the ball of mass M has abscissa $x=0$. To determine the transverse deflections $u(x,t)$ of points of the string from the position of equilibrium we have the boundary-value problem

$$u_{1tt}=a^2u_{1xx} \quad \text{if} \quad -\infty<x<0, \quad 0<t<+\infty, \tag{1}$$

$$u_{2tt}=a^2u_{2xx} \quad \text{if} \quad 0<x<+\infty, \quad 0<t<+\infty, \tag{1'}$$

$$\begin{aligned} u_1(0,t)=u_2(0,t) \quad Mu_{1tt}(0,t)&=Mu_{2tt}(0,t)\\ &=T_0u_{2x}(0,t)-T_0u_{1x}(0,t)-ku_1(0,t)\\ &=T_0u_{2x}(0,t)-T_0u_{1x}(0,t)-ku_2(0,t), \end{aligned} \tag{2}$$

$$u_1(x,0)=f(x), \quad u_{1t}(x,0)=F(x) \quad \text{if} \quad -\infty<x<0, \tag{3}$$

$$u_2(x,0)=f(x), \quad u_{2t}(x,0)=F(x) \quad \text{if} \quad 0<x<+\infty. \tag{3'}$$

If the ball is subject to a resistance proportional to the velocity, then instead of the second of the matching conditions (2) we get a condition

$$Mu_{1tt}(0,t)=Mu_{2tt}(0,t)=T_0u_{2x}(0,t)-T_0u_{1x}(0,t)-ku_1(0,t)-k^*u_{1t}(0,t)$$
$$=T_0u_{2x}(0,t)-T_0u_{1x}(0,t)-ku_2(0,t)-k^*u_{2t}(0,t) \quad \text{if} \quad 0<t<+\infty.$$

---

† Compare with the discussions on the derivation of the third boundary condition in the solution of problem 4.

41. The coordinate x of a point in the conductor is taken as the distance along the conductor from the end, earthed through a lumped resistance. In order to determine the voltage $v(x, t)$ and the current intensity $i(x, t)$ in the conductor we derive the boundary-value problem

$$v_x+Li_t = 0, \quad i_x+Cv_t = 0 \quad \text{if} \quad 0 < x < l, \quad 0 < t < +\infty, \tag{1}$$

$$-v(0, t) = R_0 i(0, t), \quad C_0 v_t(l, t) = i(l, t) \quad \text{if} \quad 0 < t < +\infty, \tag{2}$$

$$v(x, 0) = f(x), \quad i(x, 0) = \phi(x) \quad \text{if} \quad 0 < x < l \tag{3}$$

or

$$v_{tt} = a^2 v_{xx} \quad \text{if} \quad 0 < x < l_1, \quad 0 < t < +\infty, \tag{1'}$$

$$R_0 v_x(0,t) = Lv_t(0, t), \quad LC_0 v_{tt}(l, t) = -v_x(l, t) \quad \text{if} \quad 0 < t < +\infty, \tag{2'}$$

$$v(x, 0) = f(x), \quad v_x(x, 0) = -\frac{1}{C}\phi'(x) \quad \text{if} \quad 0 < x < l \tag{3'}$$

and

$$i_{tt} = a^2 i_{xx} \quad \text{if} \quad 0 < x < l, \quad 0 < t < +\infty, \tag{1''}$$

$$i_x(0, t) = C_0 R_0 i_t(0, t), \quad C_0 i_x(l, t)+Ci(l, t) = 0 \quad \text{if} \quad 0 < t < +\infty, \tag{2''}$$

$$i(x, 0) = \phi(x) \quad i_t(x, 0) = -\frac{1}{L}f'(t) \quad \text{if} \quad 0 < x < l, \tag{3''}$$

$$a^2 = \frac{1}{CL}.$$

Method. The differential equations (1) are obtained from the differential equations (1) and (2) of the answer to problem 19 with $R = G = 0$. The boundary conditions (2) are obtained from the relation

$$\Delta v = \tilde{R}_0 i+\tilde{L}_0\frac{di}{t}+\frac{1}{\tilde{C}_0}\int i\,dt, \tag{4}$$

which determines the drop in voltage in passing through lumped resistances $\tilde{R}_0$, self inductance $\tilde{L}_0$ and capacitance $\tilde{C}_0$, connected in series. Thus, for instance for the end $x = 0$ of the conductor we have:

$$0-v(0, t) = R_0 i(0, t) \quad \text{if} \quad 0 < t < +\infty, \tag{5}$$

where $0-v(0, t)$ denotes the difference between earth potential and the end of the conductor (earth potential is taken equal to zero).

Equations (1′) and (1″) are derived from equations (1) by elimination of the functions $i(x, t)$ and $v(x, t)$ respectively. The boundary conditions (2′) and (2″) are derived from the boundary conditions (2) using equations (1). The initial conditions (3′) and (3″) are derived from the initial conditions (3) using equations (1).

42. The distance along the conductor from the end, earthed through a lumped self inductance $L_0^{(1)}$, is taken as the x coordinate of a point in the conductor. To determine $v(x, t)$ and $i(x, t)$ we obtain the boundary-value problem.

$$v_x+Li_t = 0, \quad i_x+Cv_t = 0 \quad \text{if} \quad 0 < x < l, \quad 0 < t < +\infty, \tag{1}$$

$$-v(0, t) = L_0^{(1)} i_t(0, t), \quad v(l, t)-E(t) = L_0^{(2)} i_t(l, t) \quad \text{if} \quad 0 < t < +\infty, \tag{2}$$

$$v(x, 0) = f(x), \quad i(x, 0) = \phi(x) \quad \text{if} \quad 0 < x < l \tag{3}$$

or

$$v_{tt} = a^2 v_{xx} \quad \text{if} \quad 0 < x < l, \quad 0 < t < +\infty, \tag{1'}$$

$$L_0^{(1)} v_x(0, t)-Lv(0, t) = 0, \quad L_0^{(2)} v_x(l, t)+Lv(l, t) = LE(t) \quad \text{if} \quad 0 < t < +\infty, \tag{2'}$$

$$v(x, 0) = f(x), \quad v_t(x, 0) = -\frac{1}{C}\phi'(x) \quad \text{if} \quad 0 < x < l, \tag{3'}$$

$$i_{tt} = a^2 i_{xx} \quad \text{if} \quad 0 < x < l, \quad 0 < t < +\infty, \tag{1''}$$

$$CL_0^{(1)} i_{tt}(0, t) = i_x(0, t), \quad CL_0^{(2)} i_{tt}(l, t)+i_x(l, t) = E'(t) \quad \text{if} \quad 0 < t < +\infty, \tag{2''}$$

$$i(x, 0) = \phi(x), \quad i_t(x, 0) = -\frac{1}{L} f'(x) \quad \text{if} \quad 0 < x < l. \tag{3''}$$

Method. See the method for problem 41.

43. The distance along the conductor from one end of the conductor to a point is taken as the x coordinate of the point. To determine $v(x, t)$ and $i(x, t)$ we obtain the boundary-value problem

$$v_x+Li_t+Ri = 0, \quad i_x+Cv_t+Gv = 0 \quad \text{if} \quad 0 < x < l, \quad 0 < t < +\infty, \tag{1}$$

$$-v(0, t) = R_0^{(1)} i(0, t), \quad v(l, t) = R_0^{(2)} i(l, t) \quad \text{if} \quad 0 < t < +\infty, \tag{2}$$

$$v(x, 0) = f(x), \quad i(x, 0) = \phi(x) \quad \text{if} \quad 0 < x < l \tag{3}$$

or

$$v_{xx} = CLv_{tt}+(CR+GL)v_t+GRv \quad \text{if} \quad 0 < x < l, \quad 0 < t < +\infty, \tag{1'}$$

$$\left.\begin{aligned} &v_x0,(t)-\frac{L}{R_0^{(1)}}v_t(0, t)-\frac{R}{R_0^{(1)}}v(0, t) = 0,\\ &v_x(l, t)+\frac{L}{R_0^{(2)}}v_t(l, t)+\frac{R}{R_0^{(2)}}(vl, t) = 0 \end{aligned}\right\} \quad \text{if} \quad 0 < t < +\infty, \tag{2'}$$

$$v(x, 0) = f(x), \quad v_t(x, 0) = \frac{Gf(x)-\phi'(x)}{C} \quad \text{if} \quad 0 < x < l \tag{3'}$$

and

$$i_{xx} = CLi_{tt}+(CR+GL)i_t+GRi \quad \text{if} \quad 0 < x < l, \quad 0 < t < +\infty, \tag{1''}$$

$$\left.\begin{aligned} &i_x(0, t)-CR_0^{(1)} i_t(0, t)-GR_0^{(1)} i(0, t) = 0,\\ &i_x(l, t)+CR_0^{(2)} i_t(l, t)+GR_0^{(2)} i(l, t) = 0 \end{aligned}\right\} \quad \text{if} \quad 0 < t < +\infty, \tag{2''}$$

$$i(x, 0) = \phi(x), \quad i_t(x, 0) = \frac{R\phi(x)-f'(x)}{L} \quad \text{if} \quad 0 < x < l. \tag{3''}$$

44. The coordinate system is chosen as in the preceding problem. In order to determine $v(x, t)$ and $i(x, t)$ we obtain the boundary-value problem

$$v_x+Li_t+Ri = 0, \quad i_x+Cv_t+Gv = 0 \quad \text{if} \quad 0<x<l, \quad 0<t<+\infty, \tag{1}$$

$$\left.\begin{aligned} -v(0, t) &= L_0^{(1)}i_t(0, t)+R_0^{(1)}i(0, t),\\ v(l, t) &= L_0^{(2)}i_t(l, t)+R_0^{(2)}i(l, t) \end{aligned}\right\} \quad \text{if} \quad 0<t<+\infty, \tag{2}$$

$$v(x, 0) = f(x), \quad i(x, 0) = \phi(x) \quad \text{if} \quad 0<x<l. \tag{3}$$

To determine $v(x, t)$ if the conditions $R_0^{(1)}L-RL_0^{(1)}=0$ and $R_0^{(2)}L-RL_0^{(2)}=0$ hold we obtain the boundary-value problem

$$v_{xx} = CLv_{tt}+(CR+GL)v_t+GRv \quad \text{if} \quad 0<x<l, \quad 0<t<+\infty, \tag{1'}$$

$$\left.\begin{aligned} L_0^{(1)}v_x(0, t)-Lv(0, t) &= 0,\\ L_0^{(2)}v_x(l, t)+Lv(l, t) &= 0 \end{aligned}\right\} \quad \text{if} \quad 0<t<+\infty, \tag{2'}$$

$$v(x, 0) = f(x), \quad v_t(x, 0) = \frac{Gf(x)-\phi'(x)}{C} \quad \text{if} \quad 0<x<l. \tag{3'}$$

45. The origin of coordinates O is situated at the junction of the semi-infinite conductors. The distance along the conductor from the origin O to a point is taken as the x coordinate of the point.

In order to determine $v(x, t)$ and $i(x, t)$ we find the boundary-value problem

$$v_{1x}+L_1i_{1t}+R_1i_1 = 0, \quad i_{1x}+C_1v_{1t}+G_1v_1 = 0$$
$$\text{if} \quad -\infty<x<0, \quad 0<t<+\infty,$$

$$v_{2x}+L_2i_{2t}+R_2i_2 = 0, \quad i_{2x}+C_2v_{2t}+G_2v_2 = 0$$
$$\text{if} \quad 0<x<+\infty, \quad 0<t<+\infty,$$

$$\left.\begin{aligned} &i_1(0, t) = i_20(t),\\ &v_{2t}(0, t)-v_{1t}(0, t) = \frac{1}{C_0}i_1(0, t) = \frac{1}{C_0}i_2(0, t) \end{aligned}\right\} \quad \text{if} \quad 0<t<+\infty,$$

$$v_1(x, 0) = f(x), \quad i_1(x, 0) = \phi(x) \quad \text{if} \quad -\infty<x<0,$$
$$v_2(x, 0) = f(x), \quad i_2(x, 0) = \phi(x) \quad \text{if} \quad 0<x<+\infty.$$

To determine the current intensity assuming that $G_1 = G_2 = 0$ we obtain the boundary-value problem

$$i_{1xx} = C_1L_1i_{1tt}+C_1R_1i_{1t} \quad \text{if} \quad -\infty<x<0, \quad 0<t<+\infty,$$
$$i_{2xx} = C_2L_2i_{2tt}+C_2R_2i_{2t} \quad \text{if} \quad 0<x<+\infty, \quad 0<t<+\infty,$$

$$i_1(0, t) = i_2(0, t), \frac{1}{C_1}i_{1x}(0, t)-\frac{1}{C_2}i_{2x}(0, t) = \frac{1}{C_0}i_1(0, t) \quad \text{if} \quad 0<t<+\infty,$$

$$i_1(x, 0) = \phi(x), \quad i_{1t}(x, 0) = \frac{R_1\phi(x)-f'(x)}{L_1} \quad \text{if} \quad -\infty<x<0,$$

$$i_2(x, 0) = \phi(x), \quad i_{2t}(x, 0) = \frac{R_2\phi(x)-f'(x)}{L_2} \quad \text{if} \quad 0<x<+\infty.$$

46. The coordinate system and differential equations are the same as in problem 45. But the matching conditions have the form

$$i_1(0, t) = i_2(0, t), \quad v_2(0, t) - v_1(0, t) = R_0 i_1(0, t) = R_0 i_2(0, t) \quad \text{if} \quad 0 < t < +\infty$$

and, if the leakage conductance is absent

$$i_1(0, t) = i_2(0, t), \qquad \frac{1}{C_1} i_{1x}(0, t) - \frac{1}{C_2} i_{2x}(0, t) = R_0 i_{1t}(0, t).$$

47. The coordinate system is chosen in the usual way. To determine $v(x, t)$ and $i(x, t)$ we obtain the boundary-value problem

$$v_x + Li_t + Ri = 0, \quad i_x + Cv_t + Gv = 0 \quad \text{if} \quad 0 < x < l, \quad 0 < t < +\infty, \tag{1}$$

$$\left.\begin{aligned} -\frac{v_t(0, t)}{R_0} - \frac{v(0, t)}{L_0^{(1)}} &= i_t(0, t), \\ C_0 v_{tt}(l, t) + \frac{v(l, t)}{L_0^{(2)}} &= i_t(l, t) \end{aligned}\right\} \quad \text{if} \quad 0 < t < +\infty, \tag{2}$$

$$v(x, 0) = f(x), \quad i(x, 0) = \phi(x) \quad \text{if } 0 < x < l. \tag{3}$$

48. We measure the x coordinate of a point from the middle of the conductor. The system of telegraphic equations and initial conditions are written down in the usual way. The matching conditions have the form

$$v(-l, t) - v(l, t) = L_0 i_t(-l, t) = L_0 i_t(l, t), \tag{1}$$

$$v(-l, t) - v(l, t) = R_0 i(-l, t) = R_0 i(l, t), \tag{2}$$

$$v_t(-l, t) - v_t(l, t) = \frac{1}{C_0} i(l, t) = \frac{1}{C_0} i(-l, t). \tag{3}$$

5. Similarity of Boundary-value Problems

Instead of an introduction to the solutions of the problems of this section, a detailed solution of problem 49 with which this section begins, is given.

49. If $\bar{p}(x'', t'') = -p(x'', t'')$ is taken as the function characterizing the longitudinal vibrations of a rod $0 \leqslant x'' \leqslant l''$ where $p(x'', t'')$ is the tension in a cross-section with coordinate x'' (defined as in problem 1 on page 171 of the present section), then problem (II) on the longitudinal vibrations of a rod, one end of which ($x'' = 0$) is free, and the other end ($x'' = l''$) rigidly fixed, is formulated in the following way:

$$\left.\begin{aligned} &\bar{p}_{t''t''} = a''^2 \bar{p}_{x''x''}, \quad 0 < x'' < l'', \quad 0 < t'' < +\infty, \quad a''^2 = \frac{E}{\rho}, \\ &\bar{p}(0, t'') = \bar{p}_{x''}(l'', t'') = 0, \quad 0 < t'' < +\infty, \\ &\bar{p}(x'', 0) = \phi_p(x''), \quad \bar{p}_{t''}(x'', 0) = \psi_p(x''), \quad 0 < x'' < l''. \end{aligned}\right\} \tag{II}$$

If the electric voltage $v(x', t')$ is taken as the function describing the electric vibrations in a conductor $0 \leqslant x' \leqslant l'$ of negligibly small leakage conductance and capacity, then problem (I) on the electric vibrations in a conductor, one end of which ($x' = 0$) is earthed, and the other end ($x' = l'$) is insulated, is formulated in the following manner:

$$\left.\begin{aligned} &v_{t't'} = a'^2 v_{x'x'}, \quad 0 < x' < l', \quad 0 < t' < +\infty, \\ &v(0, t') = v_{x'}(l', t') = 0, \quad 0 < t' < +\infty, \\ &v(x', 0) = \phi_v(x'), \quad v_{t'}(x', 0) = \psi_v(x'), \quad 0 < x' < l'. \end{aligned}\right\} \tag{I}$$

Problem (I) is analogous to problem (II). In order that problem (I) be similar to problem (II) with given coefficients of similarity k_x, k_t, k_u, it is necessary and sufficient that the relations

$$k_x = \frac{l'}{l''}, \tag{1}$$

$$a'^2 = \frac{k_x^2}{k_t^2} a''^2, \tag{2}$$

$$\phi_v(x')\, k_u \phi_p(x''), \quad \psi_v(x') = \frac{k_u}{k_t} \psi_p(x'') \quad \text{if} \quad x' = k_x x'', \quad 0 < x'' < l'' \tag{3}$$

be fulfilled.

Solution. Let us prove the necessity and sufficiency of conditions (1), (2) and (3).

First let us prove the necessity. Let

$$v(x', t') = k_u \bar{p}(x'', t'') \quad \text{if} \quad x' = k_x x'', \quad t' = k_t t'',$$

where (x', t') passes through $D_{\mathrm{I}}[0 < x' < l',\ 0 < t' < +\infty]$, when (x'', t'') passes through $D_{\mathrm{II}}[0 < x'' < l'', 0 < t'' < +\infty]$. Then we at once obtain that $l' = k_x l''$, i.e. condition (1) is fulfilled. From $v(x', t') = k_u \bar{p}(x'', t'')$, performing a differentiation with respect to t, we obtain $v_{t'}(x', t') = (k_u/k_t)\bar{p}_t{}''(x'', t'')$, therefore for $t' = t'' = 0$ the equality

$$v(x', 0) = k_u \bar{p}(x'', 0), \quad v_{t'}(x', 0) = \frac{k_u}{k_t} \bar{p}_{t''}(x'', 0), \quad 0 < x'' < l'', \tag{4}$$

will be fulfilled, i.e. condition (3) will be fulfilled.

Differentiating the equality

$$v(x', t') = k_u \bar{p}(x'', t'') \quad \text{with respect to } x'' \text{ and } t''$$

and using the relations $x' = k_x x''$, $t' = k_t t''$, we obtain:

$$k_t^2 \frac{\partial^2 v}{\partial t'^2} = k_u \frac{\partial^2 \bar{p}}{\partial t''^2}, \quad k_x^2 \frac{\partial^2 v}{\partial x'^2} = k_u \frac{\partial^2 \bar{p}}{\partial x''^2}.$$

Since the function $\bar{p}(x'', t'')$ must satisfy the equation $\partial^2 \bar{p}/\partial t''^2 = a''^2(\partial^2 \bar{p}/\partial x''^2)$, then, consequently, the equality

$$k_t^2 \frac{\partial^2 v}{\partial t''^2} - k_x^2 a''^2 \frac{\partial^2 v}{\partial x''^2} = k_u \left(\frac{\partial^2 \bar{p}}{\partial t''^2} - a''^2 \frac{\partial^2 \bar{p}}{\partial x''^2} \right) = 0.$$

must be fulfilled.

Therefore, $v(x', t')$ is not only a solution of the equation

$$\frac{\partial^2 v}{\partial t'^2} = a'^2 \frac{\partial^2 v}{\partial x'^2}, \tag{5}$$

but it is also a solution of the equation

$$\frac{\partial^2 v}{\partial t'^2} = \frac{k_x^2}{k_t^2} a''^2 \frac{\partial^2 v}{\partial x'^2}. \tag{6}$$

Subtracting (6) from (5) we obtain:

$$\left(a'^2 - \frac{k_x^2}{k_t^2} a''^2\right) \frac{\partial^2 v}{\partial x'^2} \equiv 0,$$

which is possible only for the condition

$$a'^2 - \frac{k_x^2}{k_t^2} a''^2 = 0, \tag{7}$$

because the condition

$$\frac{\partial^2 v}{\partial x'^2} = 0 \tag{8}$$

implies $v \equiv 0$ using the equation and the boundary conditions (1), but this is impossible unless $\phi_v(x')$ and $\psi_v(x')$ are identically equal to zero. Therefore, (8) is impossible, which means that (7) holds, i.e. condition (2) is fulfilled.

Let us consider now the sufficiency. We change to the dimensionless quantities ξ, τ, U in the boundary-value problems (I) and (II) by means of the relations

$$x' = l'\xi, \quad t' = t_0'\tau, \quad v = v_0 U(\xi, \tau), \quad x'' = l''\xi, \quad t'' = t_0''\tau, \quad \bar{p} = \bar{p}_0 U(\xi, \tau),$$

where the constants t_0' and t_0'' have the dimensions of time, and v_0 and $\bar{p}_0$ have the dimensions of v and p respectively, these constants being chosen so that

$$\frac{t_0'}{t_0''} = k_t, \quad \frac{v_0}{\bar{p}_0} = k_u. \tag{9}$$

We recall that, moreover, the relation

$$k_x = \frac{l'}{l''} \tag{1}$$

is fulfilled.

The boundary-value problems (I) and (II) take respectively the form

$$\left.\begin{aligned} &\frac{\partial^2 U}{\partial \tau^2} = \frac{t_0'^2}{l'^2} a'^2 \frac{\partial^2 U}{\partial \xi^2}, \quad 0 < \xi < 1, \quad 0 < \tau < +\infty, \\ &U(0, \tau) = 0, \quad U_\xi(1, \tau) = 0, \quad 0 < \tau < +\infty, \\ &U(\xi, 0) = \frac{1}{v_0}\phi_v(l', \xi), \quad U_\tau(\xi, 0) = \frac{t_0'}{v_0}\psi_v(l', \xi), \quad 0 < \xi < 1, \end{aligned}\right\} \tag{I}$$

$$\left.\begin{aligned}
&\frac{\partial^2 U}{\partial \tau^2} = \frac{t_0''^2}{l''^2} a''^2 \frac{\partial^2 U}{\partial \xi^2}, \quad 0 < \xi < 1, \quad 0 < \tau < +\infty,\\
&U(0, \tau) = 0, \quad U_\xi(1, \tau) = 0, \quad 0 < \tau < +\infty,\\
&U(\xi, 0) = \frac{1}{\overline{p}_0} \phi_p(l''\xi), \quad U_\tau(\xi, 0) = \frac{t_0''}{\overline{p}_0} \psi_p(l''\xi), \quad 0 < \xi < 1.
\end{aligned}\right\} \quad (\mathrm{II}')$$

From (1), (2) and (9) it follows that

$$\frac{t_0'^2}{l'^2} a'^2 = \frac{t_0''^2}{l''^2} a''^2.$$

From (1), (9) and (3) it follows that

$$\frac{1}{v_0} \phi_v(l'\xi) = \frac{1}{\overline{p}_0} \phi_p(l''\xi), \quad \frac{t_0'}{v_0} \psi_v(l'\xi) = \frac{t_0''}{\overline{p}_0} \psi_p(l''\xi), \quad 0 < \xi < 1.$$

Thus, in problems (I′) and (II′) the equations, initial and boundary conditions are identical; therefore (by virtue of the uniqueness theorem) their solutions are identical.

Thus,

$$U(\xi, \tau) = \frac{1}{v_0} v(x', t') = \frac{1}{\overline{p}_0} \overline{p}(x'', t'') \quad \text{if} \quad x' = k_x x'', \quad t'' = k_t t'',$$

i.e.

$$v(x', t') = k_u \overline{p}(x'', t'') \quad \text{if} \quad x' = k_x x'', \quad t' = k_t t'',$$

which was requiring to be proved.

Note. It is possible to formulate other conditions of similarity by choosing the functions describing the physical processes in different ways. For example we could write the equations and boundary conditions for the rod in terms of longitudinal displacements instead of tensions.

50. The longitudinal displacement of cross-sections of the rod $u(x'', t'')$ is taken as the function characterizing the longitudinal vibrations of the rod $0 < x'' < l''$.

(a) If one end of the rod ($x'' = 0$) is rigidly fixed, and the other end ($x'' = l''$) is elastically attached, then to determine $u(x'', t'')$ we obtain the boundary-value problem

$$\left.\begin{aligned}
&u_{t''t''} = a''^2 u_{x''x''}, \quad 0 < x'' < l'', \quad 0 < t'' < +\infty, \quad a''^2 = \frac{E}{\rho},\\
&u(0, t'') = 0, \quad E u_{x''}(l'', t'') + k u(l'', t'') = 0, \quad 0 < t'' < +\infty,\\
&u(x'', 0) = \phi_u(x''), \quad u_{t''}(x'', 0) = \psi_u(x''), \quad 0 < x'' < l''.
\end{aligned}\right\} \quad (\mathrm{IIa})$$

If the electric voltage is taken as the function describing the electric vibrations in a conductor $0 \leqslant x' \leqslant l'$ of negligibly small resistance and leakage conductance, and if one end of the conductor ($x' = 0$) is earthed directly, and the other ($x' = l'$) through a lumped self inductance, then to determine the voltage $v(x', t')$ in the conductor we have the boundary-value problem

$$v_{t't'} = a'^2 v_{x'x'}, \quad 0 < x' < l', \quad 0 < t' < +\infty, \quad a'^2 = \frac{1}{CL},$$

where C is the capacity per unit length of the conductor and L is the self inductance per unit length of the conductor

$$\left.\begin{array}{l} v(0, t') = 0, \quad v_{x'}(l', t') + \dfrac{L}{L_0} v(l', t') = 0, \quad 0 < t' < +\infty, \\ v(x', 0) = \phi_v(x'), \quad v_{t'}(x', 0) = \psi_v(x'), \quad 0 < x' < l'. \end{array}\right\} \tag{Ia}$$

Problem (Ia) is similar to problem (IIa). In order that problem (Ia) be similar to problem (IIa) with coefficients of similarity k_x, k_t, k_u, it is necessary and sufficient that the relations

$$k_x = \frac{l'}{l''}, \tag{1a}$$

$$a'^2 = \frac{k_x^2}{k_t^2} a''^2, \tag{2a}$$

$$k_x \frac{L}{L_0} = \frac{k}{E}, \tag{3a}$$

$$\left.\begin{array}{r} \phi_v(x') = k_u \phi_u(x''), \\ \psi_v(x') = \dfrac{k_u}{k_t} \psi_u(x''), \\ x' = k_x x'', \quad 0 < x'' < l'' \end{array}\right\} \tag{4a}$$

be fulfilled.

(b) If one end of the rod ($x'' = 0$) is free, and the other ($x'' = l''$) experiences a resistance, proportional to velocity, then the boundary-value problem for determining the longitudinal displacements $u(x'', t'')$ of points of the rod has the form

$$\left.\begin{array}{l} u_{t''t''} = a''^2 u_{x''x''}, \quad 0 < x'' < l'', \quad 0 < t'' < +\infty, \\ u_{x''}(0, t'') = 0, \quad E u_{x''}(l'', t'') + r u_{t''}(l'', t'') = 0, \quad 0 < t'' < +\infty, \\ u(x'', 0) = \phi_u(x''), \quad u_{t''}(x'', 0) = \psi_u(x''), \quad 0 < x'' < l''. \end{array}\right\} \tag{IIb}$$

Here r denotes the coefficient of resistance. If one end of the conductor ($x' = 0$) is earthed directly, and the other end ($x' = l'$) is earthed through a lumped resistance R_0, then, assuming that the resistance and leakage conductance of the conductor equal zero, to determine the current intensity $i(x', t')$ we have the boundary-value problem

$$i_{t't'} = a'^2 i_{x'x'}, \quad a'^2 = \frac{1}{CL}, \quad 0 < x' < l', \quad 0 < t' < +\infty,$$

$$i_{x'}(0, t') = 0, \quad i_{x'}(l', t') + CR_0 i_{t'}(l', t') = 0, \quad 0 < t' < +\infty,$$
$$i(x', 0) = \phi_i(x'), \quad i_{t'}(x', 0) = \psi_i(x'), \quad 0 < x' < l'.$$

Problem (Ib) is similar to problem (IIb). In order that problem (Ib) be similar to problem (IIb) with coefficients of similarity k_x, k_t, k_u, it is necessary and sufficient that the relations

$$k_x = \frac{l'}{l''}, \tag{1b}$$

$$a'^2 = \frac{k_x^2}{k_t^2} a''^2, \tag{2b}$$

$$CR_0 \frac{k_x}{k_t} = \frac{r}{E}, \tag{3b}$$

$$\phi_i(x') = k_u \phi_u(x''), \quad \psi_i(x') = \frac{k_u}{k_t} \psi_u(x''), \quad x' = k_x x'', \quad 0 < x'' < l''. \tag{4b}$$

be fulfilled.

(c) If one end of the rod ($x'' = 0$) is attached elastically and the other end ($x'' = l''$) moves according to a given law, then we have:

$$\left.\begin{array}{l} u_{t''t''} = a''^2 u_{x''x''}, \quad 0 < x'' < l'', \quad 0 < t'' < +\infty, \\ E u_{x''}(0, t'') - k u(0, t'') = 0, \quad u(l'', t'') = \omega_u(t''), \quad 0 < t'' < +\infty, \\ u(x'', 0) = \phi_u(x''), \quad u_{t''}(x'', 0) = \psi_u(x''), \quad 0 < x'' < l''. \end{array}\right\} \quad \text{(IIc)}$$

If one end of a conductor ($x' = 0$) is earthed through a lumped self inductance L_0, and an e.m.f. $\omega_v(t')$ is applied to the other end ($x' = l'$), then to determine the voltage in the conductor we have the boundary-value problem

$$\left.\begin{array}{l} v_{t't'} = a'^2 v_{x'x'}, \quad 0 < x' < l', \quad 0 < t' < +\infty, \\ L_0 v_{x'}(0, t') - L v(0, t') = 0, \quad v(l', t') = \omega_v(t'), \quad 0 < t' < +\infty, \\ v(x') = \phi_v(x'), \quad v_{t'}(x', 0) = \psi_v(x'), \quad 0 < x' < l'. \end{array}\right\} \quad \text{(Ic)}$$

Problem (Ic) is similar to problem (IIc). In order that problem (Ic) be similar to problem (IIc) with coefficients of similarity k_x, k_t, k_u, it is necessary and sufficient that relations (1a), (2a), (3a), (4a) (see above) and the relation

$$\omega_v(t') = k_u \omega_u(t''), \quad t' = k_t t'', \quad 0 < t'' < +\infty.$$

be fulfilled.

Method. The problem is solved in a similar manner to the preceding one.

51. If one end of a conductor ($x'' = 0$) is earthed through a lumped resistance R_0, and the other end ($x'' = l''$) is earthed through a lumped capacitance C_0, then to determine the voltage in the conductor of negligibly small leakage conductance we obtain the boundary-value problem

$$\left.\begin{array}{l} v_{t''t''} = a''^2 v_{x''x''}, \quad a''^2 = \dfrac{1}{CL}, \quad 0 < x'' < l'', \quad 0 < t'' < +\infty, \\ R_0 v_{x''}(0, t'') - L v_{t''}(0, t'') = 0, \quad L C_0 v_{t''t''}(l'', t'') + v_{x''}(l'', t'') = 0, \\ \qquad\qquad\qquad\qquad\qquad\qquad\qquad 0 < t'' < +\infty, \\ v(x'', 0) = \phi_v(x''), \quad v_{t''}(x'', 0) = \psi_v(x''), \quad 0 < x'' < l'', \end{array}\right\} \quad \text{(IIa)}$$

and to determine the current we have the boundary-value problem

$$\left.\begin{aligned} &i_{t''t''} = a''^2 i_{x''x''}, \quad 0 < x'' < l'', \quad 0 < t'' < +\infty, \\ &i_{x''}(0,t'') - CR_0 i_{t''}(0,t'') = 0, \quad C_0 i_{x''}(l'',t'') + Ci(l'',t'') = 0, \\ &\qquad\qquad\qquad\qquad\qquad\qquad 0 < t'' < +\infty, \\ &i(x'',0) = \phi_i(x''), \quad i_{t''}(x'',0) = \psi_i(x''), \quad 0 < x'' < l''. \end{aligned}\right\} \quad \text{(IIb)}$$

If a resistance torque proportional to the angular velocity is applied to the end of a flexible cylinder ($x' = 0$), performing torsion vibrations, and a pulley with an axial moment of inertia k_0 is mounted at the other end ($x' = l'$), then to determine the deflection angles $\theta(x', t')$ of cross-sections of the rod we obtain the boundary-value problem

$$\left.\begin{aligned} &\theta_{t't'} = a'^2 \theta_{x'x'}, \quad 0 < x' < l', \quad 0 < t' < +\infty, \\ &GJ\theta_{x'}(0,t') - r_0\theta_{t'}(0,t') = 0, \quad k_0\theta_{t't'}(l',t') + GJ\theta_{x'}(l',t') = 0, \\ &\qquad\qquad\qquad\qquad\qquad\qquad 0 < t' < +\infty, \\ &\theta(x',0) = \phi_0(x'), \quad \theta_{x'}(x',0) = \psi_0(x'), \quad 0 < x' < l', \end{aligned}\right\} \quad \text{(Ia)}$$

where $a''^2 = GJ/k$ and the quantities G, J, k have the same meaning as in the answer to problem 3.

If a braking torque, proportional to the angular velocity is applied to the end of a cylinder $x' = 0$, performing torsional vibrations, and the end $x' = l'$ is fixed elastically, then to determine $\theta(x', t')$ we obtain the boundary-value problem

$$\left.\begin{aligned} &\theta_{t't'} = a'^2 \theta_{x'x'}, \quad 0 < x' < l', \quad 0 < t' < +\infty, \\ &GJ\theta_{x'}(0,t') - r_0\theta_{t'}(0,t') = 0, \quad GJ\theta_{x'}(l',t') + H_0\theta(l' t') = 0, \\ &\qquad\qquad\qquad\qquad\qquad\qquad 0 < t' < +\infty, \\ &\theta(x',0) = \phi_\theta(x'), \quad \theta_{t'}(x',0) = \psi_\theta(x'), \quad 0 < x' < l'. \end{aligned}\right\} \quad \text{(Ib)}$$

Problem (Ia) is similar to problem (IIa). Problem (Ib) is similar to problem (IIb). In order that problem (Ia) be similar to problem (IIa) with coefficients of similarity k_x, k_t, k_u, it is necessary and sufficient that the relations

$$k_x = \frac{l'}{l''}, \tag{1}$$

$$a'^2 = \frac{k_x^2}{k_t^2} a''^2, \tag{2}$$

$$\frac{R_0}{GJ}\frac{k_x}{k_t} = \frac{L}{r_0}, \tag{3}$$

$$\frac{GJ}{k_0}\frac{k_t^2}{k_x} = \frac{1}{LC_0}, \tag{4}$$

$$\left.\begin{aligned} \phi_\theta(x') &= k_u \phi_v(x''), \\ \psi_\theta(x') &= \frac{k_u}{k_t}\psi_v(x''), \\ x' &= k_x x'', \quad 0 < x'' < l'' \end{aligned}\right\} \tag{5}$$

be fulfilled. In order that (Ib) be similar to problem (IIb) with coefficients of similarity k_x, k_t, k_u, it is necessary and sufficient that relations (1), (2) and (5) and the relations

$$\frac{r_0}{GJ}\frac{k_x}{k_t} = CR_0, \qquad k_x\frac{H_0}{GJ} = \frac{C}{C_0}$$

be fulfilled.

Method. See solution 49.

§ 2. Method of Travelling Waves (D'Alembert's Method)

1. Problems for an Infinite String

The solutions of the boundary-value problems of this section, with the form

$$u_{tt} = a^2 u_{xx}, \qquad -\infty < x < +\infty, \qquad 0 < t < +\infty, \tag{1}$$

$$u(x, 0) = \phi(x), \qquad u_t(x, 0) = \psi(x), \qquad -\infty < x < +\infty, \tag{2}$$

occur in D'Alembert's formula

$$u(x, t) = \frac{\phi(x-at)+\phi(x+at)}{2} + \frac{1}{2a}\int_{x-at}^{x+at}\psi(z)\,dz. \tag{3}$$

52. In the problem under consideration $\psi(x) \equiv 0$, therefore

$$u(x, t) = \frac{\phi(x-at)+\phi(x+at)}{2} = \frac{1}{2}\phi(x-at)+\frac{1}{2}\phi(x+at), \tag{1}$$

where $\phi(x)$ is given graphically in the conditions of the problem.

The forward and backward waves $\frac{1}{2}\phi(x-at)$ and $\frac{1}{2}\phi(x+at)$ coincide at the initial time $t = 0$ having a value equal to $\frac{1}{2}\phi(x)$.

At time t ($t > 0$) the graph of the forward wave is displaced without deformation to the right by a distance at, and the graph of the backward wave to the left by at. Summing the displaced graphs of the forward and backward waves at times $t_1, t_2, \ldots$, we obtain the profile of the string at these times. The profile of the string for the times $t_k = kc/4a$, $k = 0, 1, 2, 3, 5$ (Fig. 21) is given below.

53. (a) $u(x, t) = \dfrac{\phi(x-at)+\phi(x+at)}{2}$, (1)

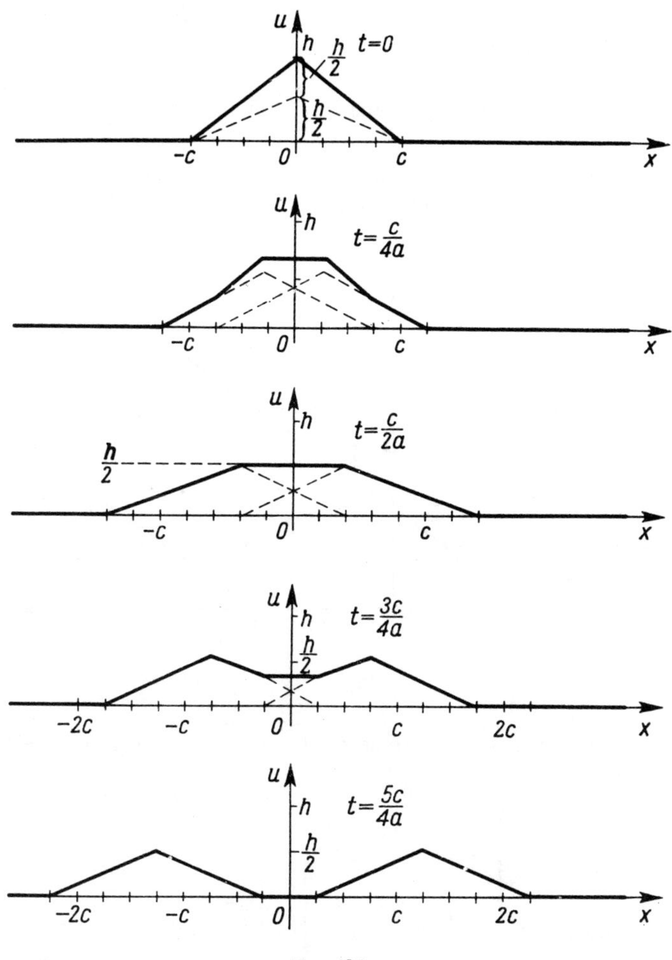

FIG. 21

where

$$\phi(x) = \left\{ \begin{array}{cc} 0, & -\infty < x < -c, \\ h\left[1-\dfrac{x^2}{c^2}\right], & -c < x < c, \\ 0, & c < x < +\infty. \end{array} \right\} \qquad (2)$$

In order to obtain the formulae required, let us consider the division of the phase plane (x, t) by the characteristics of equation (1), drawn from the ends of the interval $(-c, c)$, at which the initial deflection differs from zero (Fig. 22).

We first give formulae for the profile of the string for $t = \text{const.}$, confining ourselves to two special cases:

$$0 < t < \frac{c}{a} \quad \text{and} \quad \frac{c}{a} < t < +\infty .$$

If $t = \text{const.}$, $0 < t < c/a$, then for x, varying monotonically from $-\infty$ to $+\infty$, the point (x, t) of the phase plane in turn passes through the regions I, IV, II, VI, III.

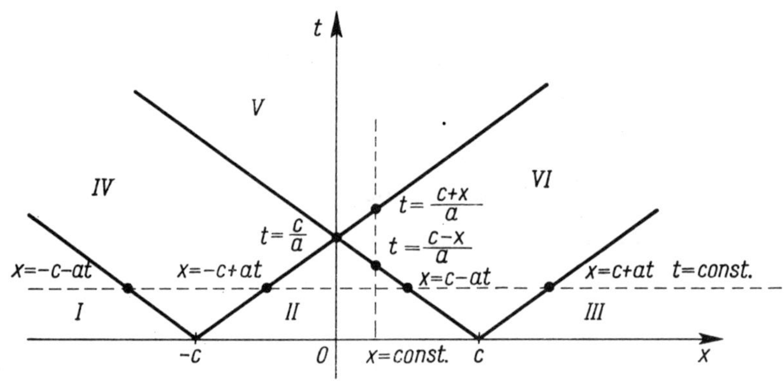

FIG. 22

Thus for $0 < t < c/a$ the profile of the string is given by the relations

$$u(x,t) = \begin{cases} 0, & -\infty < x < -c-at, \\ \frac{h}{2}\left[1-\frac{(x+at)^2}{c^2}\right], & -c-at < x < -c+at, \\ h\left[1-\frac{x^2+a^2t^2}{c^2}\right], & -c+at < x < c-at, \\ \frac{h}{2}\left[1-\frac{(x-at)^2}{c^2}\right], & c-at < x < c+at, \\ 0, & c+at < x < +\infty . \end{cases}$$

Similarly the string profile is obtained for $c/a < t < +\infty$.

(b) We give now the formulae for $u(x, t)$ for $x = \text{const.}$, representing the law of motion of a point of the string with fixed coordinate. Let us choose a fixed value of x in each of the intervals $-\infty < x < -c$, $-c < x < 0$, $0 < x < c$, $c < x < +\infty$ and investigate the expression for the solution in t, varying from 0 to $+\infty$. We obtain:

(α)

$$u(x,t)=\left\{\begin{array}{ll}0, & 0\leqslant t\leqslant -\dfrac{c+x}{a},\\ \dfrac{h}{2}\left[1-\dfrac{(x+at)^2}{c^2}\right], & -\dfrac{c+x}{a}\leqslant t\leqslant \dfrac{c-x}{a},\\ 0, & \dfrac{c-x}{a}\leqslant t<+\infty,\end{array}\right\}-\infty<x<-c,$$

(β)

$$u(x,t)=\left\{\begin{array}{ll}h\left[1-\dfrac{x^2+a^2t^2}{c^2}\right], & 0\leqslant t\leqslant \dfrac{c+x}{a},\\ \dfrac{h}{2}\left[1-\dfrac{(x+at)^2}{c^2}\right], & \dfrac{c+x}{a}\leqslant t\leqslant \dfrac{c-x}{a},\\ 0, & \dfrac{c-x}{a}\leqslant t<+\infty,\end{array}\right\}-c<x<0,$$

(γ)

$$u(x,t)=\left\{\begin{array}{ll}h\left[1-\dfrac{x^2+a^2t^2}{c^2}\right], & 0\leqslant t\leqslant \dfrac{c-x}{a},\\ \dfrac{h}{2}\left[1-\dfrac{(x-at)^2}{c^2}\right], & \dfrac{c-x}{a}\leqslant t\leqslant \dfrac{c+x}{a},\\ 0, & \dfrac{c+x}{a}\leqslant t<+\infty,\end{array}\right\}0<x<c,$$

(δ)

$$u(x,t)=\left\{\begin{array}{ll}0, & 0<t<\dfrac{-c+x}{a},\\ \dfrac{h}{2}\left[1-\dfrac{(x-at)^2}{c^2}\right], & \dfrac{-c+x}{a}\leqslant t\leqslant \dfrac{c+x}{a},\\ 0, & \dfrac{c+x}{a}<t<+\infty,\end{array}\right\}c<x<+\infty.$$

Note 1. (α) and (β) are obtained from (δ) and (γ) by the simple replacement of x by $-x$, since $u(x,t)$ is an even function of x because of the evenness of $\phi(x)$.

Note 2. A geometric method of finding the profile of the string for different moments of time is described in the solution of problem 52.

54. The deflection $u(x,t)$ reaches a maximum value at a point with abscissa

$$x=\frac{\alpha_2+\beta_2+\alpha_1+\beta_1}{2}$$

at a time

$$t=\frac{\alpha_2+\beta_2-(\alpha_1+\beta_1)}{4a};$$

this maximum value equals $(h_1+h_2)/2$.

Method. Consider the integral surface, representing the solution $u = u(x, t)$ of the boundary-value problem.

55. The solution of the boundary-value problem has the form

$$u(x, t) = \Psi(x+at) - \Psi(x-at),$$

where

$$\Psi(z) = \frac{1}{2a}\int_{z_0}^{z} \psi(\alpha)\mathrm{d}\alpha = \begin{cases} 0, & -\infty < z < -c, \\ \dfrac{v_0(z+c)}{2a}, & -c \leqslant z \leqslant c, \quad z_0 = -c, \\ \dfrac{v_0 c}{a}, & c \leqslant z < +\infty, \end{cases}$$

therefore the law of motion of points of the string with different abscissae is represented by the formulae

(a)

$$u(x, t) = \left\{\begin{array}{ll} 0, & 0 \leqslant t \leqslant -\dfrac{c+x}{a}, \\ \dfrac{v_0(x+at)}{2a} + \dfrac{v_0 c}{2a}, & -\dfrac{c+x}{a} \leqslant t \leqslant \dfrac{c-x}{a}, \\ \dfrac{v_0 c}{a}, & \dfrac{c-x}{a} \leqslant t < +\infty, \end{array}\right\} -\infty < x < -c,$$

(b)

$$u(x, t) = \left\{\begin{array}{ll} v_0 t, & 0 \leqslant t \leqslant \dfrac{c+x}{a}, \\ \dfrac{v_0(x+at)}{2a} + \dfrac{v_0 c}{2a}, & \dfrac{c+x}{a} \leqslant t \leqslant \dfrac{c-x}{a}, \\ \dfrac{v_0 c}{a}, & \dfrac{c-x}{a} \leqslant t < +\infty, \end{array}\right\} -c < x < 0,$$

(c)

$$u(x, t) = \left\{\begin{array}{ll} v_0 t, & 0 \leqslant t \leqslant \dfrac{c-x}{a}, \\ \dfrac{v_0(x-at)}{2a} + \dfrac{v_0 c}{2a}, & \dfrac{c-x}{a} \leqslant t \leqslant \dfrac{c+x}{a}, \\ \dfrac{v_0 c}{a}, & \dfrac{c+x}{a} \leqslant t < +\infty, \end{array}\right\} 0 < x < c,$$

(d)

$$u(x, t) = \left\{\begin{array}{ll} 0, & 0 \leqslant t \leqslant \dfrac{-c+x}{a}, \\ \dfrac{v_0(x-at)}{2a} + \dfrac{v_0 c}{2a}, & \dfrac{-c+x}{a} \leqslant t \leqslant \dfrac{c+x}{a}, \\ \dfrac{v_0 c}{a}, & \dfrac{c+x}{a} < t < +\infty, \end{array}\right\} c < x < +\infty.$$

The profile of the string for the times $t_1, t_2, \ldots$ may be obtained by subtraction of the graph of the forward wave $\Psi(x-at)$ from the graph of the backward wave $\Psi(x+at)$. For times $t_k = kc/4a$, $k = 0, 2, 4, 6$ it has the form (Fig. 23).

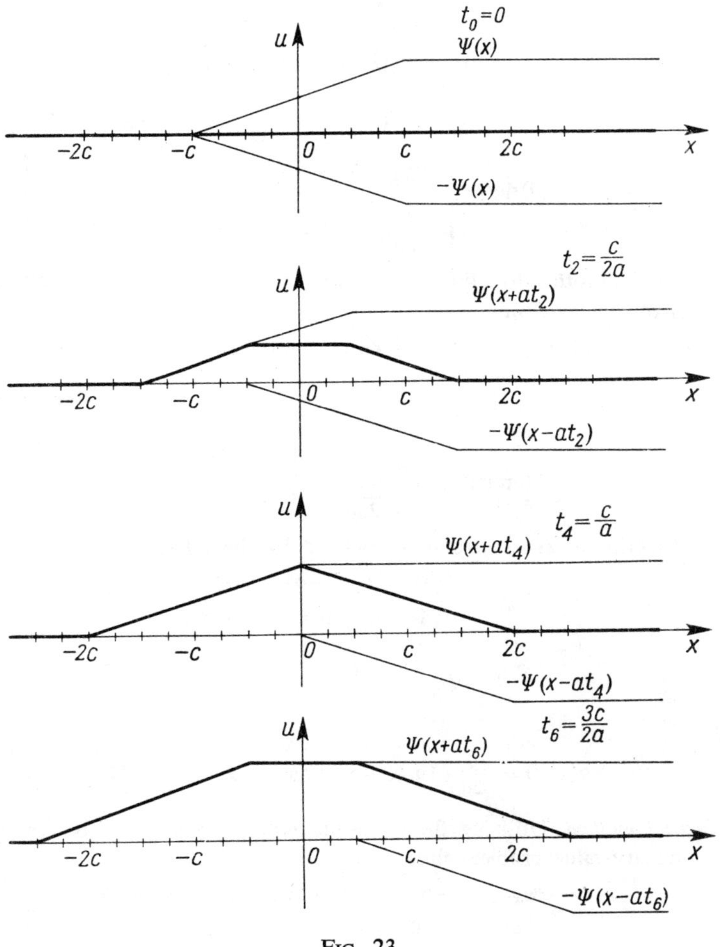

FIG. 23

56. We give two methods of solving the problem.

First method. We shall assume firstly that the impulse is uniformly distributed over the segment $x_0-\delta \leqslant x \leqslant x_0+\delta$. Then the boundary-value problem is formulated in the following way:

$$u_{tt} = a^2 u_{xx}, \qquad -\infty < x < +\infty, \quad 0 < t < +\infty, \tag{1}$$

$$\left.\begin{aligned} u(x,0) &= 0, \quad -\infty < x < +\infty, \\ u_t(x,0) = \phi_\delta(x) &= \begin{cases} 0, & -\infty < x < x_0-\delta, \\ \dfrac{I}{2\delta\rho}, & x_0-\delta < x < x_0+\delta, \\ 0, & x_0+\delta < x < +\infty, \end{cases} \end{aligned}\right\} \tag{2}$$

$$U_\delta(x,t) = \Psi_\delta(x+at) - \Psi_\delta(x-at), \tag{3}$$

where

$$\Psi_\delta(z) = \frac{1}{2a}\int_{x_0-\delta}^{z} \phi_\delta(\alpha)\,d\alpha = \begin{cases} 0, & -\infty < z < x_0-\delta, \\ \dfrac{I}{4a\delta\rho}(z-x_0+\delta), & x_0-\delta < z < x_0+\delta, \\ \dfrac{I}{2a\rho}, & x_0+\delta < z < +\infty. \end{cases} \tag{4}$$

By a formal transition to a limit as $\delta \to 0$ in the solution of (3) we obtain the solution of the problem

$$u(x,t) = \lim U_\delta(x,t) = \lim_{\delta\to 0} \Psi_\delta(x+at) - \lim_{\delta\to 0} \Psi_\delta(x-at)$$
$$= \Psi(x+at) - \Psi(x-at),$$

where

$$\Psi(z) = \lim_{\delta\to 0} \psi_\delta(z) = \begin{cases} 0 & \text{if} \quad -\infty < x < x_0, \\ \dfrac{I}{2a\rho} & \text{if} \quad x_0 < x < +\infty. \end{cases}$$

If the function $\sigma_0(z)$ is introduced, defined by the relations

$$\sigma_0(z) = \begin{cases} 0 & \text{if} \quad -\infty < z < 0, \\ 1 & \text{if} \quad 0 < z < +\infty, \end{cases}$$

then

$$\Psi(z) = \frac{I}{2a\rho}\sigma_0(z-x_0)$$

and

$$u(x,t) = \frac{I}{2a\rho}\{\sigma_0(x+at-x_0) - \sigma_0(x-at-x_0)\}.$$

Second method. Utilizing the delta-function†, it is possible to formulate the boundary-value problem thus:

$$u_{tt} = a^2 u_{xx}, \quad -\infty < x < +\infty, \quad 0 < t < +\infty,$$
$$u(x,0) = 0, \quad -\infty < x < +\infty,$$
$$u_t(x,0) = \frac{I}{\rho}\delta(x-x_0)‡ \quad -\infty < x < +\infty.$$

---

† See [7], pages 292–296.

‡ The coefficient in the delta-function $\delta(x-x_0)$ is chosen so that the total impulse, transmitted to the string at time $t = 0$, i.e. $\int_{-\infty}^{+\infty} u_t(x,0)\,dx$, is equal to 1.

Then by means of D'Alembert's formula we obtain:

$$u(x,t) = \frac{I}{2a\rho}\int\limits_{x-at}^{x+at} \delta(z-x_0)\,dz = \frac{I}{2a\rho}\int\limits_{x-x_0-at}^{x-x_0+at} \delta(\xi)\,d\xi$$

$$= \frac{I}{2a\rho}[\sigma_0(x+at-x_0)-\sigma_0(x-at-x_0)],$$

since

$$\int\limits_{z_0}^{z} \delta(\xi)\,d\xi = \begin{Bmatrix} 1, & z>0, \\ 0, & z<0 \end{Bmatrix} \text{ if } z_0<0, \text{ and } \int\limits_{z_0}^{z} \delta(\xi)\,d\xi = \begin{Bmatrix} 0, & z>0, \\ 1, & z<0 \end{Bmatrix} \text{ if } z_0>0.$$

57. In problem 52

$$u(x,0) = \phi(x) \not\equiv 0, \quad u_t(x,0) = \psi(x) \equiv 0,$$

and in the problem under consideration the travelling wave at time $t = 0$ is characterized by "initial" deflections and velocities† different from zero

$$u(x,0) = \phi(x), \quad u_t(x,0) = -a\phi'(x), \quad -\infty < x < +\infty.$$

In the case of problem 52 we had: $u(x,t) = \frac{1}{2}\phi(x-at)+\frac{1}{2}\phi(x+at)$.

In the problem under consideration D'Alembert's formula gives:

$$u(x,t) = \frac{\phi(x-at)+\phi(x+at)}{2} + \frac{1}{2a}\int\limits_{x-at}^{x+at} [-a\phi'(z)]\,dz = \phi(x-at).$$

58. The solution of the boundary-value problem

$$\left.\begin{aligned} v_x+Li_t+Ri &= 0, \\ i_x+Cv_t+Gv &= 0 \end{aligned}\right\} \text{ if } \quad -\infty<x<+\infty, \quad 0<t<+\infty, \tag{1}$$

$$v(x,0) = \phi(x), \quad i(x,0) = \sqrt{\frac{C}{L}}\,F(x) \quad \text{if} \quad -\infty<x<+\infty \tag{2}$$

for $CR = GL$ has the form

$$\left.\begin{aligned} v(x,t) &= e^{-\frac{R}{L}t}\{\phi(x-at)+\psi(x+at)\}, \\ i(x,t) &= \sqrt{\frac{C}{L}}\,e^{-\frac{R}{L}t}\{\phi(x-at)-\psi(x+at)\}, \end{aligned}\right\} \begin{aligned} &-\infty<x<+\infty, \\ &0<t<+\infty, \end{aligned} \tag{3}$$

where

$$\phi(z) = \frac{f(x)+F(z)}{2} \quad \text{and} \quad \psi(z) = \frac{f(z)-F(z)}{2} \quad \text{if} \quad -\infty<z<+\infty. \tag{4}$$

† It is assumed that a wave already exists for $t < 0$.

Method. Eliminate the current intensity from equation (1); in the equation of second order for $v(x, t)$ thus obtained eliminate the term $v_t(x, t)$ (see chapter I) then the equation takes the form

$$u_{tt} = a^2 u_{xx}.$$

Its solution will be:

$$u(x, t) = \phi(x-at)+\psi(x+at).$$

Returning to the function $v(x, t)$ and using equation (1) and the initial conditions, the answer is readily obtained.

2. Problems for a Semi-infinite Region

Let us find the solution of the boundary-value problem for the semi-infinite region

$$u_{tt} = a^2 u_{xx}; \quad 0 < x < +\infty, \quad 0 < t < +\infty, \tag{1}$$

$$\alpha_1 u_{tt}(0, t)+\alpha_2 u_t(0, t)+\alpha_3 u_x(0, t)+\alpha_4 u(0, t) = \Phi(t), \quad 0 < t < +\infty, \tag{2}$$

$$u(x, 0) = \phi(x), \quad u_t(x, 0) = \psi(x), \quad 0 < x < +\infty \tag{3}$$

in the form

$$u(x, t) = \phi_1(x-at)+\phi_2(x+at). \tag{4}$$

The functions $\phi_1(z)$ and $\phi_2(z)$ may be determined from the initial conditions only for $0 < z < +\infty$. This is sufficient for the determination of $\phi_2(z)$, since $x+at > 0$ for $0 < x < +\infty$, $0 < t < +\infty$. But the function $\phi_1(z)$ must be given for $-\infty < z < 0$. This is achieved by using the boundary condition (2).

The solution of the boundary-value problem (1), (2), (3) may be found also by using D'Alembert's formula

$$u(x, t) = \frac{\phi(x-at)+\phi(x+at)}{2}+\frac{1}{2a}\int_{x-at}^{x+at} \psi(z)\,dz \tag{5}$$

for an infinite string. To do this it is necessary to fictitiously expand the string along the negative semi-axis $-\infty < x < 0$, and then arrange the initial conditions (3) along the semi-axis so that $u(x, t)$, calculated from (5), fulfils the boundary condition (2)†. Moreover, it follows that in the case of a fixed end, the functions $\phi(x)$ and $\psi(x)$ must be expanded as odd functions along the semi-axis $-\infty < x < 0$, and in the case of a free end as even functions.

59. The profile of the string at times

$$t = \frac{c}{a}; \quad \frac{3c}{2a}; \quad \frac{2c}{a}; \quad \frac{7c}{2a}$$

is represented in Fig. 24.

---

† See [7], pages 57–58.

60. The solution of the boundary-value problem

$$u_{tt} = a^2 u_{xx}, \quad 0 < x, t < +\infty, \tag{1}$$

$$u_x(0, t) = 0, \qquad 0 < t < +\infty, \tag{2}$$

$$u(x, 0) = 0, \qquad 0 < x < +\infty, \tag{3}$$

$$u_t(x, 0) = \left\{ \begin{array}{ll} 0, & 0 < x < c, \\ v_0, & c < x < 2c, \\ 0, & 2c < x < +\infty \end{array} \right\} \tag{4}$$

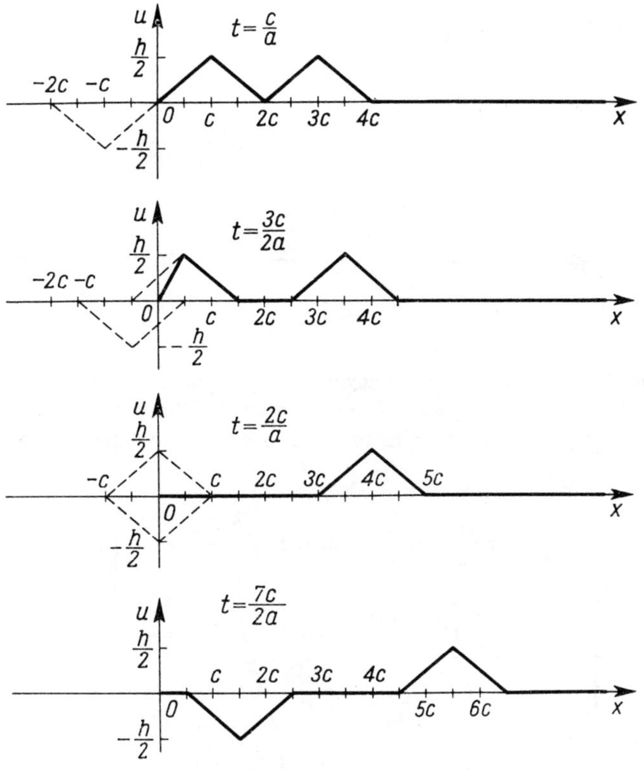

FIG. 24

may be found by means of D'Alembert's formula for an event† expansion of the initial conditions

$$u(x, t) = \Psi(x+at) - \Psi(x-at), \tag{5}$$

† See [7], pages 57–58.

where

$$\Psi(z) = \frac{1}{2a}\int_{-2c}^{z} \phi(\alpha)\,d\alpha, \tag{6}$$

$$\phi(z) = \left\{\begin{array}{ll} 0, & -\infty < z < -2c, \\ v_0, & -2c < z < -c, \\ 0, & -c < z < c, \\ v_0, & c < z < 2c, \\ 0, & 2c < z < +\infty. \end{array}\right\} \tag{7}$$

A graph of the function $\psi(z)$ has the form shown in Fig. 25.

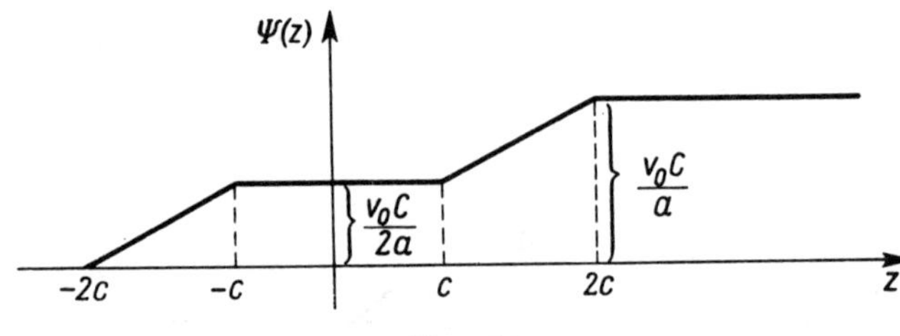

FIG. 25

A profile of the deflections at any moment of time is obtained by subtracting the graph of the forward wave from the graph of the backward wave.

For $t = 0$; c/a; $2c/a$; $3c/a$ the profile of the deflections has the form shown in Fig. 26.

61.

$$u(x,t) = \left\{\begin{array}{ll} -\frac{2Al}{\pi a}\sin\frac{\pi x}{2l}\sin\frac{at}{2l}, & 0 < t < \frac{2l-x}{a}, \\ \frac{2Al}{\pi a}\cos^2\frac{\pi}{4l}(x-at), & \frac{2l-x}{a} < t < \frac{2l+x}{a}, \\ 0, & \frac{2l+x}{a} < t < +\infty, \end{array}\right\} 0 < x < 2l,$$

$$u(x,t) = \left\{\begin{array}{ll} 0, & 0 < t < \frac{-2l+x}{a}, \\ \frac{2Al}{\pi a}\cos^2\frac{\pi}{4l}(x-at), & \frac{-2l+x}{a} < t < \frac{2l+x}{a}, \\ 0, & \frac{2l+x}{a} < t < +\infty, \end{array}\right\} 2l < x < +\infty,$$

$$A = -\frac{\pi l}{4l}.$$

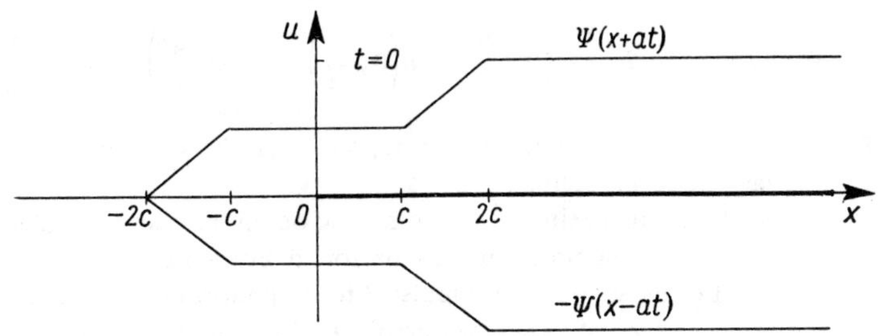
u
t=0
Ψ(x+at)
−2c
−c
0
c
2c
x
−Ψ(x−at)

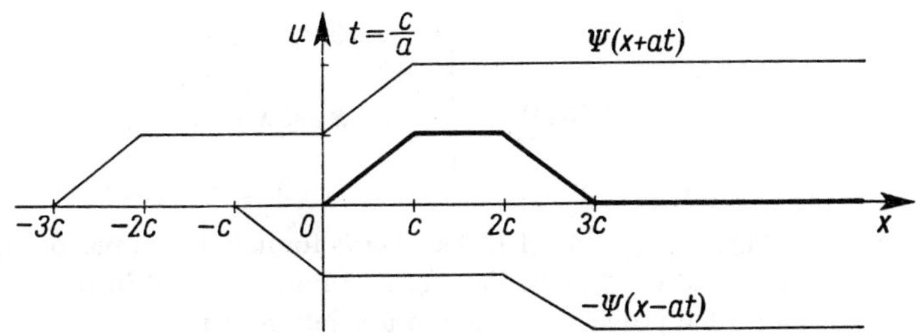
u
t=c/a
Ψ(x+at)
−3c
−2c
−c
0
c
2c
3c
x
−Ψ(x−at)

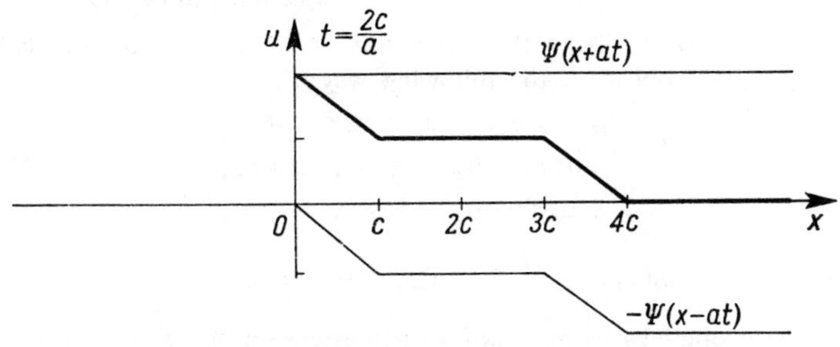
u
t=2c/a
Ψ(x+at)
0
c
2c
3c
4c
x
−Ψ(x−at)

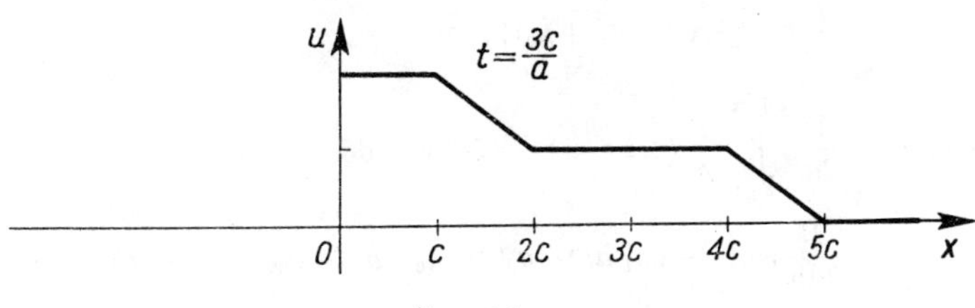
u
t=3c/a
0
c
2c
3c
4c
5c
x

Fig. 26

62.

$$\max_{\substack{-\infty<x<+\infty \\ 0<t<+\infty}} u(x,t) = h = u\left(0, \frac{2l}{a}\right) = u\left(0, \frac{6l}{a}\right) = u\left(2l, \frac{4l}{a}\right) = u\left(4l, \frac{2l}{a}\right).$$

Method. Consider the integral surface, representing the solution $u = u(x, t)$ of the boundary-value problem.

63. *Solution.* In a manner similar to that carried out in the case of problem 56, the solution of the present problem may be found by two methods.

First method. Let us assume the impulse I to be uniformly distributed over the segment $x_0 \leqslant x \leqslant x_0+\delta$. Then we arrive at the boundary-value problem

$$u_{tt} = a^2 u_{xx}, \quad 0 < x, t < +\infty, \tag{1}$$

$$u(0,t) = 0, \quad 0 < t < +\infty, \tag{2}$$

$$u(x,0) = 0, \quad u_t(x,0) = \begin{cases} 0, & 0 < x < x_0, \\ \dfrac{I}{\delta\rho}, & x_0 < x < x_0+\delta, \\ 0, & x_0+\delta < x < \infty. \end{cases} \tag{3}$$

Its solution is obtained in terms of D'Alembert's formula by means of an odd extension of the initial conditions. Passing to a limit as $\delta \to 0$ in the solution of this boundary-value problem, we obtain a solution of the original problem

$$u(x,t) = \frac{I}{2a\rho}\{\sigma_0(x-x_0+at)-\sigma_0(x-x_0-at)- -\sigma_0(x+x_0+at)+\sigma_0(x+x_0-at)\}.$$

Second method. Utilizing the δ-function, it is possible to formulate the boundary-value problem in the following way:

$$u_{tt} = a^2 u_{xx}, \quad 0 < x, t < +\infty, \tag{1}$$

$$u(0,t) = 0, \quad 0 < t < +\infty, \tag{2}$$

$$u(x,0) = 0, \quad 0 < x < +\infty, \tag{3}$$

$$u_t(x,0) = \frac{I}{\rho}\delta(x-x_0), \quad 0 < x < +\infty. \tag{3'}$$

Its solution is obtained by means of an odd extension of the initial conditions. An odd extension of the initial condition (3′) gives:

$$u_t(x,0) = \frac{I}{\rho}\{\delta(x-x_0)-\delta(x+x_0)\},$$

$$u(x,t) = \frac{1}{2a}\int_{x-at}^{x+at} \frac{I}{\rho}\{\delta(\xi-x_0)-\delta(\xi+x_0)\}\,d\xi$$

$$= \frac{I}{2a\rho}\{\sigma_0(x-x_0+at)-\sigma_0(x-x_0-at)-\sigma_0(x+x_0+at)+ +\sigma_0(x+x_0-at)\}.$$

64. By means of the δ-function the boundary-value problem is formulated in the following way:

$$u_{tt} = a^2 u_{xx}, \quad 0 < x,\ t < +\infty, \tag{1}$$

$$u(0, t) = 0, \quad 0 < t < +\infty, \tag{2}$$

$$u(x, 0) = 0, \quad 0 < x < +\infty, \tag{3}$$

$$u_t(x, 0) = \frac{I}{\rho} \sum_{k=1}^{n} \delta(x - x_k). \tag{3'}$$

Extending the initial condition oddly and applying D'Alembert's formula, we obtain:

$$u(x, t) = \frac{I}{\rho} \sum_{k=1}^{n} \{\sigma_0(x - x_k + at) - \sigma_0(x - x_k - at) - \sigma_0(x + x_k + at) + \\ + \sigma_0(x + x_k - at)\}.$$

65. The solution of the problem may be derived by passing to a limit as $x_0 \to 0+0$ in the solution of the boundary-value problem

$$u_{tt} = a^2 u_{xx}, \quad 0 < x,\ t < +\infty, \tag{1}$$

$$u_x(0, t) = 0, \quad 0 < t < +\infty, \tag{2}$$

$$u(x, 0) = 0, \quad 0 < x < +\infty, \tag{3}$$

$$u_t(x, 0) = \frac{I}{\rho} \delta(x - x_0), \quad x_0 > 0, \quad 0 < x < +\infty, \tag{3'}$$

$$u(x, t) = \frac{I}{2a\rho} \lim_{x_0 \to 0+0} \{\sigma(x - x_0 + at) - \sigma_0(x - x_0 - at) + \sigma_0(x + x_0 + at) - \\ - \sigma_0(x + x_0 - at)\} = \frac{I}{a\rho} \{\sigma_0(x + at) - \sigma_0(x - at)\}.$$

This solution may also be derived by passing to a limit as $\delta \to 0$ in the solution of the boundary-value problem

$$u_{tt} = a^2 u_{xx}, \quad 0 < x,\ t < +\infty, \tag{1}$$

$$u_x(0, t) = 0, \quad 0 < t < +\infty, \tag{2}$$

$$u(x, 0) = 0, \quad 0 < x < +\infty, \tag{3}$$

$$u_t(x, 0) = \begin{cases} \dfrac{I}{\delta\rho}, & 0 < x < \delta, \\ 0, & \delta < x < +\infty. \end{cases} \tag{3'}$$

66. *Solution.* We give two methods of solving the problem.

First method. The boundary-value problem is formulated thus:

$$u_{tt} = a^2 u_{xx}, \quad 0 < x, t < +\infty, \tag{1}$$

$$M u_{tt}(0, t) = ES u_x(0, t), \quad 0 < t < +\infty, \quad M = \frac{Q}{g}, \tag{2}$$

$$u(x,0) = 0, \quad 0 < x < +\infty, \tag{3}$$

$$u_t(x,0) = \begin{cases} v_0, & x = 0, \\ 0, & 0 < x < +\infty. \end{cases} \tag{3'}$$

We seek the solution of the boundary-value problem (1), (2), (3), (3′) in the form

$$u(x,t) = \phi\left(t - \frac{x}{a}\right) + \psi\left(t + \frac{x}{a}\right). \tag{4}$$

From the initial conditions we find:

$$\left.\begin{aligned} \phi(-z)+\psi(z) &= 0, \quad (5)\\ \phi'(-z)+\psi'(z) &= 0, \quad (5') \end{aligned}\right\} 0 < z < +\infty.$$

Integrating (5′) we obtain;

$$-\phi(-z)+\psi(z) = \text{const.}$$

The constant of integration may be assumed equal to zero. Then

$$-\phi(-z)+\psi(z) = 0, \quad 0 < z < +\infty. \tag{5''}$$

From (5) and (5″) we find:

$$\psi(z) = 0, \quad \phi(-z) = 0, \quad 0 < z < +\infty. \tag{6}$$

Therefore,

$$u(x,t) = \begin{cases} \phi\left(t - \dfrac{x}{a}\right), & t > \dfrac{x}{a}, \\ 0, & 0 < t < \dfrac{x}{a}. \end{cases} \tag{7}$$

Substituting the expression found for $u(x, t)$ in the boundary condition (2), we arrive at the differential equation for determining $\phi(z)$ for $z > 0$

$$\phi''(z) + \frac{ES}{aM}\phi'(z) = 0, \quad 0 < z < +\infty. \tag{8}$$

From (6) we find the first initial condition for equation (8)

$$\phi(0) = 0. \tag{9}$$

From the initial condition (3′) $u_t(0, 0) = 0$ and from (7) for $u(x, t)$ we find the second initial condition for (8)

$$\phi'(0) = v_0. \tag{10}$$

Integrating equation (8) with the initial conditions (9) and (10) gives:

$$\phi(z) = \frac{aMv_0}{ES}\left[1 - e^{-\frac{ES}{aM}z}\right], \quad 0 < z < +\infty. \tag{11}$$

Therefore,

$$u(x,t)=\left\{\begin{array}{ll}\dfrac{aMv_0}{ES}\left[1-e^{-\frac{ES}{aM}\left(t-\frac{x}{a}\right)}\right], & t>\dfrac{x}{a},\\[2ex] 0, & 0<t<\dfrac{x}{a}.\end{array}\right\} \tag{12}$$

Second method. The boundary-value problem is formulated by means of the one-sided δ-function† in the following way:

$$u_{tt}=a^2u_{xx},\quad 0<x,t<+\infty, \tag{1}$$

$$Mu_{tt}(0,t)=ESu_x(0,t)+I\delta(t),\quad 0\leqslant t<+\infty,\quad I=Mv_0, \tag{2'}$$

$$u(x,0)=0,\quad 0\leqslant x<+\infty, \tag{3}$$

$$u_t(x,0)=0,\quad 0\leqslant x<+\infty. \tag{3'}$$

The solution of the boundary-value problem (1), (2′), (3), (3′) has the form

$$u(x,t)=\phi\left(t-\frac{x}{a}\right)+\psi\left(t+\frac{x}{a}\right). \tag{4}$$

As before, from the initial conditions we find:

$$\phi(-z)=\psi(z)=0\quad\text{if}\quad 0\leqslant z<+\infty. \tag{6}$$

Therefore,

$$u(x,t)=\left\{\begin{array}{ll}\phi\left(t-\dfrac{x}{a}\right), & t>\dfrac{x}{a},\\[2ex] 0, & 0<t<\dfrac{x}{a}.\end{array}\right\} \tag{7}$$

Substituting this expression for $u(x,t)$ in the boundary condition (2′) and the initial conditions (3) and (3′), we obtain a differential equation for the determination of $\phi(z)$ with $z>0$

$$\phi''(z)+\frac{ES}{aM}\phi'(z)=v_0\delta(z),\quad 0\leqslant z<+\infty, \tag{13}$$

---

† The one-sided δ-function is defined for $-\infty<t<+\infty$ as the limit in the sense of weak convergence of the series of functions

$$f_n(t)=\left\{\begin{array}{lll}0 & \text{if} & -\infty<t<0,\\[1ex] n & \text{if} & 0<t<\dfrac{1}{n},\\[1ex] 0 & \text{if} & \dfrac{1}{n}<t<+\infty.\end{array}\right.$$

Compare also with the footnote on page 214.

and the initial conditions

$$\phi(0) = \phi'(0) = 0. \tag{14}$$

Integration of (13) with the initial conditions (14) gives:

$$\phi(z) = \frac{aMv_0}{ES}\left[1 - e^{-\frac{ES}{aM}z}\right], \quad 0 < z < +\infty, \tag{11}$$

and we again arrive at expression (12) for $u(x, t)$.

67. The solution of the boundary-value problem

$$u_{tt} = a^2 u_{xx}, \quad 0 < x, t < +\infty, \tag{1}$$

$$u_x(0, t) - hu(0, t) = 0, \quad 0 < t < +\infty, \tag{2}$$

$$u(x, 0) = \begin{cases} \sin\dfrac{\pi x}{l}, & 0 < x < l, \\ 0, & l < x < +\infty, \end{cases} \tag{3}$$

$$u_t(x, 0) = 0, \quad 0 \leqslant x < +\infty, \tag{3'}$$

is:

$$u(x, t) = \phi\left(t - \frac{x}{a}\right) + \psi\left(t + \frac{x}{a}\right), \tag{4}$$

where

$$-\phi(-z) = \psi(z) = \begin{cases} \dfrac{1}{2}\sin\dfrac{\pi a z}{l}, & 0 \leqslant z \leqslant \dfrac{l}{a}, \\ 0, & \dfrac{l}{a} \leqslant z < +\infty, \end{cases} \tag{5}$$

$$\phi(z) = \begin{cases} \dfrac{1}{\pi^2 + h^2 l^2}\left[\dfrac{\pi^2 - h^2 l^2}{2}\sin\dfrac{\pi a z}{l} + \pi h l\left(\cos\dfrac{\pi a z}{l} - e^{-ahz}\right)\right], & 0 \leqslant z \leqslant \dfrac{l}{a}, \\ -\dfrac{\pi h l}{\pi^2 + h^2 l^2}(1 + e^{hl})e^{-ahz}, & \dfrac{l}{a} \leqslant z < +\infty. \end{cases}$$

68. The solution of the boundary-value problem

$$\theta_{tt} = a^2 \theta_{xx}, \quad 0 < x, t < +\infty, \tag{1}$$

$$\theta_x(0, t) + \alpha\theta_t(0, t) = 0, \quad 0 < t < +\infty, \tag{2}$$

$$\theta(x, 0) = 0, \quad 0 < x < +\infty, \tag{3}$$

$$\theta_t(x, 0 = \omega, \quad 0 < x < +\infty \tag{3'}$$

has the form

$$\theta(x, t) = \begin{cases} \omega t, & 0 < at < x, \\ \dfrac{\omega(t - \alpha x)}{1 - a\alpha}, & x < at < +\infty. \end{cases} \tag{4}$$

69. We have the boundary-value problems

$$u_{tt} = a^2 u_{xx}, \quad 0 < x, t < +\infty, \tag{1}$$

$$\left.\begin{array}{l} \text{(a) or } u(0,t) = 0, \\ \text{(b) or } u_x(0,t) = 0, \\ \text{(c) or } u_x(0,t) - hu(0,t) = 0, \\ \text{(d) or } u_x(0,t) + \alpha u_t(0,t) = 0, \end{array}\right\} 0 < t < +\infty, \tag{2}$$

$$\left.\begin{array}{ll} u(x,0) = f(x), & (3) \\ u_t(x,0) = af'(x), & (3') \end{array}\right\} 0 < x < +\infty.$$

In the case of the boundary condition (a)

$$u(x,t) = \begin{cases} f(x+at), & 0 < at < x, \\ f(x+at) - f(at-x), & x < at < +\infty. \end{cases}$$

In the case of the boundary condition (b)

$$u(x,t) = \begin{cases} f(x+at), & 0 < at < x, \\ f(x+at) + f(at-x), & x \leqslant at < +\infty. \end{cases}$$

In the case of the boundary condition (c)

$$u(x,t) = \begin{cases} f(at+x), & 0 < at < x, \\ f(x+at) + f(at-x) + 2h\,e^{h(x-at)} \int\limits_0^{x-at} e^{-hs} f(-s)\,ds, & x < at < +\infty. \end{cases}$$

In the case of the boundary condition (d)

$$u(x,t) = \begin{cases} f(x+at), & 0 < at < x, \\ f(x+at) + \dfrac{1+a\alpha}{1-a\alpha} f(at-x). \end{cases}$$

70. We have the boundary-value problem

$$u_{tt} = a^2 u_{xx}, \quad 0 < x, t < +\infty, \tag{1}$$

$$M_0 u_{tt}(0,t) = -H_0 u(0,t) - R_0 u_t(0,t) - kp_0 S u_x(0,t), \quad 0 < t < +\infty, \tag{2}$$

$$u(x,0) = f(x), \quad 0 < x < +\infty, \tag{3}$$

$$u_t(x,0) = af'(x), \quad 0 < x < +\infty. \tag{3'}$$

Its solution may be represented in the following way:

$$u(x,t) = \begin{cases} f(x+at), & 0 < at < x, \\ f(x+at) + \Phi(at-x), & x < at < +\infty, \end{cases}$$

where $\Phi(z)$ is a solution of the differential equation

$$a^2 M_0 \phi''(z) + (aR_0 - kp_0 S)\phi'(z) + H_0 \phi(z) = -[a^2 M_0 f''(z) + a(R_0 + kp_0 S) f'(z) + H_0 f(z)]$$

with initial conditions

$$\phi(0) = \phi'(0) = 0.$$

Note. In the boundary condition (2) S denotes the area of the piston and R_0 denotes the coefficient of friction. We neglect the change of pressure on the outer side of the piston. The excess pressure on the piston ("pressure perturbation") equals $p-p_0 = k(p_0)/(\rho_0)(\rho-\rho_0)$. But because of the equation of continuity (see the solution of problem 4 chapter II) we have $\rho-\rho_0 = -\rho_0 u_x$. Therefore $p-p_0 = -kp_0u_x$.

71. *Solution*†. The boundary value problems (a), (b), (c) correspond to the boundary conditions (3), (3′), (3″), given below,

$$\left.\begin{aligned} &v_x+Li_t+Ri = 0, \quad (1)\\ &i_x+Cv_t+Gv = 0, \quad (1')\end{aligned}\right\} 0 < x, t < +\infty, \quad CR = GL,$$

$$\left.\begin{aligned} &v(x,0) = f(x), \quad (2)\\ &i(x,0) = -\sqrt{\frac{C}{L}}\,f(x), \quad (2')\end{aligned}\right\} 0 < x < +\infty,$$

$$\left.\begin{aligned} &-v(0,t) = R_0 i(0,t), \quad (3)\\ &-v_t(0,t) = \frac{1}{C_0}\,i(0,t), \quad (3')\\ &-v(0,t) = L_0 i_t(0,t), \quad (3'')\end{aligned}\right\} 0 < t < +\infty.$$

The solutions of these boundary-value problems have the form‡

$$v(x,t) = e^{-\frac{R}{L}t}\{\phi(x-at)+\psi(x+at)\}, \tag{4}$$

$$i(x,t) = e^{-\frac{R}{L}t}\sqrt{\frac{C}{L}}\{\phi(x-at)-\psi(x+at)\}. \tag{4'}$$

From the initial conditions (2), (2′) for the boundary-value problems (3), (3′) we obtain:

$$\phi(z) = 0, \quad \psi(z) = f(z), \quad 0 < z < +\infty. \tag{5}$$

Depending on the boundary conditions we obtain different representations of $\phi(z)$ for $-\infty < z < 0$.

In the case of the boundary condition (3)

$$\phi(z) = \frac{R_0\sqrt{C}-\sqrt{L}}{R_0\sqrt{C}+\sqrt{L}}\,f(-z), \quad -\infty < z < 0. \tag{6}$$

In the case of the boundary condition (3′)

$$\phi(z) = e^{\left(\frac{1}{CC_0}-aCR\right)z}\int_0^z e^{\left(aCR-\frac{1}{CC_0}\right)\zeta}\left[f'(-\zeta)+\left(aCR-\frac{1}{CC_0}\right)f(-\zeta)\right]d\zeta,$$
$$-\infty < z < 0. \tag{6'}$$

† See the solution of problem 58, p. 215.

‡ See the footnote on page 217.

In the case of the boundary condition (3″)

$$\phi(z) = e^{\frac{aL-L_0R}{a}z} \int_0^z e^{\left(\frac{L_0R-aL}{a}\right)\zeta} \left[-f'(-\zeta)+\frac{L_0R-aL}{a} f(-\zeta)\right] d\zeta,$$

$$-\infty < z < 0.$$

Note. In the case of the boundary condition (3) the reflected wave

$$\phi(z) = \frac{R_0\sqrt{C}-\sqrt{L}}{R_0\sqrt{C}+\sqrt{L}} f(-z), \qquad -\infty < z < 0$$

is not present at all for $R_0 = \sqrt{L/C}$ (the case of complete absorption of the incident wave). If $R_0 > \sqrt{L/C}$, then the reflected wave has the same sign as the incident wave; if $R_0 < \sqrt{L/C}$, then the opposite (preservation of phase and change of phase in the reverse). If $R_0 = 0$ (directly earthed end), then the reflected pressure wave changes sign, its amplitude being equal (at a point $x = 0$) to the amplitude of the incident wave. If $R_0 \to +\infty$ (insulated end), then the reflected pressure wave has the same sign and amplitude as the incident wave. The amplitude of the reflected wave is two times less than the amplitude of the incident wave when

$$\frac{R_0\sqrt{C}-\sqrt{L}}{R_0\sqrt{C}+\sqrt{L}} = \frac{1}{2}$$

or

$$\frac{R_0\sqrt{C}-\sqrt{L}}{R_0\sqrt{C}+\sqrt{L}} = -\frac{1}{2}.$$

72. *Solution.* In order to determine the steady-state in the conductor we seek a solution of the differential equation

$$\frac{\partial^2 v}{\partial x^2} - CL\frac{\partial^2 v}{\partial t^2} - 2CR\frac{\partial v}{\partial t} - GRv = 0, \qquad 0 < x, t < +\infty, \qquad CR = GL \tag{1}$$

which depends only on x and is independent of t,

$$v = v_0(x), \qquad 0 < x < +\infty \tag{2}$$

with a boundary condition

$$v(0) = E. \tag{3}$$

Substitution of $v_0(x)$ in (1) gives:

$$\frac{d^2 v_0}{dx^2} - GRv_0 = 0,$$

from which we obtain:

$$v_0(x) = C_1 e^{-\sqrt{GR}\,x} + C_2 e^{\sqrt{GR}\,x}.$$

Since $v_0(x)$ must be bounded for $x \to +\infty$, $C_2 = 0$. From the boundary condition (3) we find $C_1 = E$. Therefore,

$$v_0(x) = E e^{-\sqrt{GR}\,x}. \tag{4}$$

The corresponding steady-state current distribution

$$i = i_0(x) \tag{5}$$

is obtained by substituting (4) and (5) in the differential equation

$$v_x + L i_t + R i = 0. \tag{6}$$

We find:

$$i_0(x) = E\sqrt{\frac{C}{L}}\, e^{-\sqrt{GR}\,x}. \tag{7}$$

When the end $x = 0$ of the conductor is earthed (at time $t = 0$) through a lumped resistance, we obtain a boundary-value problem for the voltage and current intensity in the conductor

$$\left.\begin{aligned} v_x + L i_t + R i &= 0, \quad (8)\\ i_x + C v_t + G v &= 0, \quad (9)\end{aligned}\right\} 0 < x, t < +\infty,$$

$$-v(0, t) = R_0 i(0, t), 0 < t < +\infty, \tag{10}$$

$$\left.\begin{aligned} v(x, 0) &= E e^{-\sqrt{GR}\,x}, \quad (11)\\ i(x, 0) &= E\sqrt{\frac{C}{L}}\, e^{-\sqrt{GR}\,x}, \quad (12)\end{aligned}\right\} 0 < x < +\infty.$$

The solution of the boundary-value problem (8), (9), (10), (11), (12) has the form

$$v(x, t) = e^{-\frac{R}{L}t}[\phi(x-at)+\psi(x+at)],$$

$$i(x, t) = e^{-\frac{R}{L}t}\sqrt{\frac{C}{L}}[\phi(x-at)-\psi(x+at)].$$

We can find expressions for $\phi(z)$ and $\psi(z)$ from the initial conditions (11), (12)

$$\phi(z) = E e^{-\sqrt{GR}\,z}, \quad 0 < z < +\infty, \tag{13}$$

$$\psi(z) = 0, \quad 0 < z < +\infty. \tag{14}$$

From the boundary condition (10)

$$\phi(-z) = \frac{R_0\sqrt{C} - \sqrt{L}}{R_0\sqrt{C} + \sqrt{L}}\,\psi(z) = 0, \quad 0 < z < +\infty. \tag{15}$$

Therefore,

$$v(x, t) = E e^{-\sqrt{GR}\,x}\,\sigma_0(x-at), t < +\infty, \tag{16}$$

$$i(x, t) = E\sqrt{\frac{C}{L}}\, e^{-\sqrt{GR}\,x}\,\sigma_0(x-at), \quad 0 < x, t < +\infty. \tag{17}$$

73. The solution of the boundary-value problem

$$u_{tt} = a^2 u_{xx}, \quad 0 < x, t < +\infty, \tag{1}$$

$$u(0, t) = \mu(t), \quad 0 < t < +\infty, \tag{2}$$

$$u(x, 0) = u_t(x, 0) = 0, \quad 0 < x < +\infty \tag{3}$$

is:

$$u(x, t) = \left\{ \begin{array}{ll} \mu\left(t - \dfrac{x}{a}\right), & \dfrac{x}{a} < t < +\infty, \\ 0, & 0 < t < \dfrac{x}{a}. \end{array} \right\} \tag{4}$$

Method. The solution of the boundary-value problem (1), (2), (3) may be sought in the form

$$u(x, t) = \phi\left(t - \frac{x}{a}\right) + \psi\left(t + \frac{x}{a}\right).$$

74. The solution of the boundary-value problem

$$u_{tt} = a^2 u_{xx}, \quad 0 < x, t < +\infty, \quad a^2 = \frac{E}{\rho}, \tag{1}$$

$$ESu_x(0, t) = -F(t), \quad 0 < t < +\infty, \tag{2}$$

$$u(x, 0) = u_t(x, 0) = 0 < x < +\infty \tag{3}$$

is:

$$u(x, t) = \left\{ \begin{array}{ll} 0, & 0 < t < \dfrac{x}{a} \\ \dfrac{1}{S\sqrt{E\rho}} \displaystyle\int_0^{t - \frac{x}{a}} F(\xi)\, d\xi, & \dfrac{x}{a} < t < +\infty. \end{array} \right\} \tag{4}$$

75. The solution of the boundary-value problem

$$\left. \begin{array}{ll} -\dfrac{\partial p}{\partial x} = \dfrac{\partial w}{\partial t}, & (1) \\ -\dfrac{\partial p}{\partial t} = \lambda^2 \dfrac{\partial w}{\partial x}, & (1') \end{array} \right\} 0 < x, t < +\infty,$$

$$w_x(0, t) + \frac{P_0 S}{\lambda^2 \Omega_0 \rho_0} w(0, t) = \frac{P_0}{\lambda^2 \Omega_0 \rho_0} q(t), \quad 0 < t < +\infty, \tag{2}$$

$$w(x, 0) = 0, \quad p(x, 0) = 0, \quad 0 < x < +\infty \tag{3}$$

is:

$$w(x, t) = \left\{ \begin{array}{ll} 0, & 0 < \lambda t < x, \\ e^{h(x - \lambda t)} \displaystyle\int_0^{x - \lambda t} \Phi\left(-\frac{\zeta}{\lambda}\right) e^{h\zeta}\, d\zeta, & x < \lambda t < +\infty, \end{array} \right\} \tag{4}$$

where

$$h = \frac{P_0 S}{\lambda^2 \Omega_0 \rho_0}, \qquad \Phi(z) = -\frac{P_0}{\lambda^2 \Omega_0 \rho_0} q(z), \qquad 0 < z < +\infty, \tag{5}$$

$p(x, t)$ is derived from $w(x, t)$ by means of the relation (1) or (1′). $w(x, t) = \rho\, v(x, t)$, where ρ is the density, and v is the velocity of the liquid.

76. The solution of the boundary-value problem

$$u_{tt} = a^2 u_{xx}, \qquad 0 < x, t < +\infty, \tag{1}$$

$$u(x, 0) = u_t(x, 0) = 0, \qquad 0 < x < +\infty, \tag{2}$$

$$M u_{tt}(0, t) = ES u_x(0, t) + I \sum_{n=0}^{+\infty} \delta(t - nT), \tag{3}$$

where $\delta(t)$ is the one-sided δ-function, has the form

$$u(x, t) = \frac{aI}{ES} \sum_{n=0}^{+\infty} \left[1 - e^{-\frac{ES}{aM}\left(t - \frac{x}{a} - nT\right)}\right] \sigma_0\left(t - \frac{x}{a} - nT\right), \tag{4}$$

$$0 < x, t < +\infty.$$

77. The solution of the boundary-value problem

$$\left.\begin{aligned} v_x + L i_t + R i &= 0, \\ i_x + C v_t + G v &= 0, \end{aligned}\right\} 0 < x, t < +\infty, \qquad CR = GL, \tag{1, 1′}$$

$$v(0, t) = E \sin \omega t, \qquad 0 < t < +\infty, \tag{2}$$

$$v(x, 0) = i(x, 0) = 0, \qquad 0 < x < +\infty \tag{3}$$

has the form

$$v(x, t) = v_0(x, t) + v^*(x, t), \tag{4}$$

$$i(x, t) = i_0(x, t) + i^*(x, t), \tag{4′}$$

where

$$v_0(x, t) = E e^{-\alpha x} \sin(\omega t - \beta x) \tag{5}$$

is the voltage of the steady-state vibrations,

$$i_0(x, t) = E e^{-\alpha x} \frac{1}{R^2 + \omega^2 L^2} [(\alpha R + \beta \omega L) \sin(\omega t - \beta x) + \\ + (\beta R - \alpha \omega L) \cos(\omega t - \beta x)] \tag{6}$$

is the current intensity of the steady-state vibrations

$$\alpha = \sqrt{\frac{GR + \omega^2 CL + 2\omega CR}{2}}, \qquad \beta = \sqrt{\frac{GR + \omega^2 CL - 2\omega CR}{2}}, \tag{7}$$

and

$$v^*(x, t) = e^{-\frac{R}{L} t} [\phi(x - at) + \psi(x + at)] \tag{8}$$

is the voltage of the damped vibrations

$$i^*(x,t) = \sqrt{\frac{C}{L}}\, e^{-\frac{R}{L}t}\,[\phi(x-at)-\psi(x+at)] \tag{9}$$

is the current intensity of the damped vibrations

$$\phi(z) = -\frac{v_0(z,0)+i_0(z,0)\sqrt{\frac{L}{C}}}{2}, \quad \psi(z) = -\frac{v_0(z,0)-i_0(z,0)\sqrt{\frac{L}{C}}}{2}, \tag{10}$$

$$0 < z < \infty,$$

$$\phi(z) = -\psi(-z), \quad -\infty < z < 0. \tag{11}$$

For

$$t > \frac{1}{\frac{R}{L}+\alpha a}\left\{\ln 10\left[1+\frac{(|\beta-\alpha\omega|+\alpha R+\beta\omega L)\sqrt{L}}{(R^2+\omega^2L^2)\sqrt{C}}\right]\right\} \tag{12}$$

the amplitude of the voltage of the damped vibrations will be less than 10 per cent of the amplitude of the voltage of the steady-state vibrations.

Method. Eliminate the current intensity from (1) and (1′) and find the steady-state voltage. Substituting this into (1′) gives the steady-state current. It is best first to look for the steady-state vibrations of voltage and current in complex form $\tilde{v}(x,t) = v(x)e^{j\omega t}$, $\tilde{i}(x,t) = i(x)e^{j\omega t}$, where $j = \sqrt{-1}$, requiring a restriction for $x \to +\infty$, and then return to the real variable and satisfy the boundary condition (2).

3. Problems for an Infinite Line, Consisting of Two Homogeneous Semi-infinite Lines

If an infinite string (rod) is obtained by joining two semi-infinite homogeneous strings (rods) then, taking the point of junction as $x = 0$, we can write down equations for the displacements of points of the string

$$u_{1tt} = a_1^2 u_{1xx}, \quad -\infty < x < 0, \quad 0 < t < +\infty, \tag{1}$$

$$u_{2tt} = a_2^2 u_{2xx}, \quad 0 < x < +\infty, \quad 0 < t < +\infty \tag{1′}$$

and the initial conditions

$$u_1(x,0) = f_1(x), \quad u_{1t}(x,0) = F_1(x), \quad -\infty < x < 0, \tag{2}$$

$$u_2(x,0) = f_2(x), \quad u_{2t}(x,0) = F_2(x), \quad 0 < x < +\infty. \tag{2′}$$

To these equations and initial conditions it is necessary yet to add matching conditions at the point $x = 0$.

If, for example, the strings are joined directly (without any lumped connections), then the matching conditions have the form

$$u_1(0,t) = u_2(0,t), \tag{3}$$

$$u_{1x}(0,t) = u_{2x}(0,t). \tag{4}$$

The solution of the boundary-value problem (1), (1′), (2), (2′), (3), (4) may be sought in the form

$$u_1(x,t) = \phi_1(x-a_1t)+\psi_1(x+a_1t), \quad -\infty < x < 0, \quad 0 < t < +\infty, \tag{5}$$

$$u_2(x,t) = \phi_2(x-a_2t)+\psi_2(x+a_2t), \quad 0 < x < +\infty, \quad 0 < t < +\infty. \tag{6}$$

The functions $\phi_1(z)$, $\psi_1(z)$, $\phi_2(z)$, $\psi_2(z)$ are determined from the initial conditions (2), (2′) and the matching conditions (3) and (4).

78. The solution of the boundary-value problem†

$$\left.\begin{aligned} u_{1tt} &= a_1^2u_{1xx}, \quad -\infty < x < 0, \\ u_{2tt} &= a_2^2u_{2xx}, \quad 0 < x < +\infty, \end{aligned}\right\} \quad 0 < t < +\infty, \tag{1), (1′}$$

$$u_1(0,t) = u_2(0,t), \quad E_1\frac{\partial u_1(0,t)}{\partial x} = E_2\frac{\partial u_2(0,t)}{\partial x}, \quad 0 < t < +\infty, \tag{2}$$

$$u_1(x,0) = f\left(-\frac{x}{a_1}\right), \quad \frac{\partial u_1(x,0)}{\partial t} = f'\left(-\frac{x}{a_1}\right), \quad -\infty < x < 0, \tag{3}$$

$$u_2(x,0) = 0, \quad \frac{\partial u_2(x,0)}{\partial t} = 0, \quad 0 < t < +\infty \tag{3′}$$

is:

$$u_1x,t) = f\left(t-\frac{x}{a_1}\right)+\frac{\sqrt{E_1\rho_1}-\sqrt{E_2\rho_2}}{\sqrt{E_1\rho_1}+\sqrt{E_2\rho_2}}\,f\left(t+\frac{x}{a_1}\right), \tag{4}$$
$$-\infty < x < 0, \quad 0 < t < +\infty,$$

$$u_2(x,t) = \frac{2\sqrt{E_1\rho_1}}{\sqrt{E_1\rho_1}+\sqrt{E_2\rho_2}}\,f\left(t-\frac{x}{a_2}\right), \quad 0 < x < +\infty, \quad 0 < t < +\infty. \tag{5}$$

The reflected wave

$$\frac{\sqrt{E_1\rho_1}-\sqrt{E_2\rho_2}}{\sqrt{E_1\rho_1}+\sqrt{E_2\rho_2}}\,f\left(t+\frac{x}{a_1}\right)$$

is absent for $\sqrt{E_1\rho_1} = \sqrt{E_2\rho_2}$. For $E_2\rho_2 \to 0$, reflection will occur as from a free end, for $E_2\rho_2 \to \infty$ as from a rigidly fixed end.

The transmitted wave, for $E_2\rho_2 \to 0$, has an amplitude twice as great as the incident wave; for $E_2\rho_2 \to +\infty$ the transmitted wave vanishes.

One should note that for $E_2\rho_2 \to 0$ reflection takes place as from a free end, but a transmitted wave exists and even has an amplitude twice as great as the amplitude of the incident wave.

79. (a) for $Mk > T_0\rho$ (and $0 < x < +\infty$)

$$\left.\begin{aligned} u(x,t) &= \frac{Mv_0}{\sqrt{Mk-T_0\rho}}\, e^{-\frac{\rho}{M}(at-x)} \sinh\left[\left(t-\frac{x}{a}\right)\frac{\sqrt{Mk-T_0\rho}}{M}\right], \\ &\qquad \text{if} \quad x < at < +\infty, \\ u(x,t) &= 0, \quad \text{if} \quad 0 < at < x; \end{aligned}\right\} \tag{1}$$

† See problem 26, page 12.

(b) for $Mk = T_0\rho$ (and $0 < x < +\infty$)

$$u(x,t) = v_0 e^{-\frac{k}{M}\left(t-\frac{x}{a}\right)}\left(t-\frac{x}{a}\right)\sigma_0\left(t-\frac{x}{a}\right), \text{ if } 0 < t < +\infty; \tag{2}$$

(c) for $Mk < T_0\rho$ (and $0 < x < +\infty$)

$$\left.\begin{aligned} u(x,t) &= \frac{Mv_0}{\sqrt{T_0\rho - Mk}}\, e^{-\frac{\rho}{M}(at-x)} \sinh\left[\left(t-\frac{x}{a}\right)\frac{\sqrt{T_0\rho-Mk}}{M}\right], \\ &\qquad \text{if} \quad x < at < +\infty, \\ u(x,t) &= 0, \quad \text{if} \quad 0 < at < x. \end{aligned}\right\} \tag{3}$$

For $-\infty < x < 0$ the solution $u(x, t)$ is derived from (1), (2), (3) by replacing x by $-x$.

80. The solution of the boundary-value problem

$$\left.\begin{aligned} u_{1tt} &= a^2 u_{1xx}, \quad -\infty < x < 0, \\ u_{2tt} &= a^2 u_{2xx}, \quad 0 < x < +\infty, \end{aligned}\right\} 0 < t < +\infty, \tag{1, 1'}$$

$$\left.\begin{aligned} T_0[u_{2x}&(0,t) - u_{1x}(0,t)] \\ &= ku_1(0,t) + Mu_{1tt}(0,t) + ru_t(0,t) \\ &= ku_2(0,t) + Mu_{2tt}(0,t) + ru_{2t}(0,t), \end{aligned}\right\} 0 < t < +\infty, \tag{2}$$

$$u_1(0, t = u_2(0,t),$$

$$u_1(x,0) = f(x), \quad u_{1t}(x,0) = -af'(x), \quad -\infty < x < 0, \tag{3}$$

$$u_2(x,0), \quad u_{2t}(x,0) = 0, \quad 0 < x < +\infty \tag{3'}$$

may be represented in the form

$$u_1(x,t) = f(x-at) + \phi(x+at), \tag{4}$$

$$u_2(x,t) = \phi(x-at), \tag{4'}$$

where $\phi(z)$ is a solution of the differential equation

$$a^2M\phi''(z) + [2T_0 - ar]\phi'(z) + k\phi(z) = 2T_0 f'(z) \text{ if } -\infty < z < 0 \tag{5}$$

with zero initial conditions, and

$$\phi(z) = \phi(-z) - f(z) \text{ if } 0 < z < +\infty. \tag{6}$$

81. The solution of the boundary-value problem

$$u_{tt} = a^2 u_{xx}, \quad -\infty < x < v_0 t, \quad v_0 t < x < +\infty, \quad 0 < t < +\infty, \tag{1}$$

where

$$u = u_1(x,t), \quad -\infty < x < v_0 t,$$

$$u = u_2(x,t), \quad v_0 t < x < +\infty,$$

$$u_{1x}(v_0 t, t) = u_{2x}(v_0 t, t) = -\frac{1}{kp_0}\tilde{p}(t), \tag{2}$$

$$u(x,0) = u_t(x,0) = 0, \quad -\infty < x < 0, \quad 0 < x < +\infty \tag{3}$$

has the form

$$u_1(x,t)=\begin{cases} -\dfrac{a+v_0}{kp_0}\displaystyle\int_0^{\frac{x+at}{a+v_0}}\tilde{p}(\xi)\,d\xi, & -at<x<v_0t,\\ 0, & -\infty<x<-at,\end{cases} \quad (4)$$

$$u_2(x,t)=\begin{cases} \dfrac{a-v_0}{kp_0}\displaystyle\int_0^{\frac{at-x}{a+v_0}}\tilde{p}(\xi)\,d\xi, & v_0t<x<at,\\ 0, & at<x<+\infty.\end{cases} \quad (4')$$

In particular, if $\tilde{p}(t)=A\cos\omega t$,

$$u_1(x,t)=\begin{cases} -\dfrac{a+v_0}{kp_0\omega}A\sin\left[\dfrac{\omega}{a+v_0}(x+at)\right], & -at<x<v_0t,\\ 0, & -\infty<x<-at,\end{cases}$$

$$u_2(x,t)=\begin{cases} \dfrac{a-v_0}{kp_0\omega}A\sin\left[\dfrac{\omega}{a-v_0}(x-at)\right], & v_0t<x<at,\\ 0, & at<x<+\infty,\end{cases}$$

Thus, in the case $\tilde{p}(t)=A\cos\omega t$ a wave is propagated in a direction, opposite to the direction of motion of the source with a frequency less than the frequency of the source,

$$\omega_1=\frac{a}{a+v_0}\omega,$$

and in the direction of motion of the source a wave with a frequency, greater than the frequency of the source

$$\omega_2=\frac{a}{a-v_0}\omega$$

(Doppler effect).

82. The solution of the boundary-value problem

$$\left.\begin{aligned} &u_{tt}=a^2u_{xx}, \quad -\infty<x<v_0t, \quad v_0t<x<+\infty, \quad 0<t<+\infty,\\ &u=u_1(x,t), \quad -\infty<x<v_0t, \quad u=u_2(x,t), \quad v_0t<x<+\infty,\end{aligned}\right\} \quad (1)$$

$$u_1(v_0t,t)=u_2(v_0t,t), \quad T_0\left[\frac{\partial u_2(v_0t,t)}{\partial x}-\frac{\partial u_1(v_0t,t)}{\partial x}\right]=-F(t), \quad (2)$$

$$0<t<+\infty,$$

$$\left.\begin{aligned} &u_1(x,0)=u_{1x}(x,0)=0, \quad -\infty<x<0,\\ &u_2(x,0)=u_{2t}(x,0)=0, \quad 0<x<+\infty\end{aligned}\right\} \quad (3)$$

is:

$$u_1(x,t) = \begin{cases} \dfrac{a^2-v_0^2}{2aT_0} \displaystyle\int\limits_0^{\frac{x+at}{a+v_0}} F(\xi)\,d\xi, & -at < x < v_0 t, \\ 0, & -\infty < x < -at, \end{cases} \tag{4}$$

$$u_2(x,t) = \begin{cases} \dfrac{a^2-v_0^2}{2aT_0} \displaystyle\int\limits_0^{\frac{at-x}{a-v_0}} F(\xi)\,d\xi, & v_0 t < x < at, \\ 0, & at < x < +\infty. \end{cases} \tag{4'}$$

4. Problems for a Finite Segment

In the case of a finite homogeneous segment of length l the solution of the boundary-value problem

$$u_{tt} = a^2 u_{xx}, \quad 0 < x < l, \tag{1}$$

$$u(x,0) = \phi(x), \quad u_t(x,0) = \psi(x), \quad 0 < x < l, \tag{2}$$

$$\left.\begin{aligned} &\alpha_1 u_{tt}+\alpha_2 u_t+\alpha_3 u_x+\alpha_4 u = \mu(t), \quad x = 0, \\ &\beta_1 u_{tt}+\beta_2 u_t+\beta_3 u_x+\beta_4 u_x = \mu^*(t), \quad x = l, \end{aligned}\right\} 0 < t < +\infty \tag{3}$$

may be sought in the form

$$u(x,t) = \phi_1(x-at)+\phi_2(x+at), \tag{4}$$

where the functions $\phi_1(z)$ and $\phi_2(z)$ for $0 < z < l$ are determined from the initial conditions (2), and for other required values are extended by using the boundary conditions (3).

It is possible also to search for the solution of the boundary-value problem (1), (2), (3) by means of D'Alembert's formula

$$u(x,t) = \frac{\phi(x-at)+\phi(x+at)}{2} + \frac{1}{2a} \int\limits_{x-at}^{x+at} \psi(\xi)\,d\xi$$

for an infinite straight line, extending $\phi(z)$ and $\psi(z)$ along all the straight line $-\infty < z < +\infty$ by means of the boundary conditions (3).

83.

$$u(x,t) = A \sin\frac{\pi x}{l} \cos\frac{\pi a t}{l}, \quad 0 < x < l, \quad 0 < t < -\infty.$$

Method. The solution is derived by means of D'Alembert's formula for an odd periodic expansion of period $2l$ of the initial conditions.

84.

$$u(x,t) = \frac{\phi(x-at)+\phi(x+at)}{2}, \quad 0 < x < l, \quad 0 < t < +\infty,$$

where

$$\phi(z) = \begin{cases} Az, & -l < z < l, \\ A(2l-z), & l < z < 3l, \end{cases}$$

$$\phi(z) = \phi(z+4l), \quad -\infty < z < +\infty.$$

85.

$$u(x,t) = \frac{\phi(x-at)+\phi(x+at)}{2}, \quad 0 < x < l, \quad 0 < t < +\infty,$$

where

$$\phi(z) = \phi_n(z), \quad -l+2nl < z < l+2nl, \quad n = 0, \pm 1, \pm 2, \ldots,$$

where

$$\phi_{-n}(z) = -\phi_n(-z),$$

$$\phi_0(z) = Az, \quad \phi_1(z) = Al\mathrm{e}^{-h(z-l)} - \mathrm{e}^{-hz} \int_{-l}^{z-2l} [\phi_0'(-\zeta)+h\phi_0(-\zeta)]\, \mathrm{e}^{h(\zeta+2l)}\, \mathrm{d}\zeta,$$

$$\phi_2(z) = Al\mathrm{e}^{-h(z-l)} - \mathrm{e}^{-hz}\left\{ \int_{-l}^{l} [\phi_0'(-\zeta)+h\phi_0(-\zeta)]\, \mathrm{e}^{h(\zeta+2l)}\, \mathrm{d}\zeta + \right.$$

$$\left. + \int_{l}^{z-2l} [\phi_{-1}'(-\zeta)+h\phi_{-1}(-\zeta)]\, \mathrm{e}^{h(\zeta+2l)}\, \mathrm{d}\zeta \right\},$$

$$\phi_n(z) = Al\mathrm{e}^{-h(z-l)} - \mathrm{e}^{-hz}\left\{ \sum_{k=1}^{n-1} \int_{(2k-3)l}^{(2k-1)l} [\phi_{-k+1}'(-\zeta)+h\phi_{-k+1}(\zeta)]\, \mathrm{e}^{h(\zeta+2l)}\, \mathrm{d}\zeta + \right.$$

$$\left. + \int_{(2n-3)l}^{z-2l} [\phi_{-n+1}'(-\zeta)+h\phi_{-n+1}(-\zeta)]\, \mathrm{e}^{h(\zeta+2l)}\, \mathrm{d}\zeta \right\}.$$

86. *Solution.* Firstly let us solve the boundary-value problem

$$u_{tt} = a^2 u_{xx}, \quad 0 < x < l, \quad 0 < t < +\infty, \tag{1}$$

$$u(0,t) = 0, \quad u_x(l,t) = 0, \quad 0 < t < +\infty, \tag{2}$$

$$u(x,0) = 0 \quad u_t(x,0) = \frac{I}{\rho}\,\delta(x-x_0), \quad 0 < x_0 < l, \quad 0 < x < l, \tag{3}$$

where $\delta(x)$ is the one-sided δ-function†. Its solution is:

$$u(x,t) = \frac{1}{2a} \int_{x-at}^{x+at} \frac{I}{\rho} \sum_{k=-\infty}^{+\infty} (-1)^k \{\delta(\xi - x_0 + 2kl) - \delta(\xi + x_0 + 2kl)\}\, \mathrm{d}\xi$$

$$= \frac{I}{2a\rho} \sum_{k=-\infty}^{+\infty} (-1)^k \{\sigma_0(x+at-x_0+2kl) - \sigma_0(x+at+x_0+2kl) -$$

$$- \sigma_0(x-at-x_0+2kl) + \sigma_0(x-at+x_0+2kl)\}.$$

† See the footnote to the solution of problem 66, page 223.

Passing to a limit as $x_0 \to l$ in the solution obtained, we find the solution of the given problem

$$u(x,t) = \frac{I}{2ap} \sum_{k=-\infty}^{+\infty} (-1)^k \{\sigma_0[x+at+(2k-1)l]-\sigma_0[x+at+(2k+1)l]-$$
$$-\sigma_0[x-at+(2k-1)l]+\sigma_0[x-at+(2k+1)l]\}.$$

87. *Solution.* During the act of impact one has the boundary-value problem for the longitudinal displacements $u(x,t)$ of points of the rod

$$u_{tt} = a^2 u_{xx}, \quad 0 < x < l, \quad 0 < t < +\infty, \tag{1}$$

$$u(0,t) = 0, \quad 0 < t < +\infty, \tag{2}$$

$$Mu_{tt}(l,t) = -ESu_x(l,t), \quad 0 < t < t_0, \tag{2'}$$

where t_0 is the time of completion of the act of impact

$$u(x,0) = 0, \quad 0 < x < l, \tag{3}$$

$$u_t(x,0) = \begin{cases} 0, & 0 \leqslant x < l, \\ -v_0, & x = l. \end{cases} \tag{3'}$$

The time t_0 is characterized by the fact that for $0 < t < t_0$, $u_x(l,t)$ must be < 0, and for $t = t_0$, $u_x(l,t) = 0$.

The solution of the boundary-value problem (1), (2), (2′), (3), (3′) is:

$$u(x,t) = \phi(at-x)-\phi(at+x), \tag{4}$$

where the function $\phi(z)$ is determined in the following way:

$$\phi'(z) = 0, \quad l < z < l, \tag{5}$$

$$\phi(z) = 0, \quad -l < z < l, \tag{6}$$

$$\phi''(z)+\frac{1}{\alpha l}\phi'(z) = \phi''(z-2l)-\frac{1}{\alpha l}\phi'(z-2l), \quad l < z < +\infty, \tag{7}$$

$\alpha = M/\rho sl$ is the ratio of the mass of the load to the mass of the rod. Using the differential equation (7) and the second initial condition (3′) the function $\phi'(z)$ is determined over the segment $l < z < 3l$. Then by means of the same differential equation $\phi'(z)$ is determined successively over the intervals $3l < z < 5l$, $5l < z < 7l$, ..., etc., the constant of integration each time being determined from the condition of continuity of change in velocity of the end $u_t(l,t)$ for $t > 0$, and, in particular, for $t = 2l/a$; $4l/a$; $6l/a$;... Thus one derives the expressions

$$\phi'(z) = \frac{v_0}{a} e^{-\frac{z-l}{\alpha l}}, \quad l < z < 3l, \tag{5'}$$

$$\phi'(z) = \frac{v_0}{a} e^{-\frac{z-l}{\alpha l}} + \frac{v_0}{a}\left[1-\frac{2}{\alpha l}(z-3l)\right] e^{-\frac{z-3l}{\alpha l}}, \quad 3l < z < 5l, \tag{5''}$$

$$\phi'(z) = \frac{v_0}{a}\,e^{-\frac{z-l}{\alpha l}} + \frac{v_0}{a}\left[1-\frac{2}{\alpha l}(z-3l)\right]e^{-\frac{z-3l}{\alpha l}} +$$

$$+\frac{v_0}{a}\left[1-\frac{4}{\alpha l}(z-3l)+\frac{2}{\alpha^2 l^2}(z-5l)^2\right]e^{-\frac{z-5l}{\alpha l}}, \quad 5l < z < 7l. \qquad (5''')$$

The function $\phi(z)$ is obtained by integration of $\phi'(z)$ over the intervals $l < z < 3l$, $3l < z < 5l$, $5l < z < 7l, \ldots$ taking account of the continuity of the change of $u(l, t)$ with time.

Thus one derives the expressions

$$\left.\begin{aligned} \phi(z) &= \frac{\alpha l v_0}{a}\left[1-e^{-\frac{z-l}{\alpha l}}\right], \quad l < z < 3l, \\ \phi(z) &= -\frac{\alpha l v_0}{a}\,e^{-\frac{z-l}{a}} + \frac{\alpha l v_0}{a}\left[1+\frac{2}{\alpha l}(z-3l)\right]e^{-\frac{z-3l}{\alpha l}}, \quad 3l < z < 5l, \\ \phi(z) &< \frac{\alpha l v_0}{a}\left[e^{-\frac{z-l}{a}}-1\right] + \frac{\alpha l v_0}{a}\left[1+\frac{2}{\alpha l}(z-3l)\right]e^{-\frac{z-3l}{\alpha l}} - \\ &\quad -\frac{\alpha l v_0}{a}\left[1+\frac{2}{\alpha^2 l^2}(z-5l)^2\right]e^{-\frac{z-5l}{\alpha l}}, \quad 5l < z < 7l. \end{aligned}\right\} \qquad (6')$$

For $0 < t < l/a$ $\phi(at-x) = 0$ by virtue of (6), therefore according to (4)

$$u(x, t) = \phi(at+x) \quad \text{if} \quad 0 < t < \frac{l}{a}, \qquad (8)$$

i.e. only the "backward" wave $\phi(at+x)$, travelling from the end $x = l$, subjected to the blow, is propagated along the rod; for $t = l/a$ it reaches the fixed end and for $l/a < t < 2l/a$ a reflected wave $\phi(at-x)$ is added to it, i.e. the solution will have the form

$$u(x, t) = \phi(at-x)+\phi(at+x), \quad \frac{l}{a} < t < \frac{2l}{a}. \qquad (9)$$

For $t = 2l/a$ the wave $\phi(at-x)$ is reflected from the end $x = l$, so that the component $\phi(at+x)$ in solution (4) over the interval $2l/a < t < 3l/a$ will be changed.

Thus, $u(x, t)$ has different expressions over the intervals

$$0 < t < \frac{l}{a}, \quad \frac{l}{a} < t < \frac{2l}{a}, \quad \ldots, \quad n\frac{l}{a} < t < (n+1)\frac{l}{a}, \quad \ldots, \qquad (10)$$

$u_x(l, t)$ different expressions over the intervals

$$0 < t < \frac{2l}{a}, \quad \frac{2l}{a} < t < \frac{4l}{a}, \quad \ldots, \quad 2n\frac{l}{a} < t < (2n+2)\frac{l}{a}, \quad \ldots \qquad (11)$$

The act of impact cannot be completed for $0 < t < 2l/a$, since for these values of t, $u_x(l, t)$ will be < 0.

In order that the act of impact be completed in a time t, belonging to the interval $2l/a < t < 4l/a$, it is necessary and sufficient that the inequality

$$2+e^{-\frac{2}{\alpha}} < \frac{4}{\alpha},$$

be fulfilled, i.e.

$$\alpha < 1\cdot 73.$$

Note. Since real surfaces can have irregularities, then in order to apply this solution to real cases of impact it is necessary that the time to produce close contact of the mass with the free end of the rod should be negligibly small in comparison with the time of travel of the disturbance wave along the rod. trains arising in the load, must be negligibly small in comparison with the strains in the rod.

88. During the act of impact one has the boundary-value problem for longitudinal displacements $u(x, t)$ of points of the rod

$$u_{tt} = a^2 u_{xx}, \quad 0 < x < +\infty, \quad 0 < t < +\infty, \tag{1}$$

$$u_x(0, t) = 0, \quad 0 < t < +\infty, \tag{2}$$

$$Mu_{tt}(l, t) = -ESu_x(l, t), \quad 0 < t < t_0, \tag{2'}$$

where t_0 is the time of completion of the act of impact,

$$u(x, 0) = 0, \quad 0 \leqslant x \leqslant l, \tag{3}$$

$$u_t(x, 0) = \begin{cases} 0, & 0 \leqslant x \leqslant l, \\ -v_0, & x = l. \end{cases} \tag{3'}$$

The time of completion of the act of impact is determined in the same way as in the preceding problem.

The solution of the boundary-value problem (1), (2), (2′), (3), (3′) has the form

$$u(x, t) = \phi(at-x)+\phi(at+x), \tag{4}$$

where $\phi(z)$ is derived in the following way:

$$\phi'(z) = 0, \quad -l < z < l, \tag{5}$$

$$\phi(z) = 0, \quad -l < z < l, \tag{6}$$

$$\phi''(z)+\frac{1}{\alpha l}\phi'(z) = -\phi''(z-2l)+\frac{1}{\alpha l}\phi'(z-2l), \quad l < z < +\infty, \tag{7}$$

and $\alpha = M/\rho sl$ is the ratio of the mass of the load to the mass of the rod.

Firstly by means of the differential equation (7) $\phi'(z)$ is determined successively over the intervals $l < z < 3l$, $3l < z < 5l$, etc., account being taken of the initial condition (3′) and the continuity of $u_t(l, t)$ for $0 < t < +\infty$:

$$\phi'(z) = -\frac{v_0}{a}e^{-\frac{z-l}{\alpha l}}, \quad l < z < 3l, \tag{5'}$$

$$\phi'(z) = -\frac{v_0}{a}\,e^{-\frac{z-l}{\alpha l}} + \frac{v_0}{a}\left[1-\frac{2}{\alpha l}(z-3l)\right]e^{-\frac{z-3l}{\alpha l}}, \quad 3l < z < 5l, \quad (5'')$$

. . . .

Then by integration of $\phi'(z)$ and using the condition of continuity for $t > 0$ expressions are derived for $\phi(z)$ over these intervals.

$$\phi(z) = -\frac{v_0 \alpha l}{a}\left[1-e^{-\frac{z-l}{\alpha l}}\right], \quad l < z < 3l, \quad (6')$$

. . . .

89. The solution of the boundary-value problem†

$$\left(1-\frac{x}{H}\right)^2\frac{\partial^2 u}{\partial t^2} = a^2\frac{\partial}{\partial x}\left[\left(1-\frac{x}{H}\right)^2\frac{\partial u}{\partial x}\right], \quad 0 < x < l, \quad 0 < t < +\infty, \quad (1)$$

$$u(0, t) = 0, \quad 0 < t < +\infty, \quad (2)$$

$$Mu_{tt}(l, t) = -ESu_x(l, t) \quad 0 < t < t_0, \quad (2')$$

where t_0 is the time of completion of the act of impact,

$$u(x, 0) = 0, \quad 0 < x < l, \quad (3)$$

$$u_t(x, 0) = \left\{\begin{array}{ll} 0, & 0 \leqslant x < l, \\ -v_0, & x = l \end{array}\right\} \quad (3')$$

has the form

$$u(x, t) = \frac{\phi(at-x)-\phi(at+x)}{H-x}. \quad (4)$$

The function $\phi(x)$ is derived in the following way:

$$\phi(z) = 0, \quad -l < z < l, \quad (5)$$

$$\alpha\phi''(z)+\phi'(z)+\frac{\phi(z)}{H-l} = \alpha\phi''(z-2l)-\phi'(z-2l)+\frac{\phi(z-2l)}{H-l},$$

$$l < z < +\infty, \quad (6)$$

where $\alpha = Ma^2/Es$.

By means of this differential equation, and the initial condition (3′) and the conditions of continuity of $u_t(l, t)$ for $0 < t < +\infty$ and of the continuity of $u(l, t)$ for $0 \leqslant t < +\infty$ the function is determined successively over the intervals $l < z < 3l$, $3l < z < 5l$, etc.,

$$\phi(z) = \frac{v_0}{a}(H-l)\frac{e^{\lambda_1(z-l)}-e^{\lambda_2(z-l)}}{\lambda_1-\lambda_2}, \quad l < z < 3l, \quad (5')$$

etc., where λ_1 and λ_2 are roots of the equation

$$\lambda^2+\frac{\lambda}{\alpha}+\frac{1}{\alpha(H-l)} = 0.$$

† See problem 21, page 10.

90. The solution of the problem is similar to the solution of problems 87, 88 and 89.

91. The solutions of the boundary-value problems

$$u_{tt} = a^2 u_{xx}, \quad 0 < x < l, \quad 0 < t < +\infty, \tag{1}$$

$$u(x, 0) = u_t(x, 0) = 0, \quad 0 < x < l, \tag{2}$$

(a) $$u(0, t) = \mu(t), \quad u(l, t) = 0, \quad 0 < t < +\infty, \tag{3}$$

(b) $$u(0, t) = 0, \quad u(l, t) = \mu(t), \quad 0 < t < +\infty, \tag{3'}$$

(c) $$u(0, t) = \mu(t), \quad u_x(l, t) = 0, \quad 0 < t < +\infty \tag{3''}$$

have the corresponding forms†

$$\text{(a)}\ u(x, t) = \sum_{n=0}^{+\infty} \tilde{\mu}\left(t - \frac{x+2nl}{a}\right) - \sum_{n=1}^{+\infty} \tilde{\mu}\left(t - \frac{2nl-x}{a}\right), \tag{4}$$

$$\text{(b)}\ u(x, t) = \sum_{n=0}^{+\infty} \left[\tilde{\mu}\left(t - \frac{x-(2n+1)l}{a}\right) - \tilde{\mu}\left(t - \frac{x+(2n+1)l}{a}\right)\right], \tag{4'}$$

$$\text{(c)}\ u(x, t) = \sum_{n=1}^{+\infty} \left[\tilde{\mu}\left(t - \frac{x+2nl}{a}\right) + -\tilde{\mu}\left(t + \frac{x-2nl}{a}\right)\right](-1)^n + \tilde{\mu}\left(t - \frac{x}{a}\right), \tag{4''}$$

where

$$\tilde{\mu}(t) = \begin{Bmatrix} 0 & \text{for} & t < 0, \\ \tilde{\mu}(t) & \text{for} & t \geqslant 0. \end{Bmatrix}. \tag{5}$$

92. The solution of the boundary-value problem

$$\left.\begin{aligned} -\frac{\partial p}{\partial x} &= \frac{\partial w}{\partial t}, \\ -\frac{\partial p}{\partial t} &= \lambda^2 \frac{\partial w}{\partial x}, \end{aligned}\right\} 0 < x < l, \quad 0 < t < +\infty‡, \tag{1}$$

$$w(0, t) = \phi(t), \quad p(l, t) = 0, \quad 0 < t < +\infty, \tag{2}$$

$$w(x, 0) = 0, \quad p(x, 0) = 0, \quad 0 < x < l \tag{3}$$

has the form§

$$w(x, t) = \sum_{n=1}^{+\infty} \left[\tilde{\phi}\left(t - \frac{x+2nl}{\lambda}\right) - \tilde{\phi}\left(t + \frac{x-2nl}{\lambda}\right)\right](-1)^n + \tilde{\phi}\left(t - \frac{x}{\lambda}\right), \tag{4}$$

---

† See [7], pages 67–68.

‡ See the answer to problem 5, page 178.

§ See the solution of the preceding problem.

where

$$\tilde{\phi}(t) = \begin{cases} 0, & -\infty < t < 0, \\ \phi(t), & 0 < t < +\infty, \end{cases} \tag{5}$$

and

$$p(x,t) = \lambda\tilde{\phi}\left(t - \frac{x}{\lambda}\right) + \lambda \sum_{n=1}^{+\infty} (-1)^n \left\{\tilde{\phi}\left(t - \frac{x+2nl}{\lambda}\right) + \tilde{\phi}\left(t + \frac{x-2nl}{\lambda}\right)\right\}. \tag{6}$$

Thus

$$p(0,t) = \lambda\tilde{\phi}(t) + 2\lambda \sum_{n=1}^{+\infty} (-1)^n \tilde{\phi}\left(t - \frac{2nl}{\lambda}\right).$$

93. *Solution.* The beginning of the impact occurs when the left-hand rod reaches the right-hand; this time we take as $t = 0$, and the point of contact of the ends at this time we take as $x = 0$. The end of the impact takes place when the velocity of the striking end becomes less than the velocity of the end struck.

Let us denote by $u_1(x, t)$ and $u_2(x, t)$ the displacements of cross-sections of the striking and struck rods. Then $u_1(x, t)$ and $u_2(x, t)$ are solutions of the boundary-value problem (during the impact).

$$\left.\begin{aligned} u_{1tt} &= a^2 u_{1xx}, & -l < x < 0, \\ u_{2tt} &= a^2 u_{2xx}, & 0 < x < l, \end{aligned}\right\} 0 < t < +\infty, \tag{1}$$

$$u_{1x}(-l, t) = 0, \quad u_1(0, t) = u_2(0, t), \quad u_{1x}(0, t) = u_{2x}(0, t), \quad u_{2x}(l, t) = 0,$$
$$0 < t < +\infty\dagger, \tag{2}$$

$$\left.\begin{aligned} u_1(x, 0) &= 0, & u_{1t}(x, 0) &= v_1, & -l < x < 0, \\ u_2(x, 0) &= 0, & u_{2t}(x, 0) &= v_2, & 0 < x < l. \end{aligned}\right\} \tag{3}$$

The solution of the boundary-value problem (1), (2), (3) has the form

$$u_1(x, t) = \phi_1(x - at) + \psi_1(x + at), \quad u_2(x, t) = \phi_2(x - at) + \psi_2(x + at). \tag{4}$$

Substituting (4) in (2) and (3) we obtain:

$$\left.\begin{aligned} \phi_1'(-l-at) + \psi_1'(-l+at) = 0, \quad \phi_2'(l-at) + \phi_2'(l+at) = 0, \\ \phi_1'(-at) + \psi_1'(at) = \phi_2'(-at) + \psi_2'(at), \end{aligned}\right\} 0 < t < +\infty, \tag{5}$$

$$\phi_1(-at) + \psi_1(at) = \phi_2(-at) + \psi_2(at), \quad 0 < t < +\infty, \tag{6}$$

$$\phi_1(x) + \psi_1(x) = 0, \quad -l < x < 0, \tag{7}$$

$$-\phi_1'(x) + \psi_1'(x) = \frac{v_1}{a}, \quad -l < x < 0, \tag{8}$$

† Part of the boundary conditions (2) is fulfilled only for $0 < t < t_0$, where t_0 is the time of the end of the impact.

$$\phi_2(x)+\psi_2(x) = 0, \quad 0 < x < l, \tag{9}$$

$$-\phi_2'(x)+\psi_2'(x) = \frac{v_2}{a}, \quad 0 < x < l. \tag{10}$$

From relations (7)–(10) we find:

$$-\phi_1'(z) = \psi_1'(z) = \frac{v_1}{2a}, \quad -l < z < 0, \tag{11}$$

$$-\phi_2'(z) = \psi_2'(z) = \frac{v_2}{2a}, \quad 0 < z < l. \tag{12}$$

Relations (5), (6) give:

$$\phi_1'(-l-z) = -\psi_1'(-l+z), \tag{13}$$

$$\psi_2'(l+z) = -\phi_2'(l-z), \tag{14}$$

$$\phi_1'(-z) = \phi_2'(-z), \tag{15}$$

$$\psi_1'(z) = \psi_2'(z). \tag{16}$$

From relations (13)–(16) it follows that the functions $\phi_1'(z)$, $\psi_1'(z)$, $\phi_2'(z)$, $\psi_2'(z)$ are periodic with a period $4l$; therefore it is sufficient to define each over the interval $0 \leqslant z \leqslant 4l$; further construction is accomplished by periodic extension. Such determination of the functions $\phi_1'(z)$, $\psi_1'(z)$, $\phi_2'(z)$, $\psi_2'(z)$ by means of the relations (11)–(16) gives values for them, shown graphically in Fig. 27.

Using the functions $\phi_1'(z)$, $\psi_1'(z)$, $\phi_2'(z)$, $\psi_2'(z)$, we find expressions for

$$u_{1t}(x, t), \quad u_{2t}(x, t), \quad u_{1x}(x, t) \quad u_{2x}(x, t).$$

In Fig. 28 the distribution of velocities and strains for times $t = 0$, $t = l/2a$, $t = l/a$, $t = 3l/2a$, $t = 2l/a$ is shown graphically.

94. The solution of the boundary-value problem

$$\left.\begin{aligned} &\frac{\partial v}{\partial x}+L\frac{\partial i}{\partial t}+Ri = 0, \\ &\frac{\partial i}{\partial x}+C\frac{\partial v}{\partial t}+Gv = 0, \end{aligned}\right\} \quad \begin{aligned} &0 < x < l, \quad 0 < t < +\infty, &(1)\\ &CR = GL, &(1')\end{aligned}$$

$$v(0, t) = E, \quad v(l, t) = 0, \quad 0 < t < +\infty, \tag{2}$$

$$v(x, 0) = 0, \quad i(x, 0) = 0, \quad 0 < x < l \tag{3}$$

has the form

$$v(x, t) = v_0(x)+v^*(x, t), \tag{4}$$

$$i(x, t) = i_0(x)+i^*(x, t), \tag{4'}$$

where $v_0(x)$ and $i_0(x)$ is the steady-state solution of the system (1), (1′), satisfying the boundary conditions (2), which is also the limiting solution (4), (4′) of the boundary-value problem (1), (1′), (2), (3) for $t \to +\infty$ and $v^*(x, t)$ and $i^*(x, t)$ is the solution of the system (1), (1′) for the boundary conditions

$$v^*(0, t) = 0, \quad v^*(l, t) = 0 \tag{2'}$$

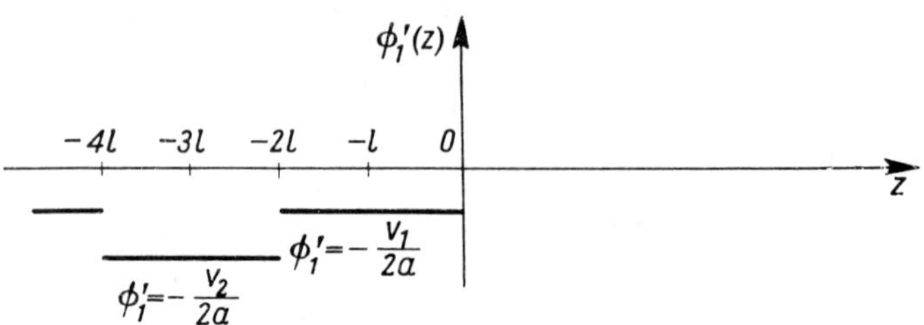

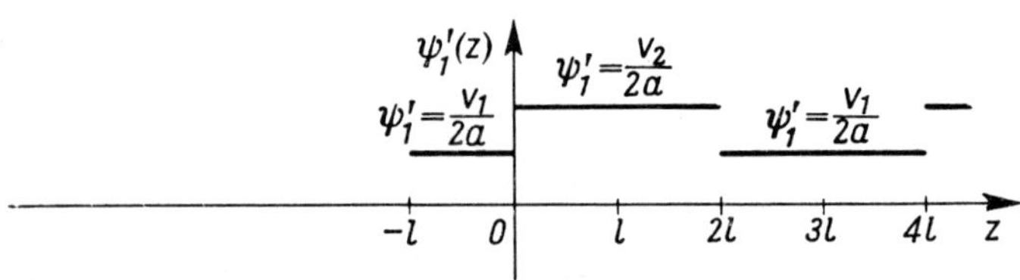

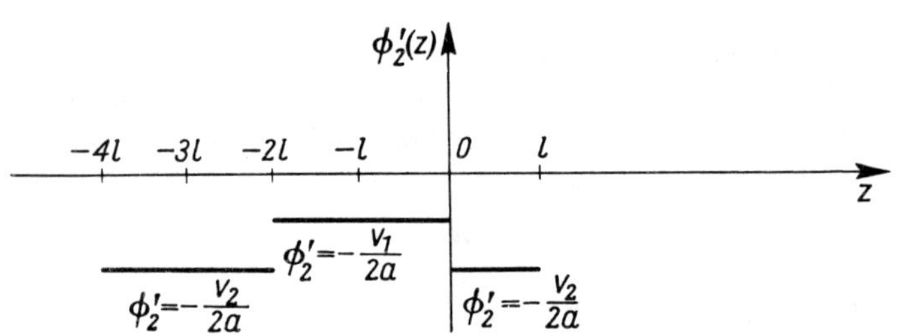

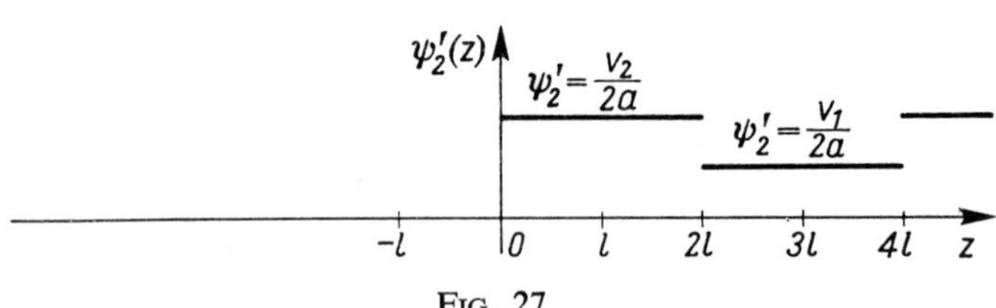

FIG. 27

and the initial conditions

$$v^*(x, 0) = -v_0(x), \qquad i^*(x, 0) = -i_0(x). \tag{3'}$$

We obtain†:

$$v_0(x) = E \frac{\sinh \sqrt{GR}(l-x)}{\sinh \sqrt{GR}l}, \tag{5}$$

---

† See the solution of problem 72, page 227.

$$i_0(x) = E\sqrt{\frac{C}{L}}\,\frac{\cosh\sqrt{GR}(l-x)}{\sinh\sqrt{GR}\,l}, \tag{5'}$$

$$v^*(x,t) = e^{-\frac{R}{L}t}[\phi(x-at)+\psi(x+at)], \tag{6}$$

$$i^*(x,t) = e^{-\frac{R}{L}t}\sqrt{\frac{C}{L}}[\phi(x-at)-\psi(x+at)], \tag{6'}$$

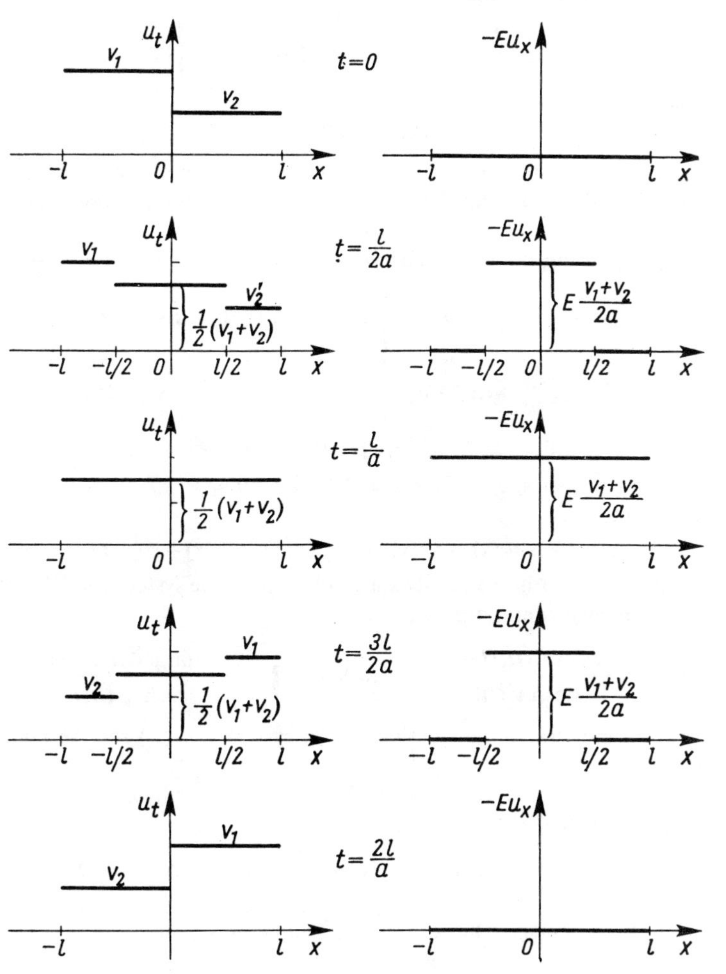

FIG. 28

where

$$\phi(x) = \frac{f(x)+F(x)}{2}, \quad \psi(x) = \frac{f(x)-F(x)}{2}, \quad 0 < x < l, \tag{7}$$

$$f(x) = -v_0(x), \quad (Fx) = -\sqrt{\frac{L}{C}}\, i_0(x), \quad 0 < x < l, \tag{8}$$

from conditions (2′) the functions $f(x)$ and $F(x)$ must be extended as even and odd functions of period $2l$.

For t, satisfying the inequality

$$t > \frac{L}{R} \ln\left\{10\left[1+\tanh \sqrt{GR}\,(l-x)\right]\right\}, \tag{9}$$

the relation

$$|i^*(x, t)| < 0{\cdot}1 i_0(x), \tag{10}$$

will be fulfilled, i.e. the current intensity at the point x of the conductor will differ from its limiting value for $t \to +\infty$ known to be not more than by 10 per cent.

95. The solution of the boundary-value problem

$$\left.\begin{aligned} &\frac{\partial v}{\partial x}+L\frac{\partial i}{\partial t}+Ri = 0, \\ &\frac{\partial i}{\partial x}+C\frac{\partial v}{\partial t}+Gv = 0, \end{aligned}\right\} \begin{aligned} &0 < x < l, \quad 0 < t < +\infty, && (1) \\ &CR = GL, && (1') \end{aligned}$$

$$v(0, t) = E, \quad i(l, t) = 0, \quad 0 < t < +\infty, \tag{2}$$

$$v(x, 0) = 0, \quad i(x, 0) = 0, \quad 0 < x < l \tag{3}$$

has the form

$$v(x, t) = v_0(x) + v^*(x, t), \quad i(x, t) = i_0(x) + i^*(x, t), \tag{4}$$

where $v_0(x)$ and $i_0(x)$ is the steady-state solution of the system (1), (1′), satisfying the boundary conditions (2),

$$v_0(x) = E\frac{\cosh\sqrt{GR}\,(l-x)}{\cosh\sqrt{GR}\,l}, \quad i_0(x) = E\sqrt{\frac{C}{L}}\,\frac{\sinh\sqrt{GR}\,(l-x)}{\cosh\sqrt{GR}\,l}, \tag{5}$$

and $v^*(x, t)$ and $i^*(x, t)$ the solution of the system (1), (1′) for the boundary conditions

$$v^*(0, t) = 0, \quad i^*(l, t) = 0, \quad 0 < t < +\infty \tag{6}$$

and initial conditions

$$v^*(x, 0) = -v_0(x), \quad i^*(x, 0) = -i_0(x), \quad 0 < x < l, \tag{7}$$

$$v^*(x, t) = e^{-\frac{R}{L}t}[\phi(x-at)+\psi(x+at)], \tag{8}$$

$$i^*(x, t) = e^{-\frac{R}{L}t}\sqrt{\frac{C}{L}}[\phi(x-at)-\psi(x+at)], \tag{9}$$

$$\phi(x) = \frac{f(x)+F(x)}{2}, \quad \psi(x) = \frac{f(x)-F(x)}{2}, \quad 0 < x < l, \tag{10}$$

$$f(x) = -v_0(x), \quad F(x) = -\sqrt{\frac{L}{C}}\, i_0(x), \quad 0 < x < l. \tag{11}$$

From the boundary conditions (6) it follows that the functions $f(x)$ and $F(x)$ are extended, respectively, oddly and evenly with respect to $x = 0$, evenly and oddly with respect to $x = l$ and periodically with period $4l$.

For t, satisfying the inequality

$$t > \frac{L}{R} \ln \{10[1+\tanh \sqrt{GR}(l-x)]\}, \tag{12}$$

the voltage at point x of the conductor will differ from its limiting value by not more than 10 per cent.

96. (a)

$$v(t,t) = \begin{cases} 0 & \text{if} \quad 0 < t < T, \\ E\left\{1-\left(\dfrac{Z-R_0}{Z+R_0}\right)^n\right\} & \text{if } (2n-1)T < t < (2n+1)T, \quad n = 1, 2, 3, \ldots, \end{cases}$$

where $Z = \sqrt{L/C}$ is the characteristic impedance, $T = l/a$, $a = 1/\sqrt{LC}$ is the velocity of propagation of the electromagnetic disturbances along the conductor;

(b)

$$v(l,t) = \begin{cases} 0 \quad \text{if} \quad 0 < t < T, \\ 2E\{1-e^{-\varkappa\left(\frac{t}{T}-1\right)}\} \quad \text{if} \quad T < t < 3T, \\ -2Ee^{-\varkappa\left(\frac{t}{T}-1\right)}+2E\left\{1-2\varkappa\left(\dfrac{t}{T}-3\right)\right\} e^{-\varkappa\left(\frac{t}{T}-3\right)} \quad \text{if} \quad 3T < t < 5T, \end{cases}$$

etc., $\varkappa = lC/C_0$.

(c)

$$v(l,t) = \begin{cases} 0 \quad \text{if} \quad 0 < t < T, \\ 2Ee^{-\varepsilon\left(\frac{t}{T}-1\right)} \quad \text{if} \quad T < t < 3T, \\ 2Ee^{-\varepsilon\left(\frac{t}{T}-1\right)}-2E\left\{1-2\varepsilon\left(\dfrac{t}{T}-3\right)\right\} e^{-\varepsilon\left(\frac{t}{T}-3\right)}, \quad \text{if} \quad 3T < t < 5T, \end{cases}$$

etc., $\varepsilon = lL/L_0$.

Method. For the laws of reflection from the end $x = l$ see the solution of problem 71, p. 226.

§ 3. Method of Separation of Variables

1. Free Vibrations in a Non-resistant Medium

97. The solution of the boundary-value problem

$$u_{tt} = a^2 u_{xx}, \quad 0 < x < l, \quad 0 < t < +\infty, \tag{1}$$

$$u(0, t) = u(l, t) = 0, \quad 0 < t < +\infty, \tag{2}$$

$$u(x, 0) = \left\{ \begin{array}{ll} \dfrac{h}{x_0} x, & 0 < x < x_0, \\ \\ \dfrac{h(l-x)}{l-x_0}, & x_0 < x < l, \end{array} \right\} \tag{3}$$

$$u_t(x, 0) = 0, \quad 0 < x < l \tag{3'}$$

is:

$$u(x, t) = \frac{2hl^2}{\pi^2 x_0(l-x_0)} \sum_{n=1}^{+\infty} \sin \frac{n\pi x_0}{l} \sin \frac{n\pi x}{l} \cos \frac{n\pi a t}{l}. \tag{4}$$

In the expression for $u(x, t)$ the terms for which $\sin (n\pi x_0/l) = 0$ disappear, i.e. the harmonics for which the point $x = x_0$ is a node are absent. The energy of the nth harmonic equals

$$E_n = Mh^2 \frac{a^2 l^2}{\pi^2 n^2 x_0^2 (l-x_0)^2} \sin^2 \frac{n\pi x_0}{l}, \quad M = \rho l.$$

98. *Solution.* Let us find the initial deflection of the string (Fig. 29).

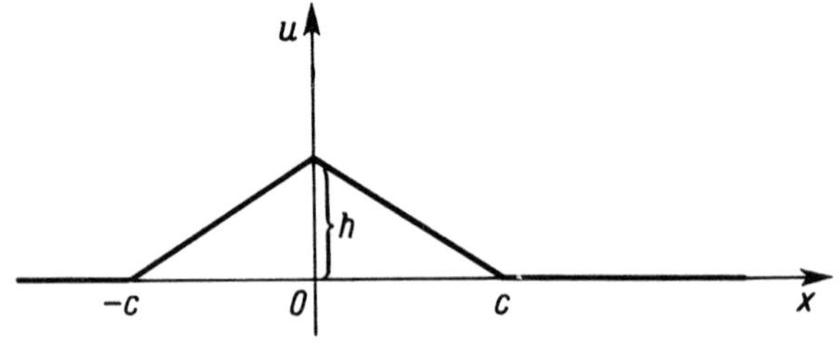

FIG. 29

To do this it is sufficient to determine the value of h. From the conditions of equilibrium (projected on the vertical axis) we find:

$$T(\sin \alpha + \sin \beta) = F_0.$$

Because of the smallness of the deflections $\sin\alpha \approx \tan\alpha$, $\sin\beta \approx \tan\beta$†, but

$$\tan\alpha = \frac{h}{x_0}, \qquad \tan\beta = \frac{h}{l-x_0}.$$

Thus

$$h = \frac{F_0 x_0(l-x_0)}{lT}\ ‡, \tag{1}$$

$$u(x,t) = -\frac{2hl^2}{\pi^2 x_0(l-x_0)} \sum_{n=1}^{+\infty} \frac{1}{n^2} \sin\frac{n\pi x_0}{l} \sin\frac{n\pi x}{l} \cos\frac{n\pi at}{l},$$

where h is given by (1).

99. $$u(x,t) = \frac{32h}{\pi^3} \sum_{n=0}^{+\infty} \frac{1}{(2n+1)^3} \sin\frac{(2n+1)\pi x}{l} \cos\frac{(2n+1)\pi at}{l},$$

where h is the maximum initial deflection of the string.

100. $$u(x,t) = \frac{4v_0 l}{\pi^2 a} \sum_{n=1}^{+\infty} \frac{1}{n^2} \sin\frac{n\pi x_0}{l} \sin\frac{n\pi\delta}{l} \sin\frac{n\pi x}{l} \sin\frac{n\pi at}{l}.$$

The energy of the nth harmonic equals

$$E_n = \frac{4Mv_0^2}{\pi^2 n^2} \sin^2\frac{n\pi x_0}{l} \sin^2\frac{n\pi\delta}{l}, \qquad M = \rho l.$$

101. *Method.* Firstly let us assume the impulse I to be uniformly distributed over the segment $x_0-\delta \leqslant x \leqslant x_0+\delta$ of the string. Then we get the expression for $u(x,t)$, deduced in the answer to the preceding problem, where

$$v_0 = \frac{I}{2\delta\rho},$$

and ρ is the linear mass density of the string. Passing to a limit as $\delta \to 0$, we obtain the solution of the given problem

$$u(x,t) = \frac{2l}{\pi a\rho} \sum_{n=1}^{+\infty} \frac{1}{n} \sin\frac{n\pi x_0}{l} \sin\frac{n\pi x}{l} \sin\frac{n\pi at}{l}.$$

---

† Because of the smallness of the deflection, T does not depend on the deflection. See [7], page 16.

‡ The initial deflection may be determined by solving the problem:

$$u''(x) = 0, \quad T\big(u'(x_0+0)-u'(x_0-0)\big) = F_0, \quad u(x_0-0) = u(x_0+0),$$
$$u(0) = u(l) = 0.$$

The energy of the nth harmonic equals

$$E_n = \frac{I^2}{M}\sin^2\frac{n\pi x_0}{l}, \qquad M = \rho l.$$

The solution of the problem may also be derived by assuming

$$u_t(x,0) = \frac{I}{\rho}\,\delta(x-x_0),$$

where $\delta(x)$ is the delta-function†.

102. $$u(x,t) = \frac{8v_0\delta}{\pi^2 a}\sum_{n=1}^{+\infty}\frac{1}{n}\,\frac{\cos\dfrac{n\pi\delta}{l}\sin\dfrac{n\pi x_0}{l}}{1-\dfrac{(2\delta n)^2}{l^2}}\sin\frac{n\pi x}{l}\sin\frac{n\pi at}{l}.$$

The energy of the nth harmonic equals

$$E_n = \frac{16v_0^2\delta^2\rho}{l\pi^2}\,\frac{1}{\left[1-\dfrac{(2\delta n)^2}{l^2}\right]^2}\cos^2\frac{n\pi\delta}{l}\sin^2\frac{n\pi\delta}{l}.$$

103. $$u(x,t) = \frac{8kl}{\pi^2}\sum_{n=0}^{+\infty}\frac{(-1)^n}{(2n+1)^2}\sin\frac{(2n+1)\pi x}{2l}\cos\frac{(2n+1)\pi at}{2l}.$$

104. The answer may be derived from the answer to the preceding problem, if one assumes

$$k = \frac{F_0}{ES},$$

where E is the modulus of elasticity, and S the cross-sectional area of the rod.

105. The solution of the boundary-value problem

$$u_{tt} = a^2 u_{xx}, \quad 0 < x < l, \quad 0 < t < -\infty, \tag{1}$$

$$u(x,0) = \phi(x), \quad u_t(x,0) = \psi(x), \quad 0 < x < l, \tag{2}$$

$$u_x(0,t) = u_x(l,t) = 0, \quad 0 < t < +\infty \tag{3}$$

is:

$$u(x,t) = \frac{1}{l}\int_0^l [\phi(z)+t\psi(z)]\,dz + \sum_{k=1}^{+\infty}\left(a_k\cos\frac{ak\pi t}{l}+b_k\sin\frac{ak\pi t}{l}\right)\cos\frac{k\pi x}{l}, \tag{4}$$

where

$$a_k = \frac{2}{l}\int_0^l \phi(z)\cos\frac{k\pi z}{l}\,dz, \qquad b_k = \frac{2}{a\pi k}\int_0^l \psi(z)\cos\frac{k\pi z}{l}\,dz.$$

---

† See the footnote to the solution of problem 56, page 213.

106. The solution of the problem may be derived from the solution of the preceding problem, if one assumes $\phi(x) = 0$,

$$\psi(x) = \begin{cases} 0, & 0 \leqslant x \leqslant l-\delta, \\ -\dfrac{I}{\delta\rho}, & l-\delta < x \leqslant l, \end{cases}$$

and then passes to a limit as $\delta \to 0$, or assumes $\phi(x) = 0$ and

$$\psi(x) = -\frac{I}{\rho}\delta(x-x_0)\dagger, \quad 0 < x_0 < l,$$

where $\delta(x)$ is the delta-function, and then passes to a limit as $x_0 \to l$

$$u(x,t) = -\frac{I}{\rho l}t - \frac{2I}{\pi a\rho}\sum_{k=1}^{\infty}\frac{(-1)^k}{k}\cos\frac{k\pi x}{l}\sin\frac{k\pi at}{l}.$$

107. $$u(x,t) = -\frac{4I}{\pi a\rho}\sum_{n=0}^{+\infty}\frac{(-1)^n}{2n+1}\sin\frac{(2n+1)\pi x}{2l}\sin\frac{(2n+1)\pi at}{2l}.$$

108. The solution of the boundary-value problem

$$u_{tt} = a^2 u_{xx}, \quad 0 < x < l, \quad 0 < t < +\infty, \tag{1}$$

$$u_x(0,t) = 0, \quad u_x(l,t)+hu(l,t) = 0, \quad 0 < t < +\infty, \tag{2}$$

$$u(x,0) = \phi(x), \quad u_t(x,0) = \psi(x), \quad 0 < x < l \tag{3}$$

has the form

$$u(x,t) = \sum_{n=1}^{+\infty}(a_n\cos\lambda_n at + b_n\sin\lambda_n at)\,\text{sos}\,\lambda_n x, \tag{4}$$

where λ_n are the eigenvalues of the boundary-value problem

$$\left.\begin{array}{l} X''(x)+\lambda^2 X(x) = 0, \quad 0 < x < l, \\ X'(0) = 0, \quad X'(l)+hX(l) = 0, \end{array}\right\} \tag{5}$$

where λ_n are the positive roots of the equation

$$\lambda\tan\lambda l = h, \tag{6}$$

$X_n(x) = \cos\lambda_n x$ are eigenfunctions of the boundary-value problem (5).

We find the square of the norm of the nth eigenfunction by means of (6)

$$\|X_n\|^2 = \int_0^l X_n^2(x)\,dx = \frac{l}{2}\left[1+hl\left(\frac{\cos\lambda_n l}{\lambda_n l}\right)^2\right] = \frac{l}{2}\left[1+\frac{h}{l(\lambda_n^2+h^2)}\right], \tag{7}$$

$$a_n = \frac{1}{\|X_n\|^2}\int_0^l \phi(z)\cos\lambda_n z\,dz, \quad b_n = \frac{1}{\|X_n\|^2 a\lambda_n}\int_0^l \psi(z)\cos\lambda_n z\,dz. \tag{8}$$

---

† In connection with the choice normalization for $\delta(x-x_0)$ see the second footnote on page 214.

109. The solution of the boundary-value problem

$$u_{tt} = a^2 u_{xx}, \quad 0 < x < l, \quad 0 < t < +\infty, \tag{1}$$

$$u_x(0, t) = 0, \quad u_x(l, t) + hu(l, t) = 0, \quad 0 < t < +\infty, \tag{2}$$

$$u(x, 0) = \frac{F_0}{ES} x, \quad u_t(x, 0) = 0, \quad 0 < x < l \tag{3}$$

is derived from the solution of the preceding boundary-value problem for

$$\phi(x) = \frac{F_0}{ES} x, \quad \psi(x) = 0,$$

$$u(x, t) = \frac{2F_0}{ElS} \sum_{n=1}^{+\infty} \frac{(1+hl)\cos\lambda_n l - 1}{\lambda_n^2 \left\{1 + hl\left(\frac{\cos\lambda_n l}{\lambda_n l}\right)^2\right\}} \cos\lambda_n x \cos\lambda_n at$$

$$= \frac{2F_0}{ElS} \sum_{n=1}^{+\infty} \frac{(1+hl) - \sqrt{1 + \frac{h^2}{\lambda_n^2}}}{\lambda_n^2 \sqrt{1 + \frac{h^2}{\lambda_n^2}} \left\{1 + \frac{h}{l(\lambda_n^2 + h^2)}\right\}} \cos\lambda_n x \cos a\lambda_n t, \tag{4}$$

where λ_n are the positive roots of the equation $\lambda \tan \lambda l = h$.

110. $u(x, t) = \dfrac{2l}{al\rho} \displaystyle\sum_{n=1}^{+\infty} \frac{\cos\lambda_n x \sin a\lambda_n t}{\lambda_n \left\{1 + hl\left(\frac{\cos\lambda_n l}{\lambda_n l}\right)^2\right\}} = \frac{2l}{al\rho} \sum_{n=1}^{+\infty} \frac{\cos\lambda_n x \sin\lambda_n at}{\lambda_n \left\{1 + \frac{h}{l(\lambda_n^2 + h^2)}\right\}},$

where λ_n are the positive roots of the equation $\lambda \tan \lambda l = h$.

111. The solution of the boundary-value problem

$$u_{tt} = a^2 u_{xx}, \quad 0 < x < l, \quad 0 < t < +\infty, \tag{1}$$

$$u_x(0, t) - hu(0, t) = 0, \quad u_x(l, t) + hu(l, t) = 0, \quad 0 < t < +\infty, \tag{2}$$

$$u(x, 0) = \phi(x), \quad u_t(x, 0) = \psi(x), \quad 0 < x < l \tag{3}$$

is:

$$u(x, t) = \sum_{n=1}^{+\infty} (a_n \cos a\lambda_n t + b_n \sin a\lambda_n t) \sin(\lambda_n x + \phi_n), \tag{4}$$

where λ_n are the eigenvalues of the boundary-value problem

$$X''(x) + \lambda^2 X(x) = 0, \quad 0 < x < l, \tag{5}$$

$$X'(0) = hX(0) = 0, \tag{6}$$

$$X'(l) + hX(l) = 0, \tag{6'}$$

and $X_n(x) = \sin(\lambda_n x + \phi_n)$ are the eigenfunctions of this boundary-value problem; λ_n are the roots of the equation

$$\cot \lambda l = \frac{1}{2}\left(\frac{\lambda}{h} - \frac{h}{\lambda}\right), \tag{7}$$

and

$$\phi_n = \text{arc tan}\,\frac{\lambda_n}{h}. \tag{8}$$

The square of the norm of the eigenfunction $X_n(x)$ equals

$$||X_n||^2 = \int_0^l \sin^2(\lambda_n x+\phi_n)\,dx = \frac{(\lambda_n^2+h^2)l+2h}{2(h^2+\lambda_n^2)}, \tag{9}$$

therefore

$$a_n = \frac{2(\lambda_n^2+h^2)}{(\lambda_n^2+h^2)l+2h}\int_0^l \phi(z)\sin(\lambda_n z+\phi_n)\,dz, \tag{10}$$

$$b_n = \frac{2(\lambda_n^2+h^2)}{a\lambda_n(\lambda_n^2+h^2)l+2a\lambda_n h}\int_0^l \psi(z)\sin(\lambda_n z+\phi_n)\,dz. \tag{11}$$

Method. 1. Equation (7) may be obtained in the following way. From the general solution of (5)

$$X_n(x) = C_1\cos\lambda x+C_2\sin\lambda x,$$

satisfying the boundary condition (6), we obtain:

$$X(x,\lambda) = C_2\left\{\frac{\lambda}{h}\cos\lambda x+\sin\lambda x\right\} = C_2\overline{X}(x,\lambda). \tag{12}$$

Substituting (12) in the boundary condition (6′) we obtain:

$$\left\{\frac{\partial X(x,\lambda)}{\partial x}+hX(x,\lambda)\right\}_{x=l} = C_2\left\{\frac{\partial \overline{X}(x,\lambda)}{\partial x}+h\overline{X}(x,\lambda)\right\}_{x=l} = 0,$$

since $C_2 \neq 0$, otherwise (12) would have been a trivial solution, then

$$\left\{\frac{\partial \overline{X}(x,\lambda)}{\partial x}+h\overline{X}(x,\lambda)\right\}_{x=l} = 0. \tag{13}$$

After substitution of the explicit relation

$$\overline{X}(x,\lambda) = \frac{\lambda}{h}\cos\lambda x+\sin\lambda x \tag{13'}$$

(13) is transformed into equation (7)

$$\cot\lambda l = \frac{1}{2}\left(\frac{\lambda}{h}-\frac{h}{\lambda}\right). \tag{7}$$

This equation may be approximately solved graphically†.

† Concerning the solution of the transcendental equation to any degree of accuracy, see [1], page 204.

Substituting in (12) the eigenvalue λ_n instead of λ we obtain the corresponding eigenfunction

$$X_n(x) = C_2\overline{X}(x, \lambda_n).$$

Thus the eigenfunction is determined correct to a constant C_2. This factor may be chosen so that the function $X_n(x)$ has the form

$$X_n(x) = X(x, \lambda_n) = \sin(\lambda_n x + \phi_n), \tag{14}$$

where

$$\phi_n = \arc\tan\frac{\lambda_n}{h}. \tag{14'}$$

Assuming $\lambda l = \xi$, we obtain:

$$\cot\xi = \frac{1}{3}\left(\frac{\xi}{lh} - \frac{lh}{\xi}\right). \tag{15}$$

Denoting by $\xi_1, \xi_2, \ldots, \xi_n, \ldots$ the abscissae of the points of intersection of the cotangent of $\eta = \cot\xi$ and the hyperbola $\eta = \frac{1}{2}(\xi/lh - lh/\xi)$, we obtain $\lambda_n = \xi_n/l$ (Fig. 30).

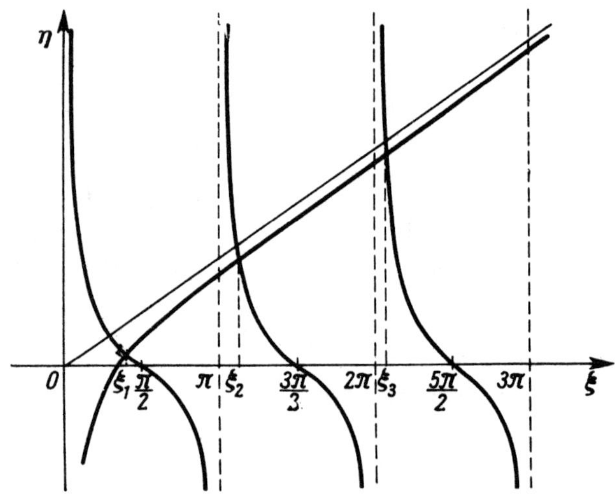

FIG. 30

Method 2. The square of the norm of the eigenfunction (9) may be found directly by integration

$$||X_n||^2 = \int_0^l \sin^2(\lambda_n x + \phi_n)\,dx \tag{16}$$

or by passing to a limit as $\lambda \to \lambda_n$ in the relation

$$\int_0^l X(x, \lambda)X(x, \lambda_n)\,dx = \frac{X'_x(l, \lambda_n)X(l, \lambda) - X'_x(l, \lambda)X(l, \lambda_n)}{\lambda^2 - \lambda_n^2}. \tag{17}$$

Expanding the right-hand side of (17) as $\lambda \to \lambda_n$ we obtain:

$$\int_0^l X^2(x, \lambda_n)\,dx = \frac{X'_x(l, \lambda_n)X'_\lambda(l, \lambda_n) - X''_{x\lambda}(l, \lambda_n)X(l, \lambda_n)}{2\lambda_n}. \tag{18}$$

The equality (17) is derived from the equations

$$X''(x, \lambda) + \lambda^2 X(x, \lambda) = 0,$$
$$X''(x, \lambda_n) + \lambda_n^2 X(x, \lambda_n) = 0$$

by multiplying the first of them by $X(x, \lambda_n)$, the second by $X(x, \lambda)$, by subtraction of the results and subsequent integration by parts.

In calculating the integral (16) or the right-hand part of the equality (18 it is necessary to use the boundary condition (6).

Note. Equation (7) may be rewritten in the form

$$\tan \lambda_n l = \frac{2\lambda_n h}{\lambda_n^2 - h^2}. \tag{19}$$

For $h \to 0$ (free ends) we obtain from (19):

$$\lim_{h\to 0} \tan \lambda_n l = 0.$$

From (14′) and (14) we find $\lim\limits_{h\to 0} \phi_n = \pi/2$, $\lim\limits_{h\to 0} X_n(x) = \sin(\lambda_n x + \pi/2)$, therefore

$$\lambda_n = \frac{n\pi}{l}, \quad n = 0, 1, 2, \ldots,$$

$$X_n(x) = \cos\frac{n\pi x}{l}, \quad n = 0, 1, 2, \ldots$$

This result was obtained directly in the solution of problem 105. For $n \to \infty$ (the ends are fixed) we derive from (19):

$$\lim_{n\to\infty} \tan \lambda_n l = 0.$$

From (14′) and (14) we find:

$$\lim_{n\to\infty} \phi_n = 0, \quad \lim_{n\to\infty} \sin(\lambda_n x + \phi_n) = \sin \lambda_n x.$$

Hence,

$$\lambda_n = \frac{n\pi}{2}, \quad n = 1, 2, 3, \ldots,$$

$$X_n(x) = \sin\frac{n\pi x}{}.$$

This result was also obtained directly in the solution of problem 97.

112. The solution of the boundary-value problem

$$u_{tt} = a^2 u_{xx}, \quad 0 < x < l, \quad 0 < t < +\infty, \tag{1}$$

$$u_x(0, t) - h_1 u(0, t) = 0, \quad u_x(l, t) + h_2 u(l, t) = 0, \quad 0 < t < +\infty, \tag{2}$$

$$u(x, 0) = \phi(x), \quad u_t(x, 0) = \psi(x), \quad 0 < x < l \tag{3}$$

is:

$$u(x, t) = \sum_{n=1}^{+\infty} (a_n \cos a\lambda_n t + b_n \sin a\lambda_n t) \sin(\lambda_n x + \phi_n), \tag{4}$$

where λ_n are the eigenvalues of the boundary-value problem

$$X''(x) + \lambda^2 X(x) = 0, \quad 0 < x < l, \tag{5}$$

$$X'(0) - h_1 X(0) = 0, \quad X'(l) + h_2 X(l) = 0. \tag{6}$$

The eigenvalues are the roots of the equation

$$\cot \lambda l = \frac{\lambda^2 - h_1 h_2}{\lambda(h_1 + h_2)}, \tag{7}$$

and $X_n(x) = \sin(\lambda_n x + \phi_n)$ the corresponding eigenfunctions, where

$$\phi_n = \arctan \frac{\lambda_n}{h_1}. \tag{8}$$

The square of the norm of the eigenfunction equals

$$\|X_n\|^2 = \int_0^l X_n^2(x)\,dx = \frac{1}{2}\left\{l + \frac{(\lambda_n^2 + h_1 h_2)(h_1 + h_2)}{(\lambda_n^2 + h_1^2)(\lambda_n^2 + h_2^2)}\right\}. \tag{9}$$

113. The solution of the boundary-value problem

$$u_{tt} = a^2 u_{xx}, \quad 0 < x < l, \quad 0 < t < +\infty, \quad l = 2\pi R, \tag{1}$$

$$u(0, t) = u(l, t), \quad u_x(0, t) = u_x(l, t), \quad 0 < t < +\infty, \tag{2}$$

$$u(x, 0) = \phi(x), \quad u_t(x, 0) = \psi(x), \quad 0 < x < l \tag{3}$$

is:

$$u(x, t) = \sum_{n=0}^{+\infty} \left(a_n' \cos \frac{2\pi nat}{l} + b_n' \sin \frac{2\pi nat}{l}\right) \cos \frac{2\pi nx}{l} +$$

$$+ \sum_{n=1}^{+\infty} \left(a_n'' \cos \frac{2\pi nat}{l} + b_n'' \sin \frac{2\pi nat}{l}\right) \sin \frac{2\pi nx}{l},$$

$$a_n' = \frac{2}{l}\int_0^l \phi(z) \cos \frac{2\pi nz}{l}\,dz, \quad a_n'' = \frac{2}{l}\int_0^l \phi(z) \sin \frac{2\pi nz}{l}\,dz, \quad n = 1, 2, 3, \ldots,$$

$$a_0' = \frac{1}{l}\int_0^l \phi(z)\,dz,$$

$$b'_n = \frac{1}{n\pi a}\int_0^l \psi(z)\cos\frac{2\pi nz}{l}\,dz, \quad b''_n = \frac{1}{n\pi a}\int_0^l \psi(z)\sin\frac{2\pi nz}{l}\,dz, \quad n = 1, 2, 3, \ldots,$$

$$b'_0 = \frac{1}{2n\pi a}\int_0^l \psi(z)\,dz.$$

Method. Substituting the general solution $X(x) = A\cos\lambda x + B\sin\lambda x$ of the equation

$$X''(x) + \lambda^2 X(x) = 0$$

in the boundary conditions

$$X(0) = X(l), \quad X'(0) = X'(l)$$

and letting the determinant of the system of equations obtained with respect to A and B equal zero, we find the transcendental equation for determining the eigenvalues. It is found that the eigenvalues $\lambda_n = 2\pi n/l$, when substituted into the equation for determining A and B transforms these equations into an identity for any A and B. Hence, to each eigenvalue λ_n there correspond two linearly independent eigenfunctions $\cos\lambda_n x$ and $\sin\lambda_n x$; since $\lambda_n = 2\pi n/l$, all the eigenfunctions are orthogonal in the segment $0 \leqslant x \leqslant l$†. In the case where k linearly independent eigenfunctions correspond to the same eigenvalue, this eigenvalue is said to be k-fold degenerate. Thus all eigenvalues of this problem are twofold degenerate.

114. *Method.* The total energy of the string $0 \leqslant x \leqslant l$ in the case of boundary conditions of the third kind $u_x(0, t) - hu(0, t) = 0$, $u_x(l, t) + hu(l, t) = 0$ is expressed in the following way (verify this):

$$E(t) = \frac{1}{2}\int_0^l \{T_0 u_x^2(z, t) + \rho u_t^2(z, t)\}\,dz + \frac{T_0 h}{2}\{u^2(l, t) + u^2(0, t)\}.$$

In the case of boundary conditions of the first and second kind

$$E(t) = \frac{1}{2}\int_0^l \{T_0 u_x^2(z, t) + \rho u_t^2(z, t)\}\,dz$$

(see [7], page 21).

Calculating the energy of the general vibration of the string

$$u(x, t) = \sum_{n=1}^{+\infty} U_n(x, t) = \sum_{n=1}^{+\infty} T_n(t) X_n(x),$$

† The orthogonal property of the eigenfunctions, corresponding to different eigenvalues, follows from the general theory. The orthogonality of $\cos 2\pi nx/l$ and $\sin 2\pi nx/l$ in the segment $0 \leqslant x \leqslant l$ may also be verified directly by evaluation of the overlap integral.

where $X_n(x)$ are the eigenfunctions of the corresponding boundary-value problem, by using the orthogonal property of the eigenfunctions, and also the boundary conditions, it is readily shown that in the case of boundary conditions of the first, second and third kinds

$$E(t) = \sum_{n=1}^{+\infty} E_n(t).$$

For boundary conditions of the first and second kind

$$E_n(t) = \frac{1}{2}\int_0^l \{T_0 U_{nx}^2(z, t) + \rho U_{nt}^2(z, t)\}\, dz,$$

and in the case of boundary conditions of the third kind

$$E_n(t) = \frac{1}{2}\int_0^l \{T_0 U_{nx}^2(z, t) + \rho U_{nt}^2(z, t)\}\, dz + \frac{T_0 h}{2}\{U_n^2(l, t) + U_n^2(0, t)\}.$$

115. The solutions of the boundary-value problems

$$u_{tt} + a^2 u_{xxxx} = 0, \quad 0 \leqslant x \leqslant l, \quad 0 < t < +\infty, \tag{1}$$

$$u(x, 0) = \phi(x), \quad u_t(x, 0) = \psi(x), \quad 0 < t < +\infty, \tag{2}$$

$$u(0, t) = u(l, t) = u_{xx}(0, t) = u_{xx}(l, t) = 0, \quad 0 < t < +\infty, \tag{3a}$$

$$u(0, t) = u(l, t) = u_x(0, t) = u_x(l, t) = 0, \quad 0 < t < +\infty, \tag{3b}$$

$$u_{xx}(0, t) = u_{xx}(l, t) = u_{xxx}(0, t) = u_{xxx}(l, t) = 0, \quad 0 < t < +\infty \tag{3c}$$

are respectively:

(a) $$u(x, t) = \sum_{n=1}^{+\infty}\left(a_n \cos\frac{n^2\pi^2 at}{l^2} + b_n \sin\frac{n^2\pi^2 at}{l^2}\right)\sin\frac{n\pi x}{l},$$

$$a_n = \frac{2}{l}\int_0^l \phi(z)\sin\frac{\pi nz}{l}\, dz, \quad b_n = \frac{2l}{n^2\pi^2 a}\int_0^l \psi(z)\sin\frac{n\pi z}{l}\, dz, \quad n = 1, 2, 3, \ldots,$$

(b) $$u(x, t) = \sum_{n=1}^{+\infty}(a_n \cos a\lambda_n^2 t + b_n \sin a\lambda_n^2 t)X_n(x),$$

where

$$X_n(x) = (\sinh\lambda_n l - \sin\lambda_n l)(\cosh\lambda_n x - \cos\lambda_n x) - (\cosh\lambda_n l - \cos\lambda_n l)(\sinh\lambda_n x - \sin\lambda_n x),$$

and λ_n are positive roots of the equation $\cosh\lambda l \cos\lambda l = 1$;

(c) $$u(x, t) = \sum_{n=1}^{+\infty}(a_n \cos a\lambda_n^2 t + b_n \sin a\lambda_n^2 t)X_n(x),$$

where

$$X_n(x) = (\sinh \lambda_n l - \sin \lambda_n l)(\cosh \lambda_n x + \cos \lambda_n x) - \\ -(\cosh \lambda_n l - \cos \lambda_n l)(\cosh \lambda_n x + \sin \lambda_n x),$$

and λ_n are positive roots of the transcendental equation

$$\cosh \lambda l \cos \lambda l = 1.$$

Note 1. The orthogonality of the eigenfunctions is established in the following way. Multiplying the equation $X_m''''(x) - \lambda_n^4 X_n(x) = 0$ by $X_m(x)$ and the equation

$$X_m''''(x) - \lambda_m^4 X_m(x) = 0$$

by $X_n(x)$, subtracting the results and integrating by parts, we obtain:

$$\int_0^l X_m(x) X_n(x)\,dx$$

$$= \frac{\{X_m'''(x)X_n(x) - X_n'''(x)X_m(x) - X_m''(x)X_n'(x) + X_n''(x)X_m'(x)\}\Big|_{x=0}^{x=l}}{\lambda_m^4 - \lambda_n^4}$$

from which the equality

$$\int_0^l X_m(x) X_n(x)\,dx = 0, \qquad m \neq n,$$

directly follows with the boundary conditions (3a), (3b), (3c) or the combination of (3a) at one end and (3b) at the other, etc.

Note 2. To calculate the square of the modulus of the eigenfunction $X_n(x)$ it is possible to treat it in the same way as was done in the method to problem III; then the following relation† is obtained (similar to formula (18) of the solution of problem III):

$$\int_0^l X_n^2(x)\,dx = \frac{l}{4}\{X_n^2(l) - 2X_n'''(l)X_n'(l) + X_n''^2(l)\},$$

from which in case (3b)

$$\int_0^l X_n^2(x)\,dx = \frac{l}{4} X_n''^2(l)$$

and in case (3c)

$$\int_0^l X_n^2(x)\,dx = \frac{l}{4} X_n^2(l).$$

† See A. N. Krylov, *Collection of Works*, vol. III, part 2, Izd. Akad. Nauk SSSR, 1949, pages 202–203.

116. If the vibrations of the rod are produced by a shock impulse I at a point $x = x_0$, then in the answer of the preceding problem we have:

(a) $a_n = 0, \quad b_n = \dfrac{2lI \sin \dfrac{n\pi x_0}{l}}{n^2\pi^2 a\rho};$

(b) $a_n = 0, \quad b_n = \dfrac{4IX_n(x_0)}{a\lambda_n^2 X_n''^2(l)};$

(c) $a_n = 0, \quad b_n = \dfrac{4IX_n(x_0)}{a\lambda_n^2 X_n^2(l)}.$

2. Free Vibrations in a Resistant Medium

If the vibrations of the string or the longitudinal vibrations of the rod occur in a medium with a resistance proportional to the velocity, then the wave equation has the form†

$$u_{tt} = a^2 u_{xx} - 2\nu u_t, \quad \nu > 0, \tag{1}$$

and the boundary conditions are the same as in the case of vibrations in a non-resistant medium. Writing the boundary conditions in the form

$$\alpha_1 u_x(0, t) + \beta_1 u(0, t) = 0, \quad 0 < t < +\infty, \tag{2}$$

$$\alpha_2 u_x(l, t) + \beta_2 u(l, t) = 0, \quad 0 < t < +\infty, \tag{2'}$$

we consider the possibility of boundary conditions of the first, second and third kinds. There are also initial conditions

$$u(x, 0) = \phi(x), \quad u_t(x, 0) = \psi(x). \tag{3}$$

Separating the variables, we arrive at the same boundary-value problem

$$X''(x) + \lambda^2 X(x) = 0, \quad 0 < x < l, \tag{4}$$

$$\alpha_1 X'(0) + \beta_1 X(0) = 0, \tag{5}$$

$$\alpha_2 X'(l) + \beta_2 X(l) = 0 \tag{5'}$$

for the eigenvalues and eigenfunctions, as in the case of a non-resistant medium. Let λ_n and $X_n(x)$ be the eigenvalues and eigenfunctions of the problem (4), (5), (5′). For $T_n(t)$ we obtain the differential equation

$$T_n''(t) + 2\nu T_n'(t) + a^2\lambda_n^2 T_n(t) = 0, \tag{6}$$

differing by the term $2\nu T_n'(t)$ from the corresponding equation in the case of vibrations in a non-resistant medium. Its general solution has the form

$$\left.\begin{aligned} T_n(t) &= (a_n \cosh \omega_n t + b_n \sinh \omega_n t)\,\mathrm{e}^{-\nu t}, \\ \omega_n &= \sqrt{\nu^2 - a^2\lambda_n^2}\,, \end{aligned}\right\} \nu^2 > a^2\lambda_n^2, \tag{7}$$

$$\left.\begin{aligned} T_n(t) &= (a_n \cos \omega_n t + b_n \sin \omega_n t)\,\mathrm{e}^{-\nu t}, \\ \omega_n &= \sqrt{a^2\lambda_n^2 - \nu^2}, \end{aligned}\right\} \nu^2 < a^2\lambda_n^2, \tag{7'}$$

$$T_n(t) = (a_n + b_n t)\mathrm{e}^{-\nu t}, \quad \nu = a\lambda_n. \tag{7''}$$

---

† See problem 15.

The solution of the boundary-value problem (1), (2), (2′), (3) has the form

$$u(x, t) = \sum_{n=1}^{+\infty} T_n(t) X_n(x). \tag{8}$$

It is readily seen that

$$\lim_{t\to+\infty} T_n(t) = 0$$

in each of the cases (7), (7′) and (7″).

The coefficients a_n and b_n are determined from the initial conditions in the following way:

$$a_n = \frac{1}{\|X_n\|^2}\int_0^l \phi(z) X_n(z)\,dz, \quad b_n\omega_n - \nu a_n = \frac{1}{\|X_n\|^2}\int_0^l \psi(z) X_n(z)\,dz, \tag{9}$$

where

$$\omega_n = 1 \quad \text{if} \quad \nu = a\lambda_n.$$

117.

$$u(x, t) = \frac{2l^2 h e^{-\nu t}}{\pi^2 x_0(l-x_0)} \sum_{n=1}^{+\infty} \frac{1}{n^2} \sin\frac{n\pi x_0}{l} \sin\frac{n\pi x}{l}\, \Theta_n(t),$$

where

$$\Theta_n(t) = \cosh\omega_n t + \frac{\nu}{\omega_n}\sinh\omega_n t, \qquad \omega_n = \sqrt{\nu^2 - \frac{a^2n^2\pi^2}{l^2}}, \qquad \frac{n\pi a}{l} < \nu,$$

$$\Theta_n(t) = 1 + \nu t, \qquad \frac{n\pi a}{l} = \nu,$$

$$\Theta_n(t) = \cos\omega_n t + \frac{\nu}{\omega_n}\sin\omega_n t, \qquad \omega_n = \sqrt{\frac{a^2n^2\pi^2}{l^2} - \nu^2}, \qquad \frac{n\pi a}{l} > \nu.$$

118.

$$u(x, t) = \frac{2Ie^{-\nu t}}{l\rho} \sum_{n=1}^{+\infty} \frac{1}{\omega_n} \sin\frac{n\pi x_0}{l} \sin\frac{n\pi x}{l}\, \Theta_n(t),$$

where

$$\Theta_n(t) = \sinh\omega_n t, \quad \omega_n = \sqrt{\nu^2 - \frac{n^2\pi^2a^2}{l^2}} \quad \text{if} \quad \frac{n\pi a}{l} < \nu,$$

$$\Theta_n(t) = t \quad \text{if} \quad \frac{n\pi a}{l} = \nu,$$

$$\Theta_n(t) = \sin\omega_n t, \quad \omega_n = \sqrt{\frac{n^2\pi^2a^2}{l^2} - \nu^2} \quad \text{if} \quad \frac{n\pi a}{l} > \nu.$$

119.

$$u(x,t) = \frac{8kl}{\pi^2} e^{-\nu t} \sum_{n=0}^{+\infty} \frac{(-1)^n}{(2n+1)^2} \sin\frac{(2n+1)\pi x}{2l}\, \Theta_n(t),$$

where $\Theta_n(t)$ has the same value as in the answer to problem 117.

120.

$$u(x,t) = a_0 + b_0 e^{-2\nu t} + e^{-\nu t} \sum_{n=1}^{+\infty} \Theta_n(t) \cos\frac{n\pi x}{l}, \tag{1}$$

$$\left.\begin{aligned} &\Theta_n(t) = a_n + b_n t, \quad \frac{n\pi a}{l} = \nu, \\ &\Theta_n(t) = a_n \cosh \omega_n t + b_n \sinh \omega_n t, \quad \frac{n\pi a}{l} < \nu, \\ &\Theta_n(t) = a_n \cos \omega_n t + b_n \sin \omega_n t, \quad \frac{n\pi a}{l} > \nu, \end{aligned}\right\} n = 1, 2, \ldots, \tag{2}$$

$$a_n = \frac{2}{l}\int_0^l \phi(z) \cos\frac{n\pi z}{l}\, dz, \quad b_n\omega_n - \nu a_n = \frac{2}{l}\int_0^l \psi(z)\cos\frac{n\pi z}{l}\, dz, \ n = 1,2,\ldots,$$

$$b_0 = \frac{1}{2\nu l}\int_0^l \psi(\xi)\, d\xi, \quad a_0 = \frac{1}{l}\int_0^l \phi(\xi)\, d\xi + \frac{1}{2\nu l}\int_0^l \psi(\xi)\, d\xi,$$

$$\omega_n = \sqrt{\nu^2 - \frac{n^2\pi^2 a^2}{l^2}} \ \text{ if } \ \frac{n\pi a}{l} < \nu, \quad \omega_n = 1 \ \text{ if } \ \frac{n\pi a}{l} = \nu,$$

$$\omega_n = \sqrt{\frac{n^2\pi^2 a^2}{l^2} - \nu^2} \ \text{ if } \ \frac{n\pi a}{l} > \nu.$$

121.

$$u(x,t) = e^{-\nu t} \sum_{n=1}^{+\infty} \Theta_n(t) \cos \lambda_n x, \tag{1}$$

$$\left.\begin{aligned} &\Theta_n(t) = a_n \cosh \omega_n t + b_n \sinh \omega_n t, \quad \omega_n = \sqrt{\nu^2 - \lambda_n^2 a^2} \ \text{ if } \ a\lambda_n < \nu, \\ &\Theta_n(t) = a_n + b_n t, \quad \omega_n = 1 \ \text{ if } \ a\lambda_n = \nu, \\ &\Theta_n(t) = a_n \cos \omega_n t + b_n \sin \omega_n t, \quad \omega_n = \sqrt{a^2\lambda_n^2 - \nu^2} \ \text{ if } \ a\lambda_n > \nu, \end{aligned}\right\} \tag{2}$$

λ_n are the positive roots of the equation $\lambda \tan \lambda l = h$,

$$a_n = \frac{2}{l\left\{1 + \frac{h}{l(\lambda_n^2 + h^2)}\right\}} \int_0^l \phi(z) \cos \lambda_n z\, dz,$$

$$b_n\omega_n - \nu a_n = \frac{2}{l\left\{1+\dfrac{h}{l(\lambda_n^2+h^2)}\right\}}\int_0^l \psi(z)\cos\lambda_n z\,dz.$$

122†.

$$u(x,t) = e^{-\nu t}\sum_{n=1}^{+\infty}\Theta_n(t)\sin(\lambda_n x+\phi_n), \tag{1}$$

where $\Theta_n(t)$ and ω_n are given by formula (2) of the answer to the preceding problem, λ_n are the positive roots of the equations

$$\left.\begin{aligned}\cot\lambda l &= \frac{\lambda^2-h_1h_2}{\lambda(h_1+h_2)},\\ \phi_n &= \arctan\frac{\lambda_n}{h_1},\end{aligned}\right\} \tag{2}$$

$$a_n = \frac{2}{\left\{l+\dfrac{(\lambda_n^2+h_1h_2)(h_1+h_2)}{(\lambda_n^2+h_1^2)(\lambda_n^2+h_2^2)}\right\}}\int_0^l \phi(z)\sin(\lambda_n z+\phi_n)\,dz, \quad n = 1, 2, 3, \ldots,$$

$$b_n\omega_n-\nu a_n = \frac{2}{\left\{l+\dfrac{(\lambda_n^2+h_1h_2)(h_1+h_2)}{(\lambda_n^2+h_1^2)(\lambda_n^2+h_2^2)}\right\}}\int_0^l \psi(z)\sin(\lambda_n z+\phi_n)\,dz, \quad n = 1,2,3\ldots$$

123. The solution of the boundary-value problem

$$v_{xx} = CLv_{tt}+CRv_t, \quad 0 < x < l, \quad 0 < t < +\infty, \tag{1}$$

$$v(0,t) = v_x(l,t) = 0, \quad 0 < t < +\infty, \tag{2}$$

$$v(x,0) = v_0, \quad v_t(x,0) = 0, \quad 0 < x < l \tag{3}$$

has the form

$$v(x,t) = e^{-\frac{R}{2L}t}\sum_{n=1}^{+\infty} a_n \sin\frac{(2n+1)\pi x}{2l}\sin(\omega_n t+\phi_n), \tag{4}$$

where

$$\omega_n = \frac{(2n+1)\pi}{2l\sqrt{CL}}\sqrt{1-\frac{C^2R^2l^2}{L\pi^2(2n+1)^2}}\ ‡,$$

$$a_n = \frac{4v_0}{\pi(2n+1)\sin\phi_n}, \quad \tan\phi_n = 2\omega_n\frac{L}{R}.$$

† See the answers to problems 111 and 112, pages 252–256.

‡ It is assumed that $L > C^2R^2l^2/\pi^2$.

124.

$$v(x,t) = -\frac{2Q\pi}{LC^2l^2(b-a)} e^{-\nu t} \sum_{n=0}^{+\infty} \frac{(2n+1)\cos(\omega_n t - \phi_n)}{\omega_n \sqrt{\omega_n^2+\sigma^2}} \sin\frac{(2n+1)\pi(a+b)}{4l} \times$$

$$\times \sin\frac{(2n+1)\pi(a-b)}{4l} \sin\frac{(2n+1)\pi x}{2l} \quad \text{if} \quad 0 < x < a,$$

$$\nu = \frac{1}{2}\left(\frac{R}{L}+\frac{G}{C}\right), \quad \sigma = \frac{1}{2}\left(\frac{R}{L}-\frac{G}{C}\right), \quad \omega_n = \sqrt{\frac{(2n+1)^2\pi^2}{4l^2CL}-\sigma^2},$$

$$\tan\phi_n = \frac{\sigma}{\omega_n}.$$

125.

$$v(x,t) = \frac{\pi^2 Q}{2LC^2l^3} e^{-\nu t} \sum_{n=0}^{+\infty} \frac{(2n+1)^2}{\omega_n \sqrt{\omega_n^2+\sigma^2}} \sin\frac{(2n+1)\pi x_0}{2l} \sin\frac{(2n+1)\pi x}{2l} \times$$

$$\times \cos(\omega_n t - \phi_n) \quad \text{if} \quad 0 < x < x_0,$$

where the values $\nu, \sigma, \omega_n, \phi_n$ are determined as in the answer to the preceding problem.

3. Forced Vibrations under the Action of Distributed and Concentrated Forces in a Non-resistant Medium and in a Resistant Medium

The differential equation of the forced vibrations of the string under the action of a continuously distributed force in a medium with a resistance, proportional to the velocity, has the form

$$u_{tt} = a^2u_{xx} - 2\nu u_t + f(x,t),$$

where

$$F(x,t) = \rho f(x,t)$$

is the applied force per unit length, ρ the linear mass density of the string, $f(x,t)$ is the acceleration, which a point of the string with coordinate x would have at time t, if no other forces except the external ones act on it. The term $-2\nu u_t$, representing the resistance, proportional to the velocity, vanishes if the vibrations occur in a non-resistant medium.

The boundary-value problem

$$u_{tt} = a^2u_{xx} - 2\nu u_t + f(x,t), \quad 0 < x < l, \quad 0 < t < +\infty, \tag{1}$$

$$\alpha_1 u_x(0,t) + \beta_1 u(0,t) = 0, \quad \alpha_2 u_x(l,t) + \beta_2 u(l,t) = 0, \quad 0 < t < +\infty, \tag{2}$$

$$u(x,0) = \phi(x), \quad u_x(x,0) = \psi(x), \quad 0 < x < l \tag{3}$$

may be reduced to a simpler problem†.

† See [7], page 104; reduction of the problem under consideration to a simpler one may be achieved similarly.

If one succeeds in finding some particular solution $w(x, t)$ of equation (1) satisfying the boundary conditions (2), then the solution of the boundary-value problem may be represented in the form

$$u(x, t) = v(x, t)+w(x, t), \tag{4}$$

where $v(x, t)$ is the solution of the boundary-value problem

$$v_{tt} = a^2 v_{xx} - 2\nu v_t, \quad 0 < x < l, \quad 0 < t < +\infty, \tag{5}$$

$$\alpha_1 v_x(0, t)+\beta_1 v(0, t) = 0, \quad \alpha_2 v_x(l, t)+\beta_2 v(l, t) = 0, \quad 0 < t < +\infty, \tag{6}$$

$$v(x, 0) = \phi(x)-w(x, 0), \quad v_t(x, 0) = \psi(x)-w_t(x, 0), \quad 0 < x < l, \tag{7}$$

which was considered in the preceding sections.

One proceeds similarly in the case of forced vibrations under the action of concentrated forces, applied to the ends or the interior points of the ktring.

126. *Solution.* We have the boundary-value problem

$$u_{tt} = a^2 u_{xx} - 2\nu u_t + g, \quad 0 < x < l, \quad 0 < t < +\infty, \tag{1}$$

$$u(0, t) = u(l, t) = 0, \quad 0 < t < +\infty, \tag{2}$$

$$u(x, 0) = \begin{cases} \dfrac{h}{x_0} x, & 0 < x < x_0, \\ \dfrac{h(l-x)}{l-x_0}, & x_0 < x < l, \end{cases} \tag{3}$$

$$u_t(x, 0) = 0, \quad 0 < x < l. \tag{3'}$$

We seek first the steady-state solution $w(x)$ of equation (1) satisfying the boundary conditions (2).

Substituting $w(x)$ in (1) we obtain:

$$0 = a^2 \frac{d^2 w}{dx^2} + g, \quad 0 < x < l,$$

from which

$$w(x) = -\frac{g}{2a^2} x^2 + C_1 x + C_2. \tag{4}$$

From the boundary conditions (2) we find:

$$C_2 = 0, \quad C_1 = +\frac{gl}{2a^2}. \tag{5}$$

Hence,

$$w(x) = -\frac{g}{2a^2} x^2 + \frac{gl}{2a^2} x. \tag{6}$$

It now remains to solve the boundary-value problem

$$v_{tt} = a^2 v_{xx} - 2\nu v_t, \quad 0 < x < l, \quad 0 < t < +\infty, \tag{7}$$

$$v(0, t) = v(l, t) = 0, \quad 0 < t < +\infty, \tag{8}$$

$$v(x, 0) = \begin{cases} \dfrac{h}{x_0} x + \dfrac{g}{2a^2}(x^2 - lx), & 0 < x < x_0, \\ \dfrac{h(l-x)}{l-x_0} + \dfrac{g}{2a^2}(x^2 - lx), & x_0 < x < l, \end{cases} \tag{9}$$

$$v_t(x, 0) = 0, \quad 0 < x < l. \tag{10}$$

$u(x, t)$ is represented in the form $u(x, t) = v(x, t) + w(x)$. (11)

The expression for $v(x, t)$ is obtained from formulae (7), (7′), (7″), (8), (9) of the introduction to the preceding section and hints of the present section.

We note that if the term $-2\nu u_t$ in equation (1) were absent, then the steady-state solution of the boundary-value problem (1), (2), and, therefore, the initial conditions (9) and (10) for finding the function $v(x, t)$ would remain as before. In this case equation (7) does not contain the term $-2\nu v_t$ and $v(x, t)$ is found without difficulty.

It is possible to find $v(x, t)$ without using the explicit expression for $w(x)$†. Let $w(x)$ be the steady-state solution of equation (1), satisfying the boundary conditions (2). Then the solution of the boundary-value problem (1), (2), (3) can be found in the form (11), where $v(x, t)$ is the solution of the boundary-value problem

$$v_{tt} = a^2 v_{xx} - 2\nu v_t, \quad 0 < x < l, \quad 0 < t < +\infty, \tag{7′}$$

$$v(0, t) = v(l, t) = 0, \quad 0 < t < +\infty, \tag{8′}$$

$$v(x, 0) = \phi(x) - w(x), \quad 0 < x < l, \quad \phi(x) = \begin{cases} \dfrac{h}{x_0} x, & 0 < x < x_0, \\ \dfrac{h(l-x)}{l-x_0}, & x_0 < x < l, \end{cases} \tag{9}$$

$$v_t(x, 0) = 0, \quad 0 < x < l. \tag{10}$$

Let $a\lambda_n > \nu$, $n = 1, 2, 3, \ldots$ Then

$$v(x, t) = e^{-\nu t} \sum_{n=1}^{+\infty} (a_n \cos \omega_n t + b_n \sin \omega_n t) X_n(t),$$

where

$$\omega_n = \sqrt{\frac{n^2\pi^2 a^2}{l^2} - \nu^2}, \quad X_n(x) = \sin \frac{n\pi x}{l}, \quad \lambda_n = \frac{n\pi}{l}.$$

† See [7], pages 105–108.

We have:

$$b_n = \frac{v}{\omega_n} a_n, \quad a_n = \frac{1}{\|X_n\|^2} \int_0^l v(z, 0) X_n(z)\,dz = \frac{2}{l} \int_0^l [\phi(z) - w(z)] X_n(z)\,dz$$

$$= \frac{2}{l} \int_0^l \phi(z) X_n(z)\,dz - \frac{2}{l} \int_0^l w(z) X_n(z)\,dz.$$

The first integral in the last equation equals

$$\frac{2}{l} \int_0^l \phi(z) X_n(z)\,dz = \frac{2l^2 h \sin \dfrac{n\pi x_0}{l}}{n^2 \pi^2 x_0 (l - x_0)}.$$

The second integral may be evaluated by using the equation

$$X_n''(x) + \lambda_n^2 X_n(x) = 0$$

and by integration by parts

$$-\frac{2}{l} \int_0^l w(z) X_n(z)\,dz = \frac{2}{l\lambda_n^2} \int_0^l w(z) X_n''(z)\,dz$$

$$= \frac{2}{l\lambda_n^2}\, w(z) X_n'(z)\Big|_0^l - w'(z) X_n(z)\Big|_0^l + \int_0^l w'' X_n\,dz.$$

Since $X_n(0) = X_n(l) = 0$, $w(0) = w(l)$, $a^2 w''(x) + g = 0$, then

$$-\frac{2}{l} \int_0^l w(z) X_n(z)\,dz = -\frac{2g}{\lambda_n^2 a^2 l} \int_0^l X_n(z)\,dz = -\frac{2g}{\pi n \lambda_n^2 a^2} [1 - (-1)^n].$$

Thus

$$v(x, t) = \frac{2l^2}{\pi^2} e^{-vt} \sum_{n=1}^{+\infty} \left\{ \frac{h \sin \dfrac{n\pi x_0}{l}}{n^2 x_0 (l - x_0)} + \frac{g[-1 + (-1)^n]}{\pi n^3 a^2} \right\} \left(\cos \omega_n t + \right.$$

$$\left. + \frac{v}{\omega_n} \sin \omega_n t \right) \sin \frac{n\pi x}{l}. \quad (12)$$

Having used the explicit expression (6) for $w(x)$ found earlier, it is possible now to write down an expression for the solution of the problem (1), (2), (3), (3′)

$$u(x, t) = -\frac{g}{2a^2} (x^2 - lx) +$$

$$+ \frac{2l^2}{\pi^2} e^{-vt} \sum_{n=1}^{+\infty} \left\{ \frac{h}{n^2 x_0 (l - x_0)} \sin \frac{n\pi x_0}{l} + \frac{g}{\pi n^3 a^2} [-1 + (-1)^n] \right\} \left(\cos \omega_n t + \right.$$

$$\left. + \frac{v}{\omega_n} \sin \omega_n t \right) \sin \frac{n\pi x}{l}. \quad (13)$$

We see that as $t \to +\infty$, $u(x, t) \to g/2a^2(x^2 - lx)$, where

$$w(x) = -\frac{g}{2a^2}(x^2 - lx),$$

is the position of equilibrium under the action of gravity.

For $\nu \to 0$ we derive from (13) the solution of the problem for the case where the vibrations occur in an non-resistant medium.

127. The solution of the boundary-value problem

$$u_{tt} = a^2 u_{xx} + g, \quad 0 < x < l, \quad 0 < t < +\infty, \tag{1}$$

$$u(x, 0) = 0, \quad u_t(x, 0) = v_0, \quad 0 < x < l, \tag{2}$$

$$u(0, t) = 0, \quad u_x(l, t) = 0, \quad 0 < t < +\infty \tag{3}$$

is:

$$u(x, t) = \frac{gx}{a^2}\left(l - \frac{x}{2}\right) + \sum_{n=0}^{+\infty} \left\{ \frac{gl^2}{(2n+1)^3 \pi^3} \cos \frac{(2n+1)\pi a t}{2l} + \frac{2 v_0 l^2}{(2n+1)^2 \pi^2 a} \sin \frac{(2n+1)\pi a t}{2l} \right\} \sin \frac{(2n+1)\pi x}{2l}.$$

128. The solution of the boundary-value problem

$$u_{tt} = a^2 u_{xx}, \quad 0 < x < l, \quad 0 < t < +\infty, \tag{1}$$

$$u(0, t) = 0, \quad u_x(l, t) = \frac{F_0}{ES}, \quad 0 < t < +\infty, \tag{2}$$

$$u(x, 0) = 0, \quad u_t(x, 0) = 0, \quad 0 < x < l \tag{3}$$

is:

$$u(x, t) = \frac{F_0}{ES} x - \frac{8 F_0 l}{ES \pi^2} \sum_{n=0}^{+\infty} \frac{(-1)^n}{(2n+1)^2} \sin \frac{(2n+1)\pi x}{2l} \cos \frac{(2n+1)\pi a t}{2l}.$$

129. *Solution.* From the boundary-value problem†

$$\left.\begin{aligned} -\frac{\partial p}{\partial x} &= \frac{\partial w}{\partial t} + 2\nu w, && (1) \\ -\frac{\partial p}{\partial t} &= a^2 \frac{\partial w}{\partial x}, && (1') \end{aligned}\right\} \quad 0 < x < l, \quad 0 < t < +\infty,$$

$$p(0, t) = 0, \quad w(l, t) = A, \quad 0 < t < +\infty, \tag{2}$$

$$w(x, 0) = 0, \quad p(x, 0) = 0, \quad 0 < x < l, \tag{3}$$

eliminating $p(x, t)$, we obtain the boundary-value problem

$$\frac{\partial^2 w}{\partial t^2} = a^2 \frac{\partial^2 w}{\partial x^2} - 2\nu \frac{\partial w}{\partial t}, \quad 0 < x < l, \quad 0 < t < +\infty, \tag{1''}$$

† See problem 5.

$$w_x(0, t) = 0, \quad w(l, t) = A, \quad 0 < t < +\infty, \tag{2'}$$

$$w(x, 0) = w_t(x, 0) = 0, \quad 0 < x < l, \tag{3'}$$

from which we find:

$$w(x, t) = A - \frac{4A}{\pi} e^{-\nu t} \sum_{n=0}^{+\infty} \frac{\cos \tilde{\omega}_n t + \dfrac{\nu}{\tilde{w}_n} \sin \tilde{\omega}_n t}{2n+1} \sin \frac{(2n+1)\pi(l-x)}{2l}, \tag{4}$$

where

$$\tilde{\omega}_n = \sqrt{\left[\frac{(2n+1)\pi a}{2l}\right]^2 - \nu^2}.$$

We find the pressure p at $x = l$ by means of (1)

$$p(l, t) = p(0, t) - \int_0^l \left(\frac{\partial w}{\partial t} - 2\nu w\right) dx$$

$$= -\left\{2\nu l A + \frac{4aA}{\pi} e^{-\nu t} \sum_{n=0}^{+\infty} \frac{\sin(\tilde{\omega}_n t - 2\phi_n)}{(2n+1)\cos\phi_n}\right\}, \tag{5}$$

where

$$\tan \phi_n = \frac{\nu}{\tilde{\omega}_n}. \tag{6}$$

130.

$$v(x, t) = E \frac{\cosh (x-l)\sqrt{GR}}{\cosh l\sqrt{GR}} -$$

$$- \frac{\pi E}{LCl^2} \sum_{n=0}^{+\infty} (-1)^n = \frac{(2n+1)e^{-\nu t}}{\tilde{\omega}_n \sqrt{\tilde{\omega}_n^2 + \nu^2}} \cos(\tilde{\omega}_n t - \phi_n) \cos \frac{(2n+1)\pi(l-x)}{2l},$$

$$\nu = \frac{1}{2}\left(\frac{R}{L} + \frac{G}{C}\right), \ \tilde{\omega} = \sqrt{\frac{(2n+1)^2\pi^2 a^2}{4l^2} - \frac{1}{4}\left(\frac{R}{L} - \frac{G}{C}\right)^2},$$

$\tan \phi_n = \nu/\tilde{\omega}_n$, where it is assumed that $\pi a/l > |R/L - G/C|$.

131. $$v(x, t) = E \frac{\sinh (l-x)\sqrt{GR}}{\sinh l\sqrt{GR}} -$$

$$-2E e^{-\nu t} \sum_{n=1}^{+\infty} \frac{n\pi}{n^2\pi^2 + RGl^2} \left\{\cosh \tilde{\omega}_n t + \frac{\nu}{\tilde{\omega}_n} \sinh \tilde{\omega}_n t\right\} \sin \frac{n\pi x}{l},$$

$$\tilde{\omega}_n = \sqrt{\frac{1}{4}\left(\frac{R}{L} - \frac{G}{C}\right)^2 - \frac{n^2\pi^2 a^2}{l^2}}, \qquad \nu = \frac{1}{2}\left(\frac{R}{L} + \frac{G}{C}\right),$$

$\tilde{\omega}_n$ can be either real or imaginary.

132. The solution of the boundary-value problem

$$u_{tt} = a^2u_{xx}, \quad 0 < x < x_0, \quad x_0 < x < l, \quad 0 < t < +\infty, \tag{1}$$

$$\left.\begin{aligned} u(0, t) = 0, \quad u(x_0-0, t) = u(x_0+0, t),\\ T_0[u'_x(x_0+0, t)-u'_x(x_0-0, t)] = -F_0,\\ u(l, t) = 0, \quad 0 < t < +\infty, \end{aligned}\right\} \tag{2}$$

$$u(x, 0) = 0, \quad u_t(x, 0) = 0, \quad 0 < x < l \tag{3}$$

is:

$$u(x, t) = \phi(x)-\frac{2lF_0}{\pi^2 T_0}\sum_{n=1}^{+\infty}\frac{1}{n^2}\sin\frac{n\pi x_0}{l}\sin\frac{n\pi x}{l}\cos\frac{n\pi at}{l}, \tag{4}$$

where

$$\phi(x) = \begin{cases} \dfrac{F_0(l-x_0)}{T_0 l}x, & 0 < x < x_0, \\ \dfrac{F_0 x_0}{T_0 l}(l-x), & x_0 < x < l. \end{cases} \tag{5}$$

Method. The steady-state solution of (1) has the form

$$w(x) = C_1x+C_2,$$

the constants C_1 and C_2 being different over the intervals

$$0 < x < x_0 \quad \text{and} \quad x_0 < x < l.$$

Their values C_1', C_2', C_1'', C_2'' in the first and second intervals are found from the conditions (2).

Notes for the solutions of problems 133–143

(1) If the inhomogeneous differential equation has the form

$$u_{tt} = a^2u_{xx}+c^2u+b\frac{\partial u}{\partial x}+\Phi(x)\sin\omega t$$

or

$$u_{tt} = a^2u_{xx}+c^2u+b\frac{\partial u}{\partial x}+\Phi(x)\cos\omega t,$$

then a particular solution may be found in the form†

$$w(x, t) = X(x)\sin\omega t$$

or respectively in the form

$$w(x, t) = X(x)\cos\omega t.$$

When one of the characteristic frequencies of the string coincides with the frequency ω of the applied force $\Phi(x)\sin\omega t$ or $\Phi(x)\cos\omega t$, then for $b = 0$

† See also the solution of problem 133, where this case is defined more precisely.

resonance may occur, in which the amplitude of the vibration which has the frequency of the applied force increases indefinitely.

(2) If the inhomogeneous differential equation contains the term $-2\nu u_t$, i.e. has the form

$$u_{tt} = a^2 u_{xx} + b\frac{\partial u}{\partial x} + cu - 2\nu\frac{\partial u}{\partial t} + \Phi(x)\sin\omega t \tag{6}$$

or the form

$$u_{tt} = a^2 u_{xx} + b\frac{\partial u}{\partial x} - 2\nu u_t + cu + \Phi(x)\cos\omega t, \tag{7}$$

i.e. the vibrations occur in a medium with a resistance proportional to the velocity, then the particular solution discussed above does not exist. In this case it is best to change to the complex representation of the applied force; more precisely, it is possible to look for the particular solution of the equation

$$U_{tt} = a^2 U_{xx} + bU_x - 2\nu U_t + cU + \Phi(x)e^{i\omega t}$$

in the form

$$U(x, t) = X(x)e^{i\omega t}. \tag{8}$$

The real part of (8) will be a particular solution of equation (7), and the imaginary part, a particular solution of equation (6).

If the particular solution (8) satisfies the boundary conditions of the problem, then it represents the forced vibrations, making up the steady-state part of the solution of the problem for $t \to +\infty$, since the characteristic vibrations, due to the initial deflections and velocities will be damped.

133. The solution of the boundary-value problem

$$u_{tt} = a^2 u_{xx} + \frac{\Phi(x)}{\rho}\sin\omega t, \quad 0 < x < l, \quad 0 < t < +\infty, \tag{1}$$

$$u(0, t) = u(l, t) = 0, \quad 0 < t < +\infty, \tag{2}$$

$$u(x, 0) = u_t(x, 0) = 0, \quad 0 < x < l \tag{3}$$

reduces to the solution of the problem on the free vibrations of a string with fixed ends for given initial conditions, if some particular solution of the inhomogeneous equation (1) is known, satisfying the boundary conditions (2) (see the introduction to the present section).

Let us find therefore the particular solution of equation (1) satisfying the boundary conditions (2).

(a) Let $\omega \neq n\pi a/l$, $n = 1, 2, 3, \ldots$ Let us look for a particular solution, in the form

$$U(x, t) = X(x)\sin\omega t. \tag{4}$$

Substitution of (4) in (1) and (2) gives:

$$X''+\frac{\omega^2}{a^2}X=-\frac{\Phi(x)}{T_0}, \quad 0<x<l, \tag{5}$$

$$X(0)=X(l)=0, \tag{6}$$

from which we find:

$$X(x)=\left\{\frac{a}{\omega T_0}\int\limits_0^l \Phi(\xi)\sin\frac{\omega}{a}(l-\xi)d\xi\right\}\frac{\sin\frac{\omega}{a}x}{\sin\frac{\omega}{a}l}-$$

$$-\frac{a}{\omega T_0}\int\limits_0^x \Phi(\xi)\sin\frac{\omega}{a}(x-\xi)d\xi, \tag{7}$$

where T_0 is the tension in the string.

(b) Let $\omega=n_0\pi a/l$.

In this case the particular solution of the boundary-value problem (1) and (2) may be found in the form (4) only if $\Phi(x)$ and $\sin n_0\pi x/b$ are orthogonal in the segment $0<x<l$. Actually, multiplying both sides of equation (5) by $\sin n_0\pi x/l$ and integrating by parts using the boundary conditions (6), we obtain:

$$-\frac{1}{T_0}\int\limits_0^l \Phi(z)\sin\frac{n_0\pi z}{l}dz=0.$$

(b_1) Let us assume firstly that $\Phi(x)$ and $\sin n_0\pi x/l$ are orthogonal in the segment $0<x<l$. Then the general solution of equation (5) has the form

$$X(x)=-\frac{a}{\omega T_0}\int\limits_0^x \Phi(z)\sin\frac{\omega}{a}(x-z)dz+C_1\sin\frac{\omega}{a}x+C_2\cos\frac{\omega}{a}x.$$

From the boundary condition $X(0)=0$ we find $C_2=0$. Since $\omega/a=n_0\pi/l$ then $\sin(\omega/a)x$ reduces to zero at the ends of the segment $0\leqslant x\leqslant l$; therefore the constant C_1 may be chosen as desired. It is readily seen that in this case the expression

$$X(x)=-\frac{a}{\omega T_0}\int\limits_0^x \Phi(z)\sin\frac{\omega}{a}(x-z)dz \tag{8}$$

is a solution of equation (5), satisfying the boundary conditions (6).

(b_2) Let us consider now the case where $\omega=n_0\pi a/l$, and Φ and $\sin n\pi x/l$ are not orthogonal in the segment $0<x<l$. In this case the particular solution of the boundary-value problem (1), (2) cannot be found in the form (4).

Let us assume

$$\psi(x) = -\frac{\Phi(x)}{T_0} + A_{n_0} \sin \frac{n_0 \pi x}{l}, \tag{9}$$

where

$$A_{n_0} = \frac{2}{l} \int_0^l \frac{\Phi(\xi)}{T_0} \sin \frac{n_0 \pi \xi}{l} \, d\xi . \tag{10}$$

The function $\psi(x)$ is orthogonal to $\sin n_0 \pi x/l$ in the segment $0 \leqslant x \leqslant l$. Now equation (1) may be rewritten in the form

$$u_{tt} = a^2 u_{xx} - \frac{T_0}{\rho} \psi(x) \sin \omega t + \frac{T_0}{\rho} A_{n_0} \sin \frac{n_0 \pi x}{l} \sin \omega t . \tag{1'}$$

The sum of the particular solutions $v(x, t)$ and $w(x, t)$ of the equations

$$v_{tt} = a^2 v_{xx} + \frac{T_0}{\rho} A_{n_0} \sin \frac{n_0 \pi x}{l} \sin \omega t, \tag{1''}$$

$$w_{tt} = a^2 w_{xx} - \frac{T_0}{\rho} \psi(x) \sin \omega t, \tag{1'''}$$

satisfying the boundary conditions (2), will be a particular solution of equation (1) satisfying the boundary conditions (2).

Since $\sin n_0 \pi x/l$ and $-(T_0/\rho)\psi(x)$ are orthogonal in the segment $0 \leqslant x \leqslant l$, then according to (8)

$$w(x, t) = \left\{ \frac{a}{\omega} \int_0^x \psi(z) \sin \frac{\omega}{a} (x-z) \, dz \right\} \sin \omega t \tag{11}$$

will be a particular solution of equation (1''') satisfying the boundary conditions (2).

If now a particular solution of equation (1'') is sought in the form

$$v(x, t) = T(t) \sin \frac{n_0 \pi x}{l}, \tag{12}$$

then the boundary conditions (2) will be satisfied for any $T(t)$. Substituting (12) in (1'') and remembering the relation $\omega = n_0 \pi a/l$, we obtain the equation

$$T''(t) + \omega^2 T(t) = \frac{T_0}{\rho} A_{n_0} \sin \omega t. \tag{13}$$

A particular solution, as is known from the theory of ordinary differential equations, has the form

$$T(t) = t(A \cos \omega t + B \sin \omega t). \tag{14}$$

Substitution of (14) into (12) gives:

$$A = -\frac{T_0 A_{n_0}}{2\omega\rho}, \quad B = 0. \tag{15}$$

Therefore

$$T(t) = -\frac{T_0 A n_0}{2\omega\rho} t \cos \omega t \tag{16}$$

and

$$w(x, t) = -\frac{T_0 A_{n_0}}{2\omega\rho} t \cos \omega t \sin \frac{n_0 \pi x}{l}. \tag{17}$$

Thus, if $\omega = n_0\pi a/l$ and the functions $\Phi(x)$ and $\sin n_0\pi x/l$ are not orthogonal in the segment $0 < x < l$, then the particular solution of equation (1) satisfying the boundary conditions (2) has the form

$$U(x, t) = \left\{ \frac{a}{\omega} \int_0^x \psi(z) \sin \frac{\omega}{a} (x-z) \mathrm{d}z \right\} \sin \omega t - \frac{T_0 A_{n_0}}{2\omega\rho} t \cos \omega t \sin \frac{n_0 \pi x}{l}. \tag{18}$$

In this case resonance occurs: the amplitude of vibrations with the frequency of the constraining force in proportional to the time and increases indefinitely.

134. The solution of the boundary-value problem

$$u_{tt} = a^2 u_{xx} + \frac{\Phi_0}{\rho} \sin \omega t, \quad 0 < x < l, \quad 0 < t < +\infty, \tag{1}$$

$$u(0, t) = u(l, t) = 0, \quad 0 < t < +\infty, \tag{2}$$

$$u(x, 0) = u_t(x, 0) = 0, \quad 0 < x < l \tag{3}$$

is:

(a) for $\omega \neq n\pi a/l$, $n = 1, 2, 3, \ldots,$

$$u(x, t) = \frac{2\Phi_0}{\omega^2\rho} \left\{ \frac{\sin \frac{\omega}{a} x}{\sin \frac{\omega}{a} l} \sin^2 \frac{\omega l}{2a} - \sin^2 \frac{\omega}{2a} x \right\} \sin \omega t +$$

$$+ \sum_{n=1}^{+\infty} b_n \sin \frac{n\pi x}{l} \sin \frac{n\pi a t}{l},$$

where

$$b_n = -\frac{4\Phi_0}{n\pi a\omega\rho} \int_0^l \left\{ \frac{\sin \frac{\omega}{a} z}{\sin \frac{\omega}{a} l} \sin^2 \frac{\omega l}{2a} - \sin^2 \frac{\omega}{2a} z \right\} \sin \frac{n\pi z}{l} \mathrm{d}z;$$

(b_1) for $\omega = n_0\pi a/l$, where n_0 is even,

$$u(x, t) = -\frac{2\Phi_0}{\omega^2\rho} \sin^2 \frac{\omega}{2a} x \sin \omega t + \sum_{n+1}^{+\infty} b_n \sin \frac{n\pi x}{l} \sin \frac{n\pi a t}{l},$$

where

$$b_n = \frac{4\Phi_0}{n\pi a\rho\omega}\int_0^l \sin^2\frac{\omega}{2a}z\cdot\sin\frac{n\pi z}{l}\,dz;$$

(b_2) for $\omega = n_0\pi a/l$, n_0 odd

$$u(x, t) = \left\{-\frac{\Phi_0}{\omega^2\rho}\sin^2\frac{\omega}{2a}x + \frac{4a\Phi_0}{n_0\pi\omega T_0}\int_0^x \sin\frac{n_0\pi z}{l}\sin\frac{n_0\pi(x-z)}{l}\,dz\right\}\sin\omega t +$$

$$+\frac{2\Phi_0\sin\frac{n_0\pi x}{l}}{n_0\pi\omega\rho}t\cos\omega t + \sum_{n=1}^{+\infty} b_n\sin\frac{n\pi x}{l}\sin\frac{n\pi at}{l},$$

where

$$b_n = \frac{a}{n\pi a}\int_0^l\left\{-\frac{2\Phi}{\omega\rho}\sin^2\frac{\omega}{2a}\xi + \frac{4a\Phi_0}{n_0\pi T_0}\int_0^\xi \sin\frac{n_0\pi z}{l}\sin\frac{n_0\pi(x-z)}{l}\,dz +\right.$$

$$\left.+\frac{2\Phi_0}{n_0\pi\omega\rho}\sin\frac{n_0\pi\xi}{l}\right\}\sin\frac{n\pi\xi}{l}\,d\xi,$$

and T_0 is the tension of the string.

In this case resonance occurs: the amplitude of vibrations with the frequency of the constraining force ω increases proportionally to t indefinitely.

135. The solution of the boundary-value problem

$$u_{tt} = a^2u_{xx}, \quad 0 < x < l, \quad 0 < t < +\infty, \tag{1}$$

$$u(0, t) = 0, \quad u(l, t) = A\sin\omega t, \quad 0 < t < +\infty, \tag{2}$$

$$u(x, 0) = u_t(x, 0) = 0, \quad 0 < x < l \tag{3}$$

is:

(a) for $\omega \neq \frac{n\pi a}{l}, \quad n = 1, 2, 3, 4, \ldots,$ (4)

$$u(x, t) = A\frac{\sin\frac{\omega}{a}x}{\sin\frac{\omega}{a}l}\sin\omega t + \sum_{n=1}^{+\infty} b_n\sin\frac{n\pi x}{l}\sin\frac{n\pi at}{l}, \tag{5}$$

where

$$b_n = -\frac{2A\omega}{n\pi a}\int_0^l \frac{\sin\frac{\omega}{a}z}{\sin\frac{\omega}{a}l}\sin\frac{n\pi z}{l}\,dz, \quad n = 1, 2, 3, \ldots, \tag{6}$$

(b) for $\omega = \dfrac{n_0\pi a}{l}$ (7)

$$u(x, t) = \left\{ \frac{A\omega}{al} \int_0^x \left(Z - \overset{*}{A}_{n_0} \sin\frac{n\pi z}{l}\right) \sin\frac{n_0\pi(x-z)}{l}\, dz \right\} \sin\omega t -$$

$$- \frac{A\overset{*}{A}_{n_0}}{2l} t \cos\omega t \sin\frac{n_0\pi x}{l} + \sum_{n=0}^{+\infty} b_n \sin\frac{n\pi x}{l} \sin\frac{n\pi at}{l}, \quad (8)$$

where

$$b_n = -\frac{2}{n\pi a}\int_0^l U_t(z, 0)\sin\frac{n\pi z}{l}\, dz, \qquad \overset{*}{A}_{n_0} = \int_0^l z \sin\frac{n_0\pi z}{l}\, dz,$$

$U(z, t)$ is the sum of the first two terms on the right hand side of equation (8).

Method 1. For $\omega \neq n\pi a/l$, $n = 1, 2, 3, \ldots$, let us seek a particular solution of the boundary-value problem (1), (2) in the form $U(x, t) = X(x) \sin\omega t$ and write the solution of the problem (1), (2), (3) in the form $u(x, t) = v(x, t) + U(x, t)$.

Method 2. For $\omega = n_0\pi a/l$ it is useful to eliminate the inhomogeneity in the boundary condition, transferring it into the equation. In order to do this we find the steady-state solution $\phi(x)$ of equation (1) satisfying the boundary conditions $\phi(0) = 0$, $\phi(l) = A$, then we seek the solution of the boundary-value problem (1), (2), (3) in the form

$$u(x, t) = v(x, t) + \phi(x) \sin\omega t.$$

136. The solution of the boundary-value problem

$$u_{tt} = a^2 u_{xx}, \quad 0 < x < l, \quad 0 < t < +\infty, \quad (1)$$

$$u(0, t) = 0, \quad u_x = (l, t) = \frac{A}{ES}\sin\omega t, \quad 0 < t < +\infty, \quad (2)$$

$$u(x, 0) = 0, \quad u_t(x, 0) = 0 \quad (3)$$

is:

(a) for $\omega \neq \dfrac{(2n+1)\pi a}{2l}$, $n = 0, 1, 2, 3, \ldots$, (4)

$$u(x, t) = U(x, t) + \sum_{n=0}^{+\infty} b_n \sin\frac{(2n+1)\pi x}{2l} \sin\frac{(2n+1)\pi at}{2l}, \quad (5)$$

where

$$U(x, t) = \frac{aA}{ES\omega} \frac{\sin\dfrac{\omega}{a}x}{\cos\dfrac{\omega}{a}l} \sin\omega t,$$

$$b_n = -\frac{4}{(2n+1)\pi a}\int_0^l U_t(z,0)\sin\frac{(2n+1)\pi z}{2l}\,dz; \tag{6}$$

(b) for $\omega = \dfrac{(2n_0+1)\pi a}{2l}$ (7)

$$u(x,t) = U(x,t)+\sum_{\substack{n=0\\ n\neq n_0}}^{+\infty} b_n \sin\frac{(2n+1)\pi x}{2l}\sin\frac{(2n+1)\pi at}{2l}, \tag{8}$$

where

$$U(x,t) = -\left\{\frac{Aa}{ES\omega}\int_0^x\left(z-A^*_{n_0}\sin\frac{(2n_0+1)\pi z}{2l}\right)\sin\frac{(2n_0+1)\pi(x-z)}{2l}\,dz\right\}\sin\omega t-$$

$$-\frac{AA^*_{n_0}}{2ES}t\cos\omega t\sin\frac{(2n+1)\pi x}{2l}, \tag{9}$$

$$b_n = -\frac{4}{(2n+1)\pi a}\int_0^l U_t(z,0)\sin\frac{(2n+1)\pi z}{2l}\,dz,$$

$$A^*_{n_0} = \int_0^l z\sin\frac{(2n_0+1)\pi z}{2l}\,dz.$$

Method. See the method for the preceding problem.

137. The solution of the boundary-value problem

$$u_{tt} = a^2u_{xx}+\omega^2(x+u)+g\sin\omega t,\quad 0<x<l,\quad 0<t<+\infty, \tag{1}$$

$$u(0,t),\quad u_x(l,t) = 0,\quad 0<t<+\infty, \tag{2}$$

$$u(x,0),\quad u_t(x,0) = 0,\quad 0<x<l \tag{3}$$

is:

$$u(x,t) = v(x)+w(x,t)+U(x,t),$$

where

$$v(x) = \frac{k\sin kx}{\cos kl}\int_0^l \xi\cos k(l-\xi)\,d\xi - k\int_0^x \xi\sin k(x-\xi)\,d\xi,$$

$$w(x,t) = X(x)\sin\omega t = \frac{g}{2\omega^2}\left\{\frac{\cos\left[\frac{\omega}{2}(l-x)\sqrt{2}\right]}{\cos\left(\frac{\omega}{a}l\sqrt{2}\right)}-1\right\}\sin\omega t,$$

$$U(x, t) = \sum_{n=0}^{+\infty} \left[A_n \cos \left(\sqrt{\omega^2 - \frac{(2n+1)\pi^2 a^2}{4l^2}} \right) t + \right.$$

$$\left. + B_n \sin \left(\sqrt{\omega^2 - \frac{(2n+1)^2 \pi^2 a^2}{4l^2}} \right) t \right] \sin \frac{(2n+1)\pi}{2l} x,$$

$$A_n = -\frac{2}{l} \int_0^l v(\xi) \sin \frac{(2n+1)\pi\xi}{2l} \, d\xi,$$

$$B_n = -\frac{2}{l\sqrt{\omega^2 - \frac{(2n+1)^2 \pi^2 a^2}{4l^2}}} \int_0^l X(\xi) \sin \frac{(2n+1)\pi\xi}{2l} \, d\xi .$$

Method. First of all find the steady-state solution, then the forced harmonic vibrations with the frequency of the applied force, then the free vibrations.

138. The solution of the boundary-value problem

$$u_{tt} = a^2 u_{xx}, \quad 0 < x < x_0, \quad x_0 < x < l, \quad 0 < t < +\infty, \tag{1}$$

$$u(0, t) = 0, \quad u(x_0 - 0, t) = u(x_0 + 0, t),$$

$$T_0[u_x(x_0+0, t) - u_x(x_0-0, t)] = A \sin \omega t, \quad u(l, t) = 0, \quad 0 < t < +\infty, \tag{2}$$

$$u(x, 0) = u_t(x, 0) = 0, \quad 0 < x < l \tag{3}$$

is†:

$$u(x, t) = U(x, t) + \sum_{n=1}^{+\infty} b_n \sin \frac{n\pi x}{l} \sin \frac{n\pi a t}{l}, \tag{4}$$

where

$$b_n = -\frac{2}{n\pi a} \int_0^l U_t(z, 0) \sin \frac{n\pi z}{l} \, dz, \tag{5}$$

$$U(x, t) = \begin{cases} \dfrac{Aa}{T_0\omega} \dfrac{\sin \dfrac{\omega(l-x_0)}{a}}{\sin \dfrac{\omega l}{a}} \sin \dfrac{\omega x}{a} \sin \omega t, & 0 \leqslant x \leqslant x_0, \quad (6) \\[2ex] \dfrac{Aa}{T_0\omega} \dfrac{\sin \dfrac{\omega x_0}{a}}{\sin \dfrac{\omega l}{a}} \sin \dfrac{\omega(l-x)}{a} \sin \omega t, & x_0 \leqslant x \leqslant l. \quad (6') \end{cases}$$

† Compare [7], pages 115–116.

139. The solution of the boundary-value problem

$$u_{tt} = a^2 u_{xx}, \quad 0 < x < x_0, \quad x_0 < x < l, \quad 0 < t < +\infty, \tag{1}$$

$$u(0, t) = 0, \quad u(x_0 - 0, t) = u(x_0 + 0, t),$$

$$T_0[u_x(x_0+0, t) - u_x(x_0-0, t)] = A \cos \omega t, \quad u(l, t) = 0, \quad 0 < t < +\infty, \tag{2}$$

$$u(x, 0) = 0, \quad u_t(x, 0) = 0, \quad 0 < x < l \tag{3}$$

is:

$$u(x, t) = U(x, t) + \sum_{n=1}^{+\infty} a_n \sin \frac{n\pi x}{l} \cos \frac{n\pi a t}{l}, \tag{4}$$

where

$$a_n = -\frac{2}{l} \int_0^l U(z, 0) \sin \frac{n\pi z}{l}\, dz \tag{5}$$

and

$$U(x, t) = \left\{ \begin{array}{ll} \dfrac{Aa}{T_0\omega} \dfrac{\sin \frac{\omega}{a}(l - x_0)}{\sin \frac{\omega}{a} l} \sin \dfrac{\omega x}{a} \cos \omega t, & (6) \\ 0 \leqslant x \leqslant x_0, & \\ \dfrac{Aa}{T_0\omega} \dfrac{\sin \frac{\omega}{a} x_0}{\sin \frac{\omega}{a} l} \sin \dfrac{\omega}{a}(l - x) \cos t, & \\ x_0 \leqslant x \leqslant l, & (6') \end{array} \right\} \omega \neq \frac{n\pi a}{l}, \ n = 1, 2, 3, \ldots †.$$

140. The solution of the boundary-value problem is:

$$u(x, t) = U(x, t) + \sum_{n=1}^{+\infty} \left(a_n \cos \frac{n\pi a t}{l} + b_n \sin \frac{n\pi a t}{l} \right) \sin \frac{n\pi x}{l}, \tag{1}$$

where

$$a_n = -\frac{2}{l} \int_0^l U(z, 0) \sin \frac{n\pi z}{l}\, dz,$$

$$b_n = -\frac{2}{n\pi a} \int_0^l U_t(z, 0) \sin \frac{n\pi z}{l}\, dz \tag{2}$$

---

† Passing to a limit as $\omega \to 0$ we obtain for $A = F_0$ the steady-state deflection found in the solution of problem 132.

and for $m\omega \neq n\pi a/l$, $m, n = 1, 2, 3, \ldots$

$$U(x,t) = \begin{cases} \dfrac{1}{T_0}\left\{\dfrac{\alpha_0 x}{2}\left(1-\dfrac{x_0}{l}\right)+ \right. \\ \left. +\sum\limits_{n=1}^{+\infty} \dfrac{a\sin\dfrac{n\omega}{a}(l-x_0)}{n\omega\sin\dfrac{n\omega}{a}l}\sin\dfrac{n\omega x}{a}(\alpha_n\cos n\omega t+\beta_n\sin n\omega t)\right\}, \\ \qquad\qquad 0 \leqslant x \leqslant x_0, \quad (3) \\ \dfrac{1}{T_0}\left\{\dfrac{\alpha_0 x_0}{2}\left(1-\dfrac{x}{l}\right)+ \right. \\ \left. +\sum\limits_{n=1}^{+\infty} \dfrac{a\sin\dfrac{n\omega}{a}x_0}{n\omega\sin\dfrac{n\omega}{a}l}\sin\dfrac{n\omega(l-x)}{a}(\alpha_n\cos n\omega t+\beta_n\sin n\omega t)\right\}, \\ \qquad\qquad x_0 \leqslant x \leqslant l. \quad (3') \end{cases}$$

Note. The first terms of the sums (3) and (3′) correspond to a steady-state deflection under the action of a force, equal to $\alpha_0/2$ and applied at the point x_0; i.e. this force produces a deflection

$$U(x) = \begin{cases} \dfrac{1}{T_0}\dfrac{\alpha_0}{2}x\left(1-\dfrac{x_0}{l}\right), & 0 \leqslant x \leqslant x_0, \\ \dfrac{1}{T_0}\dfrac{\alpha_0}{2}x_0\left(1-\dfrac{x}{l}\right), & x_0 \leqslant x \leqslant l. \end{cases}$$

141. The solution of the boundary-value problem†

$$u_{tt} = a^2u_{xx}-2\nu u_t+\Phi(x)\sin\omega t, \quad 0<x<l, \quad 0<t<+\infty, \tag{1}$$

$$u(0,t) = u(l,t) = 0, \quad 0<t<+\infty, \tag{2}$$

$$u(x,0) = 0, \quad u_t(x,0) = 0, \quad 0<x<l \tag{3}$$

is:

$$u(x,t) = U(x,t)+e^{-\nu t}\sum_{n=1}^{+\infty}\left(a_n\cos\frac{n\pi at}{l}+b_n\sin\frac{n\pi at}{l}\right)\sin\frac{n\pi x}{l}, \tag{4}$$

where

$$a_n = -\frac{2}{l}\int_0^l U(z,0)\sin\frac{n\pi z}{l}\,dz, \tag{5}$$

† See the introduction to the answers of the present section.

$$b_n = -\frac{\nu l}{n\pi a} a_n - \frac{2}{n\pi a}\int_0^l U_t(z, 0) \sin \frac{n\pi z}{2} \, dz, \tag{5}$$

$$U(x, t) = \operatorname{Im}\left[\frac{\alpha-\beta i}{(\alpha^2+\beta^2)a^2}\left\{\left(\int_0^l \Phi_0(\xi)\mathring{X}(l-\xi)\,d\xi\right)\frac{\mathring{X}(x)}{\mathring{X}(l)} - \right.\right.$$
$$\left.\left. - e^{i\omega t}\int_0^x \Phi_0(\xi)\mathring{X}(x-\xi)\,d\xi\right\}\right]^{\dagger}. \tag{6}$$

$U(x, t)$ are the steady-state vibrations

$$\mathring{X}(x) = e^{(\alpha+\beta i)x} - e^{-(\alpha+\beta i)x}, \qquad \alpha+\beta i = \frac{\sqrt{\omega^2-2\omega\nu i}}{a}$$
$$= \frac{1}{a}\sqrt{\frac{\sqrt{\omega^4+4\nu^2\omega^2}+\omega^2}{2}} - \frac{i}{a}\sqrt{\frac{\sqrt{\omega^4+4\nu^2\omega^2}-\omega^2}{2}}. \tag{7}$$

Note. Let $\mathring{Y}(x)$ be a solution of the differential equation

$$y''+Ay'+By = 0, \qquad A = \text{const.}, \qquad B = \text{const.},$$

satisfying the initial conditions

$$y(0) = 0, \qquad y'(0) = 1;$$

then

$$y = \int_0^x f(\xi)\mathring{Y}(x-\xi)\,d\xi$$

is a solution of the equation

$$y''+Ay'+By = f(x),$$

satisfying the initial conditions

$$y(0) = 0, \qquad y'(0) = 0.$$

142. The solution of the boundary-value problem

$$u_{tt} = a^2 u_{xx} - 2\nu u_t, \qquad 0 < x < l, \qquad 0 < t < +\infty, \tag{1}$$

$$u(0, t) = 0, \qquad u_x(l, t) = \frac{A}{ES}\sin \omega t, \qquad 0 < t < +\infty, \tag{2}$$

$$u(x, 0) = 0, \qquad u_t(x, 0) = 0, \qquad 0 < x < l \tag{3}$$

is:

$$u(x, t) = U(x, t) +$$
$$+ e^{-\nu t}\sum_{n=0}^{+\infty}\left(a_n \cos\frac{(2n+1)\pi a t}{2l} + b_n \sin\frac{(2n+1)\pi a t}{2l}\right)\sin\frac{(2n+1)\pi x}{2l}, \tag{4}$$

† The symbol Im indicates the imaginary part of the complex number.

$$\left.\begin{aligned} a_n &= -\frac{2}{l}\int_0^l U(z,0)\sin\frac{(2n+1)\pi z}{2l}\,dz;\\ b_n &= -\frac{4\nu}{(2n+1)\pi a}\int_0^l U_t(z,0)\sin\frac{(2n+1)\pi z}{2l}\,dz. \end{aligned}\right\} \tag{5}$$

The steady-state vibrations are given by the relation

$$U(x,t) = \mathrm{Im}\left\{\frac{A(\alpha-\beta i)}{ES(\alpha^2+\beta^2)}\,\frac{e^{(\alpha+\beta i)x}-e^{-(\alpha+\beta i)x}}{e^{(\alpha+\beta i)l}-e^{-(\alpha+\beta i)l}}\,e^{i\omega t}\right\}, \tag{6}$$

where α and β have the same meaning as in the preceding problem.

143. The solution of the boundary-value problem

$$\frac{\partial^2 v}{\partial x^2} - LC\frac{\partial^2 v}{\partial t^2} - (RC+GL)\frac{\partial v}{\partial t} - GRv = 0,\quad 0<x<l,\quad 0<t<+\infty, \tag{1}$$

$$v_x(0,t) = 0,\quad v(l,t) = E_0\sin\omega t,\quad 0<t<+\infty. \tag{2}$$

$$v(x,0) = 0,\quad v_t(x,0) = 0.\quad 0<x<l \tag{3}$$

is:

$$v(x,t) = V(x,t) + \\ + e^{-\nu t}\sum_{n=0}^{+\infty}\left(a_n\cos\frac{(2n+1)\pi at}{2l} + b_n\sin\frac{(2n+1)\pi at}{2l}\right)\cos\frac{(2n+1)\pi x}{2l}, \tag{4}$$

$$\left.\begin{aligned} a_n &= -\frac{2}{l}\int_0^l V(z,0)\cos\frac{(2n+1)\pi z}{2l}\,dz,\\ b_n &= \frac{2\nu l}{(2n+1)\pi a}a_n - \frac{4}{(2n+1)\pi a}\int_0^l V_t(z,0)\cos\frac{(2n+1)\pi z}{2l}\,dz, \end{aligned}\right\} \tag{5}$$

$$V(x,t) = \mathrm{Im}\left\{E_0\,\frac{e^{(\alpha+\beta i)x}+e^{-(\alpha+\beta i)x}}{e^{(\alpha+\beta i)l}+e^{-(\alpha+\beta i)l}}\,e^{i\omega t}\right\}, \tag{6}$$

$$\alpha+\beta i = \pm\sqrt{p\omega^2 - r - 2q\omega i},\quad p = LC,\quad 2q = RC+GL,\quad r = GR,$$

$$\nu = \frac{GL+CR}{2CL}.$$

144. The solution of the boundary-value problem

$$\frac{\partial^2 v}{\partial x^2} - CL\frac{\partial^2 v}{\partial t^2} - (CR+GL)\frac{\partial v}{\partial t} - GRv = 0,\quad 0<x<l,\quad 0<t<+\infty, \tag{1}$$

$$v(0,t) = 0\quad v(l,t) = E\sin\omega t,\quad 0<t<+\infty, \tag{2}$$

$$v(x,0) = 0,\quad v_t(x,0) = 0,\quad 0<x<l \tag{3}$$

is:

$$v(x, t) = V(x, t)+e^{-\nu t}\sum_{n=1}^{+\infty}\left(a_n \cos\frac{n\pi at}{l}+b_n \sin\frac{n\pi at}{l}\right)\sin\frac{n\pi x}{l}, \tag{4}$$

where

$$\left.\begin{aligned} a_n &= -\frac{2}{l}\int_0^l V(z, 0)\sin\frac{nz}{l}\,dz,\\ b_n &= \frac{\nu l}{n\pi a}a_n-\frac{2}{n\pi a}\int_0^l V_t(z, 0)\sin\frac{n\pi z}{l}\,dz,\end{aligned}\right\} \tag{5}$$

$$V(x, t) = \operatorname{Im}\left\{E\frac{e^{(\alpha+\beta i)x}-e^{-(\alpha+\beta i)x}}{e^{(\alpha+\beta i)l}-e^{-(\alpha+\beta i)l}}e^{i\omega t}\right\}, \tag{6}$$

$$\pm(\alpha+\beta i) = \sqrt{p\omega^2-r-2q\omega i},\quad p = CL,\quad 2q = CR+GL,\quad r = GR, \tag{7}$$

$$\nu = \frac{CR+GL}{2CL}. \tag{8}$$

145. From the boundary-value problem

$$\left.\begin{aligned} -\frac{\partial p}{\partial x} &= \left(\frac{\partial w}{\partial t}+2aw\right),\\ -\frac{\partial p}{\partial t} &= \lambda^2\frac{\partial w}{\partial x},\end{aligned}\right\} 0 < x < l,\quad 0 < t < +\infty, \tag{1}$$

$$p(0, t) = 0,\quad w(l, t)+h\frac{\partial w(l, t)}{\partial x} = Ae^{i\omega t},\quad 0 < t < +\infty \tag{2}$$

we find the steady-state vibrations of pressure of frequency ω in the section $x = l$.

$$p(l, t) = A\lambda r(\omega)R(\omega)e^{i(\omega t+\delta)}, \tag{3}$$

$$r(\omega) = \sqrt[4]{1+4\frac{a^2}{\omega^2}}, \tag{4}$$

$$R(\omega) = \frac{1}{\sqrt{\left(\dfrac{\sin 2\phi}{\cosh 2\psi-\cos 2\phi}-\beta\phi\right)^2+\left(\dfrac{\sinh^2\psi}{\cosh 2\psi-\cos 2\phi}+\beta\psi\right)^2}}, \tag{5}$$

$$\phi = \frac{l}{\lambda}\sqrt{\frac{\sqrt{\omega^4+2a^2\omega^2}+\omega^2}{2}},\quad \psi = \frac{l}{\lambda}\sqrt{\frac{\sqrt{\omega^4+2a^2\omega^2}-\omega^2}{2}},$$

$$\delta = \frac{\pi}{2}-\theta_1-\theta_2,\quad \tan\theta_1 = \frac{\psi}{\phi},\quad \tan\theta_2 = \frac{\dfrac{\sinh 2\psi}{\cosh 2\psi-\cos 2\phi}+\beta\psi}{\dfrac{\sinh 2\phi}{\cosh 2\psi-\cos 2\phi}-\beta\phi},\quad \beta = \frac{h}{\lambda}.$$

146. The solution of the boundary-value problem

$$u_{tt} = a^2 u_{xx} + \frac{1}{\rho}\Phi(x)t, \quad 0 < x < l, \quad 0 < t < +\infty, \tag{1}$$

$$u(0, t) = u(l, t), \quad 0 < t < +\infty, \tag{2}$$

$$u(x, 0) = u_t(x, 0) = 0, \quad 0 < x < l \tag{3}$$

is:

$$u(x, t) = \frac{1}{T_0 l}\left\{ l\int_0^x d\xi \int_0^\xi \Phi(z)\,dz - x\int_0^l d\xi \int_0^\xi \Phi(z)\,dz \right\} t +$$

$$+\sum_{n=1}^{+\infty} b_n \sin\frac{n\pi x}{l}\sin\frac{n\pi at}{l}, \tag{4}$$

where

$$b_n = -\frac{2}{n\pi a T_0 l}\int_0^l \left\{ l\int_0^x d\xi \int_0^\xi \Phi(z)\,dz - x\int_0^l d\xi \int_0^\xi \Phi(z)\,dz \right\} \sin\frac{n\pi x}{l}\,dx. \tag{5}$$

147. The solution of the boundary-value problem

$$u_{tt} = a^2 u_{xx}, \quad 0 < x < l, \quad 0 < t < +\infty, \tag{1}$$

$$u(0, t) = 0, \quad u_x(l, t) = \frac{A}{ES}t, \quad 0 < t < +\infty, \tag{2}$$

$$u(x, 0) = 0, \quad u_t(x, 0) = 0, \quad 0 < x < l \tag{3}$$

is:

$$u(x, t) = \frac{A}{ES}xt + \sum_{n=0}^{+\infty} b_n \sin\frac{(2n+1)\pi x}{2l}\sin\frac{(2n+1)\pi at}{2l}, \tag{4}$$

$$b_n = -\frac{4}{(2n+1)\pi a}\int_0^l \frac{Az}{ES}\sin\frac{(2n+1)\pi z}{2l}\,dz. \tag{5}$$

148. The solution of the boundary-value problem

$$u_{tt} = a^2 u_{xx} + \frac{1}{\rho}\Phi(x)t^m, \quad 0 < x < l, \quad 0 < t < +\infty, \quad m > -1, \tag{1}$$

$$u(0, t) = u(l, t) = 0, \quad 0 < t < +\infty, \tag{2}$$

$$u(x, 0) = 0, \quad u_t(x, 0) = 0, \quad 0 < x < l \tag{3}$$

is†:

$$u(x, t) = \sum_{n=1}^{+\infty} u_n(t)\sin\frac{n\pi x}{l}, \tag{4}$$

† See the method to the following problem.

$$u_n(t) = \frac{a_n}{\omega_n} \int_0^t \tau^m \sin \omega_n(t-\tau)\, d\tau, \qquad \omega_n = \frac{n\pi a}{l},$$

$$a_n = \frac{2}{l} \int_0^l \frac{\Phi(z)}{\rho} \sin \frac{n\pi z}{l}\, dz. \tag{5}$$

149. The solution of the boundary-value problem

$$u_{tt} = a^2 u_{xx}, \quad 0 < x < l, \quad 0 < t < +\infty, \tag{1}$$

$$u(0, t) = 0, \quad u_x(l, t) = \frac{A}{ES} t^m, \quad 0 < t < +\infty, \quad m > -1, \tag{2}$$

$$u(x, 0) = 0, \quad u_t(x, 0) = 0, \quad 0 < x < l \tag{3}$$

is:

$$u(x, t) = \frac{Axt^m}{ES} + \sum_{n=0}^{+\infty} u_n(t) \sin \frac{(2n+1)\pi x}{2l}, \tag{4}$$

where

$$u_n(t) = \frac{a_n}{\omega_n} \int_0^t \tau^{m-2} \sin \omega_n(t-\tau)\, d\tau, \qquad \omega_n = \frac{(2n+1)\pi a}{l}, \tag{5}$$

$$a_n = -\frac{2A}{(m-1)(m-2)ESl} \int_0^l z \sin \frac{(2n+1)\pi z}{2l}\, dz. \tag{6}$$

Method. In order to eliminate the inhomogeneity in the boundary condition we seek the solution of the boundary-value problem (1), (2) (3) in the form

$$u(x, t) = v(x, t) + \frac{Axt^{m-2}}{ES}, \tag{7}$$

which reduces to the boundary-value problem

$$v_{tt} = a^2 v_{xx} - \frac{Axt^{m-2}}{(m-1)(m-2)ES}, \quad 0 < x < l, \quad 0 < t < +\infty, \tag{8}$$

$$v(0, t) = 0, \quad v_x(l, t) = 0, \quad 0 < t < +\infty, \tag{9}$$

$$v(x, 0) = 0, \quad v_t(x, 0) = 0, \quad 0 < x < l. \tag{10}$$

The particular solution of the boundary-value problem

$$u_{tt} = a^2 u_{xx} + f(x, t), \quad 0 < x < l, \quad 0 < t < +\infty, \tag{1'}$$

$$\alpha_1 u_x(0, t) + \beta_1 u(0, t) = 0,$$
$$\alpha_2 u_x(l, t) + \beta_2 u(l, t) = 0, \quad 0 < t < +\infty \tag{2'}$$

may be sought in the form

$$u(x, t) = \sum_{n=1}^{+\infty} u_n(t) X_n(x), \tag{3'}$$

where $u_n(t)$ are functions, to be found, and $X_n(x)$ are eigenfunctions of the boundary-value problem

$$\left.\begin{aligned} X''(x)+\lambda^2 X(x) &= 0, \quad 0 < x < l, \\ \alpha_1 X'(0)+\beta_1 X(0) &= 0, \quad \alpha_2 X'(l)+\beta_2 X(l) = 0. \end{aligned}\right\} \tag{11}$$

It is also necessary to expand the term $f(x, t)$ a as series in the eigenfunctions of this problem, i.e. to represent it in the form

$$f(x, t) = \sum_{n=1}^{+\infty} f_n(t) X_n(x), \tag{12}$$

where

$$f_n(t) = \frac{1}{\|X_n\|^2} \int_0^l f(z, t) X_n(z)\, dz. \tag{13}$$

150.

(a) For $\omega \neq \omega_n = n\pi a/l$, $n = 1, 2, 3, \ldots$,

$$u(x, t) = \sum_{n=1}^{+\infty} \frac{a_n}{(\omega_n^2-\omega^2)\,\omega_n} (\omega_n \sin \omega t - \omega \sin \omega_n t) \sin \frac{n\pi x}{l}; \tag{1}$$

(b) For $\omega = \omega_{n_0} = n_0 \pi a/l$

$$u(x, t) = \sum_{n=1}^{+\infty} \frac{a_n}{(\omega_n^2-\omega^2)\,\omega_n} (\omega_n \sin \omega t - \omega \sin \omega_n t) \sin \frac{n\pi x}{l} +$$

$$+ \frac{a_{n_0}}{2\omega_{n_0}} (\sin \omega_{n_0} t - \omega_{n_0} t \cos \omega_{n_0} t) \sin \frac{n_0 \pi x}{l}, \tag{2}$$

$$a_n = \frac{2}{l} \int_0^l \frac{\Phi(z)}{\rho} \sin \frac{n\pi z}{l}\, dz. \tag{3}$$

Note. Here in contrast to the solution of problem 133 oscillations with the frequency of the applied force are given not in closed form, but in the form of a series.

151. $u(x, t) = \sum\limits_{n=1}^{+\infty} u_n(t) \sin \dfrac{n\pi x}{l},$ (1)

where

$$u_n(t) = \frac{a_n}{\tilde{\omega}_n} \int_0^t e^{-\nu(t-\tau)} \sin \omega \tau \sin \tilde{\omega}_n (t-\tau)\, d\tau, \tag{2}$$

$$a_n = \frac{2}{l} \int_0^l \frac{\Phi(x)}{\rho} \sin \frac{n\pi x}{l}\, dx,$$

$$\tilde{\omega}_n = \sqrt{\omega_n^2 - \nu^2}. \tag{3}$$

Here it is assumed that $\omega_n > \nu$. Finding an expression $u_n(t)$ for $\omega_n \leqslant \nu$ does not present any difficulty.

152. $u(x, t)$

$$= \frac{16F_0\tau\delta}{\pi^3\rho a}\sum_{n=1}^{+\infty}\frac{1}{n}\frac{\cos\frac{n\pi\delta}{l}\cos\frac{\omega_n\tau}{2}\sin\frac{n\pi x_0}{l}}{\left[1-\left(\frac{2n\delta}{l}\right)^2\right]\left[1-\left(\frac{na\tau}{l}\right)^2\right]}\sin\frac{n\pi x}{l}\sin\omega_n\left(t-\frac{\tau}{2}\right),$$

where

$$\omega_n = \frac{n\pi a}{l}.$$

153. The solution of the boundary-value problem

$$u_{tt} = a^2 u_{xx} + \frac{I}{\rho}\delta(x-x_0)\delta(t)\dagger, \quad 0 < x < l, \quad 0 < t < +\infty, \tag{1}$$

$$u(0, t) = u(l, t) = 0, \quad 0 < t < +\infty, \tag{2}$$

$$u(x, 0) = u_t(x, 0) = 0, \quad 0 < x < l \tag{3}$$

is:

$$u(x, t) = \frac{2I}{\rho l}\sum_{n=1}^{+\infty}\frac{1}{\omega_n}\sin\omega_n t\sin\frac{n\pi x_0}{l}\sin\frac{n\pi x}{l}, \quad \omega_n = \frac{n\pi a}{l}. \tag{4}$$

154. The solution of the boundary-value problem

$$u_{tt} = a^2 u_{xx} - 2\nu u_t + \frac{1}{\rho}\Phi(x)t, \quad 0 < x < l, \quad 0 < t < +\infty, \tag{1}$$

$$u(0, t) = u(l, t) = 0, \qquad 0 < t < +\infty, \tag{2}$$

$$u(x, 0) = u_t(x, 0) = 0, \quad 0 < x < +\infty \tag{3}$$

is:

$$u(x, t) = \sum_{n=1}^{+\infty} u_n(t)\sin\frac{n\pi x}{l}, \tag{4}$$

† $\delta(t)$ is the one-sided delta-function

$$\delta(t) = \lim_{n\to\infty}\phi_n(t), \quad \phi_n(t) = \begin{cases} 0, & -\infty < t < 0, \\ \frac{1}{n}, & 0 < t < n, \\ 0, & n < t < +\infty; \end{cases}$$

for more details on the delta-function see [7], page 292.

where

$$u_n(t) = \frac{a_n}{\omega_n}\int\limits_0^t \tau e^{-\nu(t-\tau)}\sin\tilde{\omega}_n(t-\tau)\,d\tau, \quad \tilde{\omega}_n = \sqrt{\omega_n^2-\nu^2}\,\dagger, \tag{5}$$

$$a_n = \frac{2}{l}\int\limits_0^l \frac{\Phi(z)}{\rho}\sin\frac{n\pi z}{l}\,dz, \quad \omega_n = \frac{n\pi a}{l}. \tag{6}$$

155. The solution of the boundary-value problem

$$u_{tt} = a^2u_{xx}-2\nu u_t+\frac{I}{\rho}\,\delta(x-x_0)\delta(t), \quad 0<x<l, \quad 0<t<+\infty, \tag{1}$$

$$u(0, t) = u(l, t) = 0, \quad 0<t<+\infty, \tag{2}$$

$$u(x, 0) = u_t(x, 0) = 0, \quad 0<x<l \tag{3}$$

is:

$$u(x, t) = \frac{2I}{\rho l}e^{-\nu t}\sum_{n=1}^{+\infty}\frac{1}{\tilde{\omega}_n}\sin\tilde{\omega}_n t\sin\frac{n\pi x_0}{l}\sin\frac{n\pi x}{l}, \tag{4}$$

where

$$\tilde{\omega}_n = \sqrt{\omega_n^2-\nu^2}, \quad \omega_n = \frac{n\pi a}{l}. \tag{5}$$

156. $$u(x, t) = -\frac{2Pl^4v_0}{\rho Sa\pi^4}\sum_{n=1}^{+\infty}\frac{1}{n^3}\,\frac{1}{a^2\pi^2n^2-v_0^2l^2}\sin\frac{n\pi x}{l}\sin\frac{n^2\pi^2at}{l^2}+$$

$$+\frac{2Pl^3}{\rho S\pi^2}\sum_{n=1}^{+\infty}\frac{1}{n^2}\,\frac{1}{a^2\pi^2n^2-v_0^2l^2}\sin\frac{n\pi x}{l}\sin\frac{n\pi v_0 t}{l}, \quad 0<t<\frac{l}{v_0},$$

$$u(x, t) = -\frac{2Pl^4v_0}{\rho Sa\pi^4}\sum_{n=1}^{+\infty}\frac{1}{n^3}\,\frac{1}{a^2\pi^2n^2-v_0^2l^2}\sin\frac{n\pi x}{l}\sin\frac{n^2\pi^2at}{l^2}, \quad \frac{l}{v_0}<t<+\infty.$$

Method. Use the delta-function.

157. $$u(x,t) = \frac{P_0}{\rho Sl\omega^2}\sum_{n=1}^{+\infty}\left\{\frac{\cos\frac{n^2\pi^2a^2}{l^2}t-\cos\left(1+\frac{n\pi v_0}{\omega l}\right)\omega t}{\left(\frac{n^2\pi^2a^2}{\omega l^2}\right)^2-\left(1+\frac{n\pi v_0}{\omega l}\right)^2}-\right.$$

$$\left.-\frac{\cos\frac{n^2\pi^2a^2}{l^2}t-\cos\left(1-\frac{n\pi v_0}{\omega l}\right)\omega t}{\left(\frac{n^2\pi^2a^2}{\omega l^2}\right)^2-\left(1-\frac{n\pi v_0}{\omega l}\right)^2}\right\}\sin\frac{n\pi x}{l}.$$

† It is assumed that $\omega_n > \nu$ for $n = 1, 2, 3, \ldots$ If for sufficiently small values of n, $\omega_n \leqslant \nu$, then the solution will contain terms with factors $\sinh\tilde{\omega}_n t$ and a term with a factor t.

158. (a) For $\omega \neq n^2\pi^2 a/l^2$, $n = 1, 2, 3, \ldots$,

$$u(x, t) = -\frac{2\omega P_0 l^5}{a\pi^2 \rho S} \sum_{n=1}^{+\infty} \frac{\sin \frac{n\pi x_0}{l}}{n^2(n^4\pi^4 a^2 - \omega^2 l^4)} \sin \frac{n\pi x}{l} \sin \frac{n^2\pi^2 at}{l^2} +$$

$$+ \sin \frac{2P_0 l^3}{\rho S} \sin \omega t \sum_{n=1}^{+\infty} \frac{\sin \frac{n\pi x_0}{l}}{n^4\pi^4 a^2 - \omega^2 l^4} \sin \frac{n\pi x}{l};$$

(b) for $\omega = n_0^2\pi^2 a/l^2$

$$u(x, t) = -\frac{2\omega P_0 l^5}{a\pi^2 \rho S} \sum_{n=1}^{+\infty} \frac{\sin \frac{n\pi x_0}{l}}{n^2(n^4\pi^4 a^2 - \omega^2 l^4)} \sin \frac{n\pi x}{l} \sin \frac{n^2\pi^2 at}{l^2} +$$

$$+ \frac{2P_0 l^3}{\rho S} \sin \omega t \sum_{\substack{n=1 \\ n \neq n_0}}^{+\infty} \frac{\sin \frac{n\pi x_0}{l}}{n^4\pi^4 a^2 - \omega^2 l^4} \sin \frac{n\pi x}{l} +$$

$$+ \frac{P_0}{\rho S l \omega^2} \sin \frac{n_0 \pi x_0}{l} \sin \omega t \sin \frac{n_0 \pi x}{l} - \frac{P_0}{\rho S l \omega} t \cos \omega t \sin \frac{n_0 \pi x_0}{l} \sin \frac{n_0 \pi x}{l}.$$

Indefinite increase of the amplitude of the forced vibrations with frequency $\omega = n_0{}^2\pi^2 a/l^2$ will occur only in the case where $\sin(n_0\pi x_0/l) \neq 0$, i.e. the point of application of the force does not coincide with any of the nodal harmonics corresponding to the number $\lambda_{n_0} = n_0\pi/l$.

Method. See the method for problem 149.

Note. The forced vibrations of frequency ω may be found in closed form in the same way as in the solution of problems 134 and 139.

For $\omega \neq n^2\pi^2 a/l$, $n = 1, 2, 3, \ldots$, for vibrations of frequency ω, thus, the following expression is obtained:

$$U(x, t) = \frac{2Pl^3}{a^2\pi^4 \rho S} \times \begin{cases} -\frac{\pi^4}{2\beta^3 l^3} \frac{\sinh \beta(l-x_0)}{\sinh \beta l} \sinh \beta x + \frac{\pi^4}{2\beta^3 l^3} \frac{\sin \beta(l-x_0)}{\sin \beta l} \sin \beta x, \\ \qquad 0 < x < x_0, \\ -\frac{\pi^4}{2\beta^3 l^3} \frac{\sinh \beta x_0}{\sinh \beta l} \sinh \beta(l-x) + \frac{\pi^4}{2\beta^3 l^3} \frac{\sin \beta x_0}{\sin \beta l} \sin \beta(l-x), \\ \qquad x_0 < x < l, \quad \beta^2 = \frac{\omega}{a}. \end{cases}$$

159. $$u(x, t) = \frac{2Pl^3}{\rho S} \sum_{n=1}^{+\infty} \frac{\sin \frac{n\pi x_0}{l} \sin(\omega t + \phi_n)}{\sqrt{(n^4\pi^4 a^2 - \omega^2 l^4)^2 + 4\nu^2\omega^2 l^8}} \sin \frac{n\pi x}{l},$$

where $\tan\phi_n = -2\nu\omega/(a^2\pi^4n^4-\omega^2l^4)$, and ν is the "coefficient of friction" in the equation

$$\frac{\partial^2 u}{\partial t^2}+a^2\frac{\partial^4 u}{\partial x^4}+2\nu\frac{\partial u}{\partial t}=0.$$

160. $$u(x,t)=\frac{2F_0l^3}{EJ}\sum_{n=1}^{+\infty}\frac{1-\cos\dfrac{\mu_n^2a^2}{l^2}t}{\mu_n^4(\sinh\mu_n+\sin\mu_n)}X_n(x), \tag{1}$$

where

$$X_n(x)=(\cosh\mu_n+\cos\mu_n)\left(\cosh\mu_n\frac{x}{l}-\sin\mu_n\frac{x}{l}\right)-$$
$$-(\sinh\mu_n+\sin\mu_n)\left(\cosh\mu_n\frac{x}{l}-\cos\mu_n\frac{x}{l}\right), \tag{2}$$

μ_n are the positive roots of the equation

$$\cosh\mu\cos\mu=-1. \tag{3}$$

161. For $t<T$ the answer coincides with the answer to the preceding problem. For $t>T$

$$u(x,t)=\frac{2F_0l^3}{EJ}\sum_{n=1}^{\infty}\frac{\cos\dfrac{\mu_n^2a^2}{l^2}(t-T)-\cos\dfrac{\mu_n^2a^2t}{l^2}}{\mu_n^4(\sinh\mu_n+\sin\mu_n)}X_n(x),$$

where μ_n and $X_n(x)$ have the same meaning as in the preceding problem.

162. $$u(x,t)=\frac{2F_0la^2}{\omega EJ}\sum_{n=1}^{+\infty}\frac{X_n(x)}{\mu_n^2(\sinh\mu_n+\sin\mu_n)}\;\frac{\sin\dfrac{\mu_n^2a^2}{l^2}t-\dfrac{\mu_n^2a^2}{\omega l^2}\sin\omega t}{1-\left(\dfrac{\mu_n^2a^2}{\omega l^2}\right)^2},$$

where μ_n and $X_n(x)$ have the same meaning as in problem 160.

163. $u(x,t)$

$$=\frac{f_0l^2a^2}{\omega EJ}\sum_{n=1}^{+\infty}\frac{\sinh\mu_n-2\cosh\mu_n\sin\mu_n+\sin\mu_n}{\mu_n^3\sinh\mu_n\sin^2\mu_n}\;\frac{\sin\dfrac{\mu_n^2a^2}{l^2}t-\dfrac{\mu_n^2a^2}{\omega l^2}\sin\omega t}{1-\left(\dfrac{\mu_n^2a^2}{\omega l^2}\right)}X_n(x),$$

where

$$X_n(x)=\sinh\mu_n\sin\frac{\mu_nx}{l}-\sin\mu_n\sinh\frac{\mu_nx}{l},$$

μ_n are the positive roots of the equation $\tan\mu=\tanh\mu$ $(\mu_1<\mu_2<\ldots)$.

4. Vibrations with Inhomogeneous Media and Other Conditions Leading to Equations with Variable Coefficients; Calculations with Concentrated Forces and Masses

164. *Solution.* The longitudinal displacement $u(x, t)$ of points of the rod is a solution of the boundary-value problem

$$\rho(x)u_{tt} = (E(x)u_x)_x, \quad 0 < x < x_0, \quad x_0 < x < l, \quad 0 < t < +\infty, \tag{1}$$

$$u(0, t) = u(l, t) = 0, \quad 0 < t < +\infty, \tag{2}$$

$$u(x_0-0, t) = u(x_0+0, t), \quad E(x_0-0)u_x(x_0-0, t) = E(x_0+0)u_x(x_0+0, t), \tag{2'}$$

$$u(x, 0) = \phi(x) = \begin{cases} \dfrac{h}{x_0}x, & 0 < x < x_0, \\ \dfrac{h(l-x)}{l-x_0}, & x_0 < x < l, \end{cases} \tag{3}$$

$$u_t(x, 0) = \psi(x) \equiv 0, \quad 0 < x < l, \tag{3'}$$

$$E(x) = \begin{cases} \overline{E}, & 0 < x < x_0, \\ \overline{\overline{E}}, & x_0 < x < l, \end{cases} \qquad \rho(x) = \begin{cases} \overline{\rho}, & 0 < x < x_0, \\ \overline{\overline{\rho}}, & x_0 < x < l, \end{cases} \tag{4}$$

$\overline{\rho}$, $\overline{\overline{\rho}}$, $\overline{E}$, $\overline{\overline{E}}$ are constants.

We seek particular solutions of the boundary-value problem (1), (2), (4) in the form

$$u(x, t) = X(x)T(t). \tag{5}$$

Substituting (5) into (1), (2) after separation of the variables, we obtain:

$$T''(t)+\omega^2 T(t) = 0, \quad 0 < t < +\infty, \tag{6}$$

$$(E(x)X'(x))'+\omega^2\rho(x)X(x) = 0, \quad 0 < x < l, \tag{7}$$

$$X(0) = X(l) = 0, \quad X(x_0-0) = X(x_0+0), \tag{7'}$$

$$\overline{E}X'(x_0-0) = \overline{\overline{E}}X'(x_0+0). \tag{7''}$$

From the general theory it is known† that the boundary-value problem (7'), (7'') has an infinite series of characteristic frequencies

$$\omega_1 < \omega_2 < \ldots < \omega_n < \ldots,$$

and corresponding to them a set of eigenfunctions

$$X_1(x), \quad X_2(x), \quad \ldots, \quad X_n(x), \quad \ldots,$$

† See [7], pages 468–469.

orthogonal with weight $\rho(x)$ in the segment $0 \leqslant x \leqslant l$. The solution of equation (7) satisfying conditions (7′) has the form

$$X(x) = \begin{cases} \dfrac{\sin \dfrac{\omega}{\bar{a}} x}{\sin \dfrac{\omega}{\bar{a}} x_0} & \text{for} \quad 0 < x < x_0, \quad \bar{a} = \sqrt{\dfrac{\bar{E}}{\bar{\rho}}}, \\ \dfrac{\sin \dfrac{\omega}{\bar{\bar{a}}} (l-x)}{\sin \dfrac{\omega}{\bar{\bar{a}}} (l-x_0)} & \text{for} \quad x_0 < x < l, \quad \bar{\bar{a}} = \sqrt{\dfrac{\bar{\bar{E}}}{\bar{\bar{\rho}}}}. \end{cases} \tag{8}$$

Satisfying condition (7″) we obtain the transcendental equation

$$\frac{1}{\sqrt{\bar{E}\rho}} \tan \frac{\omega}{\bar{a}} x_0 = \frac{1}{\sqrt{\bar{\bar{E}}\bar{\bar{\rho}}}} \tan \frac{\omega}{\bar{\bar{a}}} (x_0 - l) \tag{9}$$

for determining the characteristic frequencies ω_n.

Assuming $\omega = \omega_n$ in (8), we obtain the eigenfunctions of our boundary-value problem

$$X_n(x) = \begin{cases} \dfrac{\sin \dfrac{\omega_n}{\bar{a}} x}{\sin \dfrac{\omega_n}{\bar{a}} x_0} & \text{for} \quad 0 < x < x_0, \\ \dfrac{\sin \dfrac{\omega_n}{\bar{\bar{a}}} (l-x)}{\sin \dfrac{\omega_n}{\bar{\bar{a}}} (l-x_0)} & \text{for} \quad x_0 < x < l. \end{cases} \tag{10}$$

The square of the norm of the eigenfunction equals

$$X_n\|^2 = \int_0^l \rho(x) X_n^2(x)\, dx = \bar{\rho} \int_0^{x_0} \frac{\sin^2 \dfrac{\omega_n}{\bar{a}} x}{\sin^2 \dfrac{\omega_n}{\bar{a}} x_0}\, dx + \bar{\bar{\rho}} \int_{x_0}^l \frac{\sin^2 \dfrac{\omega_n}{\bar{\bar{a}}} (l-x)}{\sin^2 \dfrac{\omega_n}{\bar{\bar{a}}} (l-x_0)}\, dx$$

$$= \frac{\bar{\rho} x_0}{2 \sin^2 \dfrac{\omega_n}{\bar{a}} x_0} + \frac{\bar{\bar{\rho}} (l-x_0)}{2 \sin^2 \dfrac{\omega_n}{\bar{\bar{a}}} (x_0 - l)}, \tag{11}$$

$$u(x, t) = \sum_{n=1}^{+\infty} a_n X_n(x) \cos \omega_n t, \tag{12}$$

$$a_n = \frac{1}{\|X_n\|^2} \int_0^l \rho(x) \phi(x) X_n(x)\, dx = \frac{h}{\omega_n^2 \|X_n\|^2} \left(\frac{\bar{E}}{x_0} + \frac{\bar{\bar{E}}}{l-x_0)} \right). \tag{13}$$

165. $u(x,t) = X(x)\sin\omega t, \quad X(x) = \begin{cases} \dfrac{\sin\dfrac{\omega}{a}x}{\sin\dfrac{\omega}{a}x_0} & \text{for} \quad 0 < x < x_0, \\[2ex] \dfrac{\dfrac{\omega}{\bar{\bar{a}}}\cos\dfrac{\omega}{\bar{\bar{a}}}(l-x)+F_0\sin\dfrac{\omega}{\bar{\bar{a}}}(x-x_0)}{\dfrac{\omega}{\bar{\bar{a}}}\cos\dfrac{\omega}{\bar{\bar{a}}}(l-x_0)} & \\ & \text{for} \quad x_0 < x < l. \end{cases}$

166. $u(x,t) = \sum\limits_{n=1}^{+\infty} (a_n\cos\omega_n t + b_n\sin\omega_n t)X_n(x),$

$$a_n = \frac{1}{||X_n||^2}\int\limits_0^l \rho(x)\phi(x)X_n(x)\,\mathrm{d}x, \qquad b_n = \frac{1}{\omega_n||X_n||^2}\int\limits_0^l \rho(x)\psi(x)X_n(x)\,\mathrm{d}x,$$

$$X_n(x) = \begin{cases} \dfrac{\sin\dfrac{\omega_n}{a}x}{\sin\dfrac{\omega_n}{a}x_0} & \text{for} \quad 0 < x < x_0, \\[2ex] \dfrac{\dfrac{\omega_n}{\bar{\bar{a}}}\cos\dfrac{\omega_n}{\bar{\bar{a}}}(l-x)+h\sin\dfrac{\omega_n}{\bar{\bar{a}}}(l-x)}{\dfrac{\omega_n}{\bar{\bar{a}}}\cos\dfrac{\omega_n}{\bar{\bar{a}}}(l-x_0)+h\sin\dfrac{\omega_n}{\bar{\bar{a}}}(l-x_0)} & \text{for} \quad x_0 < x < l, \end{cases}$$

ω_n are the positive roots of the transcendental equation

$$p\omega = -\tan q\omega,$$

where

$$p = \frac{1}{h\bar{\bar{a}}}, \qquad q = \frac{l-x_0}{\bar{\bar{a}}} + \frac{x_0}{\bar{a}}.$$

167. $u(x,t) = \sum\limits_{n=1}^{+\infty} a_n X_n(x)\cos a\lambda_n t, \qquad 0 < x < l, \quad 0 < t < +\infty, \qquad (1)$

$$a^2 = \frac{T_0}{\rho}, \tag{2}$$

$$X_n(x) = \begin{cases} \dfrac{\sin\lambda_n x}{\sin\lambda_n x_0}, & 0 < x < x_0, \\[2ex] \dfrac{\sin\lambda_n(l-x)}{\sin\lambda_n(l-x_0)}, & x_0 < x < l, \end{cases} \tag{3}$$

λ_n are the eigenvalues of the boundary-value problem, being roots of the equation

$$\cot \lambda_n x_0 - \cot \lambda_n(l-x_0) = \frac{M}{\rho}\lambda_n. \tag{4}$$

The eigenfunctions $X_n(x)$ are orthogonal in the segment $0 < x < l$ with weight $\rho(x) = \rho + M\delta(x-x_0)$, where ρ is the linear mass density of the string, and $\delta(x-x_0)$ is the delta-function; thus,

$$\int_0^l \rho(x)X_m(x)X_n(x)\,dx = \rho\int_0^l X_m(x)X_n(x)\,dx + MX_m(x_0)X_n(x_0) = 0 \tag{5}$$

$$\text{if} \quad m \neq n,$$

the square of the norm of the eigenfunction equals

$$\|X_n\|^2 = \int_0^l \rho(x)X_n^2(x)\,dx = \frac{\rho x_0}{2\sin^2\lambda_n x_0} + \frac{\rho(l-x_0)}{2\sin^2\lambda_n(l-x_0)} + \frac{M}{2}, \tag{6}$$

$$a_n = \frac{\int_0^l \rho(x)\phi(x)X_n(x)\,dx}{\|X_n\|^2} = \frac{\rho\int_0^l \phi(x)X_n(x)\,dx + M\phi(x_0)X_n(x_0)}{\|X_n\|^2}, \tag{7}$$

from which we obtain:

$$a_n = \frac{\dfrac{\rho h}{\lambda_n^2}\left(\dfrac{1}{x_0} + \dfrac{1}{l-x_0}\right)}{\|X_n\|^2}. \tag{8}$$

Method. In calculating (6) and (8) it is necessary to use (4). Formulae (6) and (7) may be derived without using the delta-function, as was done in [7], pages 157–161.

168. $u(x,t) = \sum_{n=1}^{+\infty} \{a_n \cos\omega_n t + b_n \sin\omega_n t\} X_n(x)$,

$$X_n(x) = \begin{cases} \dfrac{\sin\dfrac{\omega_n}{\bar{a}}x}{\sin\dfrac{\omega_n}{\bar{a}}x_0}, & 0 \leqslant x \leqslant x_0, \\[2ex] \dfrac{\cos\dfrac{\omega_n}{\bar{\bar{a}}}(l-x)}{\cos\dfrac{\omega_n}{\bar{\bar{a}}}(l-x_0)}, & x_0 \leqslant x \leqslant l, \end{cases}$$

ω_n are roots of the equation

$$\bar{S}\sqrt{\bar{E}\bar{\rho}}\cot\frac{\omega_n}{\bar{a}}x_0 - \bar{\bar{S}}\sqrt{\bar{\bar{E}}\bar{\bar{\rho}}}\tan\frac{\omega_n}{\bar{\bar{a}}}(l-x_0) = M\omega_n,$$

$$a_n = \frac{\dfrac{\overline{S}\overline{\rho}}{\sin^2\dfrac{\omega_n}{\overline{a}}x_0}\displaystyle\int_0^{x_0}\phi(x)\sin^2\frac{\omega_n}{\overline{a}}x\,dx+\dfrac{\overline{\overline{S}}\overline{\overline{\rho}}}{\cos^2\dfrac{\omega_n}{\overline{\overline{a}}}(l-x_0)}\displaystyle\int_{x_0}^{l}\phi(x)\cos^2\frac{\omega_n}{\overline{\overline{a}}}(l-x)\,dx}{||X_n||^2},$$

$$b_n = \frac{\dfrac{\overline{S}\overline{\rho}}{\sin^2\dfrac{\omega_n}{\overline{a}}x_0}\displaystyle\int_0^{x_0}\psi(x)\sin^2\frac{\omega_n}{\overline{a}}x\,dx+\dfrac{\overline{\overline{S}}\overline{\overline{\rho}}}{\cos^2\dfrac{\omega_n}{\overline{\overline{a}}}(l-x_0)}\displaystyle\int_{x_0}^{l}\psi(x)\cos^2\frac{\omega_n}{\overline{\overline{a}}}(l-x)\,dx}{\omega_n||X_n||^2},$$

$$||X_n||^2 = \frac{\overline{S}\overline{\rho}}{2\sin^2\dfrac{\omega_n}{\overline{a}}x_0}+\frac{\overline{\overline{S}}\overline{\overline{\rho}}}{2\cos^2\dfrac{\omega_n}{\overline{\overline{a}}}(l-x_0)}+\frac{M}{2}.$$

169. The solution of the boundary-value problem

$$u_{tt} = a^2u_{xx}, \quad 0<x<l, \quad 0<t<+\infty, \tag{1}$$

$$u(0,t)=0, \quad u_{tt}(l,t) = -c^2u_x(l,t), \quad 0<t<+\infty, \tag{2}$$

$$u(x,0)=\phi(x), \quad u_t(x,0)=\psi(x), \quad 0<x<l, \tag{3}$$

where

$$c^2 = \frac{KG}{M}, \quad a^2 = \frac{KG}{J},$$

is:

$$u(x,t) = \sum_{n=1}^{+\infty}(a_n\cos a\lambda_n t+b_n\sin a\lambda_n t)\sin\lambda_n x. \tag{4}$$

λ_n, the eigenvalues of the boundary-value problem, are roots of the equation

$$\cot\lambda_n l = \frac{M}{J}\lambda_n, \tag{5}$$

and the eigenfunctions $X_n(x) = \sin\lambda_n x$ satisfy the orthogonality condition†

$$MX_m(l)X_n(l)+\int_0^l JX_m(x)X_n(x)\,dx = 0, \quad m\neq n, \tag{6}$$

$$a_n = \frac{M\phi(l)X_n(l)+\int_0^l J\phi(x)X_n(x)\,dx}{MX_n^2(l)+\int_0^l JX_n^2(x)\,dx}, \tag{7}$$

$$b_n = \frac{M\psi(l)X_n(l)+\int_0^l J\psi(x)X_n(x)\,dx}{\left[MX_n^2(l)+\int_0^l JX_n^2(x)\,dx\right]a\lambda_n}. \tag{8}$$

† See the method for problem 167.

170. $$u(x,t) = \frac{H-l}{H-x}\,\frac{A}{\dfrac{\omega}{a}\cos\dfrac{\omega}{a}l+\dfrac{1}{H-l}\sin\dfrac{\omega}{a}l}\sin\frac{\omega}{a}x\sin\omega t.$$

171. $$u(x,t) = \frac{1}{l}\sum_{n=1}^{+\infty}\frac{J_0\left(\mu_n\sqrt{\dfrac{x}{l}}\right)}{J_1^2(\mu_n)}\cos\frac{a\mu_n t}{2\sqrt{l}}\int_0^l \phi(\xi)J_0\left(\mu_n\sqrt{\frac{\xi}{l}}\right)\mathrm{d}\xi,$$

$\phi(x) = u(x,0)$, $J_0(x)$, $J_1(x)$ are Bessel functions of zero and first order of the first kind, μ_n are the positive roots of the equation $J_0(\mu) = 0$.

172. $$u(x,t) = \sum_{n=1}^{+\infty}(a_n\cos a\lambda_n t+b_n\sin a\lambda_n t)J_0\left(\mu_n\sqrt{\frac{x}{l}}\right),$$

$$a_n = \frac{1}{lJ_1^2(\mu_n)}\int_0^l \phi(\xi)J_0\left(\mu_n\sqrt{\frac{\xi}{l}}\right)\mathrm{d}\xi,$$

$$b_n = \frac{1}{al\lambda_n J_1^2(\mu_n)}\int_0^l \psi(\xi)J_0\left(\mu_n\sqrt{\frac{\xi}{l}}\right)\mathrm{d}\xi,$$

$$\phi(x) = u(x,0),\qquad \psi(x) = u_t(x,0),\qquad \lambda_n = \sqrt{\frac{\mu_n^2}{4l}-\frac{\omega^2}{a^2}};$$

μ_n have the same meaning as in the answer to the preceding problem.

173. $$u(x,t) = \sum_{n=1}^{+\infty}\left[a_n\cos\sqrt{2n(2n-1)}\,at+b_n\sin\sqrt{2n(2n-1)}\,at\right]P_{2n-1}\left(\frac{x}{l}\right),$$

where

$$a_n = \frac{4n-1}{l}\int_0^l \phi(\xi)P_{2n-1}\left(\frac{\xi}{l}\right)\mathrm{d}\xi,\qquad b_n = \frac{4n-1}{\sqrt{2n(2n-1)}\,al}\int_0^l \psi(\xi)P_{2n-1}\left(\frac{\xi}{l}\right)\mathrm{d}\xi,$$

$$P_n(x) = \frac{1}{2^n n!}\frac{\mathrm{d}^n}{\mathrm{d}x^n}[(x^2-1)^n]\quad\text{is a Legendre polynomial,}$$

$$\phi(x) = u(x,0)\qquad \psi(x) = u_t(x,0).$$

§ 4. Method of Integral Representations

1. The Method of the Fourier Integral

We recall that given certain restrictions on $f(x)$ the Fourier integral relation

$$f(x) = \frac{1}{2\pi}\int_{-\infty}^{+\infty}\mathrm{d}\lambda\int_{-\infty}^{+\infty}f(\xi)e^{i\lambda(x-\xi)}\,\mathrm{d}\xi, \tag{1}$$

is valid, where

$$f'(x) = \frac{1}{2\pi} \int_{-\infty}^{+\infty} i\lambda \, d\lambda \int_{-\infty}^{+\infty} f(\xi) e^{i\lambda(x-\xi)} \, d\xi, \tag{2}$$

i.e. differentiation of the integral with respect to a parameter under the integral sign is possible, and

$$F(x) = \frac{1}{2\pi} \int_{-\infty}^{+\infty} \frac{d\lambda}{i\lambda} \int_{-\infty}^{+\infty} f(\xi) e^{i\lambda(x-\xi)} \, d\xi \dagger, \tag{3}$$

where $F(x)$ is the integral of $f(x)$.

The solution of the equation

$$u_{tt} = a^2 u_{xx}, \quad -\infty < x < +\infty, \quad 0 < t < +\infty \tag{4}$$

may be sought in the form

$$u(x, t) = \frac{1}{2\pi} \int_{-\infty}^{+\infty} d\lambda \int_{-\infty}^{+\infty} U(\xi, t) e^{i\lambda(x-\xi)} \, d\xi. \tag{5}$$

Substitution of (5) into (4) gives:

$$\frac{1}{2\pi} \int_{-\infty}^{+\infty} d\lambda \int_{-\infty}^{+\infty} \left\{ \frac{d^2U}{dt^2} + a^2\lambda^2 U \right\} e^{i\lambda(x-\xi)} \, d\xi = 0. \tag{6}$$

In order to satisfy (6) it is sufficient that the equation

$$\frac{d^2U}{dt^2} + a^2\lambda^2 U = 0, \tag{7}$$

be satisfied, from which we find:

$$U(\xi, t) = A(\xi) e^{ia\lambda t} + B(\xi) e^{-ia\lambda t}, \tag{8}$$

where $A(\xi)$ and $B(\xi)$ are arbitrary functions of the parameter ξ.

Substitution of the expression obtained in (5) according to (1) gives the familiar solution as a sum of propagating waves

$$u(x, t) = \frac{1}{2\pi} \int_{-\infty}^{+\infty} d\lambda \int_{-\infty}^{+\infty} \{A(\xi) e^{i\lambda(x+at-\xi)} + B(\xi) e^{i\lambda(x-at-\xi)}\} \, d\xi$$

$$= A(x+at) + B(x-at). \tag{9}$$

Similarly the Fourier integral can be used to solve other problems, connected with the wave equation.

---

† Here the integral is understood in the sense of the principal value.

The following scheme for applying the Fourier integral to solve boundary-value problems on the infinite line $-\infty < x < +\infty$ and the semi-infinite line $0 < x < +\infty$ is widely used.

The Fourier transform of $f(x)$ over the line $-\infty < x < +\infty$ is defined by the equation

$$\bar{f}(\lambda) = \frac{1}{\sqrt{2\pi}} \int_{-\infty}^{+\infty} f(\xi) e^{-i\lambda\xi} d\xi. \tag{10}$$

Because of formula (1) the original function $f(x)$ can be expressed in terms of $\bar{f}(\lambda)$ by means of the relation

$$f(x) = \frac{1}{\sqrt{2\pi}} \int_{-\infty}^{+\infty} \bar{f}(\lambda) e^{i\lambda x} d\lambda. \tag{10'}$$

The transition from $f(x)$ to $\bar{f}(\lambda)$ in formula (10) is called the Fourier integral transform; obviously, the transformations (10) and (10′) are reciprocal.

Over the semi-infinite line $0 < x < +\infty$ it is possible to consider the Fourier cosine transform† of the function $f(x)$

$$\bar{f}^{(c)}(\lambda) = \sqrt{\frac{2}{\pi}} \int_{0}^{+\infty} f(\xi) \cos \lambda\xi \, d\xi, \tag{11}$$

The original function $f(x)$ is given by the formula

$$f(x) = \sqrt{\frac{2}{\pi}} \int_{0}^{+\infty} \bar{f}^{(c)}(\lambda) \cos \lambda x \, d\lambda. \tag{11'}$$

One may also consider the Fourier sine transform‡

$$\bar{f}^{(s)}(\lambda) = \sqrt{\frac{2}{\pi}} \int_{0}^{+\infty} f(\xi) \sin \lambda\xi \, d\xi, \tag{12}$$

in which case the original function is given by the formula

$$f(x) = \sqrt{\frac{2}{\pi}} \int_{0}^{+\infty} \bar{f}^{(s)}(\xi) \sin x\xi \, d\xi. \tag{12'}$$

It is also possible to consider transforms with other kernels. For details see the special literature.

In order to solve a boundary-value problem for $u(x, t)$ by means of the Fourier integral transform, with respect to the variable x one tries to find the Fourier

† The Fourier integral transform with kernel $\cos \lambda\xi$.

‡ The Fourier integral transform with kernel $\sin \lambda\xi$.

transform of this function. After this the original function is found by means of the inverse Fourier transform, i.e. the function $u(x, t)$ is determined by its Fourier transform. As the kernel of the Fourier integral transform for problems on the semi-infinite line it is necessary to choose a particular solution $X(x, t)$ of the equation, resulting from a separation of the variables of the fundamental equation of the given boundary-value problem, which satisfies the boundary condition of the problem, if the condition is homogeneous, or which satisfies the corresponding homogeneous boundary condition, if the boundary condition of the problem is inhomogeneous.

174. $$u(x, t) = \frac{1}{2a}\int_0^t d\tau \int_{x-a(t-\tau)}^{x+at(t-\tau)} f(\xi, \tau)\,d\xi. \qquad (1)$$

Method. Substituting in equation (1) the conditions of the problem

$$u(x, t) = \frac{1}{2\pi}\int_{-\infty}^{+\infty} d\lambda \int_{-\infty}^{+\infty} U(\xi, t)e^{i\lambda(x-\xi)}\,d\xi,$$

$$f(x, t) = \frac{1}{2\pi}\int_{-\infty}^{+\infty} d\lambda \int_{-\infty}^{+\infty} f(\xi, t)e^{i\lambda(x-\xi)}\,d\xi,$$

we arrive at the equation

$$\frac{d^2U}{dt^2} + a^2\lambda^2 U = f(\xi, t).$$

Solving it for the initial conditions

$$U(\xi, 0) = 0, \qquad \frac{dU(\xi, 0)}{dt} = 0,$$

we obtain:

$$U(\xi, t) = \frac{1}{a\lambda}\int_0^t f(\xi, \tau)\sin a(t-\tau)\,d\tau.$$

Putting $\sin a(t-\tau)$ in complex form† and substituting the expression $U(\xi, t)$ obtained in the integral

$$u(x, t) = \frac{1}{2\pi}\int_{-\infty}^{+\infty} d\lambda \int_{-\infty}^{+\infty} U(\xi, t)e^{i\lambda(x-\xi)}\,d\xi$$

by (3) of the introduction to the present section, we obtain formula (1).

† $\sin\phi = \dfrac{e^{i\phi} - e^{-i\phi}}{2i}$.

175. $u(x, t) = \dfrac{\phi(x-at)+\phi(x+at)}{2} +$

$$+\frac{ct}{2a}\int\limits_{x-at}^{x+at} \frac{I_1\left(c\sqrt{t^2-\dfrac{(x-\xi)^2}{a^2}}\right)}{\sqrt{t^2-\dfrac{(x-\xi)^2}{a^2}}}\phi(\xi)\,d\xi+$$

$$+\frac{1}{2a}\int\limits_{x-at}^{x+at} I_0\left(c\sqrt{t^2-\frac{(x-\xi)^2}{a^2}}\right)\psi(\xi)\,d\xi, \quad (1)$$

where $I_0(z)$ and $I_1(z)$ are "modified" Bessel functions of zero and first order; they can be represented by the series

$$I_0(z) = J_0(iz) = \sum_{k=0}^{+\infty} \frac{1}{(k!)^2}\left(\frac{z}{2}\right)^{2k+1}, \quad (2)$$

$$I_1(z) = -iJ_1(iz) = \sum_{k=0}^{+\infty} \frac{1}{k!(k+1)!}\left(\frac{z}{2}\right)^{2k+1}, \quad (3)$$

where

$$I_0'(z) = I_1(z). \quad (4)$$

The modified Bessel function of νth order

$$I_\nu(x) = \sum_{k=0}^{+\infty} \frac{1}{\Gamma(k+1)\Gamma(k+\nu+1)}\left(\frac{x}{2}\right)^{2k+\nu} \quad (5)$$

is the solution of the differential equation

$$y''+\frac{1}{x}y'-\left(1+\frac{\nu^2}{x^2}\right)y = 0\ \dagger \quad (6)$$

which is bounded as $x \to 0$.

Solution. The solution of the boundary-value problem

$$u_{tt} = a^2u_{xx}+c^2u, \quad -\infty < x < +\infty, \quad 0 < t < +\infty, \quad (7)$$

$$u(x, 0) = \phi(x), \quad u_t(x, 0) = \psi(x), \quad -\infty < x < +\infty \quad (8)$$

may be found in the form

$$u(x, t) = \frac{1}{2\pi}\int\limits_{-\infty}^{+\infty} d\lambda \int\limits_{-\infty}^{+\infty} U(\xi, t)e^{i\lambda(x-\xi)}\,d\xi. \quad (9)$$

---

† For more detail see [7], page 656.

Substituting (9) in (7) we obtain the equation

$$\frac{d^2U(\xi, t)}{dt^2} + (a^2\lambda^2 - c^2)U(\xi, t) = 0. \tag{10}$$

Its solution, which by (8) and (9) satisfies the initial conditions

$$U(\xi, 0) = \phi(\xi), \qquad U_t(\xi, 0) = \psi(\xi), \tag{11}$$

has the form

$$U(\xi, t) = \phi(\xi) \cos t \sqrt{a^2\lambda^2 - c^2} + \psi(\xi) \frac{\sin t \sqrt{a^2\lambda^2 - c^2}}{\sqrt{a^2\lambda^2 - c^2}}. \tag{12}$$

Substituting (12) in (9) we obtain:

$$u(x, t) = \frac{1}{2\pi} \int_{-\infty}^{+\infty} d\lambda \int_{-\infty}^{+\infty} \phi(\xi) \cos t \sqrt{a^2\lambda^2 - c^2}\, e^{i\lambda(x-\xi)} d\xi +$$

$$+ \frac{1}{2\pi} \int_{-\infty}^{+\infty} d\lambda \int_{-\infty}^{+\infty} \psi(\xi) \frac{\sin t \sqrt{a^2\lambda^2 - c^2}}{\sqrt{a^2\lambda^2 - c^2}} e^{i\lambda(x-\xi)} d\xi = u_1(x, t) + u_2(x, t). \tag{13}$$

From the theory of cylindrical functions it is known that

$$\frac{\sin r}{r} = \frac{1}{2} \int_0^{\pi} J_0(r \sin\phi \sin\theta) e^{ir\cos\phi\cos\theta} \sin\theta\, d\theta. \tag{14}$$

In this equation we make the substitution

$$r \cos\phi = -a\lambda t, \qquad r \sin\phi = ict, \qquad r^2 = t^2(a^2\lambda^2 - c^2). \tag{15}$$

Then the equation

$$\frac{\sin t \sqrt{a^2\lambda^2 - c^2}}{t \sqrt{a^2\lambda^2 - c^2}} = \frac{1}{2} \int_{-at}^{+at} J_0\left(ict \sqrt{1 - \frac{\beta^2}{a^2t^2}}\right) e^{-i\lambda\beta} \frac{d\beta}{at}$$

$$= \frac{1}{2a} \int_{-at}^{+at} I_0\left(c \sqrt{t^2 - \frac{\beta^2}{a^2}}\right) e^{-i\lambda\beta} \frac{d\beta}{t}$$

is obtained. Hence,

$$u_2 = \frac{1}{4\pi} \int_{-\infty}^{+\infty} d\lambda \int_{-\infty}^{+\infty} \left\{\psi(\xi) \int_{-at}^{+at} I_0\left(c \sqrt{t^2 - \frac{\beta^2}{a^2}}\right) e^{i\lambda(x-\xi-\beta)} d\beta\right\} d\xi. \tag{15'}$$

Let us assume

$$\Phi(\beta) = \begin{cases} 0 & \text{for} \quad \left|\dfrac{\beta}{a}\right| > |t|, \\ I_0\left(c \sqrt{t^2 - \dfrac{\beta^2}{a^2}}\right) & \text{for} \quad \left|\dfrac{\beta}{a}\right| < |t|. \end{cases} \tag{16}$$

Then

$$\int_{-at}^{at} I_0\left(c\sqrt{t^2-\frac{\beta^2}{a^2}}\right)e^{i\lambda(x-\beta)}\,d\beta = \int_{-\infty}^{+\infty} \Phi(\beta)e^{i\lambda(x-\beta)}\,d\beta. \tag{17}$$

On the right hand side of (15) we first carry out an integration with respect to λ and β. By the Fourier integral relation† we obtain:

$$\frac{1}{2\pi}\int_{-\infty}^{+\infty} d\lambda \int_{-\infty}^{+\infty} \phi(\beta)\,e^{i\lambda(z-\beta)}\,d\beta = \phi(z)$$

at points of continuity of $\phi(z)$. In our case $z = x-\xi$.

By (17)

$$\phi(x-\xi) = \begin{cases} I_0\left(c\sqrt{t^2-\dfrac{(x-\xi)^2}{a^2}}\right) & \text{for } x-at < \xi < x+at, \\ 0 & \text{for } -\infty < \xi < x-at,\ x+at < \xi < +\infty. \end{cases}$$

Therefore

$$u_2(x,t) = \frac{1}{2a}\int_{-\infty}^{+\infty} \psi(\xi)\Phi(x-\xi)\,d\xi = \frac{1}{2a}\int_{x-at}^{x+at} \psi(\xi)I_0\left(c\sqrt{t^2-\frac{(x-\xi)^2}{a^2}}\right)d\xi. \tag{18}$$

Let us recall the original expressions for

$$u_2(x,t) = \frac{1}{2\pi}\int_{-\infty}^{+\infty} d\lambda \int_{-\infty}^{+\infty} \psi(\xi)\frac{\sin t\sqrt{a^2\lambda^2-c^2}}{\sqrt{a^2\lambda^2-c^2}}\,e^{i\lambda(x-\xi)}\,d\xi, \tag{19}$$

$$u_1(x,t) = \frac{1}{2\pi}\int_{-\infty}^{+\infty} d\lambda \int_{-\infty}^{+\infty} \phi(\xi)\cos t\sqrt{a^2\lambda^2-c^2}\,e^{i\lambda(x-\xi)}\,d\xi. \tag{20}$$

Comparing (19) and (18) for $u_2(x,t)$, we have obtained, integrating (20) with respect to t,

$$\int_0^t u_1(x,\tau)\,d\tau = \frac{1}{2\pi}\int_{-\infty}^{+\infty} d\lambda \int_{-\infty}^{+\infty} \phi(\xi)\frac{\sin t\sqrt{a^2\lambda^2-c^2}}{\sqrt{a^2\lambda^2-c^2}}e^{i\lambda(x-\xi)}\,d\xi$$

$$= \frac{1}{2a}\int_{x-at}^{x+at} \phi(\xi)I_0\left(c\sqrt{t^2-\frac{(x-\xi)^2}{a^2}}\right)d\xi.$$

† See the introduction to the solutions of problems of the present section.

Differentiation of the last equation with respect to t gives:

$$u_1(x,t)=\frac{\phi(x-at)+\phi(x+at)}{2}+\frac{1}{2a}\int\limits_{x-at}^{x+at}\phi(\xi)\frac{\partial}{\partial t}I_0\left(c\sqrt{t^2-\frac{(x-\xi)^2}{a^2}}\right)\mathrm{d}\xi$$

$$=\frac{\phi(x-at)+\phi(x+at)}{2}+\frac{ct}{2a}\int\limits_{x-at}^{x+at}\phi(\xi)\frac{I_1\left(c\sqrt{t^2-\frac{(x-\xi)^2}{a^2}}\right)}{\sqrt{t^2-\frac{(x-\xi)^2}{a^2}}}\mathrm{d}\xi. \quad (21)$$

Adding (18) and (21) we obtain formula (1) of the answer.

176. $u(x,t)=\dfrac{1}{2a}\displaystyle\int\limits_0^t \mathrm{d}\tau\int\limits_{x-a(t-\tau)}^{x+a(t-\tau)} f(\xi,\tau)I_0\left(c\sqrt{(t-\tau)^2\frac{(x-\xi)^2}{a^2}}\right)\mathrm{d}\xi. \quad (1)$

Method. In order to obtain formula (1) of the answer it is possible to use the method of solving problem 175.

177. $u(x,t)=\dfrac{\phi(x+at)+\phi(|x-at|)\operatorname{sign}(x-at)}{2}+\dfrac{1}{2a}\displaystyle\int\limits_{|x-at|}^{x+at}\psi(z)\,\mathrm{d}z.$

Solution. Let us multiply the equation $u_{tt}=a^2u_{\xi\xi}$ by $\sqrt{2/\pi}\sin\lambda\xi$, integrate with respect to ξ from 0 to $+\infty$, and do the same with the initial conditions; this leads to the equation

$$\frac{\mathrm{d}^2\overline{u}^{(s)}(\lambda,t)}{\mathrm{d}t^2}+a^2\lambda^2\overline{u}^{(s)}(\lambda,t)=0 \quad (1)$$

with initial conditions

$$\overline{u}^{(s)}(\lambda,0)=\overline{f}^{(s)}(\lambda),\qquad \overline{u}_t^{(s)}(\lambda,0)=\overline{\psi}^{(s)}(\lambda), \quad (2)$$

where

$$\overline{u}^{(s)}(\lambda,t)=\sqrt{\frac{2}{\pi}}\int\limits_0^{+\infty}u(\xi,t)\sin\lambda\xi\,\mathrm{d}\xi,$$

$$\overline{f}^{(s)}(\lambda)=\sqrt{\frac{2}{\pi}}\int\limits_0^{+\infty}f(\xi)\sin\lambda\xi\,\mathrm{d}\xi,$$

$$\overline{\psi}^{(s)}(\lambda)=\sqrt{\frac{2}{\pi}}\int\limits_0^{+\infty}\psi(\xi)\sin\lambda\xi\,\mathrm{d}\xi.$$

Solving equation (1) for the initial conditions (2), we obtain:

$$\overline{u}^{(s)}(\lambda,t)=\overline{f}^{(s)}(\lambda)\cos a\lambda t+\overline{\psi}^{(s)}(\lambda)\frac{\sin a\lambda t}{a\lambda}. \quad (3)$$

Multiplying both sides of (3) by $\sqrt{2/\pi}\,\sin\lambda x$ and integrating with respect to λ from 0 to $+\infty$, we obtain:

$$u(x,t)=\sqrt{\frac{2}{\pi}}\int_0^{+\infty}\bar{u}^{(s)}(\lambda,t)\sin\lambda x\,d\lambda$$

$$=\sqrt{\frac{2}{\pi}}\int_0^{+\infty}\bar{f}^{(s)}(\lambda)\cos(a\lambda t)\sin\lambda x\,d\lambda+$$

$$+\frac{1}{a}\sqrt{\frac{2}{\pi}}\int_0^{+\infty}\bar{\psi}^{(s)}(\lambda)\,\frac{\sin(a\lambda t)\sin(\lambda x)\,d\lambda}{\lambda}$$

$$=\frac{1}{2}\sqrt{\frac{2}{\pi}}\int_0^{+\infty}\bar{f}^{(s)}(\lambda)[\sin\lambda(x+at)+\sin\lambda(x-at)]\,d\lambda+$$

$$+\frac{1}{2a}\sqrt{\frac{2}{\pi}}\int_0^{+\infty}\bar{\psi}^{(s)}(\lambda)\,\frac{\lambda}{\cos\lambda(x-at)-\cos\lambda(x+at)}\,d\lambda,$$

if $x>at$.

Taking into account the fact that

$$\int_{x-at}^{x+at}\psi(s)\,ds=\sqrt{\frac{2}{\pi}}\int_0^{+\infty}\bar{\psi}^{(s)}(\lambda)\,d\lambda\int_{x-at}^{x+at}\sin\lambda s\,ds$$

$$=\sqrt{\frac{2}{\pi}}\int_0^{+\infty}\bar{\psi}^{(s)}(\lambda)\,\frac{\cos\lambda(x-at)-\cos\lambda(x+at)}{\lambda}\,d\lambda,$$

we obtain

$$u(x,t)=\frac{f(x+at)+f(x-at)}{2}+\frac{1}{2a}\int_{x-at}^{x+at}\psi(s)\,ds\quad\text{for}\quad x>at. \tag{4}$$

If $x<at$, then in the sine and cosine it is necessary to replace $x-at$ by $at-x$, which leads to a change of sign in front of the sine and $u(x,t)$ becomes

$$u(x,t)=\frac{f(at+x)-f(at-x)}{2}+\frac{1}{2a}\int_{at-x}^{at+x}\psi(s)\,ds\quad\text{for}\quad x<at. \tag{5}$$

Combining (4) and (5) in the one relation, we obtain the answer given above

178. $$u(x,t)=\frac{\phi(x+at)+\phi(|x-at|)}{2}+$$

$$+\frac{1}{2a}\left\{\int_0^{x+at}\psi(z)\,dz-\operatorname{sign}(x-at)\int_0^{|x-at|}\psi(z)\,dz\right\}.$$

Method. Apply the Fourier cosine transform.

179. $u(x,t)=\begin{cases} 0 & \text{for} \quad 0<t<\dfrac{x}{a}, \\ \mu\left(t-\dfrac{x}{a}\right) & \text{for} \quad t>\dfrac{x}{a}. \end{cases}$

Method. Apply the Fourier sine-transform.

Solution. Let us multiply both sides of the equation $u_{tt}=a^2u_{\xi\xi}$† by $\sqrt{2/\pi}\sin\lambda\xi$ and integrate with respect to ξ from 0 to $+\infty$, applying an integration by parts‡ and using the boundary condition $u(0,t)=\mu(t)$; this gives:

$$\frac{d^2\bar{u}^{(s)}(\lambda,t)}{dt^2}=a^2\sqrt{\frac{2}{\pi}}\int\limits_0^{+\infty}\frac{\partial^2 u}{\partial\xi^2}\sin\lambda\xi\,d\xi=a^2\sqrt{\frac{2}{\pi}}\frac{\partial u}{\partial\xi}\sin\lambda\xi\Bigg|_{\xi=0}^{\xi=+\infty}-$$

$$-a^2\sqrt{\frac{2}{\pi}}\lambda(\cos\lambda\xi)u\Bigg|_{\xi=0}^{\xi=+\infty}-a^2\lambda^2\sqrt{\frac{2}{\pi}}\int\limits_0^{+\infty}u\sin\lambda\xi\,d\xi$$

$$=-a^2\lambda^2\bar{u}^{(s)}(\lambda,t)+a^2\lambda\sqrt{\frac{2}{\pi}}\mu(t).$$

We have made use of the fact that $u(\xi,t)$ and $\partial u(\xi,t)/\partial\xi$ tend to zero as $\xi\to+\infty$. Thus we arrive at the equation

$$\frac{d^2\bar{u}^{(s)}(\lambda,t)}{dt^2}+a^2\lambda^2\bar{u}^{(s)}(\lambda,t)=a^2\lambda\sqrt{\frac{2}{\pi}}\mu(t). \tag{1}$$

Since the solution

$$u(x,t)=\sqrt{\frac{2}{\pi}}\int\limits_0^{+\infty}\bar{u}^{(s)}(\lambda,t)\sin\lambda x\,d\lambda$$

must satisfy the zero initial conditions

$$u(x,0)=u_t(x,0)=0, \quad 0<x<+\infty,$$

then, solving equation (1), for $\bar{u}^{(s)}(\lambda,t)$ one must take the zero initial conditions

$$\bar{u}^{(s)}(\lambda,0)=\frac{d\bar{u}^{(s)}(\lambda,0)}{dt}=0. \tag{2}$$

The solution of equation (1) for the initial conditions (2) has the form

$$\bar{u}^{(s)}(\lambda,t)=a\sqrt{\frac{2}{\pi}}\int\limits_0^t\mu(t)\sin a\lambda(t-\tau)\,d\tau,$$

† In $u(x,t)$ we replace x by ξ.

‡ Compare with the solution by the method of travelling waves, problem 73, page 229.

hence,

$$u(x,t) = a\frac{2}{\pi}\int_0^{+\infty} d\lambda \int_0^t \mu(\tau)\sin\lambda x \sin a\lambda(t-\tau)\,d\tau.$$

Changing the order of integration, let us evaluate first of all the integral

$$\frac{2}{\pi}\int_0^{+\infty} \sin\lambda x \sin a\lambda(t-\tau)\,d\lambda = \frac{1}{\pi}\int_0^{+\infty}\cos\lambda[x-a(t-\tau)]\,d\lambda -$$

$$-\frac{1}{\pi}\int_0^{+\infty}\cos\lambda[x+a(t-\tau)]\,d\lambda = \delta(x-a[t-\tau])-\delta(x+a[t-\tau]).$$

Since $0 \leqslant \tau < t$, then $\delta(x+a[t-\tau]) \equiv 0$ for $x > 0$; hence,

$$\frac{2}{\pi}\int_0^{+\infty}\sin\lambda x\cos a\lambda(t-\tau)\,d\lambda = \delta(x-a[t-\tau]) \text{ for } 0<\tau<t,\quad 0<x<+\infty.$$

Therefore

$$u(x,t) = a\int_0^t \mu(\tau)\delta(x-a[t-\tau])\,d\tau = \int_0^{at}\mu\left(t-\frac{s}{a}\right)\delta(x-s)\,ds$$

$$= \begin{cases} 0 & \text{for} \quad t<\dfrac{x}{a}, \\ \mu\left(t-\dfrac{x}{a}\right) & \text{for} \quad t>\dfrac{x}{a}. \end{cases}$$

180. $u(x,t) = -a\int_0^{t-\frac{x}{a}} \nu(s)\,ds.$

Method. Apply the Fourier cosine transform; see the solution of the preceding problem (compare with the solution of problem 74).

181. (a) $u(x,t) = \dfrac{1}{2a}\displaystyle\int_0^t d\tau \int_{|x-a(t-\tau)|}^{x+a(t-\tau)} f(s,\tau)\,ds;$

(b) $$u(x,t) = \frac{1}{2a}\int_0^t d\tau\left\{\int_0^{x+a(t-\tau)} f(s,\tau)\,ds - \right.$$

$$\left. \operatorname{sign-}[x-a(t-\tau)]\int_0^{|x-a(t-\tau)|} f(s,\tau)\,ds\right\}.$$

Method. In case (a) use the Fourier sine-transform and in case (b) the Fourier cosine transform. In case (a) use also the equation

$$\frac{2\sin a\lambda(t-\tau)\sin\lambda x}{\lambda} = \frac{\cos\lambda[x-a(t-\tau)]-\cos[x+a(t-\tau)]}{\lambda}$$

$$= \int\limits_{x-a(t-\tau)}^{x+a(t-\tau)} \sin\lambda s\,\mathrm{d}s = \frac{\cos\lambda[a(t-\tau)-x]-\cos[a(t-\tau)+x]}{\lambda}$$

$$= \int\limits_{a(t-\tau))-x}^{a(t-\tau)+x} \sin\lambda s\,\mathrm{d}s$$

and in case (b) use similar relations.

182. $u(x,t) = -\int\limits_0^{t-x} \nu(\tau) I_0\left(c\sqrt{(t-\tau)^2-x^2}\right)\mathrm{d}\tau.$ (1)

Method. One may look for a solution of the boundary-value problem in the form

$$u(x,t) = \int\limits_0^{t-x} \phi(\tau) I_0\left(c\sqrt{(t-\tau)^2-x^2}\right)\mathrm{d}\tau, \tag{2}$$

where $\phi(\tau)$ is a function to be determined from the boundary condition.

183. $u(x,t) = \mu(t-x) - cx\int\limits_0^{t-x} \mu(\tau)\frac{I_1\left(c\sqrt{(t-\tau)^2-x^2}\right)}{\sqrt{(t-\tau)^2-x^2}}\,\mathrm{d}\tau.$

Method. Use the solution of the preceding problem.

184. The solution $u(x,t)$ of the boundary-value problem satisfies the ordinary differential equation

$$\frac{\mathrm{d}u(x,t)}{\mathrm{d}x} - hu(x,t) = \varkappa(t-x) - cx\int\limits_0^{t-x} \varkappa(\tau)\frac{I_1\left(c\sqrt{(t-\tau)^2-x^2}\right)}{\sqrt{(t-\tau)^2-x^2}}\,\mathrm{d}\tau.$$

185. $\int\limits_{-\infty}^{+\infty} \bar{f}(\lambda)\bar{g}(\lambda)\mathrm{e}^{-i\lambda x}\,\mathrm{d}\lambda = \frac{1}{\sqrt{2\pi}}\int\limits_{-\infty}^{+\infty} \bar{f}(\lambda)\mathrm{e}^{-i\lambda x}\,\mathrm{d}\lambda \int\limits_{-\infty}^{+\infty} g(s)\,\mathrm{e}^{i\lambda s}\,\mathrm{d}s$

$$= \frac{1}{\sqrt{2\pi}}\int\limits_{-\infty}^{+\infty} g(s)\,\mathrm{d}s \int\limits_{-\infty}^{+\infty} \bar{f}(\lambda)\mathrm{e}^{i\lambda(x-s)}\,\mathrm{d}\lambda = \int\limits_{-\infty}^{+\infty} g(s)f(x-s)\,\mathrm{d}s.$$

186. $$\int_0^{+\infty} \bar{f}^{(c)}(\lambda)\bar{g}^{(c)}(\lambda)\cos\lambda x\,d\lambda = \sqrt{\frac{2}{\pi}}\int_0^{+\infty}\bar{f}^{(c)}(\lambda)\cos\lambda x\,d\lambda\int_0^{+\infty} g(s)\cos\lambda s\,ds$$

$$= \frac{1}{2}\int_0^{+\infty} g(s)\,ds\sqrt{\frac{2}{\pi}}\int_0^{+\infty}\bar{f}^{(c)}(\lambda)[\cos\lambda(x-s)+\cos\lambda(x+s)]\,d\lambda$$

$$= \frac{1}{2}\int_0^{-\infty} g(s)[f(|x-s|)+f(x+s)]\,ds.$$

187. *Method.* See the solution of problem 186.

188. $$u(x,t) = \frac{1}{\sqrt{2\pi}}\int_{-\infty}^{+\infty}\phi\left(x-2\lambda\sqrt{at}\right)(\sin\lambda^2+\cos\lambda^2)\,d\lambda -$$

$$-\frac{1}{\sqrt{2\pi}}\int_{-\infty}^{+\infty}\psi\left(x-2\lambda\sqrt{at}\right)(\sin\lambda^2-\cos\lambda^2)\,d\lambda. \qquad (1)$$

For $\phi(x) = Ae^{-x^2/4k^2}$, $\psi(x)\equiv 0$ we obtain:

$$u(x,t) = \frac{Ak}{\sqrt{R}}e^{-x^2\frac{\cos\theta}{4R}}\cos\left(\frac{x^2\sin\theta}{4R}-\frac{1}{2}\theta\right), \qquad (2)$$

where

$$R\cos\theta = k^2, \qquad R\sin\theta = at.$$

Method. Apply the Fourier transform with kernel $e^{i\lambda\xi}$ over the straight line $-\infty < x < +\infty$. Use the relations

$$\frac{1}{\sqrt{2\pi}}\int_{-\infty}^{+\infty}\cos(a\xi)^2 e^{-i\xi x}\,d\xi = \frac{1}{2\sqrt{a}}\left(\cos\frac{x^2}{4a}+\sin\frac{x^2}{4a}\right), \qquad (I)$$

$$\frac{1}{\sqrt{2\pi}}\int_{-\infty}^{+\infty}\sin(a\xi)^2 e^{-i\xi x}\,d\xi = \frac{1}{2\sqrt{a}}\left(\cos\frac{x^2}{4a}-\sin\frac{x^2}{4a}\right), \qquad (II)$$

$$\int_{-\infty}^{+\infty}\cos(at\xi)^2\bar{\phi}(\xi)e^{-i\xi x}\,d\xi = \frac{1}{2\sqrt{at}}\int_{-\infty}^{+\infty}\phi(x-s)\left(\cos\frac{s^2}{4at}+\sin\frac{s^2}{4at}\right)ds, \qquad (III)$$

$$\int_{-\infty}^{+\infty}\sin(at\xi)^2\bar{\psi}(\xi)e^{-i\xi x}\,d\xi = \frac{1}{2\sqrt{at}}\int_{-\infty}^{+\infty}\psi(x-s)\left(\cos\frac{s^2}{4at}-\sin\frac{s^2}{4at}\right)ds. \qquad (IV)$$

The relations (III) and (IV) are obtained by means of relations (I) and (II) and the Parceval theorem, proved in the solution of problem 185. (I) and (II) may be derived from known integrals (see [1])

$$\int_{-\infty}^{+\infty} \cos x^2 \, dx = \sqrt{\frac{\pi}{2}} \quad \text{and} \quad \int_{-\infty}^{+\infty} \sin x^2 \, dx = \sqrt{\frac{\pi}{2}}. \tag{3}$$

First the substitution $x = y - l$ gives:

$$\cos x^2 = \cos(y^2+l^2)\cos 2ly + \sin(y^2+l^2)\sin 2ly,$$

$$\sin x^2 = \sin(y^2+l^2)\cos 2ly - \cos(y^2+l^2)\sin 2ly.$$

Representing the cos and sin of y^2+l^2 by the cos and sin of y^2 and l^2, we obtain two equations (from (3)) for the integrals

$$\int_{-\infty}^{+\infty} \cos y^2 \cos 2ly \, dy \quad \text{and} \quad \int_{-\infty}^{+\infty} \sin y^2 \cos 2ly \, dy. \tag{4}$$

Since

$$\int_{-\infty}^{+\infty} \cos y^2 \sin 2ly \, dy = 0 \quad \text{and} \quad \int_{-\infty}^{+\infty} \sin y^2 \sin 2ly \, dy = 0, \tag{5}$$

the imaginary part of the integrals (I) and (II) equals zero.

In order to derive formula (2) for the special initial conditions

$$\phi(x) = A\,e^{-\frac{x^2}{4k^2}}, \qquad \psi(x) \equiv 0$$

it is not worth using the general formula (I); it is better to use the conversion formula

$$u(x, t) = \frac{1}{\sqrt{2\pi}} \int_{-\infty}^{+\infty} \bar{u}(\lambda, t) e^{-i\lambda x} \, d\lambda,$$

having substituted the function

$$\bar{u}(\lambda, t) = \bar{\Phi}(\lambda) \cos a\lambda^2 t = \bar{\Phi}(\lambda) \frac{e^{ia\lambda^2 t} + e^{-ia\lambda^2 t}}{2},$$

where

$$\bar{\Phi}(\lambda) = \frac{A}{\sqrt{2\pi}} \int_{-\infty}^{+\infty} e^{-\frac{\xi^2}{4k^2} + i\lambda\xi} \, d\xi = Ak\sqrt{2}\, e^{-k^2\lambda^2}.$$

One should note that the latter equality holds for both real and complex k.

189. $u(x, t) = \displaystyle\int_{\frac{x}{\sqrt{2at}}}^{+\infty} \mu\left(\frac{t - x^2}{2a\lambda^2}\right) \left[\sin\frac{\lambda^2}{2} + \cos\frac{\lambda^2}{2}\right] d\lambda.$

Method. See the solution of the preceding problem. One should note also that the integral $\int_0^{+\infty} \xi \sin (a\xi)^2 \sin (\xi x) d\xi$ is obtained by differentiation with respect to x of the integral

$$\int_0^{+\infty} \sin(a\xi)^2 \cos(\xi x) d\xi .$$

190. *Method.* Make use of the fact that

(1) if $\Phi(x)$ and $\Psi(x)$ are odd functions, then

$$U(x, t) = \frac{\Phi(x-at)+\Phi(x+at)}{2} + \frac{1}{2a} \int_{x-at}^{x+at} \Psi(z) dz$$

equals zero for $x = 0$;

(2) if $u(x, t)$ is a solution of the equation $u_{tt} = a^2 u_{xx}$, then

$$U(x, t) = \sum_{k=0}^{N} A_k \frac{\partial^k u(x, t)}{\partial x^k}$$

is also a solution of this equation.

191. *Method.* Make use of the fact that

(1) if $F(x, t)$ is an odd function with respect to x, then the function

$$U(x, t) = \frac{1}{2a} \int_0^t d\tau \int_{x-a(t-\tau)}^{x+a(t-\tau)} F(\xi, \tau) d\xi$$

equals zero for $x = 0$;

(2) if $u(x, t)$ is a solution of the equation $u_{tt} = a^2 u_{xx} + f(x, t)$, then

$$U(x, t) = \sum_{k=0}^{N} \frac{\partial^k u(x, t)}{\partial x^k}$$

is a solution of the equation

$$u_{tt} = a^2 u_{xx} + \sum_{k=0}^{N} A_k \frac{\partial^k f(x, t)}{\partial x^k} .$$

192. *Method.* The proof is developed in a similar manner to that in the solution of problem 190.

193. *Method.* The proof is developed in a similar manner to that in the solution of problem 191.

1*. Transition to a Finite Interval by the Method of Images

194. $u(x, t)$

$$= \frac{\phi(x-at)+\phi(x+at)}{2}+\frac{ct}{2}\int\limits_{x-at}^{x+at}\frac{I_1\left(c\sqrt{t^2-\dfrac{(x-\xi)^2}{a^2}}\right)}{\sqrt{t^2-\dfrac{(x-\xi)^2}{a^2}}}\phi(\xi)\,d\xi+$$

$$+\frac{1}{2a}\int\limits_{x-at}^{x+at} I_0\left(c\sqrt{t^2-\frac{(x-\xi)^2}{a^2}}\right)\psi(\xi)\,d\xi, \qquad (1)$$

where $\phi(\xi)$ and $\psi(\xi)$ are obtained by an odd extension with respect to zero and then by a periodic extension of period $2l$.

195. The solution is derived by formula (1) of the answer to the preceding problem, but $\phi(x)$ and $\psi(x)$ are extended oddly with respect to $x = 0$, evenly with respect to $x = l$ and then periodically with period $4l$.

196. The solution is derived from formula (1) of the answer to problem 194; $\phi(x)$ and $\psi(x)$ are extended evenly with respect to $x = 0$ and $x = l$ and then periodically with period $2l$.

197. We shall look for a solution of the boundary-value problem

$$u_{tt} = u_{xx}+c^2u, \quad 0 < x < l, \quad 0 < t < +\infty \qquad (1)$$

$$u(0, t) = \mu_1(t), \quad u(l, t) = \mu_2(t), \quad 0 < t < +\infty, \qquad (2)$$

$$u(x, 0) = 0, \quad u_t(x, 0) = 0, \quad 0 < x < l \qquad (3)$$

in the form

$$u(x, t) = \frac{\partial}{\partial x}\int\limits_0^{t-x}\phi(\tau)I_0\left(c\sqrt{(t-\tau)^2-x^2}\right)d\tau+$$

$$+\frac{\partial}{\partial x}\int\limits_0^{t-(l-x)}\psi(\tau)I_0\left(c\sqrt{(t-\tau)^2-(l-x)^2}\right)d\tau, \qquad (4)$$

where the functions $\phi(\tau)$ and $\psi(\tau)$ are to be found from the boundary conditions (2). It is readily verified that $u(x, t)$, given by (4), is a solution of (1) for any $\phi(\tau)$ and $\psi(\tau)$. We shall assume $\phi(\tau) \equiv \psi(\tau) = 0$ for $\tau < 0$. Performing a differentiation in (4) and using the boundary conditions (2), we obtain:

$$-\phi(t)+\psi(t-l)+\int\limits_0^{t-l}\psi(\tau)\frac{clI_1\left(c\sqrt{(t-\tau)^2-l^2}\right)}{\sqrt{(t-\tau)^2-l^2}}d\tau = \mu_1(t), \qquad (5)$$

$$-\phi(t-l)+\psi(t)-\int\limits_0^{t-l}\phi(\tau)\frac{clI_1\left(c\sqrt{(t-\tau)^2-l^2}\right)}{\sqrt{(t-\tau)^2-l^2}}d\tau = \mu_2(t). \qquad (6)$$

Let us assume

$$\phi_1(t) = \psi(t)-\phi(t), \qquad \psi_1(t) = \psi(t)+\phi(t). \qquad (7)$$

From (5) and (6) we find:

$$\phi_1(t)+\phi_2(t-l)+cl\int_0^{t-l}\phi_1(\tau)\frac{I_1\left(c\sqrt{(t-\tau)^2-l^2}\right)}{\sqrt{(t-\tau)^2-l^2}}\,d\tau = \mu_1(t)+\mu_2(t), \tag{8}$$

$$-\psi_1(t)+\psi_1(t-l)+cl\int_0^{t-l}\psi_1(\tau)\frac{I_1\left(c\sqrt{(t-\tau)^2-l^2}\right)}{\sqrt{(t-\tau)^2+l^2}}\,d\tau = \mu_1(t)-\mu_2(t). \tag{9}$$

From (8) and (9) by virtue of the equality $\phi(\tau)\equiv\psi(\tau)=0$ for $\tau<0$ we find:

$$\phi_1(t)+\mu_1(t)=\mu_2(t), \qquad \psi_1(t)=\mu_2(t)-\mu_1(t), \qquad 0\leqslant t\leqslant l, \tag{10}$$

then

$$\begin{aligned}\phi_1(t) &= \mu_1(t)+\mu_2(t)-\phi_1(t-l)-cl\int_0^{t-l}\phi_1(\tau)\frac{I_1\left(c\sqrt{(t-\tau^2)-l^2}\right)}{\sqrt{(t-\tau)^2-l^2}}\,d\tau,\\ \psi_1(t) &= \mu_1(t)-\mu_2(t)+\psi_1(t-l)+cl\int_0^{t-l}\psi_1(\tau)\frac{I_1\left(c\sqrt{(t-\tau)^2-l^2}\right)}{\sqrt{(t-\tau)^2-l^2}}\,d\tau.\end{aligned} \tag{10'}$$

2. Riemann's Method

It is required to find a solution of the equation

$$L(u)\equiv u_{xx}-u_{yy}+a_1(x,y)u_x+b_1(x,y)u_y+c_1(x,y)u=f(x,y), \tag{1}$$

satisfying the initial conditions

$$u\Big|_c=\phi(x), \qquad \frac{\partial u}{\partial n}\Big|_c=\psi(x) \tag{2}$$

along the curve c, where $\partial u/\partial n$ is the derivative with respect to the normal to this curve. It is assumed that the curve c is given by the equation $y=f(x)$, where $f(x)$ is a differentiable function, with $|f'(x)|<1$.

Then the value of u at a point M (Fig. 31) is found by means of the relation

$$u(M)=\frac{(uv)_P+(uv)_Q}{2}+\frac{1}{2}\int_P^Q[v(u_\xi\,d\eta+u_\eta\,d\xi)-u(v_\xi\,d\eta+v_\eta\,d\xi)+$$

$$+uv(a\,d\eta-b\,d\xi)]+\iint_{MPQ}v(M,M')f(M')\,d\sigma_{M'}, \qquad d\sigma_{M'}=d\xi\,d\eta, \tag{3}$$

where

$$u|_c=\phi(x),$$

$$u_\xi|_c=\frac{\partial u}{\partial s}\cos(x,s)+\frac{\partial u}{\partial n}\cos(x,n)=\frac{\phi'(x)+\psi(x)f'(x)}{\sqrt{1+f'^2(x)}},$$

$$u_\eta|_c=\frac{\partial u}{\partial s}\cos(y,s)+\frac{\partial u}{\partial n}\cos(y,n)=\frac{\phi'(x)f'(x)+\psi(x)}{\sqrt{1+f'^2(x)}},$$

and the function $v(M, M') = v(x, y, \xi, \eta)$ is Riemann's function for the operator $L(u)$, defined by the relations

$$N(v) \equiv v_{xx} - v_{yy} - (a_1 v)_x - (b_1 v)_y + c_1 v = 0 \quad \text{inside} \quad PQM, \tag{4}$$

$$\frac{\partial v}{\partial s} = \frac{b_1 - a_1}{2\sqrt{2}} v \quad \text{along} \quad MP, \tag{5}$$

$$\frac{\partial v}{\partial s} = \frac{b_1 + a_1}{2\sqrt{2}} v \quad \text{along} \quad MQ, \tag{6}$$

and

$$v(M, M) = 1. \tag{7}$$

The operators $L(u)$ and $N(v)$ are said to be conjugate.

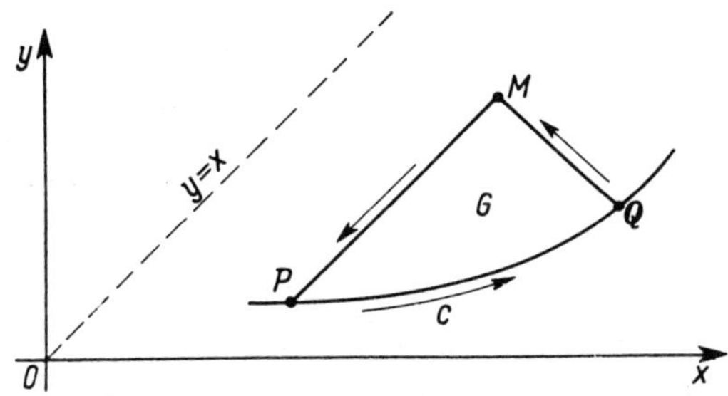

FIG. 31

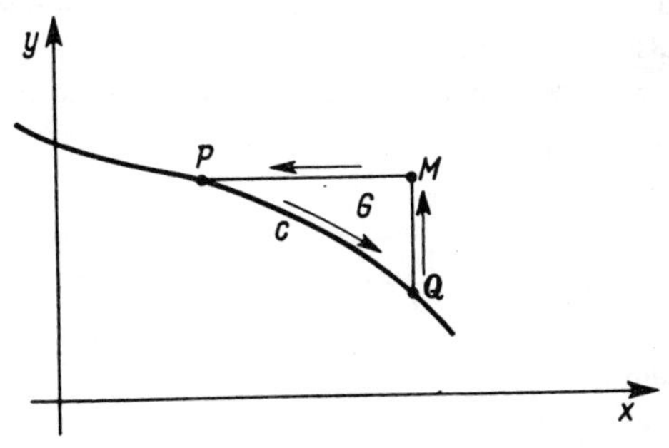

FIG. 32

If one proceeds from the other canonical form of the hyperbolic equation

$$L^*(u) = \frac{\partial^2 u}{\partial x \partial y} + a_2 \frac{\partial u}{\partial x} + b_2 \frac{\partial u}{\partial y} + c_2(u) = f(x, y), \tag{8}$$

then the solution of equation (8), satisfying the initial conditions

$$u\Big|_c = \phi(x), \quad \frac{\partial u}{\partial y}\Big|_c = \psi(x) \tag{9}$$

along the curve

$$y = f(x), \quad f'(x) < 0, \tag{10}$$

is found by means of the formula

$$u(M) = \frac{(uv)_P + (uv)_Q}{2} + \int_P^Q \left\{ \left[\frac{1}{2}\left(u\frac{\partial v}{\partial x} - v\frac{\partial u}{\partial x} \right) - b_2 uv \right] d\xi - \right.$$

$$\left. - \left[\frac{1}{2}\left(u\frac{\partial v}{\partial y} - v\frac{\partial u}{\partial y} \right) - a_2 uv \right] d\eta \right\} + \iint_{PQM} v(M, M')f(M')\,d\sigma_{M'}, \tag{11}$$

where the function v is Riemann's function for the operator $L^*(u)$ defined by the relations

$$N^*(v) \equiv \frac{\partial^2 v}{\partial x \partial y} - \frac{\partial(a_2 v)}{\partial x} - \frac{\partial(b_2 v)}{\partial y} + c_2 v = 0, \tag{12}$$

$$\frac{\partial v}{\partial x} = b_2 v \quad \text{along} \quad PM, \tag{13}$$

$$\frac{\partial v}{\partial y} = a_2 v \quad \text{along} \quad QM, \tag{14}$$

$$v(M, M) = 1. \tag{15}$$

Thus, if Riemann's function for the hyperbolic operator L or $L^*(u)$ is found, then it is possible to write down at once in integral form the solution of a wide class of boundary-value problems, connected with this hyperbolic operator.

198. The Riemann function $v \equiv 1$.

The solution of the boundary-value problem has the form

$$u(x, t) = \frac{\phi(x-at) + \phi(x+at)}{2} +$$

$$+ \frac{1}{2a}\int_{x-at}^{x+at} \psi(z)\,dz + \frac{1}{2a}\int_0^t d\tau \int_{x-a(t-\tau)}^{x+a(t-\tau)} f(z, \tau)\,dz.$$

199. Riemann's function

(a) for the operator

$$L(u) \equiv \frac{\partial^2 u}{\partial t^2} - a^2 \frac{\partial^2 u}{\partial x^2} + c^2 u$$

is:

$$v = J_0\left(c\sqrt{(t-\tau)^2 + \frac{(x-\xi)^2}{a^2}} \right);$$

(b) for the operator

$$L(u) \equiv \frac{\partial^2 u}{\partial t^2} - a^2 \frac{\partial^2 u}{\partial x^2} - c^2 u$$

is:

$$v = I_0\left(c\sqrt{(t-\tau)^2 - \frac{(x-\xi)^2}{a^2}}\right),$$

where $I_0(z) = J_0(iz)$ is the modified Bessel function of zero order.

The solution of the boundary-value problem correspondingly takes the form

(a) $$u(x,t) = \frac{\phi(x-at)+\phi(x+at)}{2} -$$

$$-\frac{ct}{2}\int_{x-at}^{x+at} \frac{J_1\left(c\sqrt{t^2-\frac{(x-\xi)^2}{a^2}}\right)}{\sqrt{t^2-\frac{(x-\xi)^2}{a^2}}}\psi(\xi)\,\mathrm{d}\xi +$$

$$+\frac{1}{2_a}\int_0^t \mathrm{d}\tau \int_{x-a(t-\tau)}^{x+a(t-\tau)} J_0\left(c\sqrt{(t-\tau)^2-\frac{(x-\xi)^2}{a^2}}\right) f(\xi,\tau)\,\mathrm{d}\xi;$$

(b) $$u(x,t)$$

$$= \frac{\phi(x-at)+\phi(x+at)}{2} + \frac{ct}{2}\int_{x-at}^{x+at} \frac{I_1\left(c\sqrt{t^2-\frac{(x-\xi)^2}{a^2}}\right)}{\sqrt{t^2-\frac{(x-\xi)^2}{a^2}}}\psi(\xi)\,\mathrm{d}\xi +$$

$$+\frac{1}{2a}\int_0^t \mathrm{d}\tau \int_{x-a(t-\tau)}^{x+a(t-\tau)} I_0\left(c\sqrt{(t-\tau)^2-\frac{(x-\xi)^2}{a^2}}\right) f(\xi,\tau)\,\mathrm{d}\xi.$$

200. $$u(x,y) = \frac{1}{2}\phi(xy) + \frac{y}{2}\phi\left(\frac{x}{y}\right) + \frac{\sqrt{xy}}{y}\int_{xy}^{\frac{x}{y}} \frac{\phi(z)}{z^{3/2}}\,\mathrm{d}z - \frac{\sqrt{xy}}{2}\int_{xy}^{\frac{x}{y}} \frac{\psi(z)}{z^{3/2}}\,\mathrm{d}z.$$

201. $$u(x,t) = \frac{\sqrt{\sqrt{l-x} - \frac{at}{2}}\,\phi\left(x+\sqrt{l-x}\,at - \frac{a^2t^2}{4}\right)}{2\sqrt[4]{l-x}} +$$

$$+\frac{\sqrt{\sqrt{l-x} + \frac{at}{2}}\,\phi\left(x-\sqrt{l-x}\,at - \frac{a^2t^2}{4}\right)}{2\sqrt[4]{l-x}} + \frac{1}{\sqrt[4]{l-x}}\int_{\sqrt{l-x}-\frac{at}{2}}^{\sqrt{l-x}+\frac{at}{2}} \Phi(x,t,z)\,\mathrm{d}z,$$

$$\Phi(x,t,z)=\frac{\sqrt{z}}{a}\psi(l-z^2)F\left(\frac{1}{2},\frac{1}{2},1,\frac{\frac{a^2t^2}{4}-(z-\sqrt{l-x})^2}{4z\sqrt{l-x}}\right)+$$

$$+\frac{at}{8\sqrt{(l-x)z}}\phi(l-z^2)F'\left(\frac{1}{2},\frac{1}{2},1,\frac{\frac{a^2t^2}{4}-(z-\sqrt{l-x})^2}{4z\sqrt{l-x}}\right),$$

where

$$F(\alpha,\beta,\gamma,x)=1+\frac{\alpha.\beta}{1.\gamma}x+\frac{\alpha(\alpha+1)\beta(\beta+1)}{1.2.\gamma(\gamma+1)}x^2+\dots$$

is a hypergeometric series†.

Method. Use the canonical form with mixed derivative for the hyperbolic operator. In characteristic coordinates Riemann's function has the form

$$G(\bar{x},\bar{y},\bar{x}_0,\bar{y}_0)=F\left(\frac{1}{2},\frac{1}{2},1,\frac{(\bar{x}-\bar{x}_0)(\bar{y}-\bar{y}_0)}{(\bar{x}_0-\bar{y}_0)(\bar{x}-\bar{y})}\right).$$

202. $$u(x,y)=\frac{\sqrt{\sin(\omega-y)}\,\phi[l\cos(\omega-y)]+\sqrt{\sin(\omega+y)}\,\phi[l\cos(\omega+y)]}{2\sqrt{\sin\omega}}+$$

$$+\frac{1}{2\sqrt{\sin\omega}}\int_{\omega-y}^{\omega+y}\Phi(\omega,y,z)dz,$$

$$\Phi(x,y,z)=\psi(l\cos z)\sqrt{\sin z}\,F\left(\frac{1}{2},\frac{1}{2},1,\frac{\cos(\omega-z)-\cos y}{2\sin\omega.\sin z}\right)+$$

$$+\frac{1}{2}\phi(l\cos z)\frac{\sin y}{\sin\omega\sqrt{\sin z}}F'\left(\frac{1}{2},\frac{1}{2},1,\frac{\cos(\omega-z)-\cos y}{2\sin\omega\sin z}\right),$$

where $\omega=\arccos x/l$.

Method. Utilize the canonical form with mixed derivative for the hyperbolic operator. Riemann's function in characteristic coordinates has the form

$$v(\bar{x},\bar{y},\bar{x}_0\bar{y}_0)=F\left(\frac{1}{2},\frac{1}{2},1,\frac{\sin(\bar{x}-\bar{x}_0)\sin(\bar{y}-\bar{y}_0)}{\sin(\bar{x}_0-\bar{y})\sin(\bar{x}-\bar{y})}\right).$$

† See the reference tables at the end of the book.

CHAPTER III

EQUATIONS OF PARABOLIC TYPE

§ 1. Physical Problems Leading to Equations of Parabolic Type; Statement of Boundary-value Problems

The equations and boundary conditions of boundary-value problems in the theory of heat conduction being considered here are derived from: (a) the law of conservation of energy, (b) the law of internal heat conduction in solids (Fourier's law) and (c) the law of convective heat exchange between the surface of the solid and the surrounding liquid or gaseous medium (Newton's law).

Fourier's law in one dimension is expressed by the relation

$$q = -\sigma\lambda \frac{\partial u}{\partial x}, \tag{1}$$

where q is the amount of heat, flowing per unit time in the direction of the x-axis through an area σ, perpendicular to the x-axis, u is the temperature in the part of the body being considered, λ the coefficient of heat conduction†.

Newton's law is expressed by the relation

$$q = \sigma\alpha(u-u_0), \tag{2}$$

where q is the amount of heat flowing per unit time across an area σ of the surface of the body into the surrounding medium, u the temperature of the surface of the body, u_0 the temperature of the surrounding medium, α the coefficient of heat exchange‡.

In boundary-value problems on diffusion the amount of diffusing substance and its concentration play the same role as the amount of heat and temperature in boundary-value problems in the theory of heat conduction.

In particular, if u is the concentration, λ the diffusion coefficient, and q the amount of substance diffusing per unit time in the direction of the x-axis

---

† λ depends on the physical properties of the body and on the temperature u, but within sufficiently wide limits one neglects the dependence of λ on the temperature, taking λ as the mean value of the temperature.

‡ All that was said in the preceding footnote about the dependence of λ on the temperature may extend within known limits to α; for more detail see [41], page 21.

through an area σ, perpendicular to the x-axis, then the law of diffusion (Nernst's law) is expressed by (1), and the law of diffusion through a semi-permeable membrane is expressed by (2).

In the case of parabolic boundary-value problems on the motion of a viscous liquid and electrodynamics, appropriate observations will be made when they are first considered.

1. Homogeneous Media; Equations with Constant Coefficients

1. The temperature of points of the rod is given by a solution of the boundary-value problem

$$u_t = a^2 u_{xx}, \quad 0 < x < l, \quad 0 < t < +\infty, \tag{1}$$

$$u(x, 0) = f(x), \quad 0 < x < l, \tag{2}$$

$$u(0, t) = \phi_1(t), \quad u(l, t) = \phi_2(t), \quad 0 < t < +\infty, \tag{3}$$

$$-\lambda\sigma u_x(0, t) = q_1(t), \quad \lambda\sigma u_x(l, t) = q_2(t), \quad 0 < t < +\infty, \tag{3'}$$

$$u_x(0, t) = h[u(0, t) - \phi_1(t)], \quad u_x(l, t) = -h[u(l, t) - \phi_2(t)], \quad 0 < t < +\infty, \tag{3''}$$

where a^2 is the coefficient of thermal conductivity, $a^2 = \lambda/c\rho$, λ the coefficient of heat conduction of the material of the rod, c the specific heat capacity, ρ the mass density, σ the cross-sectional area, $h = \alpha/\lambda$, where α is the coefficient of heat exchange, $f(x)$ the initial values of the temperature, $\phi_1(t)$ and $\phi_2(t)$ in case (3) are the temperatures of the ends of the rod, and in case (3″) are the values of the temperature of the surrounding medium at the ends of the rod; $q_1(t)$ and $q_2(t)$ the heat flows entering the rod through its ends (i.e. the amount of heat flowing in per unit time).

Method. If the lateral face of a homogeneous isotropic cylindrical rod is insulated, and the isothermal surfaces at the initial moment of time coincide with its cross-sections, the ends of the rod at all times remaining isothermal surfaces, then the isothermal surfaces in the rod will always coincide with the cross-sections, i.e. the temperature in the rod will at all times depend only on one spatial coordinate x.

Equation (1) may be derived by comparing the increase per unit of time of the amount of heat in an element $(x, x+\Delta x)$ of the rod, equal to

$$c\rho\sigma\Delta x \frac{\partial u}{\partial t}, \tag{4}$$

with the sum of the amounts of heat which have flowed into this element per unit time across the sections x and $x+\Delta x$

$$-\sigma\lambda \left.\frac{\partial u}{\partial x}\right|_x + \sigma\lambda \left.\frac{\partial u}{\partial x}\right|_{x+\Delta x}, \tag{5}$$

and then multiplying the equality obtained by Δx and passing to a limit as $\Delta x \to 0$. Let us consider in more detail the choice of sign in the terms of the sum (5). We assume $x+\Delta x > x$, which, obviously, does not disturb the general nature of the discussions. If at the end x of the element $(x, x+\Delta x)$, $\partial u/\partial x > 0$, then at points lying to the right of the end (i.e. inside the element) the temperature will be greater than at points lying to the left of the end (i.e. outside the element), which means heat will flow out from the element and, therefore, the first term of the sum (5) must be chosen with a minus sign. If $\partial u/\partial x < 0$, then the temperature to the right of the end (inside) will be less than the temperature to the left of the end (outside), therefore heat will flow into the element, the first term of (5) must be positive and, hence, it again is necessary to take a minus sign. The choice of sign in the second term is considered similarly.

To obtain the boundary conditions (3′) and (3″) it is necessary to carry out a similar argument for the boundary elements $(0, \Delta x)$ and $(l-\Delta x, l)$ using Newton's law of convective heat exchange in the case of (3″).

Note. If the coefficient of heat exchange α is considerably greater than the coefficient of internal heat conduction λ $(\alpha \to \infty)$, then the boundary conditions (3″) transform to the boundary conditions (3). If, on the other hand, α is negligibly small $(\alpha \to 0)$, then the boundary conditions (3″) become the boundary conditions (3′), where $q_1(t) = q_2(t) = 0$, i.e. we arrive at the case of heat insulation of the ends of the rod.

2. The equation of heat conduction in this case has the form

$$u_t = \frac{\lambda}{c\rho} u_{xx} - \frac{\alpha p}{c\rho\sigma}(u-u_0),$$

where p is the perimeter of a cross-section of the rod, α the coefficient of heat exchange between the surface of the rod and the surrounding medium, the temperature of which equals u_0; the remaining quantities have the same meaning as in the preceding problem; the initial and boundary conditions are written as in the preceding problem.

Method. Investigating an element $(x, x+\Delta x)$ of the rod, consider in the heat balance not only the flow of heat through the ends of the element, but also the flow of heat through its lateral face.

3. In order to determine the temperature in the ring we obtain the boundary-value problem

$$u_t = \frac{\lambda}{c\rho} u_{xx} - \frac{\alpha p}{c\rho\sigma}(u-u_0), \quad 0 < x < l, \quad 0 < t < +\infty, \tag{1}$$

$$u(0, t) = u(l, t), \quad u_x(0, t) = u_x(l, t), \quad 0 < t < +\infty, \tag{2}$$

$$u(x, 0) = f(x), \quad 0 < x < l. \tag{3}$$

Here λ, c, ρ, σ, α, p have the same meaning as in the preceding problem. The coordinate x is the length of the arc, measured along the ring. If the radius of the ring equals R, then $x = R\theta$, where θ is an angular coordinate; hence,

$l = 2\pi R$, $\partial/\partial x = (1/R)\partial/\partial\theta$ and, changing to the independent variables θ, t the boundary-value problem (1), (2), (3) may be transformed to the form

$$u_t = \frac{\lambda}{c\rho R^2}u_{\theta\theta} - \frac{\alpha p}{c\rho\sigma}(u-u_0), \quad 0<\theta<2\pi, \quad 0<t<+\infty, \tag{1'}$$

$$u(0,t) = u(2\pi,t), \quad u_\theta(0,t) = u_\theta(2\pi,t), \quad 0<t<+\infty, \tag{2'}$$

$$u(\theta,0) = F(\theta), \quad 0<\theta<2\pi. \tag{3'}$$

4. $$\frac{\partial u}{\partial t} = a^2\frac{\partial^2 u}{\partial x^2}, \quad v_0 t < x+\infty, \quad 0<t<+\infty, \tag{1}$$

$$u(v_0 t, t) = \phi(t), \quad 0<t<+\infty, \tag{2}$$

$$u(x,0) = 0, \quad 0<x<+\infty. \tag{3}$$

5. In order to determine the temperature $u(x,t)$ in the wire we must solve the boundary-value problem

$$u_t = \frac{\lambda}{c\rho}u_{xx} - \frac{\alpha p}{c\rho\sigma}(u-u_0) + \frac{\beta I^2 R}{c\rho\sigma}, \quad 0<x<l, \quad 0<t<+\infty, \tag{1}$$

$$c_1 u_t(0,t) = \lambda\sigma u_x(0,t), \quad c_2 u_t(l,t) = \lambda\sigma u_x(l,t), \quad 0<t<+\infty, \tag{2}$$

$$u(x,0) = f(x), \quad 0<x<l, \tag{3}$$

where c_1 and c_2 are the heat capacities of the terminals, I the current intensity, R the resistance per unit length of the conductor, β the coefficient of proportionality in the formula

$$q = \beta I^2 R\Delta x, \tag{4}$$

expressing the amount of heat liberated by the current I per unit time in an element $(x, x+\Delta x)$ of the conductor. The coefficients λ, c, ρ, σ, p, α have the same meaning as in problem 2.

Method. In deriving equation (1) it is necessary to use (4).

6. In order to determine the concentration $u(x,t)$ we have to solve the same equation and the same boundary conditions, as in problem 1 for determining the temperature, with this difference, however, that in the case of diffusion

$$a^2 = \lambda = D,$$

where D is the diffusion coefficient, and α the coefficient of porosity of each of the boundary planes.

7. In order to determine the concentration u of the diffusing substance we must solve the equation

$$u_t = Du_{xx} - vu_x, \tag{1}$$

where D is the diffusion coefficient, and v the velocity of motion of the medium.

Method. To obtain equation (1) it is necessary to isolate an element of constant cross-sectional area, parallel to the x-axis (Fig. 33) and consider the amount of the substance passing through the sections x and $x+\Delta x$ due to diffusion and the transfer of the moving medium.

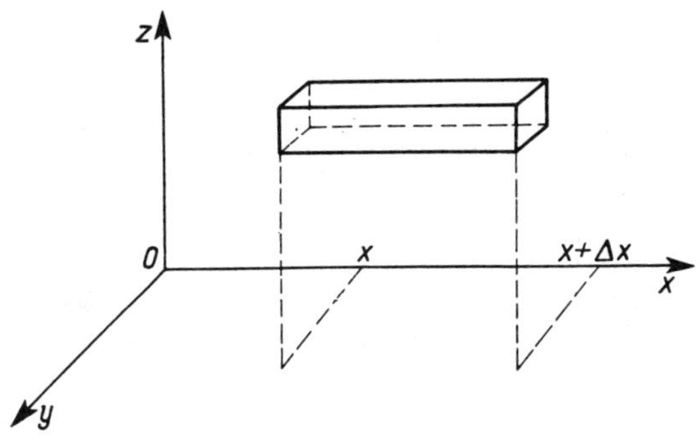

FIG. 33

8. In order to determine the concentration of the suspended particles we must solve the equation

$$\frac{\partial u}{\partial t} = D\frac{\partial^2 u}{\partial z^2} - v\frac{\partial u}{\partial z},$$

where D is the diffusion coefficient, and v the velocity of precipitation of the particles, the z-axis being directed downwards. The condition of impermeability of the plane $z = z_0$ has the form

$$D\frac{\partial u}{\partial z} - vu = 0 \quad \text{for} \quad z = z_0.$$

Method. See the method for the preceding problem. In place of the flow of the diffusing substance due to the motion of the medium it is necessary to consider the flow due to the precipitation velocity of the particles.

9. (a) $u_t = Du_{xx} - \beta_1 u, \quad \beta_1 > 0;$

(b) $u_t = Du_{xx} + \beta_2 u, \quad \beta_2 > 0,$

where D is the diffusion coefficient, β_1 the coefficient of disintegration, and β_2 the coefficient of multiplication.

Method. In the case of (a) an amount of diffusing substance equal to $\beta_1 u$ disintegrates per unit volume per unit time, and in the case of (b) an amount of diffusing substance equal to $\beta_2 u$ forms.

10. If the velocity of the moving plane maintains a constant direction, then the velocities of particles of the liquid will, obviously, be parallel to this direction. Directing the axis perpendicular to the layer and locating the origin of coordinates in the fixed plane, in order to determine the velocities of particles of the liquid we must solve the boundary-value problem

$$v_t = \nu v_{xx}, \quad 0 < x < l, \quad 0 < t < +\infty, \tag{1}$$

$$v(0, t) = 0, \quad v(l, t) = v_0(t), \quad 0 < t < +\infty, \tag{2}$$

$$v(x, 0) = 0, \quad 0 < x < l, \tag{3}$$

where l is the thickness of the layer, $v_0(t)$ is the velocity of motion of the boundary plane, $\nu = \mu/\rho$ the kinematic coefficient of viscosity, ρ the mass density, μ the dynamic coefficient of viscosity appearing in Newton's law for determining the frictional stress between layers of the viscous liquid

$$\tau = \mu \frac{\partial v}{\partial x}.$$

Method. In deriving equation (1) it is necessary to neglect the pressure gradient in comparison with the gradient of the frictional forces, which can be done, if the liquid has high viscosity.

11.

$$\frac{\partial E}{\partial t} = \frac{c^2}{4\pi\sigma\mu} \frac{\partial^2 E}{\partial \zeta^2}, \tag{1}$$

$$\frac{\partial H}{\partial t} = \frac{c^2}{4\pi\sigma\mu} \frac{\partial^2 H}{\partial \zeta^2}. \tag{2}$$

Solution. Let us write down the set of Maxwell's equations† for the condition that in the region under consideration volume charges and external electromotive forces are absent:

$$\operatorname{rot} \boldsymbol{E} + \frac{1}{c} \frac{\partial \boldsymbol{B}}{\partial t} = 0, \tag{I}$$

$$\operatorname{rot} \boldsymbol{H} - \frac{1}{c} \frac{\partial \boldsymbol{D}}{\partial t} = \frac{4\pi}{c} \boldsymbol{j}, \tag{II}$$

$$\operatorname{div} \boldsymbol{B} = 0, \tag{III}$$

$$\operatorname{div} \boldsymbol{D} = 0, \tag{IV}$$

$$\boldsymbol{j} = \sigma \boldsymbol{E}, \tag{V}$$

$$\boldsymbol{D} = \varepsilon \boldsymbol{E}, \tag{VI}$$

$$\boldsymbol{B} = \mu \boldsymbol{H}. \tag{VII}$$

Neglecting displacement currents $(1/c)\, \partial D/\partial t$ in equation (II) (conducting medium) and using (V) and (VII), we rewrite equations (I) and (II) in the form

$$\operatorname{curl} \boldsymbol{E} + \frac{\mu}{c} \frac{\partial \boldsymbol{H}}{\partial t} = 0, \tag{I'}$$

$$\operatorname{curl} \boldsymbol{H} = \frac{4\pi\sigma}{c} \boldsymbol{E}. \tag{II'}$$

† See [7], page 489.

We take the rot of both sides of equation (I′), and by differentiating the equality (II′) with respect to t, eliminate H from the results obtained. Then we make use of relations (IV) and (VI) and the known equality of vector analysis

$$\text{curl curl}\,\boldsymbol{a} = \text{grad div}\,\boldsymbol{a} - \text{div grad}\,\boldsymbol{a}\dagger;$$

obtaining the equation

$$\frac{\partial \boldsymbol{E}}{\partial t} = \frac{c^2}{4\pi\sigma\mu}\,\text{div grad}\,\boldsymbol{E}. \tag{3}$$

Similarly the equation

$$\frac{\partial \boldsymbol{H}}{\partial t} = \frac{c^2}{4\pi\sigma\mu}\,\text{div grad}\,\boldsymbol{H} \tag{4}$$

is obtained. According to the conditions of the problem $\boldsymbol{E} = \boldsymbol{E}(\zeta, t)$, $\boldsymbol{H} = \boldsymbol{H}(\zeta, t)$ where ζ is the distance read from some fixed plane. In a rectangular cartesian system of coordinates (ξ, η, ζ) the Laplacian operator is written in the form

$$\text{div grad} = \frac{\partial^2}{\partial \xi^2} + \frac{\partial^2}{\partial \eta^2} + \frac{\partial^2}{\partial \zeta^2},$$

and therefore

$$\text{div grad}\,\boldsymbol{E} = \frac{\partial^2 \boldsymbol{E}}{\partial \zeta^2},$$

$$\text{div grad}\,\boldsymbol{H} = \frac{\partial^2 \boldsymbol{H}}{\partial \zeta^2}.$$

Therefore equations (3) and (4) are transformed into (1) and (2).

2. Inhomogeneous Media; Equations with Variable Coefficients and Matching Conditions

12. If the x-axis is directed along the rod, and the origin of coordinates is chosen at the point of junction of the rods, then the boundary-value problem for determining the temperature in the composite rod may be written in the form

$$\left.\begin{aligned} &\frac{\partial u_1}{\partial t} = a_1^2 \frac{\partial^2 u_1}{\partial x^2}, \quad -\infty < x < 0, \\ &\frac{\partial u_2}{\partial t} = a_2^2 \frac{\partial^2 u_2}{\partial x^2}, \quad 0 < x < +\infty, \end{aligned}\right\} 0 < t < +\infty,$$

(a) $u_1(0, t) = u_2(0, t), \quad \lambda_1 u_{1x}(0, t) = \lambda_2 u_{2x}(0, t), \quad 0 < t < +\infty,$

(b) $u_1(0, t) = u_2(0, t), \quad \lambda_2 u_{2x}(0, t) - \lambda_1 u_{1x}(0, t) = C_0 u_{1t}(0, t) = C_0 u_{2t}(0, t),$

$$0 < t < +\infty,$$

$$u_1(x, 0) = f(x), \quad -\infty < x < 0,$$
$$u_2(x, 0) = f(x), \quad 0 < x < +\infty.$$

† This equality is valid for any two-fold continuously differentiable vector $\boldsymbol{a}$.

13. Choosing the x-axis along the axis of the cylinders and locating the origin of coordinates at the place of junction of the cylinders, we obtain the boundary-value problem

$$\left.\begin{aligned} &\frac{\partial u_1}{\partial t} = D_1 \frac{\partial^2 u_1}{\partial x^2}, \quad -l < x < 0, \\ &\frac{\partial u_2}{dt} = D_2 \frac{\partial^2 u_2}{\partial x^2}, \quad 0 < x < l_2, \end{aligned}\right\} 0 < t < +\infty,$$

$$u_{1x}(-l_1, t) = 0. \quad u_{2x}(l_2, t) = 0, \quad 0 < t < +\infty,$$

(a) $u_1(0, t) = u_2(0, t), \quad D_1 u_{1x}(0, t) = D_2 u_{2x}(0, t), \quad 0 < t < +\infty,$

(b) $$\left.\begin{aligned} -D_1 u_{1x}(0, t) &= \alpha[u_1(0, t) - u_2(0, t)], \\ -D_2 u_{2x}(0, t) &= \alpha[u_1(0, t) - u_2(0, t)], \end{aligned}\right\} 0 < t < +\infty,$$

$$u_1(x, 0) = f(x), \quad -l < x < 0,$$
$$u_2(x, 0) = f(x), \quad 0 < x < l_2.$$

14. If at time $t = 0$ the furnace was at the point $x = 0$ of the rod, then the boundary-value problem on the determination of the temperature in the rod may be written in the form

$$\frac{\partial u_1}{\partial t} = a^2 \frac{\partial^2 u_1}{\partial x^2}, \quad -\infty < x < v_0 t,$$

$$\frac{\partial u_2}{\partial t} = a^2 \frac{\partial^2 u_2}{\partial x^2}, \quad v_0 t < x < +\infty,$$

$$u_1(v_0 t, t) = u_2(v_0 t, t), \quad \lambda\sigma[u_1(v_0 t, t) - u_2(v_0 t, t)] = Q, \quad 0 < t < +\infty,$$
$$u_1(x, 0) = f(x), \quad -\infty < x < 0,$$
$$u_2(x, 0) = f(x), \quad 0 < x < +\infty,$$

where Q is the amount of heat generated by the furnace per unit time, λ the coefficient of heat conduction, and σ the cross-sectional area of the rod.

Using the delta-function the boundary-value problem may be formulated more concisely:

$$\frac{\partial u}{\partial t} = a^2 \frac{\partial^2 u}{\partial x^2} + \frac{Q}{c\rho} \delta(x - v_0 t), \quad -\infty < x < +\infty, \quad 0 < t < +\infty,$$
$$u(x, 0) = f(x), \quad -\infty < x < +\infty.$$

15. Locating the origin of coordinates on the surface of the metal and denoting by $\xi(t)$ the depth to which the solidification has spread by time t, we obtain the boundary-value problem

$$\left.\begin{aligned} &\frac{\partial u_1}{\partial t} = a_1^2 \frac{\partial^2 u_1}{\partial x^2}, \quad 0 < x < \xi(t), \\ &\frac{\partial u_2}{\partial t} = a_2^2 \frac{\partial^2 u_2}{\partial x^2}, \quad \xi(t) < x < l, \end{aligned}\right\} 0 < t < t_l, \qquad (1)$$

$$\left.\begin{aligned} &u_1(0, t) = U_1 = \text{const.},\\ &\lambda_1 \frac{\partial u_1}{\partial x}\bigg|_{x=\xi(t)} - \lambda_2 \frac{\partial u_2}{\partial x}\bigg|_{x=\xi(t)} = \varkappa\rho_2 \frac{d\xi}{dt}, \end{aligned}\right\} 0 < t < t_l,$$

$$u_1(\xi(t), t) = u_2(\xi(t), t) = 0, \quad 0 < t < t_l,$$

$$u_{2x}(l, t) = 0, \quad 0 < t < t_l,$$

$$u_2(x, 0) = U_0, \quad 0 < x < l.$$

Here the melting point (temperature of solidification) of the metal is taken as zero. λ_1 and λ_2 are the coefficients of heat conduction of the solid and liquid metal, a_1^2 and a_2^2 their coefficients of thermal conductivity; $\varkappa$ the latent heat of fusion, ρ_2 the mass density of the molten metal, t_1 the time at which $\xi(t_1) = l$.

If the temperature varies over a very wide range and it is not possible to neglect the dependence of the coefficients of heat conduction, heat capacity and mass densities on temperature, then equation (1) must be replaced by the equations

$$\left.\begin{aligned} c_1\rho_1 \frac{\partial u_1}{\partial t} &= \frac{\partial}{\partial x}\left(\lambda_1 \frac{\partial u_1}{\partial x}\right), \quad 0 < x < \xi(t),\\ c_2\rho_2 \frac{\partial u_2}{\partial t} &= \frac{\partial}{\partial x}\left(\lambda_2 \frac{\partial u_2}{\partial x}\right) \quad \xi(t) < x < l, \end{aligned}\right\} 0 < t < t_l. \tag{1'}$$

16. Placing the origin of coordinates in the plane of the lamina, directing the x-axis perpendicular to the lamina, and the u-axis vertically downwards, we obtain the following boundary-value problem for the velocities of particles of the liquid,

$$\left.\begin{aligned} &u_t = \nu u_{xx}, \quad -l_1 < x < 0,\\ &u_t = \nu u_{xx}, \quad 0 < x < l_2, \end{aligned}\right\} 0 < t < +\infty,$$

$$\left.\begin{aligned} &u(-l, t) = 0, \qquad u(l_2, t) = 0,\\ &u(0-0, t) = u(0+0, t) = w, \end{aligned}\right\} 0 < t < +\infty,$$

where w is the velocity of motion of the lamina

$$\frac{dw}{dt} = \frac{\rho\nu}{\gamma}[u_x(0+0, t) - u_x(0-0, t)] + g, \quad 0 < t < +\infty,$$

$$w(0) = 0,$$

$$u(x, 0) = 0, \quad -l_1 < x < 0, \quad 0 < x < l_2.$$

Here γ is the mass of unit area of the lamina, ρ the mass density of the liquid.

17. To determine the temperature in the rod we must solve the boundary-value problem

$$\left(1 - \frac{x}{L}\right)^2 \frac{\partial u}{\partial t} = a^2 \frac{\partial}{\partial x}\left[\left(1 - \frac{x}{L}\right)^2 \frac{\partial u}{\partial x}\right] - \frac{2\alpha\left(1 - \frac{x}{L}\right)}{c\rho r_0 \cos\gamma} u, \quad 0 < x < l,$$

$$0 < t < +\infty,$$

$$u_x(0, t) = 0, \quad u_x(l, t) = 0, \quad 0 < t < +\infty,$$

$$u(x, 0) = U_0, \quad 0 < x < l,$$

$$a^2 = \frac{\lambda}{c\rho}.$$

Here L is the height of the complete cone, resulting from an extension of the rod, γ is half the solid angle of the cone, r_0 the radius of the large base of the truncated cone, l its height, λ, c, ρ the coefficient of heat conduction, specific heat and mass density of the material of the cone, α the coefficient of convective heat exchange between the surface of the cone and the surrounding medium.

3. Similarity of Boundary-value Problems

18. The boundary-value problem on the heating of the rod thermally insulated along the surface—problem (1)

$$\frac{\partial u'}{\partial t'} = a^2 \frac{\partial^2 u'}{\partial x'^2}, \quad a^2 = \frac{\lambda}{c\rho}, \quad 0 < x' < l', \quad 0 < t' < +\infty, \tag{1}$$

$$u'(0, t') = \phi'(t') \neq 0, \quad u'(l', t') = 0, \quad 0 < t' < +\infty, \tag{2}$$

$$u'(x', 0) = 0, \quad 0 < x' < l' \tag{3}$$

is similar to boundary-value problem 10 problem (II) on the motion of a layer of viscous liquid

$$\frac{\partial u''}{\partial t''} = \nu \frac{\partial^2 u''}{\partial x''^2}. \quad 0 < x'' < l'', \quad 0 < t'' < +\infty, \tag{1'}$$

$$u''(0, t'') = \phi''(t'') \neq 0, \quad u''(l'', t'') = 0, \quad 0 < t'' < +\infty, \tag{2'}$$

$$u''(x'', 0) = 0, \quad 0 < x'' < l''. \tag{3'}$$

In order that problem (I) be similar to problem (II) with coefficients of similarity k_x, k_i, k_u, it is necessary and sufficient that the relations

$$\phi'(t') = k_u \phi''(t'') \quad \text{for} \quad 0 < t'' < +\infty, \tag{4}$$

be fulfilled, where $t' = k_t t''$, and

$$a^2 = \frac{k_x^2}{k_t} \nu, \quad k_x = \frac{l'}{l''}. \tag{5}$$

Solution. Determination of the similarity is obvious. Let us prove the necessity and sufficiency of the conditions (4) and (5).

Necessity. Let

$$u'(x', t') = k_u u''(x'', t'') \quad \text{for} \quad x' = k_x x'', \quad t' = k_t t'',$$

where (x', t') passes through $D_I[0 < x' < l', \ 0 < t' < +\infty]$, when (x'', t'') passes through

$$D_{II}[0 < x'' < l'', \quad 0 < t'' < +\infty]. \tag{6}$$

Then the equality $u'(0, t') = k_u u''(0, t'')$ must be fulfilled for $0 < t'' < +\infty$, i.e. by virtue of (2) and (2′) equality (4) must be fulfilled. Differentiating $u'(x', t') = k_u u''(x'', t'')$ with respect to x'' and t'' and utilizing the equalities $x' = k_x x''$, $t' = k_t t''$, we obtain:

$$k_t \frac{\partial u'}{\partial t'} = k_u \frac{\partial u''}{\partial t''}, \quad k_x^2 \frac{\partial^2 u''}{\partial x'^2} = k_u \frac{\partial^2 u''}{\partial x''^2}.$$

Since $u''(x'', t'')$ must satisfy equation (1′), then, therefore the equality

$$k_u \left(\frac{\partial u''}{\partial t''} - \nu \frac{\partial^2 u''}{\partial x''^2} \right) = k_t \frac{\partial u'}{\partial t'} - \nu k_x^2 \frac{\partial^2 u'}{\partial x'^2} = 0,$$

must be fulfilled, i.e. the equation for $u'(x', t')$

$$\frac{\partial u'}{\partial t'} = \nu \frac{k_x^2}{k_t} \frac{\partial^2 u'}{\partial x'^2}, \quad 0 < x' < l', \quad 0 < t' < +\infty,$$

must be fulfilled. Thus $u'(x', t')$ must be not only a solution of the boundary-value problem (1), (2), (3), but also a solution of the boundary-value problem

$$\frac{\partial u'}{\partial t'} = \nu \frac{k_x^2}{k_t} \frac{\partial^2 u'}{\partial x'^2}, \quad 0 < x' < l', \quad 0 < t' < +\infty, \tag{1''}$$

$$u'(0, t) = \phi'(t'), \quad 0 < t' < +\infty, \tag{2''}$$

$$u'(x', t') = 0, \quad 0 < x' < l'. \tag{3''}$$

Hence we conclude that the relation

$$a^2 = \nu \frac{k_x^2}{k_t} \tag{5}$$

is fulfilled. In fact, subtracting (1″) from (1) we obtain:

$$0 \equiv \left(a^2 - \nu \frac{k_x^2}{k_t} \right) \frac{\partial^2 u'}{\partial x'^2}.$$

If we assumed that $\partial^2 u'/\partial x'^2 \equiv 0$, then by virtue of equation (1″) (or (1)) $\partial u'/\partial t' \equiv 0$, but this is impossible, since $u(0, t') = \phi'(t')$, where $\phi'(t') \not\equiv 0$. Therefore,

$$a^2 - \nu \frac{k_x^2}{k_t} = 0,$$

which was requiring to be proved.

Sufficiency. Let us convert to the dimensionless quantities ξ, τ, U in the boundary-value problems (I) and (II) by means of the relations

$$x' = b'\xi, \quad t' = t_0'\tau, \quad u' = u_0' U(\xi, \tau), \quad x'' = l''\xi, \quad t'' = t_0''\tau, \quad u'' = u_0'' U(\xi, \tau),$$

where the constants t_0' and t_0'' have the dimensions of time, and u_0' and u_0'' have the dimensions of u' and u'' respectively, where these constants are chosen so that

$$\frac{t_0'}{t_0''} = k_t, \quad \frac{u_0'}{u_0''} = k_u.$$

We recall that, moreover, the relation

$$k_x = \frac{l'}{l''}$$

is fulfilled. The boundary-value problems (I) and (II) take the form

$$\left.\begin{aligned} &\frac{\partial U}{\partial \tau} = \frac{t_0'}{l'^2} a^2 \frac{\partial^2 U}{\partial \xi^2}, \quad 0 < \xi < 1, \quad 0 < \tau < +\infty, \\ &U(0, \tau) = \frac{1}{u_0'} \phi'(t_0', \tau), \quad 0 < \tau < +\infty, \\ &U(\xi, 0) = 0, \quad 0 < \xi < 1, \end{aligned}\right\} \qquad (\mathrm{I}')$$

$$\left.\begin{aligned} &\frac{\partial U}{\partial \tau} = \frac{t_0''}{l''^2} \nu \frac{\partial^2 U}{\partial \xi^2}, \quad 0 < \xi < 1, \quad 0 < \tau < +\infty, \\ &U(0, \tau) = \frac{1}{u_0''} \phi''(t_0'' \tau), \quad 0 < \tau < +\infty, \\ &U(\xi, 0) = 0, \quad 0 < \xi < 1. \end{aligned}\right\} \qquad (\mathrm{II}')$$

From (4) it follows that

$$\frac{1}{u_0'} \phi'(t_0' \tau) = \frac{1}{u_0''} \phi''(t_0'' \tau), \quad 0 < \tau < +\infty.$$

From (5) it follows that

$$\frac{t_0'}{l'^2} a^2 = \frac{t_0''}{l''^2} \nu,$$

i.e. the equations, initial conditions and boundary conditions agree identically in problems (I′) and (II′); therefore (by virtue of the uniqueness theorem) their solutions agree.

Thus

$$U(\xi, \tau) = \frac{1}{u_0'} u'(x', t') = \frac{1}{u_0''} (x'', t''),$$

i.e.

$$u'(x', t') = k_u u''(x'', t''),$$

which was requiring to be proved.

19. The boundary-value problem for determining the temperature in the rod, at the lateral surface of which a convective heat exchange takes place with a medium whose temperature equals zero,

$$\frac{\partial u'}{\partial t'} = a^2 \frac{\partial^2 u'}{\partial x'^2} - \frac{\alpha p}{c\rho\sigma} u', \quad a^2 = \frac{\lambda}{c\rho}, \quad 0 < x' < l', \quad 0 < t' < +\infty, \quad (1)$$

$$u'(0, t') = \phi'(t'), \quad 0 < t' < +\infty, \quad \left.\frac{\partial u'}{\partial x'}\right|_{x'=l'} = 0, \quad (2)$$

$$u'(x', 0) = 0, \quad 0 < x' < l' \quad (3)$$

is similar to the boundary-value problem for determining the concentration of the diffusing substance, the rate of disintegration of which is proportional to the concentration

$$\frac{\partial u''}{\partial t''} = D\frac{\partial^2 u''}{\partial x''^2} - \beta u'', \quad 0 < x'' < l'', \quad 0 < t'' < +\infty, \tag{1'}$$

$$u''(0, t'') = \phi''(t''), \quad 0 < t'' < +\infty, \quad \left.\frac{\partial u''}{\partial x''}\right|_{x''=l''} = 0, \tag{2'}$$

$$u''(x'', 0) = 0, \quad 0 < x'' < l''. \tag{3'}$$

In order that the first problem be similar to the second with given coefficients of similarity k_x, k_t, k_u, it is necessary and sufficient that the relations

$$\phi'(t') = k_u\phi''(t'') \quad \text{for} \quad 0 < t'' < +\infty, \quad \text{where} \quad t' = k_t t'', \tag{4}$$

$$a^2 = \frac{k_x^2}{k_t}D, \quad k_x = \frac{l'}{l''}, \tag{5}$$

$$\frac{\alpha p}{c\rho\sigma} = \frac{1}{k_t}\beta \tag{6}$$

be fulfilled.

Method. Proof of the necessity and sufficiency of conditions (4), (5), (6) is carried out in a manner similar to that done for conditions (4) and (5) in the solution of the preceding problem.

20. Problem (I) "Find the intensity of the electric current in a conductor of finite length of negligibly small self-inductance, if to one of its ends an electromotive force, varying according to a given law, is applied and the other end is earthed through a lumped resistance R_0" is similar to problem (II) formulated above (see the specification of the problem) for determining the temperature in the rod, since problem (I) may be written in the form†

$$\frac{\partial u'}{\partial t'} = \frac{1}{RC}\frac{\partial^2 u'}{\partial x'^2} - \frac{G}{C}u', \quad 0 < x' < l', \quad 0 < t' < +\infty, \tag{1}$$

$$u'(0, t') = \phi'(t'), \quad \left[\frac{\partial u'}{\partial x'} + \frac{R}{R_0}u'\right]_{x'=l'} = 0, \quad 0 < t' < +\infty, \tag{2}$$

$$u'(x' 0) = 0, \quad 0 < x' < l', \tag{3}$$

and problem (II) in the form

$$\frac{\partial u''}{\partial t''} = a^2\frac{\partial^2 u''}{\partial x''} - \frac{\alpha p}{c\rho\sigma}u'', \quad a^2 = \frac{\lambda}{c\rho}, \quad 0 < x'' < l'', \quad 0 < t'' < +\infty, \tag{1'}$$

† In connection with the symbols see problem 2, chapter III, and problem 19. chapter II.

$$u''(0, t'') = \phi''(t''), \quad \left[\frac{\partial u''}{\partial x''} + \frac{\alpha}{\lambda} u''\right]_{x''=l''} = 0, \quad 0 < t'' < +\infty, \tag{2'}$$

$$u''(x'',0) = 0, \quad 0 < x'' < l''. \tag{3'}$$

In order that problem (I) be similar to problem (II) with coefficients of similarity k_x, k_t, k_u, it is necessary and sufficient that the relations

$$\phi'(t') = k_u \phi''(t''), \quad 0 < t'' < +\infty, \quad \text{where} \quad t' = k_t t'', \tag{4}$$

$$\frac{1}{RC} = \frac{k_x^2}{k_t} a^2, \quad k_x = \frac{l'}{l''}, \tag{5}$$

$$\frac{G}{C} = \frac{1}{k_t} \frac{\alpha p}{c\rho\sigma}, \tag{6}$$

$$\frac{R}{R_0} = \frac{1}{k_x} \frac{\alpha}{\lambda}, \tag{7}$$

be fulfilled.

21. The boundary-value problem on the heating of the rod $0 \leqslant x \leqslant l'$ thermally insulated along the surface $x = l'$ (problem I)

$$\frac{\partial u'}{\partial t'} = a^2 \frac{\partial^2 u'}{\partial x'^2}, \quad a^2 \frac{\lambda}{c\rho}, \quad 0 < x' < l', \quad 0 < t' < +\infty, \tag{1}$$

$$u'(0, t') = U_0, \quad u'_{x'}(l', t') = 0, \quad 0 < t' < +\infty, \tag{2}$$

$$u'(x', 0) = 0, \quad 0 < x' < l' \tag{3}$$

is similar to the boundary-value problem formulated in the condition on the propagation of a plane electromagnetic field in a conducting layer $0 \leqslant x'' \leqslant l''$ (problem II)

$$\frac{\partial u''}{\partial t''} = \frac{c^2}{4\pi\sigma\mu} \frac{\partial^2 u''}{\partial x''^2}, \quad 0 < x'' < l'', \quad 0 < t'' < +\infty, \tag{1'}$$

$$u''(0, t'') = H_0, \quad u''_{x''}(l'', t'') = 0, \quad 0 < t'' < +\infty, \tag{2'}$$

$$u''(x'', 0) = 0, \quad 0 < x'' < l''. \tag{3'}$$

In order that the first problem be similar to the second with given coefficients of similarity k_x, k_t, k_u, it is necessary and sufficient that the relations

$$U_0 = k_u H_0, \tag{4}$$

$$a^2 = \frac{k_x^2}{k_t} \frac{c^2}{4\pi\sigma\mu}, \tag{5}$$

$$k_x = \frac{l'}{l''} \tag{6}$$

be fulfilled.

§ 2. Method of Separation of Variables

1. Homogeneous Isotopic Media. Equations with Constant Coefficients

(a) *Problems of heat conduction with constant boundary conditions*

22. (a) The solution of the boundary-value problem

$$u_t = a^2 u_{xx}, \quad a^2 = \frac{\lambda}{c\rho}, \quad 0 < x < l, \quad 0 < t < +\infty, \tag{1}$$

$$u(0, t) = u(l, t) = 0, \quad 0 < t < +\infty, \tag{2}$$

$$u(x, 0) = f(x), \quad 0 < x < +\infty \tag{3}$$

is:

$$u(x, t) = \sum_{n=1}^{+\infty} a_n e^{-\frac{n^2\pi^2 a^2}{l^2} t} \sin \frac{n\pi x}{l}, \quad 0 < x < l, \quad 0 < t < +\infty,$$

where

$$a_n = \frac{2}{l} \int_0^l f(\xi) \sin \frac{n\pi\xi}{l} \, d\xi .$$

(b) If $f(x) \equiv U_0 =$ const., then

$$u(x, t) = \frac{4U_0}{\pi} \sum_{k=0}^{+\infty} \frac{1}{2k+1} e^{\frac{(2k+1)^2\pi^2 a^2}{l^2} t} \sin \frac{(2k+1)\pi x}{l},$$

$$0 < x < l, \quad 0 < t < +\infty. \tag{4}$$

At the point $x = l/2$ we have:

$$u\left(\frac{l}{2}, t\right) = \frac{4U_0}{\pi} \sum_{k=0}^{+\infty} \frac{(-1)^k}{2k+1} e^{-\frac{(2k+1)^2\pi^2 a^2}{l^2} t} . \tag{5}$$

Since the series on the right hand side of the latter equality satisfies the conditions of Leibnitz's theorem on alternating series, then the remainder of series (5) does not exceed absolutely the value of the first of the residual terms, i.e.

$$\left| R_n\left(\frac{l}{2}, t\right) \right| = \left| \frac{4U_0}{\pi} \sum_{k=n+1}^{+\infty} \frac{(-1)^k}{2k+1} e^{-\frac{(2k+1)^2\pi^2 a^2}{l^2} t} \right| \leqslant \frac{4U_0}{\pi} \frac{e^{-\frac{(2n+3)^2\pi^2 a^2}{l^2} t}}{2n+3} . \tag{6}$$

Let us estimate, finally, the ratio of the sum of all the terms of series (5), starting with the second, to the first term of this series. From (6) we have:

$$\frac{\left| R_0\left(\frac{l}{2}, t\right) \right|}{\frac{4U_0}{\pi} e^{-\frac{\pi^2 a^2 t}{l^2}}} \leqslant \frac{1}{3} e^{-\frac{8\pi^2 a^2}{l^2} t} \leqslant \varepsilon \text{ for } t \geqslant t^* = -\frac{l^2}{8\pi^2 a^2} \ln 3\varepsilon, \tag{7}$$

where $\varepsilon > 0$ is an arbitrary positive quantity.

Note. To estimate the error permissible in the replacement of the sum of series (4) by its partial sum at other points $x \neq l/2$, it is possible to use Abel's theorem. But estimation of the remainder of the series by Abel's criterion of approximation at the ends of the interval $0 \leqslant x \leqslant l$ becomes useless. We indicate a method, giving a steady-state estimate of the remainder of the series over the whole interval $0 \leqslant x \leqslant l$:

$$|R_n(x,t)| = \left| \frac{4U_0}{\pi} \sum_{k=n+1}^{+\infty} \frac{1}{2k+1} e^{-\frac{(2k+1)^2\pi^2 a^2}{l^2} t} \sin \frac{(2k+1)\pi x}{l} \right|$$

$$< \frac{4U_0}{\pi} \sum_{k=n+1}^{+\infty} \frac{1}{2k+1} e^{-\frac{(2k+1)^2\pi^2 a^2}{l^2} t} < \frac{4U_0}{\pi} \int_n^{+\infty} \frac{1}{2z+1} e^{-\frac{(2z+1)^2\pi^2 a^2}{l^2} t} dz$$

$$= \frac{2U_0}{\pi} \int_{A_n}^{+\infty} \frac{e^{-\zeta^2}}{\zeta} d\zeta,$$

where $A_n = (2n+1)\pi a \sqrt{t}/l$.

Integrating by parts, we obtain:

$$\int_{A_n}^{+\infty} \frac{e^{-\zeta^2}}{\zeta} d\zeta = \frac{1}{2A_n^2} e^{-A_n^2} + \int_{A_n}^{+\infty} \frac{e^{-\zeta^2}}{\zeta^3} d\zeta.$$

But

$$\int_{A_n}^{+\infty} \frac{e^{-\zeta^2}}{\zeta^3} d\zeta < e^{-A_n^2} \int_{A_n}^{+\infty} \frac{d\zeta}{\zeta^3} = \frac{e^{-A_n^2}}{2A_n^2}.$$

Therefore

$$|R_n(x,t)| < \frac{2U_0}{\pi} \frac{e^{-A_n^2}}{A_n^2}, \quad \text{where} \quad A_n = \frac{(2n+1)\pi a \sqrt{t}}{l}.$$

23. The solution of the equation

$$\frac{\partial u}{\partial t} = a^2 \frac{\partial^2 u}{\partial x^2}, \qquad 0 < x < l, \qquad 0 < t < \infty \tag{3}$$

for the initial condition (1) and boundary conditions (2) (see the specification of the problem) is:

$$u(x,t) = U_1 + (U_2 - U_1)\frac{x}{l} +$$

$$+ \frac{2}{\pi} \sum_{n=1}^{+\infty} \frac{1}{n} \{(U_0 - U_1)[1-(-1)^n] + (-1)^{n+1}(U_1 - U_2)\} e^{-\frac{n^2\pi^2 a^2}{l^2} t} \sin \frac{n\pi x}{l}. \tag{4}$$

The steady-state temperature in the rod equals

$$\overline{u}(x) = \lim_{t \to +\infty} u(x,t) = U_1 + (U_2 - U_1)\frac{x}{l}. \tag{5}$$

Method. The solution of equation (3) for the initial condition (1) and boundary conditions (2) may be found in the form

$$u(x,t) = v(x,t)+\bar{u}(x), \tag{6}$$

where the function $\bar{u}(x)$ is defined as the steady-state solution of equation (3), satisfying the boundary conditions (2), i.e.

$$\frac{d^2\bar{u}(x)}{dx^2} = 0, \quad 0 < x < l,$$

$$\bar{u}(0) = U_1, \quad \bar{u}(l) = U_2,$$

from which

$$\bar{u}(x) = U_1+(U_2-U_1)\frac{x}{l},$$

i.e. $\bar{u}(x)$ is the limit to which the temperature in the rod tends for $t \to +\infty$.

The function $v(x,t)$ will satisfy equation (3) and the conditions

$$v(x,0) = U_0-\bar{u}(x), \tag{7}$$

$$v(0,t) = v(l,t) = 0, \tag{8}$$

i.e. $v(x,t)$ is a solution of the first boundary-value problem with zero boundary conditions. This problem has already been considered (see problem 22).

24. The solution of the boundary-value problem

$$u_t = a^2u_{xx}-h(u-u_0), \ a^2 = \frac{\lambda}{c\rho}, \quad h = \frac{\alpha p}{c\rho\sigma}, \ 0<x<l, \ 0<l<+\infty, \tag{1}$$

$$u(0,t) = U_1, \quad u(l,t) = U_2, \quad 0<t<+\infty, \tag{2}$$

$$u(x,0) = f(x), \quad 0<x<l \tag{3}$$

is:

$$u(x,t) = u_0+w(x)+v(x,t), \quad 0<x<l, \quad 0<t<+\infty \tag{4}$$

where

$$w(x) = \frac{(U_1-u_0)\sinh\frac{\sqrt{h}}{a}(l-x)+(U_2-u_0)\sinh\frac{\sqrt{h}}{a}x}{\sinh\frac{l\sqrt{h}}{a}}, \quad 0<x<l, \tag{5}$$

$$v(x,t) = \sum_{n=1}^{+\infty} A_n e^{-\left(\frac{n^2\pi^2a^2}{l^2}+h\right)t}\sin\frac{n\pi x}{l}, \quad 0<x<l, \quad 0<t<+\infty, \tag{6}$$

$$A_n = \frac{2}{l}\int_0^t [f(\xi)-w(\xi)-u_0]\sin\frac{n\pi\xi}{l}\,d\xi. \tag{7}$$

In particular, if $U_1 = U_2 = 0$ and $f(x) \equiv 0$, then

$$w(x) = -u_0 \frac{\sinh \frac{\sqrt{h}}{a}(l-x) + \sinh \frac{\sqrt{h}}{a} x}{\sinh \frac{l\sqrt{h}}{a}}, \quad 0 < x < l, \tag{5'}$$

$$vx(, t) = -\frac{4hl^2u_0}{\pi a^2} \sum_{k=1}^{+\infty} \frac{\sin \frac{(2k-1)\pi x}{l}}{(2k-1)[(2k-1)^2\pi^2a^2+hl^2]} e^{-\left[\frac{(2k-1)^2\pi^2a^2}{l^2}+h\right]t}. \tag{6'}$$

25. The solution of the boundary-value problem

$$u_t = a^2 u_{xx}, \quad 0 < x < l, \quad 0 < t < +\infty, \tag{1}$$

$$u_x(0, t) = u_x(l, t) = 0, \quad 0 < t < +\infty, \tag{2}$$

$$u(x, 0) = f(x), \quad 0 < x < l \tag{3}$$

is:

$$u(x, t) = \frac{a_0}{2} + \sum_{n=1}^{+\infty} a_n e^{-\frac{\pi^2a^2n^2}{l^2}t} \cos \frac{n\pi z}{l}, \quad 0 < x < l, \quad 0 < t < +\infty, \tag{4}$$

where

$$a_n = \frac{2}{l} \int_0^l f(z) \cos \frac{n\pi z}{l} \, dz, \quad n = 0, 1, 2, 3, \ldots \tag{5}$$

In order to derive the temperature in the case of heat exchange at the lateral surface, it is necessary to multiply the right hand side of (4) by e^{-ht}, where h has the same meaning as in the preceding problem.

26. The solution of the boundary-value problem

$$u_t = a^2 u_{xx} - hu, \quad 0 < x < l, \quad 0 < t < +\infty, \tag{1}$$

$$-\lambda\sigma u_x(0, t) = q_1, \quad \lambda\sigma u_x(l, t) = q_2, \quad 0 < t < +\infty, \tag{2}$$

$$u(x, 0) = f(x), \quad 0 < x < l \tag{3}$$

is:

$$u(x, t) = w(x) + v(x, t), \tag{4}$$

where

$$w(x) = \frac{a}{\sqrt{h}} Q_1 \sinh \frac{\sqrt{h}}{a} x + \frac{Q_2 - Q_1 \cosh \frac{\sqrt{h}}{a} l}{\frac{\sqrt{h}}{a} \sinh \frac{\sqrt{h}}{a} l} \cosh \frac{\sqrt{h}}{a} x, \quad 0 < x < l, \tag{5}$$

$$Q_1 = -\frac{q_1}{\lambda\sigma}, \quad Q_2 = -\frac{q_2}{\lambda\sigma}, \tag{6}$$

$$v(x, t) = \frac{a_0}{2} e^{-ht} + \sum_{n=1}^{+\infty} a_n e^{-\left(\frac{n^2\pi^2a^2}{l^2}+h\right)t} \cos\frac{n\pi x}{l}, \quad 0 < x < l,$$

$$0 < t < +\infty, \qquad (7)$$

$$a_n = \frac{2}{l}\int_0^t [f(z)-w(z)] \cos\frac{n\pi z}{l}\, dz, \quad n = 0, 1, 2, 3, \ldots \qquad (8)$$

Method. See the solution of problem 23.

27. The solution of the boundary-value problem

$$u_t = a^2 u_{xx}, \quad 0 < x < l, \quad 0 < t < +\infty, \qquad (1)$$

$$u(0, t) = U_0, \quad \lambda\sigma u_x(l, t) = q_0, \quad 0 < t + \infty, \qquad (2)$$

$$u(x, 0) = f(x), \quad 0 < x < l \qquad (3)$$

is:

$$u(x, t) = Q_0 x + U_0 + \sum_{n=0}^{+\infty} \left\{a_n - \frac{4}{\pi^2}\frac{(2n+1)\pi U_0 + lQ_0}{(2n+1)^2}\right\} \times$$

$$\times e^{-\frac{(2n+1)^2\pi^2a^2}{4l^2}t} \sin\frac{(2n+1)\pi x}{2l}, \qquad (4)$$

where

$$Q_0 = \frac{q_0}{\lambda\sigma}, \quad a_n = \frac{2}{l}\int_0^l f(z) \sin\frac{(2n+1)\pi z}{2l}\, dz, \qquad (5)$$

and σ is the cross-sectional area of the rod.

If $Q_0 = 0$, $f(x) = 0$, then

$$u(x, t) = U_0 - \frac{4U_0}{\pi}\sum_{k=0}^{+\infty} \frac{1}{2k+1} e^{\frac{(2k+1)^2\pi^2a^2}{4l^2}t} \sin\frac{(2k+1)\pi x}{2l},$$

$$0 < x < l, \quad 0 < t < +\infty. \qquad (6)$$

At the point $x = l$ we have:

$$u(l, t) = U_0 - \frac{4U_0}{\pi}\sum_{k=0}^{+\infty} \frac{(-1)^k}{2k+1} e^{-\frac{(2k+i)^2\pi^2a^2}{4l^2}t}, \quad 0 < t + \infty. \qquad (7)$$

According to Leibnitz' theorem on alternating series we obtain an estimate of the remainder of series (7)

$$|R_n(l, t)| = \left|\frac{4U_0}{\pi}\sum_{k=n+1}^{+\infty} \frac{(-1)^k}{2k+1} e^{-\frac{(2k+1)^2\pi^2a^2}{4l^2}t}\right| \leqslant \frac{4U_0}{\pi(2n+3)} e^{-\frac{(2n+3)^2\pi^2a^2}{4l^2}t},$$

$$0 < t < +\infty. \qquad (8)$$

Let us estimate, finally, the ratio of $R_0(l, t)$ to $\frac{4U_0}{\pi} e^{-\frac{\pi^2 a^2}{4l^2} t}$. By virtue of (8)

$$\frac{|R_0(l, t)|}{\frac{4U_0}{\pi} e^{-\frac{\pi^2 a^2}{4l^2} t}} \leqslant \frac{1}{3} e^{\frac{2\pi^2 a^2}{l^2} t} \leqslant \varepsilon \quad \text{for} \quad t \geqslant t^* = -\frac{l^2}{2\pi^2 a^2} \ln 3\varepsilon. \tag{9}$$

Note. The steady-state estimate of the remainder $R_n(x, t)$ of the series in the segment $0 \leqslant x \leqslant l$ is readily derived by the method given in the note to the answer of problem 22 of the present section.

28. The solution of the boundary-value problem

$$u_t = a^2 u_{xx}, \quad 0 < x < l, \quad 0 < t < +\infty, \tag{1}$$

$$u_x(0, t) = 0, \quad u_x(l, t) = \frac{q}{\lambda\sigma} = Q, \quad 0 < t < +\infty, \tag{2}$$

where λ is the coefficient of heat conduction, σ the cross-sectional area of the rod

$$u(x, 0) = 0, \quad 0 < x < l, \tag{3}$$

is:

$$u(x, t) = Q\left[\frac{a^2 t}{l} + \frac{3x^2 - l^2}{6l} + \frac{2l}{\pi^2}\sum_{k=1}^{+\infty} (-1)^{k+1} \frac{e^{-\frac{k^2\pi^2 a^2}{l^2} t}}{k^2} \cos\frac{k\pi x}{l}\right]. \tag{4}$$

At the point $x = 0$ we have:

$$u(0, t) = Q\left[\frac{a^2 t}{l} - \frac{l}{6} + \frac{2l}{\pi^2}\sum_{k=1}^{+\infty} \frac{(-1)^{k+1}}{k^2} e^{-\frac{k^2\pi^2 a^2}{l^2} t}\right], \quad 0 < t < +\infty. \tag{5}$$

By Leibnitz' criterion for the remainder of the series we obtain the inequality

$$|R_n(0, t)| = \left|\sum_{k=n+1}^{+\infty} \frac{2Ql}{\pi^2} \frac{(-1)^{k+1}}{k^2} e^{-\frac{k^2\pi^2 a^2}{l^2} t}\right| \leqslant \frac{2Ql}{\pi^2 (n+1)^2} e^{-\frac{(n+1)^2\pi^2 a^2}{l^2} t},$$

$$0 < t < +\infty. \tag{6}$$

Method. In order to derive (4) it is possible to reduce the boundary-value problem (1), (2), (3) to the first boundary-value problem by means of the substitution $v(x, t) = \partial u(x, t)/\partial x$, solve the boundary-value problem for v, and then integrate v with respect to x. In this integration an arbitrary function of the time appears. Calculating the amount of heat in the rod by two methods (see the method in the answer to problem 41 of the present chapter) it is possible to determine this function.

Note. In connection with the steady-state estimate of the remainder $R_n(x, t)$ over the segment $0 \leqslant x \leqslant l$ see the note to the answer of the preceding problem.

29. $$u(x, t) = U_0\left[1 + \sum_{n=1}^{+\infty} (-1)^n \frac{2hl\sqrt{(hl)^2 + \mu_n^2}}{\mu_n[(hl)^2 + (hl) + \mu_n^2]} e^{-\frac{\mu_n^2 a^2 t}{l^2}} \cos\frac{\mu_n x}{l}\right], \tag{1}$$

where h is the coefficient of heat exchange appearing in the boundary condition $u_x(l, t)+h[u(l, t)-v_0] = 0$, and μ_n are the positive roots of the transcendental equation

$$\cot \mu = \frac{1}{hl}\mu, \tag{2}$$

forming a sequence, monotonically tending to $+\infty$.

At the point $x = 0$ we have:

$$u(0, t) = U_0\left[1+\sum_{k=1}^{+\infty}(-1)^k \frac{2hl\sqrt{(hl)^2+\mu_k^2}}{\mu_k[(hl)^2+(hl)+\mu_k^2}\, e^{-\frac{\mu_k^2 a^2}{l^2}t}\right]. \tag{3}$$

It is readily verified that series (3) satisfies the conditions of Leibnitz' theorem on alternating series; therefore for the remainder of series (3) we obtain the inequality

$$|R_n(0, t)| = \left|U_0\sum_{k=n+1}^{+\infty}(-1)^k \frac{2hl\sqrt{(hl)^2+\mu_k^2}}{\mu_k[(hl)^2+(hl)+\mu_k^2]}\, e^{-\frac{\mu_k^2 a^2}{l^2}t}\right|$$

$$\leqslant \frac{2U_0 hl\sqrt{(hl)^2+\mu_{n+1}^2}}{\mu_{n+1}[(hl)^2+(hl)+\mu_{n+1}^2]}\, e^{-\frac{\mu_{n+1}^2 a^2}{l^2}t} \tag{4}$$

By virtue of (4) we have:

$$\frac{|R_1(0, t)|}{\dfrac{2U_0hl\sqrt{(hl)^2+\mu_1^2}}{\mu_1[(hl)^2+(hl)+\mu_1^2)}\, e^{-\frac{\mu_1^2 a^2}{l^2}t}}$$

$$\leqslant \frac{(hl)^2+hl+\mu_1^2}{(hl)^2+hl+\mu_2^2}\sqrt{\frac{1+\left(\dfrac{hl}{\mu_2}\right)^2}{1+\left(\dfrac{hl}{\mu_1}\right)^2}}\, e^{-\frac{(\mu_2^2-\mu_1^2)a^2}{l^2}t} \leqslant \varepsilon \tag{5}$$

for

$$t > t^* = -\frac{l^2}{(\mu_2^2-\mu_1^2)a^2}\ln\left[\varepsilon\,\frac{(hl)^2+hl+\mu_2^2}{(hl)^2+hl+\mu_1^2}\sqrt{\frac{1+\left(\dfrac{hl}{\mu_1}\right)^2}{1+\left(\dfrac{hl}{\mu_2}\right)^2}}\right].$$

Note. The steady-state estimate of the remainder of the series $R_n(x, t)$ in the segment $0 \leqslant x \leqslant l$ may be made in a manner similar to that done in the note on page 332.

Taking into account that for the roots $\mu_1 < \mu_2 < \ldots < \mu_n < \mu_{n+1} < \ldots$ of the transcendental equation (2) the inequality

$$\frac{\pi}{2} < \mu_{n+1} - \mu_n < \pi,$$

will hold, we obtain:

$$|R_n(x,t)| \leqslant 2U_0 hl \sum_{k=n+1}^{+\infty} \frac{\sqrt{(hl)^2+\mu_k^2}\, e^{-\frac{\mu_k^2 a^2}{l^2}t}}{\mu_k[(hl)^2+(hl)+\mu_k^2]}$$

$$< \frac{4U_0 hl}{\pi} \int_{\mu_n}^{+\infty} \left[\left(\frac{hl}{\mu}\right)^2+1\right]^{1/2} \frac{e^{-\frac{\mu^2 a^2 t}{l^2}}}{(hl)^2+hl+\mu^2}\, d\mu$$

$$< \frac{4U_0 hl}{\pi}\left[\left(\frac{hl}{\mu_n}\right)^2+1\right]^{1/2} \int_{\mu_n}^{+\infty} \frac{e^{-\frac{\mu^2 a^2 t}{l^2}}}{\mu^2}\, d\mu < \frac{2U_0 ah\sqrt{t}}{\pi}\left[\left(\frac{hl}{\mu_n}\right)^2+1\right]^{1/2} \frac{e^{-A_n^2}}{A_n^3},$$

where

$$A_n = \frac{\mu_n a\sqrt{t}}{l}.$$

30. (a) The solution of the boundary-value problem

$$u_t = a^2 u_{xx}, \quad 0 < x < l, \quad 0 < t < +\infty, \tag{1}$$

$$u_x(0,t) - H[u(0,t)-U_1] = 0, \quad u_x(l,t) + H[u(l,t)-U_2] = 0, \quad 0 < t < +\infty, \tag{2}$$

$$u(x,0) = f(x), \quad 0 < x < l \tag{3}$$

is:

$$u(x,t) = w(x) + v(x,t), \quad 0 < x < l, \quad 0 < t < +\infty, \tag{4}$$

where

$$w(x) = H\frac{U_2-U_1}{2+lH}x + \frac{U_2+(1+lH)U_1}{2+lH}, \quad 0 < x < l, \tag{5}$$

and

$$v(x,t) = \sum_{n=1}^{+\infty} a_n e^{-a^2\lambda_n^2 t}\left(\cos\lambda_n x + \frac{H}{\lambda_n}\sin\lambda_n x\right), 0 < x < l, 0 < t < +\infty, \tag{6}$$

$\lambda_n = z_n/l$, z_n are the positive roots of the transcendental equation

$$\cot z = \frac{1}{2}\left(\frac{z}{lH} - \frac{lH}{z}\right). \tag{7}$$

The eigenfunctions†

$$X_n(x) = \cos \lambda_n x + \frac{H}{\lambda_n} \sin \lambda_n x, \quad n = 1, 2, 3, \ldots, \tag{8}$$

are orthogonal in the segment $0 \leqslant x \leqslant l$; the square of the norm of the eigenfunction $X_n(x)$ equals

$$\|X_n\|^2 = \int_0^l X_n^2(x)\,dx = \frac{(\lambda_n^2 + H^2)l + 2H}{2\lambda_n^2}, \tag{9}$$

$$a_n = \frac{2\lambda_n^2}{(\lambda_n^2 + H^2)l + 2H} \int_0^l [f(z) - w(z)] \left(\cos \lambda_n z + \frac{H}{\lambda_n} \sin \lambda_n z\right) dz. \tag{10}$$

(b) If the temperature of the medium at both ends is the same, and the initial temperature of the rod equals zero, then, taking the middle of the rod as the origin of coordinates, we derive that the temperature in the rod is an even function of x, i.e. for $x = 0$ $\partial u/\partial x = 0$. Thus it is possible to consider instead of the whole rod only half of it, where the boundary-value problem 29 is obtained to determine the temperature (it is necessary to replace l by $l/2$).

31. The solution of the boundary-value problem

$$u_t = a^2 u_{xx} - hu, \quad 0 < x < l, \quad 0 < t < +\infty, \tag{1}$$

$$u_x(0, t) - H[u(0, t) - U_1] = 0, \quad u_x(l, t) + H[u(l, t) - U_2] = 0,$$
$$0 < t < +\infty, \tag{2}$$

$$u(x, 0) = f(x), \quad 0 < x + \infty \tag{3}$$

is:

$$u(x, t) = w(x) + v(x, t), \quad 0 < x < l, \quad 0 < t < +\infty, \tag{4}$$

where

$$w(x) = H \frac{\left[U^2 \frac{\sqrt{h}}{a} - U_1\left(H \sinh \frac{\sqrt{h}}{a} l - \frac{\sqrt{h}}{a} \cosh \frac{\sqrt{h}}{a} l\right)\right] \cosh \frac{\sqrt{h}}{a} x}{\left(H^2 + \frac{h}{a^2}\right) \sinh \frac{\sqrt{h}}{a} l} +$$
$$+ H \frac{\left[U_1\left(H \cosh \frac{\sqrt{h}}{a} l + \frac{\sqrt{h}}{a} \sinh \frac{\sqrt{h}}{a} l\right) + U_2 H\right] \sinh \frac{\sqrt{h}}{a} x}{\left(H^2 + \frac{h}{a^2}\right) \sinh \frac{\sqrt{h}}{a} l}, \tag{5}$$

---

† For more detail see the solution of problem 111, chapter II; the eigenfunctions considered there are obtained by multiplying the eigenfunctions (8) by $\lambda_n / \left(\sqrt{\lambda_n^2 + h^2}\right)$, therefore, knowing the square of the modulus of the eigenfunctions, considered in problem 11, chapter II, the square of the modulus of the eigenfunctions (8) is readily derived.

$$v(x, t) = \sum_{n=1}^{+\infty} a_n e^{-(a^2\lambda_n^2+h)t} \left(\cos \lambda_n x + \frac{H}{\lambda_n} \sin \lambda_n x\right), \tag{6}$$

λ_n, $X_n(x) = \cos \lambda_n x + H/\lambda_n \sin \lambda_n x$ and a_n are determined as in the answer to the preceding problem.

32. The solution of the boundary-value problem

$$u_t = a^2 u_{xx} - h[u-u_0], \quad -\pi < x < \pi, \quad 0 < t < +\infty, \tag{1}$$

$$u(x, 0) = f(x), \quad -\pi < x < \pi, \tag{2}$$

$$u(-\pi, t) = u(x, t), \quad u_x(-\pi, t) = u_x(\pi, t), \quad 0 < t < +\infty, \tag{3}$$

is:

$$u(x, t) = u_0 + e^{-ht} v(x, t), \tag{4}$$

$$v(x, t) = \sum_{n=0}^{+\infty} (a_n \cos nx + b_n \sin nx) e^{-n^2 a^2 t}, \tag{5}$$

where:

$$a_0 = \frac{1}{2\pi} \int_{-\pi}^{\pi} [f(x) - u_0] \, dx, \tag{6}$$

$$a_n = \frac{1}{\pi} \int_{-\pi}^{\pi} [f(x) - u_0] \cos nx \, dx, \quad b_n = \frac{1}{\pi} \int_{-\pi}^{\pi} [f(x) - u_0] \sin nx \, dx. \tag{7}$$

If the initial temperature of the ring $f(x) \equiv u_1 = \text{const.}$, then

$$u(x, t) = u_0 + e^{-ht} [u_1 - u_0].$$

(b) *Problems of heat conduction with variable boundary conditions and free terms, dependent on x and t*

33. The solution of the boundary-value problem

$$u_t = a^2 u_{xx}, \quad 0 < x < l, \quad 0 < t < +\infty, \tag{1}$$

$$u(0, t) = 0, \quad u(l, t) = At, \quad 0 < t < +\infty, \tag{2}$$

$$u(x, 0) = 0, \quad 0 < x < l \tag{3}$$

is:

$$u(x, t) = \frac{A}{l} xt + \frac{Ax}{6a^2 l} (x^2 - l^2) + v(x, t), \quad 0 < x < l, \quad 0 < t < +\infty, \tag{4}$$

$$v(x, t) = \sum_{n=1}^{+\infty} a_n e^{-\frac{n^2\pi^2 a^2}{l^2} t} \sin \frac{n\pi x}{l}, \tag{5}$$

where

$$a_n = -\frac{A}{3a^2 l^2} \int_0^l z(z^2 - l^2) \sin \frac{n\pi z}{l} \, dz. \tag{6}$$

34. $$u(x,t)=\left\{\int_0^t \Phi(\tau)e^{-\frac{\pi^2a^2}{l^2}(t-\tau)}\,d\tau\right\}\sin\frac{\pi x}{l}+$$

$$+\sum_{n-1}^{+\infty} a_n e^{-\frac{n^2\pi^2a^2}{l^2}t}\sin\frac{n\pi x}{l},\quad 0<x<l,\quad 0<t<+\infty,\tag{1}$$

where

$$a_n=\frac{2}{l}\int_0^l f(z)\sin\frac{n\pi z}{l}\,dz.\tag{2}$$

Method. A particular solution of the equation, satisfying the boundary conditions (see the specification of the problem), may be sought in the form

$$w(x,t)=\phi(t)\sin\frac{\pi x}{l},\tag{3}$$

where $\phi(t)$ is a function to be determined.

35. (a) The solution of the boundary-value problem

$$u_t=a^2u_{xx}+f(x,t),\quad 0<x<l,\quad 0<t<+\infty,\tag{1}$$

$$u(0,t)=u(l,t)=0,\quad 0<t<+\infty,\tag{2}$$

$$u(x,0)=\phi(x),\quad 0<x<l,\tag{3}$$

where $f(x,t)=F(x,t)/c\rho$ is:

$$u(x,t)=\int_0^l \phi(\xi)\left\{\frac{2}{l}\sum_{n=1}^{+\infty} e^{-\frac{n^2\pi^2a^2}{l^2}t}\sin\frac{n\pi x}{l}\sin\frac{n\pi\xi}{l}\right\}d\xi+$$

$$+\int_0^t d\tau\int_0^l f(\xi,\tau)\left\{\frac{2}{l}\sum_{n=1}^{+\infty} e^{-\frac{n^2\pi^2a^2}{l^2}(t-\tau)}\sin\frac{n\pi x}{l}\sin\frac{n\pi\xi}{l}\right\}d\xi.$$

(b) The solution of the boundary-value problem

$$u_t=a^2u_{xx}+\frac{Q}{c\rho}\delta(x-x_0),\quad 0<x<l,\quad 0<t<+\infty,\tag{1}$$

$$u(0,t)=u(l,t)=0,\quad 0<t<+\infty,\tag{2}$$

$$u(x,0)=0,\quad 0<x<l\tag{3}$$

is:

$$u(x,t)=\frac{2Ql}{c\rho\pi^2a^2}\sum_{n=1}^{+\infty}\frac{1}{n^2}\left(1-e^{-\frac{n^2\pi^2a^2}{l^2}t}\right)\sin\frac{n\pi x}{l}\sin\frac{n\pi x_0}{l}.\tag{4}$$

36. The solution of the boundary-value problem

$$u_t=a^2u_{xx}-hu+\frac{A}{c\rho\sigma}e^{-ht}\delta(x-v_0t),\quad 0<x<l,\quad 0<t<\frac{l}{v_0},$$

$$u(0, t) = u(l, t) = 0, \quad 0 < t < \frac{l}{v_0},$$

$$u(x, 0) = 0, \quad 0 < x < l$$

is:

$$u(x, t) = \frac{2A}{c\rho\sigma l} e^{-ht} \sum_{n=1}^{+\infty} \frac{\sin \frac{n\pi v}{l}}{v_0^2 + \frac{n^2\pi^2 a^2}{l^2}} \left(\sin \frac{n\pi v_0 t}{l} - \frac{v_0 l}{n\pi} \cos \frac{n\pi v_0 t}{l} + \frac{v_0 l}{n\pi} \right).$$

37. It is necessary to solve the boundary-value problem

$$u_t = a^2 u_{xx} + f(x, t), \quad 0 < x < l, \quad 0 < t < +\infty, \tag{1}$$

$$u_x(0, t) - hu(0, t) = \psi_1(t), \quad u_x(l, t) + hu(l, t) = \psi_2(t), \quad 0 < t < +\infty, \tag{2}$$

$$u(x, 0) = \phi(x), \quad 0 < x < l. \tag{3}$$

If it is required that the function

$$\psi(x, t) = (\alpha_1 x + \beta_1)\psi_1(t) + (\alpha_2 x + \beta_2)\psi_2(t), \quad 0 < x < l, \quad 0 < t < +\infty \tag{4}$$

should satisfy the boundary conditions (2) of the boundary-value problem (1), (2), (3), then the coefficients α_1, β_1, α_2, β_2 are defined uniquely:

$$\alpha_1 = \frac{1}{2+hl}, \quad \beta_1 = \frac{1+hl}{(2+hl)h}, \quad \alpha_2 = \frac{1}{2+hl}, \quad \beta_2 = \frac{1}{h(2+hl)}. \tag{5}$$

The solution of the boundary-value problem (1), (2), (3) may be sought in the form

$$u(x, t) = v(x, t) + \psi(x, t), \quad 0 < x < l, \quad 0 < t < +\infty, \tag{6}$$

where $v(x, t)$ is the new unknown function, and $\psi(x, t)$ is already defined. For the function $v(x, t)$ we obtain the boundary-value problem

$$v_t = a^2 v_{xx} + f^*(x, t), \quad 0 < x < l, \quad 0 < t < +\infty, \tag{7}$$

$$v_x(0, t) - hv(0, t) = 0, \quad v_x(l, t) + hv(l, t) = 0, \quad 0 < t < +\infty, \tag{8}$$

$$v(x, 0) = \phi^*(x), \quad 0 < x < l, \tag{9}$$

where

$$f^*(x, t) = f(x, t) - (\alpha_1 x + \beta_1)\psi_1'(t) - (\alpha_2 x + \beta_2)\psi_2'(t), \tag{10}$$

$$\phi^*(x) = \phi(x) - (\alpha_1 x + \beta_1)\psi_1(0) - (\alpha_2 x + \beta_2)\psi_2(0). \tag{11}$$

The solution of the boundary-value problem (7), (8), (9) may be found in the form

$$v(x, t) = \sum_{n=1}^{+\infty} v_n(t) X_n(x), \quad 0 < x < l, \quad 0 < t < +\infty, \tag{12}$$

where $X_n(x)$ are the eigenfunctions of the boundary-value problem

$$X''(x)+\lambda^2 X(x) = 0, \quad 0 < x < l, \tag{13}$$

$$X'(0)-hX(0) = 0, \quad X'(l)+hX(l) = 0\,\dagger. \tag{14}$$

The functions $v_n(t)$ have to be determined. The function $v(x, t)$ already satisfies the boundary conditions (8). If it is required that $v(x, t)$ should also satisfy equation (7) and the initial condition (9), then the functions $v_n(t)$ are determined. In order to do this we expand the right hand side of equation (7) and $\phi^*(x)$ in a series of the eigenfunctions $X_n(x)$

$$f^*(x,t) = \sum_{n=1}^{+\infty} \Theta_n(t)X_n(x), \quad 0 < x < l, \quad 0 < t < +\infty, \tag{15}$$

where

$$\Theta_n(t) = \frac{2\lambda_n^2}{(\lambda_n^2+h^2)l+2h}\int_0^l f^*(z, t)X_n(z)\,dz, \tag{16}$$

and

$$\phi^*(x) = \sum_{n=1}^{+\infty} a_n X_n(x), \quad 0 < x < l, \tag{17}$$

where

$$a_n = \frac{2\lambda_n^2}{(\lambda_n^2+h^2)l+2h}\int_0^l \phi^*(z)X_n(z)\,dz. \tag{18}$$

Substituting (12) and (15) in equation (7) and assuming the uniform convergence of the resulting derivatives of the series, we obtain:

$$\sum_{n=1}^{+\infty} [v_n'(t)+a^2\lambda_n^2 v_n(t)-\Theta_n(t)]X_n(x) = 0, \quad 0 < x < l, \quad 0 < t < +\infty. \tag{19}$$

In order to fulfil (19) it is sufficient that the equality

$$v_n^1(t)+a^2\lambda_n^2 v_n(t) = \Theta_n(t), \quad 0 < t < +\infty, \quad n = 1, 2, 3,\ldots \tag{20}$$

be fulfilled. Thus we obtain the differential equations for determining the functions $v_n(t)$.

Assuming in (12) $t = 0$ and comparing with (17), we obtain from (9):

$$\sum_{n=1}^{+\infty} [v_n(0)-a_n]X_n(x) = 0, \quad 0 < x < l. \tag{21}$$

To fulfil (21) it is sufficient to satisfy the equalities

$$v_n(0) = a_n, \quad n = 1, 2, 3,\ldots \tag{22}$$

---

† In connection with the determination of the eigenvalues λ_n and the norms of the eigenfunctions X_n see the answer to problem 30.

Solving the differential equations (20) for the initial conditions (22) we obtain:

$$v_n(t) = \int_0^t e^{-a^2\lambda_n^2(t-\tau)}\,\Theta_n(\tau)\,d\tau + a_n e^{-a^2\lambda_n^2 t}. \tag{23}$$

This completes the solution of the problem.

38. The solution of the boundary-value problem (1), (2), (3) is:

$$u(x,t) = v(x,t)+\psi(x,t), \quad 0<x<l, \quad 0<t<+\infty, \tag{4}$$

where $\psi(x, t)$ has the same meaning as in the answer to the preceding problem and

$$v(x,t) = \int_0^t d\tau \int_0^l f^*(z,\tau)G(x,z,t-\tau)\,dz + \int_0^l \phi^*(z)G(x,z,t)\,dz, \tag{5}$$

$$G(x,z,t-\tau) = \sum_{n=1}^{+\infty} e^{-(a^2\lambda_n^2+h)(t-\tau)}\frac{X_n(x)X_n(z)}{||X_n||^2}, \tag{6}$$

$||X_n||^2$ and λ_n have the same meaning as in the answer to problem 30,

$$f^*(x,t) = f(x,t)-h\psi(x,t)-\phi_t(x,t), \tag{7}$$

$$\phi^*(x) = \phi(x)-\psi(x,0). \tag{8}$$

39. (a) $u(x,t) \approx$

$$\approx \frac{A}{2}\left\{\frac{e^{k(x+l)}\cos[k(x-l)+\omega t]+e^{-k(x+l)}\cos[k(x-l)-\omega t]}{\cosh 2kl-\cos 2kl}-\right.$$

$$\left.-\frac{e^{k(l-x)}\cos[k(x+l)+\omega t]+e^{-k(x+l)}\cos[k(x+l)-\omega t]}{\cosh 2kl-\cos 2kl}\right\};$$

(b) $$u(x,t) \approx \frac{A}{4k}\left\{(l-i)\frac{e^{k(1+i)x+i\omega t}-e^{-k(1+i)x+i\omega t}}{e^{k(1+i)l}+e^{-k(1+i)l}}+\right.$$

$$\left.+(1+i)\frac{e^{k(1-i)x-i\omega t}-e^{-k(1-i)x-i\omega t}}{e^{k(1-i)l}+e^{-k(1-i)l}}\right\};$$

(c) $$u(x,t) \approx \frac{A}{2}\left\{\frac{e^{k(1+i)x+i\omega t}-e^{-k(1+i)x+i\omega t}}{[k(1+i)-h]\,e^{k(1+i)l}+[k(1+i)+h]\,e^{-k(1-i)l}}+\right.$$

$$\left.+\frac{e^{k(1-i)x-i\omega t}-e^{-k(1-i)x-i\omega t}}{[k(1-i)-h]\,e^{k(1-i)l}+[k(1-i)+h]\,e^{-k(1-i)l}}\right\},$$

where $k = \sqrt{\omega/2}/a$.

Method. The solution of the boundary-value problem in the case of the boundary conditions (a) for an arbitrary initial condition, i.e. the solution of the problem

$$u_t = a^2u_{xx}, \quad 0<x<l, \quad 0<t<+\infty, \tag{1}$$

$$u(0,t) = 0, \quad u(l,t) = A\cos\omega t, \quad 0<t<+\infty, \tag{2}$$

$$u(x,0) = \phi(x), \quad 0<x<l, \tag{3}$$

may be found in the form

$$u(x, t) = v(x, t)+w(x, t), \quad 0 < x < l, \quad 0 < t < +\infty, \tag{4}$$

where $v(x, t)$ is a particular solution of equation (1), satisfying the boundary conditions (2), and $w(x, t)$ is a solution of the boundary-value problem

$$w_t = a^2 w_{xx}, \quad 0 < x < l, \quad 0 < t < +\infty, \tag{1'}$$

$$w(0, t) = w(l, t) = 0, \quad 0 < t < +\infty, \tag{2'}$$

$$w(x, 0) = \phi(x) - v(x, 0), \quad 0 < x < l. \tag{3'}$$

The function $v(x, t)$ can be found as the real part of the particular solution of the boundary-value problem

$$U_t = a^2 U_{xx}, \tag{5}$$

$$U(0, t) = 0, \quad U(l, t) = A\,e^{i\omega t}, \tag{6}$$

which may readily be found in the form

$$U(x, t) = X(x)\,e^{i\omega t}. \tag{7}$$

Thus

$$v(x, t) = \frac{1}{2}\{X(x)\,e^{i\omega t} + \overline{X}(x)\,e^{-i\omega t}\}, \tag{8}$$

where the line above $X(x)$ denotes the complex conjugate.

According to (8) $v(x, t)$ does not contain terms tending to zero or to infinity for $t \to +\infty$, and since $\lim\limits_{t\to+\infty} w(x, t) = 0$, then $v(x, t)$ represents the asymptotic form of the temperature for $t \to \infty$.

In the case of boundary conditions (b) or (c) the problem is solved similarly.

40. $u(x, t) = \dfrac{Q}{\pi c\rho\sigma} e^{-ht}\left[\dfrac{1}{2} + \sum\limits_{n=1}^{+\infty} e^{-n^2a^2t}\cos nx\right].$

At a point, diametrically opposite the source†,

$$u(\pi, t) = \frac{Q}{\pi c\rho\sigma} e^{-ht}\left[\frac{1}{2} + \sum_{n=1}^{+\infty} e^{-n^2a^2t}(-1)^n\right].$$

The series on the right hand side of the latter equality satisfies the conditions of Leibnitz' theorem on alternating series; therefore the error made in replacing its sum by a partial sum does not exceed absolutely the value of the first of the residual terms.

(c) *Problems of diffusion*

41. $Q(t) = l\sigma U_0\left\{1 - \dfrac{8}{\pi^2}\sum\limits_{n=0}^{+\infty}\dfrac{e^{-\frac{(2n+1)^2\pi^2a^2}{4l^2}t}}{(2n+1)^2}\right\}.$

† In connection with the symbols see problems 3 and 32.

Method.

$$Q(t) = \sigma \int_0^l u(x, t)\, dx,$$

where $u(x, t)$ is the concentration of the diffusing substance in the cylinder at time t.

We note that $Q(t)$ may also be determined by means of the flow of the diffusing substance through the open end:

$$Q(t) = -a^2\sigma \int_0^t \frac{\partial u(0, \tau)}{\partial x}\, d\tau.$$

The equivalence of these two expressions is readily verified by means of integrating both sides of the fundamental equation

$$\int_0^l d\xi \int_0^t u_t(\xi, \tau)\, d\tau = a^2 \int_0^t d\tau \int_0^l u_{xx}(\xi, \tau)\, d\xi$$

using the boundary conditions.

An expression for $u(x, t)$ may be obtained as a particular case of the solution of problem 27.

42. $$Q(t) = U_0\sigma \left\{ l - \sum_{n=1}^{+\infty} \frac{\left[\int_0^l \left(\cos\lambda_n x + \frac{H}{\lambda_n} \sin\lambda_n x \right) dx \right]^2}{\int_0^l \left(\cos\lambda_n x + \frac{H}{\lambda_n} \sin\lambda_n x \right)^2 dx}\, e^{-a^2\lambda_n^2 t} \right\},$$

where λ_n are the roots of the trasncendental equation

$$\cot \lambda_n l = \frac{\lambda_n}{H},$$

and H the coefficient appearing in the boundary conditions

$$u_x = H(u - U_0) \quad \text{for} \quad x = 0.$$

Method. See the method to the preceding problem. An expression for $u(x, t)$ may be obtained from the solution of problem 30.

43. $$Q(t) = U_0\sigma \left\{ \frac{\frac{a}{\sqrt{\beta}} \sinh \frac{l\sqrt{\beta}}{a}}{\cosh \frac{l\sqrt{\beta}}{a}} - \frac{8l}{\pi^2} \sum_{n=0}^{\infty} \frac{e^{-\left(\frac{(2n+1)^2\pi^2a^2}{4l^2} + \beta \right) t}}{(2n+1)^2 \left(1 - \frac{4l\beta}{(2n+1)^2\pi^2a^2} \right)} \right\}.$$

Method. See the method to problem 41.

44. (a) $l_c = \dfrac{\pi a}{\sqrt{\beta}}$,

(b) $l_c = \dfrac{\pi a}{2\sqrt{\beta}}$,

(c) for any length of cylinder the process of growth of concentration is of an avalanche nature; here β is the coefficient of multiplication appearing in the equation

$$\frac{\partial u}{\partial t} = D\frac{\partial^2 u}{\partial x^2} + \beta u, \quad \beta > 0.$$

(d) *Problems of electrodynamics*

45. $v(x, t) = E_0 + \dfrac{4(E_0 - v_0)}{\pi} \displaystyle\sum_{n=1}^{+\infty} \frac{(-1)^n}{2n-1} e^{-\frac{(2n-1)^2\pi^2}{4l^2RC}t} \cos\frac{(2n-1)\pi x}{2l},$

$$0 < x < l, \quad 0 < t < +\infty,$$

where E_0 is the constant e.m.f. applied to the end $x = l$, and R and C are the resistance and capacitance per unit length of the conductor.

46.

$$v(x, t) = E_0 + 2lE_0C \sum_{n=1}^{+\infty} e^{-\frac{\alpha_n^2 t}{l^2RC}} \frac{C_0\alpha_n \sin\alpha_n\left(1-\frac{x}{l}\right) - Cl\cos\alpha_n\left(1-\frac{x}{l}\right)}{(lCC_0 + l^2C^2 + C_0^2\alpha_n^2)\alpha_n \sin\alpha_n},$$

where α_n are the roots of the equation

$$\alpha \tan\alpha = \frac{Cl}{C_0},$$

and E_0 is the constant e.m.f. applied to the end $x = 0$ of the conductor.

47.

$$v(x, t) = \frac{E_0R(l-x)}{R_0 + Rl} + 2E_0R^2 \sum_{n=1}^{+\infty} e^{-\frac{\alpha_n^2}{RC}t} \frac{\sin\alpha_n(l-x)}{\alpha_n[R(R_0+Rl) + lR_0^2\alpha_n^2]\cos\alpha_n l},$$

where R and C are the resistance and capacity per unit length of the conductor, and α_n the positive roots of the equation

$$R\tan\alpha l + \alpha R_0 = 0.$$

48. The solution of the boundary-value problem

$$H_t = a^2H_{xx}, \quad 0 < x < l, \quad 0 < t < +\infty, \quad a^2 = \frac{c^2}{4\pi\sigma\mu}, \tag{1}$$

$$H(0, t) = H(l, t) = H_0, \quad 0 < t < +\infty, \tag{2}$$

$$H(x, 0) = 0, \quad 0 < x < l \tag{3}$$

is:

$$H(x, t) = H_0 - \frac{4H_0}{\pi} \sum_{k=0}^{+\infty} \frac{e^{-\frac{(2k+1)^2\pi^2a^2}{l^2}t}}{2k+1} \sin\frac{(2k+1)\pi x}{l},$$

$$0 < x < l, \quad 0 < t < +\infty. \tag{4}$$

At the point $x = l/2$ we have:

$$H\left(\frac{l}{2}, t\right) = H_0 - \frac{4H_0}{\pi} \sum_{k=n+1}^{+\infty} \frac{(-1)^k}{2k+1} e^{-\frac{(2k+1)^2\pi^2a^2}{l^2}t}, \quad 0 < t < +\infty. \tag{5}$$

The remainder of series (5) can be estimated by Leibnitz' criterion

$$\left|R_n\left(\frac{l}{2}, t\right)\right| = \left|\frac{4H_0}{\pi} \sum_{k=n+1}^{+\infty} \frac{(-1)^k}{2k+1} e^{-\frac{(2k+1)^2\pi^2a^2}{l^2}t}\right| \leqslant \frac{4H_0}{\pi(2n+3)} e^{-\frac{(2n+3)^2\pi^2a^2}{l^2}t},$$

$$0 < t < +\infty. \tag{6}$$

From (6) we have:

$$\frac{\left|R_0\left(\frac{l}{2}, t\right)\right|}{\frac{4H_0}{\pi} e^{-\frac{\pi^2a^2}{l^2}t}} \leqslant \frac{1}{3} e^{-\frac{8\pi^2a^2}{l^2}t} \leqslant \varepsilon \quad \text{for} \quad t \geqslant t^* = -\frac{l^2}{8\pi^2a^2} \ln 3\varepsilon. \tag{7}$$

2. Inhomogeneous Media. Equations with Variable Coefficients and Matching Conditions

49. The temperature in the rod is a solution of the boundary-value problem

$$c(x)\rho(x)\frac{\partial u}{\partial t} = \frac{\partial}{\partial x}\left[\lambda(x)\frac{\partial u}{\partial x}\right], \quad 0 < x < l, \quad 0 < t < +\infty, \tag{1}$$

$$u(0, t) = u(l, t) = 0, \quad 0 < t < +\infty, \tag{2}$$

$$u(x, 0) = \phi(x), \quad 0 < x < l, \tag{3}$$

where

$$c(x) = \begin{cases} \bar{c}, 0 < x < x_0, \\ \bar{\bar{c}}, x_0 < x < l, \end{cases} \quad \rho(x) = \begin{cases} \bar{\rho}, 0 < x < x_0, \\ \bar{\bar{\rho}}, x_0 < x < l, \end{cases} \quad \lambda(x) = \begin{cases} \bar{\lambda}, 0 < x < x_0, \\ \bar{\bar{\lambda}}, x_0 < x < l, \end{cases} \tag{4}$$

$\bar{c}$, $\bar{\bar{c}}$, $\bar{\rho}$, $\bar{\bar{\rho}}$, $\bar{\lambda}$, $\bar{\bar{\lambda}}$ are constants, describing the properties of the rod,

$$u(x, t) = \sum_{n=1}^{+\infty} a_n e^{-\omega_n^2 t} X_n(x), \quad 0 < x < l, \quad 0 < t < +\infty, \tag{5}$$

where

$$X_n(x) = \begin{cases} \dfrac{\sin \dfrac{\omega_n}{\bar{a}} x}{\sin \dfrac{\omega_n}{\bar{a}} x_0}, & 0 < x < x_0, \\[2ex] \dfrac{\sin \dfrac{\omega_n}{\bar{\bar{a}}} (l-x)}{\sin \dfrac{\omega_n}{\bar{\bar{a}}} (l-x_0)}, & x_0 < x < l, \end{cases} \tag{6}$$

$$\bar{a} = \sqrt{\frac{\bar{\lambda}}{\bar{c}\bar{\rho}}}, \quad \bar{\bar{a}} = \sqrt{\frac{\bar{\bar{\lambda}}}{\bar{\bar{c}}\bar{\bar{\rho}}}}, \tag{7}$$

ω_n are the roots of the equation

$$\frac{\bar{\lambda}}{\bar{a}} \cot \frac{\omega}{\bar{a}} x_0 = \frac{\bar{\bar{\lambda}}}{\bar{\bar{a}}} \cot \frac{\omega}{\bar{\bar{a}}} (x_0 - l), \tag{8}$$

$$a_n = \frac{\int_0^l c(x)\rho(x)\phi(x)X_n(x)\,\mathrm{d}x}{||X_n||^2}, \tag{9}$$

$$||X_n||^2 = \int_0^l c(x)\rho(x)X_n^2(x)\,\mathrm{d}x = \frac{\bar{c}\,\bar{\rho}\,x_0}{2\sin^2 \frac{\omega_n}{\bar{a}} x_0} + \frac{\bar{\bar{c}}\bar{\bar{\rho}}(l-x_0)}{2\sin^2 \frac{\omega_n}{\bar{\bar{a}}}(l-x_0)}. \tag{10}$$

Method. See the solution of problem 164, § 3, chapter II.

50. $u(x, t) = \sum_{n=1}^{+\infty} a_n \mathrm{e}^{-a^2\lambda_n^2 t} X_n(x), \quad 0 < x < l, \quad 0 < t < +\infty,$ (1)

$a^2 = \lambda/c\rho$, where λ is the coefficient of heat conduction, c is the specific heat and ρ is the mass density of the material of the rod;

$$X_n(x) = \begin{cases} \dfrac{\sin \lambda_n x}{\sin \lambda_n x_0}, & 0 < x < x_0, \\ \dfrac{\sin \lambda_n (l-x)}{\sin \lambda_n (l-x_0)}, & x_0 < x < l, \end{cases} \quad n = 1, 2, 3, \ldots, \tag{2}$$

λ_n, the eigenvalues of the boundary-value problem, are the roots of the equation

$$\cot \lambda_n x_0 - \cot \lambda_n (l-x_0) = \frac{C_0}{c\rho} \lambda_n, \tag{3}$$

$$a_n = \frac{c\rho \int_0^l \phi(x)X_n(x)\,\mathrm{d}x + C_0\phi(x_0)X_n(x_0)}{\dfrac{c\rho x_0}{2\sin^2 \lambda_n x_0} + \dfrac{c\rho(l-x_0)}{2\sin^2 \lambda_n (l-x_0)} + \dfrac{C_0}{2}}, \tag{4}$$

where $u(x, 0) = \phi(x)$ are the initial values of the temperature.

Method. See the solution of problem 167, § 3, chapter II.

51.

$$u(x, t) = \frac{1}{L-x} \sum_{n=1}^{+\infty} a_n \mathrm{e}^{-\frac{n^2\pi^2 a^2}{l^2} t} \sin \frac{n\pi x}{l}, \quad 0 < x < l, \quad 0 < t < +\infty, \tag{1}$$

$$a_n = \frac{2}{l}\int_0^l (L-z)\phi(z)\sin\frac{n\pi z}{l}\,dz, \qquad n = 1, 2, 3, \ldots, \tag{2}$$

where L denotes the length of the complete cone, the truncation of which is obtained by considering a rod of length l.

52. $$u(x, t) = \sum_{n=1}^{+\infty} a_n e^{-\left(\frac{n^2\pi^2}{l^2}+\frac{2m}{c\rho}\right)a^2 t}\sin\frac{n\pi x}{l}, \quad 0 < x < l, \quad 0 < t < +\infty,$$

where

$$a_n = \frac{2}{l}\int_0^l \phi(z)\sin\frac{n\pi z}{l}\,dz.$$

53. For the velocity of liquid particles $u(x, t)$ and the velocity of motion of the plate $v(t)$ we obtain the expressions

$$u(x, t) = \frac{gl\sigma}{2\rho\nu}\left(\frac{l-x}{l}\right) - \frac{4g\rho l^3}{\sigma\nu}\sum_{n=1}^{+\infty}\frac{e^{-\frac{\lambda_n^2 \nu t}{l^2}}}{\lambda_n^2\left(\lambda_n^2+2\frac{\rho l}{\sigma}+\frac{4\rho^2 l^2}{\sigma^2}\right)}\frac{\sin\lambda_n\frac{l-x}{l}}{\sin\lambda_n}, \tag{1}$$

$$v(t) = \frac{gl\sigma}{2\rho\nu} - \frac{4g\rho l^3}{\sigma\nu}\sum_{n=1}^{+\infty}\frac{e^{-\frac{\lambda_n^2 \nu t}{l^2}}}{\lambda_n^2\left(\lambda_n^2+2\frac{\rho l}{\sigma}+4\frac{\rho^2 l^2}{\sigma^2}\right)}, \tag{2}$$

where l is half the distance between the boundary plates, ρ is the density of the liquid, ν is the kinematic coefficient of viscosity, σ is the surface density of the plate, g is the acceleration of gravity, λ_n are the positive roots of the equation

$$\lambda\tan\lambda = \frac{2\rho l}{\sigma} \tag{3}$$

(λ are the eigenvalues of the boundary-value problem, multiplied by l).

Method. For $u(x, t)$ we have the boundary-value problem

$$u_t = \nu u_{xx}, \qquad -l < x < 0, \quad 0 < x < l, \quad 0 < t < +\infty, \tag{4}$$

$$u(-l, t) = u(l, t) = 0, \quad u(0, t) = v(t), \quad 0 < t < +\infty, \tag{5}$$

$$u(x, 0) = 0, \quad 0 < x < l. \tag{6}$$

For the velocity of motion of the plate we have:

$$\frac{dv}{dt} = g + \frac{2\rho\nu}{\sigma}\left[\frac{\partial u}{\partial x}\right]_{x=0}, \tag{7}$$

$$v(0) = 0. \tag{8}$$

Since the vetocity distribution of the liquid particles is symmetrical with respect to the moving plate, then it is sufficient to determine $u(x, t)$ in the interval $0 < x < l$. The functions

$$X_n(x) = \sin \lambda_n \frac{l-x}{l}$$

are orthogonal in the segment $0 < x < l$. (See the solution of problem 167, § 3 chapter II.)

§ 3. Method of Integral Representations and Source Functions

1. Homogeneous Isotropic Media. Application of the Fourier Integral Transform to Problems on the Infinite Line and Semi-infinite Line

A definition of the Fourier integral transform and the general scheme of application to the solution of boundary-value problems is given in chapter II (pages 296–299).

54. *Solution.* Let us multiply both sides of the equation.

$$\frac{\partial u(\xi, t)}{\partial t} = a^2 \frac{\partial^2 u(\xi, t)}{\partial \xi^2} \quad \text{by} \quad \frac{1}{\sqrt{2x}} e^{-i\lambda\xi}$$

and integrate with respect to ξ from $-\infty$ to $+\infty$, assuming for example, that the function u and its derivatives tend to zero sufficiently rapidly for $\xi \to \pm\infty$. Integrating by parts, we obtain;

$$\frac{1}{\sqrt{2\pi}} \int_{-\infty}^{+\infty} \frac{\partial u}{\partial t} e^{-i\lambda\xi} d\xi = \frac{\partial}{\partial t} \frac{1}{\sqrt{2\pi}} \int_{-\infty}^{+\infty} u e^{-i\lambda\xi} d\xi = \frac{d\bar{u}(\lambda, t)}{dt}$$

$$= a^2 \frac{1}{\sqrt{2\pi}} \int_{-\infty}^{+\infty} \frac{\partial^2 u}{\partial \xi^2} e^{-i\lambda\xi} d\xi$$

$$= a^2 \frac{1}{\sqrt{2\pi}} \frac{\partial u}{\partial \xi} e^{-i\lambda\xi} \Big|_{\xi=-\infty}^{\xi=+\infty} + a^2 \frac{1}{\sqrt{2\pi}} i\lambda u e^{-i\lambda\xi} \Big|_{\xi=-\infty}^{\xi=+\infty} - a^2\lambda^2 \frac{1}{\sqrt{2\pi}} \int_{-\infty}^{+\infty} u e^{-i\lambda\xi} d\xi$$

$$= -a^2\lambda^2 \bar{u}(\lambda, t),$$

i.e.

$$\frac{d\bar{u}}{dt} + a^2\lambda^2 \bar{u} = 0. \tag{1}$$

From the equality

$$\bar{u}(\lambda, t) = \frac{1}{\sqrt{2\pi}} \int_{-\infty}^{+\infty} u(\xi, t) e^{-i\lambda\xi} d\xi$$

for $t = 0$ we obtain:

$$\bar{u}(\lambda, 0) = \frac{1}{\sqrt{2\pi}} \int_{-\infty}^{+\infty} u(\xi, 0)\, e^{-i\lambda\xi}\, d\xi = \frac{1}{\sqrt{2\pi}} \int_{-\infty}^{+\infty} f(\xi)\, e^{-i\lambda\xi}\, d\xi = \bar{f}(\lambda). \quad (2)$$

The solution of equation (1) for the initial condition (2) has the form

$$\bar{u}(\lambda) = \bar{f}(\lambda)\, e^{-a^2\lambda^2 t}.$$

The application of the inverse Fourier transform gives

$$u(x, t) = \frac{1}{\sqrt{2\pi}} \int_{-\infty}^{+\infty} \bar{u}(\lambda, t)\, e^{i\lambda x}\, d\lambda = \frac{1}{2\pi} \int_{-\infty}^{+\infty} f(\xi)\, d\xi \int_{-\infty}^{+\infty} e^{-a^2\lambda^2 t}\, e^{i\lambda(x-\xi)}\, d\lambda$$

$$= \frac{1}{\pi} \int_{-\infty}^{+\infty} f(\xi)\, d\xi \int_{0}^{+\infty} e^{-a^2\lambda^2 t} \cos \lambda(x-\xi)\, d\lambda = \frac{1}{2a\sqrt{\pi t}} \int_{-\infty}^{+\infty} f(\xi)\, e^{-\frac{(x-\xi)^2}{4a^2 t}}\, d\xi, \quad (3)$$

since

$$\int_{0}^{+\infty} e^{-\alpha^2\lambda^2} \cos \beta\lambda\, d\lambda = \frac{\sqrt{\pi}}{2\alpha}\, e^{-\frac{\beta^2}{4\alpha^2}}. \quad (4)$$

The latter integral is readily evaluated by differentiation with respect to a parameter.

55. $$u(x, t) = \frac{1}{2a\sqrt{\pi}} \int_{0}^{t} d\tau \int_{0}^{+\infty} f(\xi, \tau) \frac{e^{-\frac{(x-\xi)^2}{4a^2(t-\tau)}}}{\sqrt{t-\tau}}\, d\xi. \quad (1)$$

56. *Solution.* Multiplying both sides of the equation

$$\frac{\partial u(\xi, t)}{\partial t} = a^2 \frac{\partial^2 u(\xi, t)}{\partial \xi^2} \quad \text{by} \quad \sqrt{\frac{2}{\pi}} \sin \lambda\xi$$

and integrating with respect to ξ from 0 to $+\infty$, we obtain for the Fourier sine form of the function $u(x, t)$

$$\bar{u}^{(s)}(\lambda, t) = \sqrt{\frac{2}{\pi}} \int_{0}^{+\infty} u(\xi, t) \sin \lambda\xi\, d\xi, \quad (1)$$

the equation

$$\frac{d\bar{u}^{(s)}(\lambda, t)}{dt} + a^2\lambda^2\bar{u}^{(s)}(\lambda, t) = 0, \quad 0 < t < +\infty. \quad (2)$$

From (1) we find the initial condition

$$\bar{u}^{(s)}(\lambda, 0) = \bar{f}^{(s)}(\lambda). \quad (3)$$

The solution of equation (2) for the initial condition (3) has the form

$$\bar{u}^{(s)}(\lambda, t) = \bar{f}^{(s)}(\lambda)\, e^{-a^2\lambda^2 t}.$$

Applying the inverse Fourier sine transform to it, we find by virtue of the equality (4), deduced in the solution of problem 54

$$u(x, t) = \sqrt{\frac{2}{\pi}} \int_0^{+\infty} \bar{u}^{(s)}(\lambda, t) \sin \lambda x \, dx$$

$$= \frac{2}{\pi} \int_0^{+\infty} f(\xi) \, d\xi \int_0^{+\infty} e^{-a^2\lambda^2 t} \sin \lambda\xi \sin \lambda x \, d\lambda$$

$$= \frac{1}{\pi} \int_0^{+\infty} f(\xi) \, d\xi \int_0^{+\infty} e^{-a^2\lambda^2 t}[\cos (x-\xi) - \cos \lambda(x+\xi)] \, d\lambda$$

$$= \frac{1}{2a\sqrt{\pi t}} \int_0^{+\infty} f(\xi)\left[e^{-\frac{(x-\xi)^2}{4a^2 t}} - e^{-\frac{(x+\xi)^2}{4at^2}}\right] d\xi .$$

57. $u(x, t) = \dfrac{1}{2a\sqrt{\pi t}} \displaystyle\int_0^{+\infty} f(\xi)\left[e^{-\frac{(x-\xi)^2}{4a^2 t}} + e^{-\frac{(x+\xi)^2}{4a^2 t}}\right] d\xi .$

Method. Apply the Fourier cosine transform.

58. $u(x, t) = \dfrac{x}{2a\sqrt{\pi}} \displaystyle\int_0^{t} \frac{e^{-\frac{x^2}{4a^2(t-\tau)}}}{(t-\tau)^{3/2}} \phi(\tau) \, d\tau .$

Method. Apply the Fourier sine transform; see also the solution of the following problem.

59. Applying the Fourier cosine transform† and using the boundary condition $u_x(0, t) = \phi(t)$ we obtain:

$$\frac{d\bar{u}^{(c)}(\lambda, t)}{dt} = a^2\sqrt{\frac{2}{\pi}} \int_0^{+\infty} \frac{\partial^2 u}{\partial \xi^2} \cos \lambda\xi \, d\xi = a^2\sqrt{\frac{2}{\pi}} \frac{\partial u}{\partial \xi} \cos \lambda\xi \Bigg|_{\xi=0}^{\xi=+\infty} +$$

$$+a^2\sqrt{\frac{2}{\pi}}\lambda \int_0^{+\infty} \frac{\partial u}{\partial \xi} \sin \lambda\xi \, d\xi = -a^2\sqrt{\frac{2}{\pi}}\phi(t) + a^2\sqrt{\frac{2}{\pi}} u\lambda \sin \lambda\xi \Bigg|_{\xi=0}^{\xi=+\infty} -$$

$$-a^2\lambda^2\sqrt{\frac{2}{\pi}} \int_0^{+\infty} u(\xi, t) \cos \lambda\xi \, d\xi = -a^2\sqrt{\frac{2}{\pi}}\phi(t) - a^2\lambda^2\bar{u}^{(c)}(\lambda, t),$$

i.e.

$$\frac{d\bar{u}^{(c)}(\lambda, t)}{dt} + a^2\lambda^2\bar{u}^{(c)}(\lambda, t) = -a^2\sqrt{\frac{2}{\pi}}\phi(t), \tag{1}$$

† It is assumed that u and the derivatives of u with respect to ξ tend sufficiently rapidly to zero an $\xi \to +\infty$.

where

$$\bar{u}^{(c)}(\lambda, t) = \sqrt{\frac{2}{\pi}} \int_0^{+\infty} u(\xi, t) \cos \lambda\xi \, d\xi. \qquad (2)$$

From (2) we find:

$$\bar{u}^{(c)}(\lambda, 0) = \sqrt{\frac{2}{\pi}} \int_0^{+\infty} u(\xi, 0) \cos \lambda\xi \, d\xi. \qquad (3)$$

The solution of equation (1) for the initial condition (3) has the form

$$\bar{u}^{(c)}(\lambda, t) = -a^2 \sqrt{\frac{2}{\pi}} \int_0^t e^{-a^2\lambda^2(t-\tau)} \phi(\tau) \, d\tau.$$

Applying the inverse Fourier cosine transform (using equality (4) in the solution of problem 54) we obtain:

$$u(x, t) = \sqrt{\frac{2}{\pi}} \int_0^{+\infty} \bar{u}^{(c)}(\lambda, t) \cos \lambda x \, d\lambda$$

$$= -\frac{2a^2}{\pi} \int_0^t \phi(\tau) \, d\tau \int_0^{+\infty} e^{-a^2\lambda^2(t-\tau)} \cos \lambda x \, d\lambda$$

$$= -\frac{a}{\sqrt{\pi}} \int_0^t \frac{\phi\tau}{\sqrt{t-\tau}} e^{-\frac{x^2}{4a^2(t-\tau)}} \, d\tau.$$

60. $u(x, t) = \dfrac{1}{2a\sqrt{\pi}} \displaystyle\int_0^t d\tau \int_0^{+\infty} f(\xi, \tau) \frac{e^{-\frac{(x-\xi)^2}{4a^2(t-\tau)}} - e^{-\frac{(x+\xi)^2}{4a^2(t-\tau)}}}{\sqrt{t-\tau}} \, d\xi.$

61. $u(x, t) = \dfrac{1}{2a\sqrt{\pi}} \displaystyle\int_0^t d\tau \int_0^{+\infty} f(\xi, \tau) \frac{e^{-\frac{(x-\xi)^2}{4a^2(t-\tau)}} + e^{-\frac{(x+\xi)^2}{4a^2(t-\tau)}}}{\sqrt{t-\tau}} \, d\xi.$

62. *Method.* First establish that for the cosine transforms $\bar{f}^{(c)}(\lambda) = e^{-\alpha\lambda^2}$, $\bar{g}^{(c)}(\lambda) = 1/(\lambda^2+h^2)$ the originals are

$$f(x) = \frac{1}{\sqrt{2\alpha}} e^{-\frac{x^2}{4\alpha}}, \quad g(x) = \frac{1}{h}\sqrt{\frac{\pi}{2}} e^{-hx}.$$

63. *Method.* First establish that for the cosine forms $\bar{f}^{(c)}(\lambda) = e^{-\alpha\lambda^2}$, $\bar{g}^{(c)}(\lambda) = \lambda/(\lambda^2+h^2)$ the originals are

$$f(x) = \frac{1}{\sqrt{2\alpha}} e^{-\frac{x^2}{4\alpha}}, \quad g(x) = \sqrt{\frac{\pi}{2}} e^{-hx}.$$

64. $$u(x,t)=\frac{ah}{\sqrt{\pi}}\int_0^t\frac{\phi(\tau)}{\sqrt{t-\tau}}\left[e^{-\frac{x^2}{4a^2(t-\tau)}}-h\int_0^{+\infty}e^{-h\xi-\frac{(x+\xi)^2}{4a^2(t-\tau)}}\,d\xi\right]d\tau$$

$$=\frac{ah}{\sqrt{\pi}}\int_0^t\frac{\phi(t-\zeta)}{\sqrt{\zeta}}\left[e^{-\frac{x^2}{4a^2\zeta}}-h\int_0^{+\infty}e^{-h\xi-\frac{(x+\xi)^2}{4a^2\zeta}}\,d\xi\right]d\zeta\,.$$

Method. Use the results of problems 62 and 63.

65. $$u(x,t)=\frac{1}{2a\sqrt{\pi t}}\int_0^{+\infty}f(\xi)\left[e^{-\frac{(x-\xi)^2}{4a^2t}}+e^{-\frac{(x+\xi)^2}{4a^2t}}-\right.$$

$$\left.-2h\int_0^{+\infty}e^{-\frac{(x+\xi+\eta)^2}{4a^2t}-h\eta}\,d\eta\right]d\xi\,.$$

Method. Use the results of problems 62 and 63.

2. Homogeneous Isotropic Media. Calculation of Green's Functions

(a) *Infinite straight line*

66. $$u(x,t)=\frac{Q}{c\rho\sigma}\,G(x,\xi,t),\qquad -\infty<x,\quad \xi<+\infty,\quad 0<t<+\infty,$$

where

$$G(x,\xi,t)=\frac{1}{2a\sqrt{\pi t}}\,e^{-\frac{(x-\xi)^2}{4a^2t}}$$

is the so-called source function or Green's function for the equation $u_t=a^2u_{xx}$ in the case of an infinite region.

Method. It is possible to assume that the amount of heat Q, instantaneously liberated at a point ξ at time $t=0$, is instantaneously uniformly distributed over a small interval $(\xi-\delta,\xi+\delta)$; then the initial temperature of the rod will equal

$$u(x,0)=f_\delta(x)=\begin{cases}0, & -\infty<x<\xi-\delta,\\ \dfrac{Q}{2\delta c\rho\sigma}, & \xi-\delta<x<\xi+\delta,\\ 0, & \xi+\delta<x<+\infty\,.\end{cases}$$

Solving the problem

$$u_t=a^2u_{xx},\qquad -\infty<x<+\infty,\qquad 0<t<+\infty,\tag{1}$$

$$u(x,0)=f_\delta(x),\qquad -\infty<x<+\infty\tag{2}$$

by means of (3) in the solution of problem 54 and passing to a limit in the solution obtained as $\delta\to 0$ we obtain the answer.

One can also make use of the delta-function†, solving either the problem

$$u_t = a^2 u_{xx}, \quad -\infty < x < +\infty, \quad 0 < t < +\infty, \tag{3}$$

$$u(x, 0) = \frac{Q}{c\rho\sigma}\delta(x-\xi), \quad -\infty < x, \ \xi < +\infty \tag{4}$$

by means of formula (3) in problem 54, or the problem

$$u_t = a^2 u_{xx} + \frac{Q}{c\rho\sigma}\delta(x-\xi)\delta(t), \quad -\infty < x, \ \xi < +\infty, \quad 0 < t < +\infty, \tag{5}$$

$$u(x, 0) = 0, \quad -\infty < x < +\infty \tag{6}$$

by means of formula (1) deduced in the answer to problem 55.

For the solution of the boundary-value problems (3), (4) and (5), (6) formulae (3) and (1) were used, but one may also use the integral representation of the delta-function (see [7], pages 298–300).

The source function for the equation $u_t = a^2 u_{xx}$ over the straight line $-\infty < x < +\infty$ may also be derived from similarity (see [7], pages 243–253) or by means of a limiting transition in the expression of the source function for the segment $0 \leqslant x \leqslant l$ for $l \to +\infty$ (see [7], pages 235–240).

Note. If the instantaneous liberation of heat at the point $x = \xi$ occurred not at time $t = 0$ but at time $t = \tau$, then

$$u(x, t) = \frac{Q}{c\rho\sigma} G(x, \xi, t-\tau), \quad -\infty < x, \ \xi < +\infty, \ x \neq \xi, \ \tau < t < +\infty,$$

$$G(x, \xi, t-\tau) = \frac{1}{2a\sqrt{\pi(t-\tau)}} e^{-\frac{(x-\xi)^2}{4a^2(t-\tau)}}.$$

67. $$u(x, t) = \frac{Q}{c\rho\sigma} G(x, \xi, t), \quad -\infty < x, \quad \xi < +\infty, \quad x \neq \xi, \quad 0 < t < +\infty, \tag{1}$$

where

$$G(x, \xi, t) = \frac{e^{-ht}}{2a\sqrt{\pi t}} e^{-\frac{(x-\xi)^2}{4a^2 t}} \tag{2}$$

is the source function for the equation $u_t = a^2 u_{xx} - hu$ in the case of an infinite region.

Note. If the instantaneous liberation of an amount of heat Q took place not at time $t = 0$ but at time $t = \tau$, then

$$u(x, t) = \frac{Q}{c\rho\sigma} G(x, \xi, t-\tau), \quad -\infty < x, \xi < +\infty, \quad x \neq \xi, \quad \tau < t < +\infty, \tag{3}$$

† See the answers and hints to problems 56 and 63 § 2 chapter II and to problem 153 § 3 chapter II.

where

$$G(x, \xi, t-\tau) = \frac{e^{-h(t-\tau)}}{2a\sqrt{\pi(t-\tau)}} e^{-\frac{(x-\xi)^2}{4a^2(t-\tau)}}. \tag{4}$$

68. *Solution.* Let us replace in the solution $u(x, t)$ of the equation

$$u_t = a^2u_{xx}+f(x, t)$$

x and t by ξ and τ; let us replace, further, t by $t-\tau$ in the source function $G(x, \xi, t) = (1/2a\sqrt{\pi t})e^{-(x-\xi)^2/4a^2t}$, $0 < \tau < t$. The functions $u(\xi, \tau)$ and $G(x, \xi, t-\tau)$ satisfy the equations

$$u_\tau = a^2u_{\xi\xi}+f(\xi, \tau),$$
$$G_\tau = -a^2G_{\xi\xi},$$

therefore

$$\frac{\partial}{\partial\tau}(Gu) = a^2\left[G\frac{\partial^2 u}{\partial\xi^2} - u\frac{\partial^2 G}{\partial\xi^2}\right]+Gf. \tag{1}$$

Integrating the latter equation with respect to ξ from $-\infty$ to $+\infty$ and with respect to τ from 0 to $t-\alpha$, $0 < \alpha < t$, we obtain (if it is assumed that u and its derivatives with respect to ξ are bounded for $\xi \to \pm\infty$ or tend to ∞, but not too rapidly):

$$\int_{-\infty}^{+\infty} (Gu)_{\tau=t-\alpha} d\xi = \int_{-\infty}^{+\infty} (Gu)_{\tau=0} d\xi + \int_0^{t-\alpha} d\tau \int_{-\infty}^{+\infty} Gf d\xi. \tag{2}$$

Passing to a limit in the equality for $\alpha \to 0$ we obtain†:

$$u(x, t) = \int_{-\infty}^{+\infty} \phi(\xi)G(x, \xi, t)d\xi + \int_0^t d\tau \int_{-\infty}^{+\infty} f(\xi, \tau)G(x, \xi, t-\tau)d\xi. \tag{3}$$

69. The answer is given by formula (3) of the solution of the preceding problem, where by $G(x, \xi, t)$ one must understand the source function, found in the solution of problem 67.

Method. Problem 69 may be solved either directly or by reduction to problem 68 by means of the substitution of the unknown function $u(x, t) = e^{-ht}v(x, t)$.

70. $t = \dfrac{(x-\xi)^2}{2a^2}$, $\quad u_{\max}(x) = \dfrac{Q}{c\rho\sqrt{2\pi e}|x-\xi|}$.

71. $u(x, t) = \dfrac{Q}{c\rho}\dfrac{1}{2a\sqrt{\pi}}\displaystyle\int_0^t e^{-h\tau-\frac{x^2}{4a^2\tau}}\frac{d\tau}{\sqrt{\tau}}$, $\quad \bar{u}(x) = \dfrac{Q}{2c\rho ah} e^{-\frac{h}{a}|x|}$.

If the surface of the rod is thermally insulated, then

$$\lim_{t\to+\infty} u(x, t) = \infty.$$

† Transition to a limit on the left hand side of equality (2) for $\alpha \to 0$ is accomplished in the same way as was done in [7], on pages 248–251.

72. $u(x, t) = U_0\left[\Phi\left(\frac{x+l}{2a\sqrt{t}}\right)-\Phi\left(\frac{x-l}{2a\sqrt{t}}\right)\right],$

where

$$\Phi(z) = \frac{2}{\sqrt{\pi}}\int_0^z e^{-\xi^2}\,d\xi$$

is the so-called error integral, tables of the values of which may be found in [7], and also in Table 1 of the supplement of the present book.

73. $u(x, t) = \frac{A}{2}\,e^{-\alpha x+a^2\alpha^2 t}\left[1-\Phi\left(-\frac{x}{2a\sqrt{t}}+a\alpha\sqrt{t}\right)\right].$

74. $u(x, t) = U_0\,e^{-ht}\left[\Phi\left(\frac{x+l}{2a\sqrt{t}}\right)-\Phi\left(\frac{x-l}{2a\sqrt{t}}\right)\right].$

Method. Use the solution of problem 69, or by the substitution of the unknown function $u(x, t) = e^{-ht}\,v(x, t)$ reduce to problem 72.

75. $u(x, t) = \frac{Q}{2ac\rho\sqrt{\pi}}\int_0^t \frac{e^{-\frac{(x-v_0\tau)^2}{4a^2(t-\tau)}}}{\sqrt{t-\tau}}\,d\tau$

$$= \frac{Q}{2ac\rho\sqrt{\pi}}\int_0^t \frac{e^{-\frac{(x-v_0t+v_0\zeta)^2}{4a^2\zeta}}}{\sqrt{\zeta}}\,d\zeta,$$

in particular, the temperature of the rod at the heater equals

$$u(v_0 t, t) = \frac{Q}{c\rho v_0}\,\Phi\left(\frac{v_0\sqrt{t}}{2a}\right).$$

Note. The expression for $u(x, t)$ is derived on condition that the heat exchange at the surface of the rod, away from the heater, is negligibly small.

(b) *Semi-infinite straight line*

76. $G(x, \xi, t-\tau) = \frac{1}{2a\sqrt{\pi(t-\tau)}}\left[e^{-\frac{(x-\xi)^2}{4a^2(t-\tau)}}-e^{-\frac{(x+\xi)^2}{4a^2(t-\tau)}}\right],$

$$0 < x, \xi < +\infty, \quad x \neq \xi, \quad 0 < t < +\infty. \tag{1}$$

If a convective heat exchange takes place at the surface of the rod with a medium whose temperature equals zero, then the expression for the source function is obtained from (1) by multiplying by

$$e^{-H(t-\tau)}, \tag{2}$$

where H is the coefficient of heat exchange appearing in the equation $u_t = a^2 u_{xx} - Hu$.

Method. An expression for the temperature $u(x, t)$ and for $G(x, \xi, t-\tau)$ may be derived, by considering an infinite rod $-\infty < x < +\infty$ and assuming that at time $t = \tau$ at the point $x = \xi$ Q units of heat were instantaneously liberated, and at the point $x = -\xi$, $-Q$ units of heat were instantaneously liberated, i.e. as is sometimes said, by placing an instantaneous positive source of magnitude Q at the point $x = \xi$, and at the point $x = -\xi$ an instantaneous negative source of magnitude $-Q$†.

77. $$G(x, \xi, t-\tau) = \frac{1}{2a\sqrt{\pi(t-\tau)}}\left[e^{-\frac{(x-\xi)^2}{4a^2(t-\tau)}} + e^{-\frac{(x+\xi)^2}{4a^2(t-\tau)}}\right],$$

$$0 < x, \xi < +\infty, \quad \tau < t < +\infty, \quad x \neq \xi.$$

In the presence of a convective heat exchange at the surface of the rod the source function is derived by multiplying by $e^{-H(t-\tau)}$.

Method. See the method to the preceding problem; the present problem is solved in the same way.

78. $$G(x, \xi, t-\tau) = \frac{1}{2a\sqrt{\pi(t-\tau)}}\left\{\left[e^{-\frac{(x-\xi)^2}{4a^2(t-\tau)}} + e^{-\frac{(x+\xi)^2}{4a^2(t-\tau)}}\right] - \right.$$

$$\left. -2h\int_0^{+\infty} e^{-\frac{(x+\xi+\eta)^2}{4a^2(t-\tau)} - h\eta}\, d\eta\right\}, \quad 0 < x, \xi < +\infty, \quad x \neq \xi, \quad \tau < t < +\infty,$$

where h is the coefficient appearing in the boundary condition

$$u_x(0, t) - hu(0, t) = 0.$$

In the presence of a convective heat exchange at the surface of the rod the source function is derived by multiplying by $e^{-H(t-\tau)}$, where H is the coefficient of heat exchange appearing in the equation $u_t = a^2u_{xx} - Hu$.

Method. Use the suggestion, given in problem 82.

79. $$u(x, t) = \frac{1}{2a\sqrt{\pi t}}\int_0^{+\infty}\psi(\xi)\left[e^{-\frac{(x-\xi)^2}{4a^2t}} - e^{-\frac{(x+\xi)^2}{4a^2t}}\right]d\xi +$$

$$+ \frac{x}{2a\sqrt{\pi}}\int_0^t \frac{\phi(\tau)e^{-\frac{x^2}{4a^2(t-\tau)}}}{(t-\tau)^{3/2}}\, d\tau +$$

$$+ \frac{1}{2a\sqrt{\pi}}\int_0^t \frac{d\tau}{\sqrt{t-\tau}}\int_{-\infty}^{+\infty} f(\xi, \tau)\left[e^{-\frac{(x-\xi)^2}{4a^2(t-\tau)}} - e^{-\frac{(x+\xi)^2}{4a^2(t-\tau)}}\right]d\xi.$$

Method. Let $u(\xi, \tau)$ be the solution of the equation $u_\tau = a^2u_{\xi\xi}$, and $G(x, \xi, t-\tau)$ be the source function found in the solution of problem 76.

† The source function for the semi-infinite line is defined in the same way as the source function for a finite segment; see the introduction to the solutions of problems under section (b) of the present section.

Integrating the equality

$$\frac{\partial}{\partial \tau}(Gu) = G\frac{\partial u}{\partial \tau} + u\frac{\partial G}{\partial \tau} = a^2\left[G\frac{\partial^2 u}{\partial \xi^2} - u\frac{\partial^2 G}{\partial \xi^2}\right] + Gf\dagger$$

with respect to ξ from 0 to $+\infty$ and with respect to τ from 0 to $t-\alpha$, where $0 < \alpha < t$, we obtain:

$$\int_0^{+\infty} (Gu)_{\tau=t-\alpha} d\xi - \int_0^{+\infty} (Gu)_{\tau=0} d\xi = a^2 \int_0^{t-\alpha} d\tau \int_0^{+\infty} \left[G\frac{\partial^2 u}{\partial \xi^2} - u\frac{\partial^2 G}{\partial \xi^2}\right] d\xi +$$

$$+ \int_0^{t-\alpha} d\tau \int_0^{+\infty} Gf d\xi = a^2 \int_0^{t-\alpha} \left\{\left(G\frac{\partial u}{\partial \xi}\right)_{\xi=0}^{\xi=+\infty} - \left(u\frac{\partial G}{\partial \xi}\right)_{\xi=0}^{\xi=+\infty} -\right.$$

$$\left. - \int_0^{+\infty} \left[\frac{\partial G}{\partial \xi}\frac{\partial u}{\partial \xi} - \frac{\partial u}{\partial \xi}\frac{\partial G}{\partial \xi}\right] d\xi\right\} d\tau + \int_0^{t-\alpha} d\tau \int_0^{+\infty} Gf d\xi.$$

Imposing the correct restrictions on u and $\partial u/\partial \xi$ for $\xi \to +\infty$, we obtain:

$$\int_0^{+\infty} (Gu)_{\tau=t-\alpha} d\xi = \int_0^{+\infty} (Gu)_{\tau=0} d\xi - a^2 \int_0^{t-\alpha} \left(u\frac{\partial G}{\partial \xi}\right)_{\xi=0} d\tau + \int_0^{t-\alpha} d\tau \int_0^{+\infty} Gf d\xi$$

Passing to a limit as $\alpha \to 0$, we obtain‡:

$$\lim_{\alpha \to 0} \int_0^{+\infty} (Gu)_{\tau=t-\alpha} d\xi = u(x, t).$$

80. $$u(x, t) = \frac{1}{2a\sqrt{\pi t}} \int_0^{+\infty} \psi(\xi)\left[e^{-\frac{(x-\xi)^2}{4a^2 t}} + e^{-\frac{(x+\xi)^2}{4^2 at}}\right] d\xi -$$

$$- \frac{a}{\sqrt{\pi}} \int_0^t \frac{\phi(\tau)}{\sqrt{t-\tau}} e^{-\frac{x^2}{4a^2(t-\tau)}} d\tau +$$

$$+ \frac{1}{2a\sqrt{\pi}} \int_0^t \frac{d\tau}{\sqrt{t-\tau}} \int_{-\infty}^{+\infty} f(\xi, \tau)\left[e^{-\frac{(x-\xi)^2}{4a^2(t-\tau)}} + e^{-\frac{(x+\xi)^2}{4a^2(t-\tau)}}\right] d\xi.$$

---

† This equality is derived in the same way as (1) in the solution of problem 68.

‡ Transition to the limit is achieved in the same way as was done in [7], on pages 248–251.

Method. The problem may be solved similarly to the preceding one.

81. $$u(x,t)=\frac{1}{2a\sqrt{\pi t}}\int_0^{+\infty}\psi(\xi)\left[e^{-\frac{(x-\xi)^2}{4a^2t}}+e^{-\frac{(x+\xi)^2}{4a^2t}}-\right.$$

$$\left.-2h\int_0^{+\infty}e^{-\frac{(x+\xi+\eta)^2}{4a^2t}-h\eta}\,d\eta\right]d\xi+\frac{ha}{\sqrt{\pi}}\int_0^t\frac{\phi(\tau)}{\sqrt{t-\tau}}\left[e^{-\frac{x^2}{4a^2(t-\tau)}}-\right.$$

$$\left.-h\int_0^{+\infty}e^{-\frac{(x+\eta)^2}{4a^2(t-\tau)}-h\eta}\,d\eta\right]d\tau+\frac{1}{2a\sqrt{\pi}}\int_0^t\frac{d\tau}{\sqrt{t-\tau}}\int_{-\infty}^{+\infty}f(\xi,\tau)\left[e^{-\frac{(x-\xi)^2}{4a^2(t-\tau)}}+\right.$$

$$\left.+e^{-\frac{(x+\xi)^2}{4a^2(t-\tau)}}-2h\int_0^{+\infty}e^{-\frac{(x+\xi+\eta)^2}{4a^2(t-\tau)}-h\eta}\,d\eta\right]d\xi.$$

Method. See the method to problem 79. Problem 81 may be similarly solved.

82. *Method.* Use the fact that

(1) if $F(x)$ is an even function, then the function

$$u(x,t)=\frac{1}{2a\sqrt{\pi t}}\int_{-\infty}^{+\infty}F(\xi)e^{-\frac{(x-\xi)^2}{4a^2t}}\,d\xi$$

equals zero for $x=0$;

(2) if $u(x,t)$ is a solution of the equation $u_t=a^2u_{xx}$ then

$$U(x,t)=\sum_{k=0}^{N}A_k\frac{\partial^k u(x,t)}{\partial x^k}$$

is also a solution of this equation.

83. *Method.* Use the fact that

(1) if $F(x)$ is an even function with respect to x, then the function

$$u(x,t)=\frac{1}{2a\sqrt{\pi}}\int_0^t\frac{d\tau}{\sqrt{t-\tau}}\int_{-\infty}^{+\infty}F(\xi,\tau)e^{-\frac{(x-\xi)^2}{4a^2(t-\tau)}}d\xi$$

equals zero for $x=0$;

(2) if $u(x,t)$ is a solution of the equation

$$u_t=a^2u_{xx}+f(x,t)$$

then

$$U(x.t)=\sum^{N}A_k\frac{\partial^k u(x,t)}{\partial x^k}$$

is a solution of the equation

$$u_t = a^2 u_{xx} + \sum_{k=0}^{N} A_k \frac{\partial^k f(x, t)}{\partial x^k}.$$

84. $u(x, t) = U_0 \Phi(x/2a\sqrt{t})$. The velocity of motion of the temperature front αU_0, $\alpha = \text{const.}$, $0 < \alpha < 1$, equals

$$\frac{dx}{dt} = \frac{ak}{\sqrt{t}},$$

where k is the root of the equation $\Phi(z) = \alpha$. The graphs are given in Figs. 34 and 35.

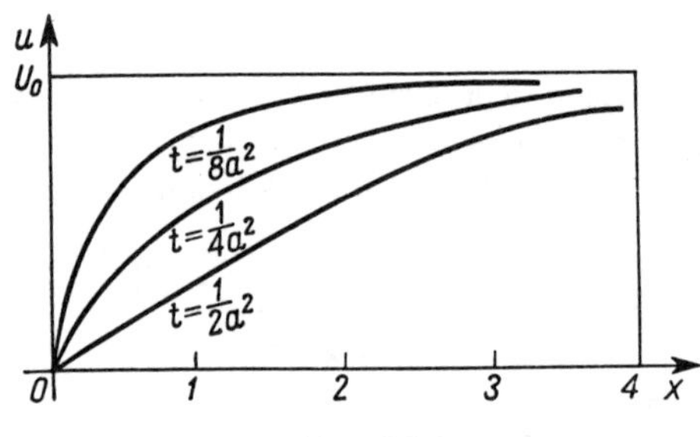

FIG. 34

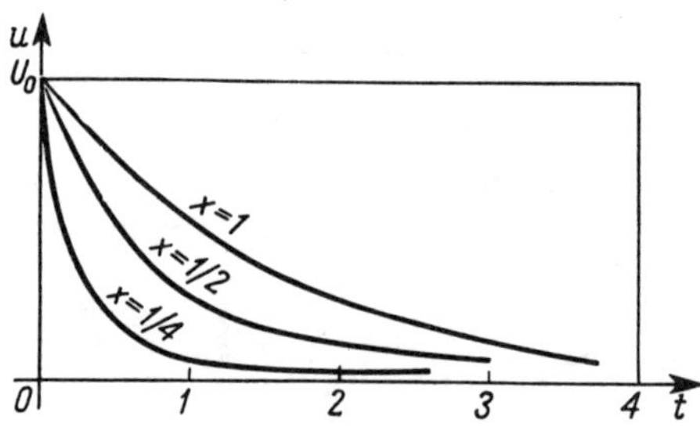

FIG. 35

85. $u(x, t) = U_0\left[1 - \Phi\left(\frac{x}{2a\sqrt{t}}\right)\right],$

$$T = \frac{x_0^2}{4a^2k^2},$$

where k is a root of the equation $\Phi(z) = 1 - \alpha$.

Method. By means of the substitution $u(x,t) = v(x,t)+U_0$ the problem reduces to the preceding one.

86. $$u(x,t) = \frac{U_0}{\sqrt{\pi}} \int\limits_{\frac{x-1}{2a\sqrt{t}}}^{\frac{x+1}{2a\sqrt{t}}} e^{-z^2}\,dz = \frac{U_0}{2}\left[\Phi\left(\frac{x+1}{2a\sqrt{t}}\right)-\Phi\left(\frac{x-1}{2a\sqrt{t}}\right)\right].$$

87. $$u(x,t) = U_0\Phi\left(\frac{x}{2a\sqrt{t}}\right)+e^{hx+h^2a^2t}\,U_0\left[1-\Phi\left(\frac{x}{2a\sqrt{t}}+ah\sqrt{t}\right)\right]. \tag{1}$$

The error made in using formula (4) of the condition does not exceed

$$\frac{1\,.\,3\,.\,5\,\ldots\,(2n-3)}{2^n}\cdot\frac{1}{z^{2n-1}}. \tag{2}$$

In order that the error made in using formula (5) should not exceed $\varepsilon > 0$, it is sufficient that the inequality

$$t = \frac{U_0^2}{4\pi a^2h^2\varepsilon^2} \tag{3}$$

be fulfilled.

Method. Integrating successively by parts, it is possible to obtain the equality

$$\int\limits_z^{+\infty} e^{-\xi^2}\,d\xi = \frac{e^{-z^2}}{2}\left\{\frac{1}{z}-\frac{1}{2z^3}+\frac{1\,.\,3}{2^2z^5}-\ldots+(-1)^{n-1}\frac{1.3\,\ldots\,(2n-3)}{2^{n-1}\,z^{2n-1}}\right\}+$$

$$+(-1)^n\,\frac{1\,.\,3\,\ldots\,(2n-1)}{2^n}\int\limits_z^{+\infty}\frac{e^{-\xi^2}}{\xi^{2n}}\,d\xi, \tag{4}$$

where, obviously

$$\int\limits_z^{+\infty} e^{-\xi^2}\frac{d\xi}{\xi^{2n}} < e^{-z^2}\int\limits_z^{+\infty}\frac{d\xi}{\xi^{2n}}. \tag{5}$$

Note. If the partial sum in the brackets of formula (4) is replaced by an infinite series, then a divergent series, called an asymptotic series, is obtained. The inequality (5) shows that the error which is made in discarding the residual term in (4)

$$(-1)^n\,\frac{1\,.\,3\,\ldots\,(2n-1)}{2^n}\int\limits_z^{+\infty}\frac{e^{-\xi^2}}{\xi^{2n}}\,d\xi,$$

tends to zero for every fixed n and $z \to +\infty$.

88. $$u(x, t) = \left(U_0 - \frac{b^2}{a^2k^2}\right)\left[1 - \Phi\left(\frac{x}{2a\sqrt{t}}\right)\right] + \frac{b^2}{a^2k^2}\,\mathrm{e}^{-kx} + \frac{b^2}{2a^2k^2}\left\{\mathrm{e}^{a^2k^2t-kx}\left[1-\Phi\left(\frac{-x}{2a\sqrt{t}} + ak\sqrt{t}\right)\right] - \mathrm{e}^{a^2k^2t+kx}\left[1-\Phi\left(\frac{x}{2a\sqrt{t}} + ak\sqrt{t}\right)\right]\right\}.$$

89. $$u(x, t) = 2aq\sqrt{\frac{t}{\pi}}\,\mathrm{e}^{-\frac{x^2}{4a^2t}} - qx\left[1-\Phi\left(\frac{x}{2a\sqrt{t}}\right)\right].$$

90. $$u(x, t) = U_2 + (U_0 - U_2)\,\mathrm{e}^{-ht}\,\Phi\left(\frac{x}{2a\sqrt{t}}\right) + \frac{U_1 - U_2}{2}\left\{\mathrm{e}^{-\frac{x\sqrt{h}}{a}}\left[1-\Phi\left(\frac{x}{2a\sqrt{t}} - \sqrt{ht}\right)\right] + \mathrm{e}^{\frac{x\sqrt{h}}{a}}\left[1-\Phi\left(\frac{x}{2a\sqrt{t}} + \sqrt{ht}\right)\right]\right\}.$$

91. $$v(x, t) = \frac{E_0}{2}\,\mathrm{e}^{-x\sqrt{RG}}\left\{1-\Phi\left(\frac{x}{2}\sqrt{\frac{RC}{t}} - \sqrt{\frac{Gt}{C}}\right)\right\} + \frac{E_0}{2}\,\mathrm{e}^{x\sqrt{RG}}\left\{1-\Phi\left(\frac{x}{2}\sqrt{\frac{RC}{t}} + \sqrt{\frac{Gt}{C}}\right)\right\},$$

where R, C, G are the resistance, capacity and leakage conductance per unit length of the conductor.

92. $$u(x, t) = \frac{Ah}{\sqrt{\left(\frac{1}{a}\sqrt{\frac{\omega}{2}}\right)^2 + \left(\frac{1}{a}\sqrt{\frac{\omega}{2}} + h\right)^2}} \times \mathrm{e}^{-\frac{x}{a}\sqrt{\frac{\omega}{2}}}\cos\left(\frac{x}{a}\sqrt{\frac{\omega}{2}} - \omega t - \gamma\right) + \frac{2Ah}{\sqrt{\pi}}\int_0^{+\infty}\mathrm{e}^{-h\eta}\,d\eta\int_0^{\frac{x+\eta}{2a\sqrt{t}}}\mathrm{e}^{-z^2}\cos\omega\left[t - \frac{(x+\eta)^2}{4a^2z^2}\right]dz, \quad (1)$$

$$\tan\gamma = \frac{\frac{1}{a}\sqrt{\frac{\omega}{2}}}{\frac{1}{a}\sqrt{\frac{\omega}{2}} + h}. \quad (2)$$

The first term on the right hand side of (1) represents a temperature wave which is damped as x increases and is periodic with respect to t. The second term vanishes for $t \to +\infty$.

93. $u(x, t) = Ae^{-\frac{x}{a}\sqrt{\frac{\omega}{2}}} \cos\left(\frac{x}{a}\sqrt{\frac{\omega}{2}} - \omega t\right).$

The velocity of propagation of the temperature wave with frequency ω equals

$$\frac{dx}{dt} = a\sqrt{2\omega}.$$

Method†. It is possible to find the steady-state temperature waves as the real part of the complex solution of the problem

$$U_t = a^2 U_{xx}, \quad 0 < x, t < +\infty, \quad U(0, t) = Ae^{i\omega t},$$

tending to zero as $x \to +\infty$. This complex solution has the form

$$U(x, t) = X(x)\, e^{i\omega t}.$$

94. $v(x, t) = E_0 e^{-x\sqrt{\frac{1}{2}RC\omega}} \cos\left(\omega t - x\sqrt{\frac{1}{2}RC\omega}\right) -$

$$-\frac{E_0}{\pi}\int_0^{+\infty} e^{-\xi t} \sin\sqrt{RC\xi}\, \frac{\xi\, d\xi}{\xi^2 + \omega^2},$$

where R and C are the resistance and capacitance per unit length of the conductor.

Method. See the method for the preceding problem.

95. $-\lambda u_x(0, t) = q(t) = \frac{\lambda}{\pi a}\frac{d}{dt}\int_0^t \frac{\mu(\tau)\, d\tau}{\sqrt{t-\tau}}.$

Method. The problem reduces to Abel's integral equation‡.

96. $\phi(t) = \frac{1}{2ah\sqrt{\pi}}\frac{d}{dt}\int_0^t \frac{\mu(\tau)\, d\tau}{\sqrt{t-\tau}},$

where h is the coefficient of heat exchange in the boundary condition

$$u_x(0, t) = h[u(0, t) - \phi(t)].$$

† For more detail on the solution of steady-state problems see [7], pages 262–266.

‡ Concerning Abel's integral equation see [2], vol II, § 79, and also the method to problem 114.

97. $\phi(t) = \frac{\lambda}{\pi a} \frac{d}{dt} \int_0^t \mu(\tau) \frac{e^{-h^*(t-\tau)}}{\sqrt{t-\tau}} d\tau,$

where h^* is the coefficient of heat exchange in the equation

$$u_t = a^2 u_{xx} - h^* u.$$

98. $\phi(t) = \frac{1}{2ah\sqrt{\pi}} \frac{d}{dt} \int_0^t \mu(\tau) \frac{e^{-h^*(t-\tau)}}{\sqrt{t-\tau}} d\tau,$

where the coefficients h and h^* have the same meaning as in problems 96 and 97.

99. $u(x, t) = \frac{e^{-\frac{v_0}{2a^2}(x-v_0t)-\frac{v_0^2}{4a^2}t}}{2a\sqrt{\pi}} \int_0^t d\tau \int_0^{+\infty} \frac{f(\xi+v_0\tau, \tau)\, e^{\frac{v_0}{2a^2}\xi+\frac{v_0^2}{4a^2}\tau}}{\sqrt{t-\tau}} \times$

$$\times \left[e^{-\frac{(x-v_0t-\xi)^2}{4a^2(t-\tau)}} - e^{-\frac{(x-v_0t+\xi)^2}{4a^2(t-\tau)}}\right] d\xi,$$

$$v_0 t < x < +\infty, \quad 0 < t < +\infty.$$

Method. Change to the new independent variables $\xi = x - v_0 t$, $t = t$ (this corresponds to a transition to a moving system of coordinates with origin at the point $x_0 = v_0 t$) and to the new unknown function by the formula $u(x, t) = e^{\alpha\xi+\beta t} v(\xi, t)$.

100. $u(x, t) = \frac{e^{-\frac{v_0}{2a^2}(x-v_0t)-\frac{v_0^2}{4a^2}t}}{2a\sqrt{\pi}} \times$

$$\times \int_0^{+\infty} e^{\frac{v_0}{2a^2}\xi} f(\xi) \left[e^{-\frac{(x-v_0t-\xi)^2}{4a^2t}} - e^{-\frac{(x-v_0t+\xi)^2}{4a^2t}}\right] d\xi.$$

Method. See the method to the preceding problem.

101.

$$u(x, t) = \frac{e^{-\frac{v_0}{2a^2}(x-v_0t)-\frac{v_0^2}{4a^2}t}}{2a\sqrt{\pi}} (x-v_0t) \int_0^t \frac{e^{-\frac{v_0^2}{4a^2}\tau} \mu(\tau)\, e^{-\frac{(x-v_0t)^2}{4a^2(t-\tau)}}}{(t-\tau)^{3/2}} d\tau.$$

Method. See the instructions to problem 99.

102. $u(x, t) = e^{-\frac{v_0}{2a^2}(x-v_0t)-\frac{v_0^2}{4a^2}t} \left\{ \frac{1}{2a\sqrt{\pi t}} \int_0^{+\infty} e^{\frac{v_0}{2a^2}\xi} f(\xi) \left[e^{-\frac{(x-v_0t-\xi)^2}{4a^2t}} + \right.\right.$

$$\left. e^{-\frac{(x-v_0t+\xi)^2}{4a^2t}} - \frac{v_0}{2a^2} \int_0^{+\infty} e^{-\frac{(x-v_0t+\xi+\eta)^2}{4a^2t} - \frac{v_0}{2a^2}\eta} d\eta \right] d\xi -$$

$$-\frac{a}{\sqrt{\pi}}\int_0^t \frac{e^{\frac{v_0^2}{4a^2}\tau}\mu(\tau)}{\sqrt{t-\tau}}\left[e^{-\frac{(x-v_0t)^2}{4a^2(t-\tau)}}-\frac{v_0}{2a^2}\int_0^{+\infty} e^{-\frac{(x-v_0t)^2}{4a^2(t-\tau)}-\frac{v_0}{2a^2}\eta}\,d\eta\right]d\tau+$$

$$+\frac{1}{2a\sqrt{\pi}}\int_0^t d\tau\int_0^{+\infty}\frac{f(\xi+v_0\tau,\tau)\,e^{\frac{v_0}{2a^2}\xi+\frac{v_0^2}{4a^2}\tau}}{\sqrt{t-\tau}}\left[e^{-\frac{(x-v_0t-\xi)^2}{4a^2(t-\tau)}}+\right.$$

$$\left.+e^{-\frac{(x-v_0t+\xi)^2}{4a^2(t-\tau)}}-\frac{v_0}{2a^2}\int_0^{+\infty} e^{-\frac{(x-v_0t+\xi+\eta)^2}{4a^2(t-\tau)}-\frac{v_0}{2a^2}\eta}\,d\eta\right]d\xi\Bigg\}.$$

Method. See the method to problem 99.

(c) *Finite segment*

The Green's function ("the source function") for a finite segment $0 < x < l$, corresponding to given boundary conditions, is the temperature $G(x, \xi, t)$ at an arbitrary point x, $0 < x < l$, at an arbitrary time $t > 0$, produced by the liberation of $Q = c\rho$† units of heat at the point ξ, $0 < \xi < l$, $\xi \neq x$ of this segment at time $t = 0$, if the ends of the segment satisfy appropriate homogeneous boundary conditions.

Thus the source function $G(x, \xi, t)$ must be (1) a solution of the equation of heat conduction, (2) satisfy the appropriate homogeneous boundary conditions, (3) reduce to zero for $t \to 0$ and $x \neq \xi$ and (4) satisfy the limiting relation

$$\lim_{\substack{t\to 0\\ t>0}}\int_{\xi-\lambda}^{\xi+\lambda} G(x, \xi, t)\,c\rho\,dx = Q,$$

or, what is the same thing

$$\lim_{\substack{t\to 0\\ >0}}\int_{\xi-\lambda}^{\xi+\lambda} G(x, \xi, t)\,dx = 1$$

or any $\lambda > 0$‡.

The source function

$$\frac{1}{2a\sqrt{\pi t}}\,e^{-\frac{(x-\xi)^2}{4a^2t}} \tag{1}$$

for the equation

$$u_t = a^2u_{xx} \tag{2}$$

along an infinite straight line satisfies requirements (1), (3) and (4).

---

† Here c is the specific heat, and ρ the linear mass density.

‡ It is assumed that $0 < \xi-\lambda < \xi+\lambda < 1$.

If we add to (1) that continuous solution $g(x, \xi, t)$ of equation (2) reducing to zero for $t = 0$ such that the sum

$$G(x, \xi, t) = \frac{1}{2a\sqrt{\pi t}} e^{-\frac{(x-\xi)^2}{4a^2 t}} + g(x, \xi, t) \tag{3}$$

satisfies the boundary conditions (2), then (3) will satisfy all the requirements (1), (2), (3), (4), i.e. will be the source function for equation (2) in the finite segment, corresponding to the boundary conditions (2).

The function $g(x, \xi, t)$ can be formed for certain types of boundary condition by the method of images; problems 103–106 are solved by this method.

103. *Solution.* Let us extend the rod $0 < x < l$ on both sides indefinitely. (We shall assume its surface is everywhere thermally insulated.) Let $Q = c\rho$, units of heat be liberated at the point ξ, $0 < \xi < l$, at time $t = 0$. The increase of temperature

$$\frac{1}{2a\sqrt{\pi t}} e^{-\frac{(x-\xi)^2}{4a^2 t}}, \tag{1}$$

produced in an infinite rod $-\infty < x < +\infty$ by this instantaneous source, is not equal to zero for $x = 0$ and $x = l$. If, however, instantaneous heat sources of magnitude $\pm Q$, distributed as indicated in Fig. 36, at the points $-\xi$, $\pm\xi \pm \pm 2nl$, $n = 1, 2, 3, \dots$† at time $t = 0$, then the temperature

$$G(x, \xi, t) = \frac{1}{2a\sqrt{\pi t}} \sum_{n=-\infty}^{+\infty} \left(e^{-\frac{(x-\xi+2nl)^2}{4a^2 t}} - e^{-\frac{(x+\xi+2nl)^2}{4a^2 t}} \right), \tag{2}$$

produced in the infinite rod $-\infty < x < +\infty$ by the action of all these sources will always be equal to zero at the point $x = 0$ and at the point $x = l$. In fact, to every source of magnitude $+Q$ according to Fig. 36 there corresponds a

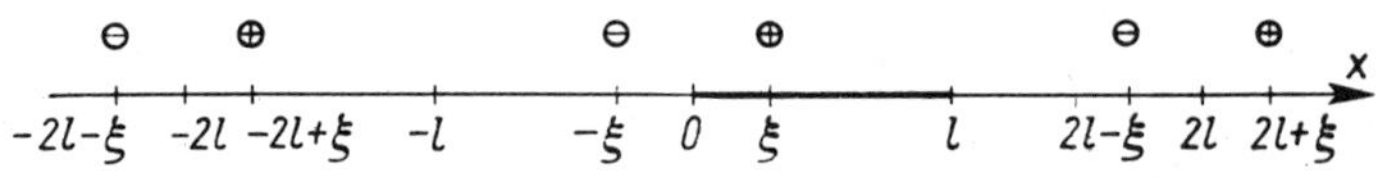

FIG. 36

source of magnitude $-Q$ symmetrical with respect to $x = 0$, and conversely, to every source of magnitude $-Q$ there corresponds a source of magnitude $+Q$ symmetrical with respect to $x = 0$, so that their actions at the point $x = 0$ cancel out. The same argument holds for the point $x = l$.

† The points $-\xi$, $\pm\xi \pm 2nl$, $n = 1, 2, 3, \dots$, are obtained from the point ξ by successive reflections with respect to $x = 0$ and $x = l$.

Let us write $G(x, \xi, t)$ in the form

$$G(x, \xi\, t) = \frac{1}{2a\sqrt{\pi t}} e^{-\frac{(x-\xi)^2}{4a^2 t}} + g(x, \xi, t), \tag{3}$$

where

$$g(x, \xi, t) = \frac{1}{2a\sqrt{\pi t}} \sum_{n=-\infty}^{+\infty}{}' \left(e^{-\frac{(x-\xi+2nl)^2}{4a^2t}} - e^{-\frac{(x+\xi+2nl)^2}{4a^2t}} \right). \tag{4}$$

In the series (2) denoted by the symbol $1/(2a\sqrt{\pi t}) \sum_{n=-\infty}^{+\infty}{}' (\ldots)$ the term (1) has been omitted. The terms of series (4) have derivatives of all orders with respect to x and t everywhere for $0 \leqslant x \leqslant l, 0 \leqslant t < +\infty$. Series (4) converges absolutely and uniformly for $0 \leqslant x \leqslant l$, $0 \leqslant t \leqslant t^*$, where t^* is an arbitrary positive quantity; the same holds for the series, resulting from (4) by termwise differentiation. For $t \to 0$, $t > 0$, each term of series (4) tends to zero.

Thus $G(x, \xi, t)$ satisfies all the requirements (1), (2), (3), (4) for the source function.

Let us estimate the error made in replacing the sum of series (4) by its partial sum $\sum_{n=-N}^{N}{}'$ for $0 \leqslant x \leqslant l$, $0 \leqslant t \leqslant t^*$. We consider first of all the series of terms with positive n. If one removes the brackets, then it becomes an alternating series, satisfying the conditions of Leibnitz' theorem. Therefore we obtain an inequality for the remainder of the series

$$|R_N^+(x, \xi, t)| = \left| \frac{1}{2a\sqrt{\pi t}} \sum_{n=N}^{+\infty} \left(e^{-\frac{(x-\xi+2nl)^2}{4a^2t}} - e^{-\frac{(x+\xi+2nl)^2}{4a^2t}} \right) \right|$$

$$\leqslant \frac{1}{2a\sqrt{\pi t}} e^{-\frac{(x-\xi+2Nl)^2}{4a^2t}} \leqslant \frac{1}{2a\sqrt{\pi t}} e^{-\frac{(N-1)^2 l^2}{a^2 t}} \quad \text{for } 0 < x < l, 0 < \xi < l. \tag{5}$$

Similarly we obtain an inequality for the remainder of the series of terms with negative n

$$|R_N^-(x, \xi, t)| \leqslant \frac{1}{2a\sqrt{\pi t}} e^{-\frac{(N-1)^2 l^2}{a^2 t}} \tag{5'}$$

Thus the inequality

$$|R_N(x, \xi, t)| \leqslant \frac{1}{a\sqrt{\pi t}} e^{-\frac{(N-1)^2 l^2}{a^2 t}}, \text{ if } 0 < x, \quad \xi < l, \quad 0 < t < +\infty, \tag{6}$$

holds for the remainder of series (4). It is readily verified that for

$$N \geqslant \frac{a}{l} \sqrt{\frac{t^*}{2}} + 1 \tag{7}$$

the inequality†

$$\frac{1}{a\sqrt{\pi t}}\,e^{-\frac{(N-1)^2 l^2}{a^2 t}} \leqslant \frac{1}{a\sqrt{\pi t^*}}\,e^{-\frac{(N-1)^2 l^2}{a^2 t^*}} \quad \text{for} \quad 0 \leqslant t \leqslant t^* \tag{8}$$

will be fulfilled. Therefore, for N, satisfying the inequality (7), the inequality

$$|R_N(x, \xi, t)| \leqslant \frac{1}{a\sqrt{\pi t^*}}\,e^{-\frac{(N-1)^2 l^2}{a^2 t^*}} \quad \text{for } 0 \leqslant t \leqslant t^*,\ 0 \leqslant x,\ \xi \leqslant l. \tag{6'}$$

will be satisfied.

Solving the boundary-value problem

$$u_t = a^2 u_{xx}, \quad 0 < x < l, \quad 0 < t < +\infty, \tag{9}$$

$$u(0, t) = u(l, t) = 0, \quad 0 < t < +\infty, \tag{10}$$

$$u(x, 0) = \delta(x), \quad 0 < x < l, \tag{11}$$

by the method of separation of variables, we obtain the expression for the source function

$$G(x, \xi, t) = \frac{2}{l}\sum_{n=1}^{+\infty} e^{-\frac{n^2\pi^2 a^2}{l^2}t} \sin\frac{n\pi x}{l}\sin\frac{n\pi\xi}{l}. \tag{12}$$

Although series (12) and (4) are formally mutually interchangeable‡, however their role in describing the source function is different; if series (4) converges more rapidly the smaller the value of t, then series (12), on the other hand, converges more rapidly, the larger the value of t.

It is easy to derive an estimate of the error made in replacing the sum of series (12) by its partial sum.

---

† In order to do this let us change to the new independent variable $\tau = (N-1)l/a\sqrt{t}$ in the function $\phi(t) = e^{-\frac{(N-1)^2 l^2}{a^2 t}}/a\sqrt{\pi t}$. We obtain:

$$\phi(t) = \frac{1}{(N-1)\,l\sqrt{\pi}} \cdot \frac{\tau}{e^{\tau^2}} = \frac{1}{(N-1)\,l\sqrt{\pi}}\,\phi(\tau),$$

where

$$\psi(\tau) = \frac{\tau}{e^{\tau^2}}.$$

Since

$$\psi'(\tau) = \frac{1-2\tau^2}{e^{\tau^2}} < 0 \quad \text{for} \quad \tau > \frac{1}{\sqrt{2}},$$

then $\psi(\tau)$ decreases monotonically in the segment $1/\sqrt{2} < \tau < +\infty$; therefore, $\phi(t)$ increases monotonically for $0 < t < 2(N-1)^2 l^2/a^2$. It means, for all N, satisfying the inequality $2(N-1)^2 l^2/a^2 \geqslant t^*$, i.e. inequality (7), inequality (8) will be fulfilled.

‡ See [7], pages 529–530.

We have:

$$|R_N(x,\xi,t)| = \left|\frac{2}{l}\sum_{n=N+1}^{+\infty} e^{-\frac{n^2\pi^2a^2}{l^2}t}\sin\frac{n\pi x}{l}\sin\frac{n\pi\xi}{l}\right| \leqslant \frac{2}{l}\sum_{n=N+1}^{+\infty} e^{-\frac{n^2\pi^2a^2}{l^2}t}$$

$$< \frac{2}{l}\int_N^{+\infty} e^{-z^2\frac{\pi^2a^2}{l^2}t}\,dz = \frac{2}{\pi a\sqrt{t}}\int_{\frac{N\pi a\sqrt{t}}{l}}^{+\infty} e^{-\xi^2}\,d\xi$$

$$= \frac{1}{a\sqrt{\pi t}}\left[1-\Phi\left(\frac{N\pi a\sqrt{t}}{l}\right)\right] \tag{13}$$

$$\text{for } 0\leqslant x, \quad \xi\leqslant l, \quad 0<t<+\infty.$$

It is better, however, to make an evaluation not of the remainder of the series representing the source function, but to make an evaluation of the remainder of the series representing the solution of the boundary-value problem, obtained by means of this function, since integration, generally speaking, improves the convergence of the series†.

104. By the method of images we obtain:

$$G(x,\ \xi,\ t) = \frac{1}{2a\sqrt{\pi t}}\sum_{n=-\infty}^{+\infty}\left\{e^{-\frac{(x-\xi+2nl)^2}{4a^2t}} + e^{-\frac{(x+\xi-2nl)^2}{4a^2t}}\right\}, \tag{1}$$

$$0\leqslant x\leqslant l, \quad 0<t<+\infty.$$

The pattern of the corresponding distribution of instantaneous sources of heat of magnitude $Q = c\rho$ is shown in Fig. 37.

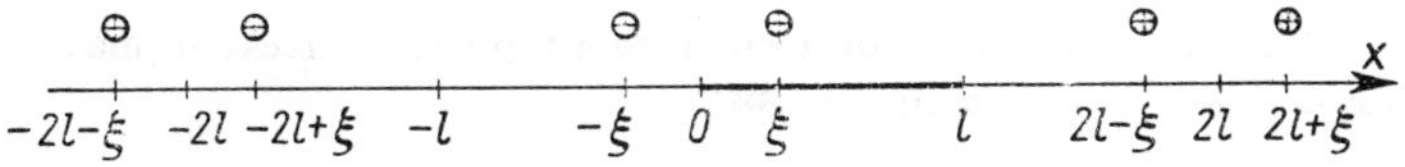

FIG. 37

By (7) and (8) of the solution of the preceding problem we have for terms of series (1):

$$\frac{1}{2a\sqrt{\pi t}}e^{-\frac{(x\pm\xi\pm 2nl)^2}{4a^2t}} \leqslant \frac{1}{2a\sqrt{\pi t^*}}e^{-\frac{(n-1)^2l^2}{a^2t^*}} \tag{2}$$

$$\text{for } 0\leqslant x, \quad \xi\leqslant l, \quad 0\leqslant t\leqslant t^*, \quad n\geqslant\frac{a}{l}\sqrt{\frac{t^*}{2}}+1. \tag{3}$$

---

† See the estimates, made in solving problems 22, 27, 28, 29, 48 of the present chapter.

Thus for the remainder of series (1) the inequality

$$|R_N(x, \xi, t)| \leqslant \frac{2}{2\sqrt{\pi t^*}} \sum_{n=N+1}^{+\infty} e^{-\frac{(n-1)^2 l^2}{a^2 t^*}} < \frac{2}{a\sqrt{\pi t^*}} \int_N^{+\infty} e^{-\frac{l^2 z^2}{a^2 t^*}} dz$$

$$= \frac{1}{l}\left[1-\Phi\left(\frac{lN}{a\sqrt{t^*}}\right)\right] \qquad (4)$$

$$\text{for } 0 \leqslant x, \quad \xi \leqslant l, \quad 0 \leqslant t \leqslant t^*, \quad N \geqslant \frac{a}{l}\sqrt{\frac{t^*}{2}}+1 \qquad (5)$$

holds.

By the method of separation of variables for this same source function the expression

$$G(x, \xi, t) = \frac{1}{l} + \frac{2}{l}\sum_{n=1}^{+\infty} e^{-\frac{n^2\pi^2a^2}{l^2}t} \cos\frac{n\pi x}{l} \cos\frac{n\pi\xi}{l} \qquad (6)$$

is obtained. The inequality

$$|R_N(x, \xi, t)| \leqslant \frac{1}{a\sqrt{\pi t}}\left\{1-\Phi\left[\frac{N\pi a\sqrt{t}}{l}\right]\right\} \qquad (7)$$

$$\text{for } 0 \leqslant x, \quad \xi \leqslant l, \quad 0 < t < +\infty \qquad (8)$$

is obtained for the remainder of series (6).

105. By the method of images we find:

$$G(x, \xi, t) = \frac{1}{2a\sqrt{\pi t}} \sum_{n=-\infty}^{+\infty} (-1)^n \left(e^{-\frac{(x-\xi+2nl)^2}{4a^2t}} - e^{-\frac{(x+\xi+2nl)^2}{4a^2t}}\right). \qquad (1)$$

The appropriate distribution of instantaneous point sources of magnitude $Q = c\rho$ and $-Q$ is shown in Fig. 38.

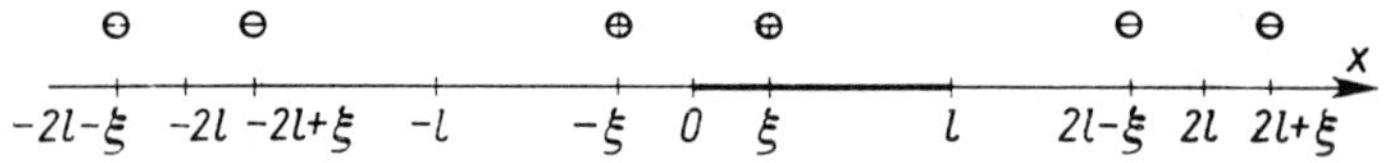

$\ominus$ $\ominus$ $\oplus$ $\oplus$ $\ominus$ $\ominus$

x

$-2l-\xi$ $-2l$ $-2l+\xi$ $-l$ $-\xi$ 0 ξ l $2l-\xi$ $2l$ $2l+\xi$

FIG. 38

The method of separation of variables gives:

$$G(x, \xi, t) = \frac{2}{l}\sum_{n=0}^{+\infty} e^{-\frac{(2n+1)^2\pi^2a^2}{4l^2}t} \cos\frac{(2n+1)\pi\xi}{2l} \cos\frac{(n+1)\pi x}{2l}. \qquad (2)$$

An estimate of the error made in replacing the sum of series (1) by its partial sum is made either by means of inequalities, similar to the inequalities (4) and (5) of the solution of the preceding problem (rough estimate), or in a manner similar to that done in the solution of problem 103 (more exact estimate). For the remainder of the series (2) we obtain the inequality

$$|R_N(x, \xi, t)| \leqslant \frac{1}{a\sqrt{\pi t}}\left\{1-\Phi\left[\frac{(2n+1)\pi a\sqrt{t}}{2l}\right]\right\}.$$

106. (a) If N satisfies the inequalities

$$N \geqslant \frac{a}{l}\sqrt{\frac{t^*}{2}}+1, \tag{1}$$

$$N > \frac{a}{l}\sqrt{t^* \ln\left[2\varepsilon a\sqrt{\pi t^*}\right]}+1, \tag{2}$$

then the remainder of series (2) of the solution of problem 103 will satisfy the inequality

$$|R_N(x, \xi, t)| \leqslant \varepsilon \quad \text{for} \quad 0 \leqslant x, \quad \xi \leqslant l, \quad 0 \leqslant t \leqslant t^*. \tag{3}$$

(b) If N satisfies the inequality

$$\Phi\left(\frac{N\pi a\sqrt{t^*}}{l}\right) \geqslant 1-\varepsilon\pi a\sqrt{t^*}, \tag{4}$$

then the remainder of series (12) of the solution of problem 103 will satisfy the inequality

$$|R_N(x, \xi, t)| \leqslant \varepsilon \quad \text{for} \quad 0 \leqslant x, \quad \xi \leqslant l, \quad t^* \leqslant t < +\infty. \tag{5}$$

Note. The inequalities (1), (2), (4) allow us for a given N to find t^* so that (3) and (5) are fulfilled.

107. (a) If N satisfies the inequalities

$$N \geqslant \frac{a}{l}\sqrt{\frac{t^*}{2}}+1, \tag{1}$$

$$\Phi\frac{lN}{a\sqrt{t^*}} \geqslant 1-\varepsilon l, \tag{2}$$

then the remainder of series (1) of problem 104 will satisfy the inequality

$$|R_N(x, \xi, t)| \leqslant \varepsilon \quad \text{for} \quad 0 \leqslant x,\ \xi \leqslant l, \quad 0 \leqslant t \leqslant t^*.$$

(b) If N satisfies the inequality

$$\Phi\left(\frac{N\pi a\sqrt{t^*}}{l}\right) \geqslant 1-\varepsilon\pi a\sqrt{t^*},$$

then the remainder of series (6) of problem 104 will satisfy the inequality

$$|R_N(x, \xi, t)| \leqslant \varepsilon \quad \text{for} \quad 0 \leqslant x,\ \xi \leqslant l, \quad t^* \leqslant t < +\infty.$$

108. An expression for the source function is obtained from expressions found in the solutions of problems 103, 104, 105 by multiplying by e^{-ht}, where h is the coefficient of heat exchange appearing in the equation $u_t = a^2 u_{xx} - hu$.

109. *Solution.* We replace in the solution $u(x, t)$ of the equation

$$u_t = a^2 u_{xx} + f(x, t), \quad 0 < x < l, \quad 0 < t < +\infty \tag{1}$$

x and t by ξ and τ; and replace t by $t-\tau$, $0 < \tau < t$ in the source function $G(x, \xi, t)$.

Integrating the equation

$$\frac{\partial}{\partial \tau}(Gu) = G\frac{\partial u}{\partial \tau} + u\frac{\partial G}{\partial \tau} = a^2\left[G\frac{\partial^2 u}{\partial \xi^2} - u\frac{\partial^2 G}{\partial \xi^2}\right] + Gf\dagger$$

with respect to ξ from zero to l and with respect to τ from zero to $t-\alpha$, $0 < \alpha < t$, we obtain:

$$\int_0^l (Gu)_{\tau=t-\alpha}\,d\xi = \int_0^l (Gu)_{\tau=0}\,d\xi + a^2\int_0^{t-\alpha}\left\{\left(G\frac{\partial u}{\partial \xi}\right)_{\xi=0}^{\xi=l} - \left(u\frac{\partial G}{\partial \xi}\right)_{\xi=0}^{\xi=l}\right\}d\tau + \int_0^{t-\alpha} d\tau \int_0^l Gf\,d\xi. \tag{2}$$

Passing to a limit in (2) as $\alpha \to 0$‡ we obtain the integral relation

$$u(x, t) = \int_0^l (Gu)_{\tau=0}\,d\xi + a^2\int_0^t\left\{\left(G\frac{\partial u}{\partial \xi}\right)_{\xi=0}^{\xi=l} - \left(u\frac{\partial G}{\partial \xi}\right)_{\xi=0}^{\xi=l}\right\}d\tau + \int_0^t d\tau\int_0^l Gf(\xi, \tau)\,d\xi. \tag{3}$$

This integral has a different value for source functions satisfying different boundary conditions. If one uses the initial and boundary conditions for u:

$$u(0, \tau) = \phi(\tau), \quad u(l, \tau) = 0, \quad 0 < \tau < +\infty, \tag{4}$$

$$u(\xi, 0) = f(\xi), \quad 0 < \xi < l, \tag{5}$$

and boundary conditions for $G(x, \xi, t-\tau)$:

$$G(x, 0, t-\tau) = 0, \quad G(x, l, t-\tau) = 0, \quad 0 < x < l, \quad 0 < \tau < t, \tag{6}$$

---

† This equality is derived in the same way as equality (1) of the solution of problem 68.

‡ The limit on the left hand side of (2) can be calculated by methods similar to those given in [7], on pages 248–251.

then the integral relation (6) reduces to:

$$u(x,\ t)=\int_0^l f(\xi)\,G(x,\ \xi,\ t)\,d\xi+a^2\int_0^t \phi(\tau)\,\frac{\partial G(x,0,\ t-\tau)}{\partial\xi}\,d\tau+$$

$$+\int_0^t d\tau\int_0^l f(\xi,\ \tau)\,G(x,\ \xi,\ t-\tau)\,d\xi.$$

Making use of the two different representations of the source function $G(x, \xi, t-\tau)$ (see the solution of problem 103), we obtain two expressions for the solution of our boundary-value problem:

$$\text{(a)}\quad u(x,\ t)=\frac{1}{2a\sqrt{\pi t}}\int_0^l f(\xi)\left\{\sum_{n=-\infty}^{+\infty}\left(e^{-\frac{(x-\xi+2nl)^2}{4a^2t}}-e^{-\frac{(x+\xi+2nl)^2}{4a^2t}}\right)\right\}d\xi+$$

$$+\frac{1}{2a\sqrt{\pi}}\int_0^t\frac{\phi(\tau)}{(t-\tau)^{3/2}}\sum_{n=-\infty}^{+\infty}\left\{(x+2nl)\,e^{-\frac{(x+2nl)^2}{4a^2(t-\tau)}}\right\}d\tau+$$

$$+\frac{1}{2a\sqrt{\pi}}\int_0^t\frac{d\tau}{\sqrt{t-\tau}}\int_0^l f(\xi,\ \tau)\left\{\sum_{n=-\infty}^{+\infty}\left(e^{-\frac{(x-\xi+2nl)^2}{4a^2(t-\tau)}}-e^{-\frac{(x+\xi+2nl)^2}{4a^2(t-\tau)}}\right)\right\}d\xi;$$

$$\text{(b)}\quad u(x,\ t)=\frac{2}{l}\int_0^l f(\xi)\left\{\sum_{n=-\infty}^{+\infty}e^{-\frac{n^2\pi^2a^2}{l^2}t}\sin\frac{n\pi\xi}{l}\sin\frac{n\pi x}{l}\right\}d\xi+$$

$$+\frac{2\pi a^2}{l^2}\sum_{n=1}^{+\infty}n\left\{\int_0^t\phi(\tau)\,e^{-\frac{n^2\tau^2a^2}{l^2}(t-\tau)}\,d\tau\right\}\sin\frac{n\pi x}{l}+$$

$$+\frac{2}{l}\int_0^t d\tau\int_0^l f(\xi,\ \tau)\left\{\sum_{n=1}^{+\infty}e^{-\frac{n^2\pi^2a^2}{l^2}(t-\tau)}\sin\frac{n\pi\xi}{l}\sin\frac{n\pi x}{l}\right\}d\xi,$$

(a) generally speaking, is more suitable for small t,
(b) for large t.

110. $$u(x,\ t)=\int_0^l f(\xi)\,G(x,\ \xi,\ t)\,d\xi-a^2\int_0^t\phi(\tau)\,G(x,\ 0,\ t-\tau)\,d\tau+$$

$$+\int_0^t d\tau\int_0^l f(\xi,\ \tau)\,G(x,\ \xi,\ t-\tau)\,d\xi, \qquad (1)$$

where $G(x, \xi, t-\tau)$ is the source function, obtained in the solution of problem 104. If the two different representations of the source function are substituted in equation (1), then two expressions for the solution of our boundary-value problem are obtained.

111. $$u(x, t) = U_0 \sum_{n=-\infty}^{+\infty} \left\{ \left[1 - \Phi\left(\frac{|x+2nl|}{2a\sqrt{t}} \right) \right] \operatorname{sign}(x+2nl) \right\}.$$

112. $$u(x, t) = q_0 \sum_{n=-\infty}^{+\infty} \left\{ 2a\sqrt{\frac{t}{\pi}}\, e^{-\frac{(x+2nl)^2}{4a^2 t}} - \right.$$
$$\left. -|x+2nl| \left[1 - \Phi\left(\frac{|x+2nl)|}{2a\sqrt{t}} \right) \right] \right\}.$$

3. Inhomogeneous Media; Equations with Piecewise Continuous Coefficients and Matching Conditions

113. $$u(x, t) = \left\{ \begin{array}{ll} U_0 + (U_0 - U_1)\, \Phi\left(-\dfrac{x}{2a_1\sqrt{t}} \right), & -\infty < x \leqslant 0, \\ U_0 + (U_0 - U_2)\, \Phi\left(\dfrac{x}{2a_2\sqrt{t}} \right), & 0 \leqslant x < +\infty, \end{array} \right\} 0 < t < +\infty,$$

$$U_0 = \frac{U_1 \dfrac{k_1}{a_1} + U_2 \dfrac{k_2}{a_2}}{\dfrac{k_1}{a_1} + \dfrac{k_2}{a_2}}.$$

Method. The problem may be solved by the following method.

It is necessary to extend the left hand rod indefinitely to the right so that an infinite homogeneous rod is obtained of the same material as the left hand semi-infinite rod. Then it is necessary to find the temperature of the infinite rod obtained for the condition that its initial temperature equals U_1 for $-\infty < < x < 0$ and U_1^* for $0 < x < +\infty$, where U_1^* is a constant, indeterminate for the present. One treats the right hand semi-infinite rod similarly. The constants U_1^* and U_2^* are found from the boundary conditions (matching conditions) at the point $x = 0$.

114. $$u(x, t) = \left\{ \begin{array}{ll} u_1(x, t), & -\infty < x < 0, \\ u_2(x, t), & 0 < x < +\infty, \end{array} \right\} 0 < t < +\infty,$$

$$u_1(x, t) = \frac{1}{2a_1\sqrt{\pi t}} \int_{-\infty}^{0} f_1(\xi) \left\{ e^{-\frac{(x-\xi)^2}{4a_1^2 t}} + e^{-\frac{(x+\xi)^2}{4a_1^2 t}} \right\} d\xi +$$
$$+ \frac{a_1}{k_1\sqrt{\pi}} \int_0^t \frac{\phi(\tau)}{\sqrt{t-\tau}}\, e^{-\frac{x^2}{4a_1^2 (t-\tau)}}\, d\tau,$$

$$u_2(x,t)=\frac{1}{2a_2\sqrt{\pi t}}\int_0^{+\infty} f_2(\xi)\left\{e^{-\frac{(x-\xi)^2}{4a_2^2 t}}+e^{-\frac{(x+\xi)^2}{4a_2^2 t}}\right\}d\xi-$$

$$-\frac{a_2}{k_2\sqrt{\pi}}\int_0^t \frac{\phi(\tau)}{\sqrt{t-\tau}}e^{-\frac{x^2}{4a_2^2(t-\tau)}}d\tau,$$

$$\phi(\tau)=\frac{1}{\pi}\frac{d}{d\tau}\int_0^\tau \frac{\Phi(z)}{\sqrt{\tau-z}}dz,$$

$$\Phi(z)=\left[\frac{1}{a_2\sqrt{z}}\int_0^{+\infty} f_2(\xi)e^{-\frac{\xi^2}{4a_2^2 z}}d\xi-\frac{1}{a_1\sqrt{z}}\int_{-\infty}^0 f_1(\xi)e^{-\frac{\xi^2}{4a_1^2 z}}d\xi\right]\frac{1}{\frac{k_1}{a_1}+\frac{k_2}{a_2}}.$$

Method. The functions $u_1(x, t)$ and $u_2(x, t)$ must be respectively the solutions of the equations of heat conduction $u_{1t}=a_1^2u_{1xx}$ and $u_{2t}=a_2^2u_{2xx}$ and must satisfy the matching conditions

$$u_1(0,t)=u_2(0,t),\qquad k_1u_{1x}(0,t)=k_2u_{2x}(0,t).$$

Assuming

$$\phi(t)=k_1u_{1x}(0,t)=k_2u_{2x}(0,t)$$

and solving the problem of heat conduction with a given boundary condition of the second kind for a semi-infinite rod $-\infty<x<0$ and for the semi-infinite rod $0<x<+\infty$, we express $u_1(x, t)$ and $u_2(x, t)$ in terms of the initial conditions and the function $\phi(t)$ unknown as yet. Using the first matching condition $u_1(0, t)=u_2(0, t)$, we obtain Abel's integral equation determining $\phi(t)$:

$$\int_0^z \frac{\phi(\tau)\,d\tau}{\sqrt{t-\tau}}=\Phi(z).$$

The solution of this equation is†:

$$\phi(\tau)=\frac{1}{\pi}\frac{d}{d\tau}\int_0^\tau \frac{\Phi(z)}{\sqrt{\tau-z}}dz.$$

If $\Phi'(z)$ exists and is continuous‡ for $0\leqslant z<+\infty$, then, performing firstly an integration by parts on the right hand side of the latter equality, and then a differentiation, we obtain:

$$\phi(\tau)=\frac{1}{\pi}\int_0^\tau \frac{\Phi'(z)}{\sqrt{\tau-z}}dz+\frac{\Phi(+0)}{\pi\sqrt{\tau}}.$$

† See, for instance, [2], vol II, § 79.

‡ For suitable restrictions on f_1 and f_2 this will be achieved.

This integral can be evaluated, in particular, if

$$\Phi(z) \equiv \text{const}.$$

In this case $\Phi'(z) = 0$ and

$$\phi(\tau) = \frac{\Phi(+0)}{\pi\sqrt{\tau}}.$$

115. The solution of the boundary-value problem

$$\frac{\partial G_1}{\partial t} = a_1^2 \frac{\partial^2 G_1}{\partial x^2}, \quad -\infty < x < 0, \quad 0 < t < +\infty, \tag{1}$$

$$\frac{\partial G_2}{\partial t} = a_2^2 \frac{\partial^2 G_2}{\partial x^2}, \quad 0 < x < +\infty, \quad 0 < t < +\infty, \tag{1'}$$

$$G_1 = G_2, \quad \lambda_1 \frac{\partial G_1}{\partial x} = \lambda_2 \frac{\partial G_2}{\partial x} \quad \text{for} \quad x = 0, \quad 0 < t < +\infty, \tag{2}$$

$$\lim_{t\to 0} G_1 = 0, \quad -\infty < x < 0, \tag{3}$$

where $\lim\limits_{t\to 0} G_2 = 0,\ 0 < x < +\infty,\ x \neq \xi$ at the point $x = \xi$ for $t \to 0$, and G_2 has a singularity

$$\frac{1}{2a_2\sqrt{\pi t}} e^{-\frac{(x-\xi)^2}{4a_2^2 t}}, \tag{3'}$$

is:

$$G_1(x, \xi, t) = \frac{2\dfrac{\lambda^2}{a_2}}{\dfrac{\lambda_1}{a_1} + \dfrac{\lambda_2}{a_2}} \frac{1}{2a_2\sqrt{\pi t}} e^{-\frac{\left(x - \frac{a_1}{a_2}\xi\right)^2}{4a_2^2 t}} \quad \text{for} \quad -\infty < x < 0, \tag{4}$$

$$G_2(x, \xi, t) = \frac{1}{2a_2\sqrt{\pi t}} e^{-\frac{(x-\xi)^2}{4a_2^2 t}} + \frac{\dfrac{\lambda_2}{a_2} - \dfrac{\lambda_1}{a_1}}{\dfrac{\lambda_2}{a_2} + \dfrac{\lambda_1}{a_1}} \cdot \frac{e^{-\frac{(x+\xi)^2}{4a_2^2 t}}}{2a_2\sqrt{\pi t}} \quad \text{for } 0 < x < +\infty. \tag{4'}$$

Solution. Let us change to dimensionless quantities (see the solution of problem 18 of the present chapter) where, so that the equation of heat conduction for the right hand and left hand rods should have the form $u_t = a^2 u_{\xi\xi}$. We have $x = l'\xi,\ -\infty < \xi < 0$,

$$x = l''\xi, \quad 0 < \xi < +\infty, \quad t = \tau, \quad l' = a_1, \quad l'' = a_2 \dagger.$$

† The emphasis is on the numerical equality, and not on the correspondence of the dimensions.

The boundary conditions (2) take the form

$$u_1(0, \tau) = u_2(0, \tau), \tag{5}$$

$$\frac{\lambda_1}{a_1}\frac{\partial u_1(0, \tau)}{\partial \xi} = \frac{\lambda_2}{a_2}\frac{\partial u_2(0, \tau)}{\partial \xi}. \tag{6}$$

We shall seek a solution for $-\infty < \xi < 0$, as the function $e^{-(\xi-\xi_0)^2/4\tau}/2\sqrt{\pi t}$ "transmitted" across the boundary of separation $\xi = 0$, i.e. as the function, having the form

$$\alpha_1 \frac{1}{2\sqrt{\pi\tau}} e^{-\frac{(\xi-\xi_0)^2}{4\tau}}, \tag{7}$$

and a solution for $0 < \xi < +\infty$ as the sum of $e^{-(\xi-\xi_0)^2/4\tau}/2\sqrt{\pi\tau}$ and a term, representing the result of "reflection" at the boundary of separation $\xi = 0$, $e^{-(\xi+\xi_0)^2/4\tau}/2\sqrt{\pi\tau}$ i.e. in the form

$$u_2(\xi, \tau) = \alpha_2 \frac{1}{2\sqrt{\pi\tau}} e^{-\frac{(\xi+\xi_0)^2}{4\pi}} + \frac{1}{2\sqrt{\pi\tau}} e^{-\frac{(\xi-\xi_0)^2}{4\tau}} \tag{8}$$

Substituting (7) and (8) in (5) and (6) we find α_1 and α_2, which leads to the answer (if one reverts to the former units).

116. The solution of the boundary-value problem

$$u_t = a^2 u_{xx}, \quad 0 < x, \ t < +\infty, \tag{1}$$

$$c_0 u_t(0, t) = \lambda S u_x(0, t), \quad 0 < t < +\infty, \tag{2}$$

$$u(x, 0) = f(x), \quad 0 < x < +\infty \tag{3}$$

is:

$$u(x, t) = \frac{1}{2a\sqrt{\pi t}} \int_{-\infty}^{+\infty} F(\xi)\, e^{-\frac{(x-\xi)^2}{4a^2 t}}\, d\xi,$$

where

$$F(x) = \begin{cases} \bar{f}(x) & \text{for } -\infty < x < 0, \\ f(x) & \text{for } 0 < x < +\infty, \end{cases}$$

$$\bar{f}(x) = \frac{e^{\alpha^2 x} - 1}{\alpha^2} f'(+0) + f(+0) + \int_0^x dz \int_0^z \{f''(-\xi) + \alpha^2 f'(-\xi)\}\, e^{-\alpha^2(\xi - z)}\, d\xi,$$

$$\alpha^2 = \frac{\lambda S}{a^2 C_0},$$

λ is the coefficient of heat conduction of the rod, S the cross-sectional area, a^2 the coefficient of thermal conductivity of the rod.

Method. Use the statement formulated in problem 82.

117. The solution of the boundary-value problem

$$\left.\begin{aligned} \frac{\partial u_1}{\partial t} &= a_1^2 \frac{\partial^2 u_1}{\partial x^2}, \quad 0 < x < \xi(t), \\ \frac{\partial u_2}{\partial t} &= a_2^2 \frac{\partial^2 u_2}{\partial x^2}, \quad \xi(t) < x < +\infty, \end{aligned}\right\} 0 < t < +\infty, \tag{1}$$

$$u_1(\xi(t), t) = u_2(\xi(t), t), \quad u_1(0, t) = U_1, \quad u_2(+\infty, t) = U_2, \tag{2}$$

where the freezing point is taken as zero, $x = \xi(t)$ the coordinate of the freezing front

$$\left(k_1 \frac{\partial u_1}{\partial x} - k_2 \frac{\partial u_2}{\partial x}\right)_{x=\xi(t)} = Q\rho \frac{d\xi}{dt}, \quad 0 < t < +\infty, \tag{2'}$$

Q the latent heat of fusion, ρ the mass density of the liquid,

$$u_2(x, 0) = U_2, \quad 0 < x < +\infty, \tag{3}$$

is:

$$u_1(x, t) = A_1 + B_1 \Phi\left(\frac{x}{2a_1\sqrt{t}}\right), \tag{4}$$

$$u_2(x, t) = A_2 + B_2 \Phi\left(\frac{x}{2a_2\sqrt{t}}\right), \tag{4'}$$

where

$$A_1 = U_1, \; B_1 = -\frac{U_1}{\Phi\left(\dfrac{\alpha}{2a_1}\right)}, \quad A_2 = -\frac{U_2\Phi\left(\dfrac{\alpha}{2a_2}\right)}{1-\Phi\left(\dfrac{\alpha}{2a_1}\right)}, \quad B_2 = \frac{U_2}{1-\Phi\left(\dfrac{\alpha}{2a_2}\right)}, \tag{5}$$

and α is the root of the transcendental equation

$$\frac{k_1 U_1 e^{-\frac{\alpha^2}{4a_1^2}}}{a_1 \Phi\left(\dfrac{\alpha}{2a_1}\right)} + \frac{k_2 U_2 e^{-\frac{\alpha^2}{4a_2^2}}}{a_2\left[1-\Phi\left(\dfrac{\alpha}{2a_2}\right)\right]} = -Q\rho\frac{\sqrt{\pi}}{2}\alpha. \tag{6}$$

CHAPTER IV

EQUATIONS OF ELLIPTIC TYPE

§ 1. Physical Problems Leading to Equations of Elliptic Type and the Statement of Boundary-value Problems

1. Boundary-value Problems for Laplace's and Poisson's Equation in a Homogeneous Medium

1. The equation for the temperature of a steady-state temperature field in a homogeneous isotropic medium has the form

$$\Delta u = -f(x, y, z), \tag{1}$$

where $f = F/k$, F the density of the sources of heat, i.e. the amount of heat, liberated per unit volume per unit time, k the coefficient of heat conduction.

A boundary condition of the first kind

$$u|_{\Sigma} = f_1$$

means that the temperature f_1 is given at the surface Σ; a condition of the second kind

$$\left.\frac{\partial u}{\partial n}\right|_{\Sigma} = f_2, \quad \text{or} \quad -k\left.\frac{\partial u}{\partial n}\right|_{\Sigma} = \bar{f}_2, \quad (\bar{f}_2 = -f_2 k),$$

means that a heat flow of magnitude $\bar{f}_2$ is given, at the surface Σ; a boundary condition of the third kind

$$\left.\frac{\partial u}{\partial n} + hu\right|_{\Sigma} = f_3, \quad \text{or} \quad -k\frac{\partial u}{\partial n} = \bar{h}(u - \bar{f}_3), \quad \bar{h} = kh, \quad \bar{f}_3 = \frac{f_3}{\bar{h}},$$

means that a heat exchange obeying Newton's law takes place with a medium of temperature $\bar{f}_3$ at the surface Σ.

A necessary condition for the existence of a steady-state temperature for the second boundary-value problem is $\int\limits_{\Sigma} f_2 d = 0$, i.e. the total heat flow through the surface Σ must equal zero. A non-uniform distribution of temperature produces a heat flow, the magnitude of which according to Fourier's law equals $Q = -k$ grad u. Its projection in the n direction equals $Q_n = -k\,\partial u/\partial n$.

Solution. In deriving equation (1) one must write down the condition of heat balance for an arbitrary volume and then use Ostrogradskii's formula.

The equation of heat balance for a volume T with boundary Σ has the form

$$\int_{\Sigma}\left(-k\frac{\partial u}{\partial n}\right)d\sigma=\int_{T}F\,d\tau; \tag{2}$$

on the left we have the total heat flowing out through Σ, on the right, the amount of heat liberated in volume T.

Ostrogradskii's formula gives:

$$\int_{T}\operatorname{div}(k\operatorname{grad}u)\,d\tau=-\int_{T}F\,d\tau, \tag{3}$$

from which by the arbitrary nature of the volume T and of the constancy of k we obtain equation (1).

2. (a) The diffusion equation in a medium at rest is

$$\Delta u=0, \tag{1}$$

where $u(x, y, z)$ is the concentration.

(b) If the medium moves with a velocity $\boldsymbol{v}=(v_x, v_y, v_z)$, where $\operatorname{div}\boldsymbol{v}=0$, then the diffusion equation takes the form

$$D\Delta u-v_x\frac{\partial u}{\partial x}-v_y\frac{\partial u}{\partial y}-v_z\frac{\partial u}{\partial z}=0, \tag{2}$$

where D is the diffusion coefficient, and v_x, v_y, v_z are the projections of the velocity $\boldsymbol{v}$ on the coordinate axes.

If $v_x=v$, $v_y=v_z=0$, then equation (2) takes the form

$$\Delta u-\frac{v}{D}\frac{\partial u}{\partial x}=0, \tag{3}$$

or

$$u_{xx}+u_{yy}+u_{zz}-\frac{v}{D}u_x=0$$

(the equation of gas attack).

Method. The diffusion flow of a substance for a non-uniform distribution of concentration equals

$$Q=-D\operatorname{grad}u. \tag{4}$$

In addition to diffusion flow one must take into account the transport flow (translational flow), equal to

$$u\boldsymbol{v},$$

so that the total flow equals

$$-D\operatorname{grad}u+u\boldsymbol{v}.$$

In deriving equations (1) and (2) one must use the principle of conservation of matter for an arbitrary volume and then apply Ostrogradskii's formula (see the solution of problem No. 1).

The principle of conservation of matter for a rigid surface Σ is written as:

$$\int\limits_{\Sigma}\left(-D\frac{\partial u}{\partial n}+v_n u\right)d\sigma = 0,$$

or

$$\int\limits_{T}[\text{div}\,(D\,\text{grad}\,u)-\text{div}\,(\boldsymbol{v}u)]\,d\tau = 0,$$

Because the volume T is arbitrary and div $\boldsymbol{v} = 0$ equation (2) follows.

3. The equation for the potential u of an electrostatic field in a vacuum has the form

$$\Delta u = -4\pi\rho,$$

where ρ is the volume density of the charges.

The physical significance of boundary conditions of first and second kind are: $u|_{\Sigma} = f$ the potential is given on the surface Σ, $\left.\frac{\partial u}{\partial n}\right|_{\Sigma} = f_1$ the density of surface charges is given.

Solution. Equations for the field of a static distributed charges, are derived from Maxwell's equations, if all the derivatives with respect to time are assumed equal to zero. For the electrostatic field ina non-conducting medium we obtain:

$$\text{curl}\,\boldsymbol{E} = 0, \tag{1}$$

$$\left.\begin{aligned}\text{div}\,\boldsymbol{D} &= 4\pi\rho,\\ \boldsymbol{D} &= \varepsilon\boldsymbol{E},\end{aligned}\right\} \tag{2}$$

where ε is the dielectric constant of the medium, $\rho = \rho(M)$ the volume charge density at the point M.

From the equation rot $\boldsymbol{E} = 0$ it follows that $\boldsymbol{E}$ is a conservative field, representable in the form

$$\boldsymbol{E} = -\text{grad}\,u,$$

where $u = u(M)$ is the potential of the field.

Equation (2) gives:

$$\text{div}\,(\varepsilon\,\text{grad}\,u) = -4\pi\rho.$$

If $\varepsilon = \text{const.}$, then we obtain the equation

$$\Delta u = -\frac{4\pi\rho}{\varepsilon};$$

in vacuum $\varepsilon = 1$, and we have

$$\Delta u = -4\pi\rho.$$

If there are conducting surfaces, then the tangential component of the electric field must be equal to zero on them:

$$E_s = -\frac{\partial u}{\partial s} = 0,$$

where $\partial/\partial s$ indicates differentiation with respect to a tangential direction on the surface. Hence on the surface of the conductor the potential is constant:

$$u = \text{const};$$

inside the conductor $u = \text{const.}$, and $\boldsymbol{E} \equiv 0$.

If the conductor is earthed, then the potential

$$u = 0.$$

The density of surface charges is evaluated by the formula

$$\sigma = -\frac{1}{4\pi} D_n = -\frac{\varepsilon}{4\pi}\frac{\partial u}{\partial n}, \tag{3}$$

where $\partial/\partial n$ means differentiation with respect to the normal to the surface. If the distribution of surface charge on the conductor is given, we obtain the condition

$$\left.\frac{\partial u}{\partial n}\right|_{\Sigma} = f, \quad f = -\frac{4\pi\sigma}{\varepsilon}.$$

But this is an unnatural statement of the problem for electrostatics; usually the total charge e on the surface is known. Therefore we require the solution of the equation $\Delta u = -4\pi\rho$ for the boundary condition $u|_{\Sigma} = u_0$, where u_0 is to be determined from the normalizing condition for the total charge

$$-\int_{\Sigma} \varepsilon \frac{\partial u}{\partial n} \mathrm{d}\sigma = 4\pi e, \quad \text{where } e = \int_{\Sigma} \rho \mathrm{d}\tau \text{ (see problem 7)}.$$

4. The magnetic field vector is $\boldsymbol{H} = -\operatorname{grad} \phi$, the potential ϕ satisfies Laplace's equation

$$\Delta\phi = 0.$$

Solution. If the magnetic field does not vary with time, then it satisfies the equations

$$\operatorname{curl} \boldsymbol{H} = 0, \tag{1}$$

$$\operatorname{div} \boldsymbol{B} = 0. \tag{2}$$

From (1) it follows:

$$\boldsymbol{H} = -\operatorname{grad}\phi;$$

substituting this expression in formula (2) and assuming the homogeneity and isotropy of the medium ($\mu = \text{const.}$) we obtain Laplace's equation.

5. Since the electric field $\boldsymbol{E}$ is conservative, then

$$\Delta u = 0,$$

and on an earthed conducting surface

$$u\,|_{\Sigma} = 0.$$

Solution. We shall start with Maxwell's equations in a conducting medium in the steady-state case

$$\left.\begin{aligned} \operatorname{curl} \boldsymbol{H} &= \frac{4\pi}{c}\,\boldsymbol{j}, \\ \operatorname{curl} \boldsymbol{E} &= 0, \\ \operatorname{div} \boldsymbol{E} &= 4\pi\rho, \\ \operatorname{div} \mu\boldsymbol{H} &= 0. \end{aligned}\right\} \tag{1}$$

Applying the operator div to the first equation, we obtain an equation for the current density $\boldsymbol{j}$

$$\operatorname{div} \boldsymbol{j} = 0. \tag{2}$$

From the equation curl $\boldsymbol{E} = 0$, it follows that

$$\boldsymbol{E} = -\operatorname{grad} u,$$

where $u = u(M)$ is a scalar potential.

Since by Ohm's law

$$\boldsymbol{j} = \sigma \boldsymbol{E} \quad (\sigma \text{ is the conductivity}) \tag{3}$$

or

$$\boldsymbol{j} = \sigma \operatorname{grad} u,$$

then for a homogeneous and isotropic medium (σ = const.) equation (2) gives:

$$\Delta u = 0.$$

From equations (1) and (2) it follows that $\rho = 0$ inside thc conductor.

(1) On an earthed ideally conducting surface the potential $u = 0$ (boundary condition of the first kind).

(2) If the conductor adjoins a dielectric, then on the boundary of separation the normal component of current density must be equal to zero:

$$j_n = -\sigma \frac{\partial u}{\partial n} = 0,$$

i.e.

$$\frac{\partial u}{\partial n} = 0$$

(boundary condition of the second kind).

6. If ϕ is the velocity potential of a steady flow of incompressible liquid, so that

$$\boldsymbol{v} = \operatorname{grad} \phi,$$

then the potential ϕ satisfies Laplace's equation

$$\Delta\phi = 0.$$

On the surface of a solid, moving with some velocity $\boldsymbol{v}$, the condition

$$\left.\frac{\partial\phi}{\partial n}\right|_{\Sigma} = v_{0n}$$

must be fulfilled. If the body is at rest, then

$$\left.\frac{\partial\phi}{\partial n}\right|_{\Sigma} = 0.$$

If the medium is of infinite extent, then the potential ϕ must satisfy a condition of regularity for $r \to \infty$.

Solution. If the liquid is incompressible, then its density $\rho = \text{const}$. From the equation of continuity (conservation of matter)

$$\frac{\partial\rho}{\partial t} + \operatorname{div}(\rho\boldsymbol{v}) = 0$$

we obtain the condition of incompressibility

$$\operatorname{div}\boldsymbol{v} = 0.$$

Since the velocity of the liquid has a potential

$$\boldsymbol{v} = \operatorname{grad}\phi,$$

then $\operatorname{div}\operatorname{grad}\phi = 0$ or $\Delta\phi = 0$.

7. The first fundamental problem of electrostatics is given by the first exterior boundary-value problem.

It is required to find the function ϕ satisfying Laplace's equation $\Delta\phi = 0$ everywhere outside a given system of conductors, which reduces to zero at infinity, and takes the given values ϕ_i on the surfaces of the conductors:

$$\phi\,|_{\Sigma_i} = \phi_i.$$

The second fundamental problem of electrostatics is stated as follows: It is required to find the function ϕ satisfying Laplace's equation $\Delta\phi = 0$ outside a given system of conductors, reducing to zero at infinity, taking constant values on the surfaces of the conductors

$$\phi\,|_{\Sigma_i} = \phi_i$$

and satisfying the integral conditions on these surfaces

$$\oint_{\Sigma_i} \frac{\partial\phi}{\partial n}\,d\sigma = -4\pi e_i,$$

where e_i is the total charge on the ith conductor.

If there is only one conductor T_0 with surface Σ_0, then the solution of the second problem of electrostatics can be written in the form

$$\phi = \phi_0 V(x, y, z),$$

where $V(x, y, z)$ is the solution of the first exterior boundary-value problem for the region, external to the conductor T_0, for the condition $V = 1$ on Σ_0: the factor ϕ_0 is determined from the normalizing condition

$$\oint_{\Sigma_0} \frac{\partial \phi}{\partial n} \, d\sigma = -4\pi e_0$$

and equals

$$\phi_0 = -\frac{e_0}{\dfrac{1}{4\pi} \displaystyle\int_{\Sigma_0} \frac{\partial V}{\partial n} \, d\sigma} = \frac{e_0}{C},$$

where $C = -\dfrac{1}{4\pi} \displaystyle\int_{\Sigma_0} \frac{\partial V}{\partial n} \, d\sigma$ is the capacity of the conductor.

2. Boundary-value Problems for Laplace's Equation in Inhomogeneous Media

8. The steady-state distribution of temperature satisfies the equation

$$\operatorname{div}(k \operatorname{grad} u) = -F(M),$$

where $k = k(M)$ is the coefficient of heat conduction, $F(M)$ the density of heat sources at the point M.

Let T be some volume with boundary Σ, on which the temperature is given

$$u\,|_{\Sigma} = f.$$

If the coefficient $k(x, y, z)$ is piecewise continuous and is discontinuous on some surface Σ_1, so that

$$k = k_1 = \text{const. in } T_1,$$

$$k = k_2 = \text{const. in } T_2 \quad (T = T_1 + T_2),$$

then on Σ_1 the matching conditions

$$\left.\begin{aligned} u_1 &= u_2, \\ k_1 \frac{\partial u_1}{\partial n} &= k_2 \frac{\partial u_2}{\partial n}, \end{aligned}\right\} \qquad (1)$$

must be satisfied, the first of which implies the continuity of temperature, and the second the continuity of heat flow on the surface of discontinuity.

The problem in this case may be stated as:

$$\Delta u_1 = -\frac{F}{k_1} \text{ in } T_1,$$

$$\Delta u_2 = -\frac{F}{k_2} \text{ in } T_2,$$

$$u|_{\Sigma} = f$$

and on Σ_1 the matching conditions for u_1 and u_2 hold.

Solution. The equation is derived as in problem 1.

The first matching condition $u_1 = u_2$ is obvious; the second condition $k_1 \partial u_1/\partial n = k_2 \partial u_2/\partial n$ may be derived by applying the equation of balance to an infinitely small cylinder T_ε of height $2h$, on an element $d\sigma$ of the surface Σ_1, and then passing to a limit for $h \to 0$.

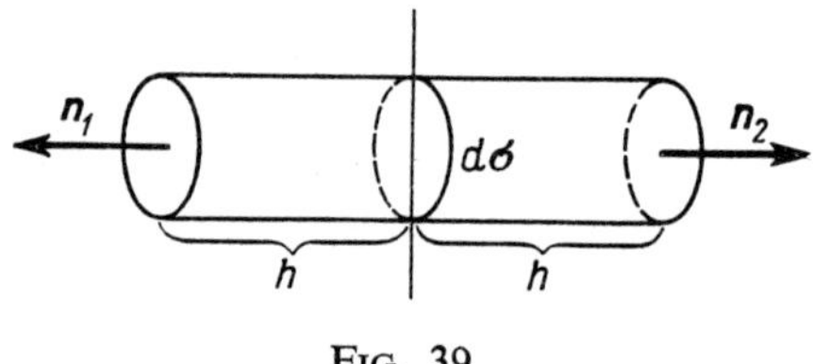

FIG. 39

As has already been noted in the solution of problem 1, the equation of heat balance has the form

$$\int_{\Sigma} \left(-k \frac{\partial u}{\partial n}\right) d\sigma = \int_{T} F d\tau, \tag{2}$$

from which the equation $\operatorname{div}(k \operatorname{grad} u) = F$ follows. Applying (2) to the cylinder T_ε, we obtain (Fig. 39):

$$-\left(-k \frac{\partial u}{\partial n}\right)_{S_1} d\sigma + \left(-k \frac{\partial u}{\partial n}\right)_{S_2} d\sigma + \int_{S_3} \left(-k \frac{\partial u}{\partial n}\right) d\sigma = \int_{T} F d\tau,$$

where S_1 is the left hand, and S_2 the right hand base of the cylinder, S_3 its lateral face. As $h \to 0$ the integrals vanish because $\partial u/\partial n$ and F are bounded everywhere. Assuming the existence of left hand and right hand limiting values of $\partial u/\partial n$ on Σ_1, we obtain:

$$k_1 \frac{\partial u_1}{\partial n_1} + k_2 \frac{\partial u_2}{\partial n_2} = 0;$$

choosing one direction of the normal $\boldsymbol{n}_2 = -\boldsymbol{n}_1 = \boldsymbol{n}$, it is possible to write:

$$k_1 \frac{\partial u_1}{\partial n} = k_2 \frac{\partial u_2}{\partial n} \quad \text{on} \quad \Sigma_1.$$

9. In an inhomogeneous dielectric the potential of the electrostatic field satisfies:

$$\operatorname{div}(\varepsilon \operatorname{grad} u) = -4\pi\rho. \tag{1}$$

If there are no surface charges on the surface of discontinuity $\varepsilon(x, y, z)$, then one may write:

$$\left.\begin{aligned} u_1 &= u_2, \\ \varepsilon_1 \frac{\partial u_1}{\partial n} &= \varepsilon_2 \frac{\partial u_2}{\partial n} \end{aligned}\right\} \text{ on the surface of discontinuity } \varepsilon,$$

where the subscripts 1 and 2 are the values of the quantities on opposite sides of the surface of discontinuity.

If

$$\varepsilon_1 = \text{const. in } T_1,$$
$$\varepsilon_2 = \text{const. in } T_2,$$

where T_1 and T_2 are regions separated by the surface Σ_1, then for the potential

$$u = \begin{cases} u_1 \text{ in } T_1, \\ u_2 \text{ in } T_2 \end{cases}$$

we have:

$$\Delta u_1 = -\frac{4\pi\rho}{\varepsilon_1} \text{ in } T_1,$$

$$\Delta u_2 = -\frac{4\pi\rho}{\varepsilon_2} \text{ in } T_2,$$

$$\left.\begin{aligned} u_1 &= u_2, \\ \varepsilon_1 \frac{\partial u_1}{\partial n} &= \varepsilon_2 \frac{\partial u_2}{\partial n} \end{aligned}\right\} \text{ on } \Sigma_1.$$

The secont matching condition implies the continuity of the normal component of the electric induction

$$\boldsymbol{D} = -\varepsilon \operatorname{grad} u, \quad D_n = -\varepsilon \frac{\partial u}{\partial n}.$$

Method. In the derivation of the equations we start from Maxwell's equations (see the solution of problem 3), assuming there that ε is a function of spatial variables. For the derivation of the matching conditions see problem 8.

In the solution of problem 3 we have:

$$\boldsymbol{E} = -\operatorname{grad} u,$$
$$\operatorname{div} \varepsilon \boldsymbol{E} = 4\pi\rho.$$

Hence equation (1) follows.

The matching conditions are derived as in problem 8. We note only that in the presence of surface charges on Σ_1

$$D_{1n}-D_{2n} = 4\pi\sigma,$$

or

$$\varepsilon_1 \frac{\partial u}{\partial n} - \varepsilon_2 \frac{\partial u_2}{\partial n} = 4\pi\sigma,$$

where σ is the density of surface charges on Σ_1.

10. If $H = -\operatorname{grad} \phi$, then in the steady-state case

$$\operatorname{div}(\mu \operatorname{grad} \phi) = 0,$$

where $\phi = \phi(P)$ is a scalar potential, $\mu = \mu(P)$ the magnetic permeability of the medium at the point P.

The matching conditions on the surface of discontinuity of the coefficient of magnetic permeability have the form

$$u_1 = u_2, \quad \mu_1 \frac{\partial u_1}{\partial n} = \mu_2 \frac{\partial u_2}{\partial n} \text{ on } \Sigma_1,$$

where the subscripts 1 and 2 correspond to the values of the quantities on opposite sides of the surface of discontinuity Σ_1.

The second condition implies the continuity of the normal component of the magnetic induction vector on Σ_1:

$$B_{1n} = B_{2n}.$$

The boundary-value problem for the piecewise continuous

$$\mu = \begin{cases} \mu_1 \text{ in } T_1, \\ \mu_2 \text{ in } T_2, \end{cases}$$

is by analogy with problems 8 and 9

$$\Delta u_1 = 0 \text{ in } T_1,$$
$$\Delta u_2 = 0 \text{ in } T_2,$$

together with matching conditions on Σ_1,

Method. See problem 9.

11. In a medium with variable conductivity $\sigma = \sigma(x, y, z)$ the equation for the potential of the electric field of a steady current

$$\operatorname{div}(\sigma \operatorname{grad} u) = 0$$

holds. If Σ is a surface of discontinuity of σ, then

$$u_1 = u_2, \quad \sigma_1 \frac{\partial u_1}{\partial n} = \sigma_2 \frac{\partial u_2}{\partial n} \text{ on } \Sigma;$$

the second condition implies the continuity of the normal component of the current density on the surface Σ $j_{1n} = j_{2n}$, since

$$j = -\sigma \operatorname{grad} u.$$

Method. See problems 5, 8, 9 and 10.

From the relations

$$\boldsymbol{E} = -\operatorname{grad} u,$$
$$\boldsymbol{j} = \sigma \boldsymbol{E},$$
$$\operatorname{div} \boldsymbol{j} = 0,$$

we obtain:

$$\operatorname{div} (\sigma \operatorname{grad} u) = 0 .$$

The matching conditions are derived by analogy with problem 9.

12. The following Table establishes a comparison of fields enumerated in the conditions of the problem.

Electric field of steady current	Potential u	Coefficient of electric conductivity σ	Current density $\boldsymbol{j} = -\sigma \operatorname{grad} u$
Heat conduction	Temperature u	Coefficient of heat conduction k	Heat flow $\boldsymbol{Q} = -k \operatorname{grad} u$
Diffusion	Concentration u	Diffusion coefficient D	Flow of matter $\boldsymbol{Q} = -D \operatorname{grad} u$
Electrostatics	Potential of the electric field u	Dielectric constant ε	Electric induction vector $\boldsymbol{D} = \varepsilon \boldsymbol{E} = -\varepsilon \operatorname{grad} u$
Magnetostatics	Potential of the magnetic field u	Magnetic permeability μ	Magnetic induction vector $\boldsymbol{B} = -\mu \operatorname{grad} u$
Potential flow of an incompressible liquid	Velocity potential u	1	$\boldsymbol{v} = -\operatorname{grad} u$

In all cases u satisfies Laplace's equation.

Method. See the preceding problem of this section and also § 1, chapter II, problem 49.

Note. If on some surface Σ_1 the constants $\sigma, k, D, \varepsilon$ or μ are discontinuous, then on the surface Σ_1 matching conditions are fulfilled, which may be represented in the form

$$u_1 = u_2, \quad p_1 \frac{\partial u_1}{\partial n} = p_2 \frac{\partial u_2}{\partial n} \text{ on } \Sigma_1,$$

where u is the unknown function, and p one of the parameters σ, k, D, ε, μ; subscripts 1 and 2 correspond to limiting values oft hese quantities on opposite sides of the surface Σ_1.

§ 2. Simplest Problems for Laplace's and Poisson's Equations

Many of the solutions of problems of this section either possess circular or spherical symmetry, or have a simple dependence on angular coordinates. Let us recall expressions for the Laplacian operator:

(1) in a polar system of coordinates

$$\Delta_2 u = \frac{1}{\rho}\frac{\partial}{\partial \rho}\left(\rho \frac{\partial u}{\partial \rho}\right) + \frac{1}{\rho^2}\frac{\partial^2 u}{\partial \phi^2};$$

(2) in a spherical system of coordinates

$$\Delta u = \frac{1}{r^2}\frac{\partial}{\partial r}\left(r^2 \frac{\partial u}{\partial r}\right) + \frac{1}{r^2 \sin\theta}\frac{\partial}{\partial \theta}\left(\sin\theta \frac{\partial u}{\partial r}\right) + \frac{1}{r^2 \sin^2\theta}\frac{\partial^2 u}{\partial \phi^2};$$

(3) in a cylindrical system of coordinates

$$\Delta u = \frac{1}{\rho}\frac{\partial}{\partial \rho}\left(\rho \frac{\partial u}{\partial \rho}\right) + \frac{1}{\rho^2}\frac{\partial^2 u}{\partial \phi^2} + \frac{\partial^2 u}{\partial z^2} = \Delta_2 u + \frac{\partial^2 u}{\partial z^2}.$$

In the solution of these problems one must remember that Laplace's equation $\Delta_2 u = 0$ is satisfied by the polynomial

$$u = A(x^2-y^2)+Bxy+Cx+Dy,$$

where A, B, C, D are arbitrary constants.

1. Boundary-value Problems for Laplace's Equation

13. (a) $u = A$;

(b) $u = \dfrac{A}{a}x$, or $u = \dfrac{A}{a}\rho\cos\phi$;

(c) $u = A+By$, or $u = A+B\rho\sin\phi$;

(d) $u = Axy$, or $u = \dfrac{A}{2}\rho^2\sin 2\phi$;

(e) $u = A+\dfrac{B}{a}y$, or $u = A+\dfrac{B}{a}\rho\sin\phi$;

(f) $u = \dfrac{A+B}{2}+\dfrac{B-A}{2a^2}(x^2-y^4)$, or $u = \dfrac{A}{2}\left(1-\dfrac{\rho^2}{a^2}\cos 2\phi\right)+$

$$+\frac{B}{2}\left(1+\frac{\rho^2}{a^2}\cos 2\phi\right).$$

Method. In order to find the solutions one must remember that x, y, xy, $x^2 - y^2$ and linear combinations of them are harmonic functions. One can verify the solution directly by substitution of the expression found for u in the equation

$$u_{xx}+u_{yy}=0, \quad \text{or} \quad \frac{1}{\rho}\frac{\partial}{\partial\rho}\left(\rho\frac{\partial u}{\partial\rho}\right)+\frac{1}{\rho^2}\frac{\partial^2 u}{\partial\phi^2}=0$$

and in the boundary condition.

Let us illustrate the method of finding the solution by example 13 (b). Changing from the variables (p, ϕ) to the variables (x, y), we rewrite the boundary condition as

$$u=\frac{A}{a}x.$$

Hence it is evident that the solution must be the harmonic function

$$u(x, y)=\frac{A}{a}x \quad \text{or} \quad u(\rho, \phi)=\frac{A}{a}\rho\cos\phi.$$

14. Problems 14 (a) and 14 (d) are inconsistent, since for the second boundary-value problem

$$\Delta u=0, \quad \left.\frac{\partial u}{\partial n}\right|_C=f$$

the condition

$$\int_C f\,ds=0$$

must be fulfilled.

(b) $u(x, y)=Aax+C$ or $u(\rho, \phi)=Aa\rho\cos\phi+C$,

(c) $u=\dfrac{A}{2}a(x^2-y^2)+C$ or $u(\rho, \phi)=\dfrac{A}{2}a\rho^2\cos^2 2\phi+C$,

(e) $u=(A+0\cdot75B)y-\dfrac{0\cdot25B}{3a^2}[3(x^2+y^2)y-4y^3]+C$,

or

$$u(\rho, \phi)=(A+0\cdot75B)\rho\sin\phi-\frac{B}{12a^2}\rho^3\sin 3\phi+C.$$

The solution of the second boundary-value problem is determined only up to an arbitrary constant C.

Method. We consider the solution of one example, for instance, 14(b), in which

$$\left.\frac{\partial u}{\partial n}\right|_{\rho=a}=Ax$$

is given.

The function $u = Dx$ or $u = D\rho \cos\phi$ is harmonic. Differentiation with respect to the normal is equivalent to differentiation with respect to ρ. Requiring that the boundary condition should be satisfied for $\rho = a$, we find $D = Aa$, so that $u(x, y) = Aax$ or $u(\rho, \phi) = Aa \cos\phi$.

In example 14 (e) one must split f into two parts: $f = f_1(\phi) + f_2(\phi)$, $f_1 = \alpha \sin\phi$, $f_2 = \beta \sin 3\phi$, and look for a solution in the form

$$u = R_1(\rho) f_1(\phi) + R_2(\rho) f_2(\phi).$$

15. (a) $u(\rho, \phi) = A$,

(b) $u(\rho, \phi) = \dfrac{Aa}{\rho} \cos\phi$,

(c) $u(\rho, \phi) = A + \dfrac{Ba^2}{\rho} \sin\phi$,

(d) $u(\rho, \phi) = \dfrac{1}{2} A \dfrac{a^4}{\rho^2} \sin 2\phi$,

(e) $u(\rho, \phi) = A + B \dfrac{a}{\rho} \sin\phi$,

(f) $u(\rho, \phi) = \dfrac{A+B}{2} - \dfrac{A-B}{2} \cdot \dfrac{a^2}{\rho^2} \cos 2\phi$.

Method. Transform to polar coordinates. If the boundary condition for $\rho = a$ has the form

$$u|_{\rho=a} = A_k \cos k\phi,$$

then look for a solution in the form

$$u(\rho, \phi) = R_k(\rho) \cos k\phi,$$

where $R(\rho)$ is a function, satisfying the equation

$$\rho^2 R'' + \rho R' - k^2 R = 0$$

and the following boundary conditions:

$$R(a) = A_k, \quad |R(\infty)| < \infty.$$

16. Problems 16(a) and 16(d) have no solution, since the condition

$$\int_C \frac{\partial u}{\partial n} ds = 0$$

is not fulfilled.

(b) $u(\rho, \phi) = -\dfrac{Aa^3}{\rho} \cos\phi + \text{const.}$,

(c) $u(\rho, \phi) = -\dfrac{Aa^5}{2\rho^2} \cos 2\phi + \text{const.}$,

(e) $u(\rho, \phi) = -(A + 0{\cdot}75B) \dfrac{a^2}{\rho} \sin\phi + 0{\cdot}25B \dfrac{a^4}{3\rho^3} \sin 3\phi + C$.

17. $u = u(\rho) = u_1 + (u_2 - u_1)\dfrac{\ln \dfrac{\rho}{a}}{\ln \dfrac{b}{a}}.$

The capacity per unit length of the cylindrical condenser equals

$$C = \frac{1}{\ln \dfrac{b}{a}}.$$

Method. Since the boundary conditions do not depend on ϕ, the solution must possess cylindrical symmetry, $u = u(\rho)$.

The capacity C of the conductor, bounded by the surface Σ, is determined by the expression

$$C = -\frac{1}{4\pi u_0} \int_{\Sigma} \frac{\partial u}{\partial n}\, d\sigma \text{ in three dmensions}$$

and

$$C = -\frac{1}{2\pi u_0} \oint_{L} \frac{\partial u}{\partial n}\, ds \text{ in two dimensions}$$

where u_0 is the potential of the conductor, and

$$-\frac{\partial u}{\partial n} = E_n$$

the normal component of the electric field.

18. $u(\rho, \phi) = \dfrac{u_0}{\alpha}\phi$

or

$$u(x, y) = \frac{u_0}{\alpha} \text{ arc tan} \frac{y}{x}.$$

Method. Writing Laplace's equation in polar coordinates

$$\Delta_2 u = \frac{1}{\rho}\frac{\partial}{\partial \rho}\left(\rho \frac{\partial u}{\partial \rho}\right) + \frac{1}{\rho^2}\frac{\partial^2 u}{\partial \phi^2} = 0,$$

we see that a function, linear in ϕ, is harmonic.

19. $u(x, y) = \phi_1 + \dfrac{\phi_2 - \phi_1}{\pi} \text{ arc tan } \dfrac{y}{x}.$

Comparison of (1) with the solution of problem 18 shows that (1) corresponds to the special case $\alpha = \pi$ of formula (1) in problem 18.

20. (a) $u = u_0$,

(b) $u = u(r) = \dfrac{a}{r} u_0.$

21. $u = u(r) = u_2 + \dfrac{u_1 - u_2}{\dfrac{1}{a} - \dfrac{1}{b}}\left(\dfrac{1}{r} - \dfrac{1}{b}\right).$

Method. The solution of the equation $\Delta u = \dfrac{1}{r^2}\dfrac{\mathrm{d}}{\mathrm{d}r}\left(r^2 \dfrac{\mathrm{d}u}{\mathrm{d}r}\right) = 0$ has the form $u = u(r) = \alpha + \beta/r$, where α and β are determined from the conditions $u(a) = u_1$, $u(b) = u_2$.

22. $C = \dfrac{\varepsilon}{\dfrac{1}{a} - \dfrac{1}{b}}.$

Method. Remember that in the presence of a dielectric the density of surface charges equals

$$\sigma = \frac{1}{4\pi} D_n = -\frac{1}{4\pi} \varepsilon \frac{\partial u}{\partial n}.$$

23. The capacity of the spherical condenser equals

$$C = \frac{\varepsilon_1}{\dfrac{1}{a} + \dfrac{1}{c} - \dfrac{\varepsilon_1}{\varepsilon_2}\left(\dfrac{1}{b} + \dfrac{1}{c}\right)}.$$

Solution. It is required to solve the boundary-value problem

$$\Delta u_1 = 0 \text{ for } a < r < c,$$
$$\Delta u_2 = 0 \text{ for } c < r < b,$$

where u_1 and u_2 satisfy boundary conditions for $r = a$ and $r = b$

$$u_1 |_{r=a} = 1, \qquad u_2 |_{r=b} = 0$$

and matching conditions for $r = c$

$$u_1 = u_2,$$

$$\varepsilon_1 \frac{\partial u_1}{\partial r} = \varepsilon_2 \frac{\partial u_2}{\partial r}.$$

The general solution has the form

$$u(r) = \begin{cases} u_1 = \dfrac{A_1}{r} + A_2 & \text{for} \quad a < r < c, \\[2ex] u_2 = \dfrac{B_1}{r} + B_2 & \text{for} \quad c < r < b. \end{cases}$$

The four coefficients A_1, A_2, B_1, B_2 are determined from the two boundary conditions at $r = a$ and $r = b$ and the two matching conditions for $r = c$. As a result, we obtain:

$$u_1 = 1 + A_1\left(\frac{1}{r} - \frac{1}{a}\right), \qquad u_2 = \frac{\varepsilon_1}{\varepsilon_2} A_1\left(\frac{1}{r} - \frac{1}{b}\right),$$

where

$$A_1 = \frac{1}{\dfrac{1}{a}+\dfrac{1}{c}-\dfrac{\varepsilon_1}{\varepsilon_2}\left(\dfrac{1}{b}+\dfrac{1}{c}\right)}.$$

The capacity is calculated from the formula

$$C = -\frac{\varepsilon_1}{4\pi}\iint\left(\frac{\partial u}{\partial n}\right)_{r=a} a^2\,\mathrm{d}\Omega = -\varepsilon_1 a^2\left(\frac{\partial u}{\partial r}\right)_{r=a} = A_1\,\varepsilon_1.$$

24. It is required to find the solution of the boundary-value problem

$$\Delta_2 u_1 = \frac{1}{\rho}\,\frac{\mathrm{d}}{\mathrm{d}\rho}\left(\rho\,\frac{\mathrm{d}u_1}{\mathrm{d}\rho}\right) = 0 \quad \text{for} \quad a<\rho<c,$$

$$\Delta_2 u_2 = \frac{1}{\rho}\,\frac{\mathrm{d}}{\mathrm{d}\rho}\left(\rho\,\frac{\mathrm{d}u_2}{\rho}\right) = 0 \quad \text{for} \quad c<\rho<b,$$

$$u_1 = 1 \quad \text{for} \quad \rho = a, \qquad u_2 = 0 \quad \text{for} \quad \rho = b,$$

$$u_1 = u_2, \qquad \varepsilon_1\frac{\partial u_1}{\partial\rho} = \varepsilon_2\frac{\partial u_2}{\partial\rho} \quad \text{for} \quad \rho = c.$$

For the capacity we obtain the expression

$$C = \frac{\varepsilon_1}{\ln\dfrac{c}{a}+\dfrac{\varepsilon_1}{\varepsilon_2}\ln\dfrac{b}{c}}.$$

When $\varepsilon_1 = \varepsilon_2 + \varepsilon$ we have:

$$C = \frac{\varepsilon}{\ln\dfrac{b}{a}}.$$

Method. The solution has the form

$$u_1 = A\ln\rho+\varepsilon; \qquad u_2 = B'\ln\rho+D.$$

25. The potential of the field equals

$$u = u_0\,\frac{\dfrac{1}{r}+\dfrac{\varepsilon_1-\varepsilon_2}{\varepsilon_2}\,\dfrac{1}{c}}{\dfrac{1}{a}+\dfrac{\varepsilon_1-\varepsilon_2}{\varepsilon_2}\,\dfrac{1}{c}} \quad \text{for} \quad a<r<c,$$

$$u = u_0\,\frac{\dfrac{\varepsilon_1}{\varepsilon_2}\,\dfrac{1}{r}}{\dfrac{1}{a}+\dfrac{\varepsilon_1-\varepsilon_2}{\varepsilon_2}\,\dfrac{1}{c}} \quad \text{for} \quad r>c.$$

Special cases:

(1) For $c \to \infty$ we obtain $u = u_0\, a/r$ for $r > a$ the potential of the field of a sphere of radius a, charged to a potential u_0 and placed in an infinite homogeneous medium;

(2) for $\varepsilon_2 \to \infty$ (medium 2 is ideally conducting)

$$u = \begin{cases} u_0 \dfrac{\dfrac{1}{r} - \dfrac{1}{c}}{\dfrac{1}{a} - \dfrac{1}{c}}, & \text{when} \quad a < r < c, \\ 0, & \text{when} \quad r > c; \end{cases}$$

(3) if $\varepsilon_1 = \varepsilon_2$, then

$$u = u_0 \frac{a}{r} \quad (r > a) \text{ (cf. part 1)}.$$

Method. See problem 22. Remember that at infinity u must go to zero.

26. The electrostatic field is

$$\boldsymbol{E} = -\operatorname{grad} u,$$

where u is the potential, equal to

$$u = u(\rho) = u_0 \frac{\ln \dfrac{b}{\rho}}{\ln \dfrac{b}{a}}.$$

27. The solution depends only on the variable z and is given by the relation

$$u = u(z) = u_1 + (u_2 - u_1)\frac{z}{h}.$$

28. The capacity per unit area of a plane condenser is

$$C = \frac{\varepsilon}{4\pi h},$$

$$C = \frac{\varepsilon_1}{4\pi\left[h_1 + \dfrac{\varepsilon_1}{\varepsilon_2}(h - h_1)\right]}.$$

29. The unknown harmonic function depends only on the variable y:

$$u = u_1 + (u_2 - u_1)\frac{y}{b}.$$

Method. Look for the solution in the form of a harmonic polynomial.

2. Boundary-value Problems for Poisson's Equation

30. $u = \frac{1}{4}(\rho^2 - a^2)$.

Method. The unknown function $u = u(\rho)$ possesses circular symmetry and is determined from the equation

$$\frac{1}{\rho}\frac{\mathrm{d}}{\mathrm{d}\rho}\left(\rho\frac{\partial u}{\partial \rho}\right) = 1$$

by the condition $u(a) = 0$.

31. A solution exists if one chooses

$$B = \frac{aA}{2},$$

and

$$u = u(\rho) = \frac{A\rho^2}{4} + \text{const.}$$

is determined up to an arbitrary constant.

32: (a) $u(\rho) = u_2 + \frac{A}{4}(\rho^2 - b^2) + \dfrac{u_1 - u_2 + \frac{A}{4}(b^2 - a_2)}{\ln \frac{b}{a}} \ln \frac{b}{\rho}$,

(b) $u(\rho) = u_1 + \frac{A}{4}(\rho^2 - a^2) + b\left(C - \frac{Ab}{2}\right)\ln\frac{\rho}{a}$,

(c) $u(\rho) = \frac{A\rho^2}{4} - a\left(\frac{aA}{2} - B\right)\ln\rho + \text{const.}$

Problem (c) has a solution only for

$$C = \frac{A(b^2 - a^2) + 2aB}{2b},$$

and the solution of problem (c) is determined up to an additive constant.

33. (a) when $\Delta u = 1$, $u(a) = 0$, then

$$u = u(r) = \frac{1}{6}(r^2 - a^2);$$

(b) when $\Delta u = Ar + B$, $u(a) = 0$, then

$$u(r) = \frac{A}{12}(r^3 - a^3) + \frac{B}{6}(r^2 - a^2).$$

Method. The unknown function possesses spherical symmetry, $u = u(r)$, and satisfies the ordinary differential equation

$$\frac{1}{r}\frac{\mathrm{d}^2}{\mathrm{d}r^2}(ru) = f(r).$$

34. (a) $u = u(r) = \frac{1}{6}(r^2-a^2) - \frac{1}{6}ab(a+b)\left(\frac{1}{a}-\frac{1}{r}\right)$,

(b) $u = u(r) = \frac{A}{6}(r^2-a^2) + \frac{B}{2}(r-a) - ab\left[\frac{A}{6}(b+a)+\frac{B}{2}\right]\left(\frac{1}{a}-\frac{1}{r}\right)$.

For $A = 1$, $B = 0$ we obtain the first expression.

Method. The solution possesses spherical symmetry $u = u(r)$.

§ 3. The Source Function

The source function $G(M, P)$ of the first boundary-value problem for the equation $\Delta u = -4\pi\rho$ is determined in the three-dimensional case† by the following conditions:

$$G(M, P) = \frac{1}{4\pi}\frac{1}{r_{MP}} + v(M, P), \tag{1}$$

where $r_{MP} = \sqrt{(x-\xi)^2+(y-\eta)^2+(z-\zeta)^2}$ is the distance between the point of observation $M(x, y, z)$ and the source at the point $P(\xi, \eta, \zeta)$ and $v(M, P)$ is a function, regular and harmonic everywhere in the region T under consideration with boundary Σ. At the boundary Σ, G satisfies the condition

$$G|_\Sigma = 0. \tag{2}$$

Thus the calculation of the source function G in some region T reduces to the solution of the first boundary-value problem for Laplace's equation

$$\Delta v = 0 \text{ in } T$$

for the special boundary condition

$$v|_\Sigma = -\frac{1}{4\pi r_{MP}}. \tag{3}$$

An electrostatic interpretation of the source function $G(M, P)$ is obvious: it is the potential at a point M of the electrostatic field, produced inside a volume T by a charge of value $e = 1/4\pi$, concentrated at a point P, if the boundary surface Σ of the region T is ideally conducting and is maintained at zero potential, i.e. is earthed; here $\frac{1}{4\pi}\frac{1}{r}$ is the potential of a charge in infinite space, and v is the potential of the field, induced by charges on Σ.

For a series of simple regions (semispace, layer, sphere and others) the induced field may be found by means of the so-called method of images. One replaces the charges induced on the boundary of the region by "fictitious" charges called image charges, which produce the same fields in the region considered as the induced surface charges. In the case of a plane boundary "the images"

† See [7], chapter IV, § 4, page 349.

are mirror images of the original charges in the plane or planes, if the region is bounded by several planes. In the case of spherical boundaries, the image is located by an inversion transformation†.

In the present section only those cases are considered which can be solved by the method of images.

If the source function $G(M, P)$ is known, then the solution of the first boundary-value problem for the equation

$$\Delta u = -F \text{ on } T \tag{4}$$

for the condition at the boundary

$$u|_{\Sigma} = f \tag{5}$$

can be found in integral form

$$u(M) = -\int_{\Sigma} f(P) \frac{\partial G}{\partial n_P} \mathrm{d}\sigma_P + \int_{T} G(M, P) F(P) \, \mathrm{d}\tau_P, \tag{6}$$

where $\partial G/\partial n$ is an outward normal derivative of G on the boundary Σ_1.

The majority of problems of the present section are taken from electrostatics. Usually besides the potential of the field one is interested in the surface density of charges, induced on conductors, and also the capacity of the conductors. Let us introduce necessary concepts.

The surface density of charges on a conductor with surface S, situated in a medium of dielectric constant ε, equals

$$\sigma = -\frac{\varepsilon}{4\pi}\left(\frac{\partial u}{\partial n}\right)_S,$$

where n is the outer normal to the surface S.

The total charge, distributed over S, is given by the integral

$$e' = \iint_S \sigma \, \mathrm{d}S.$$

The capacity of the conductor S is determined by the formula

$$S = \frac{e'}{V'},$$

where V' is the potential of the conductor S.

For the two-dimensional region D with boundary L the source function $G(M, P)$ is found similarly

$$G(M, P) = \frac{1}{2\pi} \ln \frac{1}{r_{MP}} + v(M, P), \tag{7}$$

$$G|_L = 0, \tag{8}$$

where $v(M, P)$ is a harmonic function regular in D, i.e. in this case G has a logarithmic singularity at the source.

† See [7], chapter IV, § 4, page 349.

1. The Source Function for Regions with Plane Boundaries

35. The potential of a point charge e equals

$$u = e\left(\frac{1}{r_0} - \frac{1}{r_1}\right), \tag{1}$$

where

$$r_0 = MP = \sqrt{(x-\xi)^2+(y-\eta)^2+(z-\zeta)^2},$$
$$r_1 = MP_1 = \sqrt{(x-\xi)^2+(y-\eta)^2+(z+\zeta)^2},$$

$M(x, y, z)$ is the point of observation, $P(\xi, \eta, \zeta)$ the point at which the charge is located and $P_1(\xi, \eta, -\zeta)$ its image in the plane $z = 0$.

The density of surface charges

$$\sigma = -\frac{1}{4\pi}\left(\frac{\partial u}{\partial z}\right)_{z=0} = -\frac{e}{2\pi}\,\frac{\zeta}{[(x-\xi)^2+(y-\eta)^2+\zeta^2]^{3/2}}\,.$$

The total charge, distributed over the plane $z = 0$ equals

$$e' = \iint\limits_{-\infty}^{\infty} \sigma\,\mathrm{d}x\,\mathrm{d}y = -e.$$

The source function of the first boundary-value problem for Laplace's equation in semispace is

$$G(M, P) = \frac{1}{4\pi}\left(\frac{1}{r_0} - \frac{1}{r_1}\right), \tag{2}$$

and the solution of the first boundary-value problem in the semispace $z > 0$ is given by the formula

$$u(M) = \frac{1}{4\pi}\iint\limits_{-\infty}^{\infty} \frac{2z}{[(x-\xi)^2+(y-\eta)^2+z^2]}\, f(\xi, \eta)\,\mathrm{d}\xi\,\mathrm{d}\eta. \tag{3}$$

Solution. By reflection in the plane $z = 0$ of a charge e, situated at a point $P(\xi, \eta, \zeta)$, we obtain at the point $P(\xi, \eta, -\zeta)$ a charge of value $-e$; its potential in semi-infinite space equals $-e/r_1$. We note that

$$\left.\frac{1}{r_0}\right|_{z=0} = \left.\frac{1}{r_1}\right|_{z=0} = \frac{1}{\sqrt{(x-\xi)^2+(y-\eta)^2+\zeta^2}}$$

and

$$\left(\frac{1}{r_0} - \frac{1}{r_1}\right)_{z=0} = 0,$$

i.e. the charges $e(P)$ and $-e(P_1)$ compensate one another in the plane $z = 0$. Therefore, using the principle of superposition, we have for the unknown potential:

$$u = e\left(\frac{1}{r_0} - \frac{1}{r_1}\right).$$

The source function $G(M, P)$ corresponds to $e = 1/4\pi$. Calculating then the normal derivative

$$\left(\frac{\partial G}{\partial \zeta}\right)_{\zeta=0} = \frac{2z}{4\pi[(x-\xi)^2+(y-\eta)^2+z^2]^{3/2}}$$

and using formula (6) on page 401 we obtain the solution of the first boundary-value problem

$$\Delta u = 0 \quad (z > 0), \quad u|_{z=0} = f(x, y).$$

36. The potential of the charge $e(P_0)$ equals

$$u = 4\pi e G, \quad G(M, P) = \frac{1}{4\pi} \sum_{n=-\infty}^{\infty} \left(\frac{1}{r_n} - \frac{1}{r_n'}\right), \tag{1}$$

where

$$r_n = \sqrt{(x-\xi)^2+(y-\eta)^2+[z-(2nl+\zeta)]^2},$$
$$r_n' = \sqrt{(x-\xi)^2+(y-\eta)^2+[z-(2nl-\zeta)]^2},$$

the charge is located at the point $P_0(\xi, \eta, \zeta)$, $M(x, y, z)$ is the point of observation.

Series (1) and also the series obtained by a term-by-term differentiation of series (1), converges uniformly and absolutely in the region $0 < z < l$.

Solution. In order to form series (1) it is necessary to make successive reflections in the planes $z = 0$ and $z = l$ (Fig. 40) and to find the positions of the images. Making a reflection in the plane $z = 0$, we obtain the function

$$u_0 = e\left(\frac{1}{r_0} - \frac{1}{r_0'}\right),$$

which satisfies the boundary condition $u = 0$ for $z = 0$ and does not satisfy the condition $u = 0$ for $z = l$; making then a reflection in the plane $z = l$, we obtain:

$$u_1 = e\left[\left(\frac{1}{r_0} - \frac{1}{r_0'}\right) + \left(\frac{1}{r_1} - \frac{1}{r_1'}\right)\right],$$

so that $u_1 = 0$ for $z = l$ and $u_1 \neq 0$ for $z = 0$. Continuing this process of alternate reflection in the planes $z = 0$ and $z = l$, we arrive at series (1).

Considering the fact that for each reflection the charge e changes into the charge $-e$ and conversely, it is readily established that the coordinates of the images are expressed by the formula

$$+e \qquad \zeta_n = 2nl+\zeta,$$
$$-e \qquad \zeta_n' = 2nl-\zeta,$$

where n is an integer, taking the values from $-\infty$ to $+\infty$. Using the principle of superposition and summing the action of all the images $e(P_n)$ and $-e(P'_n)$ and of the real charge $e(P)$, we obtain series (1).

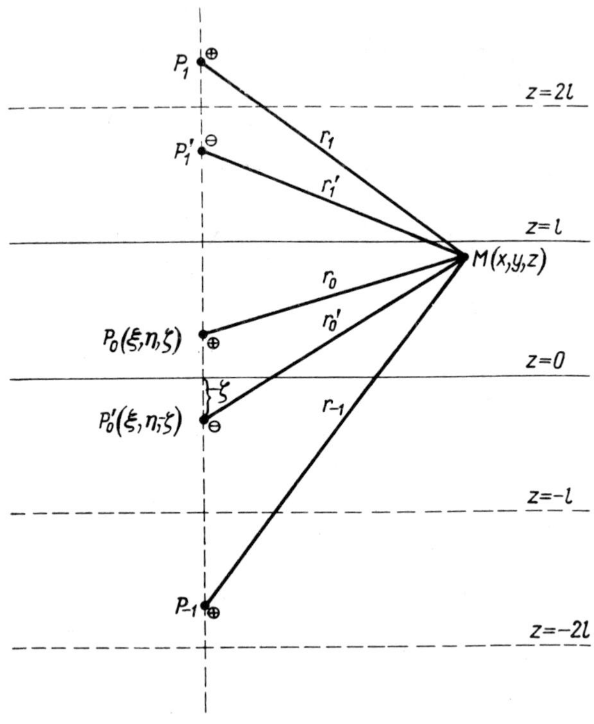

FIG. 40

Let us prove that this series converges uniformly. In order to do this we consider its nth term

$$a_n = \frac{1}{r_n} - \frac{1}{r'_n}.$$

Making use of the mean value theorem, we have:

$$a_n = 2\zeta\left[\frac{\partial}{\partial\zeta}\left(\frac{1}{r_n}\right)\right]_{\zeta=\zeta^*} = \frac{2\zeta[z-(2nl+\zeta^*)]}{r_n^{*2}\,r_n^*} \qquad (0<\zeta^*<l),$$

from which it follows:

$$|a_n| < \frac{2l}{(r_n^*)^2} < \frac{2}{(2n-1)^2 l} = b_n,$$

since $\zeta^* < l|z-\zeta^*| < l$, and, hence,

$$r_n^* = \sqrt{(x-\xi)^2+(y-\eta)^2+[z-(2nl+\zeta^*)]^2} > (2n-1)l.$$

The inequality obtained shows that the series $\sum_{n=-\infty}^{\infty} a_n$ converges uniformly and absolutely, since the majorant series $\sum_{n=-\infty}^{\infty} b_n$ converges.

Let us prove now the uniform convergence in the layer $0 < z < l$ of series obtained by one- and two-fold term-by-term differentiation of series (1).

We evaluate the derivatives

$$\left|\frac{\partial}{\partial z}\left(\frac{1}{r_n}\right)\right| \leqslant \left|-\frac{1}{r_n^2}\,\frac{z-(2nl+\zeta)}{r_n}\right| < \frac{1}{r_n^2}, \text{ because } \left|\frac{z-(2nl+\zeta)}{r_n}\right| < 1,$$

$$\left|\frac{\partial}{\partial z}\left(\frac{1}{r_n'}\right)\right| \leqslant \left|-\frac{1}{r_n'^2}\,\frac{z-(2nl-\zeta)}{r_n'}\right| < \frac{1}{r_n'^2}, \text{ because } \left|\frac{z-(2nl-\zeta)}{r_n'}\right| < 1,$$

$$\frac{\partial^2}{\partial z^2}\left(\frac{1}{r_n}\right) = \frac{\partial}{\partial z}\left\{-\frac{1}{r_n^2}\,\frac{z-(2nl+\zeta)}{r_n}\right\} = \frac{3[z-(2nl+\zeta)]^2}{r_n^5} - \frac{1}{r_n^3},$$

$$\left|\frac{\partial^2}{\partial z^2}\left(\frac{1}{r_n}\right)\right| < \frac{3}{r_n^3} + \frac{1}{r_n^3} = \frac{4}{r_n^3}.$$

Taking into account the inequalities $r_n > |n|l$, $r_n' > |n|l$, we obtain:

$$\left|\frac{\partial a_n}{\partial z}\right| < \frac{2}{n^2l^2} = b_n^{(1)}, \qquad \left|\frac{\partial^2 a_n}{\partial z^2}\right| < \frac{8}{n^3l^3} = b_n^{(2)},$$

from which follows the absolute and uniform convergence of the series $\sum_{n=-\infty}^{\infty} \frac{\partial a_n}{\partial z}$ and $\sum_{n=-\infty}^{\infty} \frac{\partial^2 a_n}{\partial z^2}$, since the series $\sum_{n=-\infty}^{\infty} b_n^{(1)}$ and $\sum_{n=-\infty}^{\infty} b_n^{(2)}$ converge.

Similarly the uniform convergence of the series

$$\sum_{n=-\infty}^{\infty} \frac{\partial a_n}{\partial x}, \quad \sum_{n=-\infty}^{\infty} \frac{\partial a_n}{\partial y}, \quad \sum_{n=-\infty}^{\infty} \frac{\partial^2 a_n}{\partial x^2}, \quad \sum_{n=-\infty}^{\infty} \frac{\partial^2 a_n}{\partial y^2}$$

is proved.

Thus series (1) may be differentiated twice.

Therefore series (1) without the term $1/r_0$ satisfies Laplace's equation everywhere in the layer $0 < z < l$, sincs all its terms satisfy this equation. The first term $1/r_0$ gives the necessary singularity at the source.

37. The rectangular components of the electric field equal

$$\left.\begin{aligned} E_x &= -\frac{I}{4\pi\sigma}\sum_{n=-\infty}^{\infty}\frac{\partial}{\partial x}\left(\frac{1}{r_n}+\frac{1}{r'_n}\right),\\ E_y &= -\frac{I}{4\pi\sigma}\sum_{n=-\infty}^{\infty}\frac{\partial}{\partial y}\left(\frac{1}{r_n}+\frac{1}{r'_n}\right),\\ E_z &= -\frac{I}{4\pi\sigma}\sum_{n=-\infty}^{\infty}\frac{\partial}{\partial z}\left(\frac{1}{r_n}+\frac{1}{r'_n}\right),\end{aligned}\right\} \tag{1}$$

where σ is the conductivity of the medium, and I is the magnitude of the current source,

$$\left.\begin{aligned} r_n &= \sqrt{(x-\xi)^2+(y-\eta)^2+[z-(2nl+\zeta)]^2},\\ r'_n &= \sqrt{(x-\xi)^2+(y-\eta)^2+[z-(2nl-\zeta)]^2}.\end{aligned}\right\} \tag{2}$$

The series for the components of the field E_x, E_y, E_z converge uniformly and absolutely and represent two-fold differentiable functions, therefore satisfying the equation $\Delta u = 0$ everywhere, except the points $r_0 = 0$ $(x = \xi, y = \eta, z = \zeta)$, at which they have a necessary singularity

$$E_x = -\frac{I}{4\pi\sigma}\frac{\partial}{\partial x}\left(\frac{1}{r_0}\right)+\dots,\qquad E_y = -\frac{I}{4\pi\sigma}\frac{\partial}{\partial y}\left(\frac{1}{r_0}\right)+\dots,$$

$$E_z = -\frac{I}{4\pi\sigma}\frac{\partial}{\partial z}\left(\frac{1}{r_0}\right)+\dots \tag{3}$$

Method. Each of the components of the field E_x, E_y, E_z satisfies Laplace's equation, so that $\Delta E = 0$, and has the required singularity (3) at the source.

For $z = 0$ the condition

$$E_z = 0 \tag{4}$$

must be fulfilled.

Placing sources of magnitude I at the points $\zeta_n = 2nl-\zeta$ and $\zeta'_n = 2nl+\zeta$, we sum the fields of these sources

$$E = -\frac{I}{4\pi\sigma}\sum_{n=-\infty}^{\infty}\operatorname{grad}\left(\frac{1}{r_n}+\frac{1}{r'_n}\right). \tag{5}$$

Boundary conditions (4) will be fulfilled, since

$$\frac{\partial}{\partial z}\left(\frac{1}{r_n}\right)\bigg|_{z=0} = -\frac{\partial}{\partial z}\left(\frac{1}{r'_n}\right)\bigg|_{z=0} = \frac{-(2nl+\zeta)}{\{(x-\xi)^2+(y-\eta)^2+(2nl+\zeta)^2\}^{3/2}}.$$

The series (5) is uniformly and absolutely convergent, since

$$\left|\frac{\partial}{\partial z}\left(\frac{1}{r_n}\right)\right| = \left|\frac{z-(2nl+\zeta)}{r_n^3}\right| \leqslant \frac{1}{r_n^2} < \frac{A}{n^2},$$

where A is some constant.

If one adds not the fields of the individual sources, but their potentials, then one obtains the series

$$\frac{I}{4\pi\sigma}\sum_{n=-\infty}^{\infty}\left(\frac{1}{r_n}+\frac{1}{r_n'}\right), \tag{6}$$

which diverges, since its terms are positive and of order $1/n$.

Term by term differentiation of series (6) is possible, since a uniformly and absolutely converging series is obtained.

38. We seek the solution of the boundary-value problem

$$\Delta u = 0 \quad \text{for} \quad 0 < z < l,$$

$$u\bigg|_{z=0} = 0, \quad \frac{\partial u}{\partial z}\bigg|_{z=l} = 0$$

for the condition that the potential u has a singularity at the point $P(z = \zeta, x = \xi, y = \eta)$

$$u \approx \frac{I}{4\pi\sigma}\frac{1}{r_0},$$

$$r_0 = \sqrt{(x-\xi)^2+(y-\eta)^2+(z-\zeta)^2}.$$

The method of images gives:

$$u(M, P) = \frac{I}{4\pi\sigma}\sum_{n=-\infty}^{\infty}\left(\frac{1}{r_n}-\frac{1}{r_n'}\right), \tag{1}$$

where

$$r_n = \sqrt{(x-\xi)^2+(y-\eta)^2+[z-(2nl+(-1)^n\zeta)]^2},$$

$$r_n' = \sqrt{(x-\xi)^2+(y-\eta)^2+[z-(2nl-(-1)^n\zeta)]^2}.$$

Method. The sources I and sinks $-I$ are located respectively at the points (Fig. 41)

$$x = \xi, \quad y = \eta, \quad z = \zeta_n = 2nl+(-1)^n\zeta,$$

$$x = \xi, \quad y = \eta, \quad z = \zeta_a' = 2nl-(-1)^n\zeta.$$

The convergence and differentiability of series (1) is proved by analogy with problem 36.

39. The source function of a point source, situated at a point $P_0(\xi, \eta, \zeta)$ with the boundary condition

$$\left(\frac{\partial u}{\partial z}+hu\right)_{z=0} = 0, \tag{1}$$

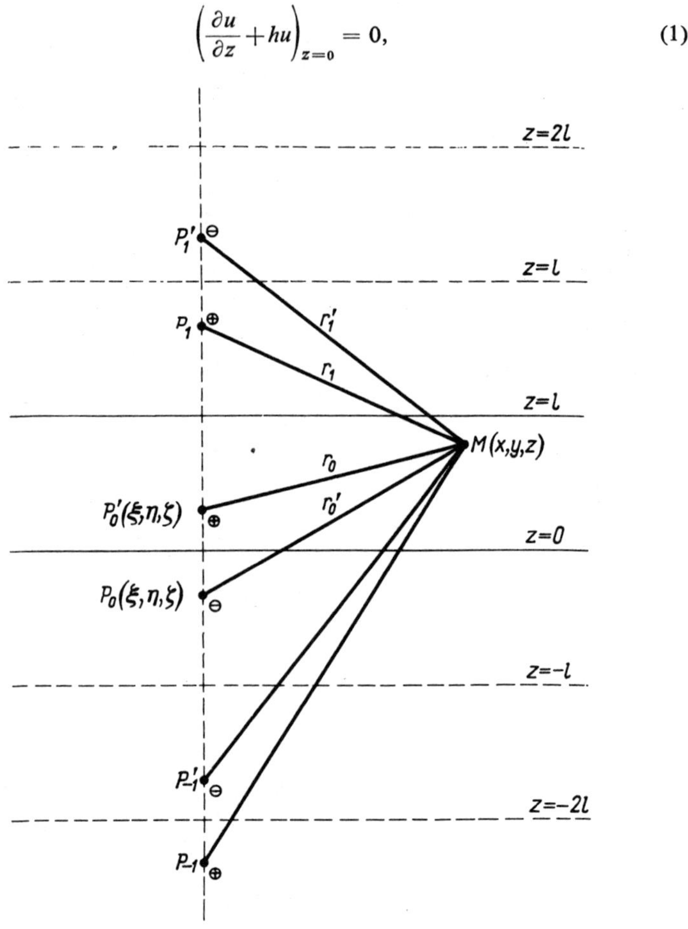

Fig. 41

is given by the relation

$$G(x, y, z;\ \xi, \eta, \zeta) = \frac{1}{4\pi}\left[\frac{1}{r_0}+\frac{1}{r_0'}-\right.$$

$$\left.-2\int\limits_{\xi}^{\infty} e^{-h(\delta-s)}\frac{ds}{\sqrt{(x-\xi)^2+(y-\eta)^2+(z+s)^2}}\right],$$

where

$$r_0 = \sqrt{(x-\xi)^2+(y-\eta)^2+(z-\zeta)^2}, \qquad r_0' = \sqrt{(x-\xi)^2+(y-\eta)^2+(z+\zeta)^2}.$$

Solution. The source $e(P_0)$ and its image $-e(P_0')$ give the condition $\partial u/\partial z = 0$ on the plane $z = 0$ for $z = 0$. In our problem $\partial u/\partial z = -hu$ for $z = 0$. Therefore we seek the solution in the form of a sum of terms, corresponding to $e(P_0)$ and $-e(P_0')$, and of an addition of the form $v(x-\xi,\ y-\eta,\ z+\zeta)$, assuming

$$G = \frac{1}{4\pi}\left(\frac{1}{r_0}+\frac{1}{r_0'}\right)+v(x-\xi, y-\eta, z+\zeta). \tag{2}$$

Substituting (2) in (1) and taking into account the fact that $\partial v/\partial z = \partial v/\partial \zeta$ we obtain:

$$\left(\frac{\partial v}{\partial \zeta}+hv\right)_{z=0} = \frac{1}{2\pi}\frac{1}{\sqrt{\rho^2+\zeta^2}}, \qquad \rho^2 = (x-\xi)^2+(y-\eta)^2.$$

Solving this equation and replacing ζ by $z+\zeta$ we obtain:

$$v(x-\xi, y-\eta, z+\zeta) = \frac{1}{2\pi}\int_{\zeta}^{\infty} \mathrm{e}^{-h(\zeta-s)}\frac{\mathrm{d}s}{\sqrt{\rho^2+(z+s)^2}}, \tag{3}$$

or

$$v = \frac{1}{2\pi}\int_{\zeta+z}^{\infty} \mathrm{e}^{-h(\zeta+z-s)}\frac{\mathrm{d}s}{\sqrt{\rho^2+s^2}}. \tag{4}$$

40. The electric field $\boldsymbol{E} = -\text{grad } u$, where $u = u(\rho, \phi, z)$ is the potential, defined by the relation

$$u = e\sum_{k=0}^{n-1}\left(\frac{1}{r_k}-\frac{1}{r_k'}\right),$$

where

$$r_k = MP_k = \sqrt{\rho^2+s^2-2\rho s\cos[\phi-(\psi+2ak)]+(z-\zeta)^2},$$

$$r_k' = MP_k' = \sqrt{\rho^2+s^2-2\rho s\cos[\phi-(2ak-\psi)]+(z-\zeta)^2},$$

$$a = \frac{\pi}{n},$$

$M = M(\rho, \phi, z)$ is the point of observation, and $P = P(s, \psi, \zeta)$ is the point at which the source is situated.

Method. Change to cylindrical coordinates ρ, ϕ, z choosing the z-axis along an edge of the bihedral angle. The mirror image of the source recurs $2n-1$ times, therefore the unknown potential can be obtained by means of summation of the potentials of $2n$ charges.

For the reflection of a charge in the sides of the bihedral angle all its image will be on a circle of radius s, lying in the plane $z = \zeta$; in absolute value they equal the unknown charge and alternate in sign.

The charges $+e$ exist at the points $P_k(s, 2\alpha k+\psi, \zeta)$; the charges $-e$ exist at the points $P_k'(s, 2\alpha k-\psi, \zeta)$, where k varies from zero to $n-1$.

It is readily seen that charges of opposite sign are symmetrically situated with respect to the planes $\phi = 0$ and $\phi = \alpha$. In fact, there corresponds to the charge $P_k(\phi = 2\alpha k+\psi)$ the charge $P'_{n-k}(\phi = 2\alpha(n-k)-\psi)$, symmetrical with respect to the plane $\phi = 0$; similarly, to the charge $P_k(\phi = 2\alpha k-\psi)$ there corresponds the charge $P'_{n-k+1}(\phi = 2\alpha(n-k+1)-\psi)$, symmetrical with respect to the plane $\phi = \alpha$.

We note that for $\alpha = \pi$ formula (1) gives the solution of problem 35.

41. $u(\rho, \phi) = V\left(-1\dfrac{\phi}{\alpha}\right).$

Method. One must find the source function inside the angle

$$G_1(\rho, \phi;\ s, \psi) = \frac{1}{2\pi}\sum_{k=0}^{n-1} \ln\frac{r_k'}{r_k}$$

(see problem 40) and use Green's formula

$$u(\rho, \phi) = -V\int_0^\infty \frac{1}{s}\left(\frac{\partial G}{\partial \psi}\right)_{\psi=\alpha} ds.$$

42. If the z-axis is directed along one of the edges, so that the perpendicular, section lies in the plane (x, y) then the potential equals

$$u = e\sum_{m=-\infty}^{\infty}\sum_{n=-\infty}^{\infty}\left(\frac{1}{r_{mn}} - \frac{1}{r'_{mn}} + \frac{1}{\bar r_{mn}} - \frac{1}{\bar r'_{mn}}\right),$$

where

$$r_{mn} = \sqrt{[x-(2ma+\xi)]^2+[y-(2nb+\eta)]^2+(z-\zeta)^2},$$

$$r'_{mn} = \sqrt{[x-(2ma-\xi)]^2+[y-(2nb-\eta)]^2+(z-\zeta)^2},$$

$$\bar r_{mn} = \sqrt{[x-(2ma+\xi)]^2+[y-(2nb-\eta)]^2+(z-\zeta)^2},$$

$$\bar r'_{mn} = \sqrt{[x-(2ma-\xi)]^2+[y-(2nb+\eta)]^2+(z-\zeta)^2},$$

where a and b are sides of the rectangle.

Method. Cover the entire plane (x, y) with rectangles, obtained from a section of the given cylinder by means of a displacement by an amount bn along the y-axis and by an amount am along the x-axis. Joining the four similar rectangles, lying inside the regions $-a \leqslant x \leqslant a$, $-b \leqslant y \leqslant b$, into one group and taking odd reflections on all sides, we obtain the first term of the sum of the series. Displacing next the whole group with respect to the x- and y-axes by $2am$ and $2bn$, we obtain the remaining terms of the series.

2. The Source Function for Regions with Spherical (Circular) and Plane Boundaries

43. If a is the radius of the sphere, e the value of the charge, O the centre of the sphere, M the point of observation, M_0 the position of the charge (Fig. 42), then the solution may be written in the form

$$u = e\left(\frac{1}{r_0} - \frac{a}{\rho_0}\frac{1}{r_1}\right),$$

where

$$r_0 = r_{MM_0}, \quad \rho_0 = OM_0, \quad r_1 = MM_1,$$

M_1 is the point lying on the extension of OM_0 and obtained from M_0 by inversion with respect to the sphere.

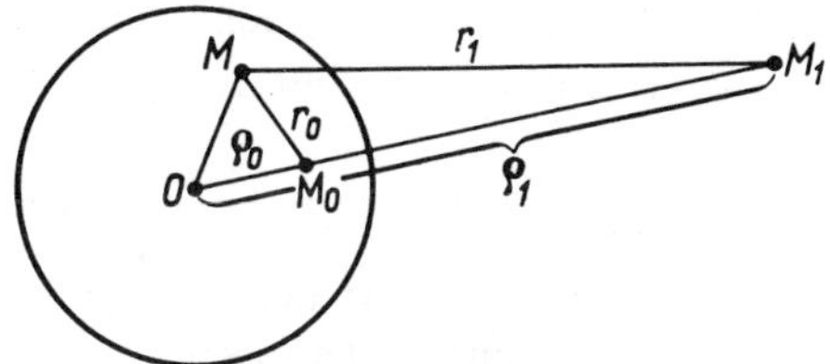

FIG. 42

Solution. The problem consists of finding the function, harmonic at all interior points of the sphere, except the point M_0, in the neighbourhood of which it is representable in the form

$$u = \frac{e}{r_0} + v(M),$$

where v is the potential of the induced field, and satisfying the boundary condition

$$u = 0.$$

To find v one must use the inversion transformation

$$OM_0 \cdot OM_1 = a^2.$$

Placing the image of the charge at M_0 at a point M_1 we find.

$$u = \frac{e}{r_0} + \frac{e_1}{r_1},$$

where e_1 is the value of the charge at M_1. The condition $u = 0$ on Σ gives:

$$e_1 = \frac{-r_1}{r_0} e.$$

Consider the triangles OMM_0 and OMM_1. They are similar, since they have a common angle MOM_0 and proportional sides $OM_0/OM = OM/OM_1$ $(OM = a)$.

Hence it follows that

$$\frac{OM_0}{OM} = \frac{OM}{OM_1} = \frac{MM_0}{MM_1}, \quad \text{or} \quad \frac{\rho_0}{a} = \frac{a}{\rho_1} = \frac{r_0}{r_1}.$$

Thus on the sphere

$$\frac{r_1}{r_0} = \frac{a}{\rho_0}.$$

Therefore the function

$$u = e\left(\frac{1}{r_0} - \frac{a}{\rho_0}\frac{1}{r_1}\right)$$

reduces to zero on the sphere; hence it also follows that $e_1 = -ae/\rho_0$.

44. The surface charge density on the sphere equals

$$\sigma = -e\frac{a^2-\rho_0^2}{4\pi a r_0^3}, \qquad r_0 = r_{MM_0}, \qquad \rho_0 = r_{OM_0},$$

where O si the origin of coordinates, M the point of observation, M_0 the position of the charge, a the radius of the sphere.

The solution of the first boundary-value problem $\Delta u = 0$ and $u|_{\rho=a} = f(\vartheta, \phi)$ is given by the relation

$$u = \frac{1}{4\pi}\iint\limits_{\Sigma} \frac{a^2-\rho_0^2}{a r_0^3} f \,\mathrm{d}S,$$

where integration is carried out over the sphere, or

$$u(\rho_0, \vartheta_0, \phi_0) = \frac{a}{4\pi}\int\limits_0^{2\pi} \mathrm{d}\phi \int\limits_0^{\pi} \frac{a^2-\rho_0^2}{(a^2-2a\rho_0\cos\gamma+\rho_0^2)^{3/2}} f(\vartheta, \phi)\sin\vartheta \,\mathrm{d}\vartheta,$$

where $\cos\gamma = \cos\vartheta\cos\vartheta_0+\sin\vartheta\sin\vartheta_0\cos(\phi-\phi_0)$.

Method. The surface charge density is

$$\sigma = D_n|_{\Sigma},$$

where D_n is the projection of the induction vector $\boldsymbol{D} = \varepsilon\boldsymbol{E}$ in the direction of the inner normal; since in this case $\varepsilon = 1$ (vacuum), then

$$4\pi\sigma = E_n|_{\Sigma} \quad \text{or} \quad 4\pi\sigma = -\left.\frac{\partial u}{\partial n_{\text{in}}}\right|_{\rho=a} = \left.\frac{\partial u}{\partial n}\right|_{\rho=a},$$

where $\partial u/\partial n$ is the derivative with respect to the direction of the outer normal. Calculating this we get

$$\left.\frac{\partial u}{\partial n}\right|_{\rho=a} = -e\frac{a^2-\rho_0^2}{4\pi a r_0^3}.$$

In order to solve the first boundary-value problem we use the formula

$$u(M) = -\iint_{\Sigma} u(P) \frac{\partial G}{\partial n_P} \, dS_p,$$

remebemring that the source function G is the potential of a point charge of value $1/4\pi$, i.e.

$$u(M) = -\iint_{\Sigma} u(P)[\sigma(M, P)]_{e=1} \, dS_p.$$

Let us find the surface charge density

$$\sigma = \frac{1}{4\pi} \frac{\partial u}{\partial n}\bigg|_{\rho=a}.$$

The derivative $\partial u/\partial n$ wtih respect to the direction n equals

$$\frac{\partial u}{\partial n} = e\left[\frac{\partial}{\partial n}\left(\frac{1}{r_0}\right) - \frac{a}{\rho_0}\frac{\partial}{\partial n}\left(\frac{1}{r_1}\right)\right].$$

Let us evaluate:

$$\frac{\partial}{\partial n}\left(\frac{1}{r_0}\right) = -\frac{1}{r_0^2}\frac{\partial r_0}{\partial n} = -\frac{1}{r_0^2}\cos(\widehat{\boldsymbol{r}_0, \boldsymbol{n}}),$$

$$\frac{\partial}{\partial n}\left(\frac{1}{r_1}\right) = -\frac{1}{r_1^2}\frac{\partial r_1}{\partial n} = -\frac{1}{r_1^2}\cos(\widehat{\boldsymbol{r}_1, \boldsymbol{n}}).$$

From the $\triangle OMM_0$ and $\triangle OMM_1$ (see Fig. 42) we find:

$$\cos(\widehat{\boldsymbol{r}_0, \boldsymbol{n}}) = \frac{a^2+r_0^2-\rho_0^2}{2ar_0}, \qquad \cos(\widehat{\boldsymbol{r}_1, \boldsymbol{n}}) = \frac{a^2+r_1^2-\rho_1^2}{2ar_1}.$$

Substituting the ratio $r_1/r_0 = a/\rho_0$, we obtain the relation for σ.

45. $u = e\left(\frac{1}{r_1} - \frac{a}{\rho_1}\frac{1}{r_1}\right)$

(in this, the symbols of problem 43 are retained) and

$$\rho_1 = r_{OM_1}.$$

Method. If the charge is placed outside the sphere at a point $M_1(\rho_1, \vartheta_1, \phi_1)$ then its image is located at the inverse point $M_0(\rho_0, \vartheta_0, \varphi_0)$ where $OM_0 \, . \, OM_1 = a^2$. Therefore

$$u = \frac{e}{r_1} + \frac{C_1}{r_0}.$$

Determination of C_1 is carried out by analogy with problem 43.

46. The surface charge density equals

$$\sigma = -e\,\frac{\rho_1^2 - a^2}{4\pi a r_1^3}.$$

The solution of the first exterior boundary-value problem for a sphere has the form

$$u = \frac{1}{4\pi} \iint_{\Sigma} \frac{\rho_1^2 - a^2}{a r_1^3} f \, ds$$

or

$$u(\rho_1, \vartheta_1, \phi_1) = \frac{a}{4\pi} \int_0^{2\pi} d\phi \int_0^{\pi} \frac{\rho_1^2 - a^2}{[a^2 - 2a\rho_1 \cos \gamma + \rho_1^2]^{3/2}} f(\vartheta, \phi) \sin \vartheta \, d\vartheta,$$

where

$$\cos \gamma = \cos \vartheta \cos \vartheta_1 + \sin \vartheta \sin \vartheta_1 \cos (\phi - \phi_1).$$

Method. Compare with problems 43 and 44.

47. (a) $u = e\left(\ln \dfrac{1}{r_0} - \ln \dfrac{a}{\rho_0} \dfrac{1}{r_1}\right).$

(b) $u = e\left(\ln \dfrac{1}{r_1} - \ln \dfrac{a}{\rho_1} \dfrac{1}{r_0}\right).$

(c) The solution of the first boundary-value problem inside a circle has the form

$$u_1(\rho_0, \phi_0) = \frac{1}{2\pi} \int_0^{2\pi} \frac{a^2 - \rho_0^2}{a^2 + \rho_0^2 - 2a\rho_0 \cos (\phi - \phi_0)} f(\phi) \, d\phi,$$

while outside the circle

$$u_2(\rho_1, \phi_1) = \frac{1}{2\pi} \int_0^{2\pi} \frac{\rho_1^2 - a^2}{a^2 + \rho_1^2 - 2a\rho_1 \cos (\phi - \phi_1)} f(\phi) \, d\phi,$$

where

$$f = u|_{\rho = a}.$$

Here the following symbols are used: a is the radius of the circle with centre at the origin of coordinates (the point O), $r_0 = MM_0$, $r_1 = MM_1$, $\rho_0 = OM_0$, $\rho_1 = OM_1$, $M_0(\rho_0, \phi_0)$ is the position of the charge, and $M_1(\rho_1, \phi_1)$ is the position of its image.

Method. In order to find the solution of problems (a) and (b) one proceeds as in problem 43, remembering that in the plane case the potential in the neighbourhood of the charge has a logarithmic singularity.

Assuming $e = 1/2\pi$, we obtain the source function G. Calculation of the normal derivatives $\partial G/\partial n$ leads to the expressions

$$\left.\frac{\partial G_1}{\partial n}\right|_{\rho = a} = -\frac{1}{2\pi a} \frac{a^2 - \rho_0^2}{r_0^2} \quad \text{(inside the circle)},$$

$$\left.\frac{\partial G_2}{\partial n}\right|_{\rho = a} = -\frac{1}{2\pi a} \frac{\rho_1^2 - a^2}{r_1^2} \quad \text{(outside the circle)}.$$

48. (a) For the hemisphere, based on the plane $z = 0$ (in the region $z \geqslant 0$) the source function has the form

$$G = G_{43}(M, M_0) - G_{43}(M, M_0'), \tag{1}$$

where

$$G_{43} = \frac{1}{4\pi}\left(\frac{1}{r_0} - \frac{a}{\rho_0}\frac{1}{r_1}\right)$$

(see problem 43), $M_0'(\rho_0, \vartheta_0, \phi_0)$ is the reflection of the point $M_0(\rho_0, \pi-\vartheta_0, \phi_0)$ in the plane $z = 0$ (Fig. 43).

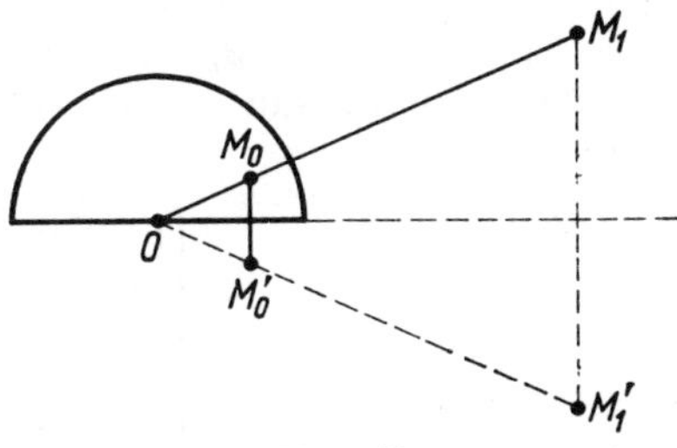

FIG. 43

(b) For the quarter sphere (Fig. 44), bounded by the planes $z = 0$, $x = 0$ and the surface of the sphere, we have:

$$G = G_{43}(M, M_0) - G_{43}(M, M_0') + G_{43}(M, M_0'') - G_{43}(M, M_0'''), \tag{2}$$

where $M_0(\rho_0, \vartheta_0, \phi_0)$, $M_0'(\rho_0, \pi-\vartheta_0, \phi_0)$, $M_0''(\rho_0, \pi-\vartheta_0, \pi+\phi_0)$, $M_0(\rho_0, \vartheta_0, \pi+$ $+\phi_0)$ are the positions of the source and its images.

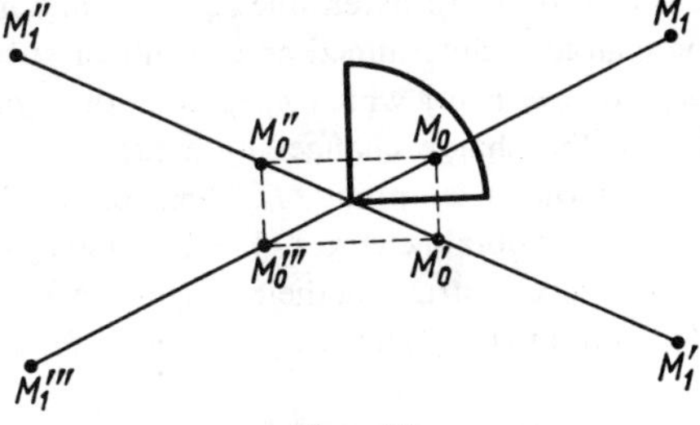

FIG. 44

Method a. Satisfying the boundary condition $u = 0$ on the sphere, we obtain $G_{43}(M, M_0)$. In order to satisfy the condition $u = 0$ for $z = 0$, it is necessary to place a charge $-\dfrac{1}{4\pi}$ at the point M_0' and a charge $+\dfrac{a}{\rho_0}\dfrac{1}{4\pi}$ at the point M_1', which gives us $G_{43}(M, M_0')$.

Method b. In order to satisfy the conditions $u = 0$ for $x = 0$ and $z = 0$ it is necessary to place sources on a sphere of radius ρ_0 at the points M_0', M_0'', M_0'''. Reflection in the sphere gives charges at M_1, M_1', M_1'', M_1'''. Adding the contributions of these we obtain (2).

49. The potential of the field, produced by a point charge e, situated inside the spherical layer $a \leqslant \rho \leqslant b$, is

$$u(M, M_0) = e \sum_{n=0}^{\infty} \left(\frac{e_n}{r_n} - \frac{e_n'}{r_n'} \right), \tag{1}$$

where $M(\rho, \vartheta, \phi)$ is the point of observation, $M_0(\rho_0, \vartheta_0, \phi_0)$ is the point at which the given charge is placed, $r_n = MM_n$, $r_n' = MM_n'$, $M_n(\rho_n, \vartheta_0 \phi_0)$ and $M_n'(\rho_n', \vartheta_0, \phi_0)$ are the points at which positive charges e_n and negative charges $-e_n'$ are situated, where

$$e_n = \begin{cases} \left(\dfrac{a}{b}\right)^k & \text{for } n = 2k, \\[2ex] \left(\dfrac{b}{a}\right)^{k+1} & \text{for } n = 2k+1, \end{cases} \qquad e_n' = \begin{cases} \left(\dfrac{a}{b}\right)^k \dfrac{a}{\rho_0} & \text{for } n = 2k, \\[2ex] \left(\dfrac{b}{a}\right)^k \dfrac{b}{\rho_0} & \text{for } n = 2k+1, \end{cases} \tag{2}$$

$$\rho_n = \begin{cases} \left(\dfrac{a^2}{b^2}\right)^k \rho_0 & \text{for } n = 2k, \\[2ex] \left(\dfrac{b^2}{a^2}\right)^{k+1} \rho_0, & \text{for } n = 2k+1, \end{cases} \qquad \rho_n' = \begin{cases} \left(\dfrac{a^2}{b^2}\right)^k \dfrac{a^2}{\rho_0} & \text{for } n = 2k, \\[2ex] \left(\dfrac{b^2}{a^2}\right)^k \dfrac{b^2}{\rho_0} & \text{for } n = 2k+1. \end{cases} \tag{3}$$

Series (1) converges uniformly and absolutely.

Solution. All the charges e_n and e_n' will, obviously, lie along the line $\phi = \phi_0$, $\theta = \theta_0, \vartheta = \vartheta_0$. Their position along the line is determined by the distances from the centres ρ_n and ρ_n'. In order to determine e_n, e_n', ρ_n and ρ_n' we notice that (1) the position of the charge is determined as a result of successive inversion at the spheres $\rho = a$ and $\rho = b$ for which $\rho_n \rho_n' = a^2$ or $\rho_n \rho_n' = b^2$, (2) for each reflection the value of the charge changes by a factor $-a/\rho_0$ or $-b/\rho_0$.

Let $e_0 = 1$ be the charge at the point M_0. Then for the first reflection at the spheres $\rho = a$ and $\rho = b$ we obtain charges $e_0' = a/\rho_0$ and $e_1' = b/\rho_0$ at the points $\rho_0' = u_2/\rho_0$ and $\rho_1' = b^2/\rho_0$. Constructing their images, we find $e_1 = b/\rho_0', e_0' = b/a$ and $e_2 = a/\rho_1'$, $e_1' = a/b$ at the points $\rho_1 = b^2/\rho_0' = b^2\rho_0/a^2$ and $\rho_2 = a^2/\rho_1' = a^2\rho_0/b^2$.

Continuing the arguments, we see that the even charges lie inside the sphere $\rho = a$, and the odd charges outside the sphere $\rho = b$. Therefore the recurrence relations are readily written down

$$e_{2k+1} = \frac{b}{a} e_{2k-1}, \qquad e_{2k} = \frac{a}{b} e_{2k-2}, \tag{4}$$

$$\rho_{2k+1} = \frac{b^2}{a^2} \rho_{2k-1}, \qquad \rho_{2k} = \frac{a^2}{b^2} \rho_{2k-2} \tag{5}$$

and similar relations for e'_{2k}, ρ'_{2k}, e'_{2k+1}, ρ'_{2k+1}. Hence we find the expressions (2), (3) for e_n, e'_n and ρ_n and ρ'_n. Summing the potentials

$$\frac{e_n}{r_n} \quad \text{and} \quad -\frac{e'_n}{r'_n},$$

we obtain series (1).

Let us consider the general term of the series

$$g_n = \frac{e_n}{r_n} - \frac{e'_n}{r'_n}$$

for sufficiently large n. We draw a plane through the points OM_1M_n; let $n = 2k$. From $\triangle OMM_{2k}$ we find:

$$r_{2k} = \sqrt{\rho^2+\rho_{2k}^2-2\rho\rho_{2k}\cos\gamma},$$

where

$$\cos\gamma = \frac{\rho^2+\rho_0^2-r_0^2}{2\rho\rho_0}.$$

Similarly we find:

$$r'_{2k} = \sqrt{\rho^2+(\rho'_{2k})^2-2\rho\rho'_{2k}\cos\gamma}.$$

Since $\rho_{2k} = (a/b)^{2k}\rho_0 \to 0$ for $k\to\infty$, then

$$\lim_{k\to\infty} r_{2k} = \rho, \quad \lim_{k\to\infty} r'_{2k} = \rho.$$

On the other hand, $e_{2k} = (a/b)^k \to 0$, $e'_{2k} = e_{2k}\,a/\rho_0 \to 0$ for $k\to\infty$. Therefore

$$|g_{2k}| < C_k = \frac{1}{b}\left(1+\frac{a}{\rho_0}\right)\left(\frac{a}{b}\right)^k. \tag{6}$$

Let $n = 2k+1$. Since $\rho_{2k+1} > b$, $\rho'_{2k+1} > b$ and for $k\to\infty$ increase infinitely, then

$$\frac{1}{r_{2k+1}} < \frac{1}{2\rho_{2k+1}} = \frac{1}{2\rho_0}\left(\frac{a}{b}\right)^{2k+2}, \quad \frac{1}{r'_{2k+1}} < \frac{1}{2\rho'_{2k+1}} = \frac{\rho_0}{2a^2}\left(\frac{a}{b}\right)^{2k+2};$$

on the other hand

$$e_{2k+1} = \left(\frac{b}{a}\right)^{k+1}, \quad e'_{2k+1} = \left(\frac{b}{a}\right)^{k+1}\frac{a}{\rho_0},$$

so that

$$|g_{2k+1}| < \frac{e_{2k+1}}{r_{2k+1}} + \frac{e'_{2k+1}}{r'_{2k+1}} \leqslant \frac{1}{b}\left(\frac{a}{b}\right)^k = C_k. \tag{7}$$

The uniform and absolute convergence of the series $\sum_{n=0}^{\infty} g_n$ follows from the inequalities (6) and (7). Its differentiability is proved similarly.

Limiting cases:

(a) for $a \to 0$ all the terms of series (1) reduce to zero except two:

$$\frac{e_0}{r_0} - \frac{e_1'}{r_1'},$$

and we obtain the solution of the interior boundary-value problem for the sphere

$$u = u_{43} = e\left(\frac{1}{r_0} - \frac{b}{\rho_0}\frac{1}{r_1'}\right)$$

(see problem 43);

(b) for $b \to \infty$ we obtain:

$$u = u_{43} = e\left(\frac{1}{r_0} - \frac{a}{\rho_0}\frac{1}{r_0'}\right)$$

the solution of the exterior problem for the sphere (see problem 45).

50. If the charge is situated at the point $M_1(\rho_1, \vartheta_0, \phi_0)$ then the potential in the presence of a charged sphere

$$u(M, M_0) = \frac{e_1}{r} + \frac{ea}{\rho_1}\frac{1}{r} + u_{45},$$

where $u_{45} = e(1/r_1 - a/\rho_1, r_0)$ is the potential of a point charge in the presence of an eathed sphere (see problem 45), $M = M(r, \vartheta, \phi)$ the point of observation, $M_0(\rho_0, \theta_0, \phi_0)$ the point at which the image of the charge is located

$$\rho_0 = \frac{a^2}{\rho_1}, \quad r_0 = MM_0, \quad r_1 = MM_1, \quad r = OM.$$

The surface charge density

$$\sigma = \frac{1}{4\pi a^2}\left(e_1 + \frac{ea}{\rho_1}\right) - \frac{e}{4\pi}\frac{\rho_1^2 - a^2}{ar_1^3} = \sigma_0 + \sigma_{\text{ind}},$$

where $\sigma_{\text{ind}} = \dfrac{e}{4\pi a}\left(\dfrac{1}{\rho_1} - \dfrac{\rho_1^2 - a^2}{r_1^3}\right)$ is the density of the induced charges.

Method. The solution must be sought in the form

$$u = U + u_{45}, \tag{1}$$

where $U = aV/r$ is the potential of the field produced by the sphere, charged to a potential V. In order to determine V one uses the equation

$$4\pi e_1 = -\iint\limits_S \frac{\partial u}{\partial r}\,dS = 4\pi a V - \iint\limits_S \frac{\partial u_{45}}{\partial r}\,dS. \tag{2}$$

By means of Green's formula

$$v(\rho_1, \vartheta_0, \phi_0) = \iint\limits_S \frac{\partial G_{45}}{\partial r}\,dS$$

and the relation

$$u_{45} = 4\pi e G_{45}$$

we obtain:

$$4\pi e v(\rho_1, \vartheta_0, \phi_0) = \iint\limits_S \frac{\partial u_{45}}{\partial r}\,dS,$$

where v is the solution of the exterior boundary-value problem for a sphere S for the condition

$$v|_S = 1,$$

equals

$$v(\rho_1, \vartheta_0, \phi_0) = \frac{a}{\rho_1}.$$

Formula (2) gives:

$$e_1 = aV - \frac{ae}{\rho_1}. \tag{3}$$

Hence we find:

$$V = \frac{e_1}{a} + \frac{e}{\rho_1}.$$

3. The Source Function in Inhomogeneous Media

If the characteristics of the medium (ε, μ, k, etc.) are discontinuous on some surface, then matching conditions must be fulfilled on this surface. In the electrostatic case we have:

$$u_1 = u_2,$$

$$\varepsilon_1\left(\frac{\partial u}{\partial n}\right)_1 - \varepsilon_2\left(\frac{\partial u}{\partial n}\right)_2 = 4\pi\eta,$$

where η is the surface density of free charges, the subscripts 1 and 2 correspond to limiting values for the outer and inner sides of the surface S, $\partial/\partial n$ indicates differentiation with respect to the direction of the normal. If $\boldsymbol{D} = \varepsilon \boldsymbol{E}$ is the electric induction vector and $\boldsymbol{E} = -\operatorname{grad} u$ then the second condition means that

$$D_{n_2} - D_{n_1} = 4\pi\eta.$$

If there are no free charges ($\eta = 0$), then

$$\varepsilon_1\left(\frac{\partial u}{\partial n}\right)_1 = \varepsilon_2\left(\frac{\partial u}{\partial n}\right)_2.$$

Let us investigate the formula for the surface density of charges on the boundary of separation of two media with dielectric constants ε_1 and ε_2 (Fig. 45).

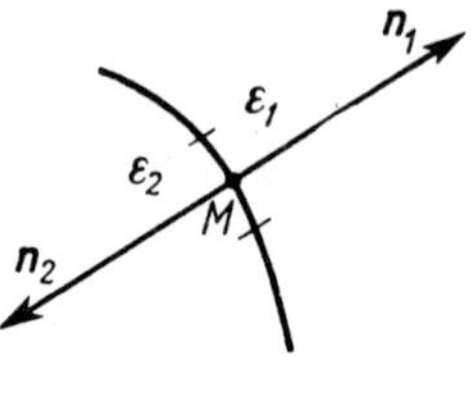

FIG. 45

From Maxwell's equations the apparent density of surface charges is

$$E_{n_1}^{(1)} - E_{n2}^{(2)} = 4\pi\sigma.$$

Considering an infinitely small element dS, we have:

$$E_{n_1}^{(1)} = 2\pi\sigma + E_{n_1}^{(0)},$$

$$E_{n_2}^{(2)} = -E_{n_1}^{(2)} = 2\pi\sigma - E_{n_1}^{(0)},$$

where $E_{n_1}^{(1)}$ and $E_{n_2}^{(2)}$ are limiting values at the point M of the boundary S of the projections of the vectors $\boldsymbol{E}^{(1)}$ and $\boldsymbol{E}^{(2)}$ in the direction of the inner normals $\boldsymbol{n}_1$ and $\boldsymbol{n}_2$, and $E_{n_1}^{(0)}$ is the value of $E_{n_1}^{(1)}$ at the point M.

From the second matching condition

$$\varepsilon_1(2\pi\sigma + E_{n_1}^{(0)}) + \varepsilon_2(2\pi\sigma - E_{n_1}^{(0)}) = 4\pi\eta$$

we obtain:

$$\sigma = \frac{2\eta}{\varepsilon_1+\varepsilon_2} + \frac{\varepsilon_2-\varepsilon_1}{2\pi(\varepsilon_1+\varepsilon_2)} E_{n_1}^{(0)}.$$

If there is no real charge on the surface, then

$$\sigma = \frac{\varepsilon_2-\varepsilon_1}{2\pi(\varepsilon_1+\varepsilon_2)} E_{n_1}^{(0)}.$$

Substituting the value $E_{n_1}^{(0)}$ on the surface S, it is possible to determine σ.

51. If the charge is placed at the point $M_0(\xi, \eta, \zeta)$ of the semispace $z > 0$ $(\zeta > 0)$, then

$$u = \begin{cases} u_1 = \dfrac{e}{\varepsilon_1}\left(\dfrac{1}{r_0} + \dfrac{\varepsilon_1-\varepsilon_2}{\varepsilon_1+\varepsilon_2}\dfrac{1}{r_0'}\right) & \text{for} \quad z > 0 \quad (\varepsilon = \varepsilon_1), \\ u_2 = \dfrac{2e}{\varepsilon_1+\varepsilon_2}\dfrac{1}{r_0} & \text{for} \quad z < 0 \quad (\varepsilon = \varepsilon_2), \end{cases} \tag{1}$$

where

$$r_0 = MM_0 = \sqrt{(x-\xi)^2+(y-\eta)^2+(z-\zeta)^2},$$

$$r_0' = MM_0' = \sqrt{(x-\xi)^2+(y-\eta)^2+(z+\zeta)^2}.$$

From formula (1) we see that the field in the region with dielectric constant ε_1 is the same as though all the space were filled with the dielectric ε_1, and an additional charge is placed at the reflected point $M_0'(\xi_1, \eta_1, -\zeta)$ with magnitude

$$e' = \frac{\varepsilon_1-\varepsilon_2}{\varepsilon_1+\varepsilon_2}\, e.$$

The field in the region ε_2 is the same as the field of a charge

$$e'' = \frac{2\varepsilon_1}{\varepsilon_1+\varepsilon_2}\, e,$$

at the point M_0 if the medium were homogeneous and $\varepsilon = \varepsilon_1$.

The density of surface charges, induced on the boundary $z = 0$, equals $\sigma = e_0\, \zeta/2\pi\bar{r}_0^3$.

Method. The solution must be sought in the form

$$u_1 = \frac{e}{\varepsilon_1}\frac{1}{r_0} + \frac{e_0'}{\varepsilon_1}\frac{1}{r_0'}, \tag{2}$$

$$u_2 = \frac{e_1}{\varepsilon_1}\frac{1}{r_0}, \tag{3}$$

where e_0' and e_1 are constants to be defined.

The matching conditions

$$u_1 = u_2, \qquad \varepsilon_1 \frac{\partial u_1}{\partial z} = \varepsilon_2 \frac{\partial u_2}{\partial z} \quad \text{for} \quad z = 0$$

give:

$$e_1 = \frac{2\varepsilon_1}{\varepsilon_1+\varepsilon_2}\, e, \qquad e_0' = \frac{\varepsilon_1-\varepsilon_2}{\varepsilon_1+\varepsilon_2}\, e.$$

The surface charge density equals

$$\sigma = \frac{\varepsilon_2-\varepsilon_1}{2\pi(\varepsilon_1+\varepsilon_2)}\, E_{n_1}^{(0)}, \tag{4}$$

where $E_{n_1}^{(0)}$ is the original field at $z = 0$ of the charge e placed at M_0, equal to

$$E_{n_1}^{(0)} = -\frac{e}{\varepsilon_1}\frac{\zeta}{\bar{r}_0^3}, \tag{5}$$

where

$$\bar{r}_0 = \sqrt{(x-\xi)^2+(y-\eta)^2+\zeta^2}.$$

From (4) and (5) it follows:

$$\sigma = \frac{\varepsilon_1-\varepsilon_2}{\varepsilon_1(\varepsilon_1+\varepsilon_2)}\, e\, \frac{\zeta}{2\pi\bar{r}_0^3} = e_0 \frac{\zeta}{2\pi\bar{r}_0^3}.$$

The total charge induced on the plane $z = 0$ equals

$$e_0 = 2\pi \int_0^\infty \sigma\rho\, d\rho = \frac{\varepsilon_1-\varepsilon_2}{\varepsilon_1(\varepsilon_1+\varepsilon_2)}\, e.$$

52. The potential of the electric field produced by a source of current I, situated at the point $M_0(0, 0, \zeta)$ is

$$u = \frac{I}{4\pi\sigma_2}\left(\frac{1}{r}+\frac{1}{r'}\right)+\sum_{\substack{n=-\infty \\ (n\neq 0)}}^{\infty}\frac{I}{4\pi\sigma_2}\left(\frac{1}{r_n}+\frac{1}{r'_n}\right) \tag{1}$$

(everywhere in the layer $0 < z < h$), where

$$r = \sqrt{x^2+y^2+(z-\zeta)^2}, \qquad r_n = \sqrt{x^2+y^2+[z-(2nh+\zeta)]^2},$$

$$r'_n = \sqrt{x^2+y^2+[z-(2nh-\zeta)]^2}, \qquad I = \varkappa^n I,$$

$$r' = \sqrt{x^2+y^2+(z+\zeta)^2},$$

$$\varkappa = \frac{\sigma_1-\sigma_2}{\sigma_1+\sigma_2}, \qquad |\varkappa| < 1.$$

If $\zeta = 0$, i.e. the source is in the plane $z = 0$, then $r_n = r'_n$ and the potential is

$$u = \frac{I}{2\pi\sigma_2 r}+\frac{1}{2\pi\sigma_2}\sum_{\substack{n=-\infty \\ (n\neq 0)}}^{\infty}\frac{I_n}{r_n},$$

where

$$r_n = \sqrt{x^2+y^2+(z-2nh)^2},$$

$$r = \sqrt{x^2+y^2+z^2}.$$

The current density for $z = 0$ equals

$$j_x = \frac{xI}{2\pi\rho^3}+\frac{1}{2\pi}\sum_{n=-\infty}^{\infty}\frac{xI_n}{[\rho^2+4n^2h^2]^{3/2}},$$

$$j_y = \frac{yI}{2\pi\rho^3}+\frac{1}{2\pi}\sum_{n=-\infty}^{\infty}\frac{yI_n}{[\rho^2+4n^2h^2]^{3/2}}, \qquad j_z = 0.$$

Method. It is required to solve the problem

$$\Delta u_1 = 0 \qquad \text{for} \quad 0 < z < h,$$

$$\Delta u_2 = 0 \qquad \text{for} \quad z > h,$$

$$\left.\begin{aligned} u_1 &= u_2 \\ \sigma_1\frac{\partial u_1}{\partial z} &= \sigma_2\frac{\partial u_2}{\partial z}\end{aligned}\right\} \qquad \text{for} \quad z = h,$$

$$u \approx \frac{I}{4\pi\sigma_2}\frac{1}{r} \qquad \text{for} \quad r \to 0,$$

$$\frac{\partial u_1}{\partial z} = 0 \qquad \text{for} \quad z = 0, \quad x \neq 0, \quad y \neq 0\,(r \neq 0).$$

The last condition means that reflection at the plane $z = 0$ will be even. For reflection in the plane $z = h$ it is necessary to utilize the method of solving problem 51. One must also take into account the fact that in order to form the solution in the layer $0 < z < h$ there is no necessity to calculate the solution in the region $z > h$.

In order to satisfy the boundary condition $(\partial u_1/\partial z|_{z=0} = 0)$, it is necessary to place a source of current I at the point $M(0, 0, -\zeta)$. In order to satisfy the matching conditions for $z = h$, it is now necessary to place sources $I_1 = \varkappa I$ at the points $M(0, 0, 2h-\zeta)$ and $M(0, 0, 2h+\zeta)$. Doing this we disturbed the condition for $z = 0$. In order to satisfy the condition for $z = 0$, it is necessary to place sources of current I_1 at the points $M(0, 0, -2h+\zeta)$, $M(0, 0, -2h-\zeta)$. Then we upset the matching conditions for $z = h$. Continuing this process, we can satisfy all the boundary conditions only by means of the series (44).

The absolute and uniform convergence of this series, and also of the derivative series, is ensured by the condition

$$|\varkappa| < 1 .$$

Making use of the formula $j = -\sigma \operatorname{grad} u$ the components of the current density j_x, j_y, j_z for $z = 0$ are readily found.

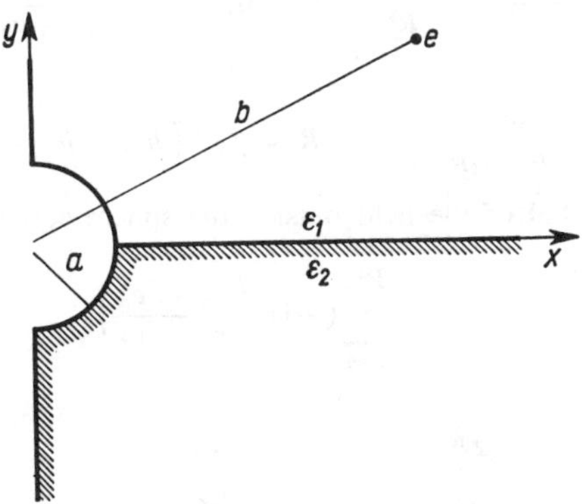

FIG. 46

53. The potential above the plane $x, z\,(y > 0)$ equals the sum of the potentials of the charge e and its seven images, distributed in the following way (Fig. 46):

e at the point $M_0(x_0, y_0, z_0)$, $-e$ at the point $M_1(-x_0, y_0, z_0)$
e' at the point $M_0'(x_0, -y_0, z_0)$ $-e'$ at the point $M_1'(-x_0, -y_0, z_0)$
$-ce$ at the point $M_2(c^2x_0, c^2y_0, c^2z_0)$ $-ce'$ at the point $M_2'(c^2x_0, -c^2y_0, c^2z_0)$
ce at the point $M_3(-c^2x_0, c^2y_0, c^2z_0)$ ce' at the point $M_3'(-c^2x_0, -c^2y_0, c^2z_0)$

where

$$c = \frac{a}{b}, \quad e' = \frac{\varepsilon_2 - \varepsilon_1}{\varepsilon_2 + \varepsilon_1} e.$$

The potential in the dielectric for $y < 0$ may be derived, using only the images in the region $y > 0$ and substituting in place of e the charge

$$e'' = \frac{2\varepsilon_2}{\varepsilon_1 + \varepsilon_2}.$$

54. The potential of the electric field produced by a point source of current existing at the point $M_0(0, -h, \zeta)$, of magnitude I_0, equals

$$u_1 = \frac{I_0}{4\pi\sigma_1}\left(\frac{1}{r_0} + \frac{1}{r_0'}\right) + \frac{\sigma_2 - \sigma_1}{\sigma_1 + \sigma_2}\frac{I_0}{4\pi\sigma_1}\left(\frac{1}{r_1} + \frac{1}{r_1'}\right) \quad \text{for} \quad y < 0,$$

where

$$r_0 = \sqrt{x^2 + (y+h)^2 + (z-\zeta)^2}, \quad r_0' = \sqrt{x^2 + (y+h)^2 + (z+\zeta)^2},$$

$$r_1 = \sqrt{x^2 + (y-h)^2 + (z-\zeta)^2}, \quad r_1' = \sqrt{x^2 + (y-h)^2 + (z+\zeta)^2}.$$

The current density for $y = 0$, $\zeta = 0$

$$j_x = \frac{\sigma_1\sigma_2 I_0}{(\sigma_1 + \sigma_2)\pi}\frac{x}{R^3}, \quad j_y = 0, \quad j_z = \frac{\sigma_1\sigma_2 I_0}{(\sigma_1 + \sigma_2)\pi}\frac{z}{R^3},$$

so that

$$|j| = \frac{\sigma_1\sigma_2 I_0}{(\sigma_1 + \sigma_2)\pi}\frac{\rho}{R^3}, \quad R = \sqrt{\rho^2 + h^2}, \quad \rho^2 = x^2 + z^2.$$

55. The potential of the field outside the sphere equals

$$u = \sum_{n=0}^{\infty} (-1)^n \left(\frac{e_n}{r_n} - \frac{e_n'}{r_n'}\right),$$

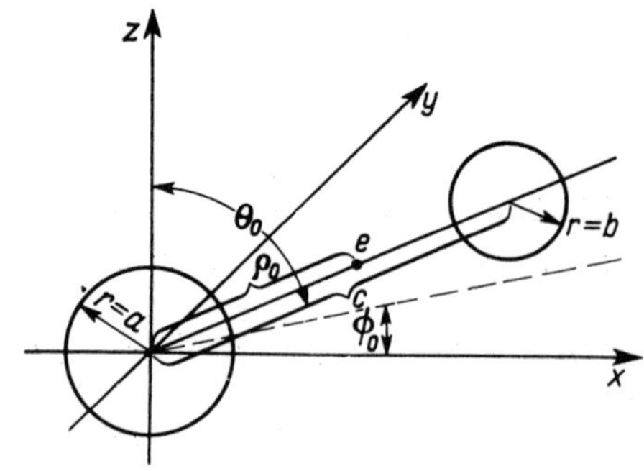

FIG. 47

where e_n and e'_n are charges, whose value is determined from the recurrence relations

$$e_{2k+1} = \frac{a}{b}\,\frac{c-\rho_{2k+1}}{\rho_{2k-1}}\,e_{2k-1}, \qquad e_{2k+2} = \frac{a}{b}\,\frac{c-\rho'_{2k+1}}{\rho'_{2k+1}}\,e_{2k},$$

$$e'_{2k+1} = \frac{a}{b}\,\frac{c-\rho'_{2k+1}}{\rho'_{2k-1}}\,e'_{2k-1}, \qquad e'_{2k+2} = \frac{a}{b}\,\frac{c-\rho_{2k+1}}{\rho_{2k+1}}\,e'_{2k}.$$

These charges exist at the points (Fig. 47) $M_n(\rho_n, \vartheta_0, \phi_0)$ and $M'_n(\rho'_n, \vartheta_0, \phi_0)$ where ρ_n and ρ'_n are determined by the recurrence relations

$$\rho_{2k+1} = \frac{(c^2-b^2)\rho_{2k-1}-a^2c}{c\rho_{2k-1}-a^2}, \qquad \rho'_{2k+1} = \frac{(c^2-b^2)\rho'_{2k-1}-a^2c}{c\rho'_{2k-1}-a^2},$$

$$\rho_{2k+2} = \frac{a^2(c-\rho_{2k})}{c(c-\rho_{2k})-b^2}, \qquad \rho'_{2k+2} = \frac{a^2(c-\rho'_{2k})}{c(c-\rho'_{2k})-b^2},$$

where

$$\rho'_0 = \frac{a^2}{\rho_0}, \qquad \rho'_1 = \frac{c(c-\rho_0)-b^2}{c-\rho_0},$$

$$\rho_1 = \frac{c(c-\rho'_0)-b^2}{c-\rho'_0}, \qquad e_0 = e, \qquad e'_0 = \frac{a}{\rho_0}\,e, \qquad e'_1 = \frac{b}{c-\rho_0}\,e,$$

$$e_1 = \frac{b}{c-\rho'_0}, \qquad e'_0 = \frac{ab}{c\rho_0-a^2}\,e,$$

$$r_n = \sqrt{\rho^2+\rho_n^2-2\rho\rho_n\cos\alpha_n},$$

$$r'_n = \sqrt{\rho^2+\rho'^2_n-2\rho\rho'_n\cos\alpha'_n},$$

where α_n is the angle between OM_i and OM_n, O is the origin of coordinates M_i is the position of the source, and M_n is the point of observation.

§ 4. The Method of Separation of Variables

1. Boundary-value Problems for a Circle, Ring and Sector

56. If on the boundary of a circle of radius a the unknown function $u|_{\rho=a} = f(\phi)$, then

$$u(\rho,\phi) = \frac{A_0}{2} + \sum_{n=1}^{\infty}\left(\frac{\rho}{a}\right)^n (A_n\cos n\phi + B_n\sin n\phi) \quad \text{for} \quad \rho < a, \tag{1}$$

where A_n, B_n are Fourier coefficients of the function $f(\phi)$, equal to

$$\left.\begin{aligned} A_n &= \frac{1}{\pi}\int_0^{2\pi} f(\phi)\cos n\phi\, d\phi \qquad (n = 0, 1, 2, \ldots), \\ B_n &= \frac{1}{\pi}\int_0^{2\pi} f(\phi)\sin n\phi\, d\phi \qquad (n = 1, 2, \ldots). \end{aligned}\right\} \tag{2}$$

From formula (1) one can derive the integral representation for the solution of the first interior boundary-value problem for Laplace's equation inside a circle (Poisson's formula)

$$u(\rho, \phi) = \frac{1}{2\pi}\int_0^{2\pi} \frac{a^2-\rho^2}{a^2+\rho^2-2a\rho\cos(\phi-\psi)} f(\psi)\,d\psi. \tag{3}$$

Solution. It is required to find the function $u(\rho, \phi)$, continuous in the circle $0 \leqslant \rho \leqslant a$, satisfying the equation

$$\frac{1}{\rho}\frac{\partial}{\partial\rho}\left(\rho\frac{\partial u}{\partial\rho}\right)+\frac{1}{\rho^2}\frac{\partial^2 u}{\partial\phi^2} = 0 \tag{4}$$

inside this circle and the boundary condition

$$u|_{\rho=a} = f(\phi), \tag{5}$$

where f is a given continuous function.

The problem is solved by the method of separation of variables (see [7], chapter IV, § 3). We seek the solution in the form of a sum

$$u(\rho, \phi) = \sum_{n=0}^{\infty} u_n(\rho, \phi),$$

where

$$u_n(\rho, \phi) = R_n(\rho)\Phi_n(\phi),$$

$$\Phi_n(\phi) = \begin{cases}\cos n\phi, \\ \sin n\phi,\end{cases} \quad R_n(\rho) = \begin{Bmatrix}\rho^n, \\ \rho^{-n}.\end{Bmatrix} \tag{6}$$

57. $$u(\rho, \phi) = \frac{A_0}{2}+\sum_{n=1}^{\infty}\left(\frac{a}{\rho}\right)^n (A_n\cos n\phi+B_n\sin n\phi), \tag{1}$$

where a is the radius of the circle, A_n and B_n are determined from formulae (2) of problem 56.

Method. It is required to find the function $u(\rho, \phi)$, satisfying the equation

$$\frac{1}{\rho}\frac{\partial}{\partial\rho}\left(\rho\frac{\partial u}{\partial\rho}\right)+\frac{1}{\rho^2}\frac{\partial^2 u}{\partial\phi^2} = 0 \quad \text{outside the circle,}$$

the boundary condition $u|_{\rho=a} = f(\phi)$ and the condition that u is bounded as $\rho \to \infty$.

We seek the solution by the method of separation of variables. From the condition at infinity it follows that $C_n = 0$, and we obtain the particular solution in the form

$$u_n(\rho, \phi) = \left(\frac{a}{\rho}\right)^n (A_n\cos n\phi+B_n\sin n\phi).$$

The general solution is given by the series

$$u(\rho, \phi) = \sum_{n=0}^{\infty} u_n(\rho, \phi).$$

Using the boundary condition for $\rho = a$ we arrive at (1).

58. (a) The solution of the second interior boundary-value problem for a circle is

$$u(\rho, \phi) = \sum_{n=1}^{\infty} \frac{\rho^n}{na^{n-1}} (A_n \cos n\phi + B_n \sin n\phi) + C_1, \tag{1}$$

(b) the solution of the exterior problem is

$$u(\rho, \phi) = \sum_{n=1}^{\infty} \frac{a^{n+1}}{n\rho^n} (A_n \cos n\phi + B_n \sin n\phi) + C_2, \tag{2}$$

where C_1 and C_2 are arbitrary constants, and a is the radius of the circle, A_n and B_n are Fourier coefficients of the function $f(\phi) = \partial u/\partial \nu|_{\rho=a}$, ν is the direction of the outer normal to the region under consideration.

Method a. It is required to find the function $u(\rho, \phi)$, continuous in the circle $0 \leqslant \rho \leqslant a$, satisfying the equation $\Delta u = 0$ inside this circle and the boundary condition

$$\left.\frac{\partial u}{\partial \nu}\right|_{\rho=a} = f(\phi)$$

on this boundary for $\rho = a$, and also the condition $\int_0^{2\pi} f(\phi)\, d\phi = 0$.

Method b. It is required to find the function $u(\rho, \phi)$, satisfying Laplace's equation outside a circle of radius $\rho = a$, the boundary condition

$$\left.\frac{\partial u}{\partial \nu}\right|_{\rho=a} = f(\phi)$$

and the condition that it is bounded for $\rho \to \infty$.

The solution of both problems is found by the method of separation of variables in a similar way to problem 56.

59. (a) $$u(\rho, \phi) = \sum_{n=1}^{\infty} \frac{\rho^n}{a^{n-1}(n-ah)} (A_n \cos n\phi + B_n \sin n\phi) - \frac{A_0}{2h}. \tag{1}$$

(b) $$u(\rho, \phi) = -\sum_{n=1}^{\infty} \frac{a^n}{(an+h)\rho^n} (A_n \cos n\phi + B_n \sin n\phi) - \frac{A_0}{2h}, \tag{2}$$

where A_n and B_n are Fourier coefficients of $f(\phi)$, determinable from formulae (2) of problem 56.

60. The qotential of the electrostatic field equals

$$u = \begin{cases} \dfrac{V_1+V_2}{2}+\dfrac{V_1-V_2}{\pi}\arctan\dfrac{2a\rho\sin\phi}{a^2-\rho^2} \text{ for } \rho < a \text{ (inside cylinder)}, \\ \dfrac{V_1+V_2}{2}+\dfrac{V_1-V_2}{\pi}\arctan\dfrac{2a\rho\sin\phi}{\rho^2-a^2} \text{ for } \rho > a \text{ (outside cylinder)}. \end{cases} \quad (1)$$

The components of the field E_ρ and E_ϕ are calculated from the relations

$$E_\rho = -\frac{\partial u}{\partial \rho}, \qquad E_\phi = -\frac{1}{\rho}\frac{\partial u}{\partial \phi}.$$

The density of the surface charges

$$\sigma = -\frac{V_1-V_2}{a\pi^2 \sin\phi}.$$

Method. The method of separation of variables gives the solution in the form of series

$$u = \begin{cases} \dfrac{V_1+V_2}{2}+\dfrac{2(V_1-V_2)}{\pi}\sum_{k=0}^{\infty}\left(\dfrac{\rho}{a}\right)^{2k+1}\dfrac{\sin(2k+1)\phi}{2k+1} \\ \qquad \text{inside cylinder} \quad (\rho < a), \\ \dfrac{V_1+V_2}{2}+\dfrac{2(V_1-V_2)}{\pi}\sum_{k=0}^{\infty}\left(\dfrac{a}{\rho}\right)^{2k+1}\dfrac{\sin(2k+1)\phi}{2k+1} \\ \qquad \text{outside cylinder} \quad (\rho > a). \end{cases} \quad (2)$$

The series on the right can be summed, if one uses the relation

$$\sum_{k=0}^{\infty}\frac{z^{2k+1}}{2k+1} = \frac{1}{2}\ln\frac{1+z}{1-z}. \quad (3)$$

In fact,

$$J = \sum_{k=0}^{\infty}\xi^{2k+1}\frac{\sin(2k+1)\phi}{2k+1} = \frac{1}{2i}\left\{\sum_{k=0}^{\infty}\frac{\xi^{2k+1}e^{i(2k+1)\phi}}{2k+1} - \sum_{k=0}^{\infty}\frac{\xi^{2k+1}e^{-i(2k+1)\phi}}{2k+1}\right\}.$$

Denoting

$$z = \xi e^{i\phi} = \xi\cos\phi + i\xi\sin\phi,$$

$$z^* = \xi e^{-i\phi} = \xi\cos\phi - i\xi\sin\phi$$

and using formula (3) we obtain:

$$J = \frac{1}{4i}\ln\frac{(1+z)(1-z^*)}{(1-z)(1+z^*)} = \frac{1}{4i}\ln\frac{1-\xi^2+i2\xi\sin\phi}{1-\xi^2-i2\xi\sin\phi} = \frac{1}{2}\arctan\frac{2\xi\sin\phi}{1-\xi^2}.$$

Hence by (2) formula (1) follows ($\xi = \rho/a$ for $\rho < a$ or $\xi = a/\rho$ for $\rho > a$).

61. (a) The solutions of the interior boundary-value problems have the form

(1) $$u(\rho,\phi) = A\frac{\rho}{a}\sin\phi,$$

(2) $$u(\rho,\phi) = B+\frac{3A}{a}\rho\sin\phi-4A\left(\frac{\rho}{a}\right)^3\sin 3\phi,$$

(3) $$u(\rho,\phi) = A\frac{\rho}{a}\sin\phi-\frac{8A}{\pi}\sum_{k=1}^{\infty}\left(\frac{\rho}{a}\right)^{2k}\frac{\cos 2k\phi}{4k^2-9}.$$

(b) The solutions of the exterior boundary-value problems are given by the expressions

(1′) $$u(\rho,\phi) = A\frac{a}{\rho}\sin\phi,$$

(2′) $$u(\rho,\phi) = B+\frac{3Aa}{\rho}\sin\phi-4A\left(\frac{a}{\rho}\right)^3\sin 3\phi,$$

(3′) $$u(\rho,\phi) = A\frac{a}{\rho}\sin\phi-\frac{8A}{\pi}\sum_{k=1}^{\infty}\left(\frac{a}{\rho}\right)^{2k}\frac{\cos 2k\phi}{4k^2-9}.$$

Method. In problems (2) and (3) use the trigonometric relation

$$\sin^3\phi = 3\sin\phi-4\sin 3\phi.$$

62. Assuming that the flow is in the negative direction of the x-axis, we introduce the cylindrical system of coordinates (ρ,ϕ,z) with the z-axis along the axis of the cylinder and the polar axis along the x-axis; then the distribution of temperature in the cylinder is given by the relation

$$u(\rho,\phi) = -\frac{q}{k}\rho\cos\phi+\text{const}.$$

The condition $\int_0^{2\pi} Q\,d\phi = 0$ is fulfilled; the problem has a solution.

63.

$$u(\rho,\phi) = \begin{cases} u_1(\rho,\phi) = \dfrac{(\varepsilon_1+\varepsilon_2)V_1+(3\varepsilon_1-\varepsilon_2)V_2}{4\varepsilon_1}+ \\ \quad +\dfrac{4(V_1-V_2)}{\pi}\displaystyle\sum_{m=0}^{\infty}\frac{\varepsilon_2 b^{2m+1}\rho^{2m+1}}{(\varepsilon_2+\varepsilon_1)b^{4m+2}+(\varepsilon_2-\varepsilon_1)a^{4m+2}}\,\frac{\sin(2m+1)\phi}{2m+1} \\ \qquad\qquad \text{for}\quad \rho<a, \\ u_2(\rho,\phi) = \dfrac{(\varepsilon_1+\varepsilon_2)(V_1-V_2)+4\varepsilon_1V_2}{4\varepsilon_1}+ \\ \quad +\dfrac{2(V_1-V_2)}{\pi}\displaystyle\sum_{m=0}^{\infty}\frac{[(\varepsilon_1+\varepsilon_2)\rho^{4m+2}+(\varepsilon_2-\varepsilon_1)a^{4m+2}]}{(\varepsilon_1+\varepsilon_2)b^{4m+2}+(\varepsilon_2-\varepsilon_1)a^{4m+2}}\,\frac{b^{2m+1}}{\rho^{2m+1}}\times \\ \qquad\qquad \times\dfrac{\sin(2m+1)\phi}{2m+1} \\ \qquad\qquad \text{for}\quad a<\rho<b. \end{cases}$$

Method. It is required to find the solution of Laplace's equation in a circle of radius $a (u = u_1)$ and the ring $a \leqslant \rho \leqslant b$ $(u = u_2)$ for the boundary condition

$$u_2(b, \phi) = \begin{cases} V_1 & \text{for} \quad 0 < \phi < \pi, \\ V_2 & \text{for} \quad \pi < \phi < 2\pi \end{cases}$$

and matching conditions

$$\left.\begin{aligned} u_1 &= u_2, \\ \varepsilon_1 \frac{\partial u_1}{\partial \rho} &= \varepsilon_2 \frac{\partial u_2}{\partial \rho} \end{aligned}\right\} \quad \text{for} \quad \rho = a.$$

Solution. We shall look for a solution $u = \begin{cases} u_1 \\ u_2 \end{cases}$ in the form of the sum

$$u_1 = V_1 + \hat{u}_1, \qquad u_2 = V_2 + \hat{u}_2,$$

where the function $\hat{u} = \begin{cases} \hat{u}_1 \\ \hat{u}_2 \end{cases}$ is harmonic, and satisfies for $\rho = b$ the boundary condition

$$\hat{u}_2 = \begin{cases} V_1 - V_2 & \text{for} \quad 0 < \phi < \pi, \\ 0 & \text{for} \quad \pi < \phi < 2\pi \end{cases}$$

and matching conditions for $\rho = a$.

Assuming next

$$\hat{u}_1 = R_1(\rho)\Phi(\phi), \qquad \hat{u}_2 = R_2(\rho)\Phi(\phi),$$

we find, as is usual, the function $\Phi(\phi)$:

$$\Phi_n(\phi) = \begin{cases} \cos n\phi, \\ \sin n\phi, \end{cases}$$

and for R_1 and R_2 we obtain the equations

$$\rho \frac{\mathrm{d}}{\mathrm{d}\rho}\left(\rho \frac{\mathrm{d}R_1}{\mathrm{d}\rho}\right) - n^2 R_1 = 0 \quad \text{for} \quad 0 < \rho < a,$$

$$\rho \frac{\mathrm{d}}{\mathrm{d}\rho}\left(\rho \frac{\mathrm{d}R_2}{\mathrm{d}\rho}\right) - n^2 R_2 = 0 \quad \text{for} \quad a < \rho < b$$

and the matching conditions

$$R_{1n}(a) = R_{2n}(a),$$
$$\varepsilon_1 R'_{1n}(a) = \varepsilon_2 R'_{2n}(a)$$

and the condition of regularity at $\rho = 0$.

Hence we find:

$$R_{1n} = A_n \rho^n, \qquad R_{2n} = B_n \rho^n + \frac{C_n}{\rho^n}.$$

The matching conditions give:

$$A_n = \frac{2\varepsilon_2}{\varepsilon_1+\varepsilon_2} B_n,$$

$$C_n = \frac{\varepsilon_2-\varepsilon_1}{\varepsilon_1+\varepsilon_2} a^{2n} B_n.$$

The general solution of the problem may be written in the form

$$\hat{u}_2(\rho, \phi) = \sum_{n=0}^{\infty} \left(\rho^n + \frac{\varepsilon_2-\varepsilon_1}{\varepsilon_1+\varepsilon_2} \frac{a^{2n}}{\rho^n}\right)(B_n \cos n\phi + \bar{B}_n \sin n\phi),$$

$$\hat{u}_1(\rho, \phi) = \sum_{n=0}^{\infty} \frac{2\varepsilon_2}{\varepsilon_1+\varepsilon_2} \rho^n (B_n \cos n\phi + \bar{B}_n \sin n\phi),$$

where B_n and $\bar{B}_n$ are coefficients obtained from the boundary condition for $\rho = b$.

64. $u = u(\rho, \phi) = \dfrac{a^2 v_0}{\rho} \cos \phi.$

Method. Introducing the coordinate system (ρ, ϕ, z), with z-axis along the axis of the cylinder we obtain for the velocity potential $u = u(\rho, \phi)$ the boundary-value problem

$$\Delta u = 0 \quad \text{for} \quad \rho > a,$$

$$-\left.\frac{\partial u}{\partial \rho}\right|_{\rho=a} = v_0 \cos \phi.$$

65. $u = u(\rho, \phi) = -v_0 \left(\rho + \dfrac{a^2}{\rho}\right) \cos \phi.$

Method. If the flow is along the x-axis, then the potential of the unperturbed motion of the liquid is

$$u_0 = -v_0 x = -v_0 \rho \cos \phi.$$

Assuming

$$u = u_0 + \hat{u},$$

we obtain the second exterior boundary-value problem for $\hat{u}$:

$$\Delta \hat{u} = 0 \quad \text{for} \quad \rho > a,$$

$$\left.\frac{\partial \hat{u}}{\partial \rho}\right|_{\rho=a} = v_0 \cos \phi.$$

66. (a) If the sphere moves in the direction of the z-axis, then in the coordinate system (r, ϑ, ϕ) with origin at the centre of the sphere the velocity potential of the liquid is

$$u = u(r, \vartheta) = \frac{1}{2} v_0 \frac{a^3}{r^2} \cos \vartheta.$$

(b) If the liquid moves in the negative direction of the z-axis, then

$$u = u(r, \vartheta) = v_0 \left(r + \frac{a^3}{2r^2}\right) \cos\vartheta .$$

Method a. It is required to find the solution of the equation

$$\Delta u = 0 \quad \text{for} \quad r > a$$

with boundary condition

$$-\left.\frac{\partial u}{\partial r}\right|_{r=a} = v_0 \cos\vartheta$$

and condition of regularity at infinity.

Method b. Assuming

$$u = v_0 + \hat{u},$$

where

$$u_0 = v_0 z = v_0 r \cos\vartheta,$$

we obtain the boundary-value problem of section (a) for u.

Solution. (a) Since the boundary condition does not depend on ϕ, the potential will be independent of ϕ, i.e. $u = u(r, \vartheta)$.

Laplace's equation for the function $u(r, \vartheta)$ has the form

$$\frac{\partial}{\partial r}\left(r^2 \frac{\partial u}{\partial r}\right) + \frac{1}{\sin\vartheta} \frac{\partial}{\partial \vartheta}\left(\sin\vartheta \frac{\partial u}{\partial \vartheta}\right) = 0.$$

We shall look for the solution in the form

$$u(r, \vartheta) = R(r) \cos\vartheta,$$

which for $R(r)$ gives:

$$\frac{\mathrm{d}}{\mathrm{d}r}\left(r^2 \frac{\mathrm{d}R}{\mathrm{d}r}\right) - 2R = 0, \qquad R'(a) = -v_0 .$$

Assuming $|R(r)| < M/r$, $R(r) = r^\sigma$ for $r \to \infty$, we find:

$$\sigma_1 = 1, \qquad \sigma_2 = -2,$$

i.e. the general solution of the equation has the form $R(r) = Ar + B/r^2$, where A and B are constants.

From the condition for $r = a$ and at infinity we obtain:

$$A = 0, \qquad B = \frac{v_0 a^3}{2},$$

so that

$$R(r) = v_0 \frac{a^3}{2r^2} .$$

Problem (b) is solved similarly.

67. Introducing a spherical system of coordinates (r, ϑ, ϕ) with origin at the centre of the sphere and the polar axis directed along the external field we obtain for the potential of the electrostatic field $\boldsymbol{E} = -\operatorname{grad} u$:

$$u_1 = -E_0 \frac{3\varepsilon_2}{2\varepsilon_2+\varepsilon_1} r \cos\vartheta \quad \text{for} \quad r < a,$$

$$u_2 = -E_0 \left(r - \frac{(\varepsilon_1-\varepsilon_2)a^3}{2\varepsilon_2+\varepsilon_1} \frac{1}{r^2} \right) \cos\vartheta \quad \text{for} \quad r > a,$$

where a is the radius of the sphere.

The polarization of the sphere is

$$\boldsymbol{P}_1 = \frac{3}{4\pi} \frac{(\varepsilon_1-\varepsilon_2)\varepsilon_2}{\varepsilon_1+2\varepsilon_2} \boldsymbol{E}_0,$$

and its dipole moment

$$\boldsymbol{p} = \frac{4}{3} \pi a^3 \boldsymbol{P}_1 = a^3 \frac{(\varepsilon_1-\varepsilon_2)\varepsilon_2}{\varepsilon_1+2\varepsilon_2} \boldsymbol{E}_0, \qquad \varepsilon_1 = 1+4\pi\varkappa_1.$$

Solution. In order to determine the field inside and outside the sphere, it is necessary to solve the following problem for the potential.

Let us assume

$$u_2 = u_0 + \hat{u}_2,$$

where

$$u_0 = -E_0 z = -Er\cos\theta.$$

To determine $\hat{u}_1$ and u_2 we solve the equations

$$\Delta u_1 = 0 \quad \text{for} \quad r < a,$$

$$\Delta u_2 = 0 \quad \text{for} \quad r > a$$

with the boundary conditions

$$\left. \begin{aligned} u_1 - \hat{u}_2 &= -E_0 a \cos\theta, \\ \varepsilon_1 \frac{\partial u}{\partial r} - \varepsilon_2 \frac{\partial \hat{u}_2}{\partial r} &= -\varepsilon_2 E_0 \cos\theta \end{aligned} \right\} \quad \text{for} \quad r = a$$

and the condition of regularity for $\hat{u}_2$ at infinity.

We seek the solution of this problem in the form

$$u_1(r, \theta) = R_1(r) \cos\theta,$$

$$\hat{u}_2(r, \theta) = R_2(r) \cos\theta.$$

Substitution in the equations and boundary conditions gives:

$$r^2 R_1'' + 2rR_1' - 2R_1 = 0, \quad r^2 R_2'' + 2rR_2' - 2R_2 = 0,$$

$$R_1(a) - R_2(a) = -aE_0,$$

$$\varepsilon_1 R_1'(a) - \varepsilon_2 R_2'(a) = -\varepsilon_2 E_0, \quad |R_2| < \frac{M}{r} \quad \text{for} \quad r \to \infty,$$

where M is some constant. Hence R_1 and R_2 are readily found. Knowing the potentials, the fields $\boldsymbol{E}_1 = -\text{grad } u_1$, $\boldsymbol{E}_2 = -\text{grad } u_2$ may be found.

The electric polarization vector $\boldsymbol{P}$ of the sphere is determined from the following equation:

$$\boldsymbol{P}_1 = \frac{(\varepsilon_1 - \varepsilon_2)\boldsymbol{E}_1}{4\pi}.$$

Indsie the sphere only the component E_z differs from zero:

$$E_z = -\frac{\partial u_1}{\partial z} = \frac{3\varepsilon_2}{2\varepsilon_2 + \varepsilon_1} E_0,$$

so that

$$E = E_z = \frac{3\varepsilon_2}{2\varepsilon_2 + \varepsilon_1} E_0.$$

Therefore the polarization of the sphere equals

$$\boldsymbol{P}_1 = \frac{3\varepsilon_2(\varepsilon_1 - \varepsilon_2)}{4\pi(\varepsilon_1 + 2\varepsilon_2)} \boldsymbol{E}_0.$$

68. Let us choose a coordinate system so that the z-axis is directed along the axis of the cylinder, and the field $\boldsymbol{E}_0$ along the x-axis.

The potential of the field inside and outside the cylinder is given by the formulae

$$u_1(\rho, \phi) = -\frac{2\varepsilon_2}{\varepsilon_1 + \varepsilon_2} E_0 \rho \cos\phi \quad \text{for} \quad \rho < a,$$

$$u_2(\rho, \phi) = -E_0\left(\rho + \frac{\varepsilon_1 - \varepsilon_2}{\varepsilon_1 + \varepsilon_2} \frac{a^2}{\rho}\right) \cos\phi \quad \text{for} \quad \rho > a,$$

where a is the radius of the cylinder.

The field inside the cylinder is

$$E_1 = \frac{2\varepsilon_2}{\varepsilon_1 + \varepsilon_2} E_0$$

and is directed along the x-axis.

The polarization is

$$\boldsymbol{P}_1 = \frac{(\varepsilon_1 - \varepsilon_2)\varepsilon_2}{2\pi(\varepsilon_1 + \varepsilon_2)} \boldsymbol{E}_0,$$

and the dipole moment per unit length is

$$\boldsymbol{p} = \pi a^2 \boldsymbol{P}_1 = \frac{(\varepsilon_1 - \varepsilon_2)\varepsilon_2}{2(\varepsilon_1 + \varepsilon_2)} a^2 \boldsymbol{E}_0.$$

Method. See problem 76.

69. The potential of the field outside the sphere is

$$u(r, \phi) = -E_0\left(r - \frac{a^3}{r^2}\right)\cos\theta \quad \text{for} \quad r > a,$$

if the origin of the spherical system of coordinates is taken at the centre of the sphere, and the polar z-axis is directed along the external field $\boldsymbol{E}_0$.

Method. The potential must be represented as a sum $u = u_0+\hat{u}$, where $u_0 = -E_0 z = -E_0 r\cos\theta$ is the potential of the external field. The following boundary-value problem is obtained for the potential of the perturbed part of the field $\hat{u}$:

$$\Delta\hat{u} = 0 \quad \text{for} \quad r > a,$$
$$\hat{u} = E_0 a\cos\theta \quad \text{for} \quad r = a.$$

70. The potential of the field

$$u = u(\rho, \phi) = -E_0\left(\rho-\frac{a^2}{\rho}\right)\cos\phi .$$

The density of the surface charges equals $\sigma = 2E_0\cos\phi$.

Method. See problem 69.

71. If

$$u|_{\rho=a} = f(\phi), \quad u|_{\rho=b} = F(\phi),$$

then

$$u(\rho, \phi) = \sum_{n=1}^{\infty}\left[\left(A_n\rho^n+\frac{B_n}{\rho^n}\right)\cos n\phi+\left(C_n\rho^n+\frac{D_n}{\rho^n}\right)\sin n\phi\right]+B_0\ln\rho+A_0, \quad (1)$$

where

$$A_n = \frac{b^n F_n^{(1)}-a^n f_n^{(1)}}{b^{2n}-a^{2n}}, \quad B_n = \frac{[b^n f_n^{(1)}-a^n F_n^{(1)}]a^n b^n}{b^{2n}-a^{2n}}, \quad A_0 = \frac{f_0^{(1)}-F_0^{(1)}}{\ln\dfrac{a}{b}},$$

$$C_n = \frac{b^n F_n^{(2)}-a^n f_n^{(2)}}{b^{2n}-a^{2n}}, \quad D_n = \frac{[b^n f_n^{(2)}-a^n F_n^{(2)}]a^n b^n}{b^{2n}-a^{2n}},$$

$$B_0 = \frac{F_0^{(1)}\ln a-f_0^{(1)}\ln b}{\ln\dfrac{a}{b}},$$

where $f_n^{(1)}, f_n^{(2)}$, and $F_n^{(1)}, F_n^{(2)}$ are Fourier coefficients of the functions $f(\phi)$ and $F(\phi)$, equal to

$$f_0^{(1)} = \frac{1}{2\pi}\int_0^{2\pi} f(\phi)\,d\phi, \quad f_n^{(1)} = \frac{1}{\pi}\int_0^{2\pi} f(\phi)\cos n\phi\,d\phi \quad (n = 1, 2, \ldots),$$

$$f_n^{(2)} = \frac{1}{\pi}\int_0^{2\pi} f(\phi)\sin n\phi\,d\phi .$$

Similar expressions hold for F_n.

Solution. It is required to find the solution of Laplace's equation inside the ring $a \leqslant \rho \leqslant b$ for the boundary conditions $u|_{\rho-a} = f(\phi)$, $u|_{\rho-b} = F(\phi)$ on its boundary. Proceeding from the method of separation of variables and assuming

$$u(\rho, \phi) = R(\rho)\Phi(\phi)$$

we obtain:

$$R_n(\rho) = A_n \rho^n + \frac{B_n}{\rho^n}, \qquad R_0(\rho) = A_0 + B_0 \ln \rho.$$

In contrast to the problem for the circle one must retain both terms, since the point $\rho = 0$ lies outside the ring.

As a result we obtain particular solutions of the form

$$u_0(\rho, \phi) = A_0 + B_0 \ln \rho,$$

$$u_n(\rho, \phi) = \left(A_n \rho^n + \frac{B_n}{\rho^n}\right) \cos n\phi + \left(C_n \rho^n + \frac{D_n}{\rho^n}\right) \sin n\phi.$$

Forming the general solution and requiring the boundary conditions for $\rho = a$ and $\rho = b$ to be satisfied, we have:

$$A_0 + B_0 \ln a + \sum_{n=1}^{\infty} \left[\left(A_n a^n + \frac{B_n}{a^n}\right) \cos n\phi + \left(C_n a^n + \frac{D_n}{a^n}\right) \sin n\phi\right] = f(\phi),$$

$$A_0 + B_0 \ln b + \sum_{n=1}^{\infty} \left[\left(A_n b^n + \frac{B_n}{b^n}\right) \cos n\phi + \left(C_n b^n + \frac{D_n}{b^n}\right) \sin n\phi\right] = F(\phi),$$

from which we obtain equations for determining A_n, B_n, C_n and D_n

$$A_n a^n + \frac{B_n}{a^n} = f_n^{(1)}, \quad C_n a^n + \frac{D_n}{a^n} = f_n^{(2)}, \quad A_0 + B_0 \ln C = f_0^{(1)},$$

$$A_n b^n + \frac{B_n}{b^n} = F_n^{(1)}, \quad C_n b^n + \frac{D_n}{b^n} = F_n^{(2)}, \quad A_0 + B_0 \ln b = F_0^{(1)}.$$

72. $$u(\rho, \phi) = u_0 - \frac{2u_0}{\pi} \sum_{m=0}^{\infty} \frac{b^{2m+1}}{b^{4m+2} - a^{4m+2}} \left(\rho^{2m+1} - \frac{a^{4m+2}}{\rho^{2m+1}}\right) \frac{\sin(2m+1)\phi}{2m+1} - \frac{u_0}{2} \frac{\ln \frac{a}{\rho}}{\ln \frac{a}{b}}. \quad (1)$$

Method. The solution is conveniently represented as a sum

$$u = u_0 + v,$$

where the function v satisfies the condition

$$v|_{\rho=a} = 0, \quad v|_{\rho=b} = \begin{cases} 0 & \text{for} \quad 0 < \phi < \pi, \\ -u_0 & \text{for} \quad \pi < \phi < 2\pi. \end{cases}$$

73. The distribution of temperature in the cable is given by the expression

$$u(\rho, \phi) = \frac{q}{4}(\rho^2 - a^2) - \frac{2\varkappa_0 b + qb^2}{2} \ln \frac{\rho}{a} + \frac{0.5 b^3 \varkappa_0}{a^4 + b^4}\left(\rho^2 - \frac{a^4}{\rho^2}\right)\cos^2\phi,$$

where $q = q_0/k$, $q_0 = 0.24 I^2 R$ is the amount of heat liberated by the passage of current per unit time per unit length of the cylinder, R the resistance per unit length of the cylinder, $\varkappa_0 = A/2k$, k the coefficient of heat conduction.

Method. It is required to find the solution of the equation $\Delta u = q$ inside the ring $a < \rho < b$ for the boundary conditions

$$u|_{\rho=a} = 0,$$

$$-k\left.\frac{\partial u}{\partial \rho}\right|_{\rho=b} = A\cos^2\phi \quad \text{or} \quad \left.\frac{\partial u}{\partial \rho}\right|_{\rho=b} = -\varkappa_0 - \varkappa_0 \cos 2\phi.$$

The function u is conveniently represented as the sum $u = u_1 + u_2$, where u_1 is the solution of the problem

$$\Delta u_1 = q, \quad u_1|_{\rho=a} = 0, \quad \left.\frac{\partial u_1}{\partial \rho}\right|_{\rho=b} = -\varkappa_0.$$

74. The temperature at the point (ρ, ϕ) equals

$$u(\rho, \phi) = \sum_{n=0}^{\infty} f_n \left(\frac{\rho}{a}\right)^{\frac{\pi n}{\alpha}} \sin \frac{\pi n}{\alpha}\phi, \tag{1}$$

where

$$f_n = \frac{2}{\alpha}\int_0^{\alpha} f(\phi) \sin \frac{\pi n}{\alpha}\phi \, d\phi.$$

In the particular case

$$f(\phi) = \begin{cases} u_1 & \text{for} \quad 0 < \phi < \dfrac{\alpha}{2}, \\ u_2 & \text{for} \quad \dfrac{\alpha}{2} < \phi < \alpha \end{cases}$$

the series may be summed (see the method to problem 60) and gives:

$$u(\rho, \phi) = \frac{u_1 - u_2}{\pi}\left[\arctan \frac{2\rho^{\frac{\pi}{\alpha}} a^{\frac{\pi}{\alpha}} \sin \frac{\pi}{\alpha}\phi}{a^{\frac{2\pi}{\alpha}} - \rho^{\frac{2\pi}{\alpha}}} + \arctan \frac{2\rho^{\frac{2\pi}{\alpha}} a^{\frac{2\pi}{\alpha}} \sin \frac{2\pi}{\alpha}\phi}{a^{\frac{4\pi}{\alpha}} - \rho^{\frac{4\pi}{\alpha}}}\right].$$

Solution. Finding the steady-state temperature reduces to solving the first boundary-value problem for Laplace's equation inside a sector for the boundary conditions

$$u|_{\rho=\alpha} = f(\phi), \quad u = 0 \quad \text{for} \quad \phi = 0 \quad \text{and} \quad \phi = \alpha.$$

Assuming

$$u = R(\rho)\Phi(\phi)$$

and performing a separation of the variables, we obtain:

$$\rho^2 R'' + \rho R' - \lambda R = 0,$$

$$\Phi' + \lambda\Phi = 0,$$

$$\Phi(0) = 0, \quad \Phi(\alpha) = 0.$$

Hence we find:

$$\Phi = A \sin\sqrt{\lambda}\phi + B\cos\sqrt{\lambda}\phi.$$

The conditions for $\phi = 0$ and $\phi = \alpha$ give:

$$B = 0, \quad \sqrt{\lambda} = \frac{\pi n}{\alpha},$$

i.e.

$$\lambda_n = \left(\frac{\pi n}{\alpha}\right)^2.$$

Thus

$$\Phi_n(\phi) = A_n' \sin\frac{\pi n}{\alpha}\phi.$$

The system of functions $\Phi_n = \sin(\pi n\phi/\alpha)$ is orthogonal in the interval $0 < \phi < \alpha$,

$$\int_0^\alpha \sin\frac{\pi n}{\alpha}\phi \sin\frac{\pi m}{\alpha}\phi \, d\phi = 0, \quad m \neq n,$$

and has the norm

$$\sqrt{\int_0^\alpha \sin^2\frac{\pi n}{\alpha}\phi \, d\phi} = \sqrt{\frac{\alpha}{2}},$$

so that the coefficient f_n of expansion of some function $f(\phi)$ as a series in the functions $\Phi_n(\phi)$,

$$f(\phi) = \sum_{n=1}^{\infty} f_n \sin\frac{\pi n}{\alpha}\phi,$$

is determined by the relation

$$f_n = \frac{2}{\alpha}\int_0^\alpha f(\phi)\sin\frac{\pi n}{\alpha}\phi\,d\phi\,.$$

Solving the equation for R and taking into account the bounded nature of the function R, we obtain a particular solution of our problem in the form

$$u_n(\rho,\phi) = A_n\rho^{\frac{\pi n}{\alpha}}\sin\frac{\pi n}{\alpha}\phi\,.$$

We seek the general solution in the form of a series

$$u(\rho,\phi) = \sum_{n=1}^{\infty} A_n\rho^{\frac{\pi n}{\alpha}}\sin\frac{\pi n}{\alpha}\phi\,.$$

Assuming $\rho = a$ and taking into account the condition for $\rho = a$, we obtain:

$$\sum_{n=1}^{\infty} A_n a^{\frac{\pi n}{\alpha}}\sin\frac{\pi n}{\alpha}\phi = f(\phi) = \sum_{n=1}^{\infty} f_n\sin\frac{\pi n}{\alpha}\phi\,,$$

from which it follows that

$$A_n = \frac{f_n}{a^{\frac{\pi n}{\alpha}}},$$

where f_n is the expansion coefficient of the function $f(\phi)$.

75. The temperature at the point (ρ,ϕ) equals

$$u(\rho,\phi) = u_1 + \frac{4(u_2-u_1)}{\pi}\sum_{m=0}^{\infty}\left(\frac{\rho}{a}\right)^{\frac{(2m+1)\pi}{\alpha}}\frac{\sin(2m+1)\frac{\pi}{\alpha}\phi}{2m+1}, \tag{1}$$

or

$$u(\rho,\phi) = \frac{2u_1}{\pi}\arctan\frac{\rho^{\frac{2\pi}{\alpha}}-a^{\frac{2\pi}{\alpha}}}{2a^{\frac{\pi}{\alpha}}\rho^{\frac{\pi}{\alpha}}\sin\frac{\pi\phi}{\alpha}} + \frac{2u_2}{\pi}\arctan\frac{2a^{\frac{\pi}{\alpha}}\rho^{\frac{\pi}{\alpha}}\sin\frac{\pi\phi}{\alpha}}{\rho^{\frac{2\pi}{\alpha}}-a^{\frac{2\pi}{\alpha}}}\,. \tag{2}$$

Method. See the method to problem 60. See also problem 74.

76. The potential of the electrostatic field equals

$$u = u(\rho,\phi) = V_1 + \frac{4(V_2-V_1)}{\pi}\sum_{m=0}^{\infty}\left(\frac{\rho}{a}\right)^{2m+1}\frac{\sin(2m+1)\phi}{2m+1}$$

or

$$u(\rho, \phi) = V_1 + \frac{2}{\pi}(V_2 - V_1) \arctan \frac{a\rho \sin \phi}{\rho^2 - a^2}.$$

The electric field vector equals

$$\boldsymbol{E} = -\operatorname{grad} u.$$

Method. See problems 60 and 75. For $\alpha = \pi$ we have $u_{75} = u_{76}$.

77. $$u(\rho, \phi) = \sum_{n=1}^{\infty} \left(A_n \rho^{\frac{\pi n}{\alpha}} + B_n \rho^{-\frac{\pi n}{\alpha}} \right) \sin \frac{\pi n}{\alpha} \phi, \tag{1}$$

where

$$A_n = \frac{b^{\frac{\pi n}{\alpha}} F_n - a^{\frac{\pi n}{\alpha}} f_n}{b^{2\frac{\pi n}{\alpha}} - a^{2\frac{\pi n}{\alpha}}}, \qquad B_n = \frac{b^{\frac{\pi n}{\alpha}} f_n - a^{\frac{\pi n}{\alpha}} F_n}{b^{2\frac{\pi n}{\alpha}} - a^{2\frac{\pi n}{\alpha}}} a^{\frac{\pi n}{\alpha}} b^{\frac{\pi n}{\alpha}},$$

$$f_n = \frac{2}{\alpha} \int_0^{\alpha} f(\phi) \sin \frac{\pi n}{\alpha} \phi \, d\phi, \qquad F_n = \frac{2}{\alpha} \int_0^{\alpha} F(\phi) \sin \frac{\pi n}{\alpha} \phi \, d\phi.$$

Special cases:
for $a \to 0$

$$B_n = 0, \qquad A_n = F_n \frac{1}{b^{\frac{\pi n}{\alpha}}}$$

and we obtain the solution of problem 74 for a circular sector;
for $b \to \infty$

$$A_n = 0, \qquad B_n = f_n a^{\frac{\pi n}{\alpha}}$$

and we obtain:

$$u(\rho, \phi) = \sum_{n=1}^{\infty} f_n \left(\frac{a}{\rho} \right)^{\frac{\pi n}{\alpha}} \sin \frac{\pi n}{\alpha} \phi \quad \text{in the region} \quad \rho > a, \quad 0 < \phi < \alpha;$$

for $\alpha = \pi$ the solution for a semi-circular ring is obtained.

Solution. It is required to find a harmonic function inside "the ring-shaped sector" $a < \rho < b$, $0 < \phi < \alpha$, satisfying the boundary conditions

$$u = 0 \quad \text{for} \quad \phi = 0, \quad \phi = \alpha, \quad u|_{\rho=a} = f(\phi), \quad u|_{\rho=b} = F(\phi).$$

Using the method of separation of variables, we obtain particular solutions of the form (see problem 74)

$$u_n(\rho, \phi) = R_n(\rho) \sin \frac{\pi n}{\alpha} \phi,$$

where $R_n(\rho)$ is determined from the equation

$$\rho^2 R_n'' + \rho R_n' - \left(\frac{\pi n}{\alpha}\right)^2 R = 0$$

and has the form

$$R_n = A_n \rho^{\frac{\pi n}{\alpha}} + B\rho^{-\frac{\pi n}{\alpha}}.$$

Forming the series

$$u(\rho, \phi) = \sum_{n=1}^{\infty} \left(A_n \rho^{\frac{\pi n}{\alpha}} + B_n \rho^{-\frac{\pi n}{\alpha}}\right) \sin \frac{\pi n}{\alpha}\,\phi$$

and satisfying the boundary conditions for $\rho = a$ and $\rho = b$, we find the coefficients A_n and B_n.

78. Let the x-axis be directed parallel to the conductors half way between them, and let the plane $z = 0$ be perpendicular to the plane, passing through the parallel conductors.

Only the x component of the vector-potential A, satisfying Laplace's equation outside the conductors, differs from zero and equals

$$A_x = \frac{2I\mu}{c} \ln \frac{R_2}{R_1}, \qquad A_y = A_z = 0,$$

where c is the velocity of light in a vacuum, μ is the magnetic permeability of the medium, I is the intensity of the current, flowing across a cross-section of each conductor,

$$R_1 = \sqrt{(y-0\cdot 5a)^2+z^2}, \qquad R_2 = \sqrt{(y+0\cdot 5a)^2+z^2},$$

and a is the distance between conductors.

The components of the induction vector $\boldsymbol{B} = \text{curl}\, \boldsymbol{A}$, determinable by the formulae

$$B_x = \frac{\partial A_z}{\partial y} - \frac{\partial A_y}{\partial z} = 0, \qquad B_y = \frac{\partial A_x}{\partial z}, \qquad B_z = -\frac{\partial A_x}{\partial y},$$

equal

$$B_x = 0, \qquad B_y = \frac{2\mu I z}{c}\left(\frac{1}{R_2^2} - \frac{1}{R_1^2}\right), \qquad B_z = -\frac{2\mu I}{c}\left[\frac{y+0\cdot 5a}{R_2^2} - \frac{y-0\cdot 5a}{R_1^2}\right].$$

Method. Use the relation for the vector-potential of the current filament I

$$A = \frac{\mu}{c} \oint_L \frac{I\,ds}{r},$$

where integration is performed over the contour of the current L. Each of the components A_x, A_y, A_z satisfies Laplace's equation outside L.

The magnetic induction vector is

$$\boldsymbol{B} = \operatorname{curl} \boldsymbol{A} = \frac{\mu I}{c} \oint_L \frac{[d\boldsymbol{s}\boldsymbol{r}]}{r^3}.$$

79. The z component of the vector-potential differs from zero and equals

$$A_z = \frac{4\pi}{c}\left\{\mu a C_0 \ln a + \frac{\mu a}{2}\sum_{n=1}^{\infty}\frac{1}{n}\left(\frac{r}{a}\right)^n (C_n \cos n\phi + D_n \sin n\phi)\right\},$$

where a is the radius of the cylinder, C_0, C_n, D_n are expansion coefficients of the surface current i_z in circular harmonics

$$i_z(a, \phi) = \sum_{n=0}^{\infty} (C_n \cos n\phi + D_n \sin n\phi).$$

Method. The vector-potential at the point $M(r, \phi)$, a distance R from the infinite conductor, carrying a current $I = i_z a d\alpha$, is parallel to it and equal to $\frac{2\mu i_z a d\alpha}{c} \ln R$. Therefore the vector-potential of the whole layer is

$$A_z = \frac{2\mu a}{c}\int_0^{2\pi} i_z \ln R \, d\alpha, \qquad R^2 = a^2 + r^2 - 2ar\cos(\phi - \alpha).$$

Expanding $\ln R$ in a series in powers of r/a, we obtain the necessary expression for A_z.

80. Let the induction vector of the external magnetic field equal $\boldsymbol{B}_0$ and be directed along the x-axis, and the z-axis be directed parallel to the current.

The components of the intensity vector of the magnetic field are determined by the formulae:

inside the cylinder

$$\left.\begin{aligned} H_r^{(1)} &= \frac{2}{\mu_1+\mu_2} B_0 \cos\phi, \\ H_\phi^{(1)} &= \frac{I}{2\pi a^2} r - \frac{2}{\mu_1+\mu_2} B_0 \sin\phi \end{aligned}\right\} \quad \text{for} \quad r < a;$$

outside the cylinder

$$\left.\begin{aligned} H_r^{(2)} &= \left(1 + \frac{\mu_1-\mu_2}{\mu_1+\mu_2}\frac{a^2}{r^2}\right) H_0 \cos\phi, \\ H_\phi^{(2)} &= \frac{I}{2\pi r} - \left(1 - \frac{\mu_1-\mu_2}{\mu_1+\mu_2}\frac{a^2}{r^2}\right) H_0 \sin\phi \end{aligned}\right\} \quad \text{for} \quad r > a.$$

Method. The resultant field is obtained as the sum

$$\boldsymbol{B} = \boldsymbol{B}_0 + \boldsymbol{B}_1,$$

where $\boldsymbol{B}_1$ is the induced field, or

$$\boldsymbol{B} = \operatorname{curl} \boldsymbol{A}, \qquad \boldsymbol{A} = \boldsymbol{A}_0 + \boldsymbol{A}_1,$$

where $\boldsymbol{A}$ is the vector-potential, $\boldsymbol{A}_1$ the vector-potential of the induced field, where $\boldsymbol{A}_0 = \boldsymbol{z}^0 B_0 r \sin\phi$ ($\boldsymbol{z}^0$is a unit vector parallel to the z-axis). On the surface of the cylinder the condition of continuity of the vector potential and the tangential components of the vector H must be fulfilled, so that

$$A^{(1)} = A^{(2)}, \quad \frac{1}{\mu^2}\frac{\partial A^{(2)}}{\partial r} = \frac{1}{\mu_1}\frac{\partial A^{(1)}}{\partial r} \quad \text{for} \quad r = a,$$

$$\Delta A^{(1)} = 0 \quad \text{for} \quad r < a, \qquad \Delta A^{(2)} = 0 \quad \text{for} \quad r > a.$$

81. Only the z-component of the vector-potential differs from zero

$$A_z = \begin{cases} A^{(1)} & \text{for} \quad r < a, \\ A^{(2)} & \text{for} \quad a < r < b, \\ A^{(3)} & \text{for} \quad r > b, \end{cases}$$

where

$$A^{(1)} = \frac{4\mu_1 I}{c}\sum_{n=0}^{\infty}\left[\alpha_{2n+1}r^{2n+1} + \frac{1}{2n+1}\left(\frac{c_0}{r}\right)^{2n+1}\right]\cos(2n+1)\phi,$$

$$A^{(2)} = \frac{4\mu_2 I}{c}\sum_{n=0}^{\infty}\left[\beta_{2n+1}r^{2n+1} + \gamma_{2n+1}\left(\frac{a}{r}\right)^{(2n+1)}\right]\cos(2n+1)\phi,$$

$$A^{(3)} = \frac{4\mu_1 I}{c}\sum_{n=0}^{\infty}\delta_{2n+1}r^{-(2n+1)}\cos(2n+1)\phi,$$

where α_n, β_n, γ_n, δ_n are coefficients, determined from the matching conditions for $r = a$ and $r = b$. In particular

$$\delta_{2n+1} = \frac{4\mu_2}{\mu_1}\frac{c_0{}^{2n+1}}{2n+1}\frac{1}{\left(\dfrac{\mu_1}{\mu_2}+1\right)^2 - \left(\dfrac{\mu_1}{\mu_2}-1\right)^2\left(\dfrac{a}{b}\right)^{4n+2}},$$

where μ_1 is the magnetic permeability of the medium. Therefore,

$$A^{(3)} = \frac{16\mu_2 I}{c}\sum_{n=0}^{\infty}\frac{1}{2n+1}\frac{1}{\left(\dfrac{\mu_1}{\mu_2}+1\right)^2 - \left(\dfrac{\mu_1}{\mu_2}-1\right)^2\left(\dfrac{a}{b}\right)^{4n+2}}\left(\frac{c_0}{r}\right)^{2n+1}\cos(2n+1)\phi.$$

The components of the vector $\boldsymbol{B}$ are given by the formulae

$$B_r = \frac{1}{r}\frac{\partial A}{\partial \phi}, \qquad B_\phi = -\frac{\partial A}{\partial r}.$$

Method. Use the expression for the vector-potential of a two-wire line

$$A_z = \frac{2\mu_1 I}{c} \ln \frac{R_2}{R_1},$$

where R_1 and R_2 are the distances of the point of observation (r, ϕ) from the conductors, and also make use of the following expansions of $\ln R_1$ and $\ln R_2$ for $r > c_0$:

$$\ln R_1 = -\sum_{n=1}^{\infty} \frac{1}{n} \left(\frac{c_0}{r}\right)^n \cos n\phi + \ln r,$$

$$\ln R_2 = -\sum_{n=1}^{\infty} \frac{1}{n} \left(\frac{c_0}{r}\right)^n (-1)^n \cos n\phi + \ln r.$$

82. The intensity vector of the resultant magnetic fiield

$$\boldsymbol{H} = -\operatorname{grad} V$$

V is the scalar potential of the field and equals

$$V = \begin{cases} V_0 + \phi \text{ in the external space} & \text{for } r > a \\ V_0 + \psi \text{ in a section of the sphere} & \text{for } b < r < a \\ V_0 + w \text{ in the internal space} & \text{for } r < b. \end{cases}$$

Here

$$V_0 = -H_0 x = -H_0 r \cos\theta.$$

$$\phi = C_1 \frac{a^3}{r^2} \cos\theta,$$

$$\psi = \left(C_2 r + C_3 \frac{a^2}{r^2}\right) \cos\theta,$$

$$w = C_4 r \cos\theta,$$

where

$$C_1 = \frac{(2\mu+1)(\mu-1)(1-\lambda)}{\Delta} H_0, \qquad C_2 = \frac{2\mu(1-\lambda)-(2+\lambda)}{\Delta} (\mu-1) H_0,$$

$$C_3 = \frac{3(\mu-1)}{\Delta} H_0, \qquad C_4 = \frac{2(1-\lambda)(\mu-1)^2}{\Delta} H_0,$$

$$\lambda = \frac{a^3}{b^3}, \qquad \Delta = 2(1-\lambda)(\mu-1)^2 - 9\mu\lambda.$$

The field intensity inside the sphere equals

$$\boldsymbol{H} = \frac{1}{1+\frac{2}{9}\left[1-\left(\frac{b}{a}\right)^3\right]\left(\sqrt{\mu}-\frac{1}{\sqrt{\mu}}\right)^2} H_0 \quad \text{for} \quad r < b.$$

Thus H is always less than H_0, i.e. screening occurs for both $\mu < 1$ and for $\mu > 1$ (for dia- and paramagnetics).

Method. The coefficients C_i must be determined from the matching conditions, for $r = a$ and for $r = b$.

2. Boundary-value Problems for Strips, Rectangles, Plane Layers and Parallelepipeds

83. If the boundary conditions

$$u|_{y=0} = f(x), \quad u|_{y=b} = \phi(x), u \quad u|_{x=0} = \psi(y), \quad u|_{x=a} = \chi(y),$$

are given along the sides of the rectangle, and they satisfy the conditions $f(0) = \psi(0)$, $f(a) = \chi(0)$, $\chi(b) = \phi(a)$, $\phi(0) = \psi(b)$, then

$$u(x, y) = \sum_{n=1}^{\infty} \left\{ \left[\bar{\phi}_n \frac{\sinh \frac{\pi n}{a} y}{\sinh \frac{\pi n}{a} b} + \bar{f}_n \frac{\sinh \frac{\pi n}{a} (b-y)}{\sinh \frac{\pi n}{a} b} \right] \sin \frac{\pi n}{a} x + \right.$$

$$\left. + \left[\bar{\chi}_n \frac{\sinh \frac{\pi n}{b} x}{\sinh \frac{\pi n}{b} a} + \bar{\psi}_n \frac{\sinh \frac{\pi n}{b} (a-x)}{\sinh \frac{\pi n}{b} a} \right] \sin \frac{\pi n}{b} y \right\} + u_0(x, y),$$

where $\bar{\phi}_n$, $\bar{f}_n$, $\bar{\chi}_n$, $\bar{\psi}_n$ are Fourier coefficients of

$$\bar{\phi}(x) = \phi(x) - u_0(x, b), \quad \bar{f}(x) = f(x) - u_0(x, 0),$$

$$\bar{\psi}(y) = \psi(y) - u_0(0, y), \quad \bar{\chi}(y) = \chi(y) - u_0(a, y),$$

equal to

$$\bar{f}_n = \frac{2}{a} \int_b^a \bar{f}(x) \sin \frac{\pi n}{a} x\, dx, \quad \bar{\phi}_n = \frac{2}{a} \int_0^a \bar{\phi}(x) \sin \frac{\pi n}{a} x\, dx,$$

$$\bar{\psi}_n = \frac{2}{b} \int_0^b \bar{\psi}(y) \sin \frac{\pi n}{b} y\, dy, \quad \bar{\chi}_n = \frac{2}{b} \int_0^b \bar{\chi}(y) \sin \frac{\pi n}{b} y\, dy.$$

The function

$$u_0(x, y) = A + Bx + Cy + Dxy,$$

where

$$A = f(0), \quad B = \frac{f(a) - f(0)}{a}, \quad C = \frac{\psi(b) - \psi(0)}{b},$$

$$D = \frac{[\phi(a) - \phi(0)] - [f(a) - f(0)]}{ab}.$$

Solution. It is required to find the solution of the equation $u_{xx}+u_{yy} = 0$ inside the rectangle $0 < x < a$, $0 < y < b$, satisfying continuous boundary conditions.

Let us represent the unknown function $u(x, y)$ as a sum

$$u(x, y) = u_0(x, y)+v(x, y),$$

where $u_0(x, y)$ is a harmonic function, chosen so that the function $v(x, y)$ reduces to zero at all vertices of the rectangle, and is in other respects completely arbitrary. Assuming

$$u_0(x, y) = A+Bx+Cy+Dxy,$$

we see that this function is harmonic; we choose the coefficients A, B, C and D so that $v(x, y)$ reduces to zero at the vertices.

The harmonic function $v(x, y)$ satisfies the boundary conditions

$$v|_{y=0} = \bar{f}(x), \quad v|_{y=b} = \bar{\varphi}(x), \quad v|_{x=0} = \bar{\psi}(y), \quad v|_{x=a} = \bar{\chi}(y),$$

where the functions $\bar{f}$, $\bar{\varphi}$, $\bar{\psi}$, $\bar{\chi}$ reduce to zero at the vertices of the rectangle.

The function $v(x, y)$ may be represented as the sum of four harmonic functions, each of which takes a given value on one of the sides and is zero on the remaining three sides. We find one such function $v(x, y)$ satisfying the equation

$$v_{1xx}+v_{1yy} = 0$$

for the boundary conditions

$$v_1|_{y=0} = 0, \quad v_1|_{y=b} = \bar{\varphi}(x), \quad v_1|_{x=0,\,a} = 0.$$

Assuming

$$v_1(x, y) = X(x)Y(y)$$

and substituting this expression in the equation, we have:

$$\frac{X''}{X} = -\frac{Y''}{Y} = -\lambda,$$

or

$$Y''-\lambda Y = 0,$$

$$X''+\lambda X = 0.$$

To the latter equation one must add the conditions

$$X(0) = 0, \quad X(a) = 0.$$

Solving this boundary-value problem for $X(x)$, we find the eigenfunctions

$$X_n(x) = \sin\frac{\pi n}{a}x,$$

corresponding to the eigenvalues

$$\lambda_n = \left(\frac{\pi n}{a}\right)^2.$$

From the equation and condition

$$Y''-\lambda Y = 0, \qquad Y(0) = 0,$$

which is a consequence of the condition $v_1(x, 0) = X(x)$, $Y(0) = 0$, we find:

$$Y_n(y) = A_n \sinh \frac{\pi n}{a} y.$$

The solution of the problem is sought in the form of a series

$$v_1(x, y) = \sum_{n=1}^{\infty} A_n \sinh \frac{\pi n}{a} y \sinh \frac{\pi n}{a} x.$$

The condition for $y = b$ gives:

$$A_n = \frac{\bar{\Phi}_n}{\sinh \dfrac{\pi n}{a} b}$$

so that

$$v_1(x,y) = \sum_{n=1}^{\infty} \bar{\Phi}_n \frac{\sinh \dfrac{\pi n}{a} y}{\sinh \dfrac{\pi n}{a} b} \sin \frac{\pi n}{a} x.$$

Now the general solution of our problem is readily written down.

84. (a) If the boundary conditions

$$u|_{y=0} = f(x), \qquad u|_{x=0} = \phi(y), \qquad u_x|_{x=a} = \psi(y), \qquad u_y|_{y=b} = \chi(x),$$

are given, with $f(0) = \phi(0)$, then

$$u(x, y) = f(0) + \sum_{n=0}^{\infty} \left\{ \frac{1}{\cosh \frac{\pi}{2a}(2n+1)b} \left[\bar{f}_n \operatorname{sohh} \frac{\pi}{2a}(2n+1)(b-y) + \right.\right.$$

$$\left. + \frac{2\pi\chi_n}{b(2n+1)} \sinh \frac{\pi}{2a}(2n+1)y \right] \sin \frac{\pi}{2a}(2n+1)x +$$

$$+ \frac{1}{\cosh \frac{\pi}{2b}(2n+1)a} \left[\bar{\Phi}_n \cosh \frac{\pi}{2b}(2n+1)(a-x) + \right.$$

$$\left.\left. + \frac{2\pi\psi_n}{a(2n+1)} \sinh \frac{\pi}{2b}(2n+1)x \right] \sin \frac{\pi}{2b}(2n+1)y \right\},$$

where $\bar{f}_n$, $\bar{\Phi}_n$, ψ_n, χ_n are Fourier coefficients of the corresponding functions, and

$$\bar{f}(x) = f(x) - f(0), \qquad \bar{\Phi}(x) = \phi(x) - f(0), \qquad \bar{\Phi}(0) = \bar{f}(0) = 0.$$

(b) If the boundary conditions

$$u|_{y=0} = f(x), \quad u|_{y=b} = \phi(x), \quad u_x|_{x=0} = \psi(y), \quad u_x|_{x=a} = \chi(y),$$

are given, then the solution of the equation $\Delta u = 0$ has the form

$$u(x, y) = \sum_{n=1}^{\infty} \left\{ \frac{1}{\sinh \frac{\pi n}{a} b} \left[f_n \sinh \frac{\pi n}{a} (b-y) + \phi_n \sinh \frac{\pi n}{a} y \right] \cos \frac{\pi n}{a} x + \right.$$

$$\left. + \frac{b}{\pi n \sinh \frac{\pi n}{b} a} \left[\chi_n \cosh \frac{\pi n}{b} x - \psi_n \cosh \frac{\pi n}{b} (a-x) \right] \sin \frac{\pi n}{b} y \right\}.$$

85. $u(x, y) = \dfrac{2V}{\pi} \arctan \left(\dfrac{\sin \frac{\pi y}{b}}{\sinh \frac{\pi x}{b}} \right)$. (1)

Solution. The method of separation of variables leads to the particular solutions

$$u_n(x, y) = \left(A_n e^{-\frac{\pi n}{b} x} + B_n e^{\frac{\pi n}{b} x} \right) \sin \frac{\pi n}{b} y.$$

From the bounded nature of the solution for $x \to \infty$ it follows that $B_n = 0$. Forming the series $\sum_{n=1}^{\infty} u_n$ and satisfying the boundary conditions for $x = 0$, we obtain:

$$u(x, y) = \frac{4V}{\pi} \sum_{m=0}^{\infty} e^{-\frac{(2m+1)\pi}{b} x} \frac{\sin \frac{(2m+1)\pi}{b} y}{2m+1}. \tag{2}$$

This series is readily summed. In fact,

$$H \equiv \sum_{m=0}^{\infty} e^{-\frac{(2m+1)\pi}{b} x} \frac{\sin \frac{2m+1}{b}}{2m+1} = \operatorname{Im} \left\{ \sum_{m=0}^{\infty} \frac{\left(e^{-\frac{\pi x}{b}} e^{i \frac{\pi y}{b}} \right)^{2m+1}}{2m+1} \right\}.$$

Assuming

$$Z = e^{-\frac{\pi x}{b}} e^{i \frac{\pi y}{b}}$$

and taking into account the fact that

$$\sum_{m=0}^{\infty} \frac{Z^{2m+1}}{2m+1} = \frac{1}{2} \ln \frac{1+Z}{1-Z}, \tag{3}$$

we have:

$$H = \operatorname{Im}\frac{1}{2}\ln\frac{\left(1-e^{-\frac{2\pi x}{b}}\right)+i2e^{-\frac{\pi x}{b}}\sin\frac{\pi y}{b}}{1-2e^{-\frac{\pi x}{b}}\cos\frac{\pi y}{b}+e^{-\frac{2\pi x}{b}}}$$

$$= \frac{1}{2}\arctan\left(\frac{2e^{-\frac{\pi x}{b}}\sin\frac{\pi y}{b}}{1-e^{-\frac{2\pi x}{b}}}\right) = \frac{1}{2}\arctan\left(\frac{\sin\frac{\pi y}{b}}{\sinh\frac{\pi x}{b}}\right),$$

from which the formula follows.

We note that by the limiting transition $b\to\infty$ in (1) the solution of Laplace's equation for the quarter plane is at once obtained for the boundary conditions

$$u|_{x=0} = V, \quad u|_{y=0} = 0.$$

In this case

$$u(x, y) = \frac{2V}{\pi}\arctan\frac{y}{x}.$$

86. $u(x, y) = \dfrac{2V}{\pi}\arctan\left(\dfrac{\sin\frac{\pi y}{b}}{\sinh\frac{\pi x}{b}}\right)+\dfrac{V_0 y}{b}-\dfrac{2V_0}{\pi}\arctan\dfrac{\sin\frac{\pi y}{b}}{e^{\frac{\pi x}{b}}+\cos\frac{\pi y}{b}}.$ (1)

A limiting transition for $b\to\infty$ gives:

$$u(x, y) = \frac{2V}{\pi}\arctan\frac{y}{x} \quad (x \geqslant 0, \quad y \geqslant 0).$$

Method. The unknown potential is conveniently represented as the sum

$$u(x, y) = \frac{V_0 y}{b}+u_1(x, y)+u_2(x, y), \tag{2}$$

where $u_1(x, y)$ is the solution of problem 85, and $u_2(x, y)$ satisfies the equation $\Delta u_2 = 0$ in the region $x > 0$, $0 < y < b$, and the conditions

$$u_2\Big|_{x=0} = -\frac{V_0 y}{b}, \quad u_2\Big|_{y=0,\,b} = 0.$$

The first term in (2) is the potential of the field in a plane condenser. Satisfying these boundary conditions for $u_2(x, y)$ we find the series

$$u_2(x, y) = -\sum_{n=1}^{\infty}(-1)^{n-1}\frac{e^{-\frac{\pi n}{b}x}\sin\frac{\pi n}{b}y}{n},$$

which is summed in a similar manner to series (2) of problem 85.

87. $$u = \frac{4V_0}{\pi}\sum_{m=0}^{\infty}\frac{\sin\frac{\pi(2m+1)}{a}x}{2m+1}\,\frac{\sinh\frac{\pi(2m+1)}{a}y}{\sinh\frac{\pi(2m+1)}{a}b}+$$

$$+\frac{4V}{\pi}\sum_{m=0}^{\infty}\frac{\sin\frac{\pi(2m+1)}{b}y}{2m+1}\,\frac{\sinh\frac{\pi(2m+1)}{b}(a-x)}{\sinh\frac{\pi(2m+1)}{b}a}. \qquad (1)$$

Limiting case

$$u \to u_{86} \quad \text{for} \quad a \to \infty.$$

88. $$u(x,y) = \begin{cases} u_1(x,y) & \text{for} \quad y<h, \\ u_2(x,y) & \text{for} \quad h<y<b, \end{cases}$$

where

$$\left.\begin{aligned} u_1(x,y) &= \sum_{n=1}^{\infty} A_n e^{-\sqrt{\lambda_n}x}\,\overline{Y}_n(y), \qquad \overline{Y}_n(y) = \frac{\sin\sqrt{\lambda_n}y}{\sin\sqrt{\lambda_n}h}, \\ u_2(x,y) &= \sum_{n=1}^{\infty} A_n e^{-\sqrt{\lambda_n}x}\,\overline{\overline{Y}}_n(y), \qquad \overline{\overline{Y}}_n(y) = \frac{\sin\sqrt{\lambda_n}(b-y)}{\sin\sqrt{\lambda_n}(b-h)}, \end{aligned}\right\} \qquad (1)$$

$$A_n = \frac{V}{||Y||^2\sqrt{\lambda_n}}\left[\frac{\varepsilon_1}{\sin\sqrt{\lambda_n}h}+\frac{\varepsilon_2}{\sin\sqrt{\lambda_n}(b-h)}\right],$$

$$||Y||^2 = \frac{\varepsilon_1 h}{2\sin^2\sqrt{\lambda_n}h}+\frac{\varepsilon_2(b-h)}{2\sin^2\sqrt{\lambda_n}(b-h)},$$

λ_n the nth root of the transcendental equation

$$\varepsilon_1\cot\sqrt{\lambda}h+\varepsilon_2\cot\sqrt{\lambda}(b-h) = 0. \qquad (2)$$

Solution. It is required to find the function $u = \begin{cases} u_1 & \text{for} \quad y<h \\ u_2 & \text{for} \quad h<y<b \end{cases}$ continuous in the region $x \geqslant 0$, $0 \leqslant y \leqslant b$, satisfying inside the region $x>0$, $0<y<b$ the equation

$$\text{div}\,(\varepsilon\,\text{grad}\,u) = 0,$$

where

$$\varepsilon = \begin{cases} \varepsilon_1 & \text{for} \quad y<h, \\ \varepsilon_2 & \text{for} \quad h<y<b, \end{cases}$$

and the boundary conditions $u_1 = 0$ for $y = 0$, $u_2 = 0$ for $y = b$, $u_1 = V$ for $x = 0$.

Because ε is piecewise continuous u_1 and u_2 satisfy the equations $\Delta u_1 = 0$, $\Delta u_2 = 0$ and matching conditions on the boundary of discontinuity $y = h$,

$$u_1 = u_2, \quad \varepsilon_1 \frac{\partial u_1}{\partial y} = \varepsilon_2 \frac{\partial u_2}{\partial y} \quad \text{for} \quad y = h.$$

Assuming

$$u(x, y) = X(x)Y(y),$$

in the equation $(\varepsilon u_x)_x + (\varepsilon u_y)_y = 0$ we obtain, after separation of the variables,

$$\frac{\mathrm{d}}{\mathrm{d}y}\left(\varepsilon \frac{\mathrm{d}Y}{\mathrm{d}y}\right) + \varepsilon\lambda Y = 0, \quad X'' - \lambda X = 0, \tag{3}$$

$$Y(0) = 0, \quad Y(b) = 0.$$

Because ε is discontinuous, we have

$$Y(y) = \begin{cases} \overline{Y}(y) & \text{for} \quad y < h, \\ \overline{\overline{Y}}(y) & \text{for} \quad h < y < b \end{cases}$$

satisfying the conditions

$$\overline{Y}'' + \lambda \overline{Y} = 0, \qquad \overline{\overline{Y}}'' + \lambda \overline{\overline{Y}} = 0,$$

$$\overline{Y}(0) = 0, \qquad \overline{\overline{Y}}(b) = 0,$$

$$\overline{Y}(h) = \overline{\overline{Y}}(h), \quad \varepsilon_1 \overline{Y}'(h) = \varepsilon_2 \overline{\overline{Y}}'(h).$$

The solution of this problem is sought in the form

$$\overline{Y}(y) = \frac{\sin\sqrt{\lambda}\, y}{\sin\sqrt{\lambda}\, h}, \qquad \overline{\overline{Y}}(y) = \frac{\sin\sqrt{\lambda}\,(b-y)}{\sin\sqrt{\lambda}\,(b-h)}.$$

Substituting these expressions in the second matching condition, we obtain the characteristic equation for determining λ:

$$\varepsilon_1 \cot\sqrt{\lambda}\, h + \varepsilon_2 \cot\sqrt{\lambda}\,(b-h) = 0. \tag{4}$$

Let $\lambda_1, \lambda_2, \ldots, \lambda_n$ be the roots of this equation, and $Y_1, Y_2, \ldots, Y_n(y)$ the corresponding eigenfunctions. From the general theory of eigenvalue problems† it follows that an infinite number of eigenvalues $\{\lambda_n\}$ exist, with corresponding eigenfunctions $\{Y_n(y)\}$. These form an orthogonal system of functions with weight $\varepsilon(y)$.

$$\int_0^b Y_m(y) Y_n(y) \varepsilon(y)\,\mathrm{d}y = 0 \quad \text{for} \quad m \neq n,$$

† See [7], chapter II, § 3, sect. 9, page 116.

or

$$\varepsilon_1 \int_0^h \bar{Y}_m(y)\bar{Y}_n(y)\,dy + \varepsilon_2 \int_h^b \bar{\bar{Y}}_m(y)\bar{\bar{Y}}_n(y)\,dy = 0 \quad \text{for} \quad m \neq n.$$

For the norm of the eigenfunction $||Y_n||$ we obtain:

$$||Y||^2 = \int_0^b Y_n^2(y)\varepsilon(y)\,dy = \varepsilon_1 \int_0^h \bar{Y}_n^2(y)\,dy + \varepsilon_2 \int_h^b \bar{\bar{Y}}_n^2(y)\,dy.$$

Evaluating these integrals and taking equation (4) for λ_n into account, we find:

$$||Y||^2 = \frac{\varepsilon_1 h}{2\sin^2\sqrt{\lambda_n}\,h} + \frac{\varepsilon_2(b-h)}{2\sin^2\sqrt{\lambda_n}\,(b-h)}. \tag{5}$$

The expansion coefficients of some function $f(y)$ in a series in the eigenfunctions $Y_n(y)$ are determined from the formula

$$f_n = \frac{1}{||Y||^2}\int_0^b f(y)Y_n(y)\varepsilon(y)\,dy.$$

From equation (3) we find:

$$X_n(x) = A_n e^{-\sqrt{\lambda_n}\,x}.$$

The general solution of the problem has the form

$$u(x, y) = \sum_{n=1}^{\infty} A_n e^{-\sqrt{\lambda_n}\,x}\, Y_n(y).$$

To determine A_n we use the condition for $x = 0$

$$V = \sum_{n=1}^{\infty} A_n Y_n(y).$$

Hence

$$A_n = \frac{V}{||Y||^2}\int_0^b Y_n(y)\varepsilon(y)\,dy = \frac{V}{\sqrt{\lambda_n}\,||Y||^2}\left[\frac{\varepsilon_1}{\sin\sqrt{\lambda_n}\,h} + \frac{\varepsilon_2}{\sin\sqrt{\lambda_n}\,(b-h)}\right].$$

89. The potential of the electrostatic field equals

$$u(x, y) = \sum_{n=1}^{\infty} A_n \frac{\sinh\sqrt{\lambda_n}\,(a-x)}{\sinh\sqrt{\lambda_n}\,a}\, Y_n(y), \tag{1}$$

where

$$Y_n(y) = \begin{cases} \bar{Y}_n(y) & \text{for} \quad y < h, \\ \bar{\bar{Y}}_n(y) & \text{for} \quad h < y < b, \end{cases}$$

expressions for $Y_n(y)$, A_n and the square of the modulus are given in the answer to the preceding problem 88, λ_n is the root of the equation

$$\varepsilon_1 \tan\sqrt{\lambda}(b-h)+\varepsilon_2 \tan\sqrt{\lambda}h = 0.$$

Going to the limit for $a \to \infty$ gives the solution of problem 88

$$u_{89}|_{a=\infty} = u_{88},$$

since

$$\lim_{a\to\infty} \frac{\sinh\sqrt{\lambda_n}(a-x)}{\sinh\sqrt{\lambda_n}a} = e^{-\sqrt{\lambda_n}x}.$$

Method. See the solution of problem 88.

90. The electric field intensity $E = -\operatorname{grad} u$, where u is the potential, equals

$$u(x,y) = \begin{cases} u_1(x,y) & \text{for} \quad y<h, \\ u_2(x,y) & \text{for} \quad h<y<b, \end{cases}$$

$$u_1(x,y) = \frac{4V}{\pi}\sum_{k=0}^{\infty} \frac{\varepsilon_2\left[\tanh\dfrac{\pi(2k+1)}{a}h+\tanh\dfrac{\pi(2k+1)}{a}(b-h)\right]}{\varepsilon_1\tanh\dfrac{\pi(2k+1)}{a}(b-h)+\varepsilon_2\tanh\dfrac{\pi(2k+1)}{a}h}\times$$

$$\times\frac{\sinh\dfrac{\pi(2k+1)}{a}y}{(2k+1)\sinh\dfrac{\pi(2k+1)}{a}b}\sin\frac{\pi(2k+1)}{a}x,$$

$$u_2(x,y) = \frac{4V}{\pi}\sum_{k=0}^{\infty}\left\{\sinh\frac{\pi(2k+1)}{a}y+\right.$$

$$\left.+\frac{\left[\varepsilon_2\tanh\dfrac{\pi(2k+1)}{a}h-\varepsilon_1\tanh\dfrac{\pi(2k+1)}{a}(b-h)\right]\sinh\dfrac{\pi(2k+1)}{a}(b-y)}{\varepsilon_1\tanh\dfrac{\pi(2k+1)}{a}(b-h)+\varepsilon_2\tanh\dfrac{\pi(2k+1)}{a}h}\right\}\times$$

$$\times\frac{\sin\dfrac{\pi(2k+1)x}{a}}{(2k+1)\sinh\dfrac{\pi(2k+1)b}{a}}.$$

Method. The solution is sought in the form $u(x,y) = \sum_{n=1}^{\infty} A_n X_n(x) Y_n(y)$, where $X_n(x)$ is the eigenfunction of the boundary-value problem

$$X''+\lambda X = 0, \quad X(0) = X(a) = 0,$$

and

$$Y_n(y) = \begin{cases} \overline{Y}_n(y) & \text{for} \quad y < h, \\ \overline{\overline{Y}}(y) & \text{for} \quad h < y < b \end{cases}$$

is the solution of the problem

$$\overline{Y}_n'' - \lambda_n \overline{Y}_n = 0, \quad \overline{\overline{Y}}_n'' - \lambda_n \overline{\overline{Y}}_n = 0, \quad \overline{Y}_n(0) = 0, \quad \overline{Y}_n(h) = \overline{\overline{Y}}_n(h),$$

$$\varepsilon_1 \overline{Y}_n'(h) = \varepsilon_2 \overline{\overline{Y}}_n'(h).$$

91. $u(x, y) = Q/2abk\,[(y-b)^2-(x-a)^2]+\text{const.}$, where k is the coefficient of heat conduction.

Method. It is required to solve the second boundary-value problem for the equation $u_{xx}+u_{yy} = 0$ inside a rectangle for the boundary conditions

$$ku_y(x, 0) = -\frac{Q}{a}, \quad ku_x(0, y) = -\frac{Q}{b}, \quad u_x(a, y) = 0, \quad u_y(x, b) = 0.$$

In this case the reduction of the problem to the two problems, in each of which zero boundary conditions are taken on three sides, is impossible since then the necessary condition of the solvability of the second boundary-value problem would be violated

$$\oint_C \frac{\partial u}{\partial n}\, ds = 0.$$

Solution. Since the boundary-values do not vary along the sides, it is possible to look for the solution in the form of a harmonic polynomial

$$u(x, y) = A+Bx+Cy+Dxy+E(x^2-y^2).$$

Satisfying the boundary conditions, we find the values of the coefficients and obtain the answer.

The solution may also be found by the method of separation of variables, assuming

$$u(x, y) = \sum_{n=0}^{\infty} u_n(x) Y_n(y),$$

where $Y_n(y) = \cos\sqrt{\lambda_n}y$ is an eigenfunction of the boundary-value problem

$$Y''+\lambda Y = 0, \quad Y'(0) = Y'(b) = 0,$$

corresponding to the eigenvalue $\lambda_n = (\pi n/b)^2$.

We find:

$$u_n(x) = \frac{2}{b}\int_0^b u(x, y) Y_n(y)\, dy \quad \text{for} \quad n > 0.$$

Substituting here

$$Y_n = -\frac{1}{\lambda_n} Y_n''$$

and integrating twice by parts, we obtain:

$$u_n(x) = -\frac{2}{b}\frac{2}{\lambda_n}\left\{[uY_n']_0^b - [u_y Y]_0^b + \int_0^b u_{yy} Y_n \,dy\right\}.$$

Taking into account the boundary conditions and the equation $u_{xx}+u_{yy} = 0$ we have:

$$u_n(x) = -\frac{2Q}{kab\lambda_n}+\frac{1}{\lambda_n}u_n''(x),$$

or

$$u_n''(x)-\lambda_n u_n(x) = \frac{2Q}{abk}.$$

Integrating the boundary conditions for $x = 0$ and $x = a$, we obtain the conditions for $u_n(x)$:

$$u_n'(0) = 0, \quad u_n'(a) = 0.$$

Hence we find:

$$u_n(x) = -\frac{2Q}{kab\lambda_n} \quad \text{for} \quad n > 0.$$

In order to find

$$u_0(x) = \frac{1}{b}\int_0^b u(x, y)\,dy,$$

let us integrate the equation $\Delta u = 0$ by parts, which gives:

$$u_0''(x) = \frac{Q}{abk}, \quad u_0'(0) = \frac{Q}{bk}, \quad u'(a) = 0,$$

from which

$$u_0(x) = -\frac{Q}{2abk}(x-a)^2+\text{const.}$$

Thus we obtain:

$$u(x, y) = -\frac{Q}{2abk}(x-a)^2+\frac{2Q}{ka\pi^2}\sum_{n=1}^{\infty}\frac{\cos\frac{\pi n}{b}y}{n^2}+\text{const.},$$

or

$$u(x, y) = \frac{Q}{2abk}[(y-b)^2-(x-a)^2]+\text{const.},$$

since

$$\frac{4b}{\pi^2}\sum_{n=1}^{\infty}\frac{\cos\frac{\pi n}{b}y}{n^2} = (y-b)^2.$$

92. If the flow q is given on the side $y = b$, then

$$u(x, y) = u_1 - \frac{4qa}{k\pi^2} \sum_{m=0}^{\infty} \frac{\sinh \frac{(2m+1)\pi}{a} y}{\cosh \frac{(2m+1)\pi}{a} b} \frac{\sin \frac{\pi(2m+1)}{a} x}{(2m+1)^2}.$$

Method. The solution is conveniently represented as the sum

$$u(x, y) = u_1 + v(x, y),$$

where $v(x, y)$ is the solution of the equation $\Delta u = 0$, satisfying the boundary conditions

$$v = 0 \quad \text{for} \quad x = 0, \quad x = a, \quad y = 0,$$

$$-kv_y(x, b) = q.$$

93. The solution of the equation $\Delta u = 0$ inside a rectangular parallelepiped for the boundary conditions

$$u|_{x=0} = f_1(y,z), \quad u|_{x=a} = f_2(y,z),$$
$$u|_{y=0} = f_3(x,z), \quad u|_{y=b} = f_4(x,z),$$
$$u|_{z=0} = f_5(x,y), \quad u|_{z=c} = f_6(x,y)$$

has the form

$$u(x, y, z) = u_1(x, y, z) + u_2(x, y, z) + u_3(x, y, z),$$

where

$$u_1(x, y, z) = \sum_{m=1}^{\infty} \sum_{n=1}^{\infty} \frac{(f_2)_{mn} \sinh \nu_{mn}^{(1)} x + (f_1)_{mn} \sinh \nu_{mn}^{(1)} (a-x)}{\sinh \nu_{mn}^{(1)} a} \sin \frac{\pi m}{b} y \sin \frac{\pi n}{c} z,$$

$$\nu_{mn}^{(1)} = \pi \sqrt{\frac{m^2}{b^2} + \frac{n^2}{c^2}},$$

$$f_{imn} = \frac{4}{ab} \int_0^b \int_0^c f_i(y, z) \sin \frac{\pi m}{b} y \sin \frac{\pi n}{c} z \, dy \, dz \quad (i = 1, 2).$$

The functions $u_2(x, y, z)$ and $u_3(x, y, z)$ are defined by similar formulae.

Solution. The unknown function $u(x, y, z)$ may be represented as the sum of three harmonic functions u_1, u_2, u_3 satisfying the boundary conditions

$$u_1|_{x=0} = f_1(y, z), \quad u_1|_{x=a} = f_2(y, z); \quad u_1 = 0 \quad \text{for} \quad y = 0, \quad b; \quad z = 0, \quad c,$$
$$u_2|_{y=0} = f_3(x, z), \quad u_2|_{y=b} = f_4(x, z); \quad u_2 = 0 \quad \text{for} \quad x = 0, \quad a; \quad z = 0, \quad c,$$
$$u_3|_{z=0} = f_5(x, y), \quad u_3|_{z=c} = f_6(x, y); \quad u_3 = 0 \quad \text{for} \quad x = 0, \quad a; \quad y = 0, \quad b.$$

Let us calculate $u_1(x, y, z)$. Assuming $u_1(x, y, z) = X(x)\, v(y, z)$, after separation of the variables, we obtain a boundary value problem for $v(x, y)$

$$v_{yy} + v_{zz} + \lambda v = 0, \quad v = 0 \quad \text{for} \quad y = 0, \ b; \quad z = 0, \ c,$$

with normalized eigenfunctions

$$v_{m,n}(y, z) = \sqrt{\frac{4}{bc}} \sin \frac{\pi m}{b} y \sin \frac{\pi n}{c} z \quad (m, n = 1, 2, \ldots),$$

and the eigenvalues equal to $\lambda_{mn} = \pi^2(m^2/b^2 + n^2/c^2)$. Determining the function $X_{mn}(x)$ from the equation $X'' - \lambda_{mn} X = 0$, we obtain:

$$u_1(x, y, z) = \sum_{m=1}^{\infty} \sum_{n=1}^{\infty} (A_{mn} \sinh \sqrt{\lambda_{mn}}\, x + B_{mn} \cosh \sqrt{\lambda_{mn}}\, x) v_{mn}(y, z).$$

The expansion coefficients A_{mn} and B_{mn} are determined from the conditions for $x = 0$ and $x = a$. Similarly we find the functions $u_2(x, y, z)$ and $u_3(x, y, z)$.

We note that in the solution of the first boundary-value problem for a rectangle we introduced an auxiliary harmonic polynomial, which reduced the values of the boundary functions at the corners to zero. In the problem for the parallelepiped the construction of such a polynomial is considerably more complicated and this was not done. Therefore the series for the solution converges non-uniformly in the neighbourhood of the corners of the parallelepiped.

94. The potential of the electrostatic field equals

$$u(x, y, z) = \frac{16V}{\pi^2} \times$$

$$\times \sum_{m=0}^{\infty} \sum_{n=0}^{\infty} \frac{\sin \frac{(2m+1)\pi}{a} x \sin \frac{2(n+1)\pi}{b} y \sinh \pi \sqrt{\frac{(2m+1)^2}{a^2} + \frac{(2n+1)^2}{b^2}} (c-z)}{(2m+1)(2n+1) \sinh \pi \sqrt{\frac{(2m+1)^2}{a^2} + \frac{(2n+1)^2}{b^2}}\, c}. \tag{1}$$

For $c \to \infty$ we obtain the solution for the semi-infinite tube

$$u(x, y, z) = \frac{16V}{\pi^2} \times$$

$$\times \sum_{m=0}^{\infty} \sum_{n=0}^{\infty} \frac{\sin \frac{(m+1)\pi}{a} x \sin \frac{(2n+1)\pi}{b} y}{(2m+1)(2n+1)} e^{-\pi \sqrt{\frac{(2m+1)^2}{a^2} + \frac{(2n+1)^2}{b^2}}\, z}. \tag{2}$$

95. $u(x, y, z) = V -$

$$-\frac{16V}{\pi^2} \sum_{m=0}^{\infty} \sum_{n=0}^{\infty} \frac{\cosh \nu_{mn}\left(\frac{c}{2} - z\right)}{\cosh \frac{\nu_{mn} c}{2}} \; \frac{\sin \frac{(2m+1)\pi}{a} x \sin \frac{(2n+1)\pi}{b} y}{(2m+1)(2n+1)},$$

where

$$\nu_{mn} = \pi \sqrt{\frac{(2m+1)^2}{a^2} + \frac{(2n+1)^2}{b^2}}\,.$$

A limiting transition for $c \to \infty$ gives:

$$u(x, y, z) = V - u_{94b}\,,$$

where u_{94b} is given by formula (2) of problem 94.

The solution may also be represented as the sum

$$u(x, y, z) = \frac{16V}{\pi^2} \sum_{m=0}^{\infty} \sum_{n=0}^{\infty} \frac{\cosh \bar{\mu}_{mn}\left(\frac{a}{2} - x\right)}{\cosh \bar{\mu}_{mn}\frac{a}{2}} \times$$

$$\times \frac{\sin \frac{\pi(2m+1)}{b} y \sin \frac{\pi(2n+1)}{c} z}{(2m+1)(2n+1)} + \frac{16V}{\pi^2} \sum_{m=0}^{\infty} \sum_{n=0}^{\infty} \frac{\cosh \bar{\bar{\mu}}_{mn}\left(\frac{b}{2} - y\right)}{\cosh \bar{\bar{\mu}}_{mn}\frac{b}{2}} \times$$

$$\times \frac{\sin \frac{\pi(2m+1)x}{a} \sin \frac{\pi(2n+1)}{c} z}{(2m+1)(2n+1)}\,,$$

$$\bar{\mu}_{mn} = \pi \sqrt{\frac{(2m+1)^2}{b^2} + \frac{(2n+1)^2}{c^2}}\,, \quad \bar{\bar{\mu}}_{mn} = \pi \sqrt{\frac{(2m+1)^2}{a^2} + \frac{(2n+1)^2}{c^2}}\,.$$

3. Problems Requiring the Application of Cylindrical Functions

96. $u(\rho, \phi, z)$

$$= \sum_{n=0}^{\infty} \sum_{m=1}^{\infty} (A_{m,n} \cos n\phi + B_{m,n} \sin n\phi) J_n\left(\frac{\mu_m^{(n)}}{a}\rho\right) \frac{\sinh \frac{\mu_m^{(n)}}{a}(l-z)}{\sinh \frac{\mu_m^{(n)}}{a} l} +$$

$$+ \sum_{n=0}^{\infty} \sum_{m=1}^{\infty} (C_{m,n} \cos n\phi + D_{m,n} \sin n\phi) J_n\left(\frac{\mu_m^{(n)}}{a}\rho\right) \frac{\sinh \frac{\mu_m^{(n)}}{a} z}{\sinh \frac{\mu_m^{(n)}}{a} l}\,, \tag{1}$$

where $\mu_m^{(n)}$ is the mth root of the equation

$$J_n(\mu) = 0,$$

$A_{m,n}$, $B_{m,n}$, $C_{m,n}$, $D_{m,n}$ are expansion coefficients of the functions $f(\rho, \phi)$ and $F(\rho, \phi)$, equal to

$$A_{m,n} = \frac{2}{a^2\pi\varepsilon_n[J_n'(\mu_m^{(n)})]^2}\int_0^{2\pi}\int_0^a f(\rho,\phi)\cos n\phi J_n\left(\frac{\mu_m^{(n)}}{a}\rho\right)\rho\,\mathrm{d}\rho\,\mathrm{d}\phi,$$

$$\varepsilon_n = \begin{Bmatrix} 2, & n = 0, \\ 1, & n \neq 0, \end{Bmatrix} \tag{2}$$

$$B_{m,n} = \frac{2}{a^2\pi[J_n'(\mu_m^{(n)})]^2}\int_0^{2\pi}\int_0^a f(\rho,\phi)\sin n\phi J_n\left(\frac{\mu_m^{(n)}}{a}\rho\right)\rho\,\mathrm{d}\rho\,\mathrm{d}\phi, \text{ etc.}$$

Solution. The problem is solved by the method of separation of variables. Substituting the expressions

$$u(\rho, \phi, z) = V(\rho, \phi)Z(z)$$

in the equation

$$\frac{1}{\rho}\frac{\partial}{\partial\rho}\left(\rho\frac{\partial u}{\partial\rho}\right)+\frac{1}{\rho^2}\frac{\partial^2 u}{\partial\phi^2}+\frac{\partial^2 u}{\partial z^2} = 0 \tag{3}$$

and separating the variables, we obtain for $V(\rho, \phi)$ the equation

$$\frac{1}{\rho}\frac{\partial}{\partial\rho}\left(\rho\frac{\partial V}{\partial\rho}\right)+\frac{1}{\rho^2}\frac{\partial^2 V}{\partial\phi^2}+\lambda V = 0 \tag{4}$$

with the boundary condition

$$V(a, \phi) = 0$$

and for $Z(z)$ the equation

$$Z''-\lambda Z = 0. \tag{5}$$

Assuming further

$$V(\rho, \phi) = R(\rho)\Phi(\phi),$$

we have:

$$\frac{1}{\rho}\frac{\mathrm{d}}{\mathrm{d}\rho}\left(\rho\frac{\mathrm{d}R}{\mathrm{d}\rho}\right)+\left(\lambda-\frac{\nu^2}{\rho^2}\right)R = 0, \tag{6}$$

$$\Phi''+\nu^2\Phi = 0, \tag{7}$$

where ν is a separation constant. From the conditions of periodicity of the function Φ with respect to the angle ϕ we find $\nu^2 = n^2$.

For $R(\rho)$ we have Bessel's equation

$$\frac{1}{\rho}\frac{\mathrm{d}}{\mathrm{d}\rho}\left(\rho\frac{\mathrm{d}R}{\mathrm{d}\rho}\right)+\left(\lambda-\frac{n^2}{\rho^2}\right)R = 0$$

with the boundary condition

$$R(a) = 0$$

and the natural condition of regularity at zero†

$$|R(0)| < \infty.$$

Hence we find:

$$R(\rho) = J_n(\sqrt{\lambda}\,\rho).$$

The boundary condition for $\rho = a$ gives:

$$J_n(\mu) = 0, \quad \text{where} \quad \mu = \sqrt{\lambda}\,a.$$

Let us denote the roots of this equation by $\mu_1^{(n)}, \mu_2^{(n)}, \ldots, \mu_m^{(n)}$. Thus the boundary-value problem for $V(\rho, \phi)$ has eigenvalues $\lambda_{mn} = (\mu_m^{(n)}/a)^2$, which correspond to the eigenfunctions

$$\overline{V}_{n,m} = J_n\left(\frac{\mu_m^{(n)}}{a}\rho\right)\cos n\phi, \quad \overline{\overline{V}}_{n,m} = J_n\left(\frac{\mu_m^{(n)}}{a}\rho\right)\sin n\phi,$$

forming two orthogonal systems of functions, for which

$$||\overline{V}_{n,m}||^2 = \frac{a^2}{2}[J_n'(\mu_m^{(n)})]^2\pi\varepsilon_n, \quad ||\overline{\overline{V}}_{n,m}||^2 = \frac{a^2}{2}[J_n'(\mu_m^{(n)})]^2\pi,$$

where

$$\varepsilon_n = \begin{cases} 2 & \text{for} \quad n = 0, \\ 1 & \text{for} \quad n \neq 0. \end{cases}$$

The general solution of our problem is represented in the form of the series

$$u(\rho, \phi, z) = \sum_{m=1}^{\infty}\sum_{n=0}^{\infty} A_{n,m}\overline{V}_{m,n}(\rho, \phi) + B_{n,m}\overline{\overline{V}}_{m,n}(\rho, \phi) Z_{n,m}(z),$$

where $Z_{n,m}(z)$ is a solution of equation (5).

Since the unknown function $u(\rho, \phi, z)$ may be represented as the sum

$$u(\rho, \phi, z) = u_1(\rho, \phi, z) + u_2(\rho, \phi, z),$$

where $u_1(\rho, \phi, z)$ and $u_2(\rho, \phi, z)$ are harmonic functions satisfying the condition

$$u_1|_{\rho=a} = 0, \quad u_1|_{z=0} = f(\rho, \phi), \quad u_1|_{z=l} = 0,$$
$$u_2|_{\rho=a} = 0, \quad u_2|_{z=0} = 0, \quad u_2|_{z=l} = F(\rho, \phi),$$

then it is sufficient to consider the function $u_1(\rho, \phi, z)$. In this case

$$Z_{m,n}(z) = \sinh\frac{\mu_m^{(n)}}{a}(l - z).$$

Satisfying the boundary condition for $z = 0$

$$\sum_{m=1}^{\infty}\sum_{n=0}^{\infty}(A_{m,n}\overline{V}_{m,n} + B_{m,n}\overline{\overline{V}}_{m,n})\sinh\frac{\mu_m^{(n)}}{a}l = f(\rho, \phi),$$

† See [7], appendix I, page 637.

we find:

$$A_{m,n} = \frac{\bar{f}_{m,n}}{\sinh \frac{\mu_m^{(n)}}{a} l}; \qquad B_{m,n} = \frac{\bar{\bar{f}}_{m,n}}{\sinh \frac{\mu_m^{(n)} l}{a}},$$

where

$$\bar{f}_{m,n} = \frac{1}{\|\bar{V}_{m,n}\|^2} \int_0^{2\pi}\int_0^a f(\rho,\phi)\bar{V}_{m,n}(\rho,\phi)\rho\, d\rho\, d\phi,$$

$$\bar{\bar{f}}_{m,n} = \frac{1}{\|\bar{\bar{V}}_{m,n}\|^2} \int_0^{2\pi}\int_0^a f(\rho,\phi)\bar{\bar{V}}_{m,n}(\rho,\phi)\rho\, d\rho\, d\phi.$$

97. $$u = u(\rho, z) = \sum_{m=1}^{\infty} \frac{A_m \sinh \frac{\mu_m^{(0)}(l-z)}{a} + B_m \sinh \frac{\mu_m^{(0)} z}{a}}{\sinh \frac{\mu_m^{(0)}}{a} l} J_0\left(\frac{\mu_m^{(0)}}{a}\rho\right),$$

where $\mu_m^{(0)}$ is the mth root of the equation

$$J_0(\mu) = 0,$$

and the coefficients A_m and B_m are given by the formulae

$$A_m = \frac{2}{a^2 J_1^2(\mu_m^{(0)})} \int_0^a f(\rho) J_0 \frac{(\mu_m^{(0)}\rho)}{a} \rho\, d\rho,$$

$$B_m = \frac{2}{a^2 J_1^2(\mu_m^{(0)})} \int_0^a F(\rho) J_0\left(\frac{\mu_m^{(0)}}{a}\rho\right) \rho\, d\rho.$$

98. $$u = u(\rho, z) = \sum_{n=1}^{\infty} f_n \frac{I_0\left(\frac{\pi n}{l}\rho\right)}{I_0\left(\frac{\pi n}{l}a\right)} \sin \frac{\pi n}{l} z,$$

where $f_n = \frac{2}{l}\int_0^b f(z) \sin \frac{\pi n}{l} z\, dz$ are Fourier coefficients, $I_0(x) = J_0(ix)$ is a Bessel function of zero order of imaginary argument, satisfying the equation

$$I_0'' x + \frac{1}{x} I_0'(x) - I_0(x) = 0.$$

Special cases:

(a) $f_n = \frac{2f_0}{\pi n}[1-(-1)^n]$,

$$u = \frac{4f_0}{\pi}\sum_{m=0}^{\infty}\frac{I_0\left(\frac{\pi(2m+1)}{l}\rho\right)}{I_0\left(\frac{\pi(2m+1)}{l}a\right)}\frac{\sin\frac{\pi(2m+1)z}{l}}{2m+1};$$

(b) $f_n = \frac{4Al}{(\pi n)^3}[1-(-1)^n]$,

$$u = \frac{8Al}{\pi^3}\sum_{m=0}^{\infty}\frac{I_0\left(\frac{\pi(2m+1)}{l}\rho\right)}{I_0\left(\frac{\pi(2m+1)}{l}a\right)}\frac{\sin\frac{\pi(2m+1)}{l}z}{(2m+1)^3}.$$

99 The solution of the general first interior boundary-value problem for a finite cylinder $0 \leqslant \rho \leqslant a$, $0 \leqslant z \leqslant l$

$$\Delta u = \frac{1}{\rho}\frac{\partial}{\partial\rho}\left(\rho\frac{\partial u}{\partial\rho}\right)+\frac{1}{\rho^2}\frac{\partial^2 u}{\partial\phi^2}+\frac{\partial^2 u}{\partial z^2} = 0,$$

$$u|_{\rho=a} = f(\phi, z)$$

$$u|_{z=0} = \psi_1(\rho, \phi), \qquad u|_{z=l} = \psi_2(\rho, \phi)$$

is represented as the sum of three solutions

$$u = u_1+u_2+u_3,$$

where u_1, u_2, u_3 are solutions of the boundary-value problems

$$\begin{array}{llll}
\Delta u_1 = 0, & u_1|_{\rho=a} = f(\phi, z), & u_1|_{z=0} = u_1|_{z=l} = 0, & \\
\Delta u_2 = 0, & u_2|_{\rho=a} = 0, & u_2|_{z=0} = \psi_1(\rho, \phi), & u_2|_{z=l} = 0, \\
\Delta u_3 = 0, & u_3|_{\rho=a} = 0, & u_3|_{z=0} = 0, & u_3|_{z=l} = \psi_2(\rho, \phi),
\end{array}$$

determinable by means of the following series:

$$u_1(\rho, \phi, z) = \sum_{n=0}^{\infty}\sum_{m=1}^{\infty}(A_{n,m}^{(1)}\cos n\phi+B_{n,m}^{(1)}\sin n\phi)\frac{I_n\left(\frac{\pi m}{l}\rho\right)}{I_n\left(\frac{\pi m}{l}a\right)}\sin\frac{\pi m}{l}z,$$

where

$$A_{n,m}^{(1)} = \frac{2}{\pi\varepsilon_n l}\int_0^{2\pi}\int_0^l f(\phi, z)\cos n\phi\sin\frac{\pi m}{l}z\,d\phi\,dz,$$

$$\varepsilon_n = \begin{cases} 2 & \text{for} \quad n = 0, \\ 1 & \text{for} \quad n \neq 0, \end{cases}$$

$$B^{(1)}_{n,m} = \frac{2}{\pi l}\int_0^{2\pi}\int_0^l f(\phi, z)\sin n\phi \sin\frac{\pi m}{l} z\,d\phi\,dz,$$

$$u_2(\rho, \phi, z) = \sum_{n=0}^{\infty}\sum_{m=1}^{\infty}(A^{(2)}_{n,m}\cos n\phi + B^{(2)}_{n,m}\sin n\phi)\frac{\sinh\dfrac{\mu^{(n)}_m(l-z)}{a}}{\sinh\dfrac{\mu^{(n)}_m}{a} l} J_n\left(\frac{\mu^{(n)}_m}{a}\rho\right),$$

and

$$A^{(2)}_{n,m} = \frac{2}{\pi\varepsilon_n a^2[J_n'(\mu^{(n)}_m)]^2}\int_0^{2\pi}\int_0^a \psi_1(\phi, \rho)\cos n\phi J_n\left(\frac{\mu^{(n)}_m}{a}\rho\right)\rho\,d\rho\,d\phi,$$

$$B^{(2)}_{n,m} = \frac{2}{\pi a^2[J_n'(\mu^{(n)}_m)]^2}\int_0^{2\pi}\int_0^a \psi_1(\phi, \rho)\sin n\phi J_n\left(\frac{\mu^{(n)}_m}{a}\rho\right)\rho\,d\rho\,d\phi.$$

$$u_3(\rho, \phi, z) = u_2(\rho, \phi, l-z),$$

if in the expression for $A^{(2)}_{n,m}$ and $B^{(2)}_{n,m}$, $\psi_1(\phi, \rho)$ is replaced by $\psi_2(\phi, \rho)$.

Method. One must look for particular solutions for u_1 in the form

$$u_1(\rho, \phi, z) = V(\rho, \phi)Z(z).$$

For $Z(z)$ one derives the boundary-value problem

$$Z''+\lambda Z = 0, \quad Z(0) = Z(l) = 0, \quad \lambda_m = \left(\frac{\pi m}{l}\right)^2, \quad Z_m = \sin\frac{\pi m}{l} z,$$

and for $V(\rho, \phi)$ the equation

$$\frac{1}{\rho}\frac{\partial}{\partial\rho}\left(\rho\frac{\partial V}{\partial\rho}\right)+\frac{1}{\rho^2}\frac{\partial^2 V}{\partial\phi^2}-\left(\frac{\pi m}{l}\right)^2 V = 0,$$

from which we find:

$$V_{m,n}(\rho, \phi) = I_n\left(\frac{\pi m}{l}\rho\right)\begin{cases}\cos n\phi,\\ \sin n\phi.\end{cases}$$

Assuming

$$u_2(\rho, \phi, z) = V(\rho, \phi)Z(z),$$

we have:

$$\frac{1}{\rho}\frac{\partial}{\partial\rho}\left(\rho\frac{\partial V}{\partial\rho}\right)+\frac{1}{\rho^2}\frac{\partial^2 V}{\partial\phi^2}+\lambda V = 0, \quad V(a, \phi) = 0,$$

so that

$$\lambda_{m,n} = \left(\frac{\mu^{(n)}_m}{a}\right)^2, \quad V_{m,n} = J_n\left(\frac{\mu^{(n)}_m}{a}\rho\right)\begin{cases}\cos n\phi,\\ \sin n\phi\end{cases}$$

and

$$Z'' - \left(\frac{\mu_m^{(n)}}{a}\right)^2 Z = 0, \qquad Z(l) = 0,$$

from which we find:

$$Z_{m,n} = C_{m,n} \sinh \frac{\mu_m^{(n)}}{a}(l-z).$$

100. The potential of the electrostatic field inside the cylindrical tank equals

$$u(\rho, z) = \frac{4V_0}{\pi} \sum_{k=0}^{\infty} \frac{I_0\left[\frac{(2k+1)\pi}{l}\rho\right]}{I_0\left[\frac{(2k+1)\pi}{l}a\right]} \frac{\sin\frac{(2k+1)\pi}{l}z}{2k+1}.$$

The field on the axis of the cylinder

$$E_z(0, z) = -\left(\frac{\partial u}{\partial z}\right)_{\rho=0} = -\frac{4V_0}{l} \sum_{k=0}^{\infty} \frac{\cos\frac{(2k+1)\pi}{l}z}{I_0\left(\frac{(2k+1)\pi}{l}a\right)},$$

$$E_z(0, 0) = -\frac{4V_0}{l} \sum_{k=0}^{\infty} \frac{1}{I_0\left(\frac{(2k+1)\pi a}{l}\right)}$$

(see the solution of problem 98).

In the limiting case for $l \to \infty$ we have:

$$u(\rho, z) = \frac{2V_0}{\pi} \int_0^{\infty} \frac{I_0(\xi, \rho)}{I_0(\xi a)} \frac{\sin \xi z}{\xi} d\xi.$$

The field on the axis of the cylinder ($\rho = 0$) equals

$$E_z = -\left(\frac{\partial u}{\partial z}\right)_{\rho=0} = -\frac{2V_0}{\pi a} \int_0^{\infty} \frac{\cos\frac{z}{a} s\, ds}{I_0(s)}.$$

In particular,

$$E_z(0, 0) = -\frac{2V_0}{\pi a} \int_0^{\infty} \frac{ds}{I_0(s)}.$$

101. $$u(\rho, z) = 2V_0 \sum_{m=1}^{\infty} \frac{\sinh \frac{\mu_m^{(0)}}{a}(l-z)}{\sinh \frac{\mu_m^{(0)}}{a} l} \frac{J_0\left(\frac{\mu_m^{(0)}\rho}{a}\right)}{\mu_m^{(0)} J_1(\mu_m^{(0)})},$$

where $\mu_m^{(0)}$ is a root of the equation $J_0(\mu_m^{(0)}) = 0$.

Noting that

$$\lim_{l\to\infty} \frac{\sinh \frac{\mu_m^{(0)}}{a}(l-z)}{\sinh \frac{\mu_m^{(0)}}{a} l} = e^{-\frac{\mu_m^{(0)}}{a} z},$$

we obtain the solution of the problem for a semi-infinite cylinder

$$u(\rho, z) = 2V_0 \sum_{m=1}^{\infty} \frac{e^{-\frac{\mu_m^{(0)}}{a} z}}{\mu_m^{(0)} J_1(\mu_m^{(0)})} J_0\left(\frac{\mu_m^{(0)}}{a}\rho\right).$$

102. The solution of the problem

$$\Delta u = 0 \quad (\rho < a,\ 0 < z < \infty), \quad u|_{\rho=a} = 0, \quad u|_{z=0} = V_0$$

has the form

$$u_{101}(\rho, z) = 2V_0 \sum_{m=1}^{\infty} \frac{e^{-\frac{\mu_m^{(0)}}{a} z}}{\mu_m^{(0)} J_1(\mu_m^{(0)})} J_0\left(\frac{\mu_m^{(0)}}{a}\rho\right),$$

and the solution of the problem

$$\Delta u = 0 \quad (\rho < a,\ 0 < z < \infty), \quad u|_{\rho=a} = V_0, \quad u|_{z=0} = 0$$

is given by the relation

$$u(\rho, z) = V_0 - u_{101}(\rho, z).$$

103. The distribution of temperature inside the cylinder is given by the formula

$$u(\rho, z) = \sum_{m=1}^{\infty} A_m \frac{\sinh \frac{\mu_m}{a}(l-z)}{\cosh \frac{\mu_m}{a} l} J_0\left(\frac{\mu_m}{a}\rho\right),$$

where

$$A_m = \frac{2aq}{k\mu_m^2 J_1(\mu_m)},$$

μ_m is a root of the equation

$$J_0(\mu) = 0,$$

and k is the coefficient of heat conduction.

Method. The problem reduces to solving Laplace's equation $\Delta u = 0$ for the boundary conditions

$$-k\left.\frac{\partial u}{\partial z}\right|_{z=0} = q, \quad u|_{\rho=a} = 0, \quad u|_{z=l} = 0.$$

104. $$u(\rho, z) = \sum_{m=1}^{\infty} A_m \frac{\sinh \frac{\nu_m(l-z)}{a}}{\cosh \frac{\nu_m}{a} l} J_0\left(\frac{\nu_m}{a}\rho\right), \tag{1}$$

$$A_m = \frac{2h^2a^3q}{k(a^2h^2+\nu_m^2)\nu_m^2 J_1(\nu_m)}, \tag{2}$$

where ν_m is the mth root of the transcendental equation

$$J_1(\nu) = \frac{ah}{\nu} J_0(\nu), \tag{3}$$

and h is the coefficient of heat exchange. Passing to a limit as $h \to \infty$, we obtain the solution of problem 103.

Method. It is required to solve the boundary-value problem

$$\Delta u = 0 \quad (0 < \rho < a,\ 0 < z < l),$$

$$-k \frac{\partial u}{\partial z}\bigg|_{z=0} = q, \quad \frac{\partial u}{\partial \rho}\bigg|_{\rho=a} + hu\bigg|_{\rho=a} = 0, \quad u\bigg|_{z=l} = 0.$$

105. (a) The solution of problem 103 for a semi-infinite cylinder ($l = \infty$) has the form

$$u(\rho, z) = \sum_{m=1}^{\infty} A_m e^{-\frac{\mu_m}{a} z} J_0\left(\frac{\mu_m}{a}\rho\right), \tag{1}$$

where μ_m is a root of the equation $J_0(\mu) = 0$,

$$A_m = \frac{2aq}{k\mu_m^2 J_1(\mu_m)}. \tag{2}$$

(b) The solution of problem 104 for $l = \infty$ is

$$u(\rho, z) = \sum_{m=1}^{\infty} A_m e^{-\frac{\nu_m z}{a}} J_0\left(\frac{\nu_m}{a}\rho\right), \tag{3}$$

where ν_m is a root of equation (3) of problem 104, and A_m is given by formula (2) of problem 104.

Method. The solution is sought in the form

$$u(\rho, z) = \sum_{m=1}^{\infty} Z_m(z) R_m(\rho),$$

where $R_m(\rho)$ agrees with the corresponding functions of problems 103 and 104, $Z_m(z)$ is determined from the equation $Z'' - \lambda_m Z = 0$ and the condition $Z(\infty) = 0$, so that $Z_m = A_m e^{-\sqrt{\lambda_m} z}$.

106. The intensity of the electrostatic field is $\boldsymbol{E} = -\text{grad } u$, where u is the potential given by

$$u(\rho, z) = \frac{4V_0}{\pi} \sum_{m=0}^{\infty} 1 \times$$

$$\times \frac{I_0\left[\frac{\pi(2m+1)}{l}\rho\right] K_0\left[\frac{\pi(2m+1)}{l}a\right] - I_0\left[\frac{\pi(2m+1)}{l}a\right] K_0\left[\frac{\pi(2m+1)}{l}\rho\right]}{I_0\left[\frac{\pi(2m+1)}{l}b\right] K_0\left[\frac{\pi(2m+1)}{l}a\right] - I_0\left[\frac{\pi(2m+1)}{l}a\right] K_0\left[\frac{\pi(2m+1)}{l}b\right]} \times$$

$$\times \frac{\sin \frac{\pi(2m+1)}{l} z}{2m+1}. \tag{1}$$

Here I_0 and K_0 are Bessel functions of zero order of imaginary argument of the first and second kind respectively.

Limiting cases:

(1) if $l = \infty$, then

$$u(\rho, z) = \frac{2V_0}{\pi} \int_0^{\infty} \frac{I_0(\rho s)K_0(as) - I_0(as)K_0(\rho s)}{I_0(bs)K_0(as) - I_0(as)K_0(bs)} \frac{\sin sz}{s} \, ds \tag{2}$$

the potential inside a semi-infinite tube;

(2) for $a = 0$ we have:

$$u(\rho, z) = \frac{4V_0}{\pi} \sum_{m=0}^{\infty} \frac{I_0\left[\frac{\pi(2m+1)}{l}\rho\right]}{I_0\left[\frac{\pi(2m+1)}{l}b\right]} \frac{\sin \frac{\pi(2m+1)}{l} z}{2m+1} = u_{100} \tag{3}$$

the potential inside a cylindrical tank.

Method. The solution is sought in the form $u = \sum R(\rho)Z(z)$, for $R(\rho)$ the problem

$$R'' + \frac{1}{\rho} R' - \lambda R = 0, \qquad R(a) = 0$$

is obtained. The general solution of this equation has the form

$$R(\rho) = AI_0(\sqrt{\lambda}\,\rho) + BK_0(\sqrt{\lambda}\,\rho).$$

The condition $R(a) = 0$ gives $B = -AI_0(\sqrt{\lambda a})/K_0(\sqrt{\lambda a})$. For $Z(z)$ we have $Z'' + \lambda Z = 0$, $Z(0) = Z(l) = 0$, $\lambda_n = (\pi n/l)^2$, $Z_n(z) = \sin(\pi nz/l)$.

For the limiting transition $l \to \infty$ let us introduce the variable $s_m = \pi(2m+1)/l$, so that $\Delta s = 2\pi/l$ and our series is transformed into an integral.

107. The temperature at the point (ρ, z) inside the toroid equals

$$u(\rho, z)=u_0+(u_1-u_0)v(\rho, z),$$

where $v(\rho, z) = v_{106}(\rho, z)+v_{106}(\rho, l-z)$, $v_{106}(\rho,z)$ is the solution of problem 106 for $V_0 = 1$.

108. If one denotes by $u(\rho, \phi, z)$ the steady-state temperature at the point (ρ, ϕ, z), then

(1) $u(\rho, \phi, z) = u_0 = \text{const.}$,

(2) $u(\rho, \phi, z) = u(z) = u_1 \dfrac{z}{l}$.

109. $$u(\rho, z) = \pi u_0 \sum_{m=1}^{\infty} \frac{J_0(\sqrt{\lambda_m} b)}{\lambda_m[J_0(\sqrt{\lambda_m} a)+J_1(\sqrt{\lambda_m} b)]} \frac{\sinh \sqrt{\lambda_m}(l-z)}{\sinh \sqrt{\lambda_m}\, l} R_m(\rho),$$

where

$$R_m(\rho) = J_0(\sqrt{\lambda_m}\rho)N_0(\sqrt{\lambda_m}a)-J_0(\sqrt{\lambda_m}a)N_0(\sqrt{\lambda_m}\rho),$$

and λ_m is the mth root of the equation

$$\frac{J_0(\sqrt{\lambda_m}a)}{N_0(\sqrt{\lambda_m}a)} = \frac{J_0(\sqrt{\lambda_m}b)}{N_0(\sqrt{\lambda_m}b)}.$$

Solution. The unknown solution of the boundary-value problem

$$\Delta u = \frac{1}{\rho}\frac{\partial}{\partial\rho}\left(\rho\frac{\partial u}{\partial\rho}\right)+\frac{\partial^2 u}{\partial z^2} = 0, \quad u|_{\rho=a} = u|_{\rho=b} = 0, \quad u|_{z=b} = 0, \quad u|_{z=0} = u_0$$

is represented in the form $u(\rho, z) = \sum R(\rho)Z(z)$, where $R(\rho)$ is determined from Bessel's equation

$$\frac{1}{\rho}\frac{\mathrm{d}}{\mathrm{d}\rho}\left(\rho\frac{\mathrm{d}R}{\mathrm{d}\rho}\right)+\lambda R = 0 \tag{1}$$

with boundary conditions $R(a) = 0$, $R(b) = 0$, and $Z(z)$ satisfies the equation $Z''-\lambda Z = 0$ and the condition

$$Z(l) = 0. \tag{2}$$

From (1) we find:

$$R(\rho) = CJ_0(\sqrt{\lambda}\,\rho)+DN_0(\sqrt{\lambda}\,\rho).$$

Using the conditions for $\rho = a$ and $\rho = b$, we have:

$$R_m(\rho) = J_0(\sqrt{\lambda_m}\rho)N_0(\sqrt{\lambda_m}a)-J_0(\sqrt{\lambda_m}a)N_0(\sqrt{\lambda_m}\rho),$$

where λ_m is determined from the equation

$$\frac{J_0(\sqrt{\lambda_m}a)}{N_0(\sqrt{\lambda_m}a)} = \frac{J_0(\sqrt{\lambda_m}b)}{N_0(\sqrt{\lambda_m}b)}.$$

Evaluating the norm of the eigenfunction $R_m(\rho)$

$$||R_m||^2 = \int_b^a \rho R_m^2(\rho)\,\mathrm{d}\rho,$$

we obtain:

$$||R_m||^2 = \frac{2}{\pi^2}\frac{J_0^2(\sqrt{\lambda_m}a) - J_0^2(\sqrt{\lambda_m}b)}{J_0^2(\sqrt{\lambda_m}b)}\dagger.$$

From equation (2) we find:

$$Z_m(z) = A_m \frac{\sinh\sqrt{\lambda_m}(l-z)}{\sinh\sqrt{\lambda_m}\,l}.$$

We seek the function $u(\rho, z)$ in the form

$$u(\rho, z) = \sum_{m=1}^{\infty} A_m R_m(\rho)\frac{\sinh\sqrt{\lambda_m}(l-z)}{\sinh\sqrt{\lambda_m}\,l}.$$

The coefficient A_m is determined from the boundary condition

$$u(\rho, 0) = u_0 = \sum_{m=1}^{\infty} A_m R_m(\rho).$$

Hence

$$A_m = \frac{u_0}{||R_m||^2}\int_a^b \rho R_m(\rho)\,\mathrm{d}\rho.$$

Replac ng $R_m(\rho)$ from the equation

$$R_m(\rho) = -\frac{1}{\lambda_m\rho}\frac{\mathrm{d}}{\mathrm{d}\rho}\left[\rho\frac{\mathrm{d}R_m}{\mathrm{d}\rho}\right],$$

we obtain:

$$A_m = -\frac{u_0}{\lambda_m||R_m||^2}\int_a^b \frac{\mathrm{d}}{\mathrm{d}\rho}\left(\rho\frac{\mathrm{d}R_m}{\mathrm{d}\rho}\right)\mathrm{d}\rho = -\left[\frac{u_0}{\lambda_m||R_m||^2}\left(\rho\frac{\mathrm{d}R_m}{\mathrm{d}\rho}\right)\right]_{\rho=a}^{\rho=b}$$

$$= \frac{u_0}{\lambda_m||R_m||^2}[aR_m'(a) - bR_m'(b)].$$

Let us evaluate

$$R_m'(a) = \sqrt{\lambda_m}[J_0'(\sqrt{\lambda_m}a)N_0(\sqrt{\lambda_m}a) - J_0(\sqrt{\lambda_m}a)N_0'(\sqrt{\lambda_m}a)] = -\frac{2}{\pi a},$$

† See chapter VII, § 2, problem 27.

$$R'_m(b) = \sqrt{\lambda_m}\,[J'_0(\sqrt{\lambda_m}\,b)N_0(\sqrt{\lambda_m}\,a) - J_0(\sqrt{\lambda_m}\,a)N'_0(\sqrt{\lambda_m}\,b)]$$

$$= \sqrt{\lambda_m}\,\frac{N_0(\sqrt{\lambda_m}\,a)}{N_0(\sqrt{\lambda_m}\,b)}\,[J'_0(\sqrt{\lambda_m}b)N_0(\sqrt{\lambda_m}\,b) - J_0(\sqrt{\lambda_m}\,b)N'_0(\sqrt{\lambda_m}\,b)]$$

$$= -\frac{z}{\pi b}\,\frac{N_0(\sqrt{\lambda_m}\,a)}{N_0(\sqrt{\lambda_m}\,b)} = -\frac{2}{\pi b}\cdot\frac{J_0(\sqrt{\lambda_m}\,a)}{J_0(\sqrt{\lambda_m}\,b)};$$

in this we made use of the expression for the Wronskian

$$J_0(x)N'_0(x) - J'_0(x)N_0(x) = \frac{2}{\pi x}.$$

After substitution of the expressions for $R'_m(a)$, $R'_m(b)$ and $||R_m||^2$, we find:

$$A_m = \frac{\pi u_0}{\lambda_m}\,\frac{J_0(\sqrt{\lambda_m}\,b)}{J_0(\sqrt{\lambda_m}\,a) + J_0(\sqrt{\lambda_m}b)}.$$

110. *Solution.* It is required to find the solution of the equation

$$\frac{1}{\rho}\frac{\partial}{\partial\rho}\left(\rho\frac{\partial u}{\partial\rho}\right) + \frac{\partial^2 u}{\partial z^2} = -4\pi f,$$

where $f)\rho, z) = e\delta(\rho)\delta(z-\zeta)/2\pi\rho$† is the volume density, corresponding to a point sicharge, tuated at the pont $\rho = 0$, $z = \zeta$, so that

$$e\int_0^{2\pi} \mathrm{d}\phi \int_0^a\int_0^h f(\rho, z)\rho\,\mathrm{d}\rho\,\mathrm{d}z = e.$$

For $z = 0$, $z = h$, $\rho = a$ the boundary conditions

$$u = 0$$

must be fulfilled. The solution is sought is the form of a series

$$u(\rho, z) = \sum_{m=1}^{\infty} R_m(\rho)\sin\frac{\pi m}{h}\,z.$$

For $R_m(\rho)$ we obtain the equation

$$\frac{1}{\rho}\frac{\mathrm{d}}{\mathrm{d}\rho}\left(\rho\frac{dR_m}{d\rho}\right) - \mu_m^2 R_m = -\frac{4e}{h\rho}\,\delta(\rho)\sin\mu_m\zeta = -f_m,$$

where $\mu_m = \pi m/h$, with boundary condition $R_m(a) = 0$. The general solution of the homogeneous equation has the form

$$R_m(\rho) = A_m I_0(\mu_m\rho) + B_m K_0(\mu_m\rho).$$

† $\delta(\rho)$ and $\delta(z-\zeta)$ are δ-functions. See [7], chap. III of the appendix, page 725.

Varying the constants A_m and B_m in the inhomogeneous equation for R_m, we obtain:

$$A_m' I_0(\mu_m\rho)+B_m' K_0(\mu_m\rho) = 0,$$

$$A_m' I_0'(\mu_m\rho)+B_m' K_0'(\mu_m\rho) = -\frac{f_m}{\mu_m}.$$

Hence we find:

$$B_m' = \frac{f_m I_0(\mu_m\rho)}{\mu_m W_m}, \qquad W_m = I_0'(x)K_0(x)-I_0(x)K_0'(x)$$

the Wronskian determinant, equal to $W = 1/x\,(x = \mu_m\rho)$, so that

$$B_m' = \rho f_m I_0(\mu_m\rho), \quad A_m' = -\rho f_m K_0(\mu_m\rho).$$

Determining A_m and B_m by integration, we have:

$$R_m(\rho) = A_m^0 I_0(\mu_m\rho)+B_m^0 K_0(\mu_m\rho)+ \\ +\int_0^\rho s[I_0(\mu_m s)K_0(\mu_m\rho)-I_0(\mu_m\rho)K_0(\mu_m s)]f_m(s)ds,$$

where A_m^0 and B_m^0 are constants.

From the condition $|R_m(0)| < \infty$ it follows $B_m^0 = 0$. The condition for $\rho = a$ gives:

$$A_m^0 = -\frac{1}{I_0(a\mu_m)}\int_0^a [I_0(s\mu_m)K_0(a\mu_m)-I_0(a\mu_m)K_0(s\mu_m)]sf_m(s)ds,$$

so that

$$R_m(\rho) = -\frac{K_0(a\mu_m)I_0(\rho\mu_m)}{I_0(a\mu_m)}\int_0^a I_0(s\mu_m)sf_m(s)ds+ \\ +K_0(\rho\mu_m)\int_0^\rho I_0(s\mu_m)sf_m(s)ds+I_0(\rho\mu_m)\int_\rho^a K_0(s\mu_m)sf_m(s)ds,$$

or

$$R_m(\rho) = \frac{4e}{h}\frac{I_0(a\mu_m)K_0(\rho\mu_m)-K_0(a\mu_m)I_0(\rho\mu_m)}{I_0(a\mu_m)}\sin\mu_m\zeta,$$

since

$$\int_\rho^a K_0(s\mu_m)sf_m(s)ds = \frac{4e}{h}\sin\mu_m\zeta\int_\rho^a K_0(s\mu_m)\delta(s)ds = 0,$$

because the point $s = 0$ lies outside the region of integration

$$\int_0^\rho I_0(s\mu_m)sf_m(s)\,\mathrm{d}s = \int_0^a I_0(s\mu_m)sf_m(s)\,\mathrm{d}s$$

$$= \frac{4e}{h}\sin\mu_m\zeta\int_0^a \delta(s)\,\mathrm{d}s = \frac{4e}{h}\sin\mu_m(\zeta).$$

Thus the potential of a point charge e, situated on the axis of the cylinder $\rho \leqslant a$, $0 \leqslant z \leqslant h$ with conducting walls, is described by means of the series

$$u(\rho, z) = \frac{4e}{h}\sum_{m=1}^{\infty}\frac{I_0\left(\frac{\pi m}{h}a\right)K_0\left(\frac{\pi m}{h}\rho\right)-K_0\left(\frac{\pi m}{h}a\right)I_0\left(\frac{\pi m}{h}\rho\right)}{I_0\left(\frac{\pi m}{h}a\right)}\times$$

$$\times\sin\frac{\pi m}{h}\zeta\sin\frac{\pi m}{h}z.$$

In the limit for $a\to\infty$ we obtain an expression for the potential of a point charge in the layer between the conducting planes $z = 0$ and $z = h$:

$$u(\rho, z) = \frac{4e}{h}\sum_{m=1}^{\infty}K_0\left(\frac{\pi m}{h}\rho\right)\sin\frac{\pi m}{h}\zeta\sin\frac{\pi m}{h}z.$$

We now make a limiting transition for $h\to\infty$, and obtain:

$$u = \frac{4e}{\pi}\sum_{m=1}^{\infty}K_0(\mu_m\rho)\sin\mu_m\zeta\sin\mu_m z\Delta\mu \to \frac{4e}{\pi}\int_0^\infty K_0(\mu\rho)\sin\mu\zeta\sin\mu z\,\mathrm{d}\mu$$

$$= \frac{2e}{\pi}\int_0^\infty K_0(\mu\rho)[\cos\mu(z-\zeta)-\cos\mu(z+\zeta)]\,\mathrm{d}\mu. \qquad (1)$$

Taking into consideration the relation

$$\int_0^\infty K_0(\mu\rho)\cos\mu z\,\mathrm{d}\mu = \frac{\pi}{2(\rho^2+z^2)^{1/2}},$$

we obtain the following expression for the potential:

$$u = \frac{e}{\sqrt{\rho^2+(z-\zeta)^2}}-\frac{e}{\sqrt{\rho^2+(z+\zeta)^2}},$$

which agrees with the familiar expression for the potential of a point charge in semispace. In order to obtain the potential of the charge in infinite space from (1), we introduce a new variable

$$z' = z-\zeta,$$

which gives:

$$u = \frac{2e}{\pi}\int_0^\infty K_0(\mu, \rho)\cos\mu z'\,d\mu - \frac{2e}{\pi}\int_0^\infty K_0(\mu,\rho)\cos\mu(z+2\zeta)\,d\mu$$

$$= \frac{e}{\sqrt{\rho^2+(z')^2}} - \frac{e}{\sqrt{\rho^2+(z'+2\zeta)^2}} \to \frac{e}{\sqrt{\rho^2+(z')^2}} \quad \text{for} \quad \zeta\to\infty,$$

111. The potential of a point charge, situated at the point $(0, \zeta)$ inside a semi-infinite tube $\rho \leqslant a$, $z \geqslant 0$ with conducting walls kept at zero potential, is determined by the expression

$$u(\rho, z) = \frac{e}{\sqrt{\rho^2+(z-\zeta)^2}} - \frac{e}{\sqrt{\rho^2+(z+\zeta)^2}} - \frac{4e}{\pi}\int_0^\infty \frac{K_0(a\mu)I_0(\rho\mu)}{I_0(a\mu)} \times$$

$$\times \sin\mu\zeta\sin\mu z\,d\mu.$$

112. A charge e is placed at the point (r', ϕ', ζ) inside an infinite circular cylinder with conducting walls; the potential of the electric field, produced by this charge, is given by the series

$$u(r,\phi,z) = \frac{4e}{a}\sum_{n=0}^\infty\sum_{m=1}^\infty \frac{J_n\left(\frac{\mu_m^{(n)}}{a}r\right)J_n\left(\frac{\mu_m^{(n)}}{a}r'\right)}{\mu_m^{(n)}[J_n'(\mu_m^{(n)})]^2}\,e^{-\frac{\mu_m^{(n)}|z-\zeta|}{a}}\cos n(\phi-\phi'), \quad (1)$$

where $\mu_m^{(n)}$ is a root of the equation $J_n(\mu) = 0$, and a is the radius of the cylinder.

If the charge is placed on the axis of the cylinder ($r' = 0$), then

$$u(r,\phi,z) = \frac{4e}{a}\sum_{m=1}^\infty \frac{J_0\left(\frac{\mu_m^{(0)}}{a}r\right)}{\mu_m^{(0)}J_1^2(\mu_m^{(0)})}\,e^{-\frac{\mu_m^{(0)}}{a}|z-\zeta|}$$

Solution. In order to find the source function we solve the inhomogeneous equation

$$\Delta u = -4\pi\rho \qquad (2)$$

with the boundary condition

$$u|_\Sigma = 0,$$

where Σ is the surface of the cylinder by the method of separation of variables. In order to do this we assume that the cross-section of the cylinder is an arbitrary region S with boundary C.

Let $\{\psi_n(M)\}$ and $\{\lambda_n\}$ be the eigenfunctions and eigenvalues of the problem

$$\Delta_2\psi+\lambda\psi = 0 \quad \text{in} \quad S, \quad \psi = 0 \quad \text{on} \quad C.$$

We shall look for a solution in the form

$$u(M, z) = \sum_{n=1}^\infty u_n(z)\psi_n(M). \qquad (3)$$

Expanding $\rho(M)$ also in a series, we obtain:

$$\rho(M, z) = \sum_{n=1}^{\infty} \rho_n(z)\psi_n(M),$$

where

$$u_n(z) = \frac{1}{||\psi_n||^2} \int_S u(M', z)\psi_n(M')\,d\sigma_{M'},$$

$$\rho_n(z) = \frac{1}{||\psi_n||^2} \int_S \rho(M', z)\psi_n(M')\,d\sigma_{M'},$$

where $|| \psi_n ||^2 = \int_S \psi_n^2(M')\,d\sigma_{M'}$.

From equation (2) an equation for $u_n(z)$

$$u_n'' - \lambda_n u_n = -4\pi\rho_n (\lambda_n > 0),$$

follows, where $u_n \to 0$ for $z \to \pm\,\infty$.

Hence we find:

$$u_n(z) = 4\pi \int_{-\infty}^{+\infty} \frac{e^{-\sqrt{\lambda_n}|z-\zeta|}}{2\sqrt{\lambda_n}} \rho_n(\zeta)\,d\zeta,$$

or

$$u_n(z) = 4\pi \int_{-\infty}^{+\infty} \int_S \rho(M', \zeta) e^{-\sqrt{\lambda_n}|z-\zeta|} \frac{\psi_n(M')}{2\sqrt{\lambda_n}}\,d\sigma_{M'}\,d\zeta.$$

Substituting this expression in (3) and formally changing the order of summation and integration, we obtain:

$$u(M, z) = 4\pi \int_{-\infty}^{+\infty} \int_S \rho(M', \zeta) \left\{ \sum_{n=1}^{\infty} \frac{\psi_n(M)\psi_n(M')}{2\sqrt{\lambda_n}||\psi_n||^2} e^{-\sqrt{\lambda_n}|z-\zeta|} \right\} d\sigma_{M'}\,d\zeta.$$

Hence it follows that the Green's function equals

$$G(M, M', z, \zeta) = \sum_{n=1}^{\infty} \frac{\psi_n(M)\psi_n(M')}{2\sqrt{\lambda_n}||\psi_n||^2} e^{-\sqrt{\lambda_n}|z-\zeta|}.$$

The potential of a charge e equals $u = 4\pi eG$. If also the cylinder is circular (S a circle), then

$$\psi_n = \psi_{m,n}(r, \phi) = J_n\left(\frac{\mu_m^{(n)}}{a} r\right) \begin{cases} \cos n\phi, \\ \sin n\phi, \end{cases}$$

$$\lambda_n = \lambda_{m,n} = \left(\frac{\mu_m^{(n)}}{a}\right)^2,$$

$$||\psi_n||^2 = ||\psi_{m,n}||^2 = \frac{\pi}{2} \varepsilon_n a^2 [J_n'(\mu_{m,n})]^2,$$

where

$$\varepsilon_n = \begin{cases} 2 & \text{for} \quad n = 0, \\ 1 & \text{for} \quad n \neq 0. \end{cases}$$

Substituting these expressions in the formula for G, we obtain the solution of the problem in the form (1).

4. Problems Requiring the Application of Spherical and Cylindrical Functions

113. The solution of the first interior boundary-value problem for a sphere

$$\Delta u = 0 \quad \text{for} \quad r < a, \quad u|_{r=a} = f(\theta, \phi)$$

may be written as a series

$$u(r, \theta, \phi) = \sum_{n=0}^{\infty} \left(\frac{r}{a}\right)^n Y_n(\theta, \phi), \tag{1}$$

where

$$Y_n = \sum_{k=0}^{n} (A_{nk} \cos k\phi + B_{nk} \sin k\phi) P_n^{(k)}(\cos\theta), \tag{2}$$

$$\left.\begin{aligned} A_{0,0} &= \frac{1}{4\pi} \int_0^{2\pi}\int_0^{\pi} f(\theta, \phi) \sin\theta \, d\theta \, d\phi \\ A_{n,k} &= \frac{(2n+1)(n-k)!}{2\pi(n+k)!} \int_0^{2\pi}\int_0^{\pi} f(\theta, \phi) P_n^{(k)}(\cos\theta) \cos k\phi \sin\theta \, d\theta \, d\phi, \qquad n > 0, \\ B_{nk} &= \frac{(2n+1)(n-k)!}{2\pi(n+k)!} \int_0^{2\pi}\int_0^{\pi} f(\theta, \phi) P_n^{(k)}(\cos\theta) \sin k\phi \sin\theta \, d\theta \, d\phi. \end{aligned}\right\} \tag{3}$$

114. The solution of the first exterior boundary-value problem for a sphere

$$\Delta u = 0 \quad \text{for} \quad r > a, \quad u|_{r=} = f(\theta, \phi)$$

is represented by the series

$$u(r, \theta, \phi) = \sum_{n=0}^{\infty} \left(\frac{a}{r}\right)^{n+1} Y_n(\theta, \phi)$$

$$= \sum_{n=0}^{\infty} \sum_{k=0}^{n} \left(\frac{a}{r}\right)^{n+1} (A_{nk} \cos k\phi + B_{nk} \sin k\phi) P_n^{(k)}(\cos\theta),$$

where A_{nk} and B_{nk} are coefficients, given by formulae (3) of problem 113.

115. (a) The solution of the second interior boundary-value problem for a sphere

$$\Delta u = 0 \quad \text{for} \quad r < a, \quad \left.\frac{\partial u}{\partial n}\right|_{r=a} = f(\theta, \phi),$$

where $f(\theta, \phi)$ is a function satisfying the condition

$$\int_0^{2\pi}\int_0^{\pi} f(\theta, \phi) \sin\theta \, d\theta \, d\phi = 0,$$

has the form

$$n(r, \theta, \phi) = \sum_{n=1}^{\infty} \frac{r^n}{na^{n-1}} Y_n(\theta, \phi) + \text{const.}$$

$$= \sum_{n=1}^{\infty}\sum_{k=0}^{n} \frac{r^n}{na^{n-1}} (A_{nk}\cos k\phi + B_{nk}\sin k\phi) P_n^{(k)}(\cos\theta) + \text{const.},$$

where A_{nk} and B_{nk} are given by formulae (3) of problem 123.

(b) The solution of the second exterior boundary-value problem for a sphere

$$\Delta u = 0 \quad \text{for} \quad r > a, \quad \left.\frac{\partial u}{\partial n}\right|_{r=a} = f(\theta, \phi) \quad \text{or} \quad -\left.\frac{\partial u}{\partial r}\right|_{r=a} = f(\theta, \phi)$$

(n the outer normal) has the form

$$u(r, \theta, \phi) = \sum_{n=1}^{\infty} \frac{a^{n+2}}{(n+1)r^{n+1}} Y_n(\theta, \phi) + \text{const.}$$

For the special case

$$f(\theta, \phi) = A\cos\theta$$

we obtain:

$$Y_n(\theta, \phi) = 0 \quad \text{for} \quad n > 1,$$

$$Y_1(\theta, \phi) = A\cos\theta,$$

$$u(r, \theta) = Ar\cos\theta \quad (r < a),$$

$$u(r, \theta) = \frac{Aa^3}{2r^2}\cos\theta \quad (r > a).$$

116. The intensity of the electrostatic field, as is usual, is expressed in terms of the potential u,

$$\boldsymbol{E} = \text{grad}\, u,$$

equal to

$$u = V_2 + \frac{V_1 - V_2}{2}\sum_{n=1}^{\infty}\left(\frac{r}{a}\right)^n \frac{2n+1}{n+1} P_{n-1}(0)P_n(\cos\theta), \qquad r < a,$$

where a is the radius of the sphere, and P_n is the Legendre polynomial of nth order

$$P_n(0) = \begin{cases} \dfrac{(-1)^\nu \, 1.3.5 \ldots (2\nu-1)}{\nu! \, 2^\nu} & \text{for} \quad n = 2\nu, \\ 0 & \text{for} \quad n = 2\nu+1. \end{cases}$$

Method. In evaluating the integral

$$\int_0^1 P_n(x)\,\mathrm{d}x$$

one uses:

(1) the recurrence relation

$$P_n(x) = \frac{1}{2n+1}[P'_{n+1}(x)-P'_{n-1}(x)],$$

(2) the formulae

$$P_{n-1}(0)-P_{n+1}(0) = P_{n-1}(0)\frac{2n+1}{n+1},$$

$$P_{2\nu}(0) = (-1)^\nu \frac{1.3.5 \ldots (2\nu-1)}{\nu! \, 2^\nu},$$

$$P_{2\nu+1}(0) = 0.$$

117. A charge is placed at the point $r = r_0$, $\theta = 0$, where r, θ are spherical coordinates, and the origin of coordinates is at the point $r = 0$.

(a) If $r_0 < a$, where a is the radius of the sphere then the potential

$$u(r, \theta) = \begin{cases} e\displaystyle\sum_{n=0}^{\infty}\left(\frac{r^n}{r_0^{n+1}} - \frac{r_0^n r^n}{a^{2n+1}}\right)P_n(\cos\theta) & \text{for} \quad r < r_0, \\ e\displaystyle\sum_{n+0}^{\infty}\left(\frac{r_0^n}{r^{n+1}} - \frac{r_0^n r^n}{a^{2n+1}}\right)P_n(\cos\theta) & \text{for} \quad r > r_0, \end{cases}$$

and the density of surface charges on the sphere

$$\sigma = -\frac{e}{4\pi}\sum_{n=0}^{\infty}(2n+1)\frac{r_0^n}{a^{n+2}}P_n\cos\theta).$$

(b) If $r_0 > a$, then the potential

$$u(r, \theta) = \begin{cases} e\displaystyle\sum_{n=0}^{\infty}\left(\frac{r^n}{r_0^{n+1}} - \frac{a^{2n+1}}{r^{n+1} r_0^{n+1}}\right)P_n(\cos\theta) & \text{for} \quad r < r_0, \\ e\displaystyle\sum_{n=0}^{\infty}\left(\frac{r_0^n}{r^{n+1}} - \frac{a^{2n+1}}{r^{n+1} r_0^{n+1}}\right)P_n(\cos\theta) & \text{for} \quad r > r_0, \end{cases}$$

and the density of surface charges on the sphere equals

$$\sigma = -\frac{e}{4\pi}\sum_{n=0}^{\infty}(2n+1)\frac{a^{n-1}}{r_0^{n+1}}P_n(\cos\theta).$$

Method. One must use the expansion

$$\frac{1}{R} = \begin{cases} \dfrac{1}{r_0}\displaystyle\sum_{n=0}^{\infty}\left(\frac{r}{r_0}\right)^n P_n(\cos\theta) & \text{for} \quad r < r_0, \\ \dfrac{1}{r}\displaystyle\sum_{n=0}^{\infty}\left(\frac{r_0}{r}\right)^n P_n(\cos\theta) & \text{for} \quad r > r_0, \end{cases}$$

where R is the distance of the point of observation (r, θ) from the charge $(r_0, 0)$.

The potential of the point charge is sought in the form of the sum

$$u(r,\theta) = \frac{e}{R} + \sum_{n=0}^{\infty} A_n\left(\frac{r}{a}\right)^n P_n(\cos\theta) \qquad \text{for} \quad r < a, \tag{1}$$

$$u(r,\theta) = \frac{e}{R} + \sum_{n=0}^{\infty} B_n\left(\frac{a}{r}\right)^{n+1} P_n(\cos\theta) \qquad \text{for} \quad r > a. \tag{2}$$

The coefficients A_n and B_n are determined from the boundary condition $u\,|_{r=a} = 0$:

$$A_n = -e\frac{r_0^n}{a^{n+1}}, \qquad B_n = -e\frac{a^n}{r_0^{n+1}}.$$

The surface charge density is found from the formula

$$\sigma = -\frac{1}{4\pi}\left(\frac{\partial u}{\partial n}\right)_{r=a} \tag{3}$$

One must use formula (2), in the case of (a), and formula (1) in the case of (b).

118. Let the sphere of radius a, on which a charge e_1 is distributed, be situated in the field of a point charge e, existing at the point $(r_0, 0)$.

The potential of the field equals

$$u = \frac{e}{R} + \frac{aV_0}{r} - e\sum_{n=0}^{\infty}\frac{a^{2n+1}}{r_0^{n+1}r^{n+1}}P_n(\cos\theta),$$

and the surface density of charges induced in the sphere is given by the expression

$$\sigma = -\frac{e}{4\pi}\sum_{n=0}^{\infty}(2n+1)\frac{a^{n-1}}{r_0^{n+1}}P_n(\cos\theta) + \frac{V_0}{4\pi a},$$

$$V_0 = \frac{e_1}{a} + \frac{e}{r_0}.$$

Method. The solution as before is sought in the form

$$u(r, \theta) = \frac{e}{R} + \sum_{n=0}^{\infty} B_n \left(\frac{a}{r}\right)^{n+1} P_n(\cos\theta).$$

The boundary condition $u|_{r=a} = V_0$ gives:

$$u(a, \theta) = \sum_{n=0}^{\infty} \left\{ e \frac{a^n}{r_0^{n+1}} + B_n \right\} P_n(\cos\theta) = V_0 = \text{const.}$$

Hence we find:

$$B_0 = V_0 - e\frac{1}{r_0}, \qquad B_n = -e\frac{a^n}{r_0^{n+1}}.$$

After making use of the expansion

$$\frac{1}{R} = \frac{1}{r_0} \sum_{n=0}^{\infty} \left(\frac{r}{r_0}\right)^n P_n(\cos\theta),$$

we find:

$$\sigma = -\frac{1}{4\pi}\left(\frac{\partial u}{\partial n}\right)_{r=a}.$$

In order to determine V_0 one uses the normalization condition

$$\int_0^{2\pi}\int_0^{\pi} \sigma a^2 \sin\theta \,d\theta \,d\phi = e_1.$$

119. Let (r, θ, ϕ) be spherical coordinates, $r = 0$ the centre of the sphere, a its radius, $u = u(r, \theta, \phi)$ the velocity potential.

(a) $u = u(r, \theta) = \dfrac{v_0 a^3}{2r^2} P_1(\cos\theta) = \dfrac{v_0 a^3}{2r^2} \cos\theta,$

(b) $u = u(r, \theta) = -v_0\left(r + \dfrac{a^3}{2r^2}\right) P_1(\cos\theta).$

Method. The solution is sought in the form

$$u = \sum_{n=0}^{\infty} \frac{a^{n+2}}{r^{n+1}} Y_n(\theta, \phi),$$

where $Y_n(\theta, \phi)$ is a spherical function. From the boundary condition for $r = a$ we find:

$$Y_n(\theta, \phi) = \begin{cases} 0 & \text{for} \quad n \neq 1, \\ \dfrac{v_0}{2} P_1 \cos\theta . \end{cases}$$

Compare the solution of problems 65 and 66 of this chapter.

120. The potential of the perturbed field equals

$$u = \begin{cases} u_1 = -E_0 \dfrac{3\varepsilon_2}{2\varepsilon_2+\varepsilon_1} r\cos\theta & \text{inside sphere} \quad (r < a), \\ u_2 = -E_0 \left[1 + \dfrac{\varepsilon_2-\varepsilon_1}{2\varepsilon_2+\varepsilon_1}\left(\dfrac{a}{r}\right)^3\right] r\cos\theta & \text{outside sphere.} \end{cases}$$

where a is the radius of the sphere, $u = u_0 = -E_0 z = -E_0 r \cos\theta$ is the potential of the external field in the absence of the sphere. The z-component of the field E_z is

$$E_z = -\frac{\partial u}{\partial z} = \begin{cases} -\dfrac{3\varepsilon_2}{2\varepsilon_2+\varepsilon_1} E_0 & \text{inside sphere,} \\ \left[1 - \dfrac{\varepsilon_2-\varepsilon_1}{2\varepsilon_2+\varepsilon_1} \dfrac{2a^3}{r^3}\right] E_0 & \text{outside sphere.} \end{cases}$$

Method. See [7], supplement, part II, § 3, page 716.

121. (a) If the charge is placed outside the sphere at the point (r_0, θ), $r_0 > a$, then the potential of the electric field equals

$$u(r, \theta) = \begin{cases} u_1(r, \theta) & \text{for} \quad r < a, \\ u_2(r, \theta) & \text{for} \quad r > a, \end{cases}$$

where

$$u_1(r, \theta) = e \sum_{n=0}^{\infty} \frac{2n+1}{n\varepsilon_1+(n+1)\varepsilon_2} \frac{r^n}{r_0^{n+1}} P_n(\cos\theta),$$

$$u_2(r, \theta) = e\, \frac{\varepsilon_2-\varepsilon_1}{\varepsilon_2} \sum_{n=0}^{\infty} \frac{n}{n\varepsilon_1+(n+1)\varepsilon_2} \frac{a^{2n+1}}{r_0^{n+1} r^{n+1}} P_n(\cos\theta) + \frac{e}{\varepsilon_2} \frac{1}{R}.$$

Method. The solution is sought in the form

$$u_1(r, \theta) = \sum_{n=0}^{\infty} A_n \left(\frac{r}{a}\right)^n P_n(\cos\theta) \qquad (r < a),$$

$$u_2(r, \theta) = \frac{e}{\varepsilon_2} \frac{1}{R} + \sum_{n=0}^{\infty} B_n \left(\frac{a}{r}\right)^{n+1} P_n(\cos\theta) \qquad (r > a),$$

where A_n and B_n are coefficients, determined from the matching conditions

$$u_1 = u_2, \quad \varepsilon_1 \frac{\partial u_1}{\partial r} = \varepsilon_2 \frac{\partial u_2}{\partial r} \quad \text{for} \quad r = a.$$

(b) If $r_0 < a$, then

$$u_1(r, \theta) = e \frac{\varepsilon_1 - \varepsilon_2}{\varepsilon_1} \sum_{n=0}^{\infty} \frac{n+1}{n\varepsilon_1 + (n+1)\varepsilon_2} \cdot \frac{r_0^n r^n}{a^{2n+1}} P_n(\cos\theta) + \frac{e}{\varepsilon_1 R} \quad \text{for} \quad r < a,$$

$$u_2(r, \theta) = e \sum_{n=0}^{\infty} \frac{2n+1}{n\varepsilon_1 + (n+1)\varepsilon_2} \frac{r_0^n}{r^{n+1}} P_n(\cos\theta) \quad \text{for} \quad r > a.$$

Method. See problem 131(a).

122. The current density is

$$j_1 = -\sigma_1 \operatorname{grad} u_1 \quad \text{for} \quad r < a,$$

$$j_2 = -\sigma_2 \operatorname{grad} u_2 \quad \text{for} \quad r > a.$$

A source of current exists at the point $(r_0, 0)$.

(a) If $r_0 < a$, then

$$u_1 = \frac{I}{4\pi} \frac{\sigma_1 - \sigma_2}{\sigma_1} \sum_{n=0}^{\infty} \frac{n+1}{n\sigma_1 + (n+1)\sigma_2} \frac{r_0^n r^n}{r^{2n+1}} P_n(\cos\theta) + \frac{I}{4\pi\sigma_1 R} \quad (r < a),$$

$$u_2 = \frac{I}{4\pi} \sum_{n=0}^{\infty} \frac{2n+1}{n\sigma_1 + (n+1)\sigma_2} \frac{r_0^n}{r^{n+1}} P_n(\cos\theta) \quad (r > a).$$

(b) If $r_0 > a$, then

$$u_1(r, \theta) = \frac{I}{4\pi} \sum_{n=0}^{\infty} \frac{2n+1}{n\sigma_1 + (n+1)\sigma_2} \frac{r^n}{r_0^{n+1}} P_n(\cos\theta) \quad (r < a),$$

$$u_2(r, \theta) = \frac{I}{4\pi} \frac{\sigma_2 - \sigma_1}{\sigma_2} \sum_{n=0}^{\infty} \frac{n}{n\sigma_1 + (n+1)\sigma_2} \frac{a^{2n+1}}{r_0^{n+1} r^{n+1}} P_n(\cos\theta) + \frac{I}{4\pi\sigma_2 R} \quad (r > a).$$

123. If the sphere is ideally conducting ($\sigma_1 = \infty$), and the source of current of magnitude I exists at the point $(r_0, 0)$ $r_0 > a$ of a medium, possessing conductivity σ_2, then

$$j_1 = 0, \quad j_2 = -\sigma_2 \operatorname{grad} u_2,$$

where

$$u_2(r, \theta) = -\frac{I}{4\pi\sigma_2} \sum_{n=0}^{\infty} \frac{a^{2n+1}}{r_0^{n+1} r^{n+1}} P_n(\cos\theta) + \frac{I}{4\pi\sigma_2 R} \quad (r > a).$$

124. The temperature $u(r, \theta)$ outside a sphere equals (the source at the point $(r_0, 0)$)

$$u(r, \theta) = \frac{Q_0}{4\pi kR} + \frac{Q_0}{4\pi k} \sum_{n=0}^{\infty} \frac{n}{n+1} - \frac{a^{2n+1}}{r_0^{n+1} r^{n+1}} P_n(\cos\theta),$$

where k is the coefficient of heat conduction of the medium,

$$R = \sqrt{r^2 + r_0^2 - 2r_0 r \cos\theta},$$

and a is the radius of the sphere.

125. The temperature inside the sphere $r < a$ equals

$$u(r, \theta) = \frac{Q_0}{4\pi kR} + \frac{Q_0}{4\pi k} \sum_{n=0}^{\infty} \frac{(n+1)-ah}{n+ah} \frac{r_0^n r^n}{a^{2n+1}} P_n(\cos\theta),$$

where h is the coefficient of heat exchange, and the source is at the point $(r_0, 0)$.

Method. For $r = a$ the condition

$$k\frac{\partial u}{\partial r} + hu = 0$$

holds. The solution must, as is usual, be sought in the form

$$u(r, \theta) = \frac{Q_0}{4\pi kR} + \sum_{n=0}^{\infty} A_n \left(\frac{r}{a}\right)^n P_n(\cos\theta).$$

Use the expansion of $1/R$ and determine A_n from the boundary condition.

126. The potential of a point charge e between concentric spheres $(a \leqslant r \leqslant b)$ is

$$u(r, \theta) = \frac{e}{R} - e\sum_{n=0}^{\infty} \left[\frac{r_0^{2n+1} - a^{2n+1}}{b^{2n+1} - a^{2n+1}} \frac{r^n}{r_0^{n+1}} + \frac{b^{2n+1} - r_0^{2n+1}}{b^{2n+1} - a^{2n+1}} \frac{a^{2n+1}}{r_0^{n+1} r^{n+1}}\right] P_n(\cos\theta).$$

For $a \to 0$ therefore we obtain the solution of problem 117(a) for $b \to \infty$, the solution of problem 117(b).

The density of induced charges

$$\sigma_1 = \sigma|_{r=a} = -\frac{e}{4\pi} \sum_{n=0}^{\infty} \frac{(2n+1)(b^{2n+1} - r_0^{2n+1})}{b^{2n+1} - a^{2n+1}} \frac{a^{n-1}}{r_0^{n+1}} P_n(\cos\theta),$$

$$\sigma_2 = \sigma|_{r=b} = \frac{e}{4\pi} \sum_{n=0}^{\infty} \frac{(2n+1)(r_0^{2n+1} - a^{2n+1})}{b^{2n+1} - a^{2n+1}} \frac{b^{n-1}}{r_0^{n+1}} P_n(\cos\theta).$$

Method. The solution must be sought in the form

$$u(r,\theta)=\frac{e}{R}+\sum_{n=0}^{\infty}\left(A_n r^n+\frac{B_n}{r^{n+1}}\right)P_n(\cos\theta).$$

A_n and B_n are determined from the conditions $u=0$ for $r=a$ and $r=b$.

127. (1) The field intensity of a point charge e, situated at the point $r=r_0$, $\theta=0$ outside the sphere $r_0>b$, $E=-\operatorname{grad} u$, where the potential

$$u=\begin{cases} u_1 & \text{for} \quad r<a, \\ u_2 & \text{for} \quad a<r<b, \\ u_3 & \text{for} \quad r>b, \end{cases}$$

is given by the formulae

$$u_1(r,\theta)=\sum_{n=0}^{\infty}A_n\left(\frac{r}{a}\right)^n P_n(\cos\theta) \qquad \text{for} \quad r<a,$$

$$u_2(r,\theta)=\sum_{n=0}^{\infty}\left[B_n\left(\frac{r}{a}\right)^n+C_n\left(\frac{b}{r}\right)^{n+1}\right]P_n(\cos\theta) \qquad \text{for} \quad a<r<b,$$

$$u_3(r,\theta)=\sum_{n=0}^{\infty}D_n\left(\frac{b}{r}\right)^{n+1}P_n(\cos\theta)+\frac{e}{\varepsilon_3 R} \qquad \text{for} \quad r>b,$$

where $R=\sqrt{r^2+r_0^2-2rr_0\cos\theta}$ is the distance between the point of observation (r,θ) and the charge, and

$$B_n=\frac{ea^n\delta_{23}[n\varepsilon_3+(n+1)]}{r_0^{n+1}[1+n(n+1)]\left(\frac{a}{b}\right)^{2n+1}(\varepsilon_2-\varepsilon_1)(\varepsilon_3-\varepsilon_2)\delta_{12}\delta_{23}},$$

$$A_n=(2n+1)\varepsilon_2\delta_{12}B_n,$$

$$C_n=\left(\frac{a}{b}\right)^{n+1}(\varepsilon_2-\varepsilon_1)\delta_{12}nB_n,$$

$$D_n=\left(\frac{b}{a}\right)^n B_n+C_n-\frac{eb^n}{\varepsilon_3 r_0^{n+1}},$$

$$\delta_{i,i+1}=\frac{1}{n\varepsilon_i+(n+1)\varepsilon_{i+1}} \qquad (i=1,2).$$

The problem is solved similarly, if the point charge is situated inside the sphere $r_0<a$ or $a<r_0<b$. We do not quote the corresponding expressions for the potential.

128. If the external field E_0 is directed along the polar axis z, so that its potential

$$V_0 = -E_0 z = -E_0 r P_1(\cos\theta) \quad [P_1(x) = x],$$

then the solution of the problem has the form

$$V = V_1 = ArP_1(\cos\theta) \qquad \text{for} \quad r \leqslant a,$$

$$V = V_2 = \left(Br + \frac{C}{r^2}\right) P_1(\cos\theta) \qquad \text{for} \quad a \leqslant r \leqslant b \quad \text{inside the case,}$$

$$V = V_3 = \left(-E_0 r + \frac{A}{r^2}\right) P_1(\cos\theta) \quad \text{for} \quad r \geqslant b,$$

where A, B, C, D are coefficients equal to

$$A = -\frac{9E_0 H}{\Delta}, \quad B = -\frac{3(2H+1)}{\Delta} E_0, \quad C = -\frac{3(H-1)E_0}{\Delta} a^3,$$

$$D = E_0 b^3 + \left[1 + \frac{(2H+1)b^3}{(H-1)a^3}\right], \quad H = \frac{\varepsilon_2}{\varepsilon_1}, \quad \Delta = 9H - 2(H-1)^2 \left[\left(\frac{a}{b}\right)^3 - 1\right].$$

129. Introducing the symbols: a is the radius of the inner surface, b is the radius of the outer surface of the spherical condenser, δ is the distance between the centres of the spheres, and neglecting the terms δ^2, δ^3, δ^4, etc., we obtain for the density of the surface charges on the inner surface

$$\sigma = -\frac{\varepsilon}{4\pi}\left(\frac{\partial V}{\partial r}\right)_{r=a} = \frac{\varepsilon ab(V_1 - V_2)}{4\pi(b-a)}\left(\frac{1}{a^2} - \frac{3\delta}{(b^3 - a^3)}\cos\theta\right),$$

where $V_1 - V_2$ is the potential difference between the surfaces. It is assumed that the centre of a spherical system of coordinates is placed at the centre of centre of the inner sphere.

Method. Use the expansion of the inverse distance between points in a Legendre polynomial. If $r = a$ is the inner surface, then the equation (correct to terms of order δ^2 and above) of the outer surface may be written in the form

$$r = b + \delta P_1(\cos\theta).$$

130. If points of the ring have spherical coordinates $r = c$, $\theta = \alpha$, then the potential at the point (r, θ) equals

$$V = \frac{e}{\varepsilon c}\sum_{n=0}^{\infty}\left(\frac{c}{r}\right)^{n+1} P_n(\cos\alpha) P_n(\cos\theta) \quad \text{for} \quad r > c \quad \text{or} \quad 0 \neq \alpha, \quad r = c,$$

and

$$V = \frac{e}{\varepsilon c}\sum_{n=0}^{\infty}\left(\frac{r}{c}\right)^{n} P_n(\cos\alpha) P_n(\cos\theta) \quad \text{for} \quad r < c \quad \text{or} \quad 0 \neq \alpha, \quad r = c.$$

Method. First the potential is sought at a point $(r_0, 0)$ on the polar axis z

$$V(r_0, 0) = \frac{e}{\varepsilon} \frac{1}{\sqrt{r_0^2 + c^2 - 2r_0 c \cos \alpha}} \begin{cases} \dfrac{e}{\varepsilon c} \displaystyle\sum_{n=0}^{\infty} \left(\frac{c}{r_0}\right)^{n+1} P_n(\cos \alpha), \\ \dfrac{e}{\varepsilon c} \displaystyle\sum_{n=0}^{\infty} \left(\frac{r_0}{c}\right)^{n} P_n(\cos \alpha). \end{cases}$$

131. If $a \cos \alpha > b$, then the potential of the field between the ring and sphere, correct in the region $b < r < a$, is given by the relation

$$V(r, \theta) = \frac{2\pi\varkappa}{\varepsilon_1} \sum_{n=0}^{\infty} P_n(\cos \alpha) \sin \alpha \left\{ \left(\frac{r}{a}\right)^n - \frac{n(\varepsilon_2 - \varepsilon_1) b^{2n+1}}{a_n[n(\varepsilon_2 + \varepsilon_1) + \varepsilon_1] r^{n+1}} \right\} P_n(\cos \theta),$$

where $\varkappa = e/2\pi a \sin \alpha$, e is the total charge on the ring. The origin of coordinates is at the centre of the sphere.

Method. Use the solution of problem 130 on the potential of a charged ring, points of which have spherical coordinates $r = a$, $\theta = a$.

The solution is sought in the form of the sum

$$V(r, \theta) = \begin{cases} \displaystyle\sum_{n=0}^{\infty} A_n \left(\frac{r}{b}\right)^n P_n(\cos \theta) & \text{for} \quad r < b, \\ V_k + \displaystyle\sum_{n=0}^{\infty} B_n \left(\frac{b}{r}\right)^{n+1} P_n(\cos \theta) & \text{for} \quad r > b, \end{cases}$$

where A_n and B_n are coefficients, determinable from the matching condition for $r = b$, and V_K is the potential of the ring from problem 130.

132. The origin of coordinates is placed at the centre of the ring and sphere, the polar axis is directed perpendicularly to the plane of the ring. Then we shall have for the potential the expressions (a the radius of the ring)

$$V(r, \theta) = \begin{cases} V_1(r, \theta) & \text{for} \quad r < a \quad \text{or} \quad r = a, \quad 0 \neq \dfrac{\pi}{2}, \\ V_2(r, \theta) & \text{for} \quad b > r > a \quad \text{or} \quad r = a, \quad 0 \neq \dfrac{\pi}{2}, \end{cases}$$

where

$$V_1(r, \theta) = \frac{e}{\varepsilon a} \sum_{n=0}^{\infty} (-1)^n \frac{(2n-1)!!}{(2n)!!} \left[\left(\frac{r}{a}\right)^{2n} - \left(\frac{a}{b}\right)^{2n+1} \left(\frac{r}{b}\right)^{2n} \right] P_{2n}(\cos \theta),$$

$$V_2(r, \theta) = \frac{e}{\varepsilon a} \sum_{n=0}^{\infty} (-1)^n \frac{(2n-1)!!}{(2n)!!} \left[\left(\frac{a}{r}\right)^{2n+1} - \left(\frac{a}{b}\right)^{2n+1} \left(\frac{r}{b}\right)^{2n} \right] P_{2n}(\cos \theta).$$

The normal component of the electric field intensity on the sphere $r = b$ equals

$$E_r = -\left.\frac{\partial V_2}{\partial r}\right|_{r=b} = \frac{e}{ab\varepsilon}\sum_{n=0}^{\infty}(-1)^n\frac{(2n-1)!!}{(2n)!!}(4n+1)P_{2n}(\cos\theta).$$

Here ε is the dielectric constant of the substance filling the sphere,

$$(2n)!! = 2.4\ldots(2n-2).2n, \qquad (2n-1)!! = 1.3.5\ldots(2n-1).$$

Method. One must use the solution of problem 130 on the potential of a charged ring, namely: if the origin of the spherical system of coordinates is at the centre of the ring, on which a charge e is distributed, then the potential at any point (r, θ) equals (see problem 130 for $\alpha = \pi/2$)

$$V_{\mathrm{r}}(r,\theta) = \begin{cases} \dfrac{e}{\varepsilon a}\sum\limits_{n=0}^{\infty}\left(\dfrac{a}{r}\right)^{n+1}P_n(0)P_n(\cos\theta) & \text{for } r > a \text{ or } r = a, \ \theta \neq \dfrac{\pi}{2}, \\ \dfrac{e}{\varepsilon a}\sum\limits_{n=0}^{\infty}\left(\dfrac{r}{a}\right)^{n}P_n(0)P_n(\cos\theta) & \text{for } r < a \text{ or } r = a, \ \theta \neq \dfrac{\pi}{2}, \end{cases}$$

where

$$P_n(0) = \begin{cases} 0, & \text{for } n \text{ odd}, \\ (-1)^{\frac{n}{2}}\dfrac{1.3.5\ldots(n-1)}{2.4\ldots n}, & \text{for } n \text{ even}. \end{cases}$$

The solution of the given problem may be sought in the form of a sum

$$V = V_{\mathrm{r}} + V_{\mathrm{i}},$$

where V_{i} is the potential of charges, induced on the sphere, and equals

$$V_{\mathrm{i}} = \frac{e}{\varepsilon a}\sum_{n=0}^{\infty}A_{2n}\left(\frac{r}{b}\right)^{2n}P_{2n}(\cos\theta),$$

A_{2n} is found from the condition $V = 0$ for $r = b$.

133. If a is the radius of the sphere, with the origin of coordinates at the centre, then the potential of the current at all interior points of the sphere is expressed by the formula

$$V(r,\theta) = \frac{I}{2\pi a\sigma}\sum_{n=0}^{\infty}\frac{4n+3}{2n+1}\left(\frac{r}{a}\right)^{2n+1}P_{2n+1}(\cos\theta).$$

Method. By the symmetry of the problem $V = 0$ for $\theta = \pi/2$ and therefore it is possible to solve the problem for $0 \leqslant \theta \leqslant \pi/2$, assuming

$$V(r,\theta) = \frac{I}{2\pi a\sigma}\sum_{n=0}^{\infty}A^{2n+1}\left(\frac{r}{a}\right)^{2n+1}P_{2n+1}(\cos\theta).$$

The coefficients A_{2n+1} must be determined from the condition

$$\lim_{\delta\to 0} \sigma \int_0^\delta \left(-\frac{\partial V_\delta}{\partial r}\right) 2\pi a^2 \sin\theta \, d\theta = I, \qquad \left.\frac{\partial V_\delta}{\partial r}\right|_{\substack{r=a\\ \theta>\delta}} = 0.$$

134. We assume that the point charge exists at the origin of coordinates, and $z = a$ and $z = b$ are planes, bounding the plate, ε_2 is the dielectric constant of the plate, ε_1 is the dielectric constant of space.

The potential of the field in the region $z > b$ equals

$$V(\rho, z) = \frac{e(1-\beta^2)}{\varepsilon_1} \sum_{n=0}^{\infty} \frac{\beta^{2n}}{\sqrt{(z+2nh)^2+\rho^2}},$$

where

$$\beta = \frac{\varepsilon_2-\varepsilon_1}{\varepsilon+\varepsilon_1}, \quad \rho^2 = x^2+y^2, \quad h = b-a.$$

Solution. It is required to find the function

$$V(\rho, z) = \begin{cases} V_1(\rho, z) & \text{for } -\infty < z < a \text{ (under the plate)}, \\ V_2(\rho, z) & \text{for } a < z < b \text{ (in the plate)}, \\ V_3(\rho, z) & \text{for } b < z < \infty \text{ (above the plate)}, \end{cases}$$

harmonic everywhere, except the point $\rho = 0$, $z = 0$, at which V_1 has a singularity of the form $e/\varepsilon_1 r$, and satisfying for $z = a$ and $z = b$ the usual matching conditions

$$V_1 = V_2, \quad \varepsilon_1 \frac{\partial V_1}{\partial z} = \varepsilon_2 \frac{\partial V_2}{\partial z} \quad \text{for} \quad z = a;$$

$$V_2 = V_3, \quad \varepsilon_2 \frac{\partial V_2}{\partial z} = \varepsilon_1 \frac{\partial V_3}{\partial z} \quad \text{for} \quad z = b.$$

Since the region is infinite, the solution must be sought in integral form, proceeding from the expansion

$$\frac{1}{r} = \frac{1}{\sqrt{\rho^2+z^2}} \int_0^\infty J_0(\lambda\rho) e^{-\lambda|z|} d\lambda\dagger, \tag{1}$$

assuming

$$V_1(\rho, z) = \frac{e}{\varepsilon_1}\left[\int_0^\infty J_0(\lambda\rho) e^{-\lambda|z|} d\lambda + \int_0^\infty A(\lambda) J_0(\lambda\rho) e^{\lambda z} d\lambda\right],$$

$$V_2(\rho, z) = \frac{e}{\varepsilon_1}\left[\int_0^\infty B(\lambda) J_0(\lambda\rho) e^{-\lambda z} d\lambda + \int_0^\infty C(\lambda) J_0(\lambda\rho) e^{\lambda z} d\lambda\right],$$

$$V_3(\rho, z) = \frac{e}{\varepsilon_1} \int_0^\infty D(\lambda) J_0(\lambda\rho) e^{-\lambda z} d\lambda.$$

† See [7] page 672.

All the integrals, except the first, must remain finite for $\rho \to 0$ and $z \to 0$; moreover,

$$\lim_{z\to-\infty} V_1 = 0, \qquad \lim_{z\to\infty} V_3 = 0.$$

The matching conditions for $z = a$ and $z = b$ give:

$$A(\lambda) = (\beta+1)C(\lambda)-\beta e^{-2a\lambda}, \qquad B(\lambda) = \beta e^{2a\lambda}C(\lambda)+\beta-1,$$

$$C(\lambda) = \frac{\beta e^{-2\lambda b}}{1+\beta} D(\lambda), \qquad D(\lambda) = \frac{1-\beta^2}{1-\beta^2 e^{-2h\lambda}},$$

so that

$$V_3 = \frac{e(1-\beta^2)}{\varepsilon_1} \int_0^\infty \frac{J_0(\lambda\rho)e^{-\lambda z}}{1-\beta^2 e^{-2\lambda h}} d\lambda.$$

In order to evaluate this integral we expand $1/(1-\beta^2 e^{-2\lambda h})$ in a series in powers of $\beta^2 e^{-2\lambda h}$:

$$V_3 = \frac{e(1-\beta^2)}{\varepsilon_1} \left\{ \int_0^\infty J_0(\lambda\rho)e^{-\lambda z} d\lambda + \beta^2 \int_0^\infty J_0(\lambda\rho)e^{-\lambda(z+2h)} d\lambda + \dots \right\},$$

from which, using formula (1), we obtain a result in the form of the series deduced above.

135. The potential on the earth's surface for $z = 0$ may be represented in two forms:

$$V(\rho, 0) = \frac{I}{2\pi\sigma_1} \int_0^\infty \frac{1-\beta e^{-2\lambda h}}{1+\beta e^{-2\lambda h}} J_0(\lambda\rho)\, d\lambda$$

or

$$V(\rho, 0) = \frac{I}{2\pi\sigma_1} \left[\frac{1}{\rho} + 2 \sum_{n=1}^\infty \frac{(-1)^n \beta^n}{\sqrt{4h^n n^2+\rho^2}} \right], \tag{1}$$

where

$$\beta = \frac{\sigma_2-\sigma_1}{\sigma_2+\sigma_1}.$$

If at the points $x = a$ and $x = -a$ two electrodes are situated, a current I flowing out through the first electrode, and a current $-I$ flowing out through the second, then

$$V(\rho, 0) = \frac{I}{2\pi\sigma_1} \left\{ \frac{1}{R_1} - \frac{1}{R_2} + 2 \sum_{n=1}^\infty (-\beta)^n \left[\frac{1}{\sqrt{4h^2n^2+R_1^2}} - \frac{1}{\sqrt{4h^2n^2+R_2^2}} \right] \right\}, \tag{2}$$

where

$$R_1 = \sqrt{(x-a)^2+y^2}, \qquad R_2 = \sqrt{(x+a)^2+y^2}.$$

Method. The solution is sought in the form

$$V_1 = \frac{I}{2\pi\sigma_1}\left[\int_0^\infty J_0(\lambda\rho)e^{-\lambda z}d\lambda + \int_0^\infty A(\lambda)J_0(\lambda\rho)e^{-\lambda z}d\lambda + \int_0^\infty B(\lambda)J_0(\lambda\rho)e^{\lambda z}d\lambda\right]$$

in the region $0 \leqslant z \leqslant a$

$$V_2 = \frac{I}{2\pi\sigma_2}\int_0^\infty C(\lambda)J_0(\lambda\rho)^{-\lambda z}d\lambda \quad \text{in the region} \quad z \geqslant a.$$

The expansion coefficients $A(\lambda)$, $B(\lambda)$ and $C(\lambda)$ are determined from the two matching conditions for $z = 0$ and the condition $\partial V/\partial z = 0$ for $\rho \neq 0$. Transition from the integral to a sum is carried out by analogy with problem 134.

Formula (1) is obtained from formula (2) by means of the principle of superposition.

136. On the earth's surface the potential equals

$$V(x, y, 0) = V_0 \frac{\operatorname{arc\,sinh} a\sqrt{\dfrac{\alpha_2-1}{x^2+y^2}}}{\operatorname{arc\,cosh} \alpha}, \qquad \alpha = \sqrt{\frac{\sigma_g}{\sigma_w}} > 5.$$

Solution. It is required to solve the equation

$$\sigma_g(V_{xx}+V_{yy})+\sigma_w V_{zz} = 0$$

in the semispace $z \geqslant 0$ for the boundary condition

$$V = V_0 \quad \text{on the sphere} \quad x^2+y^2+z^2 = a^2.$$

Assuming $t = \alpha z$, we obtain $V_{xx}+V_{yy}+V_{tt} = 0$ and $V = V_0$ on the surface of the ellipsoid of revolution

$$\frac{x^2+y^2}{a^2}+\frac{t^2}{c^2} = 1,$$

where $a^2 = b^2$, $c^2 = \alpha^2 a^2$. Outside the ellipsoid on the plane $z = 0$, obviously, $\partial V/\partial z = 0$. Therefore the problem reduces to the evaluation of the potential of the field of a charged ellipsoid of revolution. Its solution (see problem 154) has the form

$$V = V_0 \frac{\displaystyle\int_\lambda^\infty \frac{ds}{(a^2+s)\sqrt{c^2+s}}}{\displaystyle\int_0^\infty \frac{ds}{(a^2+s)\sqrt{c^2+s}}},$$

where λ is an elliptic coordinate. In the given case the ellipsoid is elongated, therefore the relations

$$c^2+\lambda = (c^2-a^2)\eta^2, \quad 1 < \eta^2 < \infty,$$
$$c^2+\lambda = (c^2-a^2)\xi^2, \quad 0 < \xi^2 < 1,$$

$$\frac{\rho^2}{(c^2-a^2)(\eta^2-1)}+\frac{t^2}{(c^2-a^2)\eta^2}=1 \quad \text{and} \quad \frac{\rho^2}{-(c^2-a^2)(1-\xi^2)}+\frac{t^2}{(c^2-a^2)\xi^2}=1,$$

hold, hence

$$t=\sqrt{c^2-b^2}\,\eta\xi, \qquad \rho=\sqrt{(c^2-a^2)(1-\xi^2)(\eta^2-1)}.$$

Calculating the integrals, we obtain:

$$V=V_0\,\frac{\operatorname{arc\,tanh}\sqrt{(c^2-a^2)(c^2+\lambda)^{-2}}}{\operatorname{arc\,tanh}\sqrt{(c^2-a^2)c^{-2}}},$$

or

$$V=V_0\,\frac{\operatorname{arc\,tanh}\dfrac{1}{\eta}}{\operatorname{arc\,tanh}\sqrt{\dfrac{\alpha^2-1}{\alpha^2}}}=V_0\,\frac{\operatorname{arc\,sinh}\dfrac{1}{\sqrt{\eta^2-1}}}{\operatorname{arc\,cosech}\,\alpha}\,.$$

From the equations

$$t=\alpha z=a\sqrt{\alpha^2-1}\,\xi\eta, \qquad \rho^2=a^2(\alpha^2-1)(1-\xi^2)(\eta^2-1),$$

eliminating ξ, we obtain:

$$\frac{\rho^2}{\eta^2-1}+\frac{\alpha^2z^2}{\eta^2}=a^2(\alpha^2-1).$$

Hence for $z=0$ it follows:

$$\eta^2-1=\frac{\rho^2}{a^2(\alpha^2-1)},$$

so that

$$V(\rho,0)=V_0\,\frac{\operatorname{arc\,sinh}\dfrac{a\sqrt{\alpha^2-1}}{\rho}}{\operatorname{arc\,cosech}\,\alpha}\,.$$

§ 5. Potentials and Their Application

137. The volume potential of a homogeneous sphere

$$V=u(r)=\left\{\begin{array}{ll}2\pi\rho_0\left(a^2-\dfrac{r^2}{3}\right) & \text{for} \quad r<a,\\[2ex] \dfrac{M}{r} & \text{for} \quad r>a,\end{array}\right\} \tag{1}$$

where a is the radius of the sphere, and $M=4\pi\rho_0a^3/3$ is its mass.

Method. The volume potential

$$V(M)=\int\limits_T \frac{\rho_0}{r_{MP}}\,d\tau_P, \tag{2}$$

where T is the volume of the sphere, is a function, harmonic outside the sphere (for $r > a$), satisfying the equation

$$\Delta V = -4\pi\rho \tag{3}$$

inside the sphere and continuous together with the normal derivative on its boundary. Since $\rho_0 = \text{const.}$, the potential possesses spherical symmetry.

138. *Solution.* The problem reduces to the evaluation of the volume integral

$$V(r) = 2\pi\rho_0 \int_0^a \int_0^\pi \frac{\xi^2 d\xi \sin\theta d\theta}{R},$$

where

$$R^2 = \xi^2 + r^2 - 2\xi r \cos\theta .$$

Introducing the new variable of integration R in place of θ and taking into account the fact that

$$R dR = r\xi \sin\theta d\theta,$$

we obtain:

$$V(r) = \frac{2\pi\rho_0}{r} \int_0^a \xi \left[\int_{|r-\xi|}^{r+\xi} dR \right] d\xi .$$

If $r > a$, then $r > \xi$ always and

$$V(r) = \frac{2\pi\rho_0}{r} \int_0^a \xi \left[\int_{r-\xi}^{r+\xi} dR \right] d\xi = \frac{4\pi\rho_0 a^3}{3r} = \frac{M}{r}.$$

If $r < a$, then

$$V(r) = \frac{2\pi\rho_0}{r} \left[\int_0^r \xi(r+\xi-r+\xi) d\xi + \int_r^a \xi(r+\xi+r-\xi) d\xi \right] = 2\pi\rho_0 \left(a^2 - \frac{r^2}{3} \right).$$

139. (a)

$$V = \begin{cases} 2\pi\rho_0(b^2-a^2) & \text{for } r < a, \\ 2\pi\rho_0 b^2 - \dfrac{2\pi\rho_0}{3}\left(r^2 + \dfrac{2a^3}{r}\right) & \text{for } a < r < b, \\ \dfrac{4\pi\rho_0}{3}(b^3-a^3)\dfrac{1}{r} & \text{for } r > b; \end{cases}$$

(b)

$$V = \begin{cases} 2\pi \left[\rho_1 \left(a^2 - \dfrac{r^2}{3} \right) + \rho_2(c^2-b^2) \right] & \text{for } r < a, \\ 2\pi\rho_2(c^2-b^2) + \dfrac{4\pi}{3}\rho_1 a^3 \dfrac{1}{r} & \text{for } a < r < b, \\ \dfrac{4\pi[\rho_2(c^3-b^3)+a^3\rho_1]}{3r} = \dfrac{M_1+M_2}{r} & \text{for } r > c. \end{cases}$$

Method. By the principle of the superposition of the solutions of the linear equation the unknown potential is represented as the sum

$$V = V_{137}+V_{139a},$$

where V_{137} and V_{139a} are solutions of problems 137 and 139(a);

(c) the potential

$$V = \begin{cases} \dfrac{M(c)}{r} & \text{for} \quad r > c, \\ \dfrac{M(r)}{r}+4\pi\displaystyle\int_r^c \xi\rho(\xi)\,d\xi & \text{for} \quad r < c, \end{cases}$$

where

$$M(c) = 4\pi\int_0^c \rho\xi^2\,d\xi, \qquad M(r) = 4\pi\int_0^r \rho(\xi)\xi^2\,d\xi$$

is the mass, distributed with volume density $\rho(r)$ inside the sphere of radius c (or radius r).

If

$$\rho = \begin{cases} 0 & \text{for} \quad r < a, \\ \rho_0 & \text{for} \quad a < r < b, \end{cases}$$

then we at once obtain the solution of problem 139(a)

$$V = \begin{cases} \dfrac{M}{r} & \text{for} \quad r > b, \\ \dfrac{4\pi\rho_0}{3r}(r^3-a^3)+2\pi\rho_0(b^2-r^2) & \text{for} \quad a < r < b, \end{cases}$$

where

$$M = \frac{4\pi}{3}(b^3-a^3)\rho_0.$$

For $\rho = \rho_0$ inside a sphere of radius a $(c = a)$ we obtain from the general formula:

$$V = \begin{cases} \dfrac{M}{r} & \text{for} \quad r > a, \\ 2\pi\rho_0\left(a^2-\dfrac{r^2}{3}\right) & \text{for} \quad r < a, \end{cases}$$

where $M = 4\pi a^3\rho_0/3$ etc.

140. The potential of a homogeneous spherical single layer is

$$u = \begin{cases} 4\pi a\nu_0 & \text{for} \quad r < a, \\ \dfrac{M}{r} & \text{for} \quad r > a, \end{cases}$$

where $M = 4\pi a^2\nu_0$ is the total mass of the single layer, distributed over the sphere.

Method. The potential of a single layer

$$u(r) = \int_0^{2\pi}\int_0^{\pi} \frac{\nu_0}{R} \sin\theta \, d\theta \, d\phi,$$

where $R = \sqrt{r^2+a^2-2ra\cos\theta}$, may be calculated by direct integration, or found as the solution of the equation

$$\Delta u = 0 \quad \text{for} \quad r \neq a,$$

which is everywhere continuous, and has discontinuous normal derivatives for $r = a$

$$\left.\frac{du_2}{dr}\right|_{r=a} - \left.\frac{du_1}{dr}\right|_{r=a} = 4\pi\nu_0,$$

where u_1 is the solution of the equation $\Delta u = 0$ outside the sphere $(r > a)$, u_2 the solution inside the sphere $(r < a)$.

141. The centre of the phere of radius a is situated at the point $x = 0$, $y = 0$, $z = b$ and $\rho = \rho_0$ is the density of the volume charges. The potential of the electrostatic field is

$$V = \begin{cases} 2\pi\rho_0\left(a^2 - \dfrac{r^2}{3}\right) - \dfrac{M}{r_1} & \text{for} \quad r < a, \\ M\left(\dfrac{1}{r} - \dfrac{1}{r_1}\right) & \text{for} \quad r > a, \end{cases}$$

where

$$M = \frac{4\pi}{3}\rho_0 a^3, \qquad r = \sqrt{x^2+y^2+(z-b)^2}, \qquad r_1 = \sqrt{x^2+y^2+(z+b)^2}.$$

Method. In order to evaluate the effect of an ideally conducting plane $z = 0$ one must reflect the original sphere with centre at the point $(0, 0, b)$ with respect to the plane $z = 0$. The solution in this case is represented in the form of the sum

$$V = \begin{cases} C_1 - \dfrac{2}{3}\pi\rho_0 r^2 - \dfrac{M}{r_1} & \text{for} \quad r < a, \\ M\left(\dfrac{1}{r} - \dfrac{1}{r_1}\right) & \text{for} \quad r > a. \end{cases}$$

The constant C_1 is determined from the matching condition of the solutions for $r = a$.

142. The potential in two dimensions of a circle at the point (r, ϕ)

$$V = V(r) = \rho_0 \int_0^{2\pi}\int_0^{a} \ln \frac{1}{\sqrt{\lambda^2+r^2-2\lambda r\cos\varphi}} \lambda \, d\lambda \, d\psi$$

is calculated directly and equals

$$V(r)=\begin{cases} M\left(\frac{1}{2}-\ln a-\frac{1}{2}\frac{r^2}{a^2}\right) & \text{for} \quad r<a, \\ M\ln\frac{1}{r} & \text{for} \quad r>a. \end{cases}$$

Method. In evaluating the integrals one must expand the function under the integral sign

$$\ln\frac{1}{R}=\ln\frac{1}{\sqrt{\lambda^2+r^2-2\lambda r\cos\psi}}$$

in a series

$$\ln\frac{1}{R}=\begin{cases} \ln\frac{1}{r}+\sum_{n=1}^{+\infty}\frac{1}{n}\left(\frac{\lambda}{r}\right)^n\cos n\psi & \text{outside the circle} \quad r>a, \\ \ln\frac{1}{r}+\sum_{n=1}^{\infty}\frac{1}{n}\left(\frac{\lambda}{r}\right)^n\cos n\psi & \text{for} \quad \lambda<r<a, \\ \ln\frac{1}{\lambda}+\sum_{n=1}^{\infty}\frac{1}{n}\left(\frac{r}{\lambda}\right)^n\cos n\psi & \text{for} \quad r<\lambda<a. \end{cases}$$

143. The logarithmic potential of a single layer of the segment $-a\leqslant x\leqslant a$ with constant density $\rho=\rho_0$

$$V(x,y)=\rho_0\int_{-a}^{+a}\ln\frac{1}{\sqrt{(\xi-x)^2+y^2}}\,d\xi$$

is calculated directly and equals

$$V=2a-y\arctan\frac{2ay}{y^2+x^2-a^2}-\frac{a-x}{2}\ln[y^2+(a-x)^2]-\frac{a+x}{2}\ln[y^2+(a+x)^2].$$

Method. Integrate by parts.

144. Let $M(x,y)$ be the point of observation, ϕ the angle subtended at M by the segment $(-a,a)$.

The logarithmic potential of the double layer of the segment

$$W(M)=\nu\int_{-a}^{+a}\frac{\cos\theta_{MP}}{R_{MP}}\,d\xi_P=\nu y\int_{-a}^{+a}\frac{d\xi}{R^2}$$

(R the distance between M and the point of integration P) equals

$$W(M)=\nu\left[\arctan\frac{x+a}{y}-\arctan\frac{x-a}{y}\right]=\pm\nu\phi,$$

where

$$W = \begin{cases} \nu\phi, & \text{when} \quad y > 0, \\ -\nu\phi, & \text{when} \quad y < 0. \end{cases}$$

145. The potential of a single layer, uniformly distributed over a circular disc, has two analytical representations:

(1) the representation of the potential in the form of an expansion in spherical functions

$$V(r,\theta,\phi) = \begin{cases} \dfrac{2e}{a}\sum_{n=0}^{\infty}\left(\dfrac{r}{a}\right)^n [P_n(0)+P_{n-2}(0)]P_n(\cos\theta) - \\ \qquad\qquad -\dfrac{2er}{a^2}P_1(\cos\theta) \quad \text{for} \quad r < a, \\ \dfrac{2e}{a}\sum_{n=0}^{\infty}\left(\dfrac{a}{r}\right)^{n+1} [P_n(0)+P_{n+2}(0)]P_n(\cos\theta) \\ \qquad\qquad \text{for} \quad r > a; \end{cases} \tag{1}$$

(2) the representation of the potential in the form of an elliptic integral

$$V = \frac{\dfrac{2e}{\pi}}{\sqrt{r^2-2ar\sin\theta+a^2}} K\left(\sqrt{\frac{4ar\sin\theta}{r^2-2ar\sin\theta+a^2}}\right), \tag{2}$$

where $K(x) = \int\limits_0^{\pi/2} \dfrac{dx}{\sqrt{1-x^2\cos^2\alpha}}$ is an elliptic integral. Here $e = \pi a^2\sigma$ is the total charge.

Method. In deriving formulae (1) the value of the potential on the z-axis is evaluated, perpendicular to the plane in which the disc lies,

$$V(0,0,z) = \frac{2e}{a^2}[\sqrt{z^2+a^2}-z],$$

and then its expansion in zonal spherical functions is found. The latter argument is developed by analogy with problem 130.

146. Let us choose a system of coordinates (ρ, ϕ, z) with origin at the centre of the circle and the z-axis, perpendicular to the plane in which the circular loop of current lies. The vector-potential has only one component A_ϕ:

$$A_\phi = \frac{\mu I}{c}\oint\frac{ds}{R} = \frac{2\mu I}{c}\int\limits_0^{\pi}\frac{a\cos\phi\, d\phi}{\sqrt{a^2+\rho^2+z^2+2a\rho\cos\phi}},$$

which equals

$$A_\phi = \frac{4\mu I}{ck}\sqrt{\frac{a}{\rho}}\left[\left(1-\frac{1}{2}k^2\right)K-E\right],$$

where μ is the magnetic permeability of the medium, I the total current, flowing through the loop,

$$k^2 = \frac{4a\rho}{(a+\rho)^2+z^2},$$

K and E the complete elliptic integrals of first and second kind:

$$K(k) = \int_0^{\frac{\pi}{2}} \frac{d\theta}{\sqrt{1-k^2 \sin^2\theta}}, \qquad E(k) = \int_0^{\frac{\pi}{2}} \sqrt{1-k^2\sin^2\theta}\, d\theta.$$

At large distances from the current ($k \ll 1$) we have:

$$A_\phi = \frac{\pi\mu I}{8c}\sqrt{\frac{a}{\rho}}\, k^3\left(1+\frac{3}{4}k^2+\frac{75}{128}k^4+\dots\right).$$

For very small loops $\sqrt{\rho^2+z^2} \gg a$ we have:

$$A_\phi = \frac{\pi a^2 \mu I \sin\theta}{cr^2}.$$

147. In polar coordinates (ρ, ϕ) we find:
(a) the solution of the interior first boundary-value problem for a circle

$$u = W(\rho, \phi) = \frac{1}{2\pi}\int_0^{2\pi} \frac{(a^2-\rho^2)f(\psi)\,d\psi}{a^2+\rho^2-2a\rho\cos(\phi-\psi)},$$

(b) the solution of the exterior problem

$$u = W(\rho, \phi) = \frac{1}{2\pi}\int_0^{2\pi} \frac{(\rho^2-a^2)f(\psi)\,d\psi}{a^2+\rho^2-2a\rho\cos(\phi-\psi)},$$

where a is the radius of the circle.

The solutions of the corresponding integral equations have the form

$$\text{(a)}\quad \nu(s) = \frac{1}{\pi}f(s) - \frac{1}{4\pi^2 a}\int_C f(s)\,ds,$$

where C is a circle of radius a,

$$\text{(b)}\quad \nu(s) = -\frac{1}{\pi}f(s) + \frac{1}{4\pi^2 a}\int_C f(s)\,ds.$$

Method a. If the contour C is a circle of radius a, then

$$\frac{\cos\phi}{r} = \frac{1}{2a}$$

and the equation for $\nu(s_0)$ takes the form

$$\nu(s_0)+\frac{1}{2\pi a}\int\limits_C \nu(s)\,\mathrm{d}s = \frac{1}{\pi}f(s_0), \tag{1}$$

i.e.

$$\nu(s) = \frac{1}{\pi}f(s)+A. \tag{2}$$

Substituting (2) in (1) we find:

$$A = -\frac{1}{4\pi^2 a}\int\limits_C f(c)\,\mathrm{d}s.$$

Knowing $\nu(s)$, after simple transformations we arrive at Poisson's integral.

Method b. For the exterior boundary-value problem

$$\nu(s) = -\frac{1}{\pi}f(s)+A,$$

where

$$A = \frac{1}{4\pi^2 a}\int\limits_C f(s)\,\mathrm{d}s.$$

148. The solution of the second boundary-value problem

$$\frac{1}{\rho}\frac{\partial}{\partial\rho}\left(\rho\frac{\partial u}{\partial\rho}\right)+\frac{1}{\rho^2}\frac{\partial^2 u}{\partial\phi^2} = 0 \quad \text{for} \quad \rho < a,$$

$$\left.\frac{\partial u}{\partial\rho}\right|_{\rho=a} = f(\phi)$$

is sought in the form of the expression for the potential of a single layer

$$u = V(\rho,\phi) = a\int\limits_0^{2\pi}\ln\frac{1}{\sqrt{a^2+\rho^2-2a\rho\cos(\phi-\psi)}}\,\nu(\psi)\,\mathrm{d}\psi+\text{const}.$$

The solution of the integral equation for $\nu(\phi)$ gives:

$$\nu(\phi) = \frac{1}{\pi}f(\phi).$$

149. (a) The solution of the first boundary-value problem $u_{xx}+u_{yy}+u_{zz}=0$ in semispace $z>0$, $u|_{z=0}=f$ is sought in the form of the potential of a double layer

$$u(x,y,z) = W = \iint\limits_{-\infty}^{+\infty}\frac{\cos\phi}{r^2}\,\nu(\xi,\eta)\,\mathrm{d}\xi\,\mathrm{d}\eta, \qquad r^2 = (x-\xi)^2+(y-\eta)^2+z^2$$

and is given by the formula

$$u(x,y,z) = \frac{1}{2\pi}\iint\limits_{-\infty}^{+\infty}\frac{zf(\xi,\eta)\,\mathrm{d}\xi\,\mathrm{d}\eta}{[(x-\xi)^2+(y-\eta)^2+z^2]^{3/2}} \qquad \left(\nu = \frac{1}{2\pi}f\right).$$

(b) The solution of the second boundary-value problem

$$u_{xx}+u_{yy}+u_{zz}=0 \quad \text{for} \quad z>0, \quad \left.\frac{\partial u}{\partial z}\right|_{z=0}=f$$

is sought in the form of the potential of a single layer

$$u(x, y, z)=V(x, y, z)=\iint_{-\infty}^{+\infty} \frac{\mu(\xi, \eta)\,d\xi\,d\eta}{\sqrt{(x-\xi)^2+(y-\eta)^2+z^2}}$$

and is given by the formula

$$u(x, y, z)=\frac{1}{2\pi}\iint_{-\infty}^{+\infty} \frac{f(\xi, \eta)\,d\xi\,d\eta}{\sqrt{(x-\xi)^2+(y-\eta)^2+z^2}}+\text{const.} \qquad \left(\mu=\frac{1}{2\pi}f\right).$$

150. The first boundary-value problem

$$\Delta_2 u=u_{xx}+u_{yy}=0 \quad \text{for} \quad y>0, \quad u|_{y=0}=f(x)$$

has the solution

$$u(x, y)=\frac{1}{\pi}\int_{-\infty}^{+\infty} \frac{y f(\xi)\,d\xi}{(x-\xi)^2+y^2}.$$

Method. The solution is sought in the form of the potential of a double layer

$$u=W(M)=\int_{-\infty}^{+\infty} \nu(\xi)\frac{\cos\phi_{MP}}{r_{MP}}\,d\xi_P.$$

The kernel of the integral equation for the density ν is identically equal to zero, so that

$$\nu(\xi)=\frac{1}{\pi}f(\xi).$$

151. *Solution.* If the surface Σ is equipotential, then to each value of the parameter s there must correspond a definite potential

$$V=f(s),$$

satisfying Laplace's equation. Differentiation gives:

$$V_x=f'(s)s_x, \quad V_{xx}=f''(s)(s_x)^2+f'(s)s_{xx}, \;\dots,$$

so that

$$\Delta V=V_{xx}+V_{yy}+V_{zz}=f''(s)(\operatorname{grad} s)^2+f'(s)\Delta s.$$

Hence it follows:

$$\frac{\Delta s}{(\operatorname{grad} s)^2}=-\frac{f''(s)}{f'(s)}\,\phi(s),$$

i.e. the surface Σ is equipotential, if the ratio $\Delta s/(\operatorname{grad} s)^2$ is a function of s only.

Denoting

$$q_n = \frac{x^2}{(a^2+s)^n}+\frac{y^2}{(b^2+s)^n}+\frac{z^2}{(c^2+s)^n}, \qquad p = \frac{1}{a^2+s}+\frac{1}{b^2+s}+\frac{1}{c^2+s},$$

we see that the equation

$$\frac{x^2}{a^2+s}+\frac{y^2}{b^2+s}+\frac{z^2}{c^2+s} = 1 \tag{1}$$

reduces to $q_1 = 1$. Differentiating it with respect to x, we obtain:

$$s_x = \frac{2x}{(s+a^2)q_2}, \quad s_y = \frac{2y}{(s+b^2)q_2}, \quad s_z = \frac{2z}{(s+c^2)q_2}, \quad (\text{grad } s)^2 = \frac{4}{q^2}.$$

Calculations give:

$$s_{xx} = \frac{2}{q_2(a^2+s)} - \frac{8x^2}{q_2^2(a^2+s)^3} - \frac{8x^2q_3}{q_2^3(a^2+s)^2}, \ldots,$$

$\Delta s = 2p/q_2$ and therefore, $\phi(s) = p/2$. After integrating the equation

$$\frac{f''(s)}{f'(s)} = -\phi(s) = -\frac{1}{2}\left(\frac{1}{s+a^2}+\frac{1}{s+b^2}+\frac{1}{s+c^2}\right),$$

we obtain:

$$V = f(s) = A\int_0^s \frac{ds}{R(s)}+B,$$

where

$$R(s) = \sqrt{(s+a^2)(s+b^2)(s+c^2)}.$$

At infinity for $s \to \infty$ the potential must be equal to zero; hence it follows that

$$V = -A\int_s^\infty \frac{ds}{R(s)}.$$

152. If the ellipsoid, given by the equation

$$\frac{x^2}{a^2}+\frac{y^2}{b^2}+\frac{z^2}{c^2} = 1,$$

is conducting and carries a charge e, then

$$V = \frac{e}{2\varepsilon}\int_s^\infty \frac{ds}{R(s)}.$$

On the surface of the ellipsoid $s = 0$ the potential V equals

$$V_0 = \frac{e}{2\varepsilon}\int_0^\infty \frac{ds}{R(s)}.$$

The capacity of the ellipsoid equals

$$C = \frac{e}{V_0} = 2\varepsilon\left[\int_0^\infty \frac{ds}{R(s)}\right]^{-1}.$$

The surface density of the charge is given by the expression

$$\sigma = -\frac{\varepsilon}{4\pi}|\text{grad } V|_{s=0} = -\frac{\varepsilon}{4\pi}(V_s|\text{grad } s|)_{s=0},$$

from which by virtue of the equalities $(V_s)_{s=0} = -e/2\varepsilon abc$, $|\text{grad } s| = 2/\sqrt{q_2}$ it follows:

$$\sigma = \frac{e}{4\pi abc}\left(\frac{x^2}{a^4}+\frac{y^2}{b^4}+\frac{z^2}{c^4}\right)^{-\frac{1}{2}}.$$

If $a = b > c$ (oblate spheroid or oblate ellipsoid of revolution), then we obtain:

$$V = \frac{e}{2\varepsilon}\int_\lambda^\infty \frac{ds}{(s+a^2)\sqrt{s+c^2}} = \frac{e}{\varepsilon\sqrt{a^2-c^2}} \text{arc tan} \sqrt{\frac{a^2-c^2}{\lambda+c^2}}.$$

If $a > b = c$ (prolate spheroid or prolate ellipsoid of revolution), then we obtain:

$$V = \frac{e}{2\varepsilon}\frac{1}{\sqrt{a^2-b^2}}\ln\frac{\sqrt{\lambda+a^2}+\sqrt{b^2-a^2}}{\sqrt{\lambda+a^2}-\sqrt{b^2-a^2}}.$$

Here λ is a positive root of equation (1).

153. The surface charge density on an elliptic disc

$$\sigma = \frac{e}{4\pi bc}\left(1-\frac{y^2}{b^2}-\frac{z^2}{c^2}\right)^{-\frac{1}{2}},$$

where e is the total charge of the disc.

The capacity of a circular disc ($a = 0$, $b = c$) is

$$C = 8\varepsilon b.$$

The charge density on each side of the circular disc is

$$\sigma = \frac{e}{4\pi b\sqrt{b^2-\rho^2}}.$$

The potential produced by the circular disc is expressed by the relation

$$V = 4V_0 \text{ arc tan}\frac{b}{\sqrt{\lambda}},$$

or

$$V = 4V_0 \arc\tan \frac{\sqrt{2}b}{\sqrt{r^2-b^2+\sqrt{(r^2-b^2)^2+4b^2x^2}}} \quad (x^2+y^2+z^2 = r^2).$$

Method. In order to evaluate σ for an elliptic disc one uses the formula for σ from the solution of problem 152

$$\sigma = \frac{e}{4\pi bc}\left(\frac{x^2}{b^2}+\frac{a^2y^2}{b^4}+\frac{a^2z^2}{c^4}\right)^{-\frac{1}{2}}.$$

Eliminating x^2/a_2 by using the equation $x^2/a_2+y^2/b^2+z^2/c^2 = 1$, we obtain:

$$\sigma = \frac{e}{4\pi bc}\left(1-\frac{y^2}{b^2}-\frac{z^2}{c^2}+\frac{a^2y^2}{b^4}+\frac{a^2z^2}{c^4}\right)^{-\frac{1}{2}}.$$

The limiting case for $a \to 0$ gives the necessary relation for σ.

For a circular disc $a = 0$, $c^2 = b^2$, $\rho^2 = y^2+z^2$. The parameter λ is defined as the positive root of the equation

$$\frac{x^2}{s}+\frac{y^2+z^2}{b^2+s} = 1,$$

equal to

$$\lambda = \tfrac{1}{2}[r^2-b^2+\sqrt{(r^2-b^2)^2+4b^2x^2}].$$

154. *Method.* It is required to prove that

$$\Delta V = -4\pi\rho_0 \text{ inside an ellipsoid,}$$
$$\Delta V = 0 \qquad \text{outside an ellipsoid.}$$

Proof of the first equality does not present difficulty. In the proof of the second equality one must use the relations grad $\lambda\cdot$grad $f = 4$, div grad $f = 4R'(s)/R(s)$.

155. (a) The gravitational potential of any prolate ellipsoid of revolution $(b = c < a)$ is

$$V(x,y,z) = 2\pi(1-\varepsilon^2)\rho_0\left\{\frac{a^2}{2\varepsilon}\ln\frac{1+\varepsilon}{1-\varepsilon}-\frac{1}{\varepsilon^3}\left(\frac{1}{2}\ln\frac{1+\varepsilon}{1-\varepsilon}-\varepsilon\right)x^2-\right.$$
$$\left.-\frac{1}{2\varepsilon^3}\left(\frac{\varepsilon}{1-\varepsilon^2}-\frac{1}{2}\ln\frac{1+\varepsilon}{1-\varepsilon}\right)(y^2+z^2)\right\} \quad \text{inside the ellipsoid,}$$

$$V(x,y,z) = 2\pi(1-\varepsilon^2)\rho_0\left\{\frac{a^2}{2\varepsilon}\ln\frac{\sqrt{a^2+\lambda}+\varepsilon a}{\sqrt{a^2+\lambda}-\varepsilon a}-\right.$$
$$-\frac{1}{\varepsilon^3}\left(\frac{1}{2}\ln\frac{\sqrt{a^2+\lambda}+\varepsilon a}{\sqrt{a^2+\lambda}-\varepsilon a}-\frac{\varepsilon a}{\sqrt{a^2+\lambda}}\right)x^2-$$
$$\left.-\frac{1}{2\varepsilon^3}\left[\frac{\varepsilon a\sqrt{a^2+\lambda}}{(1-\varepsilon^2)a^2+\lambda}-\frac{1}{2}\ln\frac{\sqrt{a^2+\lambda}+\varepsilon a}{\sqrt{a^2+\lambda}-\varepsilon a}\right](y^2+z^2)\right\} \quad \text{outside the ellipsoid,}$$

where $\varepsilon^2 = 1 - c^2/a_2$, λ is a positive root of the equation

$$\frac{x^2}{a^2+s} + \frac{y^2+z^2}{c^2+s} = 1.$$

The gravitational potential of an oblate ellipsoid of revolution ($b = a > c$) is

$$V(x, y, z) = 2\pi(1+\varepsilon^2)\rho_0 \left\{\frac{c^2}{\varepsilon} \arctan \varepsilon - \frac{1}{2\varepsilon^3}\left(\arctan \varepsilon - \frac{\varepsilon}{1+\varepsilon^2}\right)(x^2+y^2) - \right.$$

$$\left. - \frac{1}{\varepsilon^3}(\varepsilon - \arctan \varepsilon)z^2\right\} \quad \text{inside the ellipsoid,}$$

$$V(x, y, z) = 2\pi(1+\varepsilon^2)\rho_0 \left\{\frac{c^2}{\varepsilon} \arctan \frac{\varepsilon c}{\sqrt{c^2+\lambda}} - \frac{1}{2\varepsilon^3}\left[\arctan \frac{\varepsilon c}{\sqrt{c^2+\lambda}} - \right.\right.$$

$$\left.\left. - \frac{\varepsilon c\sqrt{c^2+\lambda}}{(1+\varepsilon^2)c^2+\lambda}\right](x^2+y^2) - \frac{1}{\varepsilon^3}\left(\frac{\varepsilon c}{\sqrt{c^2+\lambda}} - \arctan \frac{\varepsilon c}{\sqrt{c^2+\lambda}}\right)z^2\right\} \quad \text{outside the ellipsoid,}$$

where $\varepsilon^2 = a^2/c_2 - 1$, and λ is a positive root of the equation

$$\frac{x^2+y^2}{a^2+s} + \frac{z^2}{c^2+s} = 1.$$

The limiting case for $\varepsilon \to 0$ leads to the potential of a homogeneous sphere of radius a:

$$V = \begin{cases} 2\pi\rho_0\left(a - \frac{1}{3}r^2\right) & \text{for} \quad r < a, \\ \frac{M}{r} & \text{for} \quad r > a\left(M = \frac{4\pi}{3}a^3\rho_0\right). \end{cases}$$

156. The logarithmic potential of a homogeneous elliptic region

$$\frac{x^2}{a^2} + \frac{y^2}{b^2} - 1 \leqslant 0$$

is given by the formulae

$$V(x, y) = \pi ab\rho_0\left(\frac{1}{2} - \ln\frac{a+b}{2}\right) - \frac{\pi ab\rho_0}{a+b}\left(\frac{x^2}{a^2} + \frac{y^2}{b^2}\right) \quad \text{inside the ellipse,}$$

$$V(x, y) = \pi ab\rho_0\left(\frac{1}{2} - \ln\frac{a+b}{2}\right) - \pi ab\rho_0 \frac{\dfrac{x^2}{\sqrt{a^2+\lambda}} + \dfrac{y^2}{\sqrt{b^2+\lambda}}}{\sqrt{a^2+\lambda} + \sqrt{b^2+\lambda}} -$$

$$- \pi ab\rho_0 \ln\frac{\sqrt{a^2+\lambda} + \sqrt{b^2+\lambda}}{a+b} \quad \text{outside the ellipse,}$$

where λ is the positive root of the equation

$$\frac{x^2}{a^2+s}+\frac{y^2}{b^2+s}=1.$$

157. The potential outside an ellipse equals

$$V(x,y)=V_0\left[1-\frac{1}{2\ln\frac{2}{a+b}}\int_0^\lambda\frac{ds}{\sqrt{(s+a^2)(s+b^2)}}\right]\quad(a\geqslant b>0),$$

or

$$V(x,y)=V_0\left[1-\frac{1}{2\ln\frac{2}{a+b}}\ln\frac{2\lambda+a^2+b^2+\sqrt{(\lambda+a^2)(\lambda+b^2)}}{(a+b)^2}\right],$$

where λ is the positive root of the equation

$$\frac{x^2}{a^2+s}+\frac{y^2}{b^2+s}=1.$$

The density of the charge, distributed on the ellipse, equals

$$\sigma=\frac{V_0}{2\pi ab\ln\frac{2}{a+b}}\frac{1}{\sqrt{\frac{x^2}{a^4}+\frac{y^2}{b^4}}}.$$

The limiting case for $b\to 0$ gives the potential of the segment $0\leqslant x\leqslant a$ in the plane (x,y)

$$V(x,y)=V_0\left[1-\frac{1}{2\ln\frac{2}{a}}\times\right.$$

$$\left.\times\ln\frac{\rho^2+\sqrt{(\rho^2-a^2)^2+4a^2y^2}+\sqrt{0\cdot 5\rho^2[\rho^2-a^2+\sqrt{(\rho^2-a^2)^2+4a^2y^2}]+a^2y^2}}{a^2}\right],$$

where $\rho^2=x^2+y^2$.

Method. The derivation of the formula for $V(x,y)$ is entirely analogous to the derivation deduced in the solution of problem 153.

At infinity we lay down the condition

$$u=V-A\ln\frac{1}{\rho}\to 0\quad\text{for}\quad\rho=\sqrt{x^2+y^2}\to\infty,\quad|\text{grad}\,u|<\frac{B}{\rho^2},$$

where A and $B>0$ are some constants.

158. Let I be the current flowing through the loop C_a with centre at the point $z=0$, $\rho=0$ and of radius a, I' the current flowing through the loop C_b of radius b with centre at the point $z=d$, $\rho=0$.

For the force of interaction between C_a and C_b one of the representations is possible:

$$\text{(1)}\quad F = \frac{\mu II' dk}{2c^2\sqrt{ab}}\left[-K(k)+\frac{a^2+b^2+d^2}{(a-b)^2+d^2}E(k)\right],$$

where

$$k^2 = \frac{4ab}{(a+b)^2+d^2},\qquad K(k) = \int_0^{\frac{\pi}{2}} \frac{d\phi}{\sqrt{1-k^2\sin^2\phi}},$$

$$E(k) = \int_0^{\frac{\pi}{2}} \sqrt{1-k^2\sin^2\phi}\, d\phi$$

are elliptic integrals of the first and second kind;

$$\text{(2)}\quad F = -\frac{\pi\mu II'\sin\alpha}{c^2}\sum_{n=1}^{\infty}\left(\frac{a^2}{b^2+d^2}\right)^{\frac{n}{2}} P_n^1(\cos\alpha)P_n(0)\qquad (d^2+b^2 > a^2),$$

where $\alpha = \theta_b$ is a coordinate of points of the ring C_b, if the origin of coordinates is at the centre of the ring C_a;

$$\text{(3)}\quad F = -\frac{\pi\mu II'\sin^2\beta}{c^2}\sum_{n=2}^{\infty}\frac{1}{n+1}\left(\frac{a}{b}\right)^{n+1} P_{n+1}^1(\cos\beta)P_n^1(\cos\beta)\qquad (b > a);$$

in this the origin of coordinates is situated at the apex of the circular cone passing through C_a and C_b ($a \neq b$) and having solid angle β; if $a < b$, then the series on the right converges rapidly.

Method. The force, acting on the contour, situated in a magnetic field, through which a current I flows, equals

$$\boldsymbol{F} = \frac{I}{c}\oint [d\boldsymbol{s}\boldsymbol{B}],$$

where $\boldsymbol{B}$ is the magnetic induction of the external field, and integration is performed over the given contour. In our case

$$\boldsymbol{B} = \frac{\mu I'}{c}\oint_2 \frac{[d\boldsymbol{s}\boldsymbol{r}]}{r^3}.$$

To evaluate B on the contour C_b one must use the solution of problem 146.

159. Let the ring C_a of radius a and the ring C_b of radius b lie in the parallel planes Σ_a and Σ_b, and their centres are situated on one straight line, perpendicular to the planes Σ_a and Σ_b; the coefficient of mutual inductance can be represented in the following way:

(1) $M_{12} = \dfrac{2\mu\sqrt{ab}}{k}\left[\left(1-\dfrac{1}{2}k^2\right)K-E\right]$, where $k^2 = \dfrac{4ab}{(a+b)^2+d^2}$, $K(k)$ and $E(k)$ are elliptic integrals, d is the distance between centres of the rings.

(2) If the origin of coordinates is situated at the centre of C_a, then the ring C_b will have coordinates $r_b = \sqrt{b^2+d^2}$, $\theta_b = \beta$, and

$$M_{12} = \pi\mu b \sum_{m=0}^{\infty} (-1)^m \frac{(2m-1)!!}{(2m-1)(2m)!!}\left(\frac{b^2+d^2}{a^2}\right)^{\frac{2m-1}{2}} P^1_{2m-1}(\cos\beta),$$

$$\left[\frac{b^2+d^2}{a^2} < 1\right].$$

If $\dfrac{b^2+d^2}{a^2} > 1$, then instead of $\left[\dfrac{b^2+d^2}{a^2}\right]^{\frac{2m-1}{2}}$ it is necessary to write $\left[\dfrac{a^2}{b^2+d^2}\right]^m$.

The expression for the mutual inductance of two arbitrarily oriented rings has a similar form, if their axes intersect.

Method. The coefficient of mutual inductance of contours 1 and 2 is given by the formula

$$M_{12} = \oint_1 A_2\,\mathrm{d}s_1,$$

where A_2 is the vector-potential of the field produced by unit current in contour 2.

In our case

$$M_{ba} = \oint_{C_b} A_a\,\mathrm{d}s_b = 2\pi b|A_a|_{\rho=b}^{z=d},$$

where $|A_a|$ is calculated on the basis of the solution of problem 146.

CHAPTER V

EQUATIONS OF PARABOLIC TYPE

§ 1. Physical Problems Leading to Equations of Parabolic Type; Statement of Boundary-value Problems

1. For the temperature of a liquid in the non-steady case we have:

$$\frac{\partial u}{\partial t} = a^2\left(\frac{\partial^2 u}{\partial x^2}+\frac{\partial^2 u}{\partial y^2}+\frac{\partial^2 u}{\partial z^2}\right)-v_0\frac{\partial u}{\partial x}, \quad -\infty < x, y < +\infty,$$

$$0 < z, t < +\infty, \tag{1}$$

a^2 is the coefficient of thermal conductivity;

$$\lambda\frac{\partial u}{\partial z} = \alpha(u-f) \quad \text{for} \quad z = 0, \tag{2}$$

where $f(x, y, t)$ is the temperature of the plane $z = 0$,

$$u|_{t=0} = \phi(x, y, z), \quad -\infty < x, y < +\infty, \quad 0 < z < +\infty. \tag{3}$$

In the steady-state case (with "negligibly small" heat conduction in the direction of the x-axis)

$$\frac{\partial u}{\partial x} = \frac{a^2}{v_0}\left(\frac{\partial^2 u}{\partial y^2}+\frac{\partial^2 u}{\partial z^2}\right), \quad -\infty < y < +\infty, \quad 0 < x, z < +\infty. \tag{1'}$$

$$\lambda\frac{\partial u}{\partial z} = \alpha(u-f) \quad \text{for} \quad z = 0, \tag{2'}$$

where $f(x, y)$ is the tepmerature of the plane $z = 0$,

$$u|_{x=0} = \phi(y, z), \quad -\infty < y < +\infty, \quad 0 < z < +\infty. \tag{3'}$$

2. We obtain the following equations for the concentration of a substance, diffusing into a moving medium, which fills the semispace $z > 0$ and moves with constant velocity in the direction of the x-axis with the condition that the plane $z = 0$ is impermeable

$$\frac{\partial u}{\partial t} = D\left(\frac{\partial^2 u}{\partial x^2}+\frac{\partial^2 u}{\partial y^2}+\frac{\partial^2 u}{\partial z^2}\right)-v_0\frac{\partial u}{\partial x}, \quad -\infty < x, y < +\infty,$$

$$0 < z, t < +\infty \tag{1}$$

where D is the diffusion coefficient.

$$\frac{\partial u}{\partial z} = 0 \quad \text{for} \quad z = 0.$$

In the steady-state case

$$\frac{\partial u}{\partial x} = \frac{D}{v_0}\left(\frac{\partial^2 u}{\partial y^2} + \frac{\partial^2 u}{\partial z^2}\right), \quad -\infty < y < +\infty, \quad 0 < x, z < +\infty, \tag{1'}$$

$$\frac{\partial u}{\partial z} = 0 \quad \text{for} \quad z = 0, \tag{2'}$$

$$u|_{x=0} = \phi(y, z), \quad -\infty < y < +\infty, \quad 0 < z < +\infty. \tag{3'}$$

3. (a) $$\frac{\partial u}{\partial t} = D\left(\frac{\partial^2 u}{\partial x^2} + \frac{\partial^2 u}{\partial y^2} + \frac{\partial^2 u}{\partial z^2}\right) - \beta u, \quad \beta > 0,$$

$$-\infty < x, y, z < +\infty, \quad 0 < t < +\infty, \tag{1}$$

$$u|_{t-0} = \phi(x, y, z,) \quad -\infty < x, y, z < +\infty, \tag{2}$$

(b) $$\frac{\partial u}{\partial t} = D\left(\frac{\partial^2 u}{\partial x^2} + \frac{\partial^2 u}{\partial y^2} + \frac{\partial^2 u}{\partial z^2}\right) + \beta u, \quad \beta > 0,$$

$$-\infty < x, y, z < +\infty, \quad 0 < t < +\infty, \tag{1'}$$

$$u|_{t=0} = \phi(x, y, z), \quad -\infty < x, y, z < +\infty. \tag{2'}$$

4. $$\left.\begin{aligned} \frac{\partial \boldsymbol{E}}{\partial t} &= \frac{c^2}{4\pi\mu\sigma}\left(\frac{\partial^2 \boldsymbol{E}}{\partial x^2} + \frac{\partial^2 \boldsymbol{E}}{\partial y^2} + \frac{\partial^2 \boldsymbol{E}}{\partial z^2}\right), & (1)\\ \frac{\partial \boldsymbol{H}}{\partial t} &= \frac{c^2}{4\pi\mu\sigma}\left(\frac{\partial^2 \boldsymbol{H}}{\partial x^2} + \frac{\partial^2 \boldsymbol{H}}{\partial y^2} + \frac{\partial^2 \boldsymbol{H}}{\partial z^2}\right), & (1') \end{aligned}\right\} \begin{aligned} &-\infty < x, y, z < +\infty, \\ &0 < t < +\infty, \end{aligned}$$

where $\boldsymbol{E}$ and $\boldsymbol{H}$ are the electric and magnetic field vectors, c is the velocity of light in vacuum, μ is the magnetic permeability, and σ is conductivity.

$$\left.\begin{aligned} \boldsymbol{E}|_{t=0} &= \boldsymbol{i}\phi_1(x, y, z) + \boldsymbol{j}\phi_2(x, y, z) + \boldsymbol{k}\phi_3(x, y, z), & (2)\\ \boldsymbol{H}|_{t=0} &= \boldsymbol{i}\psi_1(x, y, z) + \boldsymbol{j}\psi_2(x, y, z) + \boldsymbol{k}\psi_3(x, y, z), & (2') \end{aligned}\right\} -\infty < x, y, z < +\infty,$$

where $\boldsymbol{i}, \boldsymbol{j}, \boldsymbol{k}$ are unit vectors along the x, y, z axes, and $\phi_1, \phi_2, \phi_3, \psi_1, \psi_2, \psi_3$ are given functions.

Method. We consider the system of Maxwell's equations

$$\operatorname{curl} \boldsymbol{H} = \frac{1}{c}\frac{\partial \boldsymbol{D}}{\partial t} + \frac{4\pi}{c}\boldsymbol{j}, \tag{3}$$

$$\operatorname{curl} \boldsymbol{E} = -\frac{1}{c}\frac{\partial \boldsymbol{B}}{\partial t}, \tag{4}$$

$$\operatorname{div} \boldsymbol{B} = 0, \tag{5}$$

$$\operatorname{div} \boldsymbol{D} = 0, \tag{6}$$

written on the assumption that there are no volume charges.

Using the so-called physical field equations

$$D = \varepsilon E, \quad B = \mu H, \quad j = \sigma E \tag{7}$$

and the condition of constancy of ε, μ, σ and neglecting the displacement currents $\frac{1}{c}\frac{\partial D}{\partial t}$ in comparison with conduction currents $\frac{4\pi}{c} j = \frac{4\pi\sigma}{c} E$, we obtain the equations

$$\operatorname{curl} E = -\frac{\mu}{c}\frac{\partial H}{\partial t}, \tag{8}$$

$$\operatorname{curl} H = \frac{4\pi\sigma}{c} E. \tag{9}$$

If one takes the rot of both sides of equation (8) and uses the familiar equality of vector analysis

$$\operatorname{curl}\operatorname{curl} a = \operatorname{grad}\operatorname{div} a - \operatorname{div}\operatorname{grad} a,$$

then equation (1) may be derived from equations (6), (7) and (9). Equation (1′) is derived similarly.

5. $$\frac{\partial u}{\partial t} = a^2\left(\frac{\partial^2 u}{\partial x^2} + \frac{\partial^2 u}{\partial y^2} + \frac{\partial^2 u}{\partial z^2}\right), \quad 0 \leqslant x \leqslant l, \quad -\infty < y, z < +\infty,$$

$$0 < t < +\infty, \tag{1}$$

$$\lambda u_x(0, y, z, t) - hu(0, y, z, t) = 0, \quad \lambda u_x(l, y, z, t) + hu(l, y, z, t) = 0, \tag{2}$$

$$u(x, y, z, 0) = f(x, y, z),$$

where l is the thickness of the plate, and λ is the coefficient of heat conduction. If the temperature is independent of x, then

$$u = u(y, z, t)$$

and

$$\frac{\partial u}{\partial t} = a^2\left(\frac{\partial^2 u}{\partial y^2} + \frac{\partial^2 u}{\partial z^2}\right) - 2h_1 u, \quad -\infty < y, z < +\infty, \quad 0 < t < +\infty, \tag{1}$$

$$h_1 = \frac{h}{c\rho_1},$$

where ρ_1 is the mass per unit area of the plate.

6. $$\frac{\partial u}{\partial t} = a^2\left\{\frac{1}{r}\frac{\partial}{\partial r}\left(z\frac{\partial u}{\partial r}\right) + \frac{1}{r_2}\frac{\partial^2 u}{\partial \phi^2}\right\}, \quad r_1 \leqslant r \leqslant r_2, \quad 0 \leqslant \phi \leqslant 2\pi,$$

$$0 < t < +\infty, \tag{1}$$

$$\lambda u_r(r_1, \phi, t) - h[u(r_1, \phi, t) - U(t)] = 0, \quad 0 < t < +\infty, \tag{2}$$

$$\lambda u_r(r_2, \phi, t) + h[u(r_2, \phi, t) - U]_0 = 0, \quad 0 < t < +\infty, \tag{2′}$$

$$\pi r_1 c^* \rho^* \frac{dU(t)}{dt} = -h\left[2\pi U(t) - \int_0^{2\pi} u(r_1, \phi, t)\, d\phi\right], \quad 0 < t < +\infty, \quad (2'')$$

where $U(t)$, ρ^*, c^* are the temperature, mass density and specific heat of the liquid inside the tube,

$$u(r, \phi, 0) = f(r, \phi), \quad r_1 \leqslant r \leqslant r_2, \quad 0 \leqslant \phi \leqslant 2\pi. \quad (3)$$

7. In order to determine the velocity $v(r, t)$ of the liquid† and the angular velocity $\omega(t)$ of the cylinder we obtain the boundary-value problem

$$\frac{\partial v}{\partial t} = \nu\left\{\frac{\partial^2 v}{\partial r^2} + \frac{1}{r}\frac{\partial v}{\partial r} - \frac{v}{r^2}\right\}, \quad r_0 \leqslant r \leqslant \infty, \quad 0 < t < +\infty, \quad (1)$$

$$v|_{r=r_0} = r_0\omega(t), \quad v \to 0 \quad \text{for} \quad r \to +\infty, \quad 0 < t < +\infty, \quad (2)$$

$$K\frac{d\omega}{dt} = M + 2\pi r_0^2 \rho\nu\left[\frac{\partial v}{\partial r} - \frac{v}{r}\right]_{r=r_0}. \quad (3)$$

Method. In cylindrical coordinates

(1) the equations of motion of an incompressible viscous liquid are

$$\frac{\partial v_r}{\partial t} + v_r\frac{\partial v_r}{\partial r} + \frac{v_\phi}{r}\frac{\partial v_r}{\partial \phi} + v_z\frac{\partial v_r}{\partial z} - \frac{v_\phi^2}{r}$$
$$= -\frac{1}{\rho}\frac{\partial p}{\partial r} + \nu\left(\frac{\partial^2 v_r}{\partial r^2} + \frac{1}{r^2}\frac{\partial^2 v}{\partial \phi^2} + \frac{\partial^2 v_r}{\partial z^2} + \frac{1}{r}\frac{\partial v_r}{\partial r} - \frac{2}{r^2}\frac{\partial v_\phi}{\partial \phi} - \frac{v_r}{r^2}\right),$$

$$\frac{\partial v_\phi}{\partial t} + v_r\frac{\partial v_\phi}{\partial r} + \frac{v_\phi}{r}\frac{\partial v_\phi}{\partial \phi} + v_z\frac{\partial v_\phi}{\partial z} + \frac{v_r v_\phi}{r}$$
$$= -\frac{1}{\rho r}\frac{\partial p}{\partial \phi} + \nu\left(\frac{\partial^2 v_\phi}{\partial r^2} + \frac{1}{r^2}\frac{\partial^2 v_\phi}{\partial \phi^2} + \frac{\partial^2 v_\phi}{\partial z^2} + \frac{1}{r}\frac{\partial v_\phi}{\partial r} + \frac{2}{r^2}\frac{\partial^2 v_r}{\partial \phi^2} - \frac{v_\phi}{r^2}\right),$$

$$\frac{\partial v_z}{\partial t} + v_r\frac{\partial v_z}{\partial r} + \frac{v_\phi}{r}\frac{\partial v_z}{\partial \phi} + v_z\frac{\partial v_z}{\partial z}$$
$$= -\frac{1}{\rho}\frac{\partial p}{\partial z} + \nu\left(\frac{\partial^2 v_z}{\partial r^2} + \frac{1}{r^2}\frac{\partial^2 v_z}{\partial \phi^2} + \frac{\partial^2 v_z}{\partial z^2} + \frac{1}{r}\frac{\partial v_z}{\partial r}\right);$$

(2) the equation of continuity is

$$\frac{\partial v_r}{\partial r} + \frac{1}{r}\frac{\partial v_\phi}{\partial \phi} + \frac{\partial v_z}{\partial z} + \frac{v_r}{r} = 0,$$

where v_r, v_ϕ, v_z, are components of the velocity vector in the direction of the unit vectors of the cylindrical system of coordinates;

† $v(r, t) = v_\phi(r, t)$; see the method to the present problem.

(3) the components of the stress tensor are

$$\sigma_r = -p+2\nu\frac{\partial v_r}{\partial r}, \qquad \tau_{r\phi} = \nu\left(\frac{1}{r}\frac{\partial v_r}{\partial \phi}+\frac{\partial v_\phi}{\partial r}-\frac{v_\phi}{r}\right),$$

$$\sigma_\phi = -p+2\nu\left(\frac{1}{r}\frac{\partial v_\phi}{\partial \phi}+\frac{v_r}{r}\right), \qquad \tau_{\phi z} = \nu\left(\frac{\partial v_\phi}{\partial z}+\frac{1}{r}\frac{\partial v_z}{\partial \phi}\right),$$

$$\sigma_z = -p+2\nu\frac{\partial v_z}{\partial z}, \qquad \tau_{zr} = \nu\left(\frac{\partial v_z}{\partial r}+\frac{\partial v_r}{\partial z}\right).$$

8. *Solution.* We choose the origin of coordinates on the waterproof base and direct the z-axis vertically upwards. The components of the vectors $\boldsymbol{f}, \boldsymbol{V}, \boldsymbol{U}$ are $\boldsymbol{f} = (f_x, f_y, f_z)$, $\boldsymbol{V} = \{V_x, V_y, V_z\}$, $\boldsymbol{U} = \{u, v, w\}$. Then the equations of motion of the subsurface waters may be written in the form

$$\rho\frac{dV_x}{dt} = -\frac{\partial p}{\partial x}+g\rho f_x, \quad \rho\frac{dV_y}{dt} = -\frac{\partial p}{\partial y}+g\rho f_y, \quad \rho\frac{dV_z}{dt} = -\frac{\partial p}{\partial z}+g\rho f_z-g\rho,$$

where p is the pressure. Neglecting inertia forces and using $\boldsymbol{f} = -\frac{1}{k}\boldsymbol{U}$, we obtain the approximate equations

$$u = -\frac{k}{g\rho}\frac{\partial p}{\partial x}, \quad v = -\frac{k}{g\rho}\frac{\partial p}{\partial y}, \quad w = -k\left(\frac{1}{g\rho}\frac{\partial p}{\partial z}+1\right), \tag{1}$$

which may be written in vector form in the following way:

$$\boldsymbol{U} = -k \operatorname{grad} H, \tag{2}$$

where

$$H(x, y, z, t) = \frac{p-p_0}{g\rho}+z; \tag{3}$$

p_0 is the pressure at the free surface (not dependent on x, y, z).

Let p_1 denote the hydrostatic pressure at a point lying at a height z above the waterproof base, and $z = H_0(x, y, t)$ be the equation of the free surface; then we obtain the following expression for the hydrostatic pressure:

$$p_1-p_0 = g\rho[H_0(x, y, t)-z], \quad 0 \leqslant z \leqslant H(x, y, t),$$

i.e.

$$\frac{p_1-p_0}{g\rho}+z = H_0(x, y, t). \tag{4}$$

From (3) and (4) we find the following expression for the excess pressure:

$$\frac{p-p_1}{g\rho} = H(x, y, z, t)-H_0(x, y, t). \tag{5}$$

By assumption (1) of the conditions of the problem it follows from (2), (3) and (5):

$$u = -k\frac{\partial H_0}{\partial x}, \quad v = -k\frac{\partial H_0}{\partial y}, \tag{6}$$

i.e. particles of subsurface water lying on the one vertical line have the same horizontal velocities.

Considering a thin vertical prism with base $\Delta x \Delta y$ and of height $H_0(x, y, t)$ and using (6), the equation of continuity may be written in the form

$$\frac{\partial H_0}{\partial t} = \frac{k}{m}\left\{\frac{\partial}{\partial x}\left(H_0 \frac{\partial H_0}{\partial x}\right) + \frac{\partial}{\partial y}\left(H_0 \frac{\partial H_0}{\partial y}\right)\right\}. \qquad (7)$$

If the subsurface layer extends indefinitely, then the boundary-value problem to determine the motion of the free surface of the subsurface waters may be formulated in the following way:

$$\frac{\partial H_0}{\partial t} = \frac{k}{m}\left\{\frac{\partial}{\partial x}\left(H_0 \frac{\partial H_0}{\partial x}\right) + \frac{\partial}{\partial y}\left(H_0 \frac{\partial H_0}{\partial y}\right)\right\}, \quad -\infty < x, y < +\infty, \quad 0 < t < +\infty, \qquad (8)$$

$$H_0(x, y, 0) = \phi(x, y), \quad -\infty < x, y < +\infty. \qquad (9)$$

Note. Often one changes from the non-linear equation (7) to the linear equation

$$\frac{\partial H_0}{\partial t} = a^2\left(\frac{\partial^2 H_0}{\partial x^2} + \frac{\partial^2 H_0}{\partial y^2}\right), \quad a^2 = \frac{kh_0}{m}, \qquad (7')$$

by replacing the factor H_0 in the curly brackets on the right-hand side of equation (7) by the mean height $h_0 = \text{const.}$ of the free surface.

§ 2. The Method of Separation of Variables

1. Boundary-value Problems not Requiring the Application of Special Functions

(a) *Homogeneous media*

9. The solution of the boundary-value problem

$$\frac{\partial u}{\partial t} = a^2\left\{\frac{\partial^2 u}{\partial x^2} + \frac{\partial^2 u}{\partial y^2} + \frac{\partial^2 u}{\partial z^2}\right\}, \quad 0 < x < l_1, \quad 0 < y < l_2, \quad 0 < z < l_3, \quad 0 < t < +\infty, \qquad (1)$$

$$u|_{x=0} = u|_{x=l_1} = u|_{y=0} = u|_{y=l_2} = u|_{z=0} = u_{z=l_3} = 0, \quad 0 < t < +\infty, \qquad (2)$$

$$u|_{t=0} = f(x, y, z), \quad 0 < x < l_1, \quad 0 < y < l_2, \quad 0 < z < l_3, \qquad (3)$$

is:

$$u(x, y, z, t) = \sum_{k,m,n=1}^{+\infty} A_{k,m,n}\, e^{-a^2\pi^2\left(\frac{k^2}{l_1^2} + \frac{m^2}{l_2^2} + \frac{n^2}{l_3^2}\right)t} \sin\frac{k\pi x}{l_1} \sin\frac{m\pi y}{l_2} \sin\frac{n\pi z}{l_3}, \qquad (4)$$

where

$$A_{k,m,n} = \frac{8}{l_1 l_2 l_3}\int_0^{l_1} d\xi \int_0^{l_2} d\eta \int_0^{l_3} f(\xi, \eta, \zeta) \sin\frac{k\pi\xi}{l_1} \sin\frac{m\pi\eta}{l_2} \sin\frac{n\pi\zeta}{l_3} d\zeta. \qquad (5)$$

10. $u(x, y, z, t)$

$$=\left(\frac{4}{\pi}\right)^3 U_0 \sum_{k,m,n=0}^{+\infty} \frac{e^{-\frac{a^2\pi^2}{l^2}[(2k+1)^2+(2m+1)^2+(2n+1)^2]t}}{(2k+1)(2m+1)(2n+1)} \times$$

$$\times \sin\frac{(2k+1)\pi x}{l} \sin\frac{(2m+1)\pi y}{l} \sin\frac{(2n+1)\pi z}{l}. \quad (1)$$

At the centre of the cube

$$u\left(\frac{l}{2}, \frac{l}{2}, \frac{l}{2}, t\right) = U_0\left(\frac{4}{\pi}\right)^3 \left\{\sum_{k=0}^{+\infty} (-1)^k \frac{e^{-\frac{a^2\pi^2(2k+1)^2}{l^2}t}}{2k+1}\right\}^3. \quad (2)$$

For all t satisfying the inequality

$$t \geqslant t^* = -\frac{l^2}{8\pi^2 a^2} \ln 3\tilde{\varepsilon}, \quad (3)$$

where $\tilde{\varepsilon}$ is less than the least of the numbers 1 and $\varepsilon/9$, at the centre of the cube it is known a steady-state will occur with relative accuracy ε.

Method. Let us denote the first term of the series in brackets of equality (2) by α, and the sum of all the remaining terms of it by S. For all t satisfying the inequality (3)†:

$$\left|\frac{S}{\alpha}\right| < \tilde{\varepsilon}; \quad (4)$$

since $\tilde{\varepsilon} < 1$ and $\tilde{\varepsilon} < \varepsilon/9$ then moreover:

$$\left|\frac{3\alpha^2 S + 3\alpha S^2 + S^3}{\alpha^3}\right| \leqslant 3\left|\frac{S}{\alpha}\right|\left\{1+\left|\frac{S}{\alpha}\right|+\frac{1}{3}\left|\frac{S}{\alpha}\right|^2\right\} < 9\left|\frac{S}{\alpha}\right| < \varepsilon, \quad (5)$$

i.e. at the centre of the cube a steady-state will occur with relative accuracy ε.

11. The solution of the boundary-value problem

$$\frac{\partial u}{\partial t} = a^2\left\{\frac{\partial^2 u}{\partial x^2}+\frac{\partial^2 u}{\partial y^2}+\frac{\partial^2 u}{\partial z^2}\right\}, \quad 0 < x < l_1, \quad 0 < y < l_2,$$

$$0 < z < l_3, \quad 0 < t < +\infty, \quad (1)$$

$$\left(\frac{\partial u}{\partial x} - hu\right)\bigg|_{x=0} = \left(\frac{\partial u}{\partial x} + hu\right)_{x=l_1} = \left(\frac{\partial u}{\partial y} - hu\right)\bigg|_{y=0} = \left(\frac{\partial u}{\partial y} + hu\right)_{y=l_2}$$

$$= \left(\frac{\partial u}{\partial z} - hu\right)_{z=0} = \left(\frac{\partial u}{\partial z} + hu\right)_{z=l_3} = 0, \quad 0 < t < +\infty, \quad (2)$$

$$u(x, y, z, 0) = f(x, y, z), \quad 0 < x < l_1, \quad 0 < y < l_2, \quad 0 < z < l_3 \quad (3)$$

---

† For more detail see chapter III, § 2, problem 22.

is:

$$u(x,y,z,t) = \sum_{k,m,n=1}^{+\infty} A_{k,m,n}\, e^{-a^2\left(\lambda_k^2+\mu_m^2+\nu_n^2\right)} X_k(x)Y_m(y)Z_n(z), \tag{4}$$

where

$$A_{k,m,n} = \frac{8\lambda_k^2\mu_m^2\nu_n^2 \int_0^{l_1}\int_0^{l_2}\int_0^{l_3} f(x,y,z)X_k(x)Y_m(y)Z_n(z)\,dx\,dy\,dz}{[l_1(\lambda_k^2+h^2)+2h]\,[l_2(\mu_m^2+h^2)+2h]\,[l_3(\nu_n^2+h^2)+2h]}, \tag{5}$$

$\lambda_1, \lambda_2, \ldots; \mu_1, \mu_2, \ldots; \nu_1, \nu_2, \ldots$ are respectively the positive roots of the equations

$$\cot l_1\lambda = \frac{1}{2}\left(\frac{\lambda}{h}-\frac{h}{\lambda}\right), \quad \cot l_2\mu = \frac{1}{2}\left(\frac{\mu}{h}-\frac{h}{\mu}\right), \quad \cot l_3\nu = \frac{1}{2}\left(\frac{\nu}{h}-\frac{h}{\nu}\right), \tag{6}$$

$$X_k(x) = \cos\lambda_k x + \frac{h}{\lambda_k}\sin\lambda_k x, \qquad Y_m(y) = \cos\mu_m y + \frac{h}{\mu_m}\sin\mu_m y,$$

$$Z_n(z) = \cos\nu_n z + \frac{h}{\nu_n}\sin\nu_n z. \tag{7}$$

In particular, if $f(x, y, z) = U_0 \equiv$ const., then

$$u(x,y,z,t) = 4^3h^3U_0 \sum_{k,m,n=0}^{+\infty} e^{-a^2\left(\lambda_{2k+1}^2+\mu_{2m+1}^2+\nu_{2n+1}^2\right)t} \times$$

$$\times \frac{X_{2k+1}(x)Y_{2m+1}(y)Z_{2n+1}(z)}{[l_1(\lambda_{2k+1}^2+h^2)+2h]\,[l_2(\mu_m^2+h^2)+2h]\,[l_3(\nu_n^2+h^2)+2h]}. \tag{8}$$

Method. The roots λ_k of the equation $\cot l_1\lambda = (\lambda/h-h/\lambda)/2$ satisfy the inequalities $0 < \lambda_1 l_1 < \pi$, $\pi < \lambda_2 l_1 < 2\pi$, $2\pi < \lambda_3 l_1 < 3\pi, \ldots$, i.e. the inequalities $0 < \lambda_1 l_1/2 < \pi/2$,

$$\frac{\pi}{2} < \frac{\lambda_2 l_1}{2} < \pi, \qquad \pi < \frac{\lambda_3 l_1}{2} < \frac{3\pi}{2}, \quad \ldots \tag{9}$$

Substituting λ_k in equation $\cot l_1\lambda = (\lambda/h-h/\lambda)/2$, we rewrite the result in the form

$$\tan\lambda_k l_1 = \frac{2\lambda_k h}{\lambda_k^2-h} = \frac{2\dfrac{h}{\lambda_k}}{1-\dfrac{h^2}{\lambda_k^2}} = -\frac{2\dfrac{\lambda_k}{h}}{1-\dfrac{\lambda_k^2}{h^2}}. \tag{10}$$

By (9) $\tan\lambda_k l_1/2 > 0$ for k odd and is less than zero for k even. But

$$\tan\beta = \frac{2\tan\dfrac{\beta}{2}}{1-\tan^2\dfrac{\beta}{2}},$$

therefore from (10) it follows that

$$\tan\frac{\lambda_k l_1}{2} = \begin{cases} \dfrac{h}{\lambda_k} & \text{for } k \text{ odd,} \\[2ex] -\dfrac{\lambda_k}{h} & \text{for } k \text{ even.} \end{cases}$$

Hence,

$$\int_0^{l_1} X_k(x)\,dx = \frac{1}{\lambda_k}\left[\sin\lambda_k l_1 + \frac{h}{\lambda_k}\left(1-\cos\lambda_k l_1\right)\right]$$

$$= \frac{1}{\lambda_k}2\sin\frac{\lambda_k l_1}{2}\cos\frac{\lambda_k l_1}{2}\left[1+\frac{h}{\lambda_k}\tan\frac{\lambda_k l_1}{2}\right]$$

$$= \frac{2}{\lambda_k}\cdot\frac{\tan\dfrac{\lambda_k l_1}{2}}{1+\tan^2\dfrac{\lambda_k l_1}{2}}\left[1+\frac{h}{\lambda_k}\tan\frac{\lambda_k l_1}{2}\right] = \begin{cases} 0 & \text{for } k \text{ even,} \\[1ex] \dfrac{2h}{\lambda_k} & \text{for } k \text{ odd.} \end{cases}$$

Similarly $\int_0^{l_2} Y_m(y)\,dy$, $\int_0^{l} Z_n(z)\,dz$ are calculated.

Note. If the parallelepiped is uniformly heated at the initial moment of time (i.e. $f(x, y, z) = U_0 \equiv \text{const.}$), then obviously, the distribution of temperature in it will be symmetrical with respect to the planes $x = l_1/2$, $y = l_2/2$, $z = l_3/2$, therefore it is possible to confine oneself to determining the temperature in one of the eight parallelepipeds, into which the original parallelepiped is divided by these planes.

12. The temperature at the centre of the cube $-l \leqslant x, y, z \leqslant l$ equals

$$U = 8U_0h^3\left\{\sum_{k=0}^{+\infty} e^{-a^2\lambda_k^{2}t}(-1)^k\,\frac{\sqrt{1+\dfrac{h^2}{\lambda_k^2}}}{l(\lambda_k^2+h^2)+h}\right\}^3, \tag{1}$$

where $\lambda_0, \lambda_1, \lambda_2, \ldots$ are positive roots of the equation

$$\tan\lambda l = \frac{h}{\lambda}. \tag{2}$$

For all values of time t satisfying the inequality

$$t \geqslant t^* = -\frac{1}{a^2(\lambda_1^2-\lambda_0^2)}\ln\left[\tilde{\varepsilon}\,\frac{(hl)^2+hl+(l\lambda_1)^2}{(hl)^2+hl+(l\lambda_0)^2}\sqrt{\frac{1+\left(\dfrac{h}{\lambda_0}\right)^2}{1+\left(\dfrac{h}{\lambda_1}\right)^2}}\right], \tag{3}$$

where $\tilde{\varepsilon}$ equals the smaller of the numbers 1 and $\varepsilon/9$, a steady-state will occur at the centre of the cube correct to ε.

Method. In order to derive equation (1) for the temperature at the centre of the cube $-l \leqslant x, y, z \leqslant l$ it is sufficient, according to the note to the preceding problem, to find firstly the temperature of part $0 \leqslant x, y, z \leqslant l$ of this cube, assuming the planes $x = 0$, $y = 0$, $z = 0$ to be thermally insulated.

In connection with the determination of the time to reach a steady-state correct to $\varepsilon > 0$, see the answer and instructions to problem 10 of the present chapter and the answers and instruction to problem 29 of chapter III.

13. The solution of the boundary-value problem

$$\frac{\partial u}{\partial x} = \frac{a^2}{v_0}\left(\frac{\partial^2 u}{\partial y^2} + \frac{\partial^2 u}{\partial z^2}\right), \quad 0 < x < +\infty, \quad 0 \leqslant y \leqslant l_1, \quad 0 \leqslant z \leqslant l_2, \tag{1}$$

$$\left(\frac{\partial u}{\partial y} - hu\right)\bigg|_{y=0} = \left(\frac{\partial u}{\partial y} + hu\right)\bigg|_{y=l} = \left(\frac{\partial u}{\partial z} - hu\right)\bigg|_{z=0} = \left(\frac{du}{dz} + hu\right)\bigg|_{z=l_2} = 0,$$
$$0 < x < +\infty, \tag{2}$$

$$u|_{x=0} = U_0, \quad 0 \leqslant y \leqslant l_1, \quad 0 \leqslant z \leqslant l_2 \tag{3}$$

is:

$$u(x, y, z) = 16U_0h^2 \sum_{m,n=0}^{+\infty} e^{-\frac{a^2}{v_0^2}\left(\mu_{2m+1}^2 + \nu_{2n+1}^2\right)x} \times$$

$$\times \frac{\left(\cos \mu_{2m+1}y + \frac{h}{\mu_{2m+1}} \sin \mu_{2m+1}y\right)\left(\cos \nu_{2n+1}z + \frac{h}{\nu_{2n+1}} \sin \nu_{2n+1}z\right)}{[l_1(\mu_{2m+1}^2 + h^2) + 2h]\,[l_2(\nu_{2n+1}^2 + h^2) + 2h]},$$

where $\mu_1, \mu_2, \ldots; \nu_1, \nu_2, \ldots$ are respectively the roots of the equations

$$\cot l_1\mu = \frac{1}{2}\left(\frac{\mu}{h} - \frac{h}{\mu}\right), \quad \cot l_2\nu = \frac{1}{2}\left(\frac{\nu}{h} - \frac{h}{\nu}\right).$$

Method. See the method to problem 11.

14. (a) $l_{cr} = a\pi\sqrt{\frac{3}{\beta}}$;

(b) the avalanche process of multiplication of particles occurs for a cube of any dimensions.

(c) $l_{cr} = a\sqrt{3}/\sqrt{\beta}$ arc cot $1/2(\sqrt{\beta}/ah\sqrt{3} - ah\sqrt{3}/\sqrt{\beta})$, if $\beta > 3a^2h^2$; the avalanche process will occur for any dimensions, if $\beta < 3a^2h^2$; β is the coefficient of multiplication in the equation

$$\frac{\partial u}{\partial t} = a^2\left(\frac{\partial^2 u}{\partial x^2} + \frac{\partial^2 u}{\partial y^2} + \frac{\partial^2 u}{\partial z^2}\right) + \beta u.$$

15. $$u(r, t) = \sum_{n=1}^{+\infty} A_n e^{-\frac{n^2\pi^2 a^2}{r_0^2} t} \frac{\sin \frac{n\pi r}{r_0}}{r}, \qquad (1)$$

$$A_n = \frac{2}{r_0} \int_0^{r_0} r f(r) \sin \frac{n\pi r}{r_0} dr. \qquad (2)$$

Method. Because of the radial symmetry the equation of heat conduction has the form

$$\frac{\partial u}{\partial t} = a^2 \left\{ \frac{\partial^2 u}{\partial r^2} + \frac{2}{r} \frac{\partial u}{\partial r} \right\}.$$

Transition to the new unknown function $v(r, t) = ru(r, t)$ leads to the boundary-value problem on the cooling of a rod

$$\frac{\partial v}{\partial t} = a^2 \frac{\partial^2 v}{\partial r^2}, \quad 0 < r < r_0, \quad 0 < t < +\infty, \qquad (3)$$

$$v(0, t) = 0, \quad v(r_0, t) = 0, \quad 0 < t < +\infty, \qquad (4)$$

$$v(r, 0) = rf(r), \quad 0 < r < r_0. \qquad (5)$$

The first of the boundary conditions (4) follows because the temperature $u(r, t)$ is bounded at the centre of the sphere

$$v(+0, t) = \lim_{r \to +0} ru(r, t) = 0.$$

16. $$u(r, t) = U_1 + 2\frac{r_0}{\pi}(U_0 - U_1) \sum_{n=1}^{+\infty} \frac{(-1)^{n+1}}{n} e^{-\frac{n^2\pi^2 a^2}{r_0^2} t} \frac{\sin \frac{n\pi r}{r_0}}{r}.$$

For all values of the time t satisfying the inequality

$$t \geqslant t^* = -\frac{r_0^2}{3\pi^2 a^2} \ln \varepsilon,$$

a steady-state will occur at the centre of the sphere with relative accuracy $\varepsilon > 0$.

Method. See the solution of problem 22 § 2 chapter III.

17. The solution of the boundary-value problem

$$\frac{\partial u}{\partial t} = a^2 \left\{ \frac{\partial^2 u}{\partial r^2} + \frac{2}{r} \frac{\partial u}{\partial r} \right\}, \quad 0 \leqslant r < r_0, \quad 0 < t < +\infty, \qquad (1)$$

$$\lambda \frac{\partial u}{\partial r} = q, \quad r = r_0, \quad 0 < t < +\infty, \qquad (2)$$

$$u(r, 0) = U_0, \quad 0 \leqslant r < r_0 \qquad (3)$$

is:

$$u(r,t) = U_0 + \frac{qr_0}{\lambda}\left\{\frac{3a^2t}{r_0^2} - \frac{3r_0^2-5r^2}{10r_0^2} - \sum_{n=1}^{+\infty} \frac{2r_0 e^{-\frac{a^2\mu_n^2 t}{r_0^2}}}{\mu_n^2 \cos\mu_n} \cdot \frac{\sin\frac{\mu_n r}{r_0}}{r}\right\}, \tag{4}$$

where μ_n are positive roots of the equation

$$\tan\mu = \mu. \tag{5}$$

18. The solution of the boundary-value problem

$$\frac{\partial u}{\partial t} = a^2\left\{\frac{\partial^2 u}{\partial r^2} + \frac{2}{r}\frac{\partial u}{\partial r}\right\}, \quad 0 \leqslant r < r_0, \quad 0 < t < +\infty, \tag{1}$$

$$\frac{\partial u}{\partial r} + hu = 0 \quad \text{for} \quad r = r_0, \quad 0 < t < +\infty, \tag{2}$$

$$u(r, 0) = f(r), \quad 0 < r < r_0 \tag{3}$$

is:

$$u(r,t) = \sum_{n=1}^{+\infty} A_n e^{-a^2\lambda_n^2 t} \frac{\sin\lambda_n r}{r}, \tag{4}$$

where

$$A_n = \frac{2}{r_0}\,\frac{r_0^2\lambda_n^2 + (r_0h-1)^2}{r_0^2\lambda_n^2 + (r_0h-1)r_0h}\int_0^{r_0} rf(r)\sin\lambda_n r\,dr, \tag{5}$$

λ_n are the positive roots of the equation

$$\tan\lambda_n r_0 = \frac{\lambda_n r_0}{1-r_0h}. \tag{6}$$

19. $u(r,t) =$

$$U_1 + 2(U_1-U_0)hr_0^2 \sum_{n+1}^{+\infty} (-1)^{n+1} \frac{\sqrt{\mu_n^2+(hr_0-1)^2}}{\mu_n(\mu_n^2+h^2r_0^2-hr_0)}\, e^{-\frac{a^2\mu_n^2 t}{r_0^2}}\, \frac{\sin\frac{\mu_n r}{r_0}}{r}, \tag{1}$$

where μ_n are the positive roots of the equation

$$\tan\mu = -\frac{\mu}{r_0h-1}, \tag{2}$$

and h is the coefficient of heat exchange appearing in the boundary condition

$$\frac{\partial u}{\partial r} = h[U_1-u] \quad \text{for} \quad r = r_0, \quad 0 < t < +\infty. \tag{3}$$

At the centre of the sphere

$$u(0, t) = U_1+2(U_1-U_0)hr_0 \sum_{n=1}^{+\infty} (-1)^{n+1} \frac{\sqrt{\mu_n^2+(hr_0-1)^2}}{\mu_n^2+h^2r_0^2-hr_0} e^{-\frac{a^2\mu_n^2 t}{r_0^2}} \tag{4}$$

If $hr_0 < 1$, then, obviously, series (4) satisfies the conditions of Leibnitz's theorem on alternating series. Using this, we find that for all values of the time t satisfying the inequality

$$t \geqslant t^* = -\frac{r_0^2}{a^2(\mu_1^2-\mu_2^2)} \ln\left\{\varepsilon \frac{\mu_2^2+h^2r_0^2-hr_0}{\mu_1^2+h^2r_0^2-hr_0} \sqrt{\frac{\mu_1^2+(hr_0-1)^2}{\mu_2^2+(hr_0-1)^2}}\right\}, \tag{5}$$

a steady-state will occur at the centre of the sphere with relative accuracy $\varepsilon > 0$.

20. The solution of the boundary-value problem

$$\frac{\partial u}{\partial t} = a^2\left\{\frac{\partial^2 u}{\partial r^2}+\frac{2}{r}\frac{\partial u}{\partial r}\right\}, \quad 0 \leqslant r < r_0, \quad 0 < t < +\infty, \tag{1}$$

$$\left.\frac{\partial u}{\partial r}\right|_{r=r_0} = h[U_1+at-u]\Big|_{r=r_0}, \quad 0 < t < +\infty, \tag{2}$$

$$u(r, 0) = U_0, \quad 0 \leqslant r < r_0 \tag{3}$$

is:

$$u(r, t) = U_0+a\left[t+\frac{r^2-r_0^2-2\dfrac{r_0}{h}}{6a^2}\right]+$$

$$+\frac{2hr_0^4 a}{a^2}\sum_{n=1}^{+\infty}(-1)^{n+1}\frac{\sqrt{\mu_n^2+(hr_0-1)^2}}{\mu_n^3(\mu_n^2+h^2r_0^2-hr_0)} e^{-\frac{a^2\mu_n^2 t}{r_0^2}} \frac{\sin\dfrac{\mu_n r}{r_0}}{r}, \tag{4}$$

where μ_n are positive roots of the equation

$$\tan\mu = \frac{\mu}{hr_0-1}. \tag{5}$$

Method. First one must find the particular solution of equation (1) satisfying the inhomogeneous boundary condition (2). The particular solution may be sought in the form $U(r, t) = U_1+at+F(r)$, where $F(r)$ is an unknown function.

21. The solution of the boundary-value problem

$$\frac{\partial u}{\partial t} = a^2\left\{\frac{\partial^2 u}{\partial r^2}+\frac{2}{r}\frac{\partial u}{\partial r}\right\}, \quad r_1 < r < r_2, \quad 0 < t < +\infty, \tag{1}$$

$$\left.\left(\frac{\partial u}{\partial r}-h_1 u\right)\right|_{r=r_1} = 0, \quad \left.\left(\frac{\partial u}{\partial r}+h_2 u\right)\right|_{r=r_2} = 0, \quad 0 < t < +\infty, \tag{2}$$

$$u(r, 0) = f(r), \quad r_1 < r < r_2 \tag{3}$$

is:

$$u(r,t) = \sum_{n=1}^{+\infty} A_n \, e^{-a^2\lambda_n^2 t} \frac{\sin[\lambda_n(r-r_1)+\nu_n]}{r}, \tag{4}$$

where

$$A_n = \frac{2\int\limits_{r_1}^{r_2} rf(r)\sin[\lambda_n(r-r_1)+\nu_n]\,dr}{r_2-r_1+\dfrac{\left[\left(h_1+\dfrac{1}{r_1}\right)\left(h_2-\dfrac{1}{r_2}\right)+\lambda_n^2\right]\left[h_1+h_2+\dfrac{1}{r_1}-\dfrac{1}{r_2}\right]}{\left[\left(h_1+\dfrac{1}{r_1}\right)^2+\lambda_n^2\right]\left[\left(h_2-\dfrac{1}{r_2}\right)^2+\lambda_n^2\right]}}, \tag{5}$$

λ_n are the positive roots of the equation

$$\cot\lambda_n(r_2-r_1) = \frac{\lambda_n^2-\left(h_1+\dfrac{1}{r_1}\right)\left(h_2-\dfrac{1}{r_2}\right)}{\lambda_n\left(h_1+h_2+\dfrac{1}{r_1}-\dfrac{1}{r_2}\right)}, \tag{6}$$

$$\nu_n = \arctan\frac{\lambda_n}{h_1+\dfrac{1}{r_1}}. \tag{7}$$

22. (a) $R_{cr} = \dfrac{\pi a}{\sqrt{\beta}}$,

(b) the process will have an avalanche character for any size of sphere,

(c) R_{cr} is the smaller positive root of the equation

$$\tan\left(\frac{\sqrt{\beta}}{a}R\right) = \frac{\dfrac{\sqrt{\beta}}{a}R}{1-hR}.$$

(b) *Inhomogeneous media*

23. The solution of the boundary-value problem

$$c_1\rho_1\frac{\partial u}{\partial t} = k_1\left\{\frac{\partial^2 u}{\partial x^2}+\frac{\partial^2 u}{\partial y^2}\right\}, \quad 0 \leqslant x \leqslant x_0, \quad 0 \leqslant y \leqslant l_2, \quad 0 < t < +\infty, \tag{1}$$

$$c_2\rho_2\frac{\partial u}{\partial t} = k_2\left\{\frac{\partial^2 u}{\partial x^2}+\frac{\partial^2 u}{\partial y^2}\right\}, \quad x_0 \leqslant x \leqslant l_1, \quad 0 \leqslant y \leqslant l_2, \quad 0 < t < +\infty, \tag{1'}$$

$$u(x_0-0, y, t) = u(x_0+0, y, t), \quad 0 < y < l_2, \quad 0 < t < +\infty, \tag{2}$$

$$k_1 u_x(x_0-0, y, t) = k_2 u_x(x_0+0, y, t), \quad 0 < y < l_2, \quad 0 < t < +\infty, \tag{2'}$$

$$u|_{x=0} = u|_{x=l_1} = u|_{y=0} = u|_{y=l_2} = 0, \tag{2''}$$

$$u(x, y, 0) = f(x, y), \quad 0 \leqslant x \leqslant l_1, \quad 0 \leqslant y \leqslant l_2 \tag{3}$$

is:

$$u(x, y, t) = \sum_{m,n=1}^{+\infty} A_{m,n} \, e^{-\lambda_{m,n}^2 t} \, v_{m,n}(x, y), \tag{4}$$

where

$$v_{m,n}(x, y) = \begin{cases} \dfrac{\sin \overline{\omega}_{mn} x}{\sin \overline{\omega}_{mn} x_0} \sin \dfrac{n\pi y}{l_2}, & 0 \leqslant x \leqslant x_0, \quad 0 \leqslant y \leqslant l_2, \\ \dfrac{\sin \overline{\overline{\omega}}_{mn}(l_1 - x)}{\sin \overline{\overline{\omega}}_{mn}(l_1 - x_0)} \sin \dfrac{n\pi y}{l_2}, & x_0 \leqslant x \leqslant l, \quad 0 \leqslant y \leqslant l_2, \end{cases} \tag{5}$$

$$\overline{\omega}_{mn}^2 = \frac{c_1\rho_1}{k_1} \lambda_{mn}^2 - \frac{n^2\pi^2}{l_2^2}, \qquad \overline{\overline{\omega}}_{mn}^2 = \frac{c_2\rho_2}{k_2} \lambda_{mn}^2 - \frac{n^2\pi^2}{l_2^2}, \tag{6}$$

λ_{mn} $(m = 1, 2, 3, \ldots;\ n = 1, 2, 3, \ldots)$ are roots of the transcendental equation

$$k_1 \sqrt{\frac{c_1\rho_1}{k_1} \lambda_{mn}^2 - \frac{n^2\pi^2}{l_2^2}} \cot \left\{ x_0 \sqrt{\frac{c_1\rho_1}{k_1} \lambda_{mn}^2 - \frac{n^2\pi^2}{l_2^2}} \right\}$$

$$= k_2 \sqrt{\frac{c_2\rho_2}{k_2} \lambda_{mn}^2 - \frac{n^2\pi^2}{l_2^2}} \cot \left\{ (x_0 - l_1) \sqrt{\frac{c_2\rho_2}{k_2} \lambda_{mn}^2 - \frac{n^2\pi^2}{l_2^2}} \right\} \tag{7}$$

$$A_{m,n} = \frac{\int\limits_0^{l_1} \int\limits_0^{l_2} \mu(x, y) f(x, y) \, v_{m,n}(x, y) \, dx \, dy}{\| v_{m,n} \|^2}, \tag{8}$$

$$\mu(x, y = \begin{cases} c_1\rho_1 & \text{for} \quad 0 \leqslant x < x_0, \quad 0 \leqslant y \leqslant l_2, \\ c_2\rho_2 & \text{for} \quad x_0 < x < l_1, \quad 0 \leqslant y \leqslant l_2, \end{cases} \tag{9}$$

$$\| v_{m,n} \|^2 = \int\limits_0^{l} \int\limits_0^{l_2} \mu(x, y) \, v_{m,n}^2(x, y) \, dx \, dy$$

$$= \frac{l_2}{4} \left\{ \frac{c_1\rho_1 x_0}{\sin^2 \overline{\omega}_{mn} x_0} + \frac{c_2\rho_2 (l - x_0)}{\sin^2 \overline{\overline{\omega}}_{mn}(l_1 - x_0)} \right\}. \tag{10}$$

The functions $v_{m,n}$ are orthogonal with weight $\mu(x, y)$ in the rectangle $0 \leqslant x \leqslant l_1$, $0 \leqslant y \leqslant l_2$.

24. The solution of the boundary-value problem

$$c_1\rho_1 \frac{\partial u}{\partial t} = k_1 \left\{ \frac{\partial^2 u}{\partial x^2} + \frac{\partial^2 u}{\partial y^2} + \frac{\partial^2 u}{\partial z^2} \right\}, \quad 0 \leqslant x \leqslant x_0, \quad 0 \leqslant y \leqslant l_2, \quad 0 \leqslant z \leqslant l_3,$$
$$0 < t < +\infty, \tag{1}$$

$$c_2\rho_2 \frac{\partial u}{\partial t} = k_2 \left\{ \frac{\partial^2 u}{\partial x^2} + \frac{\partial^2 u}{\partial y^2} + \frac{\partial^2 u}{\partial z^2} \right\}, \quad x_0 \leqslant x \leqslant l_1, \quad 0 \leqslant y \leqslant l_2, \quad 0 \leqslant z \leqslant l_3,$$
$$0 < t < +\infty, \tag{1'}$$

$$u(x_0 - 0, y, z, t) = u(x_0 + 0, y, z, t), \quad 0 \leqslant y \leqslant l_2, \quad 0 \leqslant z \leqslant l_3,$$
$$0 < t < +\infty, \tag{2}$$

$$k_1u(x_0-0, y, z, t) = k_2u_x(x_0+0, y, z, t), \quad 0 \leqslant y \leqslant l_2, \quad 0 \leqslant z \leqslant l_3, \quad 0 < t < +\infty, \tag{2'}$$

$$u|_{x=0} = u|_{x=l_1} = u|_{y=0} = u|_{y=l_2} = u|_{z=0} = u|_{z=l_3} = 0, \tag{2''}$$

$$u|_{t=0} = f(x, y, z), \quad 0 \leqslant x \leqslant l_1, \quad 0 \leqslant y \leqslant l_2, \quad 0 \leqslant z \leqslant l_3 \tag{3}$$

is:

$$u(x, y, z, t) = \sum_{m,\, n,\, p=1}^{+\infty} A_{m,\, n,\, p}\, e^{-\lambda^2_{m,\, n,\, p} t}\, v_{m,\, n,\, p}(x, y, z), \tag{4}$$

where

$$v_{mnp}(x, y, z) = \begin{cases} \dfrac{\sin \overline{\omega}_{mnp} x}{\sin \overline{\omega}_{mnp} x_0} \sin \dfrac{n\pi y}{l_2} \sin \dfrac{p\pi z}{l_3}, & 0 \leqslant x \leqslant x_0, \\ & 0 \leqslant y \leqslant l_2, \quad 0 \leqslant z \leqslant l_3, \\ \dfrac{\sin \overline{\overline{\omega}}_{mnp}(l_1 - x)}{\sin \overline{\overline{\omega}}_{mnp}(l_1 - x_0)} \sin \dfrac{n\pi y}{l_2} \sin \dfrac{p\pi z}{l_3}, & x_0 \leqslant x \leqslant l_1, \\ & 0 \leqslant y \leqslant l_2, \quad 0 \leqslant z \leqslant l_3. \end{cases} \tag{5}$$

$$\overline{\omega}^2_{m,n,p} = \frac{c_1\rho_1}{k_1}\lambda^2_{m,n,p} - \frac{n^2\pi^2}{l_2^2} - \frac{p^2\pi^2}{l_3^2}, \qquad \overline{\overline{\omega}}^2_{m,n,p} = \frac{c_2\rho_2}{k_2}\lambda^2_{m,n,p} - \frac{n^2\pi^2}{l_2^2} - \frac{p^2\pi^2}{l_3^2}, \tag{6}$$

$\lambda_{m,n,p}$ $(m = 1, 2, 3, \ldots;\ n = 1, 2, 3, \ldots;\ p = 1, 2, 3, \ldots)$ are roots of the transcendental equation

$$k_1 \sqrt{\frac{c_1\rho_1}{k_1}\lambda^2_{mnp} - \frac{n^2\pi^2}{l_2^2} - \frac{p^2\pi^2}{l_3^2}}\, \cot\left\{x_0 \sqrt{\frac{c_1\rho_1}{k_1}\lambda^2_{mnp} - \frac{n^2\pi^2}{l_2^2} - \frac{p^2\pi^2}{l_3^2}}\right\}$$

$$= k_2 \sqrt{\frac{c_2\rho_2}{k_2}\lambda^2_{mnp} - \frac{n^2\pi^2}{l_2^2} - \frac{p^2\pi^2}{l_3^2}}\, \cot\left\{(x_0 - l_1) \sqrt{\frac{c_2\rho_2}{k_2}\lambda^2_{mnp} - \frac{n^2\pi^2}{l_2^2} - \frac{p^2\pi^2}{l_3^2}}\right\} \tag{7}$$

$$A_{m,n,p} = \frac{\int_0^{l_1}\int_0^{l_2}\int_0^{l_3} \mu(x, y, z) f(x, y, z)\, v_{m,n,p}(x, y, z)\, dx\, dy\, dz}{||v_{m,n,p}||^2}, \tag{8}$$

$$\mu(x, y, z) = \begin{cases} c_1\rho_1 & \text{for} \quad 0 \leqslant x \leqslant x_0, \quad 0 \leqslant y \leqslant l_2, \quad 0 \leqslant z \leqslant l_3, \\ c_2\rho_2 & \text{for} \quad x_0 \leqslant x \leqslant l_1, \quad 0 \leqslant y \leqslant l_2, \quad 0 \leqslant z \leqslant l_3, \end{cases} \tag{9}$$

$$||v_{m,n,p}||^2 = \int_0^{l_1}\int_0^{l_2}\int_0^{l_3} \mu(x, y, z)\, v^2_{m,n,p}(x, y, z)\, dx\, dy\, dz$$

$$= \frac{l_2 l_3}{8}\left\{\frac{c_1\rho_1 x_0}{\sin^2 \overline{\omega}_{mnp} x_0} + \frac{c_2\rho_2(l_1 - x_0)}{\sin^2 \overline{\overline{\omega}}_{mnp}(l_1 - x_0)}\right\}. \tag{10}$$

The functions $v_{m,n,p}(x, y, z)$ are orthogonal with weight $\mu(x, y, z)$ in the parallelepiped $0 \leqslant x \leqslant l_1$, $0 \leqslant y \leqslant l_2$, $0 \leqslant z \leqslant l_3$.

25. The solution of the boundary-value problem

$$c_1\rho_1\frac{\partial u}{\partial t}=k_1\left\{\frac{\partial^2 u}{\partial r^2}+\frac{2}{r}\frac{\partial u}{\partial r}\right\},\quad 0\leqslant r\leqslant r_0,\quad 0<t<+\infty, \tag{1}$$

$$c_2\rho_2\frac{\partial u}{\partial t}=k_2\left\{\frac{\partial^2 u}{\partial r^2}+\frac{2}{r}\frac{\partial u}{\partial r}\right\},\quad r_0\leqslant r\leqslant r_1,\quad 0<t<+\infty \tag{1'}$$

$$u(r_0-0,t)=u(r_0+0,t),\quad 0<t<+\infty, \tag{2}$$

$$k_1u_r(r_0-0,t)=k_2u_r(r_0+0,t),\quad 0<t<+\infty, \tag{2'}$$

$$u(r_1,t)=0,\quad 0<t<+\infty, \tag{2''}$$

$$u(r,0)=f(r),\quad 0\leqslant r<r_1 \tag{3}$$

is:

$$u(r,t)=\sum_{n=1}^{+\infty}A_n e^{-\lambda_n^2 t}v_n(r), \tag{4}$$

where λ_n $(n=1,2,3,\ldots)$ are roots of the transcendental equation

$$\sqrt{k_1c_1\rho_1}\cot\left\{r_0\lambda_n\sqrt{\frac{c_1\rho_1}{k_1}}\right\}-\sqrt{k_2c_2\rho_2}\cot\left\{(r_0-r_1)\lambda_n\sqrt{\frac{c_2\rho_2}{k}}\right\}=\frac{k_1-k_2}{\lambda_n r_0}, \tag{5}$$

$$v_n(r)=\begin{cases}\dfrac{\sin\overline{\omega}_n r}{r\sin\overline{\omega}_n r_0}, & 0\leqslant r\leqslant r_0,\\[2ex] \dfrac{\sin\overline{\overline{\omega}}_n(r_1-r)}{r\sin\overline{\overline{\omega}}_n(r_1-r_0)}, & r_0\leqslant r\leqslant r_1,\end{cases} \tag{6}$$

$$\overline{\omega}_n=\lambda_n\sqrt{\frac{c_1\rho_1}{k_1}},\qquad \overline{\overline{\omega}}_n=\lambda_n\sqrt{\frac{c_2\rho_2}{k_2}}, \tag{7}$$

$$A_n=\frac{\int\limits_0^{r_1}\mu(r)f(r)v_n(r)\,dr}{\|v_n\|^2}, \tag{8}$$

$$\mu(r)=\begin{cases}c_1\rho_1r^2 & \text{for}\quad 0\leqslant r<r_0,\\ c_2\rho_2r^2 & \text{for}\quad r_0<r\leqslant r_1,\end{cases} \tag{9}$$

$$\|v_n\|^2=\int\limits_0^{r_1}\mu(r)v_n^2(r)\,dr=\frac{c_1\rho_1r_0}{2\sin^2\overline{\omega}_nr_0}+\frac{c_2\rho_2(r_1-r_0)}{2\sin^2\overline{\overline{\omega}}_n(r_1-r_0)}+\frac{k_2-k_1}{\lambda_nr_0}.$$

The functions $v_n(r)$ are orthogonal in the segment $0\leqslant r\leqslant r_1$ with weight $\mu(r)$.

26. The solution of the boundary-value problem

$$\frac{\partial u}{\partial t}=a^2\left\{\frac{\partial^2 u}{\partial r^2}+\frac{2}{r}\frac{\partial u}{\partial r}\right\},\quad r_1<r<r_2,\quad 0<t<+\infty, \tag{1}$$

$$\frac{4}{3}\pi r_1^3\rho^*c^*\frac{dU}{dt}=4\pi r_1^2\lambda\left.\frac{\partial u}{\partial r}\right|_{r=r_1},\quad \left.u\right|_{r=r_1}=U(t),\quad \left.u\right|_{r=r_2}=0,$$
$$0<t<+\infty, \tag{2}$$

$$u(r,0)=f(r),\quad r_1<r<r_2, \tag{3}$$

where ρ^* and c^* are the mass density and specific heat of the liquid is:

$$u(r,t)=\sum_{n=1}^{+\infty} A_n \mathrm{e}^{-a^2\lambda_n^2 t}\frac{\sin\lambda_n(r-r_2)}{r}, \quad r_1<r<r_2, \quad 0<t<+\infty, \tag{4}$$

where λ_n are positive roots of the equation

$$\cot\lambda_n(r_2-r_1)=\frac{a^2r_1\lambda_n^2-3\lambda}{3\lambda\lambda_n}, \tag{5}$$

$$A_n=\frac{2\int\limits_{r_1}^{r_2} rf(r)\sin\lambda_n(r-r_2)\,\mathrm{d}r}{r_2-\left[r_1\left(\frac{a^2\lambda_n r_1\rho^*c^*}{3\lambda}-\frac{1}{\lambda_n r_1}\right)^2+\lambda_n^2+\frac{2a^2r_1\rho^*c^*}{3\lambda}\right]\sin^2\lambda_n(r_1-r_2)}. \tag{6}$$

Method. By means of the substitution $v(r,t)=ru(r,t)$ problem (1), (2), (3) reduces to the problem for the cooling of a rod with lumped heat capacity at its end. This problem is solved in the same way as was done in chapter III (see problem 45).

2. Boundary-value Problems Requiring the Application of Special Functions

(a) *Homogeneous media*

27. $$u(r,t)=U_0\left[1-2\sum_{n=1}^{+\infty}\mathrm{e}^{-\frac{\mu_n^2a^2}{r_0^2}t}\frac{J_0\left(\frac{\mu_n r}{r_0}\right)}{\mu_nJ_1(\mu_n)}\right],$$

$$0\leqslant r\leqslant r_0, \quad 0<t<+\infty, \tag{1}$$

where r_0 is the radius of the cylinder, and μ_n are the positive roots of the equation $J_0(\mu)=0$.

In the steady-state, i.e. for values of t so large that the sum of the second and higher terms of series (1) is negligibly small in comparison with the first term†

$$u(r,t)\approx U_0\left[1-\frac{2J_0\left(\frac{\mu_1 r}{r_0}\right)}{\mu_1J_1(\mu_1)}\mathrm{e}^{-\frac{\mu_1^2a^2}{r_0^2}t}\right], \quad 0\leqslant r\leqslant r_0, \tag{2}$$

† We recall the approximate expression for a root of the equation $J_0(\mu)=0$

$$\mu_n=\pi\left(n-\frac{1}{4}+\frac{0\cdot05661}{4n-1}-\frac{0\cdot053041}{(4n-1)^2}+\dots\right),$$

so that

$$\mu_1\approx 2\cdot4048, \quad \mu_2\approx 5\cdot5201, \quad \mu_3\approx 8\cdot6537,\dots$$

The values $J_1(\mu_n)$ see [7], page 679.

the average tepemrature over a cross-section is

$$U(t) \approx U_0 \left[1 - \frac{4}{\mu_1^2} e^{-\frac{\mu_1^2 a^2}{r_0^2}} \right]. \tag{3}$$

Note. At points with coordinate $r_1 = \mu_1 r_0/\mu_2$ the steady-state occurs sooner, since at these points the second term of series (1) reduces to zero.

28. $u(r, t) = 8U_0 \sum_{n=1}^{+\infty} e^{-\frac{\mu_n^2 a^2}{r_0^2} t} \frac{J_0\left(\frac{\mu_n r}{r_0}\right)}{\mu_n^3 J_1(\mu_n)}$, where μ_n are positive roots of the equation $J_0(\mu) = 0$. In the steady-state

$$u(r, t) \approx 8U_0 \frac{J_0\left(\frac{\mu_1 r}{r_0}\right)}{\mu_1^3 J_1(\mu_1)} e^{-\frac{\mu_1^2 a^2}{r_0^2} t}$$

the average temperature over a cross-section

$$U(t) \approx \frac{16U_0}{\mu_1^4} e^{-\frac{\mu_1^2 a^2}{r_0^2} t}.$$

Note. The steady-state comes sooner at the same points as in the preceding problem.

29. The solution of the boundary-value problem

$$\frac{\partial u}{\partial t} = a^2 \left\{ \frac{\partial^2 u}{\partial r^2} + \frac{1}{r} \frac{\partial u}{\partial r} \right\}, \quad 0 \leqslant r < r_0, \quad 0 < t < +\infty \tag{1}$$

$$\lambda \frac{\partial u}{\partial r} = q \quad \text{for} \quad r = r_0, \quad 0 < t < +\infty, \tag{2}$$

$$u(r, 0) = U_0, \quad 0 \leqslant r \leqslant r_0 \tag{3}$$

is:

$$u(r, t) = U_0 + \frac{q r_0}{\lambda} \left[2 \frac{a^2 t}{r_0^2} - \frac{1}{4} \left(1 - 2 \frac{r^2}{r_0^2} \right) - \sum_{n=1}^{+\infty} \frac{2 e^{-\frac{a^2 \mu_n^2}{r_0^2} t}}{\mu_n^2 J_0(\mu_n)} J_0\left(\frac{\mu_n r}{r_0} \right) \right], \tag{4}$$

where μ_n are positive roots of the transcendental equation

$$J_0'(\mu) = 0. \tag{5}$$

30. The solution of the boundary-value problem

$$\frac{\partial u}{\partial t} = a^2 \left\{ \frac{\partial^2 u}{\partial r^2} + \frac{1}{r} \frac{\partial u}{\partial r} \right\}, \quad 0 \leqslant r < r_0, \quad 0 < t < +\infty, \tag{1}$$

$$u|_{t=0} = f(r), \quad 0 \leqslant r \leqslant r_0, \tag{2}$$

$$\left[\frac{\partial u}{\partial r}+hu\right]_{r=r_0}=0, \quad 0<t<+\infty \tag{3}$$

is:

$$u(r,t)=\sum_{n=1}^{+\infty}A_n e^{-\frac{\mu_n^2 a^2 t}{r_0^2}} J_0\left(\frac{\mu_n r}{r_0}\right), \tag{4}$$

where

$$A_n=\frac{2\mu_n^2}{r_0^2[\mu_n^2+h^2r_0^2]\,J_0^2(\mu_n)}\int_0^{r_0} rf(r)J_0\left(\frac{\mu_n r}{r_0}\right)\mathrm{d}r, \tag{5}$$

μ_n are positive roots of the equation

$$\mu J_0'(\mu)+hr_0J_0(\mu)=0. \tag{6}$$

For a steady-state

$$u(r,t)\approx\frac{2\mu_1^2\int_0^{r_0} rf(r)J_0\left(\frac{\mu_1 r}{r_0}\right)\mathrm{d}r}{r_0^2[\mu_1^2+h^2r_0^2]J_0(\mu_1)}J_0\left(\frac{\mu_1 r}{r_0}\right)e^{-\frac{\mu_1^2a^2}{r_0^2}t}.$$

31. The solution of the boundary-value problem

$$\frac{\partial u}{\partial t}=a^2\left\{\frac{\partial_2 u}{\partial r_2}+\frac{1}{r}\frac{\partial u}{\partial r}\right\}, \quad 0\leqslant r<r_0, \quad 0<t<+\infty, \tag{1}$$

$$\frac{\partial u}{ur}=h[U_1-u] \quad \text{for} \quad r=r_0, \quad 0<t<+\infty, \tag{2}$$

$$u(r,0)=U_0, \quad 0\leqslant r<r_0 \tag{3}$$

is:

$$u(r,t)=U_1+2(U_1-U_0)\sum_{n=1}^{+\infty}\frac{J_1(\mu_n)\,e^{-\frac{a^2\mu_n^2t}{r_0^2}}}{\mu_n[J_0^2(\mu_n)+J_1^2(\mu_n)]}J_0\left(\frac{\mu_n r}{r_0}\right), \tag{4}$$

where μ_n are the positive roots of the equation

$$\mu J_0'(\mu)+hr_0J_0(\mu)=0. \tag{5}$$

Note. By (5)

$$\frac{J_1(\mu_n)}{\mu_n[J_0^2(\mu_n)+J_1^2(\mu_n)]}=\frac{hr_0}{J_0(\mu_n)\,[\mu_n^2+h^2r_0^2]}.$$

Thus, expression (4) for $u(r,t)$ may be written in the form

$$u(r,t)=U_1+2(U_1-U_0)hr_0\sum_{n=1}^{+\infty}\frac{e^{-\frac{a^2\mu_n^2t}{r_0^2}}}{J_0(\mu_n)\,[\mu_n^2+h^2r_0^2]}J_0\left(\frac{\mu_n r}{r_0}\right). \tag{6}$$

32. $$u(r,t) = U_0 + \alpha\left[t + \frac{r^2 - r_0^2 - 2\frac{r_0}{h}}{4a^2}\right] + \frac{2hr_0^3\alpha}{a^2}\sum_{n=1}^{+\infty}\frac{e^{-\frac{a^2\mu_n t}{r_0^2}}J_0\left(\frac{\mu_n r}{r_0}\right)}{\mu_n^2 J_0(\mu_n)\,[\mu_n^2 + h^2 r_0^2]}, \qquad (1)$$

where μ_n have the same meaning as in the answer to the preceding problem.

33. The magnetic field intensity

$$H = H_0\left\{1 - 2\sum_{n=1}^{+\infty} e^{-\frac{\mu_n^2 a^2}{r_0^2}t}\frac{J_0\left(\frac{\mu_n r}{r_0}\right)}{\mu_n J_1(\mu_n)}\right\},$$

where μ_n are the positive roots of the equation $J_0(\mu) = 0$. The flux of magnetic induction across a cross-section of the cylinder

$$\Phi = \int_0^{2\pi}\int_0^{r_0} Br\,\mathrm{d}r\,\mathrm{d}\phi = \mu\int_0^{2\pi}\int_0^{r_0} H_0\left\{1 - 2\sum_{n=1}^{+\infty} e^{-\frac{\mu_n^2 a^2}{r_0^2}t}\frac{J_0\left(\frac{\mu_n r}{r_0}\right)}{\mu_n J_1(\mu_n)}\right\} r\,\mathrm{d}r\,\mathrm{d}\phi$$

$$= \pi r_0^2\mu H_0\left\{1 - 4\sum_{n=1}^{+\infty}\frac{e^{-\frac{\mu_n^2 a^2}{r_0^2}t}}{\mu_n^2}\right\}.$$

Method. In the equation for the magnetic field†

$$\Delta \boldsymbol{H} = \frac{\varepsilon\mu}{c^2}\frac{\partial^2 \boldsymbol{H}}{\partial t^2} + \frac{4\pi\mu\sigma}{c^2}\frac{\partial \boldsymbol{H}}{\partial t}$$

for a conducting medium of high conductivity it is possible to neglect the term $\frac{\varepsilon\mu}{c^2}\frac{\partial^2 \boldsymbol{H}}{\partial t^2}$ in comparison with the term $\frac{4\pi\mu\sigma}{c^2}\frac{\partial \boldsymbol{H}}{\partial t}$ which leads to the equation

$$\frac{\partial \boldsymbol{H}}{\partial t} = a^2\Delta\boldsymbol{H}, \qquad a^2 = \frac{c^2}{4\pi\mu\sigma}.$$

Let us expand $\boldsymbol{H}$ in the unit vectors $\boldsymbol{e}_r$, $\boldsymbol{e}_\phi$, $\boldsymbol{e}_z$ of the cylindrical system, whose axis coincides with the axis of the cylinder,

$$\boldsymbol{H} = H_r\boldsymbol{e}_r + H_\phi\boldsymbol{e}_\phi - H_z\boldsymbol{e}_z.$$

Since the external field does not depend on ϕ and is parallel to the z-axis, then it is natural to assume that $H_r = H_\phi = 0$, and $H_z = H(r,t)$. This hypothesis

† See [7], pages 489–493.

is justified by the uniqueness theorem for the solution of the boundary-value problem. For $H(r, t)$ we obtain the boundary-value problem

$$\frac{\partial H}{\partial t} = a_2\left\{\frac{\partial^2 H}{\partial r^2}+\frac{1}{r}\frac{\partial H}{\partial r}\right\}, \quad 0 \leqslant r < r_0, \quad 0 < t < +\infty, \tag{1}$$

$$H(r, 0) = 0, \quad 0 < r < r_0, \tag{2}$$

$$H(r_0, t) = H_0, \quad 0 < t < +\infty.$$

34. In a cylindrical system of coordinates, the z-axis of which coincides with the axis of the cylinder,

$$\boldsymbol{H} = \boldsymbol{e}_r H_r + \boldsymbol{e}_\phi H_\phi + \boldsymbol{e}_z H_z, \quad H_r = H_\phi \equiv 0, \quad H_z = H(r, t),$$

$$H(r, t) = H_0 \frac{\text{ber}\,\omega' r\,\text{bei}\,\omega' r_0 - \text{bei}\,\omega' r\,\text{ber}\,\omega' a}{\text{ber}^2\,\omega' r_0 + \text{bei}^2\,\omega' r_0}\cos\omega t +$$

$$+H_0 \frac{\text{ber}\,\omega' r\,\text{bei}\,\omega' r_0 - \text{bei}\,\omega' r\,\text{ber}\,\omega' r_0}{\text{ber}^2\,\omega' r_0 + \text{bei}^2\,\omega' r_0}\sin\omega t +$$

$$+2H_0 \sum_{n=1}^{+\infty} \mathrm{e}^{-\frac{a^2\mu_n^2 t}{r_0^2}} \frac{\mu_n^3}{\mu_n^4+\omega'^4 r_0^4} \frac{J_0\left(\frac{\mu_n r}{r_0}\right)}{J_1(\mu_n)},$$

where μ_n are the positive roots of the equation $J_0(\mu) = 0$, and $\omega' = \sqrt{\omega}/a$.

Solution. The solution of the boundary-value problem

$$\frac{\partial H}{\partial t} = a^2\left\{\frac{\partial^2 H}{\partial r^2}+\frac{1}{r}\frac{\partial H}{\partial r}\right\}, \quad 0 \leqslant r \leqslant r_0, \quad 0 < t < +\infty, \tag{1}$$

$$H(r, 0) = 0, \quad 0 \leqslant r < r_0, \tag{2}$$

$$H(r_0, t) = H_0 \cos\omega t, \quad 0 < t < +\infty \tag{3}$$

is the real part of the solution of the boundary-value problem

$$\frac{\partial U}{\partial t} = a^2\left\{\frac{\partial^2 U}{\partial r^2}+\frac{1}{r}\frac{\partial U}{\partial r}\right\}, \quad 0 \leqslant r < r_0, \quad 0 < t < +\infty, \tag{1'}$$

$$U(r, 0) = 0, \quad 0 \leqslant r < r_0, \tag{2'}$$

$$U(r_0, t) = H_0\,\mathrm{e}^{i\omega t}, \quad 0 < t < +\infty. \tag{3'}$$

We look for the solution of the boundary-value problem (1′), (2′), (3′) in the form

$$U(r, t) = V(r, t)+W(r, t), \tag{4}$$

where $V(r, t)$ is a particular solution of equation (1′) satisfying the boundary condition (3′) and having the form

$$V(r, t) = R(r)\,\mathrm{e}^{i\omega t}, \tag{5}$$

and $W(r, t)$ is a solution of equation (1′) satisfying the initial condition

$$W(r, 0) = -V(r, 0) = -R(r) \tag{6}$$

and boundary condition

$$W(r_0, t) = 0. \tag{7}$$

Substituting (5) into (1′) and (3′) we find:

$$V(r, t) = H_0 \frac{I_0(r\omega' \sqrt{i}\,)}{I_0(r_0\omega' \sqrt{i}\,)} e^{i\omega t} = H_0 \frac{\text{ber } \omega' r + i \text{ bei } \omega' r}{\text{ber } \omega' r_0 + i \text{ bei } \omega' r_0} e^{i\omega t}\dagger, \tag{8}$$

where

$$\omega' + \frac{\sqrt{\omega}}{a}$$

$$R(r) = H_0 \frac{I_0(r\omega' \sqrt{i}\,)}{I_0(r_0\omega' \sqrt{i}\,)}, \tag{9}$$

$$W(r, t) = \sum_{n=1}^{+\infty} A_n e^{-\frac{a^2\mu_n^2 t}{r_0^2}} J_0\left(\frac{\mu_n r}{r_0}\right), \tag{10}$$

$$A_n = \frac{\int\limits_0^{r_0} rR(r) J_0\left(\frac{\mu_n r}{r_0}\right) \mathrm{d}r}{\frac{r_0^2}{2}[J_1(\mu_n)]^2} = 2H_0 \frac{\mu_n^3 - \mu_n\omega'^2 i}{(\mu_n^4 + \omega'^4 r_0^4) J_1(\mu_n)}. \tag{11}$$

Evaluation of the integral in the numerator of equality (11) is carried out by means of the following general method.

Let $Z_\nu(\lambda^* x)$ and $Z_\nu(\lambda x)$ be arbitrary cylindrical functions of νth order, λ and λ^* real or complex numbers. We have:

$$\frac{\mathrm{d}}{\mathrm{d}x}\left[x \frac{\mathrm{d}Z_\nu(\lambda x)}{\mathrm{d}x}\right] + \left(\lambda^2 x - \frac{\nu^2}{x}\right) Z_\nu(\lambda x) = 0, \tag{12}$$

$$\frac{\mathrm{d}}{\mathrm{d}x}\left[x \frac{\mathrm{d}Z_\nu(\lambda^* x)}{\mathrm{d}x}\right] + \left(\lambda^{*2} x - \frac{\nu^2}{x}\right) Z_\nu(\lambda^* x) = 0. \tag{13}$$

---

† We recall that $J_0(xi\sqrt{i}) = I_0(x\sqrt{i}) = \text{ber } x + i \text{ bei } x$; $\text{ber } x = 1 - \frac{x^4}{2^2 4^2} + \frac{x^8}{2^2 4^2 6^2 8^2} - \ldots$; $\text{bei } x = \frac{x^2}{2^2} - \frac{x^6}{2^2 4^2 6^2} + \frac{x^{10}}{2^2 4^2 6^2 8^2 10^2} - \ldots$

Multiplying the first of these by $Z_\nu(\lambda^* x)$ and the second by $Z_\nu(\lambda x)$, subtracting the results and performing an integration, we obtain:

$$\int xZ_\nu(\lambda^* x)Z_\nu(\lambda x)\,dx = \frac{x[\lambda Z_\nu(\lambda^* x)Z_\nu'(\lambda x)-\lambda^* Z_\nu(\lambda x)Z_\nu'(\lambda^* x)]}{\lambda^{*2}-\lambda^2}. \quad (14)$$

Assuming $\lambda^* = \omega'\sqrt{i}$, $\lambda = \mu_n/r_0$, we obtain:

$$\int_0^{r_0} rI_0(r\omega'\sqrt{i})J_0\left(\frac{\mu_n r}{r_0}\right)dr = -H_0\frac{r_0^2\mu_n J_1(\mu_n)}{\mu_n^2+\omega'^2 r_0^2 i}, \quad (15)$$

from which (11) at once follows.

Separating the real part of $V(r, t)$ and $W(r, t)$ and adding, we obtain equation (1).

Note. Passing to a limit in (14) as $\lambda^* \to \lambda$ and using equation (12), the relation, important for the evaluation of the norm of the eigenfunctions, is readily derived,

$$\int xZ_\nu^2(\lambda x)\,dx = \frac{(\lambda x)^2[Z_\nu'(\lambda x)]^2+[(\lambda x)^2-\nu^2][Z_\nu(\lambda x)]^2}{2\lambda^2}. \quad (16)$$

35. $$u(r, t) = \frac{\pi^2}{2r_0^2}\sum_{n=1}^{+\infty}\frac{\mu_n^2 J_0^2(\mu_n)\int_{r_1}^{r_2} rf(r)Z_0\left(\frac{\mu_n r}{r_1}\right)dr}{J_0^2(\mu_n)-J_0^2(\mu_n k)}e^{-\frac{\mu_n^2 a^2 t}{r_1^2}}Z_0\left(\frac{\mu_n r}{r_1}\right)-$$

$$-\pi\sum_{n=1}^{+\infty}\frac{U_2 J_0(\mu_n)-U_1 J_0(\mu_n k)}{J_0^2(\mu_n)-J_0^2(\mu_n k)}J_0(\mu_n)Z_0\left(\frac{\mu_n r}{r_1}\right)e^{-\frac{\mu_n^2 a^2 t}{r_1^2}}+$$

$$+\left[U_1\ln\frac{r_2}{r}+U_2\ln\frac{r}{r_1}\right]\ln k,$$

where $k = r_2/r_1$, μ_n are the positive roots of the equation

$$J_0(\mu)N_0(\mu k)-J_0(\mu k)N_0(\mu) = 0,$$

and

$$Z_0\left(\frac{\mu_n r}{r_1}\right) = N_0(\mu_n k)J_0\left(\frac{\mu_n r}{r_1}\right)-J_0(\mu_n k)N_0\left(\frac{\mu_n r}{r_1}\right).$$

For $U_1 = U_2 = U^* = \text{const.}$, $f(r) = U_0 = \text{const.}$

$$u(r, t) = U^*+\pi(U_0-U^*)\sum_{n=1}^{+\infty}\frac{J_0(\mu_n)Z_0\left(\frac{\mu_n r}{r_1}\right)}{J_0(\mu_n)+J_0(\mu_n k)}e^{-\frac{\mu_n^2 a^2 t}{r_0^2}}.$$

Method. In order to evaluate the modulus of the eigenfunctions

$$Z_1(\lambda_k r) = J_1(\lambda_k r_1)N_1(\lambda_k r)-N_1(\lambda_k r_1)J_1(\lambda_k r)$$

it is necessary to use equation (16) of the note to the solution of problem 34 and the expression for the Wronskian of the cylindrical functions $J_\nu(z)$, $N_\nu(z)$

$$\begin{vmatrix} J_\nu(z) & N_\nu(z) \\ J_\nu'(z) & N_\nu'(z) \end{vmatrix} = \frac{2}{\pi z}.$$

36. The solution of the boundary-value problem

$$\frac{\partial u}{\partial t} = a^2 \left\{\frac{\partial^2 u}{\partial r^2} + \frac{1}{r}\frac{\partial u}{\partial r}\right\}, \quad r_1 < r < r_2, \quad 0 < t < +\infty, \tag{1}$$

$$u(r_1, t) = 0, \quad u_r(r_2, t) = \frac{q_0}{\lambda}, \quad 0 < t < +\infty, \tag{2}$$

$$u(r, 0) = 0, \quad r_1 < r < r_2 \tag{3}$$

is:

$$u(r, t) = U(r) + \sum_{k=1}^{+\infty} A_k e^{-a^2\lambda_k^2 t} [J_0(\lambda_k r_1) N_0(\lambda_k r) - N_0(\lambda_k r_1) J_0(\lambda_k r)], \tag{4}$$

where $U(r) = q_0 r_2/\lambda \ln r/r_1$ is the steady-state solution of equation (1) satisfying the boundary conditions (2) (the limit to which the temperature tends for $t \to +\infty$), and the coefficients A_k are found from the formula

$$A_k = \frac{\pi^2\lambda_n^2}{2} \frac{J_1^2(\lambda_k r_2)}{J_0^2(\lambda_k r_1) - J_1^2(\lambda_k r_2)} \int_{r_1}^{r_2} r U(r) [J_0(\lambda_k r_1) N_0(\lambda_k r) - N_0(\lambda_k r_1) J_0(\lambda_k r)] \, dr, \tag{5}$$

λ_k are positive roots of the equation

$$J_0(\lambda r_1) N_0'(\lambda r_2) - N_0(\lambda r_1) J_0'(\lambda r_2) = 0. \tag{6}$$

37. The solution of the boundary-value problem

$$\frac{\partial u}{\partial t} = a^2 \left\{\frac{\partial^2 u}{\partial r^2} + \frac{1}{r}\frac{\partial u}{\partial r}\right\}, \quad r_1 < r < r_2, \quad 0 < t < +\infty, \tag{1}$$

$$u_r(r_1, t) - h_1 u(r_1, t) = 0, \quad u_r(r_2, t) + h_2 u(r_2, t) = 0, \quad 0 < t < +\infty, \tag{2}$$

$$u(r, 0) = U_0, \quad r_1 < r < r_2 \tag{3}$$

is:

$$u(r, t) = \sum_{k=1}^{+\infty} A_k e^{-a^2\lambda_k^2 t} \{[\lambda_k J_0'(\lambda_k r_1) - h_1 J_0(\lambda_k r_1)] N_0(\lambda_k r) - [\lambda_k N_0'(\lambda_k r_1) - h_1 N_0(\lambda_k r_1)] J_0(\lambda_k r)\}, \tag{4}$$

where

$$A_k = \frac{\pi^2\lambda_k^2}{2} \times$$

$$\times \frac{[\lambda_k J_0'(\lambda_k r_2) + h_2 J_0(\lambda_k r_2)]^2}{(h_2^2 + \lambda_k^2)[\lambda_k J_0'(\lambda_k r_1) - h_1 J_0(\lambda_k r_1)]^2 - (h_1^2 + \lambda_k^2)[\lambda_k J_0'(\lambda_k r_2) + h_2 J_0(\lambda_k r_2)]^2} \times$$

$$\times U_0 \frac{1}{\lambda_k} \{[\lambda_k J_0'(\lambda_k r_1) - h_1 J_0(\lambda_k r_1)] [r_2 N_1(\lambda_k r_2) - r_1 N_1(\lambda_k r_1)] - [\lambda_k N_0'(\lambda_k r_1) - h_1 N_0(\lambda_k r_1)] [r_2 J_1(\lambda_k r_2) - r_1 J_1(\lambda_k r_1)]\},$$

λ_k are positive roots of the equation

$$\begin{vmatrix} \lambda J_0'(\lambda r_1)-h_1 J_0(\lambda r_1) & \lambda N_0'(\lambda r_1)-h_1 N_0(\lambda r_1) \\ \lambda J_0'(\lambda r_2)+h_2 J_0(\lambda r_2) & \lambda N_0'(\lambda r_2)+h_2 N_0(\lambda r_2) \end{vmatrix} = 0.$$

38. The solution of the boundary-value problem

$$\frac{\partial v}{\partial t} = \nu \left\{ \frac{\partial^2 v}{\partial r^2} + \frac{1}{r}\frac{\partial v}{\partial r} - \frac{v}{r^2} \right\}, \quad r_1 < r < r_2, \quad 0 < t < +\infty, \tag{1}$$

$$v(r_1 t) = 0, \quad v(r_2, t) = \omega r_2, \quad 0 < t < +\infty, \tag{2}$$

$$v(r, 0) = 0, \quad r_1 < r < r_2, \tag{3}$$

where $v(r, t) = v\phi(r, t)$†, is:

$$v(r, t) = \frac{\omega r^2}{r}\frac{r_2^2 - r_1^2}{r_2^2 - r_1^2} - \pi\omega r_2 \sum_{k=1}^{+\infty} \frac{J_1(\lambda_k r_1) J_1(\lambda_k r_2)\, e^{-\nu\lambda_k^2 t}}{J_1^2(\lambda_k r_2) - J_1^2(\lambda_k r_1)} v_k(r), \tag{4}$$

$$v_k(r) = J_1(\lambda_k r_1) N_1(\lambda_k r) - N_1(\lambda_k r_1) J_1(\lambda_k r), \tag{5}$$

where λ_k are positive roots of the equation

$$J_1(\lambda_k r_1) N_1(\lambda_k r_2) - N_1(\lambda_k r_1) J_1(\lambda_k r_2) = 0. \tag{6}$$

39. $$u(r, \phi, t) = \sum_{n,k=0}^{+\infty} J_n\left(\frac{\mu_k^{(n)} r}{r_0}\right) [A_{n,k} \cos n\phi + B_{n,k} \sin n\phi]\, e^{-\frac{a^2 \mu_k^{(n)2}}{r_0^2} t}, \tag{1}$$

where

$$A_{n,k} = \frac{\varepsilon_n}{\pi r_0^2 [J_n'(\mu_k^{(n)})]^2} \int_0^{r_0}\int_0^{2\pi} f(r, \phi) J_n\left(\frac{\mu_k^{(n)} r}{r_0}\right) \cos n\phi r \, dr \, d\phi, \tag{2}$$

$$\varepsilon_n = \begin{Bmatrix} 1 & \text{for} & n = 0, \\ 2 & \text{for} & n \neq 0, \end{Bmatrix} \tag{3}$$

$$B_{n,k} = \frac{2}{\pi r_0^2 [J_n'(\mu_k^{(n)})]^2} \int_0^{r_0}\int_0^{2\pi} f(r, \phi) J_n\left(\frac{\mu_k^{(n)} r}{r_0}\right) \sin n\phi r \, dr \, d\phi, \tag{4}$$

$\mu_k^{(n)}$ are positive roots of the equation $J_n(\mu) = 0$. (5)

40. $$u(r, \phi, t) = \sum_{n,k=0}^{+\infty} J_n\left(\frac{\mu_k^{(n)} r}{r_0}\right) [A_{n,k} \cos n\phi + B_{n,k} \sin n\phi]\, e^{-\frac{a^2 \mu_k^{(n)2}}{r_0^2} t}, \tag{1}$$

† See the method to the answer of problem 7.

$$A_{n,k} = \frac{\varepsilon_n}{\pi r_0^2 J_n^2(\mu_k^{(n)}) \left[1+\frac{r_0^2h^2-n^2}{\mu_k^{(n)2}}\right]} \int_0^{r_0}\int_0^{2\pi} f(r,\phi) J_n\left(\frac{\mu_k^{(n)}r}{r_0}\right) \cos n\phi r \,dr\,d\phi, \tag{2}$$

$$\varepsilon_n = \begin{cases} 1 & \text{for} \quad n = 0, \\ 2 & \text{for} \quad n \neq 0, \end{cases} \tag{3}$$

$$B_{n,k} = \frac{2}{\pi r_0^2 J_0^2(\mu_k^{(n)}) \left[1+\frac{r_0^2h^2-n^2}{\mu_k^{(n)2}}\right]} \int_0^{r_0}\int_0^{2\pi} f(r,\phi) J_n\left(\frac{\mu_k^{(n)}r}{r_0}\right) \sin n\phi r \,dr\,d\phi, \tag{4}$$

$\mu_k^{(n)}$ positive roots of the equation

$$\mu J_n'(\mu)+r_0hJ_n(\mu) = 0, \tag{5}$$

and h is the coefficient of heat exchange appearing in the boundary condition

$$\left[\frac{\partial u}{\partial r}+hu\right]_{r=r_0} = 0, \quad 0<t<+\infty. \tag{6}$$

41. $$u(r,\phi,t) = \sum_{n,k=0}^{+\infty} e^{-a^2\lambda_k^2 t} Z_n(\lambda_k^{(n)}r)\{A_{n,k}\cos n\phi + B_{n,k}\sin n\phi\}, \tag{1}$$

$$Z_n(\lambda_k^{(n)}r) = J_n(\lambda_k^{(n)}r_1)N_n(\lambda_k^{(n)}r) - N_n(\lambda_k^{(n)}r_1)J_n(\lambda_k^{(n)}r), \tag{2}$$

where $\lambda_k^{(n)}$ are positive roots of the equation

$$J_n(\lambda_k^{(n)}r_1)N_n(\lambda_k^{(n)}r_2) - N_n(\lambda_k^{(n)}r_1)J_n(\lambda_k^{(n)}r_2) = 0, \tag{3}$$

$$A_{n,k} = \frac{\pi\lambda_k^{(n)2}}{2\varepsilon_n} \frac{J_n^2(\lambda_k^{(n)}r_2)}{J_n^2(\lambda_k^{(n)}r_1) - J_n^2(\lambda_k^{(n)}r_2)} \int_0^{r_0}\int_0^{2\pi} f(r,\phi) Z_n(\lambda_k^{(n)}r) \cos n\phi r\,dr\,d\phi, \tag{4}$$

$$\varepsilon_n = \begin{cases} 2 & \text{for} \quad n = 0, \\ 1 & \text{for} \quad n \neq 0, \end{cases} \tag{5}$$

$$B_{n,k} = \frac{\pi\lambda_k^{(n)2}}{2\varepsilon_n} \frac{J_n(\lambda_k^{(n)}r_2)}{J_n(\lambda_k^{(n)}r_1) - J_n^2(\lambda_k^{(n)}r_2)} \int_0^{r_0}\int_0^{2\pi} f(r,\phi) Z_n(\lambda_k^{(n)}r) \sin n\phi r\,dr\,d\phi. \tag{6}$$

Note. If the solution is represented by the eigenfunctions

$$\tilde{Z}_n(\lambda_k^{(n)}r) = J_n(\lambda_k^{(n)}r_2)N_n(\lambda_k^{(n)}r) - N_n(\lambda_k^{(n)}r_2)J_n(\lambda_k^{(n)}r) = \frac{J_n(\lambda_k^{(n)}r_2)}{J_n(\lambda_k^{(n)}r_1)} Z_n(\lambda_k^{(n)}r), \tag{7}$$

(this relation between $\tilde{Z}_n$ and Z_n is established by (3)), then

$$u(r,\phi,t) = \sum_{n,k=0}^{+\infty} e^{-a^2\lambda_k^{(n)2}t} \tilde{Z}_n(\lambda_k^{(n)}r)\{\tilde{A}_{n,k}\cos n\phi + \tilde{B}_{n,k}\sin n\phi\}. \tag{8}$$

The formulae for $A_{n,k}$ and $B_{n,k}$ are obtained from formulae (4) and (5), if the fraction $J_n^2(\lambda_k^{(n)}r_2)/[J_n^2(\lambda_k^{(n)}r_1)-J_n^2(\lambda_k^{(n)}r_2)]$ replaces the fraction

$$\frac{J_n^2(\lambda_k^{(n)}r_1)}{J_n^2(\lambda_k^{(n)}r_1)-J_n^2(\lambda_k^{(n)}r_2)}.$$

42. The solution of the boundary-value problem

$$\frac{\partial u}{\partial t}=a^2\left\{\frac{\partial^2 u}{\partial r^2}+\frac{1}{r}\frac{\partial u}{\partial r}+\frac{1}{r^2}\frac{\partial^2 u}{\partial \phi^2}\right\},\quad r_1<r<r_2,\quad 0\leqslant\phi\leqslant 2\pi,\qquad(1)$$

$$0<t<+\infty,$$

$$\left[\frac{\partial u}{\partial r}-h_1 u\right]_{r=r_1}=0,\quad \left[\frac{\partial u}{\partial r}+h_2 u\right]_{r=r_2}=0,\quad 0<t<+\infty,\qquad(2)$$

$$u|_{t=0}=f(r,\phi),\quad r_1<r<r_2,\quad 0\leqslant\phi\leqslant 2\pi\qquad(3)$$

is:

$$u(r,\phi,t)=\sum_{n,k=0}^{+\infty}\mathrm{e}^{-a^2\lambda_k^{(n)^2}t}Z_n(\lambda_k^{(n)}r)\{A_{n,k}\cos n\phi+B_{n,k}\sin n\phi\},\qquad(4)$$

where

$$Z_n(\lambda_k^{(n)}r)=[\lambda_k^{(n)}J_n'(\lambda_k^{(n)}r_1)-h_1J_n(\lambda_k^{(n)}r_1)]N_n(\lambda_k^{(n)}r)-$$
$$-[\lambda_k^{(n)}N_n(\lambda_k^{(n)}r_1)-h_1N_n(\lambda_k^{(n)}r_1)]J_n(\lambda_k^{(n)}r),\qquad(5)$$

where $\lambda_k^{(n)}$ are positive roots of the equation

$$\begin{vmatrix}\lambda_k^{(n)}J_n'(\lambda_k^{(n)}r_1)-h_1J_n(\lambda_k^{(n)}r_1) & \lambda_k^{(n)}N_n'(\lambda_k^{(n)}r_1)-h_1N_n(\lambda_k^{(n)}r_1)\\ \lambda_k^{(n)}J_n'(\lambda_k^{(n)}r_2)+h_2J_n(\lambda_k^{(n)}r_2) & \lambda_k^{(n)}N_n'(\lambda_k^{(n)}r_2)+h_2N_n(\lambda_k^{(n)}r_2)\end{vmatrix}=0,\qquad(6)$$

$$A_{n,k}=\frac{2\lambda_k^{(n)^2}\int\limits_0^{2\pi}\int\limits_{r_1}^{r_2}f(r,\phi)Z_n(\lambda_k^{(n)}r)\cos n\phi r\,\mathrm{d}r\,\mathrm{d}\phi}{\pi\varepsilon_n[h_2^2r_2^2+\lambda_k^{(n)^2}r_2^2-n^2]Z_n^2(\lambda_k^{(n)}r_2)-[h_1^2r_1^2+\lambda_k^{(n)^2}r_1^2-n^2]Z_n^2(\lambda_k^{(n)}r_1)},\qquad(7)$$

$$\varepsilon_n=\begin{cases}2 & \text{for}\quad n=0,\\ 1 & \text{for}\quad n\neq 0,\end{cases}\qquad(8)$$

$$B_{n,k}=\frac{2\lambda_k^{(n)^2}\int\limits_0^{2\pi}\int\limits_{r_1}^{r_2}f(r,\phi)Z_n(\lambda_k^{(n)}r)\sin n\phi r\,\mathrm{d}r\,\mathrm{d}\phi}{\pi[h_2^2r_2^2+\lambda_k^{(n)^2}r_2^2-n^2]Z_n^2(\lambda_k^{(n)}r_2)-[h_1^2r_1^2+\lambda_k^{(n)^2}r_1^2-n^2]Z_n^2(\lambda_k^{(n)}r_1)}.\qquad(9)$$

Note. It is possible to represent the solution by means of the eigenfunctions

$$\tilde{Z}_n(\lambda_k^{(n)}r)=[\lambda_k^{(n)}J_n'(\lambda_k^{(n)}r_2)+h_2J_n(\lambda_k^{(n)}r_2)]N_n(\lambda_k^{(n)}r)-$$
$$-[\lambda_k^{(n)}N_n'(\lambda_k^{(n)}r_2)+h_2N_n(\lambda_k^{(n)}r_2)]J_n(\lambda_k^{(n)}r),\qquad(10)$$

related to the functions $Z_n(\lambda_k^{(n)}r)$ by the relations

$$\tilde{Z}_n(\lambda_k^{(n)}r) = \frac{\lambda_k^{(n)}J_n'(\lambda_k^{(n)}r_2)+h_2J_n(\lambda_k^{(n)}r_2)}{\lambda_k^{(n)}J_n'(\lambda_k^{(n)}r_1)-h_1J_n(\lambda_k^{(n)}r_1)}\, Z_n(\lambda_k^{(n)}r), \tag{11}$$

$$u(r,\phi,t) = \sum_{n,k=0}^{+\infty} e^{-a^2\lambda_k^{(n)^2}t}\,\tilde{Z}_n(\lambda_k^{(n)}r)\{\tilde{A}_{n,n}\cos n\phi + B_{n,k}\sin n\phi\}. \tag{12}$$

The formulae for $\tilde{A}_{n,k}$ and $\tilde{B}_{n,k}$ are obtained from formulae (7) and (9) by replacing Z_n by $\tilde{Z}_n$.

43. $$u(r,\phi,t) = \sum_{n,k=1}^{+\infty} A_{n,k}\, e^{-a^2\lambda_k^{(n)^2}t}\, J_{\frac{n\pi}{\phi_0}}(\lambda_k^{(n)}r)\sin\frac{n\pi\phi}{\phi_0}, \tag{1}$$

where $\lambda_k^{(n)}$ are positive roots of the equation

$$J_{\frac{n\pi}{\phi_0}}(\lambda_k^{(n)}r_0) = 0, \tag{2}$$

$$A_{n,k} = \frac{4}{r_0^2\phi_0[J'_{\frac{n\pi}{\phi_0}}(\lambda_k^{(n)}r_0)]^2}\int_0^{r_0}\int_0^{\phi_0} f(r,\phi)J_{\frac{n\pi}{\phi_0}}(\lambda_k^{(n)}r)\sin\frac{n\pi\phi}{\phi_0}\, r\,dr\,d\phi. \tag{3}$$

44. The solution of the boundary-value problem

$$\frac{\partial u}{\partial t} = a^2\left\{\frac{\partial^2 u}{\partial r^2}+\frac{2}{r}\frac{\partial u}{\partial r}+\frac{1}{r^2\sin\theta}\frac{\partial}{\partial\theta}\left(\sin\theta\frac{\partial u}{\partial\theta}\right)+\frac{1}{r\ \sin^2\theta}\frac{\partial^2 u}{\partial\phi^2}\right\}, \tag{1}$$

$$u|_{r=r_0} = 0, \tag{2}$$

$$u|_{t=0} = f(r,\theta,\phi) \tag{3}$$

is

$$u(r,\theta,\phi,t) = \sum_{m,n=0}^{+\infty}\sum_{k=0}^{n} e^{-\left(\frac{a\mu_m^{(n)}}{r_0}\right)^2 t}\,\frac{J_{n+\frac{1}{2}}\left(\frac{\mu_m^{(n)}r}{r_0}\right)}{\sqrt{r}}P_{n,k}(\cos\theta)\times$$
$$\times\{A_{m,n,k}\cos k\phi + B_{m,n,k}\sin k\phi\}, \tag{4}$$

where $\mu_m^{(n)}$ are positive roots of the equation

$$J_{n+\frac{1}{2}}(\mu_m^{(n)}) = 0, \tag{5}$$

$$A_{m,n,k} = \frac{\int_0^{r_0}\int_0^{\pi}\int_0^{2\pi} f(r,\theta,\phi)\,r^{\frac{3}{2}}J_{n+\frac{1}{2}}\left(\frac{\mu_m^{(n)}r}{r_0}\right)\sin\theta P_{n,k}(\cos\theta)\cos k\phi\,dr\,d\theta\,d\phi}{\varepsilon_k\dfrac{\pi r_0^2(n+k)!}{(2n+1)(n-k)!}\left[J'_{n+\frac{1}{2}}(\mu_m^{(n)})\right]^2}, \tag{6}$$

$$\varepsilon_k = \begin{cases} 2 & \text{for} \quad k = 0, \\ 1 & \text{for} \quad k \neq 0, \end{cases}$$

$B_{m,n,k}$

$$= \frac{\int_0^{r_0}\int_0^{\pi}\int_0^{2\pi} f(r,\theta,\phi) r^{\frac{3}{2}} J_{n+\frac{1}{2}}\left(\frac{\mu_m^{(n)} r}{r_0}\right) \sin\theta P_{n,k}(\cos\theta) \sin k\phi \, dr \, d\theta \, d\phi}{\frac{\pi r_0^2 (n+k)!}{(2n+1)(n-k)!} \left[J'_{n+\frac{1}{2}}(\mu_m^{(n)})\right]^2}. \tag{7}$$

45. The solution of the boundary-value problem

$$\frac{\partial u}{\partial t} = a^2 \left\{\frac{\partial^2 u}{\partial r^2} + \frac{2}{r}\frac{\partial u}{\partial r} + \frac{1}{r^2 \sin\theta}\frac{\partial}{\partial\theta}\left(\sin\theta \frac{\partial u}{\partial\theta}\right) + \frac{1}{r^2\sin^2\theta}\frac{\partial^2 u}{\partial\phi^2}\right\}, \tag{1}$$

$$\left[\frac{\partial u}{\partial r} + hu\right]_{r=r_0} = 0, \tag{2}$$

$$u|_{t=0} = f(r,\theta,\phi) \tag{3}$$

is:

$$u(r,\theta,\phi,t) = \sum_{m,n=0}^{+\infty} \sum_{k=0}^{n} e^{-\left(\frac{a\mu_m^{(n)}}{r_0}\right)^2 t} \frac{J_{n+\frac{1}{2}}\left(\frac{\mu_m^{(n)} r}{r_0}\right)}{\sqrt{r}} P_{n,k}(\cos\theta) \times$$
$$\times \{A_{m,n,k}\cos k\phi + B_{m,n,k}\sin k\phi\}, \tag{4}$$

where $\mu_m^{(n)}$ are positive roots of the equation

$$\mu_m^{(n)} J'_{n+\frac{1}{2}}(\mu_m^{(n)}) + \left(r_0 h - \frac{1}{2}\right) J_{n+\frac{1}{2}}(\mu_m^{(n)}) = 0, \tag{5}$$

$A_{m,n,k}$

$$= \frac{\int_0^{r_0}\int_0^{\pi}\int_0^{2\pi} f(r,\theta,\phi) r^{\frac{3}{2}} J_{n+\frac{1}{2}}\left(\frac{\mu_m^{(n)} r}{r_0}\right) \sin\theta P_{n,k}(\cos\theta) \cos k\phi \, dr \, d\theta \, d\phi}{\varepsilon_k \frac{\pi r_0^2 (n+k)!}{(2n+1)(n-k)!} \left[1 + \frac{(r_0 h + n)(r_0 h - n - 1)}{(\mu_m^{(n)})^2}\right] J^2_{n+\frac{1}{2}}(\mu_m^{(n)})}. \tag{6}$$

$$\varepsilon_k = \left\{\begin{matrix} 2 & \text{for} \quad k = 0, \\ 1 & \text{for} \quad k \neq 0, \end{matrix}\right\} \tag{7}$$

$B_{m,n,k}$

$$= \frac{\int_0^{r_0}\int_0^{\pi}\int_0^{2\pi} f(r,\theta,\phi) r^{\frac{3}{2}} J_{n+\frac{1}{2}}\left(\frac{\mu_m^{(n)} r}{r_0}\right) \sin\theta P_{n,k}(\cos\theta) \sin k\phi \, dr \, d\theta \, d\phi}{\frac{\pi r_0^2 (n+k)!}{(2n+1)(n-k)!} \left[1 + \frac{(r_0 h + n)(r_0 h - n - 1)}{(\mu_m^{(n)})^2}\right] J_{n+\frac{1}{2}}(\mu_m^{(n)})}. \tag{8}$$

46. $$u(r,\theta,\phi,t)=\sum_{m,\,n=0}^{+\infty}\sum_{k=0}^{n}\mathrm{e}^{-a^2\lambda_m^{(n)2}t}\,\frac{Z_{n+\frac{1}{2}}(\lambda_m^{(n)}r)}{\sqrt{r}}P_{n,\,k}(\cos\theta)\times$$
$$\times\{A_{m,n,k}\cos k\phi+B_{m,n,k}\sin k\phi\},\qquad(1)$$

where

$$Z_{n+\frac{1}{2}}(\lambda_m^{(n)}r)=J_{n+\frac{1}{2}}(\lambda_m^{(n)}r_1)N_{n+\frac{1}{2}}(\lambda_m^{(n)}r)-N_{n+\frac{1}{2}}(\lambda_m^{(n)}r_1)J_{n+\frac{1}{2}}(\lambda_m^{(n)}r),\qquad(2)$$

$\lambda_m^{(n)}$ are positive roots of the equation

$$Z_{n+\frac{1}{2}}(\lambda_m^{(n)}r_2)=0,\qquad(3)$$

$$A_{m,n,k}=\frac{\int_0^{r_0}\int_0^{\pi}\int_0^{2\pi}f(r,\theta,\phi)r^{\frac{3}{2}}Z_{n+\frac{1}{2}}(\lambda_m^{(n)}r)\sin\theta\,P_{n,k}(\cos\theta)\cos k\phi\,\mathrm{d}r\,\mathrm{d}\theta\,\mathrm{d}\phi}{\varepsilon_k\dfrac{4\pi(n+k)!}{(2n+1)(n-k)!\,\pi^2\lambda_m^{(n)2}}\,\dfrac{J^2_{n+\frac{1}{2}}(\lambda_m^{(n)}r_1)-J^2_{n+\frac{1}{2}}(\lambda_m^{(n)}r_1)}{J^2_{n+\frac{1}{2}}(\lambda_m^{(n)}r_2)}},\qquad(4)$$

$$\varepsilon_k=\begin{cases}2 & \text{for}\quad k=0,\\ 1 & \text{for}\quad k\neq 0,\end{cases}\qquad(5)$$

$$B_{m,n,k}=\frac{\int_0^{r_0}\int_0^{\pi}\int_0^{2\pi}f(r,\theta,\phi)r^{\frac{3}{2}}Z_{n+\frac{1}{2}}(\lambda_m^{(n)}r)\sin\theta P_{n,k}(\cos\theta)\cos k\phi\,\mathrm{d}r\,\mathrm{d}\theta\,\mathrm{d}\phi}{\dfrac{4\pi(n+k)!}{(2n+1)(n-k)!\,\pi^2\lambda_m^{(n)2}}\,\dfrac{J^2_{n+\frac{1}{2}}(\lambda_m^{(n)}r_1)-J^2_{n+\frac{1}{2}}(\lambda_m^{(n)}r_2)}{J^2_{n+\frac{1}{2}}(\lambda_m^{(n)}r_2)}}.\qquad(6)$$

47. The solution of the boundary-value problem

$$\frac{\partial u}{\partial t}=a^2\left\{\frac{\partial^2u}{\partial r^2}+\frac{2}{r}\frac{\partial u}{\partial r}+\frac{1}{r^2\sin\theta}\frac{\partial}{\partial\theta}\left(\sin\theta\frac{\partial u}{\partial\theta}\right)+\frac{1}{r^2\sin^2\theta}\frac{\partial^2u}{\partial\phi^2}\right\},\ r_1<r<r_2,\qquad(1)$$

$$\left[\frac{\partial u}{\partial r}-h_1u\right]_{r=r_1}=0,\quad\left[\frac{\partial u}{\partial r}+h_2u\right]_{r=r_2}=0,\quad 0<t<+\infty,\qquad(2)$$

$$u|_{t=0}=f(r,\theta,\phi),\qquad r_1<r<r_2\qquad(3)$$

is:

$$u(r,\theta,\phi,t)=\sum_{m,n=0}^{+\infty}\sum_{k=0}^{+\infty}\mathrm{e}^{-a^2\lambda_m^{(n)2}t}\,\frac{Z_{n+\frac{1}{2}}(\lambda_m^{(n)}r)}{\sqrt{r}}P_{n,k}(\cos\theta)\times$$
$$\times\{A_{m,n,k}\cos k\phi+B_{m,n,k}\sin k\phi\},\qquad(1')$$

$$Z_{n+\frac{1}{2}}(\lambda_m^{(n)} r)$$

$$= \left[\lambda_m^{(n)} J'_{n+\frac{1}{2}}(\lambda_m^{(n)} r_1) - \left(h_1 + \frac{1}{2r_1}\right) J_{n+\frac{1}{2}}(\lambda_m^{(n)} r_1)\right] N_{n+\frac{1}{2}}(\lambda_m^{(n)} r) -$$

$$- \left[\lambda_m^{(n)} N'_{n+\frac{1}{2}}(\lambda_m^{(u)} r_1) - \left(h_1 + \frac{1}{2r_1}\right) N_{n+\frac{1}{2}}(\lambda_m^{(n)} r_1)\right] J_{n+\frac{1}{2}}(\lambda_m^{(n)} r), \quad (2')$$

$$A_{m,n,k,} = \frac{\int\limits_0^{r_0}\int\limits_0^{\pi}\int\limits_0^{2\pi} f(r, \theta, \phi) r^{\frac{3}{2}} Z_{n+\frac{1}{2}}(\lambda_m^{(n)} r) \sin\theta P_{n,k}(\cos\theta) \cos k\phi \, dr\, d\theta\, d\phi}{\varepsilon_k \dfrac{2\pi}{2n+1} \cdot \dfrac{(n+k)!}{(n-k)!} \int\limits_{r_1}^{r_2} r Z^2_{n+\frac{1}{2}}(\lambda_m^{(n)} r)\, dr}, \quad (3')$$

$$\varepsilon_k = \begin{cases} 2 & \text{for} \quad k = 0 \\ 1 & \text{for} \quad k \neq 0 \end{cases}$$

$$B_{m,n,k} = \frac{\int\limits_0^{r_0}\int\limits_0^{\pi}\int\limits_0^{2\pi} f(r, \theta, \phi) r^{\frac{3}{2}} Z_{n+\frac{1}{2}}(\lambda_m^{(n)} r) \sin\theta P_{n,k}(\cos\theta) \sin k\phi \, dr\, d\theta\, d\phi}{\dfrac{2\pi}{2n+1} \cdot \dfrac{(n+k)!}{(n-k)!} \int\limits_{r_1}^{r_2} r Z^2_{n+\frac{1}{2}}(\lambda_m^{(n)} r)\, dr}, \quad (4)$$

$$\int\limits_{r_1}^{r_2} r Z^2_{n+\frac{1}{2}}(\lambda_m^{(n)} r)\, dr$$

$$= \frac{2}{\pi^2 (\lambda_m^{(n)})^2 \left[\lambda_m^{(n)} J'_{n+\frac{1}{2}}(\lambda_m^{(n)} r_2) + \left(h_2 - \frac{1}{2r_2}\right) J_{n+\frac{1}{2}}(\lambda_m^{(n)} r_2)\right]^2} \times$$

$$\times \left\{ \left[\left(h_2 - \frac{1}{2r_2}\right)^2 + (\lambda_m^{(n)})^2 - \frac{\left(n+\frac{1}{2}\right)^2}{r_2^2}\right] \times \right.$$

$$\times \left[\lambda_m^{(n)} J'_{n+\frac{1}{2}}(\lambda_m^{(n)} r_1) - \left(h_1 + \frac{1}{2r_1}\right) J_{n+\frac{1}{2}}(\lambda_m^{(n)} r_1)\right] -$$

$$- \left[\left(h_1 + \frac{1}{2r_1}\right)^2 + (\lambda_m^{(n)})^2 - \frac{\left(n+\frac{1}{2}\right)^2}{r_1^2}\right] \times$$

$$\left. \times \left[\lambda_m^{(n)} J'_{n+\frac{1}{2}}(\lambda_m^{(n)} r_2) + \left(h_2 - \frac{1}{2r_2}\right) J_{n+\frac{1}{2}}(\lambda_m^{(n)} r_2)\right] \right\}, \quad (5)$$

where $\lambda_m^{(n)}$ are positive roots of the equation

$$\lambda_m^{(n)} Z'_{n+\frac{1}{2}}(\lambda_m^{(n)} r_2) + \left(h_2 - \frac{1}{2r_2}\right) Z_{n+\frac{1}{2}}(\lambda_m^{(n)} r_2) = 0, \quad (6)$$

(b) *Inhomogeneous media; central factors*

48. The solution of the boundary-value problem

$$c_1\rho_1\frac{\partial u}{\partial t}=k_1\left\{\frac{\partial^2 u}{\partial r^2}+\frac{1}{r}\frac{\partial u}{\partial r}+\frac{1}{r^2}\frac{\partial^2 u}{\partial\phi^2}+\frac{\partial^2 u}{\partial z^2}\right\},\quad 0\leqslant r\leqslant r_0,\qquad 0\leqslant\phi\leqslant 2\pi,\tag{1}$$

$$0<z<l,$$

$$c_2\rho_2\frac{\partial u}{\partial t}=k_2\left\{\frac{\partial^2 u}{\partial r^2}+\frac{1}{r}\frac{\partial u}{\partial r}+\frac{1}{r^2}\frac{\partial^2 u}{\partial\phi^2}+\frac{\partial^2 u}{\partial z^2}\right\},\quad r_0<r<r_1,\qquad 0<t<+\infty\tag{1'}$$

$$u(r_0-0,\phi,z,t)=u(r_0+0,\phi,z,t),\tag{2}$$

$$k_1u_r(r_0-0,\phi,z,t)=k_2u_r(r_0+0,\phi,z,t),\quad 0\leqslant\phi\leqslant 2\pi,\ 0<z<l,\ 0<t<+\infty\tag{2'}$$

$$u(r_1,\phi,z,t)=0,\tag{2''}$$

$$u(r,\phi,z,0)=f(r,\phi,z)\tag{3}$$

is:

$$u(r,\phi,z,t)=\sum_{m,n,p=0}^{+\infty}R_{m,n,p}(r)e^{-\lambda^2_{m,n,p}t}\times$$

$$\times\{A_{m,n,p}\cos n\phi+B_{m,n,p}\sin n\phi\}\sin\frac{m\pi z}{l},\tag{4}$$

where $\lambda_{m,n,p}$ are roots of the transcendental equation

$$\begin{vmatrix} J_n(\overline{\omega}r_0) & -N_n(\overline{\overline{\omega}}r_0) & -J_n(\overline{\overline{\omega}}r_0)\\ k_1\overline{\omega}J_n'(\overline{\omega}r_0) & k_2\overline{\overline{\omega}}N_n'(\overline{\overline{\omega}}r_0) & k_2\overline{\overline{\omega}}J_n'(\overline{\overline{\omega}}r_0)\\ 0 & N_n(\overline{\overline{\omega}}r_1) & J_n(\overline{\overline{\omega}}r_1)\end{vmatrix}=0,\tag{5}$$

$$\overline{\omega}=\sqrt{\frac{c_1\rho_1}{k_1}\lambda^2-\frac{m^2\pi^2}{l^2}},\quad \overline{\overline{\omega}}=\sqrt{\frac{c_1\rho_2}{k_2}\lambda^2-\frac{m^2\pi^2}{l^2}},\tag{6}$$

$$R_{m,n,p}(r)=\begin{cases}[J_n(\overline{\overline{\omega}}_{mnp}r_0)N_n(\overline{\overline{\omega}}_{mnp}r_1)-\\ \quad -N_n(\overline{\overline{\omega}}_{mnp}r_0)J_n(\overline{\overline{\omega}}_{mnp}r_1)]J_n(\overline{\omega}_{mnp}r), & 0\leqslant r\leqslant r_0,\\ [J_n(\overline{\overline{\omega}}_{mnp}r)N_n(\overline{\overline{\omega}}_{mnp}r_1)-\\ \quad -N_n(\overline{\overline{\omega}}_{mnp}r)J_n(\overline{\overline{\omega}}_{mnp}r_1)]J_n(\overline{\omega}_{mnp}r_0), & r_0\leqslant r\leqslant r_1,\end{cases}\tag{7}$$

$$A_{mnp}=\frac{\int\limits_0^{r_1}r\mu\,dr\int\limits_1^{2\pi}d\phi\int\limits_0^{l}f(r,\phi,z)R_{mnp}(r)\cos n\phi\sin\frac{m\pi z}{l}\,dz}{\varepsilon_n\frac{\pi l}{2}\int\limits_0^{r_1}r\mu R^2_{mnp}(r)\,dr},$$

$$\varepsilon_n=\begin{cases}1, & n\neq 0,\\ 2, & n=0,\end{cases}\tag{8}$$

$$B_{mnp} = \frac{\int\limits_0^{r_1} r\mu \,\mathrm{d}r \int\limits_0^{2\pi} \mathrm{d}\phi \int\limits_0^{l} f(r, \phi, z) R_{mnp}(r) \sin n\phi \sin \frac{m\pi z}{l} \mathrm{d}z}{\frac{\pi l}{2} \int\limits_0^{r_1} r\mu R^2_{mnp}(r) \,\mathrm{d}r}, \tag{9}$$

$$\mu = \begin{cases} c_1\rho_1 & \text{for} \quad 0 \leqslant r < r_0, \\ c_2\rho_2 & \text{for} \quad r_0 < r < r_1. \end{cases} \tag{10}$$

The functions $R_{mnp_1}(r)$ and $R_{mnp_2}(r)$ for different p_1 and p_2 are orthogonal in the segment $0 \leqslant r \leqslant r_1$ with weight $r\mu$.

49. The solution of the boundary-value problem

$$\frac{\partial u}{\partial t} = a^2 \left\{ \frac{\partial^2 u}{\partial^2 r} + \frac{1}{r} \frac{\partial u}{\partial r} \right\}, \quad r_1 < r < r_2, \quad 0 < t < +\infty, \tag{1}$$

$$\pi r_1^2 c^* \rho^* \frac{\mathrm{d}U}{\mathrm{d}t} = 2\pi r_1 \lambda \left.\frac{\partial u}{\partial r}\right|_{r = r_1}, \quad U(t) = u\bigg|_{r = r_1}, \quad \left.\frac{\partial u}{\partial r}\right|_{r = r_2} = 0,$$
$$0 < t < +\infty, \tag{2}$$

$$u(r, 0) = f(r), \quad r_1 < r < r_2, \tag{3}$$

where λ is the thermal conductivity of the material of the tube, c^* and ρ^* are the specific heat and mass density of the liquid, is:

$$u(r, t) = \sum_{n=1}^{+\infty} A_n e^{-a^2 \lambda_n^2 t} Z_0(\lambda_n r), \tag{4}$$

where

$$Z_0(\lambda_n r) = J'(\lambda_n r_2) N_0(\lambda_n r) - N_0'(\lambda_n r_2) J_0(\lambda_n r). \tag{5}$$

λ_n are the roots of the transcendental equation

$$Z_0'(\lambda_k r_1) = -\lambda_k \frac{a^2 r_1 c^* \rho^*}{2\lambda} Z_0(\lambda_k r_1), \tag{6}$$

$$A_n = \frac{\int\limits_{r_1}^{r_2} r f(r) Z_0(\lambda_n r) \,\mathrm{d}r - \frac{a^2 r_1^2 c^* \rho^*}{2\lambda} f(r_1) Z_0(\lambda_n r_1)}{\int\limits_{r_1}^{r_2} r [Z_0(\lambda_n r)]^2 \,\mathrm{d}r - \frac{a^2 r_1^2 c^* \rho^*}{2\lambda} [Z_0(\lambda_n r_1)]^2} \dagger. \tag{7}$$

50. The solution of the boundary-value problem

$$\frac{\partial u}{\partial t} = a^2 \left\{ \frac{\partial^2 u}{\partial r^2} + \frac{1}{r} \frac{\partial u}{\partial r} \right\}, \quad r_1 < r < r_2, \quad 0 < t < +\infty, \tag{1}$$

† See (21) and (27) in problem 52, page 544.

$$\pi r_1^2 c^* \rho^* \frac{dU}{dt} = 2\pi r_1 \lambda \frac{\partial u}{\partial r}\bigg|_{r=r_1}, \quad U(t) = u\bigg|_{r=r_1}, \quad \frac{\partial u}{\partial r}\bigg|_{r=r_2} = -hu\bigg|_{r=r_2},$$

$$0 < t < +\infty, \quad (2)$$

$$u(r, 0) = f(r), \quad r_1 < r < r_2 \quad (3)$$

is obtained from the solution of the preceding problem if it is assumed

$$Z_0(\lambda_k r) = [\lambda_k J_0'(\lambda_k r_2) + hJ_0(\lambda_k r_2)]N_0(\lambda_k r) -$$
$$-[\lambda_k N_0'(\lambda_k r_2) + hN_0(\lambda_k r_2))]J_0(\lambda_k r). \quad (4)$$

51. To determine the velocity $v(r, t)$ of the liquid† and the angular velocity $\omega(t)$ of the cylinder we obtain the boundary-value problem

$$\frac{\partial v}{\partial t} = \nu \left\{ \frac{\partial^2 v}{\partial r^2} + \frac{1}{r}\frac{\partial v}{\partial r} - \frac{v}{r^2} \right\}, \quad r_1 < r < r_2, \quad 0 < t < +\infty, \quad (1)$$

$$v|_{r=r_1} = r_1 \omega(t), \quad v|_{r=r_2} = 0, \quad K\frac{d\omega}{dt} = M + 2\pi r_1^2 \rho \nu \left[\frac{\partial v}{\partial r} - \frac{v}{r} \right]_{r=r_1},$$

$$0 < t < +\infty, \quad (2)$$

$$v(r, 0) = 0, \quad r_1 < r < r_2. \quad (3)$$

Eliminating $\omega(t)$ from the boundary conditions (2), we obtain:

$$K\frac{\partial v}{\partial t}\bigg|_{r=r_1} = M + 2\pi r_1^2 \rho \nu \left[\frac{\partial v}{\partial r} - \frac{v}{r} \right]_{r=r_1}, \quad v\bigg|_{r=r_2} = 0, \quad 0 < t < +\infty. \ (2')$$

We look for the steady-state solution of equation (1)

$$V = V(r) \quad (4)$$

satisfying the inhomogeneous boundary conditions (2′)‡. If next it is assumed

$$v(r, t) = u(r, t) + V(r), \quad (5)$$

then for $u(r, t)$ we obtain the boundary-value problem

$$\frac{\partial u}{\partial t} = \nu \left\{ \frac{\partial^2 u}{\partial r^2} + \frac{1}{r}\frac{\partial u}{\partial r} - \frac{u}{r^2} \right\}, \quad r_1 < r < r_2, \quad 0 < t < +\infty, \quad (6)$$

$$K\frac{\partial u}{\partial t}\bigg|_{r=r_1} = 2\pi r_1^2 \rho \nu \left[\frac{\partial u}{\partial r} - \frac{u}{r} \right]_{r=r_1}, \quad u\bigg|_{r=r_2} = 0, \quad 0 < t < +\infty, \quad (7)$$

$$u(r, 0) = -V(r), \quad r_1 < r < r_2. \quad (8)$$

Having solved the boundary-value problem (6), (7), (8) and determined $v(r, t)$ by means of (5), we also find $\omega(t)$ from the boundary condition $v|_{r=r_1} = r_1\omega(t)$.

---

† $v(r, t) = v_\phi(r, t)$ (see the solution of problem 7).

‡ $V(r)$ is the limit to which the velocity of the liquid particles tends for $t \to +\infty$.

The solution of the boundary-value problem (6), (7), (8) can be accomplished in the same way as the preceding problem was solved. Let us look for particular solutions of equation (6) satisfying the boundary conditions (7) in the form

$$U_k(r, t) = e^{-\nu\lambda_k^2 t} R_k(r) \tag{9}$$

For $R_k(r)$ we obtain the equation

$$\frac{d^2R}{dr^2} + \frac{1}{r}\frac{dR}{dr} + \left(\lambda_k^2 - \frac{1}{r^2}\right)R = 0, \tag{10}$$

$R_k(r) = Z_1(\lambda_k r)$, where $Z_1(z)$ is the general solution of the equation of cylindrical functions of first order, in which the arbitrary constants are chosen so that the boundary condition $R_k(r_2) = 0$ would be fulfilled for any λ_k:

$$Z_1(\lambda_k r) = N_1(\lambda_k r_2)J_1(\lambda_k r) - J_1(\lambda_k r_2)N_1(\lambda_k r). \tag{11}$$

Requiring the fulfilment of the first of the boundary conditions (7) we obtain the equation for determining the eigenvalues

$$-\lambda_k^2 K Z_1(\lambda_k r_1) = 2\pi r_1^2 \rho\nu \left[\lambda_k Z_1'(\lambda_k r_1) - \frac{Z_1(\lambda_k r_1)}{r_1}\right]. \tag{12}$$

By relation (14) of problem 34 and of equality (12) we find a relation expressing the generalized orthogonality of the eigenfunctions $Z_1(\lambda_k r)$,

$$\int_{r_1}^{r_2} r Z_1(\lambda_k r) Z_1(\lambda_n r)\, dr + \frac{K}{2\pi r_1 \rho\nu} Z_1(\lambda_k r_1) Z_1(\lambda_n r_1) = 0\dagger \tag{13}$$

for $k \neq n$

$$u(r, t) = \sum_{n=1}^{+\infty} A_n e^{-a^2\lambda_n^2} Z_1(\lambda_n r), \tag{14}$$

where

$$A_n = \frac{\int_{r_1}^{r_2} rV(r) Z_1(\lambda_n r)\, dr + \frac{K}{2\pi r_1 \rho\nu} V(r_1) Z_1(\lambda_n r_1)}{\int_{r_1}^{r_2} r[Z_1(\lambda_n r)]^2\, dr + \frac{K}{2\pi r_1 \rho\nu}[Z_1(\lambda_n r_1)]^2}. \tag{15}$$

52. *Solution.* As in problem 33 we obtain:

$$\frac{\partial H}{\partial t} = a^2 \left\{\frac{\partial^2 H}{\partial r^2} + \frac{1}{r}\frac{\partial H}{\partial r}\right\}, \quad r_1 < r < r_2, \quad 0 < t < +\infty, \quad a^2 = \frac{c^2}{4\pi\mu\sigma}, \tag{1}$$

$$H(r, 0) = 0, \quad r_1 < r < r_2, \tag{2}$$

$$H(r_2, t) = H_0, \quad 0 < t < +\infty \tag{3}$$

---

† See (21) and (27) in problem 52.

where H is the component of the magnetic field along the z-axis which coincides with the axis of the cylinder (the other components of the magnetic field equal zero).

Let us find the boundary condition for $r = r_1$. We write down Maxwell's equations

$$\operatorname{curl} \boldsymbol{H} = \frac{4\pi\sigma}{c}\boldsymbol{E} + \frac{\varepsilon}{c}\frac{\partial \boldsymbol{E}}{\partial t}, \tag{4}$$

$$\operatorname{curl} \boldsymbol{E} = -\frac{\mu}{c}\frac{\partial \boldsymbol{H}}{\partial t}; \tag{5}$$

in cylindrical coordinates

$$\frac{1}{r}\frac{\partial H_z}{\partial \phi} - \frac{\partial H_\phi}{\partial z} = \left(\frac{4\pi\sigma}{c} + \frac{\varepsilon}{c}\frac{\partial}{\partial t}\right)E_r, \tag{6}$$

$$\frac{\partial H_r}{\partial z} - \frac{\partial H_z}{\partial r} = \left(\frac{4\pi\sigma}{c} + \frac{\varepsilon}{c}\frac{\partial}{\partial t}\right)E_\phi, \tag{6'}$$

$$\frac{1}{r}\frac{\partial (rH_\phi)}{\partial r} - \frac{1}{r}\frac{\partial H_r}{\partial \phi} = \left(\frac{4\pi\sigma}{c} + \frac{\varepsilon}{c}\frac{\partial}{\partial t}\right)E_z, \tag{6''}$$

$$\frac{1}{r}\frac{\partial E_z}{\partial \phi} - \frac{\partial E_\phi}{\partial z} = -\frac{\mu}{c}\frac{\partial H_r}{\partial t}, \tag{7}$$

$$\frac{\partial E_r}{\partial z} - \frac{\partial E_z}{\partial r} = -\frac{\mu}{c}\frac{\partial H_\phi}{\partial t}, \tag{7'}$$

$$\frac{1}{r}\frac{\partial (rE_\phi)}{\partial r} - \frac{1}{r}\frac{\partial E_r}{\partial \phi} = -\frac{\mu}{c}\frac{\partial H_z}{\partial t}. \tag{7''}$$

Since we neglect displacement currents and since $H_z = H_\phi = 0$ (see the solution of problem 33), then we obtain from (6′):

$$-\frac{\partial H_z}{\partial r} = \frac{4\pi\sigma}{c}E_\phi. \tag{8}$$

Integrating (5) over a cross-section of the inner cavity, applying Stokes' law and using the condition that everywhere in the cavity $\boldsymbol{H}$ is equal to the value of $\boldsymbol{H}$ on the inner surface of the tube, we obtain:

$$2E_\phi\Big|_{r=r_1} = -\frac{r_1}{c}\frac{\partial H_z}{\partial r}\Big|_{r=r_1}. \tag{9}$$

From (8) and (9) we obtain, finally, the unknown boundary condition

$$\frac{\partial H_z}{\partial t}\Big|_{r=r_1} = \frac{\mu}{r_1}a^2\frac{\partial H_z}{\partial r}\Big|_{r=r_1},$$

i.e.

$$\frac{\partial H}{\partial t}\Big|_{r=r_1} = \frac{\mu}{r_1}a^2\frac{\partial H}{\partial r}\Big|_{r=r_1}, \quad 0 < t < +\infty. \tag{3'}$$

In order to eliminate the inhomogeneity in the boundary condition (3), we look for the solution of the boundary-value problem (1), (2), (3), (3′) in the form

$$H(r, t) = H_0 + u(r, t). \tag{10}$$

For $u(r, t)$ we obtain the boundary-value problem

$$\frac{\partial u}{\partial t} = a^2 \left\{ \frac{\partial^2 u}{\partial r^2} + \frac{1}{r}\frac{\partial u}{\partial r} \right\}, \quad r_1 < r < r_2, \quad 0 < t < +\infty, \tag{11}$$

$$u(r, 0) = -H_0, \quad r_1 < r < r_2, \tag{12}$$

$$u(r_2, t) = 0, \quad 0 < t < +\infty \tag{13}$$

$$\left.\frac{\partial u}{\partial t}\right|_{r=r_1} = \frac{\mu}{r_1} a^2 \left.\frac{\partial u}{\partial r}\right|_{r=r_1}. \tag{13′}$$

We look for particular solutions of equation (11) satisfying the boundary conditions (13) in the form

$$U_k(r, t) = \mathrm{e}^{-a^2\lambda_k^2 t} R_k(r). \tag{14}$$

Substituting (14) in (11) we obtain:

$$\frac{\mathrm{d}^2 R_k}{\mathrm{d}r^2} + \frac{1}{r}\frac{\mathrm{d}R_k}{\mathrm{d}r} + \lambda_k^2 R_k = 0\,. \tag{15}$$

Hence,

$$R_k(r) = Z_0(\lambda_k r), \tag{15′}$$

where $Z_0(z) = AN_0(z) + BJ_0(z)$ is the general solution of the equation of cylindrical functions of zero order. Let us choose the constants A and B so that condition (13) for $Z_0(\lambda_k r)$ is fulfilled for any value of λ_k; for instance, let us assume

$$Z_0(\lambda_k r) = N_0(\lambda_k r_2) J_0(\lambda_k r) - J_0(\lambda_k r_2) N_0(\lambda_k r). \tag{16}$$

Substituting (14) into (13′) we find:

$$\left.\frac{\mathrm{d}R_k(r)}{\mathrm{d}r}\right|_{r=r_1} = -\lambda_k^2 \frac{r_1}{\mu} \left. R_k(r)\right|_{r=r_1}, \tag{17}$$

or

$$Z_0'(\lambda_k r_1) = -\lambda_k \frac{r_1}{\mu} Z_0(\lambda_k r_1). \tag{18}$$

This is the equation from which the eigenvalues $\lambda_1, \lambda_2, \lambda_3 \ldots$ of the boundary-value problem are found. From equation (15) and from the equation obtained by replacing k by n in (15), multiplying them respectively by $R_n(r)$ and $R_k(r)$, subtracting the results and integrating, we obtain:

$$(\lambda_k^2 - \lambda_n^2) \int_{r_1}^{r_2} r R_k(r) R_n(r) \mathrm{d}r = \left\{ r \left[R_k(r) \frac{\mathrm{d}R_n(r)}{\mathrm{d}r} - R_n(r) \frac{\mathrm{d}R_k(r)}{\mathrm{d}r} \right] \right\}_{r=r_1}^{r=r_2}. \tag{19}$$

By virtue of the boundary conditions (13) and (17) we obtain:

$$(\lambda_k^2-\lambda_n^2)\left\{\int_{r_1}^{r_2} rR_k(r)R_n(r)\,\mathrm{d}r-\frac{r_1^2}{\mu}R_k(r_1)R_n(r_1)\right\}=0, \tag{20}$$

from which for $n\neq k$ we find:

$$\int_{r_1}^{r_2} rR_k(r)R_n(r)\,\mathrm{d}r-\frac{r_1^2}{\mu}R_k(r_1)R_n(r_1)=0\,. \tag{21}$$

Thus the functions $R_k(r)$ and $R_n(r)$ are generally orthogonal (relation (21) is an expression of the generalized orthogonality).

Let us look for the solution of the boundary-value problem (11), (12), (13) as the sum of the series

$$u(r,t)=\sum_{k=1}^{+\infty} A_k e^{-t^2\lambda_k^2}R_k(r)\,, \tag{22}$$

$u(r,t)$ satisfies equation (11) (if the series converges sufficiently well) and the boundary conditions (13), (13′). We now investigate the initial conditions, assuming first for the generality that $u(r,0)=f(r)$. Assuming $t=0$ in (22) we obtain:

$$f(r)=\sum_{k=1}^{+\infty} A_kR_k(r); \tag{23}$$

for $r=r_1$

$$f(r_1)=\sum_{k=1}^{+\infty} A_kR_k(r_1)\,. \tag{24}$$

Let us multiply (23) by $rR_n(r)$ and integrate with respect to r from r_1 to r_2:

$$\int_{r_1}^{r_2} rf(r)R_n(r)\,\mathrm{d}r=\sum_{k=1}^{+\infty} A_k\int_{r_1}^{r_2} rR_k(r)R_n(r)\,\mathrm{d}r\,. \tag{25}$$

Let us multiply (24) by $r_1^2R_n(r_1)/\mu$:

$$\frac{r_1^2}{\mu}f(r_1)R_n(r_1)=\sum_{k=1}^{+\infty} A_k\frac{r_1^2}{\mu}R_k(r_1)R_n(r_1)\,. \tag{26}$$

Subtracting (25) and (26) we obtain by virtue of (21)

$$\int_{r_1}^{r_2} rf(r)R_n(r)\,\mathrm{d}r-\frac{r_1^2}{\mu}f(r_1)R_n(r_1)=A_n\left\{\int_{r_1}^{r_2} rR_n^2(r)\,\mathrm{d}r-\frac{r_1^2}{\mu}R_n^2(r_1)\right\}. \tag{27}$$

Hence,

$$A_n = \frac{\int\limits_{r_1}^{r_2} rf(r)R_n(r)\,\mathrm{d}r - \frac{r_1^2}{\mu}f(r_1)R_n(r_1)}{\int\limits_{r_1}^{r_2} rR_n^2(r)\,\mathrm{d}r - \frac{r_1^2}{\mu}R_n^2(r_1)}. \tag{28}$$

By equality (16) of problem 34, for the Wronskian of the cylindrical functions and of the boundary condition (18) we obtain:

$$\int\limits_{r_1}^{r_2} rR_n^2(r)\,\mathrm{d}r = \frac{2}{\pi^2\lambda_n^2} - \frac{r_1^2}{2}\left(1+\frac{r_1^2\lambda_n^2}{\mu^2}\right)Z_0^2(\lambda_n r_1). \tag{29}$$

Substituting $f(r) = -H_0$ in the numerator of (28) and using the Wronskian of the cylindrical functions, we obtain the value

$$H_0\left\{\left(\frac{\mu}{\lambda_n^2}+\frac{r_1^2}{\mu}\right)Z_0(\lambda_n r_1) - \frac{2}{\pi\lambda_n^2}\right\}. \tag{30}$$

By virtue of (29) and (30), (28) takes the form

$$A_n = H_0\,\frac{\left(\frac{\mu}{\lambda_n^2}+\frac{r_1^2}{\mu}\right)Z_0(\lambda_1 r_1) - \frac{2}{\pi\lambda_n^2}}{\frac{2}{{}^2\lambda_n^2}+\frac{r_1^2}{2}\left(\frac{2}{\mu}-1-\frac{r_1^2\lambda_k^2}{\mu^2}\right)Z_0^2(\lambda_k r_1)} \tag{31}$$

where

$$Z_0(\lambda_n r_1) = N_0(\lambda_n r_2)J_0(\lambda_n r_1) - J_0(\lambda_n r_2)N_0(\lambda_n r_1). \tag{32}$$

53. The solution of the boundary-value problem

$$c_1\rho_1\frac{\partial u}{\partial t} = k_1\left\{\frac{1}{r^2}\frac{\partial}{\partial r}\left(r^2\frac{\partial u}{\partial r}\right)+\frac{1}{r^2\sin\theta}\frac{\partial}{\partial\theta}\left(\sin\theta\frac{\partial u}{\partial\theta}\right)+\frac{1}{r^2\sin^2\theta}\frac{\partial^2 u}{\partial\phi^2}\right\},$$
$$0\leqslant r<r_0, \tag{1}$$

$$c_2\rho_2\frac{\partial u}{\partial t} = k_2\left\{\frac{1}{r^2}\frac{\partial}{\partial r}\left(r^2\frac{\partial u}{\partial r}\right)+\frac{1}{r^2\sin\theta}\frac{\partial}{\partial\theta}\left(\sin\theta\frac{\partial u}{\partial\theta}\right)+\frac{1}{r^2\sin^2\theta}\frac{\partial^2 u}{\partial\phi^2}\right\},$$
$$r_0<r\leqslant r_1, \tag{1'}$$

$$\left.\begin{aligned} u(r_0-0,\theta,\phi,t) &= u(r_0+0,\theta,\phi,t),\\ k_1u_r(r_0-0,\theta,\phi,t) &= k_2u_r(r_0+0,\theta,\phi,t),\end{aligned}\right\}\begin{aligned}&0\leqslant\theta\leqslant\pi,\quad 0\leqslant\phi\leqslant 2\pi, &(2)\\ &0<t<+\infty, &(2')\end{aligned}$$

$$u(r_1,\theta,\phi,t) = 0, \tag{2''}$$

$$u(r,\theta,\phi,0) = f(r,\theta,\phi) \tag{3}$$

is:

$$u(r, \theta, \phi, t)$$
$$= \sum_{n,p=0}^{+\infty} \sum_{m=0}^{+\infty} R_{mnp}(r) P_n^{(m)}(\cos\theta) e^{-\lambda_{mnp}^2 t} \{A_{mnp}(\cos m\phi + B_{mnp} \sin m\phi\},$$

where the eigenvalues λ_{mnp} and eigenfunctions $R_{mnp}(r)$ are found in the same way as in problems 48 and 24.

§ 3. The Method of Integral Representations

1. The Application of the Fourier Integral

54. The solution of the boundary-value problem

$$u_t = a^2 \Delta_3 u, \quad -\infty < x, y, z < +\infty, \quad 0 < t < +\infty, \tag{1}$$

$$u|_{t=0} = f(x, y, z), \quad -\infty < x, y, z < +\infty, \tag{2}$$

where $\Delta_3 = \partial^2/\partial x^2 + \partial^2/\partial y^2 + \partial^2/\partial z^2$, is:

$$u(x, y, z, t) = \frac{1}{(2a\sqrt{\pi t})^3} \iiint_{-\infty}^{+\infty} f(\xi, \eta, \zeta) e^{-\frac{(x-\xi)^2 + (y-\eta)^2 + (z-)\xi^2}{4a^2 t}} d\xi\, d\eta\, d\zeta. \tag{3}$$

If f does not depend on z, then

$$u(x, y, t) = \frac{1}{(2a\sqrt{\pi t})^2} \iint_{-\infty}^{+\infty} f(\xi, \eta) e^{-\frac{(x-\xi)^2 + (y-\eta)^2}{4a^2 t}} d\xi\, d\eta. \tag{3'}$$

Method. The Fourier transform of an arbitrary† function $F(x, y, z)$, defined for $-\infty < x, y, z < +\infty$, is defined by

$$\overline{F}(\lambda, \mu, \nu) = \frac{1}{(2\pi)^{3/2}} \iiint_{-\infty}^{+\infty} F(\xi, \eta, \zeta) e^{i[\lambda\xi + \mu\eta + \nu\zeta]} d\xi\, d\eta\, d\zeta. \tag{4}$$

Transition from F to $\overline{F}$ in formula (4) is called a Fourier transform with kernel $e^{i[\lambda\xi + \mu\eta + \nu\xi]}$. Transition from the form $\overline{F}$ to the original F is accomplished through the formula

$$F(x, y, z) = \frac{1}{(2\pi)^{3/2}} \iiint_{-\infty}^{+\infty} \overline{F}(\lambda, \mu, \nu) e^{-i[\lambda x + \mu y + \nu z]} d\lambda\, d\mu\, d\nu. \tag{5}$$

Multiplying both sides of the equalities (1) and (2) by $e^{i[\lambda\xi + \mu\eta + \nu\zeta]}$ and integrating with respect to ξ, η, ζ from $-\infty$ to $+\infty$, we obtain an ordinary

† We do not consider the restrictions on $F(x, y, z)$, for $\overline{F}(\lambda, \mu, \nu)$ to exist and the inversion formula (5) to hold, referring to the special literature for this.

differential equation and the initial condition for the Fourier transform $\overline{u}$ of the solution u of the boundary-value problem (1), (2). Finding $\overline{u}$ and applying an inverse Fourier transformation we obtain u.

In the case where f does not depend on z, the boundary-value problem (1), (2) reduces to the boundary-value problem

$$u_t = a^2\Delta_2 u, \quad -\infty < x, y < +\infty, \quad 0 < t < +\infty, \tag{1'}$$

$$u|_{u=0} = f(x, y), \quad -\infty < x, y < +\infty, \tag{2'}$$

where

$$\Delta_2 = \frac{\partial^2}{\partial x^2} + \frac{\partial^2}{\partial y^2}.$$

In order to solve the problem (1′) and (2′) it is necessary to apply the Fourier transform for a function of two variables

$$\overline{F}(\lambda, \mu) = \frac{1}{2\pi} \iint\limits_{-\infty}^{+\infty} F(\xi, \eta) e^{i[\lambda\xi + \mu\eta]} \, d\xi \, d\eta. \tag{4'}$$

The inversion formula has the form

$$F(x, y) = \frac{1}{2\pi} \iint\limits_{-\infty}^{+\infty} \overline{F}(\lambda, \mu) e^{-i[\lambda x + \mu y]} \, d\lambda \, d\mu. \tag{5'}$$

55. The solution of the boundary-value problem

$$u_t = a^2\Delta_3 u + g(x, y, z, t), \quad -\infty < x, y, z < +\infty, \quad 0 < t < +\infty, \tag{1}$$

$$u|_{t=0} = 0, \quad -\infty < x, y, z < +\infty \tag{2}$$

is:

$$u(x, y, z, t) = \frac{1}{(2a\sqrt{\pi})^3} \int\limits_0^t \frac{d\tau}{(t-\tau)^{3/2}} \iiint\limits_{-\infty}^{+\infty} e^{-\frac{(x-\xi)^2 + (y-\eta)^2 + (z-\zeta)^2}{4a^2(t-\tau)}} g(\xi, \eta, \zeta, \tau) \, d\xi \, d\eta \, d\zeta. \tag{3}$$

If $g(x, y, z, t)$ does not depend on t, i.e. $g = g(x, y, z)$, then the expression (3) for the solution may be transformed to the form

$$u(x, y, z, t) = \frac{1}{4\pi a^2} \iiint\limits_{-\infty}^{+\infty} \frac{g(\xi, \eta, \zeta)}{r} \left\{1 - \Phi\left(\frac{r}{2a\sqrt{t}}\right)\right\} d\xi \, d\eta \, d\zeta, \tag{4}$$

where $\Phi(\alpha) = \dfrac{2}{\sqrt{\pi}} \int\limits_0^\alpha e^{-\omega^2} d\omega$, and $r = \sqrt{(x-\xi)^2 + (y-\eta)^2 + (z-\zeta)^2}$. If $g(x, y, z, t)$ does not depend on z, then

$$u(x, y, t) = \frac{1}{4\pi a^2} \iint\limits_{-\infty}^{+\infty} \left\{ \int\limits_0^t g(\xi, \eta, \tau) \frac{e^{-\frac{(x-\xi)^2 + (y-\eta)^2}{4a^2(t-\tau)}}}{t-\tau} d\tau \right\} d\xi \, d\eta.$$

Method. For $g = g(x, y, z)$ expression (3) is transformed into (4) by means of the substitution

$$\mu = \frac{r}{2a\sqrt{t-\tau}}.$$

56. $$u(x, y, z, t) = \frac{1}{(2a\sqrt{\pi t})^3} \int_{-\infty}^{+\infty} d\xi \int_{-\infty}^{+\infty} d\eta \int_{0}^{+\infty} \left[e^{-\frac{(x-\xi)^2+(y-\eta)^2+(z-\zeta)^2}{4a^2t}} - \right.$$
$$\left. -e^{-\frac{(x-\xi)^2+(y-\eta)^2+(z+\zeta)^2}{4a^2t}} \right] f(\xi, \eta, \zeta)\, d\zeta. \qquad (1)$$

If $f(x, y, z)$ does not depend on y, then

$$u(x, z, t) = \frac{1}{(2a\sqrt{\pi t})^2} \int_{-\infty}^{+\infty} d\xi \int_{0}^{+\infty} \left[e^{-\frac{(x-\xi)^2+(z-\zeta)^2}{4a^2t}} = e^{-\frac{(x-\xi)^2+(z+\zeta)^2}{4a^2t}} \right] f(\xi, \zeta)\, d\zeta. \qquad (2)$$

Method. Use the Fourier transform with kernel

$$\frac{1}{2^{1/2}\pi^{3/2}} e^{i[\lambda\xi+\mu\eta]} \sin \nu\zeta$$

in the semispace $-\infty < \xi, \eta < +\infty$, $0 < \zeta < +\infty$. If f does not depend on y, then it is necessary to use the Fourier transform with kernel

$$\frac{1}{\pi} e^{i\lambda\xi} \sin \nu\zeta$$

for $-\infty < \xi < +\infty$, $0 < \zeta < +\infty$.

Method. See also the solution of problem 59 chapter III.

57. $$u(x, y, z, t) = \frac{z}{(2a\sqrt{\pi})^3} \int_0^t \frac{d\tau}{(t-\tau)^{5/2}} \iint_{-\infty}^{+\infty} e^{-\frac{(x-\xi)^2+(y-\eta)^2+z^2}{4a^2(t-\tau)}} f(\xi, \eta, \tau)\, d\xi\, d\eta. \qquad (1)$$

If $f(x, y, t)$ does not depend on y, then

$$u(x, z, t) = \frac{z}{4\pi a^2} \int_0^t \frac{d\tau}{(t-\tau)^2} \int_{-\infty}^{+\infty} e^{-\frac{(x-\xi)^2+z^2}{4a^2(t-\tau)}} f(\xi, \tau)\, d\xi. \qquad (2)$$

Method. Use the Fourier transform, suggested in the method to the preceding problem.

58. $$u(x, y, z, t) = \frac{1}{(2a\sqrt{\pi t})^3} \int_{-\infty}^{+\infty} d\xi \int_{-\infty}^{+\infty} d\eta \times$$
$$\times \int_0^{+\infty} \left[e^{-\frac{(x-\xi)^2+(y-\eta)^2+(z-\zeta)^2}{4a^2t}} + e^{-\frac{(x-\xi)^2+(y-\eta)^2+(z+\zeta)^2}{4a^2t}} \right] f(\xi, \eta, \zeta)\, d\zeta.$$

Method. Use the Fourier transform with kernel

$$\frac{1}{2^{1/2}\pi^{3/2}} e^{i[\nu\xi+\mu\eta]} \cos \nu\zeta$$

in the semispace $-\infty < \xi,\ \eta < +\infty,\ 0 < \zeta < +\infty$.

59.

$$u(x, y, z, t) = \frac{1}{(2a\sqrt{\pi})^3} \int_0^t \frac{d\tau}{(t-\tau)^{3/2}} \iint_{-\infty}^{+\infty} e^{-\frac{(x-\xi)^2+(y-\eta)^2+z^2}{4a^2(t-\tau)}} f(\xi, \eta, \tau)\, d\xi\, d\eta .$$

Method. Use the Fourier transform, suggested in the method to the preceding problem.

60.

$$u(x, y, z, t) = \frac{1}{(2a\sqrt{\pi t})^3} \int_{-\infty}^{+\infty} d\xi \int_{-\infty}^{+\infty} d\eta \int_0^{+\infty} \Big[e^{-\frac{(x-\xi)^2+(y-\eta)^2+(z-\zeta)^2}{4a^2t}} + e^{-\frac{(x-\xi)^2+(y-\eta)^2+(z+\zeta)^2}{4a^2t}} - 2h \int_0^{-\infty} e^{-h\omega-\frac{(x-\xi)^2+(y-\eta)^2+(z+\zeta+\omega)^2}{4a^2(t-\tau)}} d\omega \Big] f(\xi, \eta, \zeta)\, d\zeta.$$

If f does not depend on y, then

$$u(x, z, t) = \frac{1}{4\pi a^2 t} \iint_{-\infty}^{+\infty} f(\xi, \zeta) \Big[e^{-\frac{(x-\xi)^2+(z-\zeta)^2}{4a^2t}} + e^{-\frac{(x-\xi)^2+(z+\zeta)^2}{4a^2t}} - 2h \int_0^{+\infty} e^{-h\omega-\frac{(x-\xi)^2+(z+\zeta+\omega)^2}{4a^2t}} d\omega \Big] d\xi\, d\zeta .$$

Method. Use the Fourier transform with kernel

$$\frac{1}{2^{1/2}\pi^{3/2}} e^{i[\lambda\xi+\mu\eta]} \frac{\nu \cos \nu\zeta + h \sin \nu\zeta}{\nu^2+h^2}$$

in the semispace $-\infty < \xi,\ \eta < +\infty,\ 0 < \zeta < +\infty$. If f does not depend on y, then it is necessary to use the Fourier transform with kernel

$$\frac{1}{\pi} e^{i\lambda\xi} \frac{\nu \cos \nu\zeta + h \sin \nu\zeta}{\nu^2+h^2}$$

for $-\infty < \xi < +\infty,\ 0 < \zeta < +\infty$.

See also the solution of problem 65, chapter III.

61. $$u(x, y, z, t) = \frac{h}{(2a\sqrt{\pi})^3}\int_0^t \frac{d\tau}{(t-\tau)^{5/2}}\int_{-\infty}^{+\infty} d\xi \int_{-\infty}^{+\infty} d\eta \int_0^{+\infty} (z+\zeta)f(\xi, \eta, t)\times$$
$$\times e^{-h\zeta-\frac{(x-\xi)^2+(y-\eta)^2+(z+\zeta)^2}{4a^2t}}\,d\zeta\,.$$

If f does not depend on y, then

$$u(x, z, t) = \frac{h}{4\pi a^2}\int_0^t \frac{d\tau}{(t-\tau)^2}\int_{-\infty}^{+\infty} d\xi \int_0^{+\infty} (z+\zeta)f(\xi, \tau)e^{-h\zeta-\frac{(x-\xi)^2+(z+\zeta)^2}{4a^2t}}\,d\zeta.$$

Method. Use the Fourier transform, suggested in the preceding problem.

62.

$$u(x, y, z, t) = \frac{1}{(2a\sqrt{\pi})^3}\int_0^t \frac{d\tau}{(t-\tau)^{3/2}}\int_0^{+\infty} d\zeta \iint_{-\infty}^{+\infty}\Big[e^{-\frac{(x-\xi)^2+(y-\eta)^2+(z-\zeta)^2}{4a^2(t-\tau)}} -$$
$$- e^{-\frac{(x-\xi)^2+(y-\eta)^2+(z+\zeta)^2}{4a^2(t-\tau)}}\Big]f(\xi, \eta, \zeta, \tau)\,d\xi\,d\eta.$$

63. $u(x, y, z, t)$

$$= \int_{-\infty}^{+\infty} d\zeta \int_0^{l_2} d\eta \int_0^{l_1} f(\xi, \eta, \zeta)G_i(x, y, z, \xi, \eta, \zeta, t)d\xi, \qquad i = 1,\ 2,\ 3,$$

where in the case of (a) $i = 1$ in case (b) $i = 2$, in case (c) $i = 3$, where

$G_1(x, y, z, \xi, \eta, \zeta, t)$

$$= \frac{2}{l_1 l_2 a\sqrt{\pi t}}e^{-\frac{(z-\zeta)^2}{4a^2t}}\sum_{k, n=1}^{+\infty} e^{-\pi^2 a^2\left(\frac{k^2}{l_1^2}+\frac{n^2}{l_2^2}\right)t}\times$$
$$\times \sin\frac{k\pi x}{l_1}\sin\frac{k\pi\xi}{l_1}\sin\frac{n\pi y}{l_2}\sin\frac{n\pi\eta}{l_2},$$

$G_2(x, y, z, \xi, \eta, \zeta, t)$

$$= \frac{2}{l_1 l_2 a\sqrt{\pi t}}e^{-\frac{(z-\zeta)^2}{4a^2t}}\sum_{k,n=0}^{+\infty} e^{-a^2\pi^2\left(\frac{k^2}{l_1^2}+\frac{n^2}{l_2^2}\right)t}\times$$
$$\times \varepsilon_k\varepsilon_n\cos\frac{k\pi x}{l_1}\cos\frac{k\pi\xi}{l_1}\cos\frac{n\pi y}{l_2}\cos\frac{n\pi\eta}{l_2},$$

$$\varepsilon_k = \begin{cases}\frac{1}{2} & \text{for } k = 0,\\ 1 & \text{for } k \neq 0,\end{cases} \qquad \varepsilon_n = \begin{cases}\frac{1}{2} & \text{for } n = 0,\\ 1 & \text{for } n \neq 0,\end{cases}$$

$$G_3(x, y, z, \xi, \eta, \zeta, t)$$

$$= \frac{2e^{-\frac{(z-\zeta)^2}{4a^2t}}}{a\sqrt{\pi t}} \sum_{k,n=1}^{+\infty} \frac{e^{-a^2(\lambda_k^2+\mu_n^2)t}}{[(\lambda_k^2+h^2)l_1+2h][(\mu_n^2+h^2)l_2+2h]}(\lambda_k \cos\lambda_k x + h\sin\lambda_k x)\times$$

$$\times(\lambda_k\cos\lambda_k\xi + h\sin\lambda_k\xi)(\mu_n\cos\mu_n y + h\sin\mu_n y)(\mu_n\cos\mu_n\eta + h\sin\mu_n\eta)$$

$$= \frac{2e^{-\frac{(z-\zeta)^2}{4a^2t}}}{a\sqrt{\pi t}} \sum_{k,n=1}^{+\infty} \frac{(\lambda_k^2+h^2)(\mu_n^2+h^2)e^{-a^2(\lambda_k^2+\mu_n^2)t}}{[(\lambda_k^2+h^2)l_1+2h][(\mu_n^2+h^2)l_2^2+2h]}\times$$

$$\times \sin(\lambda_k x+\phi_k)\sin(\lambda_k\xi+\phi_k)\sin(\mu_n y+\psi_n)\sin(\mu_n\eta+\psi_k),$$

where λ_k and μ_n are respectively the positive roots of the equations

$$\cot l_1\lambda = \frac{\lambda^2-h^2}{2\lambda h} \quad \text{and} \quad \cot l_2\mu = \frac{\mu^2-h^2}{2\mu h},$$

$$\phi_k = \arctan\frac{\lambda_k}{l_1}, \quad \psi_n = \arctan\frac{\mu_n}{l_2},$$

and h is the coefficient of heat exchange.

Method. Applying the Fourier transform with respect to z

$$\bar{u}(x, y, \nu, t) = \frac{1}{\sqrt{2\pi}} \int_{-\infty}^{+\infty} u(x, y, \zeta, t)e^{i\nu\zeta}d\zeta, \tag{1}$$

$$\bar{f}(x, y, \nu) = \frac{1}{\sqrt{2\pi}} \int_{-\infty}^{+\infty} f(x, y, \zeta)e^{i\nu\zeta}d\zeta, \tag{2}$$

we arrive at the equation

$$\frac{\partial\bar{u}}{\partial t} = a^2\left\{\frac{\partial^2\bar{u}}{\partial x^2}+\frac{\partial^2\bar{u}}{\partial y^2}-\lambda^2\bar{u}\right\} \tag{3}$$

and the initial condition

$$\bar{u}|_{t=0} = f(x, y, \nu). \tag{4}$$

The substitution

$$\bar{u} = e^{-a^2\lambda^2t}\,\bar{v}(x, y, \nu, t)$$

leads to the equation

$$\frac{\partial\bar{v}}{\partial t} = a^2\left\{\frac{\partial^2\bar{v}}{\partial x^2}+\frac{\partial^2\bar{v}}{\partial y^2}\right\} \tag{3'}$$

and the initial condition

$$v|_{t=0} = f(x, y, \nu). \tag{4'}$$

The boundary conditions for $\bar{v}$ will be the same as for u. We find $\bar{v}$ by the method of separation of variables, and then, having substituted its expression in (5), we apply an inverse Fourier transformation to $\bar{u}$. Integrating this with respect to ν gives the answer.

64. $$u(x, y, z, t) = \int_0^{+\infty} d\zeta \int_0^{l_2} d\eta \int_0^{l_1} f(\xi, \eta, \zeta)\tilde{G}_i(x, y, z, \xi, \eta, \zeta, t)d\xi, \quad i = 1, 2,$$

where to the boundary conditions (a) there corresponds a function $\tilde{G}_1$, obtained from the function G_1 of the answer to the preceding problem by replacing the factor $e^{-\frac{(z-\zeta)^2}{4a^2t}}$ by the factor $\left[e^{-\frac{(z-\zeta)^2}{4a^2t}} - e^{-\frac{(z+\zeta)^2}{4a^2t}}\right]$; similarly in case (b) $\tilde{G}_2$ is obtained from G_2 by the substitution

$$e^{-\frac{(z-\zeta)^2}{4a^2t}} \quad \text{for} \quad \left[e^{-\frac{(z-\zeta)^2}{4a^2t}} + e^{-\frac{(z+\zeta)^2}{4a^2t}}\right].$$

Method. In case (a) one must apply the Fourier sine-transform with respect to z, and in the case of (b) the cosine-transform with respect to z. The problem is then solved similarly to the preceding one.

65. $u(r, \phi, z, t)$

$$= \int_{-\infty}^{+\infty} d\zeta \int_0^{r_0} r'dr' \int_0^{2\pi} f(r', \phi', \zeta)G_i(r, \phi, z, r', \phi', \zeta, t)d\phi', \qquad i = 1, 2, 3.$$

In case (a) $i = 1$,

$$G_1(r, \phi, z, r', \phi', \zeta, t) = \frac{e^{-\frac{(z-\zeta)^2}{4a^2t}}}{ar_0^2\pi\sqrt{\pi t}} \sum_{n,k=0}^{+\infty} \frac{J_n\left(\frac{\mu_k^{(n)}r}{r_0}\right) J_n\left(\frac{\mu_k^{(n)}r'}{r_0}\right)}{\varepsilon_n[J_n'(\mu_k^{(n)})]^2} \cos n(\phi-\phi'),$$

$$\varepsilon_n = \begin{cases} 2 & \text{for} \quad n = 0, \\ 1 & \text{for} \quad n \neq 0; \end{cases}$$

$\mu_k^{(n)}$ are positive roots of the equation $J_n(\mu) = 0$.

In case (b) $i = 2$,

$$G_2(r, \phi, z, r', \phi', \zeta, t) = \frac{e^{-\frac{(z-\zeta)^2}{4a^2t}}}{ar_0^2\pi\sqrt{t\pi}} \sum_{k,n=0}^{+\infty} \frac{J_n\left(\frac{\mu_k^{(n)}r}{r_0}\right) J_n\left(\frac{\mu_k^{(n)}r'}{r_0}\right)}{\varepsilon_n J_n^2(\mu_k^{(n)})\left[1-\frac{n^2}{\mu_k^{(n)2}}\right]} \cos n(\phi-\phi'),$$

$$\varepsilon_n = \begin{cases} 2 & \text{for} \quad n = 0, \\ 1 & \text{for} \quad n \neq 0; \end{cases}$$

$\mu_k^{(n)}$ are roots of the equation $J_n'(\mu) = 0$, $\mu_k^{(0)} \geqslant 0$, $\mu_k^{(n)} > 0$ for $n \neq 0$.

In case (c) $i = 3$,

$G_3(r, \phi, z, r', \phi', \zeta, t)$

$$= \frac{e^{-\frac{(z-\zeta)^2}{4a^2t}}}{ar_0^2\pi\sqrt{\pi t}} \sum_{k,\,n=0}^{+\infty} \frac{J_n\left(\frac{\mu_k^{(n)} r}{r_0}\right) J_n\left(\frac{\mu_k^{(n)} r'}{r_0}\right)}{\varepsilon_n J_n^2(\mu_k^{(n)})\left[1+\frac{r_0^2h^2-n^2}{\mu_k^{(n)2}}\right]} \cos n(\phi-\phi'),$$

$$\varepsilon_n = \begin{cases} 2 & \text{for} \quad n = 0, \\ 1 & \text{for} \quad n \neq 0; \end{cases}$$

$\mu_k^{(n)}$ are positive roots of the equation $\mu J_n'(\mu)/r_0+hJ_n(\mu) = 0$.

Note. In case (b) the eigenfunction, identically equal to a constant, corresponds to the root $\mu_0^{(0)} = 0$.

Method. The problem is solved similarly to problem 63.

66.

$$u(r,\phi,z,t) = \int_0^{+\infty} d\zeta \int_0^{r_0} r'dr' \int_0^{2\pi} f(r',\phi',\zeta)\, \tilde{G}_i(r,\phi,z,r',\phi',\zeta,t)\, d\phi', \quad i = 1, 2;$$

in case (a) $i = 1$, $\tilde{G}_1$ is obtained from G_1 of the preceding problem by replacing the factor $e^{-\frac{(z-\zeta)^2}{4a^2t}}$ by the factor $\left[e^{-\frac{(z-\zeta)^2}{4a^2t}} - e^{-\frac{(z+\zeta)^2}{4a^2t}}\right]$: in case (b) $i = 2$, $\tilde{G}_2$ is obtained from G_2 of the preceding problem by replacing the factor $e^{-\frac{(z-\zeta)^2}{4a^2t}}$ by the factor $\left[e^{-\frac{(z-\zeta)^2}{4a^2t}} + e^{-\frac{(z+\zeta)^2}{4a^2t}}\right]$.

67. $u(r,\phi,t) = \int_0^{+\infty} \rho\, d\rho \int_0^{\phi_0} f(\rho,\phi') G_i(r,\phi,\rho,\phi',t)\, d\phi', \; i = 1, 2;$

in case (a) $i = 1$,

$$G_1(r,\phi,\rho,\phi',t) = \frac{2}{\phi_0} \sum_{n=1}^{+\infty} \left\{ \int_0^{+\infty} e^{-a^2\lambda^2t} J_{\frac{n\pi}{\phi_0}}(\lambda\rho) J_{\frac{n\pi}{\phi_0}}(\lambda r)\lambda d\lambda \right\} \sin\frac{n\pi\phi'}{\phi_0} \sin\frac{n\pi\phi}{\phi_0};$$

in case (b) $i = 2$,

$$G_2(r,\phi,\rho,\phi',t) = \frac{2}{\phi_0} \sum_{n=0}^{+\infty} \varepsilon_n \left\{ \int_0^{+\infty} e^{-a^2\lambda^2t} J_{\frac{n\pi}{\phi_0}}(\lambda\rho) J_{\frac{n\pi}{\phi_0}}(\lambda r)\lambda d\lambda \right\} \cos\frac{n\pi\phi'}{\phi_0} \cos\frac{n\pi\phi}{\phi_0},$$

$$\varepsilon_n = \begin{cases} \frac{1}{2} & \text{for} \quad n \neq 0, \\ 1 & \text{for} \quad n = 0. \end{cases}$$

If one uses the relation for Bessel functions

$$\int_0^{+\infty} e^{-\beta^2\tau^2} J_\nu(\alpha\tau) J_\nu(\gamma\tau)\tau \, d\tau = \frac{1}{2\beta^2} e^{-\frac{\alpha^2+\gamma^2}{4\beta^2}} I_\nu\left(\frac{\alpha\gamma}{2\beta^2}\right),$$

$$\operatorname{Re}(\nu) > -1, \quad |\arg\beta| < \frac{\pi}{2},$$

we obtain:

$$\int_0^{+\infty} e^{-a^2\lambda^2 t} J_{\frac{n\pi}{\phi_0}}(\lambda\rho) J_{\frac{n\pi}{\phi_0}}(\lambda r)\lambda \, d\lambda = \frac{1}{2a^2 t} e^{-\frac{\rho^2+r^2}{4a^2 t}} I_{\frac{n\pi}{\phi_0}}\left(\frac{\rho r}{2a^2 t}\right).$$

Therefore G_1 and G_2 can be represented in the form

$$G_1(r, \phi, \rho, \phi', t) = \frac{e^{-\frac{r^2+\rho^2}{4a^2 t}}}{a^2\phi_0 t} \sum_{n=1}^{+\infty} I_{\frac{n\pi}{\phi_0}}\left(\frac{r\rho}{2a^2 t}\right) \sin\frac{n\pi\phi}{\phi_0} \sin\frac{n\pi\phi'}{\phi_0},$$

$$G_2(r, \phi, \rho, \phi', t) = \frac{e^{-\frac{r^2+\rho^2}{4a^2 t}}}{a^2\phi_0 t} \sum_{n=0}^{+\infty} \varepsilon_n I_{\frac{n\pi}{\phi_0}}\left(\frac{r\rho}{2a^2 t}\right) \cos\frac{n\pi\phi}{\phi_0} \cos\frac{n\pi\phi'}{\phi_0},$$

$$\varepsilon_n = \begin{cases} \dfrac{1}{2} & \text{for} \quad n = 0, \\ 1 & \text{for} \quad n \neq 0. \end{cases}$$

Method. Let us look for particular solutions of the equation

$$\frac{\partial u}{\partial t} = a^2\left\{\frac{\partial^2 u}{\partial r^2} + \frac{1}{r}\frac{\partial u}{\partial r} + \frac{1}{r^2}\frac{\partial^2 u}{\partial \phi^2}\right\}$$

in the form $U(r, \phi, t) = W(r, t)\phi(\phi)$, requiring that in case (a) and (b) the appropriate boundary conditions are fulfilled.

In the case of (a) this leads to particular solutions $u_n(r, t) \neq \sin n\pi\phi/\phi_0$, $n = 1, 2, 3, \ldots$, and in the case of (b) to particular solutions $u_n(r, t) \neq \cos n\pi\phi/\phi_0$, $n = 0, 1, 2, 3, \ldots$ In both cases $u_n(r, t)$ is a solution of the equation

$$\frac{\partial u_n}{\partial t} = a^2\left\{\frac{\partial^2 u_n}{\partial r^2} + \frac{1}{r}\frac{\partial u_n}{\partial r} - \frac{\left(\frac{n\pi}{\phi_0}\right)^2}{r^2} u_n\right\}, \qquad 0 < r, \ t < +\infty. \qquad (1)$$

We seek the solution of the original boundary-value problem in the form of the sum of these particular solutions:

in case (a)

$$u(r, \phi, t) = \sum_{n=1}^{+\infty} u_n(r, t) \sin \frac{n\pi\phi}{\phi_0}; \tag{2}$$

in case (b)

$$u(r, \phi, t) = \sum_{n=0}^{+\infty} u_n(r, t) \cos \frac{n\pi\phi}{\phi_0}. \tag{3}$$

Expanding $f(r, \phi) = u|_{t=0}$ in a series in sin $n\pi\phi/\phi_0$ in the first case and in a series in cos $n\pi\phi/\phi_0$ in the second case, we find the initial conditions for $u_n(r, t)$:

in case (a)

$$u_n(r, 0) = f_n(r) = \frac{2}{\phi_0} \int_0^{\phi_0} f(r, \phi') \sin \frac{n\pi\phi'}{\phi_0} \mathrm{d}\phi'; \tag{4}$$

in case (b)

$$\left. \begin{aligned} u_n(r, 0) &= f_n(r) = \frac{2}{\phi_0} \int_0^{\phi_0} f(r, \phi') \cos \frac{k\pi\phi'}{\phi_0} \mathrm{d}\phi', \qquad n \neq 0, \\ u_0(r, 0) &= f_0(r) = \frac{1}{\phi_0} \int_0^{\phi_0} f(r, \phi') \mathrm{d}\phi'. \end{aligned} \right\} \tag{4'}$$

We seek the solution of equation (1) for the initial condition (4) or (4′), bounded for $r \to 0$, in the form

$$u_n(r, t) = \int_0^{+\infty} \int_0^{+\infty} U_n(\rho, t) J_{\frac{n\pi}{\phi_0}}(\lambda\rho) J_{\frac{n\pi}{\phi_0}}(\lambda r) \lambda \, \mathrm{d}\lambda \, \rho \, \mathrm{d}\rho,$$

using the Fourier–Bessel–Hankel† integral

$$F(r) = \int_0^{+\infty} \int_0^{+\infty} F(\rho) J_\nu(\lambda\rho) J_\nu(\lambda r) \lambda \, \mathrm{d}\lambda \rho \, \mathrm{d}\rho, \quad \nu \geqslant -\frac{1}{2}.$$

† See [42], pages 459–500.

68. The solution of the boundary-value problem

$$\frac{\partial u}{\partial t}=a^2\left\{\frac{\partial^2 u}{\partial r^2}+\frac{1}{r}\frac{\partial u}{\partial r}+\frac{1}{r^2}\frac{\partial^2 u}{\partial \phi^2}\right\},\quad 0<\phi<\phi_0,\quad 0\leqslant r<+\infty,$$

$$0<t<+\infty, \tag{1}$$

$$u(r,0,t)=\frac{\partial u(r,\phi_0,t)}{\partial \phi}=0,\quad 0<r<+\infty,\quad 0<t<+\infty, \tag{2}$$

$$u(r,\phi,0)=f(r,\phi),\quad 0<\phi<\phi_0,\quad 0<r<+\infty \tag{3}$$

is:

$$u(r,\phi,t)=\int_0^{\phi_0}d\phi'\int_0^{+\infty}f(\rho,\phi')\,G(\rho,r,\phi',\phi,t)\,\rho\,d\rho, \tag{4}$$

where

$$G(\rho,r,\phi',\phi,t) \tag{5}$$

$$=\frac{2}{\phi_0}\sum_{n=0}^{+\infty}\left\{\int_0^{\infty}e^{-a^2\lambda^2 t}J_{\frac{(2n+1)\pi}{2\phi_0}}(\lambda\rho)\,J_{\frac{(2n+1)\pi}{2\phi_0}}(\lambda r)\,\lambda\,d\lambda\right\}\sin\frac{(2n+1)\pi\phi'}{2\phi_0}\times$$

$$\times\sin\frac{(2n+1)\pi\phi}{2\phi_0}.$$

69. $u(r,\phi,z,t)$

$$=\int_{-\infty}^{+\infty}d\zeta\int_0^{r_0}r'dr'\int_0^{\phi_0}f(r',\phi',\zeta)\tilde{G}_i(r,\phi,z,r',\phi',\zeta,t)d\phi',\quad i=1,2;$$

in case (a) $i=1$,

$$\tilde{G}_1=\frac{e^{-\frac{(z-\zeta)^2}{4a^2t}}}{2a\sqrt{\pi t}}G_1,$$

where G_1 was found in problem 67;

in case (b) $i=2$,

$$\tilde{G}_2=\frac{e^{-\frac{(z-\zeta)^2}{4a^2t}}}{2a\sqrt{\pi t}}G_2,$$

where G_2 was found in problem 67.

Method. If the Fourier transform with respect to z is applied, then the problem reduces to problem 67.

70. The solution of the boundary-value problem

$$\frac{\partial u}{\partial t}=a^2\left\{\frac{\partial^2 u}{\partial r^2}+\frac{1}{r}\frac{\partial u}{\partial r}\right\},\quad r_0\leqslant r<+\infty,\quad 0<t<+\infty, \tag{1}$$

$$u(r_0,t)=U_0=\text{const.}\quad 0<t<+\infty, \tag{2}$$

$$u(r,0)=0,\quad r_0<r<+\infty \tag{3}$$

is:

$$u(r, t) = \frac{2U_0}{\pi} \int_{r_0}^{+\infty} \frac{[1-e^{-a^2\lambda^2 t}]K(r, \lambda)}{J_0^2(r_0\lambda)+N_0^2(r_0\lambda)} \frac{d\lambda}{\lambda}, \tag{4}$$

where

$$K(r, \lambda) = J_0(r_0\lambda)N_0(r\lambda)-N_0(r_0\lambda)J_0(r\lambda). \tag{5}$$

Method. Use Weber's integral transform with kernel $rK(r, \lambda)$ in the interval $r_0 \leqslant r < +\infty$, namely: firstly, applying this transformation to equation (1), an equation is obtained for the Weber form of the unknown function

$$u(\lambda, t) = \int_{r_0}^{+\infty} u(r, t) rK(r, \lambda) dr, \tag{6}$$

and then, finding $\bar{u}(\lambda, t)$, apply Weber's inversion formula

$$u(r, t) = \int_{r_0}^{+\infty} \frac{\bar{u}(\lambda, t)\lambda K(r, \lambda) d\lambda}{J_0^2(r_0\lambda)+N_0^2(r_0\lambda)}. \tag{7}$$

2. Formation and Application of Green's Functions

71. *Method.* The validity of the statement is verified directly by substituting the function $u(x, y, z, t) = u_1(x, t)u_2(y, t)u_3(z, t)$ in equation (1) and the initial condition (2).

72. $$G(x, y, z) = \frac{1}{(2a\sqrt{\pi t})^3} e^{-\frac{(x-\xi)^2+(y-\eta)^2+(z-\zeta)^2}{4a^2t}}.$$

Method. If in problem 71 it is assumed that $f_1(x) = \delta(x)$, $f_2(y) = \delta(y)$, $f_3(z) = \delta(z)$, then it at once follows that the Green's function for the space $-\infty < x, y, z < +\infty$ is the product of Green's functions for the straight lines $-\infty < x < +\infty$, $-\infty < y < +\infty$, $-\infty < z < +\infty$.

73. $u(x, y, z, t)$

$$= \frac{1}{(2a\sqrt{\pi t})^3} \iiint_{-\infty}^{+\infty} e^{-\frac{(x-\xi)^2+(y-\eta)^2+(z-\xi)^2}{4a^2t}} f(\xi, \eta, \zeta) d\xi \, d\eta \, d\zeta +$$

$$+ \frac{1}{(2a\sqrt{\pi t})^3} \int_0^t \frac{d\tau}{(t-\tau)^{3/2}} \iiint_{-\infty}^{+\infty} e^{-\frac{(x-\xi)^2+(y-\eta)^2+(z-\zeta)^2}{4a^2(t-\tau)}} F(\xi, \eta, \zeta, \tau) d\xi \, d\eta \, d\zeta. \tag{1}$$

Method. Formula (1) may be derived quite simply, but not rigorously, using the physical significance of the source function, obtained in the solution of problem 72, and considering the unknown temperature $u(x, y, z, t)$ as the result of an addition of the effects of instantaneous elementary sources, distributed at the initial time with density $f(x, y, z)$, and of continuously acting sources, distributed with density $F(x, y, z, t)$.

Formula (1) may also be derived by means of Green's formula, as was done in the solution of problem 68 chapter III.

74. (a) $G_1(x, y, z, \xi, \eta, \zeta, t)$

$$=\frac{1}{(2a\sqrt{\pi t})^3}\left[e^{-\frac{(x-\xi)^2+(y-\eta)^2+(z-\zeta)^2}{4a^2t}}-e^{-\frac{(x-\xi)^2+(y-\eta)^2+(z+\zeta)^2}{4a^2t}}\right],$$

(b) $G_2(x, y, z, \xi, \eta, \zeta, t)$

$$=\frac{1}{(2a\sqrt{\pi t})^3}\left[e^{-\frac{(x-\xi)^2+(y-\eta)^2+(z-\zeta)^2}{4a^2t}}+e^{-\frac{(x-\xi)^2+(y-\eta)^2+(z+\zeta)^2}{4a^2t}}\right],$$

(c) $$G_3(x, y, z, \xi, \eta, \zeta, t)=\frac{1}{(2a\sqrt{\pi t})^3}\left[e^{-\frac{(x-\xi)^2+(y-\eta)^2+(z-\zeta)^2}{4a^2t}}+\right.$$

$$\left.+e^{-\frac{(x-\xi)^2+(y-\eta)^2+(z+\zeta)^2}{4a^2t}}-2h\int_0^{+\infty}e^{-h\omega}e^{-\frac{(x-\xi)^2+(y-\eta)^2(z+\zeta+\omega)^2}{4a^2t}}d\omega\right].$$

75. (a) $$u(x, y, z, t)=\int_{-\infty}^{+\infty}d\xi\int_{-\infty}^{+\infty}d\eta\int_0^{+\infty}f(\xi, \eta, \zeta)G_1(x, y, z, \xi, \eta, \zeta, t)d\zeta+$$

$$+a^2\int_0^t d\tau\iint_{-\infty}^{+\infty}\Phi(\xi, \eta, \tau)G_{1\zeta}(x, y, z, \xi, \eta, 0, t-\tau)d\xi d\eta+$$

$$+\int_0^t d\tau\int_0^{+\infty}d\zeta\iint_{-\infty}^{+\infty}F(\xi, \eta, \zeta, \tau)G_1(x, y, z, \xi, \eta, \zeta, t-\tau)d\xi d\eta,$$

(b) $$u(x, y, z, t)=\int_0^{+\infty}d\xi\iint_{-\infty}^{+\infty}f(\xi, \eta, \zeta)G_2(x, y, z, \xi, \eta, \zeta, t)d\xi d\eta-$$

$$-a^2\int_0^t d\tau\iint_{-\infty}^{+\infty}\Phi(\xi, \eta, \tau)G_2(x, y, z, \xi, \eta, 0, t-\tau)d\xi d\eta+$$

$$+\int_0^t d\tau\int_0^{+\infty}d\zeta\iint_{-\infty}^{+\infty}F(\xi, \eta, \zeta, \tau)G_2(\xi, \eta, \zeta, x, y, z, t-\tau)d\xi d\eta,$$

(c) $$u(x, y, z, t)=\int_0^{+\infty}d\zeta\iint_{-\infty}^{+\infty}f(\xi, \eta, \zeta)G_3(x, y, z, \xi, \eta, \zeta, t)d\xi d\eta+$$

$$+ha^2\int_0^t d\tau\iint_{-\infty}^{+\infty}\Phi(\xi, \eta, \tau)G_3(x, y, z, \xi, \eta, 0, t-\tau)d\xi d\eta+$$

$$+\int_0^t d\tau\int_0^{+\infty}d\zeta\iint_{-\infty}^{+\infty}F(\xi, \eta, \zeta, \tau)G_3(x, y, z, \xi, \eta, \zeta, t-\tau)d\xi d\eta.$$

76. *Method.* Use the method, suggested in the instructions to problem 72.

77. (a) $G(x, y, z, \xi, \eta, \zeta, t)$

$$= \frac{e^{-\frac{(x-\xi)^2+(y-\eta)^2}{4a^2t}}}{(2\sqrt{\pi t})^3} \sum_{n=-\infty}^{+\infty} \left(e^{-\frac{(z-\zeta+2nl)^2}{4a^2t}} - e^{-\frac{(z+\zeta+2nl)^2}{4a^2t}} \right)$$

$$= \frac{e^{-\frac{(x-\xi)^2+(y-\eta^2)}{4a^2t}}}{(2a\sqrt{\pi t})^2} \frac{2}{l} \sum_{n=1}^{+\infty} e^{-\frac{n^2\pi^2a^2}{l^2}t} \sin\frac{n\pi z}{l} \sin\frac{n\pi\zeta}{l},$$

(b) $G(x, y, z, \xi, \eta, \zeta, t)$

$$= \frac{e^{-\frac{(x-\xi)^2+(y-\eta)^2}{4a^2t}}}{(2a\sqrt{\pi t})^3} \sum_{n=-\infty}^{+\infty} \left(e^{-\frac{(z-\zeta+2nl)^2}{4a^2t}} + e^{-\frac{(z-\zeta+2nl)^2}{4a^2t}} \right)$$

$$= \frac{e^{-\frac{(x-\xi)^2+(y-\eta)^2}{4a^2t}}}{(2a\sqrt{\pi t})^2} \frac{2}{l} \left\{ \frac{1}{2} + \sum_{n=0}^{+\infty} e^{-\frac{n^2\pi^2a^2}{l^2}t} \cos\frac{n\pi z}{l} \cos\frac{n\pi\zeta}{l} \right\},$$

(c) $G(x, y, z, \xi, \eta, \zeta, t)$

$$= \frac{e^{-\frac{(x-\xi)+(y-\eta)^2}{4a^2t}}}{(2a\sqrt{\pi t})^3} \sum_{n=-\infty}^{+\infty} (-1)^n \left(e^{-\frac{(x-\zeta+2nl)^2}{4a^2t}} - e^{-\frac{(z+\zeta+2nl)^2}{4a^2t}} \right)$$

$$= \frac{e^{-\frac{(x-\xi)^2+(y-\eta)^2}{4a^2t}}}{(2a\sqrt{\pi t})^2} \frac{2}{l} \sum_{n=0}^{+\infty} e^{-\frac{(2n+1)^2\pi^2a^2}{4l^2}t} \cos\frac{(2n+1)\pi z}{2l} \cos\frac{(2n+1)\pi\zeta}{2l}.$$

$$\text{(d) } G(x, y, z, \xi, \eta, \zeta, t) = \frac{e^{-\frac{(x-\xi)^2+(y-\eta)^2}{4a^2t}}}{(2a\sqrt{\pi t})^2} 2\sum_{n=1}^{+\infty} \frac{e^{-a^2\lambda_n^2t}}{(\lambda_n^2+h^2)l+2h} \times$$

$$\times (\lambda_n \cos\lambda_n x + h\sin\lambda_n x)(\lambda_n\cos\lambda_n\xi + h\sin\lambda_n\xi);$$

λ_n are positive roots of the equation $\cot l\lambda = (\lambda^2-h^2)/2\lambda h$.

78. Placing the origin of a spherical system of coordinates at the centre of the sphere, we obtain:

$$u = \frac{Q}{c\rho} G(r, r', t), \tag{1}$$

where

$$G(r, r', t) = \frac{1}{8\pi a r r' \sqrt{\pi t}} \left[e^{-\frac{(r-r')^2}{4a^2 t}} - e^{-\frac{(r+r')^2}{4a^2 t}} \right] \tag{2}$$

is called the source function of the effect of an instantaneous spherical source of heat.

Method. We solve the equation

$$\frac{\partial u}{\partial t} = a^2 \left\{ \frac{\partial^2 u}{\partial r^2} + \frac{2}{r} \frac{\partial u}{\partial r} \right\}, \quad 0 < r < +\infty, \quad 0 < t < +\infty \tag{3}$$

for the initial condition

$$u(r, 0) = \begin{cases} 0 & \text{for} \quad 0 < r < r', \\ \dfrac{Q}{c\rho 4 r'^2 \, dr'} & \text{for} \quad r' < r < r' + dr', \\ 0 & \text{for} \quad r' + dr' < r < +\infty, \end{cases} \tag{4}$$

and then in the solution obtained let us pass to a limit for $dr' \to 0$. The solution of equation (3) with initial condition (4) by the substitution $v(r, t) = ru(r, t)$ reduces to the one-dimensional case, in which $v(0, t) = 0$, since $u(0, t)$ is a finite quantity.

79. $$u(r, t) = \frac{1}{2ar\sqrt{\pi t}} \int_0^{+\infty} \xi^2 F(\xi) \left[e^{-\frac{(r-\xi)^2}{4a^2 t}} - e^{\frac{(r+\xi)^2}{4a^2 t}} \right] d\xi +$$

$$+ \frac{1}{2a^2 \sqrt{\pi}} \int_0^t \frac{d\tau}{\sqrt{t-\tau}} \int_0^{+\infty} \xi^2 f(\xi_1 \tau) \left[e^{-\frac{(r-\xi)^2}{4a^2 (t-\tau)}} - e^{-\frac{(r+\xi)^2}{4a^2 (t-\tau)}} \right] d\xi.$$

80. $$u = \frac{Q}{c\rho} G(r, r', t), \tag{1}$$

where

$$G(r, r', t) = \frac{1}{2\pi} \int_0^{+\infty} e^{-a^2 \lambda^2 t} J_0(\lambda r) J_0(\lambda r') \lambda \, d\lambda = \frac{1}{4\pi a^2 t} e^{-\frac{r^2 + r'^2}{4a^2 t}} I_0 \left(\frac{r r'}{2a^2 t} \right) \tag{2}$$

is called the source function of an instantaneous cylindrical source of heat.

Method. We solve the equation

$$\frac{\partial u}{\partial t} = a^2 \left\{ \frac{\partial^2 u}{\partial r^2} + \frac{1}{r} \frac{\partial u}{\partial r} \right\}, \ 0 < r < +\infty, \ 0 < t < +\infty \tag{3}$$

for the initial condition

$$u(r, 0) = \begin{cases} 0 & \text{for} \quad 0 < r < r', \\ \dfrac{Q}{2\pi r' \, dr' \, c\rho} & \text{for} \quad r' < r < r' + dr', \\ 0 & \text{for} \quad r' + dr' < r < +\infty, \end{cases} \tag{4}$$

and then in the solution obtained pass to a limit for $dr' \to 0$. For $r = 0$, $u(r, t)$ must be bounded. We look for the solution of equation (3), satisfying the initial condition (4) and bounded for $r = 0$, in the form

$$u(r, t) = \int_0^{+\infty} \int_0^{+\infty} U(\rho, t) J_0(\lambda\rho) J_0(\lambda r) \lambda \, d\lambda \rho \, d\rho . \qquad (5)$$

See also the method to problem 74.

81. $$u(r, t) = \frac{1}{2a^2 t} e^{-\frac{r^2}{4a^2 t}} \int_0^{+\infty} \xi F(\xi) e^{-\frac{\xi^2}{4a^2 t}} I_0\left(\frac{r\xi}{2a^2 t}\right) d\xi +$$

$$+ \int_0^t \frac{e^{-\frac{r^2}{4a^2(t-\tau)}}}{2a^2(t-\tau)} d\tau \int_0^{+\infty} \xi f(\xi, \tau) e^{-\frac{\xi^2}{4a^2(t-\tau)}} I_0\left(\frac{r\xi}{2a^2(t-\tau)}\right) d\xi .$$

82. The source function for the equation

$$\frac{\partial u}{\partial t} = D\Delta u - \boldsymbol{v} \operatorname{grad} u,$$

is:

$$G(x, y, z, x', y', z', t) = \frac{1}{(2\sqrt{Dt})^3} e^{-\frac{(x-v_1 t-x')^2+(y-v_2 t-y')^2+(z-v_3 t-z)'^2}{4Dt}}, \qquad (1)$$

where v_1, v_2, v_2 are components of the vector $\boldsymbol{v}$ with respect to the x, y, z axes, and x', y', z' are the coordinates of the point at which the source acted at time $t = 0$.

Method. In a coordinate system moving along with the medium, the diffusion equation takes the form $\partial u/\partial t = D\Delta u$. After finding the expression for the source function in a moving system of coordinates and reverting to a stationary system, we obtain (1).

83. For a source with coordinates $(0, y', z')$ we have:

$$G(x, y, z, y', z') = \frac{v}{4D\pi x} e^{-\frac{(y-y')^2+(z-z')^2}{4\frac{D}{v}x}}$$

84.

(a) $$G(x, y, y', z, z') = \frac{v}{4D\pi x}\left[e^{-\frac{(y-y')^2+(z-z')^2}{4\frac{D}{v}x}} + e^{-\frac{(y-y')^2+(z+z')^2}{4\frac{D}{v}x}}\right],$$

(b) $$G(x, y, z, y', z') = \frac{v}{4D\pi x}\left[e^{-\frac{(y-y')^2+(z-z')^2}{4\frac{D}{v}x}} - e^{-\frac{(y-y')^2+(z+z')^2}{4\frac{D}{v}x}}\right],$$

(c) $$G(x,y,z,y',z') = \frac{v}{4D\pi x}\left[e^{-\frac{(y-y')^2+(z-z')^2}{4\frac{D}{v}x}} + e^{-\frac{(y-y')^2+(z+z')^2}{4\frac{D}{v}x}} - \right.$$

$$\left. -2h\int_0^{+\infty} e^{-h\omega-\frac{(z+z'+\omega)^2}{4\frac{D}{v}x}}\,d\omega\right].$$

85. $u(x,y,z,t)$

$$= \frac{1}{(2\sqrt{\pi D})^3}\int_0^t \frac{f(\tau)}{(t-\tau)^{3/2}}\, e^{-\frac{[x-\phi(\tau)]^2+[y-\psi(\tau)]^2+[z-\varkappa(\tau)]^2}{4D(t-\tau)}}\,d\tau.$$

Method. Let us look for the solution of the equation

$$\frac{\partial u}{\partial t} = a^2\left\{\frac{\partial^2 u}{\partial x^2}+\frac{\partial^2 u}{\partial y^2}+\frac{\partial^2 u}{\partial z^2}\right\} + f(t)\delta(x-\phi(t))\delta(y-\psi(t))\delta(z-\varkappa(t)), \quad (1)$$

with initial condition

$$u|_{t=0} = 0; \quad (2)$$

δ is the delta-function.

86. $$u(r,t) = \frac{U_0}{2}\left[\Phi\left(\frac{r+r_0}{2\sqrt{Dt}}\right) - \Phi\left(\frac{r-r_0}{2\sqrt{Dt}}\right)\right] +$$

$$+\frac{U_0}{r}\sqrt{\frac{Dt}{\pi}}\left(e^{-\frac{(r-r_0)^2}{4a^2t}} - e^{-\frac{(r+r_0)^2}{4a^2t}}\right), \quad (1)$$

where

$$\Phi(z) = \frac{2}{\sqrt{\pi}}\int_0^z e^{-\zeta^2}d\zeta. \quad (2)$$

Method. If one uses the source function of the instantaneous spherical source, found in the solution of problem 78, having in mind the similarity of the heat and diffusion problems, then the solution of the equation

$$\frac{\partial u}{\partial t} = D\left\{\frac{\partial^2 u}{\partial r^2}+\frac{2}{r}\frac{\partial u}{\partial r}\right\}, \quad 0 < r, t < +\infty, \quad (3)$$

satisfying the initial condition

$$u(r,0) = f(r), \quad 0 < r < +\infty, \quad (4)$$

may be represented in the form

$$u(r,t) = \int_0^{+\infty} f(r')G(r,r',t)4\pi r'^2 dr', \quad (5)$$

where

$$G(r, r', t) = \frac{1}{8\pi r r' \sqrt{\pi D t}} \left[e^{-\frac{(r-r')^2}{4Dt}} - e^{-\frac{(r+r')^2}{4Dt}} \right]. \tag{6}$$

The problem may also be solved by reduction to a semi-infinite rod by means of the substitution $v(r, t) = ru(r, t)$.

87. (a) $u(x, y, z, t)$

$$= u\left(\sqrt{x^2+y^2+(z-z_0)^2},\ t\right) + u\left(\sqrt{x^2+y^2+(z+z_0)^2},\ t\right),$$

(b) $u(x, y, z, t)$

$$= u\left(\sqrt{x^2+y^2+(z-z_0)^2},\ t\right) - u\left(\sqrt{x^2+y^2+(z+z_0)^2},\ t\right),$$

where $u(r, t)$ is the solution of the preceding problem.

88. $$u(r, t) = \frac{U_0}{2Dt} \int_0^{r_0} e^{-\frac{r^2+r'^2}{4Dt}} I_0\left(\frac{rr'}{2Dt}\right) r' \, dr'. \tag{1}$$

Method. If one uses the source function for an instantaneous cylindrical source, obtained in the solution of problem 80, having in view the similarity of the heat and diffusion problems, then the solution of the equation

$$\frac{\partial u}{\partial t} = a^2 \left\{ \frac{\partial^2 u}{\partial r^2} + \frac{1}{r} \frac{\partial u}{\partial r} \right\}, \quad 0 \leqslant r < \infty, \quad 0 < t < +\infty, \tag{2}$$

satisfying the initial condition

$$u(r, 0) = f(r), \quad 0 < r < +\infty, \tag{3}$$

may be represented in the form

$$u(r, t) = \int_0^{+\infty} f(r') G(r, r', t) \, 2\pi r' \, dr', \tag{4}$$

where

$$G(r, r', t) = \frac{1}{4\pi D t} e^{-\frac{r^2+r'^2}{4Dt}} I_0\left(\frac{rr'}{2Dt}\right). \tag{5}$$

89. (a) $u(x, y, t) = u\left(\sqrt{(x-x_0)^2+y^2},\ t\right) + u\left(\sqrt{(x+x_0)^2+y^2},\ t\right)$,

(b) $u(x, y, t) = u\left(\sqrt{(x-x_0)^2+y^2},\ t\right) - u\left(\sqrt{(x+x_0)^2+y^2},\ t\right)$,

where $u(r, t)$ is the solution of the preceding problem.

90. The solution of the boundary-value problem (Fig. 48)

$$\frac{\partial H}{\partial t} = a^2 \left(\frac{\partial^2 H}{\partial x^2} + \frac{\partial^2 H}{\partial y^2} \right), \quad a^2 = \frac{k\bar{h}}{m}\dagger; \quad -\infty < x < +\infty,$$
$$0 < y, t < +\infty, \tag{1}$$

† See the solution of problem 8.

$$H(x, y, 0) = H_0 = \text{const.}, \quad -\infty < x < +\infty, \quad 0 < y < +\infty, \tag{2}$$

$$H(x, 0, t) = \left\{\begin{matrix} H_1, & -\infty < x < 0, \\ H_2, & 0 < x < +\infty, \end{matrix}\right\} 0 < t < +\infty \tag{3}$$

is:

$$H(x, y, t) = H_0 \Phi\left(\frac{y}{2a\sqrt{t}}\right) + \frac{H_1+H_2}{2}\left[1 - \Phi\left(\frac{y}{2a\sqrt{t}}\right)\right] - \\ - \frac{(H_1 - H_2)y}{\pi} \int_0^x e^{-\frac{\eta^2+y^2}{4a^2t}} \frac{d\eta}{\eta^2 + y^2}.$$

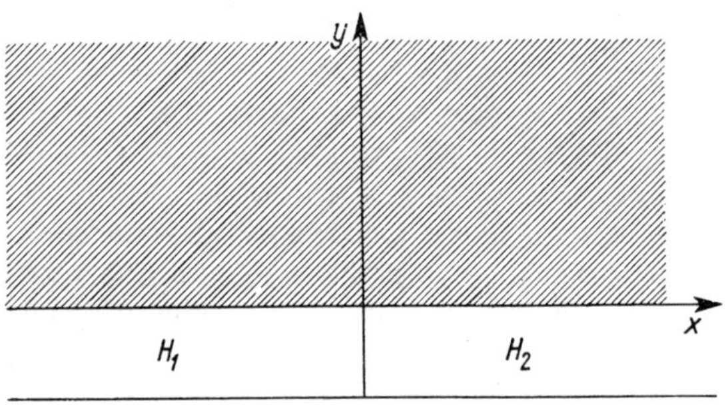

FIG. 48

Method. Find the source function for the semiplane $y \leqslant 0$ with homogeneous boundary condition of the first kind for equation (1), and then represent the solution of the problem (1), (2), (3) by means of this source function.

91. For the unknown flow $q(t)$ we obtain the expression

$$-\lambda u_r(r_0, t) = q(t) = \frac{\lambda}{\pi}\frac{d}{dt}\int_0^t \left\{\frac{\sqrt{\pi}}{a}\phi(\tau) + \frac{1}{r_0}\int_0^t \frac{\phi(\xi)d\xi}{\sqrt{\tau-\xi}}\right\}\frac{d\tau}{\sqrt{t-\tau}}.$$

Method. By means of the substitution $v(r, t) = ru(r, t)$, where $u(r, t)$ is the temperature of space, we arrive at the problem:

$$\frac{\partial v}{\partial t} = a^2 \frac{\partial^2 v}{\partial r^2}, \quad r_0 \leqslant r < +\infty, \quad 0 < t < +\infty,$$

$$v(r, 0) = 0, \qquad r_0 < r < +\infty,$$

$$v(r_0, t) = r_0 \phi(t), \quad 0 < t < +\infty,$$

$$\left.\frac{\partial v}{\partial r}\right|_{r=r_0} = -\frac{r_0}{\lambda} q(t) + \phi(t), \quad 0 < t < +\infty,$$

where $q(t)$ is an unknown function. Then, as in problems 95 and 96, § 2, chapter III, solving Abel's integral equation, we find $q(t)$.

CHAPTER VI

EQUATIONS OF HYPERBOLIC TYPE

§ 1. Physical Problems Leading to Equations of Hyperbolic Type; Statement of Boundary-value Problems

1. As the Lagrangian coordinates† of a particle we take its cartesian coordinates x, y, z in the unperturbed state. Let the cartesian coordinates of the particle in the perturbed state be

$$\xi = x+u^{(1)}(x, y, z, t),$$
$$\eta = y+u^{(2)}(x, y, z, t),$$
$$\zeta = z+u^{(3)}(x, y, z, t).$$

The vector $\boldsymbol{u} = \boldsymbol{i}u^{(1)}+\boldsymbol{j}u^{(2)}+\boldsymbol{k}u^{(3)}$ describes the displacement of the particle from the unperturbed state x, y, z. The velocity of the particle is

$$\boldsymbol{v} = \frac{d\boldsymbol{u}}{dt} = \boldsymbol{i}\dot{u}^{(1)}+\boldsymbol{j}\dot{u}^{(2)}+\boldsymbol{k}\dot{u}^{(3)} = \boldsymbol{i}v^{(1)}+\boldsymbol{j}v^{(2)}+\boldsymbol{k}v^{(3)},$$

where the dot above indicates the derivative with respect to time. The velocity potential and the displacement potential are defined by the equalities

$$\operatorname{grad} U = \boldsymbol{v}, \quad \operatorname{grad} \Phi = \boldsymbol{u}.$$

Each is determined up to an arbitrary function of the time. A perturbation of the density $\tilde{\rho}$ and a perturbation of the pressure p are defined as before‡. Each of the quantities

$$\tilde{\rho}, \tilde{p}, \rho, p, U, \Phi, u^{(i)}, v^{(i)}; \quad i = 1, 2, 3$$

assuming small perturbations satisfies the equation

$$u_{tt} = a^2(u_{xx}+u_{yy}+u_{zz}), \tag{1}$$

where $a^2 = k\,p_0/\rho_0$, $k = c_p/c_v$ is the ratio of the specific heat at constant pressure to the specific heat at constant volume; p_0 = const. and ρ_0 = const. are the unperturbed pressure and unperturbed density. The initial conditions are written in the form

$$u(x, y, z, 0) = f(x, y, z), \quad u_t(x, y, z, 0) = F(x, y, z),$$
$$-\infty < x, y, z < +\infty. \tag{2}$$

† For more detail on Lagrangian coordinates see problem 4, § 1, chapter II.

‡ See problem 4, § 1, chapter II.

Each of the quantities $\tilde{\rho}$, $\tilde{p}$, U, Φ, $\boldsymbol{v}$, $\boldsymbol{u}$ can be expressed in terms of any other of these quantities by means of the relations

$$\tilde{p} = a^2 \tilde{\rho}, \tag{3}$$

$$\rho_0 U_t + \tilde{p} = 0, \tag{4}$$

$$\rho_0 \Phi_{tt} + \tilde{p} = 0, \tag{5}$$

$$\boldsymbol{v} = \operatorname{grad} U, \tag{6}$$

$$\boldsymbol{u} = \operatorname{grad} \Phi, \tag{7}$$

$$\boldsymbol{v} = \frac{d\boldsymbol{u}}{dt}. \tag{8}$$

Method. The equation of continuity in Lagrangian coordinates can be derived, by considering the deformation of an elementary volume $\Delta x \Delta y \Delta z$ and taking into account the fact that its mass remains constant; the coefficient of volume deformation is the Ostrogradskii determinant ("Jacobian"). The linearized adiabatic equation and equations (4) and (5) are derived in the same way as the corresponding equations in the solution of problem 4, § 1, chapter II.

2. At the plane, bounding the semispace under consideration, the boundary conditions must be fulfilled

(a) $\dfrac{\partial \rho}{\partial n} = \dfrac{\partial p}{\partial n} = \dfrac{\partial U}{\partial n} = \dfrac{\partial \Phi}{\partial n} = 0$, where $\dfrac{\partial}{\partial n}$ is the derivative with respect to the normal to the plane;

(b) $\dfrac{\partial U}{\partial n} = V$, $\dfrac{\partial \Phi}{\partial n} = \displaystyle\int_0^t V\, dt$, $\dfrac{\partial \tilde{p}}{\partial n} = -\rho_0 \dot{V}$, $\dfrac{\partial \tilde{\rho}}{\partial n} = -\dfrac{\rho_0}{a^2}\dot{V}$, where $V(t)$ is the projection of the velocity of the plane on the direction of the normal.

3. Quantities on one side of the surface Σ are denoted by the subscript 1, and on the other side by the subscript 2. On the surface Σ the boundary condition

$$\rho_{01} U_1 = \rho_{02} U_2, \tag{1}$$

$$\frac{\partial U_1}{\partial n} = \frac{\partial U_2}{\partial n}, \tag{2}$$

must be fulfilled, where $\partial/\partial n$ indicates the derivative with respect to the normal to the surface Σ, and ρ_{01} and ρ_{02} are the unperturbed densities of the gases.

Method. Boundary condition (1) is obtained by means of equality (4) of the answer to problem 1. Boundary condition (2) expresses the conservation of fluid at the boundary of separation of the gases (equality of the normal components of the velocity of particles of both gases, adjoining the surface of separation Σ at the same spot).

4. For the deflection $u(x, y, t)$ of particles of the membrane from the plane of the unperturbed state (the plane XOY) we obtain:

$$\frac{\partial^2 u}{\partial t^2} = a^2\left(\frac{\partial^2 u}{\partial x^2} + \frac{\partial^2 u}{\partial y^2}\right), \quad 0 < t < +\infty, \quad (x, y) \in G, \tag{1}$$

where G is the region in the plane x, y bounded by the contour Γ,

$$u(x, y, 0) = f(x, y), \quad u_t(x, y, 0) = F(x, y), \quad (x, y) \in G, \tag{2}$$

$$u\,|_\Gamma = 0, \quad 0 < t < +\infty\dagger. \tag{3}$$

5. Equation (1) in the answer to the preceding problem must be replaced by the equation

$$\frac{\partial^2 u}{\partial t^2} = a_1^2\left\{\frac{\partial^2 u}{\partial x^2} + \frac{\partial^2 u}{\partial y^2}\right\} - \frac{\rho_0 a_0^2}{\Omega_0 \rho_1}\iint\limits_G u \,\mathrm{d}x\,\mathrm{d}y, \tag{1}$$

where a_1 is the velocity of propagation of transverse waves in the membrane, ρ_1 the surface density of the membrane, Ω_0 the volume of the vessel, ρ_0 the unperturbed density of air, a_0 the velocity of propagation of small disturbances in the air.

Method. Because $a_0 \gg a_1$ the pressure of the air, enclosed in the vessel acting on an element of the membrane, can be assumed to be independent of the coordinates of the element, and determined only by the total change of volume of the vessel due to deflection of the membrane.

Note. If the velocity of propagation of small disturbances in the surrounding medium is considerably less than the velocity of propagation of disturbances in the membrane, i.e. if $a_0 \ll a_1$, then the reaction of the medium on each element of the membrane is determined by the state of the medium in direct proximity to this element. In this case the wave equation of the membrane‡ can be written in the form

$$\frac{\partial^2 u}{\partial t^2} = a_1^2\left\{\frac{\partial^2 u}{\partial x^2} + \frac{\partial^2 u}{\partial y^2}\right\} - \frac{\rho_0}{\rho_1}\frac{\partial u}{\partial t}.$$

6.

$$\left\{\frac{\partial}{\partial t} + (\boldsymbol{v}_0, \nabla)\right\}^2 U = a^2 \Delta U, \tag{1}$$

where U is the velocity potential of particles of the gas, produced by small disturbances, $\boldsymbol{v}_0 = \boldsymbol{i}v_1^{(0)} + \boldsymbol{j}v_2^{(0)} + \boldsymbol{k}v_3^{(0)}$ is the velocity of motion of the medium; the operator $(\boldsymbol{v}_0, \nabla)$ is given by the relation

$$(\boldsymbol{v}_0, \nabla) = v_1^{(0)}\frac{\partial}{\partial x} + v_2^{(0)}\frac{\partial}{\partial y} + v_3^{(0)}\frac{\partial}{\partial z}, \tag{2}$$

† For a more detailed derivation of equation (1) see [7], pages 24–27.

‡ See [38], page 224.

where the potential U is considered as a function of the coordinates (x, y, z) of a geometric point and of time t in a stationary system of coordinates, with respect to which the medium moves with velocity $\boldsymbol{v}_0$; in other words, U is investigated in Eulerian coordinates†.

If the x-axis coincides in direction with the vector $\boldsymbol{v}_0$, then

$$(\boldsymbol{v}_0, \nabla) = \boldsymbol{v}_0 \frac{\partial}{\partial x}$$

and equation (1) takes the form

$$\frac{\partial^2 U}{\partial t^2} + 2\boldsymbol{v}_0 \frac{\partial^2 U}{\partial x \partial t} + \boldsymbol{v}_0^2 \frac{\partial^2 U}{\partial x^2} = a^2 \left\{ \frac{\partial^2 U}{\partial x^2} + \frac{\partial^2 U}{\partial y^2} + \frac{\partial^2 U}{\partial z^2} \right\}. \tag{1'}$$

Similar equations hold for the density and pressure.

Method. Firstly it is necessary to derive the fundamental equations of hydrodynamics in Eulerian coordinates

$$\frac{\partial \boldsymbol{v}^*}{\partial t} + (\boldsymbol{v}^*, \nabla)\boldsymbol{v}^* = -\frac{1}{\rho} \operatorname{grad} p, \tag{3}$$

$$\frac{\partial \rho}{\partial t} + \operatorname{div}(\rho \boldsymbol{v}^*) = 0, \tag{4}$$

$$p = f(\rho), \quad f(\rho) = p_0 \frac{\rho^k}{\rho_0^k}, \quad k = \frac{c_p}{c_v}, \tag{5}$$

$\boldsymbol{v}^* = \boldsymbol{v}_0 + \boldsymbol{v}$, $\rho = \rho_0 + \tilde{\rho}$, $p = p_0 + \tilde{p}$, where $\boldsymbol{v}^*$ is the total (absolute) velocity of the particles, $\boldsymbol{v}_0$ the transfer velocity, $\boldsymbol{v}$ the relative velocity, and the quantities ρ_0, p_0, $\tilde{\rho}$, $\tilde{p}$ are given as in problem 1. The linearization of equation (3), (4), (5) and the elimination of p and ρ leads to equation (1) of the answer.

Equation (1) may also be obtained in the following way. In the coordinate system (O', x', y', z'), moving along with the medium and coinciding at time $t = 0$ with the stationary system (O, x, y, z), for the potential $U = U(x', y', z', t)$ the equation

$$\frac{\partial^2 U}{\partial t^2} = a^2 \left(\frac{\partial^2 U}{\partial x^2} + \frac{\partial^2 U}{\partial y^2} + \frac{\partial^2 U}{\partial z^2} \right), \tag{6}$$

will hold. Transition from the Eulerian coordinates (x', y', z', t) to the Eulerian coordinates (x, y, z, t) transforms equation (6) into equation (1) of the answer.

7. Let us choose the axis Oz of a cartesian rectangular system along the edge of the wedge so that the wedge is symmetrical with respect to the plane

---

† For more detail on Lagrangian and Eulerian coordinates see problem 4, § 1, chapter II.

xOz and so that the direction of the velocity of the forward flow v_0 coincides with the direction of the axis Ox (Fig. 49).

The angle of inclination of the wedge is denoted by 2ε. Since the velocity potential U, $\boldsymbol{v} = \operatorname{grad} U$, will not depend on z and t, equation (1′) of the answer to the preceding problem reduces to the form

$$\frac{\partial^2 U}{\partial x^2} = \frac{1}{M^2 - 1} \frac{\partial^2 U}{\partial y^2}, \tag{1}$$

where $M = v_0/a > 1$ by the conditions of the problem (the velocity of the forward flow is greater than the velocity of sound). Equation (1) holds between

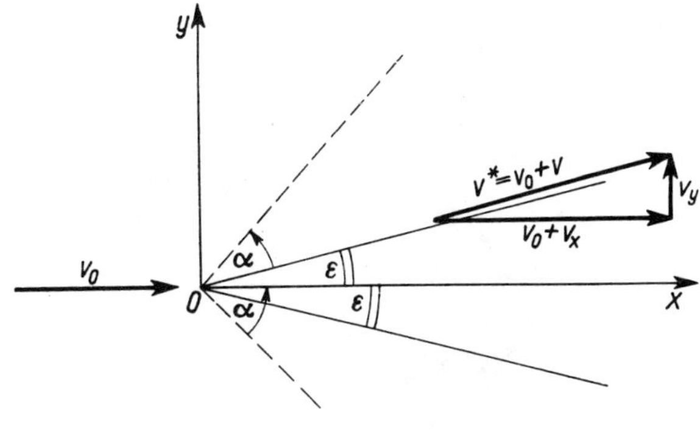

FIG. 49

the surface of the wedge and the wave front of the weak shock†. On the surface of the wedge we have:

$$\frac{\partial U}{\partial y} = \left(v_0 + \frac{\partial U}{\partial x}\right) \tan \varepsilon \quad \text{for} \quad y = x \tan \varepsilon. \tag{2}$$

On the wave front of a weak shock

$$U = 0 \quad \text{for} \quad y = x \tan \alpha, \tag{3}$$

where $\tan \alpha = \dfrac{1}{\sqrt{M^2 - 1}}$.

† The wave front of a weak shock separates the perturbed region from the unperturbed; at the surface of a wave front of weak shock the potential U and its derivatives of first order are continuous. For more detail see [15].

8. In a cylindrical system of coordinates, the axis Oz of which coincides with the axis of the cone (Fig. 50), we obtain a boundary-value problem for the velocity potential $U = U(r, z)$

$$\frac{\partial^2 U}{\partial z^2} = \frac{1}{M^2-1}\left\{\frac{\partial^2 U}{\partial r^2} + \frac{1}{r}\frac{\partial U}{\partial r}\right\} \tag{1}$$

between the surface of the cone and the surface of the wave front of a weak shock

$$\frac{\partial U}{\partial r} = \left\{v_0 + \frac{\partial U}{\partial z}\right\} \tan \varepsilon \tag{2}$$

on the surface of the cone, i.e. for $r = z \tan \alpha$; on the surface of the wave front of the weak shock

$$U = 0. \tag{3}$$

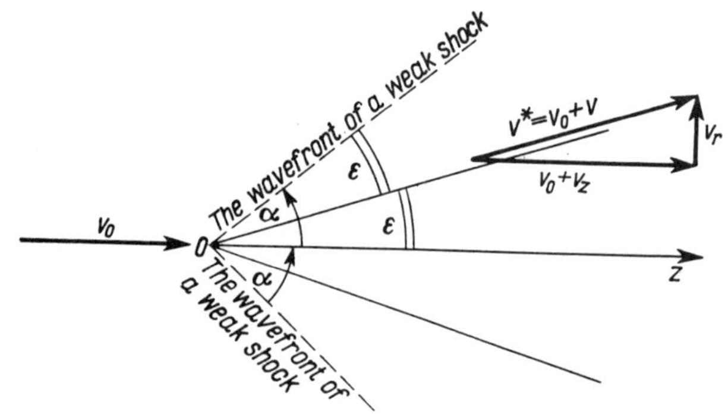

FIG. 50

9. For $\zeta(x, y, t)$ we obtain the boundary-value problem

$$\frac{\partial^2 \zeta}{\partial t^2} = a^2\left(\frac{\partial^2 \zeta}{\partial x^2} + \frac{\partial^2 \zeta}{\partial y^2}\right), \tag{1}$$

$a^2 = gh$, g is the acceleration of gravity;

$$\zeta(x, y, 0) = f(x, y), \quad \zeta_t(x, y, 0) = F(x, y), \tag{2}$$

$$\frac{\partial \zeta}{\partial n} = 0 \quad \text{at the walls}, \tag{3}$$

where $\partial/\partial n$ is the derivative with respect to the normal to the wall. For the horizontal velocity potential $U(x, y, t)$ we obtain the boundary-value problem.

$$\frac{\partial^2 U}{\partial t^2} = a^2\left(\frac{\partial^2 U}{\partial x^2} + \frac{\partial^2 U}{\partial y^2}\right), \quad a^2 = gh, \tag{1'}$$

$$U(x, y, 0) = f_1(x, y), \quad U_t(x, y, 0) = F_1(x, y), \tag{2'}$$

$$\frac{\partial U}{\partial n} = 0 \quad \text{at the walls.} \tag{3'}$$

Method. Derive:

(1) the equation of continuity

$$\frac{\partial \zeta}{\partial t} = -\operatorname{div} \boldsymbol{w},$$

where $\boldsymbol{w}$ is the horizontal velocity;

(2) the equation of motion

$$\rho \frac{\partial \boldsymbol{w}}{\partial t} = -\operatorname{grad}_{xy} p = -\boldsymbol{i}\frac{\partial p}{\partial x} - \boldsymbol{j}\frac{\partial p}{\partial y};$$

(3) the equation, expressing the pressure in the liquid at a distance z from the bottom of the tank,

$$p - p_0 = g\rho(h + \zeta - z).$$

10. The equation for the horizontal velocity potential takes the form

$$\frac{\partial^2 U}{\partial t^2} = a^2 \left\{ \frac{\partial^2 U}{\partial x^2} + \frac{\partial^2 U}{\partial y^2} \right\} + \frac{1}{\rho}\frac{\partial p_0}{\partial t}, \quad a^2 = gh. \tag{1}$$

The initial and boundary conditions are formulated as in the answer to the preceding problem.

11.

$$\rho \frac{\partial^2 u}{\partial t^2} = \frac{\partial \sigma_x}{\partial x} + \frac{\partial \tau_{xy}}{\partial y} + \frac{\partial \tau_{xz}}{\partial z} + X,$$

$$\rho \frac{\partial^2 v}{\partial t^2} = \frac{\partial \tau_{yx}}{\partial x} + \frac{\partial \sigma_y}{\partial y} + \frac{\partial \tau_{yz}}{\partial z} + Y,$$

$$\rho \frac{\partial^2 w}{\partial t^2} = \frac{\partial \tau_{zx}}{\partial x} + \frac{\partial \tau_{zy}}{\partial y} + \frac{\partial \sigma_z}{\partial z} + Z,$$

where σ_x, τ_{xy}, τ_{xz} are the projections on the coordinate axis of the stress, acting on an area, perpendicular to the x-axis; similarly $\tau_{yx}, \sigma_y, \tau_{yz}$ and $\tau_{zx}, \tau_{zy}, \sigma_z$ are defined. Here $\sigma_x, \sigma_y, \sigma_z$ are called normal stresses, and τ_{xy}, τ_{xz}, τ_{yz} the tangential or shear stresses; X, Y, Z are the projections on the coordinate axes of the density of the volume forces.

12. *Method.* From the equations of motion, obtained in the answer to problem 11, and Hooke's law, deduced in note (2) to the present problem, the following equations for the components of the vector $\boldsymbol{U}$ are readily derived:

$$\left.\begin{aligned} \rho \frac{\partial^2 u}{\partial t^2} &= \mu \Delta u + (\lambda + \mu)\frac{\partial \Theta}{\partial x} + X, \\ \rho \frac{\partial^2 v}{\partial t^2} &= \mu \Delta v + (\lambda + \mu)\frac{\partial \Theta}{\partial y} + Y, \\ \rho \frac{\partial^2 w}{\partial t^2} &= \mu \Delta w + (\lambda + \mu)\frac{\partial \Theta}{\partial z} + Z, \end{aligned}\right\} \text{where } \Theta = \operatorname{div} \boldsymbol{U}.$$

14. $U = \boldsymbol{i}\left(\frac{\partial \phi}{\partial x}+\frac{\partial \psi}{\partial y}\right)+\boldsymbol{j}\left(\frac{\partial \phi}{\partial y}-\frac{\partial \psi}{\partial x}\right), \quad \phi = \phi(x, y, t), \quad \psi = \psi(x, y, t),$

$$\rho \frac{\partial^2 \phi}{\partial t^2} = (\lambda+2\mu)\left\{\frac{\partial^2 \phi}{\partial x^2}+\frac{\partial^2 \phi}{\partial y^2}\right\}+F_1(x, y, t),$$

$$\rho \frac{\partial^2 \psi}{\partial t^2} = \mu\left\{\frac{\partial^2 \psi}{\partial x^2}+\frac{\partial^2 \psi}{\partial y^2}\right\}+F_2(x, y, t).$$

$F_1(x, y, t)$ and $F_2(x, y, t)$ are terms, obtained from the density of the volume forces.

15. (a)
$$\left.\begin{aligned} \sigma_x \cos(n, x)+\tau_{xy} \cos(n, y)+\tau_{xz} \cos(n, z) &= 0,\\ \tau_{yx} \cos(n, x)+\sigma_y \cos(n, y)+\tau_{yz} \cos(n, z) &= 0,\\ \tau_{zx} \cos(n, x)+\tau_{zy} \cos(n, y)+\sigma_z \cos(n, z) &= 0, \end{aligned}\right\} \quad (1)$$

where $\cos(n, x)$, $\cos(n, y)$, $\cos(n, z)$ are direction cosines of the normal to the element of boundary being considered.

(b) $U = 0$, i.e. $u = 0$, $v = 0$, $w = 0$. (2)

Taking the plane xz as the boundary plane and directing the y-axis into the body, in the case of a plane problem† we obtain the following expression for the boundary conditions:

$$\text{(a}'\text{)} \quad \left[a^2 \frac{\partial^2 \phi}{\partial y^2}+(a^2-2b^2)\frac{\partial^2 \phi}{\partial x^2}-2b^2\frac{\partial^2 \phi}{\partial x \partial y}\right]_{y=0} = 0, \quad (1')$$

$$\left[2\frac{\partial^2 \phi}{\partial x \partial y}+\frac{\partial^2 \psi}{\partial y^2}-\frac{\partial^2 \psi}{\partial x^2}\right]_{y=0} = 0;$$

$$\text{(b}'\text{)} \quad \left[\frac{\partial \phi}{\partial x}+\frac{\partial \psi}{\partial y}\right]_{y=0} = 0, \quad \left[\frac{\partial \phi}{\partial y}-\frac{\partial \psi}{\partial x}\right]_{y=0} = 0,$$

where ϕ and ψ are potentials, appearing in the answer to the preceding problem,

Method. The left-hand sides of equation (1) are the projections on the co-ordinate axes of the vector of the stress applied to an area with normal n‡.

16. For a radial displacement $u(r, t)$ of particles of the tube, at a distance r from the axis of the tube, we obtain:

$$\frac{\partial^2 u}{\partial t^2} = a^2\left(\frac{\partial^2 u}{\partial r^2}+\frac{1}{r}\frac{\partial u}{\partial r}-\frac{u}{r^2}\right)+F(r, t), \quad r_1 r \leqslant r \leqslant_2, \quad 0<t<+\infty, \quad (1)$$

where r_1 and r_2 are the inner and outer radii of the tube, $a^2=(\lambda+2\mu)/\rho$, a is the velocity of propagation of longitudinal deformations,

$$\left[r\frac{\partial u}{\partial r}+hu\right]_{r=r_1} = 0, \quad \left[r\frac{\partial u}{\partial r}+hu\right]_{r=r_2} = 0, \quad 0<r<+\infty, \quad (2)$$

† See problem 14.

‡ For more detail see [26], pages 17–18.

where $h = \lambda/(\lambda+2\mu)$,

$$\left.\begin{aligned} u(r,0) &= \phi(r), \quad 0 \leqslant r \leqslant r_0, \\ u_t(r,0) &= \psi(r), \quad 0 \leqslant r \leqslant r_0. \end{aligned}\right\} \tag{3}$$

17. For a radial displacement $u(r, t)$ of particles of a spherical shell for the conditions of the problem we obtain:

$$\frac{\partial^2 u}{\partial t^2} = a^2 \left\{\frac{\partial^2 u}{\partial r^2} + \frac{2}{r}\frac{\partial u}{\partial r} - \frac{2u}{r^2}\right\}, \quad r_1 \leqslant r \leqslant r_2, \quad 0 < t < +\infty. \tag{1}$$

a^2 has the same meaning as in the preceding problem,

$$(\lambda+2\mu)\frac{\partial u}{\partial r} + \lambda\frac{u}{r} = \left\{\begin{aligned} &-p(t) \quad \text{for} \quad r = r_1, \\ &\ 0 \qquad\ \ \text{for} \quad r = r_2, \end{aligned}\right\} \tag{2}$$

$$\left.\begin{aligned} u(r,0) &= 0, \\ u_t(r,0) &= 0, \end{aligned}\right\} \quad r_1 < r < r_2. \tag{3}$$

18. For transverse deflections from the unperturbed position of points of the lamina we obtain the equation

$$\frac{\partial^2 u}{\partial t^2} + c^2\left(\frac{\partial^4 u}{\partial x^4} + 2\frac{\partial^4 u}{\partial x^2 \partial y^2} + \frac{\partial^4 u}{\partial y^4}\right) = \frac{1}{2\rho h} p(x, y, t), \tag{1}$$

where $c^2 = Eh^2/3\rho(1-m^2)$, E is Young's modulus, m is Poisson's coefficient, $2h$ is the thickness of the lamina, ρ is the mass density of the lamina, $p(x, y, t)$ is the transverse force, acting on unit area of the lamina.

If the lamina lies on a flexible base, then

$$\frac{\partial^2 u}{\partial t^2} + c^2\left(\frac{\partial^4 u}{\partial x^4} + 2\frac{\partial^4 u}{\partial x^2 \partial y^2} + \frac{\partial^4 u}{\partial y^4}\right) + \frac{k}{2h\rho} u = \frac{1}{2h\rho} p(x, y, t); \tag{1'}$$

k is the coefficient of elasticity of the base†.

Note. The combination of terms in the curly brackets is conveniently written in the form $\Delta_2 \Delta_2 u$, where $\Delta_2 = \text{div grad}$ is the Laplacian operator in the plane.

19. $$\frac{\partial^2 u}{\partial t^2} + c^2\left(\frac{\partial^2 u}{\partial r^2} + \frac{1}{r}\frac{\partial}{\partial r} + \frac{1}{r^2}\frac{\partial^2}{\partial \phi^2}\right)^2 u = 0, \quad 0 \leqslant r < r_0,$$

$$0 \leqslant \phi \leqslant 2\pi, \quad 0 < t < +\infty, \tag{1}$$

$$u(r, \phi, 0) = f(r, \phi), \quad u_t(r, \phi, 0) = F(r, \phi), \quad 0 \leqslant r \leqslant r_0, \quad 0 \leqslant \phi \leqslant 2\pi, \tag{2}$$

$$u(r_0, \phi, t) = u_r(r_0, \phi, t) = 0, \quad 0 \leqslant \phi \leqslant 2\pi, \quad 0 < t < +\infty. \tag{3}$$

20. In spherical coordinates with origin at the dipole and with the axis $\theta = 0$, directed along the dipole, we obtain the boundary-value problem

$$\frac{\partial^2 H_\phi}{\partial t^2} = a^2 \left\{\frac{1}{r}\frac{\partial^2 (rH_\phi)}{\partial r^2} + \frac{1}{r^2}\frac{\partial}{\partial \theta}\left[\frac{1}{\sin\theta}\frac{\partial}{\partial \theta}\sin\theta H_\phi)\right]\right\}, \quad r > 0, \quad t > 0, \tag{1}$$

---

† See problem 10, § 1, chapter II.

$$H_\phi|_{t=0} = 0 \quad \text{for} \quad r > 0, \tag{2}$$

$$\left.\frac{\partial H_\phi}{\partial t}\right|_{t\to 0} = -\frac{3aM_0 \sin\theta}{r^4} \quad \text{for} \quad r > 0, \tag{2'}$$

$$H_\phi|_{r\to 0} = -\frac{\omega M_0}{ar^2} \sin\omega t \sin\theta \quad \text{for} \quad t > 0. \tag{3}$$

Method. Use Maxwell's equations in spherical coordinates. By virtue of the cylindrical symmetry and by virtue of elementary electrodynamic considerations $H_r = H_\theta = E_\phi = 0$ for $t \geqslant 0$.

For $t = 0$ there is an electrostatic field, produced by the electrostatic dipole, such that $H_\phi|_{t=0} = 0$ and

$$E_r|_{t=0} = \frac{2M_0 \cos\theta}{r^3}, \quad E_\theta|_{t=0} = \frac{M_0 \sin\theta}{r^3}.$$

We obtain the initial condition (2′) from these relations by means of the Maxwellian equation

$$\frac{1}{r}\left[\frac{\partial(rE_\theta)}{\partial r} - \frac{\partial E_r}{\partial\theta}\right] = -\frac{1}{a}\frac{\partial H_\phi}{\partial t}.$$

Finally, the boundary condition (3) expresses the intensity of the magnetic field at points so close to the dipole that it is possible to neglect the time of propagation of the disturbances (see [17]).

§ 2. The Simplest Problems; Different Methods of Solution

21. (a) $u(r, t)$

$$= \frac{(r-at)\phi(r-at)+(r+at)\phi(r+at)}{2r} + \frac{1}{2ar}\int_{r-at}^{r+at} \xi\psi(\xi)\, d\xi, \tag{1}$$

where the functions $\phi(\xi)$ and $\psi(\xi)$ are expanded evenly for negative ξ;

(*b*) $$\lim_{r\to 0} u(r, t) = at\,\phi'(at)+\phi(at)+t\psi(at). \tag{2}$$

Method. Formula (1) is obtained on the assumption that $u(r, t)$ remanis bounded for $r \to 0$.

22. $$u(r, t) = \frac{1}{2ar}\int_0^t d\tau \int_{r-a(t-\tau)}^{r+a(t-\tau)} \xi f(\xi, \tau)\, d\xi,$$

where $f(\xi, \tau)$ is expanded evenly for negative values of ξ.

23. For the initial conditions (a):

$$u(r,t)=\begin{cases} U_0 & \text{for} \quad 0 \leqslant t < \dfrac{r_0-r}{a}, \\ U_0\dfrac{r-at}{2r} & \text{for} \quad \dfrac{r_0-r}{a} < t < \dfrac{r_0+r}{a}, \\ 0 & \text{for} \quad \dfrac{r_0+r}{a} < t < +\infty, \end{cases} \quad \text{for} \quad 0 < r < r_0,$$

$$u(r,t)=\begin{cases} 0 & \text{for} \quad 0 \leqslant t < \dfrac{r-r_0}{a}, \\ U_0\dfrac{r-at}{2r} & \text{for} \quad \dfrac{r-r_0}{a} < t < \dfrac{r+r_0}{a}, \\ 0 & \text{for} \quad \dfrac{r+r_0}{a} < t < +\infty, \end{cases} \quad \text{for} \quad r_0 < r < +\infty.$$

For the initial conditions (b):

$$u(r,t)=\begin{cases} U_0 t & \text{for} \quad 0 \leqslant t < \dfrac{r_0-r}{a}, \\ U_0\dfrac{r_0^2-(r-at)^2}{4ar} & \text{for} \quad \dfrac{r_0-r}{a} < t < \dfrac{r_0+r}{a}, \\ 0 & \text{for} \quad \dfrac{r_0+r}{a} < t < +\infty, \end{cases} \quad \text{for} \quad 0 < r < r_0,$$

$$u(r,t)=\begin{cases} 0 & \text{for} \quad 0 \leqslant t < \dfrac{r_0-r}{a}, \\ U_0\dfrac{r_0^2-(r-at)^2}{4ar} & \text{for} \quad \dfrac{r_0-r}{a} < t < \dfrac{r_0+r}{a}, \\ 0 & \text{for} \quad \dfrac{r_0+r}{a} < t < +\infty, \end{cases} \quad \text{for} \quad r_0 < r < +\infty.$$

24. The velocity potential of particles of the gas equals $u(r, t)$ from the answer to the preceding problem for initial conditions (b), if it is assumed that $U_0 = -a^2\rho_1/\rho_0$, where $a^2 = kp_0/\rho_0$.

25. Let $U(r, t)$ denote the solution of problem 23(b) for infinite space (see the answer to problem 23(b)); then

(a) $u(x, y, z, t) = U(\tilde{r}, t) - U(\tilde{\tilde{r}}, t)$,

(b) $u(x, y, z, t) = U(\tilde{r}, t) + U(\tilde{\tilde{r}}, t)$,

where $\tilde{r} = \sqrt{x^2+y^2+(z-z_0)^2}$, $\tilde{\tilde{r}} = \sqrt{x^2+y^2+z(+z_0)^2}$.

26. Let $U(r, t)$ denote the same function as in the answer to the preceding problem; then

(a) $u(x, y, z, t) = U(r_1, t) - U(r_2, t) + U(r_3, t) - U(r_4, t)$,

(b) $u(x, y, z, t) = U(r_1, t) + U(r_2 t) - U(r_3 t) - U(r_4, t)$,

where

$$r_1 = \sqrt{x^2+(y-y_0)^2+(z-z_0)^2}, \qquad r_2 = \sqrt{x^2+(y+y_0)+(z-z_0)^2},$$

$$r_3 = \sqrt{x^2+(y+y_0)^2+(z+z_0)^2}, \qquad r_4 = \sqrt{x^2+(y-y_0)^2+(z+z_0)^2}.$$

27. $\phi(r, t) = \dfrac{-q\left(t-\dfrac{r}{a}\right)}{4\pi r}, \qquad q(t) = 0 \quad \text{for} \quad t < 0.$

Method. $\phi(r, t)$ is a solution of the boundary-value problem

$$\phi_{tt} = a^2\Delta\phi, \tag{1}$$

$$\phi|_{t=0} = \phi_t|_{t=0} = 0, \tag{2}$$

$$\lim_{r\to 0} 4\pi r^2\phi_r = q(t). \tag{3}$$

28. (a) Let the source lie in the plane $z = z_0$ and have polar coordinates $r_0, \theta_0, 0 < \theta_0 < \pi/n$. Then, denoting the solution of the preceding problem by $\phi(r, t)$, we obtain:

$$\phi(x, y, z, t) = \sum_{k=0}^{n-1} \{\phi(r_k^+, t) + \phi(r_k^-, t)\}, \tag{1}$$

where

$$r_k^+ = \sqrt{r^2+r_0^2-2rr_0\cos\left(\theta+\theta_0+\frac{2k\pi}{n}\right)+(z-z_0)^2}, \tag{2}$$

$$r_k^- = \sqrt{r^2+r_0^2-2rr_0\cos\left(\theta-\theta_0-\frac{2k\pi}{n}\right)+(z-z_0)^2}. \tag{3}$$

(b) Let the source lie in the layer $0 \leqslant z \leqslant l$ and have coordinates $x_0, y_0, z_0)$ $0 < z_0 < l$. Denoting the solution of problem 27 by $\phi(r, t)$ we obtain:

$$\phi(x, y, z, t) = \sum_{k=-\infty}^{+\infty} \{\phi(r_k^+, t) + \phi(r_k^-, t)\}, \tag{1'}$$

where

$$r_k^+ = \sqrt{r^2+(z+z_0-2kl)^2}, \tag{2'}$$

$$r_k^- = \sqrt{r^2+(z-z_0-2kl)^2}, \tag{3'}$$

here $r = \sqrt{(x-x_0)^2+(y-y_0)^2}$.

We note that for every value of $t > 0$ the formally infinite series (1′) reduces to the sum of a finite number of terms, since

$$\phi(r_k^+, t) = 0 \quad \text{for} \quad t < \frac{r_k^+}{a}, \qquad \phi(r_k^-, t) = 0 \quad \text{for} \quad t < \frac{r_k^-}{a}.$$

29. $$u^*(x, y, t) = \frac{1}{2\pi a}\left\{\frac{\partial}{\partial t}\iint\limits_{\rho \leqslant at} \frac{\phi^*(\xi, \eta)\, d\xi\, d\eta}{\sqrt{a^2 t^2 - \rho^2}} + \iint\limits_{\rho \leqslant at} \frac{\psi^*(\xi, \eta)\, d\xi\, d\eta}{\sqrt{a^2 t^2 - \rho^2}} + \int_0^t d\tau \iint\limits_{\rho \leqslant a(t-\tau)} \frac{f^*(\xi, \eta, \tau)\, d\xi\, d\eta}{\sqrt{a^2 (t-\tau)^2 - \rho^2}}\right\},$$

where $\rho = \sqrt{(x-\xi)^2 + (y-\eta)^2}$.

30. $$u^*(x, y, t) = \frac{1}{2\pi a}\left\{\frac{\partial}{\partial t}\iint\limits_{\rho \leqslant at} \phi^*(\xi, \eta) \frac{\cosh c\sqrt{a^2 t^2 - \rho^2}}{\sqrt{a^2 t^2 - \rho^2}}\, d\xi\, d\eta + \iint\limits_{\rho \leqslant at} \psi^*(\xi, \eta) \frac{\cosh c\sqrt{a^2 t^2 - \rho^2}}{\sqrt{a^2 t^2 - \rho^2}}\, d\xi\, d\eta + \int_0^t d\tau \iint\limits_{\rho \leqslant a(t-\tau)} f^*(\xi, \eta, \tau) \frac{\cosh c\sqrt{a^2 (t-\tau)^2 - \rho^2}}{\sqrt{a^2 (t-\tau)^2 - \rho^2}}\, d\xi\, d\eta\right\},$$

where $\rho = \sqrt{(x-\xi)^2 + (y-\eta)^2}$, if the plus sign stands in front of the term $c^2 u$ in the equation; if a minus sign stands in front of this term, then it is necessary to replace everywhere cosh by cos in the answer deduced.

Method. The solution of the equation

$$u_{tt} = a^2 (u_{xx} + u_{zz}), \tag{1}$$

satisfying the initial conditions

$$u|_{t=0} = 0, \qquad u_t|_{t=0} = F(x, y)\, e^{cz}, \tag{2}$$

is connected by the relation

$$u(x, y, z, t) = e^{cz} u^*(x, y, t)$$

to the solution of the equation

$$u^*_{tt} = a^2 (u^*_{xx} + u^*_{yy}) + c^2 u^*,$$

satisfying the initial conditions

$$u^*|_{t=0} = 0, \qquad u^*_t|_{t=0} = F(x, y),$$

which is readily obtained by representing the solution of problem (1), (2) in terms of Poisson's integral†. If a minus sign stands in front of $c^2 u$, then it is necessary to make the substitution $u(x, y, z, t) = e^{icz} u^*(x, y, t)$.

† See [2], vol. II, pages 553–554.

31. For the velocity potential $u(r, t)$ we obtain the expression

$$u(\rho, t) = \begin{cases} 0 & \text{for} \quad t < \dfrac{\rho}{a}, \\ -\dfrac{1}{2\pi a} \displaystyle\int_0^{t-\frac{\rho}{a}} \dfrac{q(\tau)\,d\tau}{\sqrt{a^2(t-\tau)^2-\rho^2}} & \text{for} \quad t > \dfrac{\rho}{a} \end{cases} \tag{1}$$

or the equivalent expression

$$u(\rho, t) = \begin{cases} 0 & \text{for} \quad t < \dfrac{\rho}{a}, \\ -\dfrac{1}{2\pi a} \displaystyle\int_0^{\operatorname{arc\,cosh}\frac{at}{\rho}} q\left(t - \dfrac{\rho}{a}\cosh\zeta\right) d\zeta & \text{for} \quad t > \dfrac{\rho}{a}, \end{cases} \tag{2}$$

or, if it is assumed $q(t) = 0$ under the integral for $t < 0$,

$$u(\rho, t) = -\frac{1}{2\pi} \int_0^{+\infty} q\left(t - \frac{\rho}{a}\cosh\zeta\right) d\zeta, \tag{3}$$

where $\rho = \sqrt{x^2+y^2}$, if the straight line, on which the sources are situated, is taken as the z-axis.

Method. $u(\rho, t)$ is a solution of the boundary-value problem

$$\frac{\partial^2 u}{\partial t^2} = a^2\left(\frac{\partial^2 u}{\partial \rho^2} + \frac{1}{\rho}\frac{\partial u}{\partial \rho}\right), \quad 0 \leqslant \rho < +\infty, \quad 0 < t < +\infty,$$

$$\lim_{\rho \to 0}\left(2\pi\rho\frac{\partial u}{\partial \rho}\right) = q(t), \quad 0 < t < +\infty,$$

$$u(\rho, 0) = u_t(\rho, 0) = 0, \quad 0 < \rho < +\infty.$$

Formally $u(\rho, t)$ in the form (3) can be derived by the method of "descent" (by integration with respect to z from $-\infty$ to $+\infty$) from the solution of problem 27; then it is readily verified that if $q'(t)$ is regular the function thus obtained satisfies all the conditions of the problem.

Note. At the origin of coordinates $u(\rho, t)$ has a logarithmic singularity with respect to ρ. Utilizing form (1) for $u(\rho, t)$ and applying an integration by parts to Taylor's formula, it is possible to represent $u(\rho, t)$ in the form

$$u(\rho,t) = \frac{1}{a^2} q\left(t - \frac{\rho}{a}\right)\ln\rho - \frac{1}{a^2} q(0)\ln 2t - \frac{1}{a^2}\int_0^t q'(\tau)\ln 2(t-\tau)\,d\tau + \varepsilon(\rho, t),$$

where $\varepsilon(\rho, t) \to 0$ for $\rho \to 0$.

32. $u(x, y, t) = -\dfrac{1}{2\pi}\displaystyle\sum_{k=1}^{4}\int_0^{+\infty} q\left(t - \frac{\rho_k}{a}\cosh\zeta\right) d\zeta,$

where

$$\rho_1 = \sqrt{(x-x_0)^2+(y-y_0)^2}, \qquad \rho_2 = \sqrt{(x+x_0)^2+(y-y_0)^2},$$

$$\rho_3 = \sqrt{(x+x_0)^2+(y+y_0)^2}, \qquad \rho_4 = \sqrt{(x-x_0)^2+(y+y_0)^2}.$$

33. For the velocity potential of particles of gas outisde the sphere we obtain the expression

$$U(r, t) = \left\{\begin{array}{ll} -a\dfrac{r_0}{r}\,e^{-\frac{a}{r_0}\left(t-\frac{r-r_0}{a}\right)}\displaystyle\int_0^t \mu(\tau)e^{\frac{a\tau}{r_0}}\,d\tau, & t > \dfrac{r-r_0}{a}, \\ 0, & t < \dfrac{r-r_0}{a}, \end{array}\right\} r_0 < r < +\infty,$$

and inside the sphere the expression

$$U(r, t) = \frac{r_0 A}{\dfrac{\omega}{a}\cos\dfrac{\omega}{a}r_0 + \dfrac{a}{r_0}\sin\dfrac{\omega}{a}r_0}\,\frac{\sin\dfrac{\omega}{a}r}{r}\sin\omega t +$$

$$+\sum_{n=1}^{+\infty} B_n \frac{\sin\lambda_n r}{r}\sin a\lambda_n t, \qquad 0 \leqslant r < r_0, \qquad 0 < t < +\infty,$$

where λ_n are positive roots of the equation

$$\tan(r_0\lambda) = -\frac{r_0\lambda}{a},$$

and

$$B_n = \frac{\displaystyle\int_0^{r_0} f(r)\sin(\lambda_n r)dr}{\displaystyle\int_0^{r_0} \sin^2(\lambda_n r)dr}$$

and

$$f(r) = \frac{\omega r_0 A}{\dfrac{\omega}{a}\cos\dfrac{\omega}{a}r_0 + \dfrac{a}{r_0}\sin\dfrac{\omega}{a}r_0}\sin\frac{\omega}{a}r.$$

Note. The expression for $U(r, t)$ for $0 \leqslant r \leqslant r_0$ is obtained on the assumption that there is no resonance, i.e. that $\tilde{\lambda} = \omega/a$ does not coincide with any of the eigenvalues λ_n†.

34. Let the centre of the sphere lie on the axis Oz at a point $z_0 > r_0 > 0$, and the plane $z = 0$ is a boundary plane for the semispace under consideration.

---

† For an investigation of the solution in the case of resonance see problem 134. § 3, chapter II.

Then, denoting the solution of the preceding problem by $U(r, t)$ we obtain the solution of problem 34 in the form

$$u(x, y, z, t) = U(r_1, t)+U(r_2, t) \quad \text{for} \quad r_1 > r_0 \text{ and } z > 0,$$
$$u(x, y, z, t) = U(r_1, t) \quad \text{for} \quad 0 < r_1 < r_0,$$

where $r_1 = \sqrt{x^2+y^2+(z-z_0)^2}$, $r_2 = \sqrt{x^2+y^2+(z+z_0)^2}$.

35. For the velocity potential we obtain the expression

$$U = \begin{cases} \operatorname{div} \dfrac{f\left(t-\dfrac{r-r_0}{a}\right)}{r} & \text{for} \quad t > \dfrac{r-r_0}{a}, \\ 0 & \text{for} \quad t < \dfrac{r-r_0}{a}, \end{cases}$$

where the vector

$$f(t) = ar_0^2 e^{-\frac{at}{r_0}} \int_0^t V(\tau) \sin \frac{a(t-\tau)}{r_0} e^{\frac{a\tau}{r_0}} d\tau .$$

Method. The solution of the problem may be sought in the form

$$U = \operatorname{div} \frac{f\left(t-\dfrac{r-r_0}{a}\right)}{r}.$$

The velocity of particles of the gas

$$\boldsymbol{v} = \operatorname{grad} U = \frac{3(\boldsymbol{f}\boldsymbol{n})\boldsymbol{n}-\boldsymbol{f}}{r^3} + \frac{3(\boldsymbol{f}'\boldsymbol{n})\boldsymbol{n}-\boldsymbol{f}'}{ar^2} + \frac{\boldsymbol{n}(\boldsymbol{n}\boldsymbol{f}'')}{a^2r}$$

($\boldsymbol{n}$ is the unit vector in the direction $\boldsymbol{r}$; the dash denotes differentiation of $\boldsymbol{f}$ with respect to its argument) satisfies the boundary condition $v_r = \boldsymbol{V}\boldsymbol{n}$ for $r = r_0$. Hence we obtain the equation for $\boldsymbol{f}$

$$\boldsymbol{f}''(t)+\frac{2a}{r_0}\boldsymbol{f}'(t)+\frac{2a^2}{r_0^2}\boldsymbol{f}(t) = r_0a^2\boldsymbol{V}(t).$$

36. For the velocity potential U produced by a small perturbation, and for the pressure perturbation $\tilde{p}$ we obtain the expressions

$$U(x, y) = \frac{v_0(y-x\tan\alpha)}{\cot\varepsilon+\tan\alpha}, \quad 0 < x < +\infty, \quad x\tan\varepsilon \leqslant y \leqslant x\tan\alpha,$$

$$\tilde{p} = \rho_0 v_0^2 \frac{\tan\alpha}{\cot\varepsilon+\tan\alpha}.$$

Method. To determine $\tilde{p}$ it is necessary to use relation (4) of the answer to problem 1†.

† It is necessary to change to Eulerian coordinates in this relation and to use the steady-state nature of the process and the smallness of the perturbations. In connection with the symbols see the answer to problem 7.

37. $U(r,x)$

$$= -v_0 r \frac{-\sqrt{\frac{x^2}{r^2}\tan^2\alpha - 1} + \frac{1}{2}\frac{x}{r}\tan\alpha \ln \frac{\frac{x}{r}\tan\alpha + \sqrt{\frac{x^2}{r^2}\tan^2\alpha-1}}{\frac{x}{r}\tan\alpha - \sqrt{\frac{x^2}{r^2}\tan^2\alpha-1}}}{\sqrt{\nu^2-1}+\frac{1}{2}\tan\varepsilon\tan\alpha\ln\frac{\nu+\sqrt{\nu^2-1}}{\nu-\sqrt{\nu^2-1}}}\tan\varepsilon,$$

$$0 < x < +\infty, \qquad \nu\cot\alpha \leqslant \frac{x}{r} \leqslant \cot\alpha, \qquad \nu = \frac{\tan\alpha}{\tan\varepsilon} > 1,$$

$$\tilde{p}\,|_{r=x\tan\varepsilon} = \rho_0 v_0^2 \frac{\ln\frac{\nu+\sqrt{\nu^2+1}}{\nu-\sqrt{\nu^2-1}}}{\sqrt{\nu^2-1}+\frac{1}{2}\tan\alpha\tan\varepsilon\ln\frac{\nu+\sqrt{\nu^2-1}}{\nu-\sqrt{\nu^2-1}}}\tan\alpha.$$

Method. See the answer to problem 8; the solution of equation (1) with boundary conditions (2) and (3) may be sought in the form

$$U(r,x) = r\psi\left(\frac{x}{r}\right) = r\psi(\zeta), \qquad \zeta = \frac{x}{r}.$$

To determine $\tilde{p}$ it is necessary to use (4) of the answer to problem 1.

39. To determine the potential of the flow produced by the action of the wall, we obtain the boundary-value problem (in Lagrangian coordinates)

$$(1-M^2)u_{xx}+u_{yy} = 0, \quad -\infty < x < +\infty, \quad 0 < y < +\infty, \quad M = \frac{U}{a}, \tag{1}$$

where a is the velocity of sound in the gas,

$$u_y(x,0) = U\varepsilon\omega\cos\omega x, \qquad -\infty < x < +\infty. \tag{2}$$

(a) In the first case $1-M^2 > 0$ and equation (1) is elliptic, and

$$u(x,y) = -\frac{U\varepsilon}{\sqrt{1-M^2}}\, e^{-\omega y\sqrt{1-M^2}}\cos\omega x.$$

(b) In the case of supersonic flow $1-M^2 < 0$ and

$$u(x,y) = -\frac{U\varepsilon}{\sqrt{M^2-1}}\sin\omega(x-y\sqrt{M^2-1}).$$

Method. In the elliptic case the solution must be sought in the form

$$u(x,y) = u_1(x)u_2(y),$$

and in the hyperbolic case in the form of expanding waves, taking into account the fact that in the hyperbolic (supersonic) case small disturbances are propagated to the right of the sources of disturbance. The boundary condition (2) is obtained from the exact boundary condition

$$\left(\frac{u_y}{u+u_x}\right)_{\text{bound}} = \left(\frac{\mathrm{d}y}{\mathrm{d}x}\right)_{\text{bound}}$$

by neglecting small quantities of higher order.

Note. Comparing the solutions in the elliptic and hyperbolic case, we see that disturbances produced by an oscillating wall are rapidly attenuated with distance, in the elliptic case, and maintain their amplitude in the hyperbolic case.

40.

$$\psi(r,t) = \left\{\begin{array}{lll} \pi U_0 & \text{for} & 0 < t < \dfrac{r_0-r}{a}, \\ U_0\left(\dfrac{\pi}{2}+\arcsin\dfrac{r_0-at}{r}\right) & \text{for} & \dfrac{r_0-r}{a} < t < \dfrac{r_0+r}{a}, \\ 0 & \text{for} & \dfrac{r_0+r}{a} < t < +\infty, \end{array}\right\} \quad 0 < r < r_0,$$

$$\psi(r,t) = \left\{\begin{array}{lll} 0 & \text{for} & 0 < t < \dfrac{r-r_0}{a}, \\ U_0\left(\dfrac{\pi}{2}+\arcsin\dfrac{r_0-at}{r}\right) & \text{for} & \dfrac{r-r_0}{a} < t < \dfrac{r+r_0}{a}, \\ 0 & \text{for} & \dfrac{r+r_0}{a} < t < +\infty, \end{array}\right\} \quad r_0 < r +\infty.$$

Method. Assuming $x = r\cos\phi$, $y = r\sin\phi$, $\alpha = \cos\theta$, $\beta = \sin\theta$, integrate firstly with respect to θ from 0 to 2π, and then change the variable of integration; this reduces to expression (1) of the conditions of the problem.

41. *Method.* Integrate the spherically symmetrical waves $f_1(at-r)/r$ and $f_2(at+r)/r$ with respect to z from $-\infty$ to $+\infty$, and then make a suitable substitution of the variable of integration.

42. *Solution.* Let us look for the solution of the equation

$$\frac{\partial^2 u}{\partial t^2} = a^2\left(\frac{\partial^2 u}{\partial r^2}+\frac{1}{r}\frac{\partial u}{\partial r}\right), \quad r^2 = x^2+y^2$$

in the form $u(r,t) = \mathrm{e}^{-i\omega t}f(r)$; this gives:

$$u(r,t) = A\mathrm{e}^{-i\omega t}J_0(kr)+BH_0^{(1)}(kr)\mathrm{e}^{-i\omega t}, \quad k = \frac{\omega}{a};$$

A and B are arbitrary constants†, $u_1(r,t) = A\mathrm{e}^{-i\omega t}J_0(kr)$ is a monochromatic cylindrical standing wave, not having a singularity at $r = 0$, for large r

$$u_1(r,t) \approx A\sqrt{\frac{2}{\pi}}\,\frac{\cos\left(kr-\dfrac{\pi}{4}\right)}{\sqrt{kr}}\,\mathrm{e}^{-i\omega t},$$

† Concerning the functions $J_0^{(1)}$ and $H_0^{(1)}$ see [7], pages 566, 642, 653, 676 and others.

$u_2(r, t) = Be^{-i\omega t} H_0^{(1)}(kr)$ is an expanding, monochromatic cylindrical wave, having a singularity at $r = 0$. For small r

$$u_2(r, t) \approx B\frac{2i}{\pi} \ln (kr) e^{-i\omega t},$$

for large r

$$u_2(r, t) \approx B\sqrt{\frac{2}{\pi}} \frac{e^{i\left(kr-\omega t-\frac{\pi}{4}\right)}}{\sqrt{kr}}.$$

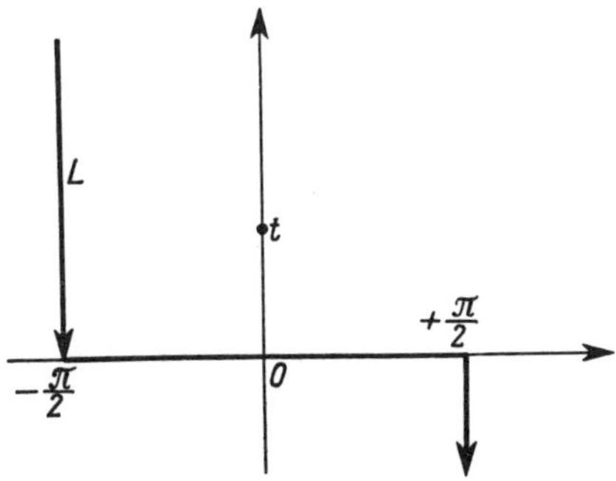

FIG. 51

Integrating the plane monochromatic wave

$$e^{-i\omega\left(t-\frac{x\cos\theta+y\sin\theta}{a}\right)}$$

with respect to the angle θ from 0 to π, we obtain:

$$\tilde{u}_1(r, t) = e^{-i\omega t}\int_0^\pi e^{ikr\cos(\theta-\phi)}\, d\theta = 2\pi e^{-i\omega t} J_0(kr), \qquad k = \frac{\omega}{a}.$$

If one integrates in the plane of the complex variable θ over the path L (Fig. 51), then we obtain:

$$\tilde{u}_2(r, t) = e^{-i\omega t}\int_L e^{ikr\cos\theta}\, d\theta = \pi e^{-i\omega t} H_0^{(1)}(kr).$$

44. *Solution.* Let us take as the plane of separation of the two media the plane $z = 0$ (Fig. 52). Quantities in the semispace $z < 0$ are denoted by the subscript 1, and in the semispace $z > 0$ by the subscript 2. We denote the incident, reflected and refracted waves respectively by

$$\phi_1 = A_1 e^{i(\omega_1 t - k_1 n_1 r)},$$

$$\phi_1^* = A_1^* e^{i(\omega_1^* t - k_1^* n_1^* r)},$$

$$\phi_2 = A_2 e^{i(\omega_2 t - k_2 n_2 r)}.$$

Here $k_1 = \omega_1/a_1$, $k_1^* = \omega_1^*/a_1$ and $k_2 = \omega_2/a_2$ are the wave numbers, ω_1, ω_1^*, ω_2 the frequencies of the incident, reflected and refracted waves, a_1 and a_2 the velocity of propagation of the wave in the first and second medium; $\boldsymbol{n}_1$, $\boldsymbol{n}_1^*$, $\boldsymbol{n}_2$ unit vectors in the direction of propagation of the corresponding waves; the vector $\boldsymbol{r} = [x, y, z]$. On the plane $z = 0$ the boundary conditions†

$$\rho_1\{\phi_1+\phi_1^*\} = \rho_2\phi_2 \quad \text{for} \quad z = 0, \tag{1}$$

$$\frac{\partial\phi_1}{\partial z}+\frac{\partial\phi_1^*}{\partial z} = \frac{\partial\phi_2}{\partial z} \quad \text{for} \quad z = 0, \tag{2}$$

must be fulfilled. We shall assume the vector $\boldsymbol{n}_1$ to be parallel to the plane xOz i.e.

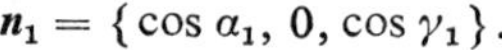

$$\boldsymbol{n}_1 = \{\cos\alpha_1, 0, \cos\gamma_1\}.$$

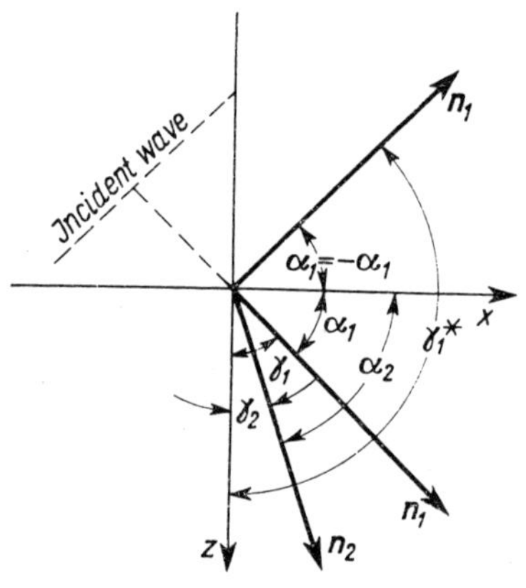

FIG. 52

Let us write now the vectors $\boldsymbol{n}_1^*$ and $\boldsymbol{n}_2$ in coordinate form:

$$\boldsymbol{n}_1^* = \{\cos\alpha_1^*, \cos\beta_1^*, \cos\gamma_1^*\}, \qquad \boldsymbol{n}_2 = \{\cos\alpha_2, \cos\beta_2, \cos\gamma_2\}.$$

Since the functions of τ: $e^{\nu_1\tau}$, $e^{\nu_2\tau}$, $e^{\nu_3\tau}$, when ν_1, ν_2, ν_2 are different, are linearly independent, then the substitution of ϕ_1, ϕ_1^*, ϕ_2 in the boundary conditions (1) and (2) leads to the equalities

$$\left.\begin{aligned} \omega_1^* = \omega_2 = \omega_1, \\ k_1^* = \frac{\omega_1^*}{a_1} = k_1 = \frac{\omega_1}{a_1}, \end{aligned}\right\} \tag{3}$$

$$\cos\beta_1^* = \cos\beta_2 = 0, \tag{4}$$

† See the answer to problem 3.

i.e. the unit vectors $\boldsymbol{n}_1^*$ and $\boldsymbol{n}_2$ are also parallel to plane xOz

$$k_1 \cos \alpha_1 = k_1 \cos \alpha_1^* = k_2 \cos \alpha_2, \tag{5}$$

from which the familiar relations between the angles of incidence, reflection and refraction are obtained: $\alpha_1 = -\alpha_1^*$, since the reflected wave, like the incident, lies in the semispace $z < 0$, and

$$\frac{\cos \alpha_1}{\cos \alpha_2} = \frac{k_2}{k_2} = \frac{\dfrac{\omega}{a_2}}{\dfrac{\omega}{a_1}} = \frac{a_1}{a_2}.$$

If the equalities obtained as the result of the substitution of ϕ_1, ϕ_1^* and ϕ_2 in the boundary conditions (1) and (2), are divided by the common variable factor, then relations for determining the amplitudes of the reflected and refracted waves

$$\rho_1 A_1 + \rho_1 A_1^* = \rho_2 A_2,$$
$$k_1 \cos \gamma_1 A_1 + k_1 \cos \gamma_1^* A_1^* = k_2 \cos \gamma_2 A_2;$$

are obtained; from these equations, using the equality

$$\cos \gamma_1^* = -\cos \gamma$$

we obtain:

$$A_1^* = \frac{\rho_2 k_1 \cos \gamma_1 - \rho_1 k_2 \cos \gamma_2}{\rho_2 k_1 \cos \gamma_1 + \rho_1 k_2 \cos \gamma_2} A_1,$$

$$A_2 = \frac{2\rho_1 k_1 \cos \gamma_1}{\rho_2 k_1 \cos \gamma_1 + \rho_1 k_2 \cos \gamma_2} A_1.$$

45. Denoting by $\boldsymbol{n}_1$, $\boldsymbol{n}_1^*$, $\boldsymbol{n}_2$ the unit vectors in the direction of the incident reflected and refracted waves, we obtain (see Fig. 52)

$$\alpha_1^* = -\alpha_1,$$

$$\frac{\cos \alpha_1}{\cos \alpha_2} = \frac{v_1}{v_2} = \nu_{12} = \sqrt{\frac{\varepsilon_2}{\varepsilon_1}},$$

$$v_1 = \frac{c}{\sqrt{\varepsilon_1}}, \qquad v_2 = \frac{c}{\sqrt{\varepsilon_2}},$$

where c is the velocity of light in vacuum, ε_1 and ε_2 are the dielectric constants of the first and second medium (we assume $\mu_1 = \mu_2 = 1$).

Method. A plane electromagnetic monochromatic wave may be represented in the form

$$\boldsymbol{E} = \boldsymbol{E}^{(0)} e^{i(\omega t - k\boldsymbol{n}\cdot\boldsymbol{r})}, \quad \boldsymbol{H} = \boldsymbol{H}^{(0)} e^{i(\omega t - k\boldsymbol{n}\cdot\boldsymbol{r})}.$$

Then it is necessary to use the conditions at the boundary of separation of the two dielectrics†.

46. Representing the incident wave in the form‡

$$E_1 = \{E_1 e^{i(\omega t - k_1 z)};\ 0;\ 0\},$$
$$H_1^* = \{0;\ \sqrt{\varepsilon_1}\, E_1 e^{i(\omega t - k_1 z)};\ 0\},$$

we obtain:

$$E_1^* = \{E_1^* e^{i(\omega t - k_1 z)};\ 0;\ 0\}, \qquad H_1^* = \{0;\ -\sqrt{\varepsilon_1}\, E_1^* e^{i(\omega t + k_1 z)};\ 0\},$$
$$E_2 = \{E_2 e^{i(\omega t - k_2 z)};\ 0;\ 0\}, \qquad H_2 = \{0;\ -\sqrt{\varepsilon_2}\, E_2 e^{i(\omega t - k_2 z)};\ 0\},$$

where

$$E_1^* = \frac{1-\nu_{12}}{1+\nu_{12}} E_1, \qquad E_2 = \frac{2}{1+\nu_{12}} E_1, \qquad \nu_{12} = \sqrt{\frac{\varepsilon_2}{\varepsilon_1}}.$$

§ 3. The Method of Separation of Variables

1. Boundary-value Problems not Requiring the Application of Special Functions

(a) *Homogeneous media*

47. The solution of the boundary-value problem

$$u_{tt} = a^2 \{u_{xx} + u_{yy}\}, \quad 0 < x < l_1, \quad 0 < y < l_2, \quad 0 < t < +\infty, \tag{1}$$

$$u|_{x=0} = u|_{x=l_1} = u|_{y=0} = u|_{y=l_2} = 0, \tag{2}$$

$$u(x, y, 0) = Axy(l_1 - x)(l_2 - y), \quad u_t(x, y, 0) = 0, \quad 0 < x < l_1, \quad 0 < y < l_2 \tag{3}$$

is:

$$u(x, y, t) = \frac{16 A l_1^2 l_2^2}{\pi^6} \sum_{m, n=0}^{+\infty} \frac{\sin \frac{(2m+1)\pi x}{l_1} \sin \frac{(2n+1)\pi y}{l_2}}{(2m+1)^3 (2n+1)^3} \times$$
$$\times \cos\left\{\pi a t \sqrt{\frac{(2m+1)^2}{l_1^2} + \frac{(2m+1)^2}{l_2^2}}\right\}.$$

48. The solution of the boundary-value problem

$$u_{tt} = a^2 \{u_{xx} + u_{yy}\}, \quad 0 < x < l_1, \quad 0 < y < l_2, \quad 0 < t < +\infty, \tag{1}$$

$$u|_{x=0} = u|_{x=l_1} = u|_{y=0} = u|_{y=l_2} = 0, \tag{2}$$

$$u(x, y, 0) = 0, \quad u_t(x, y, 0) = Axy(l_1 - x)(l_2 - y), \quad 0 < x < l_1, \quad 0 < y < l_2 \tag{3}$$

† See [7], page 490.

‡ See [17], pages 551–563.

is:

$$u(x, y, t) = \frac{16Al_1^2l_2^2}{\pi^7 a} \times$$

$$\times \sum_{m,n=0}^{+\infty} \frac{\sin\frac{(2m+1)\pi x}{l_1} \sin\frac{(2n+1)\pi y}{l_2} \sin\left\{\pi at\sqrt{\frac{(2m+1)^2}{l_1^2}+\frac{(2n+1)^2}{l_2^2}}\right\}}{(2m+1)^3(2n+1)^3\sqrt{\frac{(2m+1)^2}{l_1^2}+\frac{(2n+1)^2}{l_2^2}}}.$$

49. $u(x, y, t)$

$$= \frac{4K}{\pi a\rho l_1 l_2} \sum_{m,n=1}^{+\infty} \frac{\sin\frac{m\pi x_0}{l_1}\sin\frac{m\pi x}{l_1}\sin\frac{n\pi y_0}{l_2}\sin\frac{n\pi y}{l_2}}{\sqrt{\frac{m^2}{l_1^2}+\frac{n^2}{l_2^2}}} \sin\left\{\pi at\sqrt{\frac{m^2}{l_1^2}+\frac{n^2}{l_2^2}}\right\},$$

where ρ is the surface density.

Method. It is possible to find firstly a solution, assuming the impulse K to be uniformly distributed over the neighbourhood $x_0-\varepsilon < x < x_0+\varepsilon$ $y_0-\varepsilon < y < y_0+\varepsilon$ of the point (x_0, y_0), and then pass to a limit as $\varepsilon \to 0$†.

It is possible also to use Dirac's impulse delta-function and formulate the initial conditions in the following way:

$$u(x, y, 0)=0, \quad u_t(x, y, 0) = \frac{K}{\rho}\delta(x-x_0)\,\delta(y-y_0), \quad 0<x<l_1, \quad 0<y<l_2.$$

The second method leads to the result much more rapidly.

50. The solution of the boundary-value problem

$$u_{tt} = a^2\{u_{xx}+u_{yy}\}+A^{(0)}(x, y)\sin\omega t \qquad A^{(0)}(x, y) = \frac{1}{\rho}A(x, y), \tag{1}$$

$$u|_{x=0} = u|_{x=l_1} = u|_{y=0} = u|_{y=l_2} = 0, \tag{2}$$

$$u(x, y, 0) = 0, \qquad u_t(u, y, 0) = 0, \qquad 0<x<l_1, \qquad 0<y<l_2 \tag{3}$$

is:

$$u(x, y, t) = \sum_{m,n=0}^{+\infty} A_{mn}\left(\sin\omega t - \frac{\omega}{\omega_{mn}}\sin\omega_{mn}t\right)\sin\frac{m\pi x}{l_1}\sin\frac{n\pi y}{l_2}, \tag{4}$$

where

$$A_{mn} = \frac{4}{l_1 l_2(\omega_{mn}^2-\omega^2)}\int_0^{l_1} dx \int_0^{l_2} A^{(0)}(x, y)\sin\frac{m\pi x}{l_1}\sin\frac{n\pi y}{l_2}\,dy, \tag{5}$$

$$\omega_{mn} = \pi a\sqrt{\frac{m^2}{l_1^2}+\frac{n^2}{l_2^2}}, \tag{6}$$

† See the solution of problem 101, § 3, chapter II.

for the condition that the frequency of the constraining force does not coincide with any of the natural frequencies $\omega \neq \omega_{mn}$. If $\omega = \omega_{m_0 n_0}$ (resonance), then

$$u(x, y, t) = \sum_{\substack{m, n=0 \\ m \neq m_0,\, n \neq n_0}}^{+\infty} A_{mn} \left(\sin \omega t - \frac{\omega}{\omega_{mn}} \sin \omega_{mn} t \right) \sin \frac{m\pi x}{l_1} \sin \frac{n\pi y}{l_2} +$$

$$+ A_{m_0 n_0} (\sin \omega t - \omega t \cos \omega t) \sin \frac{m_0 \pi x}{l_1} \sin \frac{n_0 \pi y}{l_2}, \tag{7}$$

where A_{mn} is given by formula (5), and

$$A_{m_0 n_0} = \frac{2}{l_1 l_2 \omega} \int_0^{l_1} dx \int_0^{l_2} A^{(0)}(x, y) \sin \frac{m_0 \pi x}{l_1} \sin \frac{n_0 \pi y}{l_2} dy. \tag{8}$$

Note. If the frequency $\omega_{m_0 n_0}$ is multiple, i.e. corresponds to a multiple eigenvalue, then instead of one resonance term a group of resonance terms of specified form appears.

51. If the frequency of the constraining force does not coincide with any of the natural frequencies of the membrane, i.e. $\omega \neq \omega_{mn}$, $m, n = 1, 2, 3, \ldots$, then

$$u(x, y, t)$$

$$= \frac{4A}{\rho l_1 l_2} \sum_{m, n=1}^{+\infty} \frac{\sin \omega t - \dfrac{\omega}{\omega_{mn}} \sin \omega_{mn} t}{\omega_{mn}^2 - \omega^2} \sin \frac{m\pi x_0}{l_1} \sin \frac{n\pi y_0}{l_2} \sin \frac{m\pi x}{l_1} \sin \frac{n\pi y}{l_2}.$$

If $\omega = \omega_{m_0 n_0}$ (resonance), then

$$u(x, y, t)$$

$$= \frac{4A}{\rho l_1 l_2} \sum_{\substack{m, n=1 \\ m \neq m_0;\, n \neq n_0}}^{-\infty} \frac{\sin \omega - \dfrac{\omega}{\omega_{mn}} \sin \omega_{mn} t}{\omega_{mn}^2 - \omega^2} \sin \frac{m\pi x_0}{l_1} \sin \frac{n\pi y_0}{l_2} \sin \frac{m\pi x}{t_1} \sin \frac{n\pi y}{l_2} +$$

$$+ \frac{2A}{\rho l_1 l_2 \omega} (\sin \omega t - \omega t \cos \omega t) \sin \frac{m_0 \pi x_0}{l_1} \sin \frac{n_0 \pi y_0}{l_2} \sin \frac{m_0 \pi x}{l_1} \sin \frac{n_0 \pi y}{l_2}.$$

Note. If the frequency $\omega_{m_0 n_0}$ is multiple, then instead of one resonance term a group of resonance terms of specified form appears.

52. The horizontal velocity potential of the water is a solution of the boundary-value problem

$$\frac{\partial^2 U}{\partial t^2} = a^2 \left\{ \frac{\partial^2 U}{\partial x^2} + \frac{\partial^2 U}{\partial y^2} \right\} + \frac{A}{\rho} \cos \frac{\pi x}{l_1} \cos \frac{\pi y}{l_2} \phi'(t), \quad 0 < t < +\infty, \ a^2 = gh, \tag{1}$$

$$\left. \frac{\partial U}{\partial x} \right|_{x=0} = \left. \frac{\partial U}{\partial x} \right|_{x=l_1} = \left. \frac{\partial U}{\partial y} \right|_{y=0} = \left. \frac{\partial U}{\partial y} \right|_{y=l_2} = 0, \tag{2}$$

$$U(x, y, 0) = 0, \quad U_t(x, y, 0) = 0. \tag{3}$$

It can be represented in the form

$$U(x, y, t) = \frac{A}{k_0 \rho} \left\{ \int_0^t \phi'(\tau) \sin k(t-\tau)\, d\tau \right\} \cos \frac{\pi x}{l_1} \cos \frac{\pi y}{l_2}. \tag{4}$$

53.

$$u(x, y, t) = \frac{4k}{\rho l_1 l_2} e^{-\nu^2 t} \sum_{m, n=1}^{+\infty} \frac{\sin \frac{m\pi x_0}{l_1} \sin \frac{n\pi y_0}{l_2} \sin \frac{m\pi x}{l_1} \sin \frac{n\pi y}{l_2}}{\tilde{\omega}_{mn}} \sin \tilde{\omega}_{mn} t,$$

where

$$\tilde{\omega}_{mn}^2 = \pi^2 a^2 \left(\frac{m^2}{l_1^2} + \frac{n^2}{l_2^2} \right) - \nu^4,$$

and ν^2 is the coefficient of resistance in the equation

$$\frac{\partial^2 u}{\partial t^2} = a^2 \left\{ \frac{\partial^2 u}{\partial x^2} + \frac{\partial^2 u}{\partial y^2} \right\} - 2\nu^2 \frac{\partial u}{\partial t}.$$

54. $u(x, y, t)$

$$= -\frac{16A}{\pi^2 \rho} \sum_{m, n=1}^{+\infty} \frac{(\dot{\omega^2} + \omega_{mn}^2) \sin \omega t + 2\nu^2 \omega \cos \omega t}{(2m+1)(2n+1)[(\omega^2 - \omega_{mn}^2)^2 + 4\omega^2 \nu^4]} \sin \frac{m\pi x}{l_1} \sin \frac{n\pi y}{l_2},$$

$$\omega_{mn} = \pi a \sqrt{\frac{m^2}{l_1^2} + \frac{n^2}{l_2^2}}.$$

Method. Let us look for the solution of the equation

$$\frac{\partial^2 U}{\partial t^2} = a^2 \left\{ \frac{\partial^2 U}{\partial x^2} + \frac{\partial^2 U}{\partial y^2} \right\} - 2\nu^2 \frac{\partial U}{\partial t} + \frac{A}{\rho} e^{-i\omega t},$$

reducing to zero for $x = 0$, $x = l_1$, $y = 0$, $y = l_2$, in the form

$$U(x, y, t) = V(x, y) e^{i\omega t};$$

then $u(x, y, t) = \text{Im}\,(U(x, y, t))$. To determine $V(x, y)$ we obtain the boundary-value problem

$$\Delta V + \frac{\omega^2 - 2\nu^2 \omega i}{a^2} V = -\frac{A}{\rho a^2},$$

$$V|_{x=0} = V|_{x=l_1} = V|_{y=0} = V|_{y=l_2} = 0.$$

We look for its solution in the form

$$V(x, y) = \sum_{m, n=0}^{+\infty} A_{mn} \sin \frac{m\pi x}{l_1} \sin \frac{n\pi y}{l_2}.$$

55. The solution of the boundary-value problem

$$\frac{\partial^2 U}{\partial t^2} = a^2\left(\frac{\partial^2 U}{\partial r^2} + \frac{2}{r}\frac{\partial U}{\partial r}\right), \quad r_1 < r < r_2, \quad 0 < t < +\infty,$$

$$\left.\frac{\partial U}{\partial r}\right|_{r=r_1} = \varepsilon\omega\cos\omega t, \quad \left.\frac{\partial U}{\partial r}\right|_{r=r_2} = 0, \quad 0 < t < +\infty$$

($U(r, t)$ is the velocity potential of the gas), representing steady-state harmonic oscillations of frequency ω, is:

$$U(r,t) = \left\{\frac{2a\varepsilon\left(\frac{\omega}{a}\cos\frac{\omega}{a}r_2 - \frac{1}{r_2}\sin\frac{\omega}{a}r_2\right)}{\left(\frac{1}{r_1}-\frac{1}{r_2}\right)\cos\frac{\omega}{a}(r_1-r_2)}\cdot\frac{\cos\frac{\omega}{a}r}{r} + \right.$$

$$\left. + \frac{2a\varepsilon\left(\frac{\omega}{a}\sin\frac{\omega}{a}r_2 + \frac{1}{r_2}\cos\frac{\omega}{a}r_2\right)}{\left(\frac{1}{r_1}-\frac{1}{r_2}\right)\cos\frac{\omega}{a}(r_1-r_2)}\cdot\frac{\sin\frac{\omega}{a}r}{r}\right\}\cos\omega t.$$

Method. Look for the solution in the form

$$U(r, t) = R(r)\cos\omega t.$$

56. The solution of the boundary-value problem†

$$\frac{\partial^2 U}{\partial t^2} = a^2\left(\frac{\partial^2 U}{\partial r^2} + \frac{2}{r}\frac{\partial U}{\partial r}\right), \quad r_1 < r < r_2, \quad 0 < t < +\infty, \tag{1}$$

$$\left.\frac{\partial U}{\partial r}\right|_{r=r_1} = 0, \quad \left.\frac{\partial U}{\partial r}\right|_{r=r_2} = 0, \quad 0 < t < +\infty, \tag{2}$$

$$U(r, 0) = 0, \quad U_t(r, 0) = -\frac{a^2}{\rho_0}f(r), \quad r_1 < r < r_2 \tag{3}$$

is:

$$U(r,t) = \sum_{n=1}^{+\infty} A_n \frac{\cos\lambda_n r + \gamma_n \sin\lambda_n r}{r}\sin a\lambda_n t,$$

$$\lambda_n = (2n+1)\frac{\pi}{2(r_2-r_1)}, \quad n = 0, 1, 2, \dots,$$

$$\gamma_n = \frac{\lambda_n\sin\lambda_n r_2 + \frac{1}{r_2}\cos\lambda_n r_2}{\lambda_n\cos\lambda_n r_2 - \frac{1}{r_2}\sin\lambda_n r_2},$$

† $U(r, t)$ is the velocity potential of gas particles.

$$A_n = -\frac{a^2}{\rho_0}\int_{r_1}^{r_2} rf(r)\,[\cos\lambda_n r+\gamma_n\sin\lambda_n r]\,\mathrm{d}r.$$

Method. Change to the new unknown function

$$V(r,t) = rU(r,t).$$

(b) *Inhomogeneous media*

57. The solution of the boundary-value problem

$$\left.\begin{aligned} &\rho_1\frac{\partial^2 u_1}{\partial t^2} = T_0\left\{\frac{\partial^2 u_1}{\partial x^2}+\frac{\partial^2 u_1}{\partial y^2}\right\}, \quad 0\leqslant x\leqslant x_0,\quad 0\leqslant y\leqslant l_2,\ 0<t<+\infty,\\ &\rho_2\frac{\partial^2 u_2}{\partial t^2} = T_0\left\{\frac{\partial^2 u_2}{\partial x^2}+\frac{\partial^2 u_2}{\partial y^2}\right\}, \quad x_0\leqslant x\leqslant l_1,\quad 0\leqslant y\leqslant l_2,\end{aligned}\right\} \tag{1}$$

$$\left.\begin{aligned} &\left.\begin{aligned} u(x_0-0,y,t) &= u(x_0+0,y,t),\\ u_x(x_0-0,y,t) &= u_x(x_0+0,y,t),\end{aligned}\right\}\ 0\leqslant y\leqslant l_2,\quad 0<t<+\infty,\\ &u|_{x=0} = u|_{x=l_1} = u|_{y=0} = u|_{y=l_2} = 0,\end{aligned}\right\} \tag{2}$$

$$u(x,y,0) = f(x,y),\quad u_t(x,y,0) = F(x,y),\quad 0\leqslant x\leqslant l_1,\quad 0\leqslant y\leqslant l_2 \tag{3}$$

is:

$$u(x,y,t) = \sum_{m,n=1}^{+\infty}(A_{mn}\cos\lambda_{mn}t+B_{mn}\sin\lambda_{mn}t)\,v_{mn}(x,y), \tag{4}$$

where

$$v_{mn}(x,y) = \left\{\begin{aligned} &\frac{\sin\overline{\omega}_{mn}x}{\sin\overline{\omega}_{mn}x_0}\sin\frac{n\pi y}{l_2}, \quad 0\leqslant x\leqslant x_0,\quad 0\leqslant y\leqslant l_2,\\ &\frac{\sin\overline{\overline{\omega}}_{mn}(l_1-x)}{\sin\overline{\overline{\omega}}_{mn}(l_1-x_0)}\sin\frac{n\pi y}{l_2}, \quad x_0\leqslant x\leqslant l_1,\quad 0\leqslant y\leqslant l_2,\end{aligned}\right\} \tag{5}$$

$$\overline{\omega}^2_{mn} = \frac{\rho_1}{T_0}\lambda^2_{mn}-\frac{n^2\pi^2}{l_2^2},\qquad \overline{\overline{\omega}}^2_{mn} = \frac{\rho_2}{T_0}\lambda^2_{mn}-\frac{n^2\pi^2}{l_2^2}, \tag{6}$$

where $\lambda_{mn}(m, n = 1, 2, 3, \ldots)$ are roots of the transcendental equation

$$\sqrt{\frac{\rho_1}{T_0}\lambda^2_{mn}-\frac{n^2\pi^2}{l_2^2}}\cot\left\{x_0\sqrt{\frac{\rho_1}{T_0}\lambda^2_{mn}-\frac{n^2\pi}{l_2^2}}\right\}$$
$$= \sqrt{\frac{\rho_2}{T_0}\lambda^2_{mn}-\frac{n^2\pi^2}{l_2^2}}\cot\left\{(x_0-l_1)\sqrt{\frac{\rho_2}{T_0}\lambda^2_{mn}-\frac{n^2\pi^2}{l_2^2}}\right\}, \tag{7}$$

$$A_{mn} = \frac{\int_0^{l_1}\int_0^{l_2}\mu(x,y)f(x,y)\,v_{mn}(x,y)\,\mathrm{d}x\,\mathrm{d}y}{\|v_{mn}\|^2}, \tag{8}$$

$$B_{mm} = \frac{\int_0^{l_1}\int_0^{l_2} \mu(x, y) F(x, y) v_{mn}(x, y)\,dxdy}{\lambda_{mn}\|v_{mn}\|^2}, \tag{9}$$

$$\mu(x, y) = \begin{cases} \rho_1, & 0 \leqslant x < x_0, \quad 0 \leqslant y \leqslant l_2, \\ \rho_2, & x_0 < x \leqslant l_1, \quad 0 \leqslant y \leqslant l_2, \end{cases} \tag{10}$$

$$\|v_{mn}\|^2 = \int_0^{l_1}\int_0^{l_2} \mu(x, y) v_{mn}^2(x, y)\,dxdy$$

$$= \frac{l_2}{4}\left\{\frac{\rho_1 x_0}{\sin^2 \overline{\omega}_{mn} x_0} + \frac{\rho_2 (l_1 - x_0)}{\sin^2 \overline{\overline{\omega}}_{mn}(l_1 - x_0)}\right\}. \tag{11}$$

58. The solution of the boundary-value problem

$$\left.\begin{aligned} \rho_{01}\frac{\partial^2 u}{\partial t^2} &= K_1 p_0\left\{\frac{\partial^2 u}{\partial r^2} + \frac{2}{r}\frac{\partial u}{\partial r}\right\}, \quad 0 \leqslant r \leqslant r_1, \quad 0 < t < +\infty, \\ \rho_{02}\frac{\partial^2 u}{\partial t^2} &= K_2 p_0\left\{\frac{\partial^2 u}{\partial r^2} + \frac{2}{r}\frac{\partial u}{\partial r}\right\}, \quad r_1 < r < r_2, \end{aligned}\right\} \tag{1}$$

$$\left.\frac{\partial u}{\partial r}\right|_{r=r_2} = 0, \quad \rho_{01} u\,|_{r=r_1-0} = \rho_{02} u\,|_{r=r_1+0},$$

$$\left.\frac{\partial u}{\partial r}\right|_{r=r_1-0} = \left.\frac{\partial u}{\partial r}\right|_{r=r_1+0}, \quad 0 < t < +\infty, \tag{2}$$

$$u(r, 0) = f(r), \quad u_t(r, 0) = 0, \quad 0 \leqslant r \leqslant r_2 \tag{3}$$

is:

$$u(r, t) = \sum_{n=1}^{+\infty} A_n \frac{v_n(r)}{r}\cos\lambda_n t, \quad 0 \leqslant r < r_2, \quad 0 < t < +\infty, \tag{4}$$

where $\lambda_n (n = 1, 2, 3, \ldots)$ are roots of the transcendental equation

$$\begin{vmatrix} \rho_{01}\sin\frac{\lambda r_1}{a_1} & -\rho_{02}\cos\frac{\lambda}{a_2}r_1 & -\rho_{02}\sin\frac{\lambda}{a_2}r_1 \\ \frac{\lambda}{a_1}\cos\frac{\lambda}{a_1}r_1 - \frac{1}{r_1}\sin\frac{\lambda}{a_1}r_1 & \frac{\lambda}{a_2}\sin\frac{\lambda}{a_2}r_1 + \frac{1}{r_1}\cos\frac{\lambda}{a_2}r_1 & -\frac{\lambda}{a_2}\cos\frac{\lambda}{a_2}r_1 + \frac{1}{r_1}\sin\frac{\lambda}{a_2}r_1 \\ 0 & \frac{\lambda}{a_2}\sin\frac{\lambda}{a_2}r_2 + \frac{1}{r_2}\cos\frac{\lambda}{a_2}r_2 & -\frac{\lambda}{a_2}\cos\frac{\lambda}{a_2}r_2 + \frac{1}{r_2}\sin\frac{\lambda}{a_2}r_2 \end{vmatrix} = 0, \tag{5}$$

$$v_n(r) = \begin{cases} \alpha_n \sin\frac{\lambda_n}{a_2}r, & 0 \leqslant r \leqslant r_1, \\ \beta_n \cos\frac{\lambda_n}{a_2}r + \gamma_n \sin\frac{\lambda_n}{a_2}r, & r_1 \leqslant r \leqslant r_2. \end{cases} \tag{6}$$

The constants α_n, β_n and γ_n are determined up to an arbitrary constant factor from the system of equations

$$\left.\begin{aligned} &\left(\rho_{01} \sin \frac{\lambda_n}{a_1} r_1\right) \alpha_n - \left(\rho_{02} \cos \frac{\lambda_n}{a_2} r_1\right) \beta_n - \left(\rho_{02} \sin \frac{\lambda_n}{a_2} r_1\right) \gamma_n = 0, \\ &\left(\frac{\lambda_n}{a_1} \cos \frac{\lambda_n}{a_1} r_1 - \frac{1}{r_1} \sin \frac{\lambda_n}{a_1} r_1\right) \alpha_n + \left(\frac{\lambda_n}{a_2} \sin \frac{\lambda_n}{a_2} r_1 + \frac{1}{r_1} \cos \frac{\lambda_n}{a_2} r_1\right) \beta_n + \\ &\qquad + \left(-\frac{\lambda_n}{a_2} \cos \frac{\lambda_n}{a_2} r_1 + \frac{1}{r_1} \sin \frac{\lambda_n}{a_2} r_1\right) \gamma_n = 0, \\ &\left(\frac{1}{r_2} \cos \frac{\lambda_n}{a_2} r_2 + \frac{\lambda_n}{a_2} \sin \frac{\lambda_n}{a_2} r_2\right) \beta_n + \left(\frac{1}{r_2} \sin \frac{\lambda_n}{a_2} r_2 - \frac{\lambda_n}{a_2} \cos \frac{\lambda_n}{a_2} r_2\right) \gamma_n = 0, \end{aligned}\right\} \tag{7}$$

$$A_n = \frac{\int_0^{r_2} r\mu(r) f(r) v_n(r)\, \mathrm{d}r}{\|v_n\|^2}, \quad n = 1, 2, 3, \ldots, \tag{8}$$

$$\mu(r) = \left\{\begin{aligned} &\frac{\rho_{01}}{k_1 p_0}, \quad 0 \leqslant r < r_1, \\ &\frac{\rho_{02}}{k_2 p_0}, \quad r_1 < r \leqslant r_2, \end{aligned}\right\} \tag{9}$$

$$\|v_n\|^2 = \int_0^{r_2} \mu(r) v_n^2(r)\, \mathrm{d}r. \tag{10}$$

2. Boundary-value Problems Requiring the Application of Special Functions

(a) *Homogeneous media*

59. The solution of the boundary-value problem

$$\frac{\partial^2 u}{\partial t^2} = a^2 \left(\frac{\partial^2 u}{\partial r^2} + \frac{1}{r} \frac{\partial u}{\partial r}\right), \quad 0 \leqslant r < r_0, \quad 0 < t < +\infty, \tag{1}$$

$$u(r_0, t) = 0, \quad 0 < t < +\infty, \tag{2}$$

$$u(r, 0) = \phi(r), \quad u_t(r, 0) = \psi(r), \quad 0 < r < r_0 \tag{3}$$

is:

$$u(r, t) = \sum_{n=1}^{+\infty} \left\{A_n \cos\left(a \frac{\mu_n}{r_0} t\right) + B_n \sin\left(a \frac{\mu_n}{r_0} t\right)\right\} J_0\left(\frac{\mu_n r}{r_0}\right), \tag{4}$$

where

$$\left.\begin{aligned} A_n &= \frac{2}{r_0^2 [J_1(\mu_n)]^2} \int_0^{r_0} r\phi(r) J_0\left(\frac{\mu_n r}{r_0}\right) \mathrm{d}r, \\ B_n &= \frac{2}{a\mu_n r_0 [J_1(\mu_n)]^2} \int_0^{r_0} r\psi(r) J_0\left(\frac{\mu_n r}{r_0}\right) \mathrm{d}r, \end{aligned}\right\} \tag{5}$$

and μ_n are positive roots of the equation $J_0(\mu) = 0$.

60. The solution of the boundary-value problem

$$\frac{\partial^2 u}{\partial t^2} = a^2\left\{\frac{\partial^2 u}{\partial r^2}+\frac{1}{r}\frac{\partial u}{\partial r}\right\}, \quad 0 \leqslant r \leqslant r_0, \quad 0 < t < +\infty, \tag{1}$$

$$u(r_0, t) = 0, \tag{2}$$

$$u(r, 0) = A\left(1-\frac{r^2}{r_0^2}\right), \quad u_t(r, 0) = 0, \quad 0 \leqslant r \leqslant r_0, \tag{3}$$

is:

$$u(r, t) = 8A\sum_{n=1}^{+\infty}\frac{J_0\left(\frac{\mu_n r}{r_0}\right)}{\mu_n^3 J_1(\mu_n)}\cos\frac{a\mu_n t}{r_0}; \tag{4}$$

μ_n are positive roots of the equation $J_0(\mu) = 0$.

Method. To calculate the coefficients of the series (4) use the formula $\int\limits_0^x J_0(x)\mathrm{d}x = xJ_1(x)$, to prove the validity of the formula

$$\int\limits_0^x x^3 J_0(x)\mathrm{d}x = 2x^2 J_0(x)+(x^3-4x)J_1(x). \tag{5}$$

61. The horizontal velocity potential of the water is a solution of the boundary-value problem

$$\frac{\partial^2 U}{\partial t^2} = a^2\left\{\frac{\partial^2 U}{\partial r^2}+\frac{1}{r}\frac{\partial U}{\partial r}\right\}, \quad 0 \leqslant r \leqslant r_0, \quad 0 < t < +\infty, \tag{1}$$

$$|U(0, t)| < +\infty, \quad \frac{\partial U(r_0, t)}{\partial r} = 0, \quad 0 < t < +\infty, \tag{2}$$

$$U(r, 0) = \phi(r), \quad U_t(r, 0) = \psi(r), \quad 0 \leqslant r \leqslant r_0. \tag{3}$$

The solution has the form

$$U(r, t) = \frac{2}{r_0^2}\int\limits_0^{r_0} r\,\{\phi(r)+t\psi(r)\}\,\mathrm{d}r + \sum_{n=1}^{+\infty}\left(A_n\cos\frac{a\mu_n t}{r_0}+B_n\sin\frac{a\mu_n t}{r_0}\right)J_0\left(\frac{\mu_n r}{r_0}\right), \tag{4}$$

where

$$\left.\begin{aligned} A_n &= \frac{2}{r_0^2[J_0(\mu_n)]^2}\int\limits_0^{r_0} r\phi(r)J_0\left(\frac{\mu_n r}{r_0}\right)\mathrm{d}r,\\ B_n &= \frac{2}{a\mu_n r_0[J_0(\mu_n)]^2}\int\limits_0^{r_0} r\psi(r)J_0\left(\frac{\mu_n r}{r_0}\right)\mathrm{d}r,\end{aligned}\right\} \tag{5}$$

and μ_n are positive roots of the equation $J_1(\mu) = 0$.

62. The solution of the boundary-value problem

$$\frac{\partial^2 u}{\partial t^2} = a^2\left\{\frac{\partial^2 u}{\partial r^2}+\frac{1}{r}\frac{\partial u}{\partial r}\right\}+\frac{p_0}{\rho}, \quad 0 \leqslant r \leqslant r_0, \quad 0 < t < +\infty, \tag{1}$$

$$u(r_0, t) = 0, \quad 0 < t < +\infty, \tag{2}$$

$$u(r, 0) = 0, \quad u_t(r, 0) = 0, \quad 0 \leqslant r \leqslant r_0 \tag{3}$$

is:

$$u(r, t) = \frac{p_0}{\rho a^2}\left\{\frac{r_0^2-r^2}{4}-2r_0^2\sum_{k=1}^{+\infty}\frac{J_0\left(\frac{\mu_k r}{r_0}\right)}{\mu_k^2 J_1(\mu_k)}\cos\frac{a\mu_k t}{r_0}\right\}, \tag{4}$$

where μ_k are positive roots of the equation $J_0(\mu) = 0$, and ρ is the surface density of the membrane.

63. The solution of the boundary-value problem

$$\frac{\partial^2 u}{\partial t^2} = a^2\left\{\frac{\partial^2 u}{\partial r^2}+\frac{1}{r}\frac{\partial u}{\partial r}\right\}+\frac{1}{\rho}f(r, t), \quad 0 \leqslant r < r_0, \quad 0 < t < +\infty, \tag{1}$$

$$u(r_0, t) = 0, \quad 0 < t < +\infty, \tag{2}$$

$$u(r, 0) = 0, \quad u_t(r, 0) = 0, \quad 0 \leqslant r < r_0 \tag{3}$$

is:

$$u(r, t) = \sum_{n=1}^{+\infty} A_n(t) J_0\left(\frac{\mu_n r}{r_0}\right),$$

$$A_n(t) = \frac{1}{\omega_n}\int_0^t d\tau\int_0^{r_0} f(\xi, \tau) J_0\left(\frac{\mu_n \xi}{r_0}\right)\sin\omega_n(t-\tau)\,d\xi, \tag{4}$$

where $\omega_n = a\mu_n/r_0$, and μ_n are roots of the equation $J_0(\mu) = 0$.

64. The solution of the boundary-value problem

$$\frac{\partial^2 u}{\partial t^2} = a^2\left\{\frac{\partial^2 u}{\partial r^2}+\frac{1}{r}\frac{\partial u}{\partial r}\right\}+\frac{p_0}{\rho}\sin\omega t, \quad 0 \leqslant r < r_0, \quad 0 < t < +\infty, \tag{1}$$

$$u(r_0, t) = 0, \quad 0 < t < +\infty, \tag{2}$$

$$u(r, 0) = 0, \quad u_t(r, 0) = 0, \quad 0 \leqslant r \leqslant r_0 \tag{3}$$

is:

$$u(r, t) = \frac{p_0}{\omega^2 \rho}\left\{\frac{J_0\left(\frac{\omega r}{a}\right)}{J_0\left(\frac{\omega r_0}{a}\right)}-1\right\}\sin\omega t+\frac{2p_0\omega r_0^3}{a\rho}\sum_{n=1}^{+\infty}\frac{J_0\left(\frac{\mu_n r}{r_0}\right)\sin\frac{a\mu_n t}{r_0}}{\mu_n^2(\omega^2 r_0^2-a^2\mu_n^2)J_1(\mu_n)} \tag{4}$$

where μ_n are positive roots of the equation $J_0(\mu) = 0$, provided the frequency ω of the constraining force does not coincide with any of the characteristic frequencies of the membrane $\omega_n = a\mu_n/r_0$ (no resonance). In the case of resonance the solution is sought in the same way as was done in the solution of problem 133, §3, chapter II.

65. $$u(r, t) = A\frac{J_0\left(\frac{\omega r}{a}\right)}{J_0\left(\frac{\omega r_0}{a}\right)}\sin \omega t - \sum_{n=1}^{+\infty} A_n J_0\left(\frac{\mu_n r}{r_0}\right)\sin\frac{a\mu_n t}{r_0},$$

where

$$A_n = \frac{2A\omega}{a\mu_n r_0 J_0\left(\frac{\omega_0 r}{a}\right)[J_1(\mu_n)]^2}\int_0^{r_0} rJ_0\left(\frac{\omega r}{a}\right)J_0\left(\frac{\mu_n r}{r_0}\right)dr;$$

μ_n are roots of the equation $J_0(\mu) = 0$.

Method. First it is necessary to find the forced oscillations with the frequency of the constraining force in the form $U(r, t) = R(r)\sin \omega t$.

Note. The solution is written on the assumption that there is no resonance, i.e. that $\omega \neq \omega_n = a\mu_n/r_0$, $n = 1, 2, 3, \ldots$

66. The solution of the boundary-value problem

$$\frac{\partial^2 u}{\partial t^2} = a^2\left(\frac{\partial^2 u}{\partial r^2}+\frac{1}{r}\frac{\partial u}{\partial r}\right)-2\nu^2\frac{\partial u}{\partial t}, \quad 0 \leqslant r < r_0, \quad 0 < t < +\infty, \tag{1}$$

$$u(r_0, t) = 0, \quad 0 < t < +\infty, \tag{2}$$

$$u(r, 0) = \phi(r), \quad u_t(r, 0) = \psi(r), \quad 0 < r < r_0 \tag{3}$$

is:

$$u(r, t) = \sum_{n=1}^{+\infty} e^{-\nu^2 t}\{A_n \cos\tilde{\omega}_n t + B_n \sin\tilde{\omega}_n t\}\, J_0\left(\frac{\mu_n r}{r_0}\right), \tag{4}$$

where

$$A_n = \frac{2}{r_0^2[J_1(\mu_n)]^2}\int_0^{r_0} r\phi(r)J_0\left(\frac{\mu_n r}{r_0}\right)dr, \tag{5}$$

$$B_n = \frac{\nu^2}{\tilde{\omega}_n}A_n + \frac{2}{\tilde{\omega}_n r_0^2[J_1(\mu_n)]^2}\int_0^{r_0} r\psi(r)J_0\left(\frac{\mu_n r}{r_0}\right)dr, \tag{6}$$

and μ_n are positive roots of the equation $J_0(\mu) = 0$,

$$\tilde{\omega}_n = \sqrt{\frac{a^2\mu_n^2}{r_0^2} - \nu^4}\,\dagger.$$

† We assume $\tilde{\omega}_n$ real for $n = 1, 2, 3, \ldots$ If for $n = 1, 2, \ldots$ $\tilde{\omega}_n$ is imaginary, then in the corresponding terms cos and sin are replaced by cosh and sinh and the sign in front of the first term in formula (6) is replaced by the opposite sign.

67. (a) $$u(r,t) = 2\frac{p_0}{\rho} r_0^2 \sum_{n=1}^{+\infty} \frac{J_0\left(\frac{\mu_n r}{r_0}\right)}{\mu_n J_1(\mu_n)} \cdot \frac{[(a^2\mu_n^2 - r_0^2\omega^2)\sin\omega t - 2\nu^2\omega\cos\omega t]}{[(a^2\mu_n^2 - r_0^2\omega^2)^2 + 4\nu^4\omega^2]},$$

(b) $$u(r,t) = 2\frac{p_0}{\rho} r_0^2 \sum_{n=1}^{+\infty} \frac{J_0\left(\frac{\mu_n r}{r_0}\right)}{\mu_n J_1(\mu_n)} \cdot \frac{[(a^2\mu_n^2 - r_0^2\omega^2)\cos\omega t + 2\nu^2\omega\sin\omega t]}{[(a^2\mu_n^2 - r_0^2\omega^2)^2 + 4\nu^4\omega^2]},$$

where μ_n are positive roots of the equation $J_0(\mu) = 0$.

Method. See the method to problem 50.

68. It is necessary to find the solution of the equation

$$\frac{\partial^2 U}{\partial t^2} = a^2\left\{\frac{\partial^2 U}{\partial r^2} + \frac{1}{r}\frac{\partial U}{\partial r}\right\} - 2\nu^2\frac{\partial U}{\partial t},$$

satisfying the boundary conditions

$$|U(0,t)| < +\infty, \qquad U(r_0,t) = Ae^{i\omega t},$$

and then take its imaginary part. With this object we eliminate the inhomogeneity in the boundary condition, transferring it into the right-hand part of the differential equation; namely, we shall look for the solution of the problem in the form

$$U(r,t) = v(r,t) + A\frac{r^2}{r_0^2}e^{i\omega t},$$

where

$$v(r,t) = R(r)e^{i\omega t}, \qquad |R(0)| < +\infty, \qquad R(r_0) = 0. \tag{1}$$

For $R(r)$ we obtain the differential equation

$$\frac{d^2R}{dr^2} + \frac{1}{r}\frac{dR}{dr} + \frac{\omega^2 - 2\nu^2\omega i}{a^2}R = -\left\{A\left(\frac{\omega^2 r^2}{a^2 r_0^2} + \frac{4}{r_0^2}\right) - 2\omega\frac{\nu^2}{a^2}i\right\}.$$

The solution, satisfying the boundary conditions (1), is sought in the form

$$R(r) = \sum_{n=1}^{+\infty} A_n J_0\left(\frac{\mu_n r}{r_0}\right),$$

where μ_n are positive roots of the equation $J_0(\mu) = 0$.

69. The solution of the boundary-value problem

$$\frac{\partial^2 u}{\partial t^2} = a_1^2\left\{\frac{\partial^2 u}{\partial r^2} + \frac{1}{r}\frac{\partial u}{\partial r}\right\} - \alpha^2\int_0^{r_0} ru(r,t)\,dr,$$

$$0 < r < r_0, \qquad 0 < t < +\infty, \qquad \alpha^2 = \frac{2\pi\rho_0 a_0^2}{\Omega_0\rho_1}, \tag{1}$$

$$u(r_0,t) = 0, \qquad 0 < t < +\infty, \tag{2}$$

$$u(r,0) = \phi(r), \qquad u_t(r,0) = \psi(r), \qquad 0 < r < r_0 \tag{3}$$

is:

$$u(r,t)=\sum_{n=1}^{+\infty}\left\{A_n\cos\frac{a_1\mu_n t}{r_0}+B_n\sin\frac{a_1\mu_n t}{r_0}\right\}\left\{J_0\left(\mu_n\frac{r}{r_0}\right)-J_0(\mu_n)\right\},\tag{4}$$

where μ_n are positive roots of the equation

$$J_0(\mu)+\varkappa J_2(\mu)=0,\qquad \varkappa=\frac{\pi\rho_0 a_0^2}{\Omega_0\rho_1 a_1^2}r_0^2\,\dagger,\tag{5}$$

$$A_n=\frac{2}{r_0^2\left[J_1^2(\mu_n)+\dfrac{2\varkappa}{\mu_n^2}J_2^2(\mu_n)\right]}\int_0^{r_0}r\phi(r)\left[J_0\left(\mu_n\frac{r}{r_0}\right)-J_0(\mu_n)\right]\mathrm{d}r,$$

$$B_n=\frac{2}{a\mu_n r_0\left[J_1^2(\mu_n)+\dfrac{2\varkappa}{\mu_n^2}J_2^2(\mu_n)\right]}\int_0^{r_0}r\psi(r)\left[J_0\left(\mu_n\frac{r}{r_0}\right)-J_0(\mu_n)\right]\mathrm{d}r.$$

Method. We seek particular solutions of the equation

$$\frac{\partial^2 u}{\partial t^2}=a_1^2\left\{\frac{\partial^2 u}{\partial r^2}+\frac{1}{r}\frac{\partial u}{\partial r}\right\}-\frac{\rho_0 a_0^2}{\rho_1\Omega_0}\int_0^{2\pi}\mathrm{d}\phi\int_0^{r_0}ru(r,t)\,\mathrm{d}r,$$

satisfying the conditions

$$|u(0,t)|<+\infty,\qquad u(r_0,t)=0,$$

in the form

$$U(r,t)=R(r)T(t).$$

After separation of the variables this leads to the equations

$$T''+a^2\lambda^2T=0,\qquad \frac{\mathrm{d}^2R}{\mathrm{d}r^2}+\frac{1}{r}\frac{\mathrm{d}R}{\mathrm{d}r}+\lambda^2R=2\pi\frac{\rho_0 a_0^2}{\rho_1 a_1^2}\frac{1}{\Omega_0}\int_0^{r_0}rR(r)\,\mathrm{d}r.$$

Before looking for the solution of the latter equation, satisfying the conditions

$$|R(0)|<+\infty,\qquad R(r_0)=0,\tag{1'}$$

we make a change of variables: $\lambda r=x$, $R(r)=R(\frac{x}{\lambda})=y(x)$; this leads to the equation

$$y''+\frac{1}{x}y'+y(x)=\frac{2\pi}{\Omega_0}\frac{\rho_0 a_0^2}{\rho_1 a_1^2}\frac{r_0^2}{\mu^2}\int_0^{\mu}xy(x)\,\mathrm{d}x,\tag{2'}$$

where $\mu=\lambda r_0$.

In addition conditions (1) take the form

$$|y(0)|<+\infty,\qquad y(\mu)=0.\tag{3'}$$

† In connection with the symbols see problem 5.

We look for the solutions of equation (2), satisfying the boundary conditions (3), in the form

$$y(x) = J_0(x) - J_0(\mu). \tag{4'}$$

Substitution of (4) into (2) gives:

$$-J_0(\mu) = \frac{2\mu}{\Omega_0}\frac{\rho_0 a_0^2}{\rho_1 a_1^2}\frac{r_0^2}{\mu^2}\int_0^{\mu} x\{J_0(x) - J_0(\mu)\}\,dx, \tag{5'}$$

which leads to the following equation for determining the values of μ corresponding to the eigenvalues $\lambda = \mu/r_0$ of the original boundary-value problem

$$J_0(\mu) + \varkappa J_2(\mu) = 0\,\dagger, \tag{6}$$

where

$$\varkappa = \pi\frac{\rho_0 a_0^2}{\rho_1 a_1^2}\frac{r_0^2}{\Omega_0}.$$

Assuming

$$R_n(r) = J_0\left(\frac{\mu_n r}{r_0}\right) - J_0(\mu_n),$$

where μ_n are positive roots of the transcendental equation (6), the following orthogonality relations‡ are readily determined for the eigenfunctions $R_n(r)$

$$\int_0^{r_0} rR_n(r)R_m(r)\,dr = \begin{cases} \dfrac{r_0^2}{2}\left[J_1^2(\mu_n) + \dfrac{2\varkappa}{\mu_n^2}J_2^2(\mu_n)\right] & \text{for} \quad m = n, \\ 0 & \text{for} \quad m \neq n. \end{cases} \tag{7}$$

70. $$u(r, t) = \frac{\dfrac{\Pi_0}{T_0}\left\{J_0\left(\dfrac{\omega r_0}{a_1}\right) - J_0\left(\dfrac{\omega r}{a_1}\right)\right\}}{\varkappa J_2\left(\dfrac{\omega r_0}{a_1}\right) + \dfrac{\omega^2}{a_1^2}J_0\left(\dfrac{\omega r_0}{a_1}\right)}\sin\omega t + \sum_{n=1}^{+\infty} A_n B_n(r)\sin\frac{a_1\mu_n t}{r_0}, \tag{1}$$

where $\varkappa$ and $R(r)$ have the same meaning as in the preceding problem,

$$A_n = \frac{2}{a\mu_n r_0\left[J_1^2(\mu_n) + \dfrac{2\varkappa}{\mu_n^2}J_2^2(\mu_n)\right]}\int_0^{r_0} r\psi(r)R_n(r)\,dr, \tag{2}$$

where

$$\psi(r) = \frac{\dfrac{\Pi_0}{T_0}\left\{J_0\left(\dfrac{\omega r_0}{a_1}\right) - J_0\left(\dfrac{\omega r}{a_1}\right)\right\}}{\varkappa J_2\left(\dfrac{\omega r_0}{a_1}\right) + \dfrac{\omega^2}{a_1^2}J_0\left(\dfrac{\omega r}{a_1}\right)}; \tag{3}$$

T_0 is the tension of the membrane.

---

† First integrate on the right-hand side of (5) by means of relation (20), page 643 [7], then use the first of the relations (21), page 644 [7] assuming $\nu = 1$.

‡ See [38].

Method. See the method to the preceding problem.

71. We seek the solution of the boundary-value problem

$$\frac{\partial^2 u}{\partial t^2}+c^4\Delta_2\Delta_2 u=0, \qquad 0\leqslant r<r_0, \qquad 0<t<+\infty, \tag{1}$$

$$u(r_0, t)=u_r(r_0, t)=0, \qquad 0<t<+\infty, \tag{2}$$

$$u(r, 0)=f(r), \qquad u_t(r, 0)=F(r), \qquad 0\leqslant r<r_0 \tag{3}$$

by the method of separation of variables. We note that by the conditions of radial symmetry

$$\Delta_2=\frac{\partial^2}{\partial r^2}+\frac{1}{r}\frac{\partial}{\partial r}. \tag{4}$$

We look for particular solutions of equation (2) in the form

$$U(r, t)=R(r)\,T(r). \tag{5}$$

We obtain:

$$T''+\omega^2 T=0, \qquad T(t)=A\cos\omega t+B\sin\omega t, \tag{6}$$

$$\Delta\Delta R-k^4 R=0, \qquad k^4=\frac{\omega^4}{c^4}. \tag{7}$$

The latter equation may be written thus:

$$(\Delta+k^2)(\Delta-k^2)R(r)=0. \tag{8}$$

Thus $R(r)$ may be a solution of the equation

$$(\Delta+k^2)R(r)=0, \qquad \text{i.e.} \quad \frac{d^2R}{dr^2}+\frac{1}{r}\frac{dR}{dr}+k^2R=0, \tag{9}$$

or of the equation

$$(\Delta-k^2)R(r)=0, \qquad \text{i.e.} \quad \frac{d^2R}{dr^2}+\frac{1}{r}\frac{dR}{dr}+k^2R=0. \tag{9'}$$

Since only the solutions of equation (8) bounded at $r=0$ are of interest to us, then

$$R(r)=CJ_0(kr)+DI_0(kr). \tag{10}$$

In order to satisfy the boundary conditions (2) $R(r)$ must satisfy the boundary conditions

$$R(r_0)=R'(r_0)=0. \tag{11}$$

Substituting (10) into (11) we obtain the equations

$$\left.\begin{aligned} CJ_0(kr_0)+DI_0(kr_0)&=0,\\ CJ_0'(kr_0)+DI_0'(kr_0)&=0.\end{aligned}\right\} \tag{12}$$

We look for non-trivial solutions of equation (8) satisfying the boundary conditions (11), therefore the constants C and D must not simultaneously reduce to zero, hence, the determinant of system (12) must be equal to zero. Thus we arrive at the transcendental equation

$$J_0(kr_0)I_0'(kr_0)-I_0(kr_0)J_0'(kr_0) = 0 \tag{13}$$

for determining the eigenvalues of our boundary-value problem $k_1, k_2, \ldots$ $\ldots, k_n, \ldots$ As eigenfunctions one may take

$$R_n(r) = R(k_n r) = I_0(k_n r_0)J_0(k_n r)-J_0(k_n r_0)I_0(k_n r). \tag{14}$$

These functions are orthogonal in the segment $0 \leqslant r < r_0$ with weight r. To prove this statement we note that equation (7) may be rewritten in the form

$$\frac{1}{r}\frac{\mathrm{d}}{\mathrm{d}r}\left\{r\frac{\mathrm{d}}{\mathrm{d}r}\left[\frac{1}{r}\frac{\mathrm{d}}{\mathrm{d}r}\left(r\frac{\mathrm{d}R}{\mathrm{d}r}\right)\right]\right\}-k^4R = 0. \tag{15}$$

We assume $R = R_n(r)$, $k = k_n$ in equation (15) and then $R = R_m(r)$ and $k = k_m$, and multiply the first of the equalities obtained by $rR_m(r)$ and the second by $rR_n(r)$, subtract the results and integrate with respect to r from zero to r_0; this gives:

$$(k_m^4-k_n^4)\int_0^{r_0} rR_mR_n\,\mathrm{d}r = \left\{R_n\frac{\mathrm{d}}{\mathrm{d}r}\left[\frac{1}{r}\frac{\mathrm{d}}{\mathrm{d}r}\left(r\frac{\mathrm{d}R_m}{\mathrm{d}r}\right)\right]-\right.$$

$$\left.-R_m\frac{\mathrm{d}}{\mathrm{d}r}\left[\frac{1}{r}\frac{\mathrm{d}}{\mathrm{d}r}\left(r\frac{\mathrm{d}R_n}{\mathrm{d}r}\right)\right]+R_m'R_n''-R_n'R_m''\right\}_{r=0}^{r=r_0}. \tag{16}$$

The right-hand part of the latter equality reduces to zero identically for $r = 0$, because of the structure of the series for the Bessel functions $J_0(x)$ and $I_0(x)$, and for $r = r_0$ by virtue of the boundary conditions (11). Therefore for $k_m \neq k_n$, $k_m \geqslant 0$, $k_n \geqslant 0$,

$$\int_0^{r_0} rR_mR_n\,\mathrm{d}r = 0\,\dagger. \tag{16'}$$

† The orthogonality of R_n and R_m can be proved without detailed investigation of their behaviour at $r = 0$. Let us consider the equations

$$\Delta\Delta R_n(r)-k_n^4R_n(r) = 0,$$
$$\Delta\Delta R_m(r)-k_m^4R_m(r) = 0;$$

let us multiply the first by $R_m(r)$ and the second by $R_n(r)$ and subtract; we arrive at the equality

$$R_m\Delta\Delta R_n-R_n\Delta\Delta R_m = (k_n^4-k_m^4)R_mR_n.$$

Let us integrate this equality over the circle K with boundary Γ, $0 \leqslant r \leqslant r_0$, $0 \leqslant \phi \leqslant 2\pi$ and let us make use of Green's theorem

$$(k_n^4-k_m^4)\iint_K R_mR_n\,\mathrm{d}\sigma = \iint_K [R_m\Delta\Delta R_n-R_n\Delta\Delta R_m]\,\mathrm{d}\sigma =$$

If in (16) one replaces k_m by k and passes to a limit as $k \to k_n$, then, expanding according to Hospital's rule, we find†:

$$\|R_n\|^2 = \int_0^{r_0} rR_n^2(r)\,\mathrm{d}r$$

$$= \frac{r_0^6}{4}\left\{R_n''^2 - R_n' \frac{\mathrm{d}}{\mathrm{d}r}\left[\frac{1}{r}\frac{\mathrm{d}}{\mathrm{d}r}\left(r\frac{\mathrm{d}R_n}{\mathrm{d}r}\right)\right] - \frac{R_n'R_n''}{k_n^4 r_0^4} - R_n'R_n'''\right\}_{r=r_0}$$

$$= \frac{r_0^6}{4} R_n''^2(r_0) = r_0^2 J_0^2(k_n r_0) I_0^2(k_0 r_n). \tag{18}$$

We obtain the solution of the boundary-value problem (1), (2), (3) as the sum of the series

$$u(r, t) = \sum_{n=1}^{+\infty} \{A_n \cos(a^2k_n^2 t) + B_n \sin(a^2 k_n^2 t)\}\, R_n(r),$$

where

$$A_n = \frac{\int_0^{r_0} rf(r)R_n(r)\,\mathrm{d}r}{\|R_n\|^2}, \quad B_n = \frac{\int_0^{r_0} rF(r)R_n(r)\,\mathrm{d}r}{a^2k_n^2\|R_n\|^2}.$$

72. $u(r, t) = \dfrac{c^2 J}{2\pi D} \displaystyle\sum_{n=1}^{+\infty} \frac{[I_0(k_n r_0) - J_0(k_n r_0)] R_n(r)}{k_n^2 r_0^2 I_0^2(k_n r_0) J_0^2(k_n r_0)} \sin(k_n^2 c^2 t),$

$$= \int_\Gamma \left(R_m \frac{\mathrm{d}\Delta R_n}{\mathrm{d}r} - R_n \frac{\mathrm{d}\Delta R_m}{\mathrm{d}r}\right)\mathrm{d}s + \int_\Gamma \left(\frac{\mathrm{d}R_n}{\mathrm{d}r}\Delta R_m - \frac{\mathrm{d}R_m}{\mathrm{d}r}\Delta R_n\right)\mathrm{d}s = 0, \tag{17}$$

since on the circuit Γ, $r = r_0$,

$$R_m(r_0) = R_n(r_0) = 0, \quad R_m'(r_0) = R_n'(r_0) = 0$$

holds.

If $k_m \neq k_n (k_m, k_n \geqslant 0)$, then from the equality

$$(k_n^4 - k_m^4)\iint_K R_m R_n\,\mathrm{d}\sigma = 0$$

there follows the equality

$$\iint_K R_m R_n\,\mathrm{d}\sigma = 0,$$

i.e.

$$\int_0^{r_0} rR_m(r)R_n(r)\,\mathrm{d}r = 0.$$

† It is also possible to use relation (17), obtained in the preceding footnote.

where the cylindrical rigidity D equals

$$D = \frac{2Eh^3}{3(1-m^2)}\,^{\dagger},$$

and $R_n(r)$ has the same meaning as in the preceding problem.

73. $$u(r, t) = \frac{2p_0 c^2 r_0^2}{\omega D} \sum_{n=1}^{+\infty} \frac{\sin(k_n^2 c^2 t) - \frac{k_n^2 c^2}{\omega} \sin \omega t}{1 - \left(\frac{k_n^2 c^2}{\omega}\right)^2} \frac{J_1(k_n r_0) R_n(r)}{k_n^2 r_0^2 J_0^2(k_n r_0) I_0^2(k_n r_0)},$$

where R_n and D have the same meaning as in the preceding problem.

74. $$u(r, t) = \frac{c^2 p_0}{2\pi D} \sum_{n=1}^{+\infty} \frac{\sin(k_n^2 c^2 t) - \frac{k_n^2 c^2}{\omega} \sin \omega t}{1 - \left(\frac{k_n^2 c^2}{\omega}\right)^2} \frac{[I_0(k_n r_0) - J_0(k_n r_0)] R_n(r)}{k_n^2 r_0^2 J_0^2(k_n r_0) I_0^2(k_n r_0)}.$$

75. The solution of the boundary-value problem

$$\frac{\partial^2 u}{\partial t^2} = a^2 \left\{ \frac{\partial^2 u}{\partial r^2} + \frac{1}{r} \frac{\partial u}{\partial r} \right\}, \quad r^* < r < r^{**}, \quad 0 < t < +\infty, \tag{1}$$

$$u(r^*, t) = u(r^{**}, t) = 0, \quad 0 < t < +\infty, \tag{2}$$

$$u(r, 0) = \phi(r), \quad u_t(r, 0) = \psi(r), \quad r^* \leqslant r \leqslant r^{**} \tag{3}$$

is:

$$u(r, t) = \sum_{n=1}^{+\infty} \{A_n \cos(a\lambda_n t) + B_n \sin(a\lambda_n t)\} R_n(r), \tag{4}$$

where

$$R_n(r) = J_0(r\lambda_n) H_0^{(1)}(r^{**}\lambda_n) - J_0(r^{**}\lambda_n) H_0^{(1)}(r\lambda_n); \tag{5}$$

λ_n are positive roots of the equation

$$J_0(r^*\lambda) H_0^{(1)}(r^{**}\lambda) - J_0(r^{**}\lambda) H_0^{(1)}(r^*\lambda) = 0, \tag{6}$$

$$A_n = \frac{\pi^2 \lambda_n^2}{2} \cdot \frac{J_0^2(\lambda_n r^*)}{J_0^2(\lambda_n r^{**}) - J_0^2(\lambda_n r^*)} \int_{r^*}^{r^{**}} r\phi(r) R_n(r) \, dr, \tag{7}$$

$$B_n = \frac{\pi^2 \lambda_n}{2a} \cdot \frac{J_0^2(\lambda_n r^*)}{J_0^2(\lambda_n r^{**}) - J_0^2(\lambda_n r^*)} \int_{r^*}^{r^{**}} r\psi(r) R_n(r) \, dr. \tag{7'}$$

76. $$u(r, t) = R(r) \sin \omega t + \sum_{n=1}^{+\infty} B_n R_n(r) \sin a\lambda_n t, \tag{1}$$

---

† See the solution of problem 18 of the present chapter.

$$R(r) = \frac{p_0}{\omega^2\rho}\left\{\frac{\left[H_0^{(1)}\left(\frac{\omega r^{**}}{a}\right) - H_0^{(1)}\left(\frac{\omega r^*}{a}\right)\right]J_0\left(\frac{\omega r}{a}\right)}{J_0\left(\frac{\omega r^*}{a}\right)H_0^{(1)}\left(\frac{\omega r^{**}}{a}\right) - H_0^{(1)}\left(\frac{\omega r^*}{a}\right)J_0\left(\frac{\omega r^{**}}{a}\right)} + \right.$$

$$\left. + \frac{\left[J_0\left(\frac{\omega r^*}{a}\right) - J_0\left(\frac{\omega r^{**}}{a}\right)\right]H_0^{(1)}\left(\frac{\omega r}{a}\right)}{J_0\left(\frac{\omega r^*}{a}\right)H_0^{(1)}\left(\frac{\omega r^{**}}{a}\right) - H_0^{(1)}\left(\frac{\omega r^*}{a}\right)J_0\left(\frac{\omega r^{**}}{a}\right)} - 1\right\}, \quad (2)$$

$$B_n = \frac{\pi^2\lambda_n}{2a} \cdot \frac{J_0^2(\lambda_n r^*)}{J_0^2(_n r^{**}) - J_0^2(\lambda_n r^*)} \int_{r^*}^{r^{**}} rR(r)R_n(r)\,dr, \quad (3)$$

$$R_n(r) = J_0(\lambda_n r)H_0^{(1)}(\lambda_n r^{**}) - J_0(\lambda_n r^{**})H_0^{(1)}(\lambda_n r); \quad (4)$$

λ_n are positive roots of the equation

$$J_0(\lambda r^*)H_0^{(1)}(\lambda r^{**}) - J_0(\lambda r^{**})H_0^{(1)}(\lambda r^*) = 0. \quad (5)$$

77.

$$u(r, t) = \sum_{n=1}^{+\infty} \{A_n \cos a\lambda_n t + B_n \sin a\lambda_n t\}\, R_n^*(r), \quad (1)$$

$$R_n^* = J_0(\lambda_n r)\, H_0^{(1)\prime}(\lambda_n r^{**}) - J_0'(\lambda_n r^{**})\, H_0^{(1)}(\lambda_n r), \quad (2)$$

λ_n are positive roots of the equation

$$J_0'(\lambda r^*)\, H_0^{(1)\prime}(\lambda r^{**}) - J_0'(\lambda r^{**})\, H_0^{(1)\prime}(\lambda r^*) = 0, \quad (3)$$

$$A_n = \frac{\int_{r^*}^{r^{**}} r\phi(r)\, R_n^*(r)\,dr}{\int_{r^*}^{r^{**}} rR_n^{*2}(r)\,dr}, \qquad B_n = \frac{\int_{r^*}^{r^{**}} r\psi(r)\, R_n^*(r)\,dr}{a\lambda_n \int_{r^*}^{r^{**}} rR_n^{*2}(r)\,dr}.$$

78.

$$u(r, z, t) = \left\{\sum_{n=1}^{+\infty} A_n I_0\left(r\sqrt{\frac{n^2\pi^2}{l^2} - \frac{\omega^2}{a^2}}\right) \cos\frac{n\pi z}{l}\right\} \cos\omega t +$$

$$+ \sum_{m,\, n=0}^{+\infty} B_{mn} J_0\left(\frac{\mu_m r}{r_0}\right) \cos\frac{n\pi z}{l} \cos ta\sqrt{\frac{\mu_m^2}{r_0^2} + \frac{n^2\pi^2}{l^2}}, \quad (1)$$

$$A_n = -\frac{2}{l\sqrt{\frac{n^2\pi^2}{l^2} - \frac{\omega^2}{a^2}}\, I_1\left(r_0\sqrt{\frac{n^2\pi^2}{l^2} - \frac{\omega^2}{a^2}}\right)} \times$$

$$\times \int_1^l f(z) \cos\frac{n\pi z}{l}\,dz, \qquad n = 1, 2, 3, \ldots, \quad (2)$$

$$A_0 = \frac{a^2}{l\omega^2 J_1\left(\dfrac{\omega r_0}{a}\right)} \int_0^l f(z)\, dz, \qquad (2')$$

$$B_{nm} = -\frac{2A_n}{r_0^2 J_0^2(\mu_m)} \int_0^{r_0} r I_0\left(r\sqrt{\frac{n^2\pi^2}{l^2} - \frac{\omega^2}{a^2}}\right) J_0\left(\frac{\mu_m r}{r_0}\right) dr, \quad n = 1, 2, 3, \ldots, \qquad (3)$$

$$B_{0m} = -\frac{2A_0}{r_0^2 J_0^2(\mu_m)} \int_0^{r_0} r J_0\left(\frac{\omega r}{a}\right) J_0\left(\frac{\mu_m r}{r_0}\right) dr; \qquad (3')$$

$\mu_m = 0, 1, 2, 3, \ldots$ are positive roots of the equation $J_1(\mu) = 0$.

Note. The term with factor $\cos \omega t$ in equation (1) is a particular solution of the equation

$$\frac{\partial^2 u}{\partial t^2} = a^2 \left\{\frac{\partial^2 u}{\partial r^2} + \frac{1}{r}\frac{\partial u}{\partial r} + \frac{\partial^2 u}{\partial z^2}\right\}, \qquad (3a')$$

satisfying the inhomogeneous boundary conditions of the problem.

79. The velocity potential equals

$$u(r, z, t) = \left\{\sum_{m=0}^{+\infty} A_m \cosh z\sqrt{\frac{\mu_m^2}{r_0^2} - \frac{\omega^2}{a^2}}\, J_0\left(\frac{\mu_m r}{r_0}\right)\right\} \cos \omega t +$$

$$+ \sum_{n,\, m=0}^{+\infty} B_{nm} \cos \frac{n\pi z}{l} J_0\left(\frac{\mu_m r}{r_0}\right) \cos ta\sqrt{\frac{\mu_m^2}{r_0^2} + \frac{n^2\pi^2}{l^2}},$$

$$A_m = \frac{2}{r_2^2 J_0^2(\mu_m)\sqrt{\dfrac{\mu_m^2}{r_0^2} - \dfrac{\omega^2}{a^2}}\, \sinh l\sqrt{\dfrac{\mu_m^2}{r_0^2} - \dfrac{\omega^2}{a^2}}} \int_0^{r_0} r f(r) J_0\left(\frac{\mu_m r}{r_0}\right) dr,$$

$$B_{nm} = -\frac{2A_m}{l}\int_0^l \cosh z\sqrt{\frac{\mu_m^2}{r_0^2} - \frac{\omega^2}{a^2}} \cos\frac{n\pi z}{l}\, dz,$$

$$B_{0m} = -\frac{A_m}{l}\int_0^l \cosh z\sqrt{\frac{\mu_m^2}{r_0^2} - \frac{\omega^2}{a^2}}\, dz,$$

μ_m, $m = 0, 1, 2, \ldots$ are positive roots of the equation $J_1(\mu) = 0$

80. The velocity potential equals

$$u(r, z, t) = \left\{\sum_{n=0}^{+\infty} A_n R_n(r) \cos \frac{n\pi z}{l}\right\} \cos \omega t +$$

$$+ \sum_{m=0}^{+\infty} B_{nm} R_m^*(r) \cos \frac{n\pi z}{l} \cos ta\sqrt{\lambda_m^2 + \frac{n^2\pi^2}{l^2}}, \qquad (1)$$

$$R_n(r) = K_0'(\varkappa_n r^*) I_0(\varkappa_n r) - K_0(\varkappa_n r) I_0'(\varkappa_n r^*),$$

$$x_n = \sqrt{\frac{n^2\pi^2}{l^2} - \frac{\omega^2}{a^2}}, \quad n = 1, 2, \ldots\dagger \tag{2}$$

$$R_0(r) = N_0'\left(\frac{\omega r^*}{a}\right) J_0\left(\frac{\omega r}{a}\right) - N_0\left(\frac{\omega r}{a}\right) J_0'\left(\frac{\omega r^*}{a}\right), \tag{3}$$

$$A_n = \frac{2}{lR_n'(r^{**})}\int_0^l f(z)\cos\frac{n\pi z}{l}\,dz, \quad n = 1, 2, \ldots,$$

$$A_0 = \frac{1}{lR_0'(r^{**})}\int_0^l f(z)\,dz, \tag{4}$$

$$R_m^*(r) = J_0(\lambda_m r) H_0^{(1)'}(\lambda_m r^{**}) - J_0'(\lambda_m r^{**}) H_0^{(1)}(\lambda_m r),$$

λ_m are positive roots of the equation

$$J_0'(\lambda r^*) H_0^{(1)'}(\lambda r^{**}) - J_0'(\lambda r^{**}) H_0^{(1)'}(\lambda r^*) - 0,$$

$$B_{nm} = -\frac{A_n \int_{r^*}^{r^{**}} rR_n(r)R_m^*(r)\,dr}{\int_{r^*}^{r^{**}} rR_m^{*2}(r)\,dr}.$$

81. The velocity potential equals

$$u(r, z, t) = \left\{\sum_{n=0}^{+\infty} A_n R_n^*(r)\cosh z\sqrt{\lambda_n^2 - \frac{\omega^2}{a^2}}\right\}\cos\omega t +$$

$$+ \sum_{n,m=0}^{+\infty} B_{nm} R_n^*(r)\cos\frac{m\pi z}{l}\cos ta\sqrt{\lambda_n^2 + \frac{m^2n^2}{l^2}},$$

$$R_n^*(r) = J_0(\lambda_n r) H_0^{(1)'}(\lambda_n r^{**}) - J_0'(\lambda_n r^{**}) H_0^{(1)}(\lambda_n r),$$

$\lambda_n (n = 0, 1, 2, \ldots)$ are positive roots of the equation

$$J_0'(\lambda r^*) H_0^{(1)'}(\lambda r^{**}) - J_0'(\lambda r^{**}) H_0^{(1)'}(\lambda r^*) = 0\ddagger,$$

$$A_n = \frac{\int_{r^{**}}^{r^{**}} rf(r)R_n^*(r)\,dr}{\sqrt{\lambda_n^2 - \frac{\omega^2}{a^2}}\sinh l\sqrt{\lambda_n^2 - \frac{\omega^2}{a^2}}\int_{r^*}^{r^{**}} rR_n^{*2}(r)\,dr},$$

† Here it is assumed that $\varkappa_n$ is real; otherwise K_0 and I_0 are replaced by N_0 and J_0.

‡ Here it is assumed that $\lambda_n > \omega/a$ for all $n = 0, 1, 2, \ldots$

$$B_{nm} = -\frac{2A_n}{l}\int_0^l \cosh z\sqrt{\lambda_n^2-\frac{\omega^2}{a^2}}\cos\frac{m\pi z}{l}\,dz, \quad m = 1, 2, 3, \ldots,$$

$$B_{n0} = -\frac{A_n}{l}\int_0^l \cosh z\sqrt{\lambda_n^2-\frac{\omega^2}{a^2}}\,dz.$$

82. $$u(r, \phi, t) = \frac{2K}{\pi a r_0}\sum_{n,k=0}^{+\infty}\frac{J_n\left(\frac{\mu_k^{(n)}r}{r_0}\right)J_n\left(\frac{\mu_k^{(n)}r_1}{r_0}\right)}{\varepsilon_n\mu_k^{(n)}J_n'^2(\mu_k^{(n)})}\cos n(\phi-\phi_1)\sin\frac{a\mu_k^{(n)}t}{r_0},$$

where $\mu_k^{(n)}$ are positive roots of the equation $J_n(\mu) = 0$,

$$\varepsilon_n = \begin{cases} 2 & \text{for} \quad n = 0, \\ 1 & \text{for} \quad n \neq 0, \end{cases}$$

r_0 is the radius of the membrane, and (r_1, ϕ_1) is the point of impact.

Method. One may assume first that the impulse K is uniformly distributed at time $t = 0$ over the elementary area $\phi_1 \leqslant \phi \leqslant \phi_1+\Delta\phi$, $r_1 \leqslant r \leqslant r_1+\Delta r$ i.e. that the initial conditions have the form

$$u(r, \phi, 0) = 0, \quad 0 \leqslant \phi \leqslant \phi_0, \quad 0 \leqslant r \leqslant r_0,$$

$$u_t(r, \phi, 0) = \begin{cases} \dfrac{K}{\rho r_1\Delta\phi\Delta r} & \text{inside the area shown,} \\ 0 & \text{outside the area shown} \end{cases}$$

and then pass to a limit as $\Delta\phi \to 0$ and $\Delta r \to 0$.

One may also use the delta-function to formulate the initial conditions assuming

$$u(r, \phi, 0) = 0, \quad 0 \leqslant \phi \leqslant \phi_0, \quad 0 \leqslant r \leqslant r_0,$$

$$u_t(r, \phi, 0) = \frac{K}{\rho}\delta^*(r-r_1)\delta(\phi-\phi_1),$$

where the delta-function $\delta(\phi-\phi_1)$ is defined in the usual way, and the function $\delta^*(r-r_1)$ is given by the equalities

$$\int_{r_0'}^{r_0''} r\delta^*(r-r_1)f(r)\,dr = f(r_1), \quad \text{when} \quad r_0' < r_1 < r_0'',$$

$$\int_{r_0'}^{r_0''} r\delta^*(r-r_1)f(r)\,dr = 0, \quad \text{when } r_1 \text{ lies outside the interval } [r_0', r_0''],$$

for any function $f(r)$. Thus, the product $\delta^*(r-r_1)\delta(\phi-\phi_1)$ is the ordinary delta-function for a plane region; multiplying it by an element of area in polar coordinates $r\,dr\,d\phi$ and integrating over the region under consideration, we obtain 1 or 0, according to whether the point (r_1, ϕ_1) belongs to this region or not†.

83. The horizontal velocity potential of the water is a solution of the boundary-value problem

$$\frac{\partial^2 u}{\partial t^2} = a^2 \left\{ \frac{\partial^2 u}{\partial r^2} + \frac{1}{r}\frac{\partial u}{\partial r} + \frac{1}{r^2}\frac{\partial^2 u}{\partial \phi^2} \right\}, \quad 0 \leqslant r \leqslant r_0, \quad 0 \leqslant \phi \leqslant 2\pi, \quad 0 < t < +\infty, \tag{1}$$

$$u_r(r_0, \phi, t) = 0, \quad 0 \leqslant \phi \leqslant 2\pi, \quad 0 < t < +\infty, \tag{2}$$

$$u(r, \phi, 0) = v_0 r \cos\phi, \quad u_t(r, \phi, 0) = 0, \quad 0 \leqslant r \leqslant r_0, \quad 0 \leqslant \phi \leqslant 2\pi. \tag{3}$$

The solution is

$$u(r, \phi, t) = v_0 \cos\phi \sum_{n=1}^{+\infty} A_n J_1\left(\frac{\mu_n r}{r_0}\right) \cos\frac{a\mu_n t}{r_0}, \tag{4}$$

where μ_n are positive roots of the equation $J_1'(\mu) = 0$, and

$$A_n = \frac{2\mu_n^2 \int_0^{r_0} r^2 J_1\left(\frac{\mu_n r}{r_0}\right) dr}{r_0[\mu_n^2 - 1] J_1^2(\mu_n)}. \tag{5}$$

84. $$u(r, \phi, t) = \frac{r_0}{a\rho} \sum_{n=1}^{+\infty} \frac{A t}{\mu_n} J_1\left(\frac{\mu_n r}{r_0}\right) \int_0^t \cos(\phi - \omega\tau) \sin\frac{a\mu_n}{r_0}(t-\tau)\, d\tau, \tag{1}$$

where μ_n are positive roots of the equation $J_1(\mu) = 0$,

$$A_n = \frac{\int_0^{r_0} r f(r) J_1\left(\frac{\mu_n r}{r_0}\right) dr}{\int_0^{r_0} r J_1^2\left(\frac{\mu_n r}{r_0}\right) dr} = \frac{2\int_0^{r_0} r f(r) J_1\left(\frac{\mu_n r}{r_0}\right) dr}{r_0^2 J_1'^2(\mu_n)}. \tag{2}$$

Note. The solution may be obtained in another form. For this it is necessary first to find a particular solution of the inhomogeneous equation

$$\frac{\partial^2 u}{\partial t^2} = a^2 \left\{ \frac{\partial^2 u}{\partial r^2} + \frac{1}{r}\frac{\partial u}{\partial r} + \frac{1}{r^2}\frac{\partial^2 u}{\partial \phi^2} \right\} + \frac{f(r)}{\rho}\cos(\phi - \omega t),$$

reducing to zero for $r = r_0$ in the form

$$U(r, \phi, t) = R(r)\cos(\phi - \omega t).$$

† See also [7], page 292.

Applying the method of variation of parameters using the Wronskian of the cylindrical functions $W\{J_1(z), N_1(z)\} = 2/\pi z$, an expression for $R(r)$ is readily obtained

$$R(r) = \frac{\pi\omega}{2a\rho} \frac{J_1\left(\frac{\omega r}{a}\right)}{J_1\left(\frac{\omega r_0}{a}\right)} \left\{ J_1\left(\frac{\omega r_0}{a}\right) \int_0^{r_0} r f(r) N_1\left(\frac{\omega r}{a}\right) \mathrm{d}r - \right.$$

$$\left. - N_1\left(\frac{\omega r_0}{a}\right) \int_0^{r_0} r f(r) J_1\left(\frac{\omega r}{a}\right) \mathrm{d}r \right\} +$$

$$+ \frac{\pi\omega}{2a\rho} J_1\left(\frac{\omega r}{a}\right) \int_0^{r} r f(r) N_1\left(\frac{\omega r}{a}\right) \mathrm{d}r + \frac{\pi\omega}{2a\rho} N_1\left(\frac{\omega r}{a}\right) \int_0^{r} r f(r) J_1\left(\frac{\omega r}{a}\right) \mathrm{d}r .$$

Then it is necessary to find the solution of the homogeneous equation with the corresponding inhomogeneous initial conditions.

85. $$u(r, \phi, t) = \left\{ \sum_{n=1}^{+\infty} C_n (a^2\mu_n^2 - r_0^2\omega^2) J_1\left(\frac{\mu_n r}{r_0}\right) \right\} \cos(\phi - \omega t) -$$

$$- 2r_0^2 \varkappa\omega \left\{ \sum_{n=1}^{+\infty} C_n J_1\left(\frac{\mu_n r}{r_0}\right) \right\} \sin \phi - \omega t),$$

$$C_n = \frac{2\int_0^{r_0} r f(r) J_1\left(\frac{\mu_n r}{r_0}\right) \mathrm{d}r}{\rho J_1'^2(\mu_n)[(r_0^2\omega^2 - a^2\mu_n^2)^2 + 4r_0^4\varkappa^2\omega^2]},$$

μ_n are positive roots of the equation $J_1(\mu) = 0$.

Method. The solution of the problem may be obtained as the real part of the solution of the equation

$$\frac{\partial^2 u}{\partial t^2} - a^2 \left\{ \frac{\partial^2 u}{\partial r^2} + \frac{1}{r}\frac{\partial u}{\partial r} + \frac{1}{r^2}\frac{\partial^2 u}{\partial \phi^2} \right\} - 2\varkappa \frac{\partial u}{\partial t} + \frac{f(r)}{\rho} \mathrm{e}^{i(\phi - \omega t)}, \qquad (1)$$

reducing to zero for $r = r_0$. Such a solution of equation (1) may be sought in the form

$$U(r, \phi, t) = R(r) \mathrm{e}^{i(\phi - \omega t)}.$$

The solution of the differential equation obtained for $R(r)$ may be sought in the form

$$R(r) = \sum_{n=1}^{+\infty} A_n J_1\left(\frac{\mu_n r}{r_0}\right).$$

86. $u(r, \phi, t) = \left\{v(r, t) + \frac{r^2}{r_0^2} f(t)\right\} \cos n\phi,$

$$v(r, t) = \frac{r_0}{a} \sum_{m=1}^{+\infty} \frac{J_n\left(\frac{\mu_m r}{r_0}\right)}{\mu_m} \int_0^t \psi_m(\tau) \sin \frac{a\mu_m}{r_0}(t-\tau)\,d\tau,$$

$$\psi_m(t) = \frac{2\int_0^{r_0} r[3a^2 f(t) - r^2 f''(t)] J_m\left(\frac{\mu_m r}{r_0}\right) dr}{r_0^4 J_n'^2(\mu_m)};$$

μ_m are positive roots of the equation $J_n(\mu) = 0$.

87. $u(r, \phi, t) = \sin \omega t \sum_{n=0}^{+\infty} J_n\left(\frac{\omega r}{a}\right)(A_n \cos n\phi + B_n \sin n\phi) +$

$$+ \sum_{n, m=0}^{+\infty} J_n\left(\frac{\mu_m^{(n)} r}{r_0}\right)(A_{nm} \cos n\phi + B_{nm} \sin n\phi) \sin \frac{a\mu_m^{(n)} t}{r_0},$$

$\mu_m^{(n)}$ are positive roots of the equation $J_n(\mu) = 0$,

$$A_0 = \frac{1}{2\pi J_0\left(\frac{\omega r_0}{a}\right)} \int_0^{2\pi} F(\phi)\,d\phi, \quad A_n = \frac{1}{\pi J_n\left(\frac{\omega r_0}{a}\right)} \int_0^{2\pi} F(\phi) \cos n\phi\, d\phi,$$

$$n = 1, 2, 3, \ldots,$$

$$B_n = \frac{1}{\pi J_n\left(\frac{\omega r_0}{a}\right)} \int_0^{2\pi} F(\phi) \sin n\phi\, d\phi, \quad n = 1, 2, 3, \ldots,$$

$$A_{nm} = \frac{2\omega A_n \int_0^{r_0} r J_n\left(\frac{\omega r}{a}\right) J_n\left(\frac{\mu_m^{(n)} r}{r_0}\right) dr}{a r_0 \mu_m^{(n)} J_n'^2\left(\mu_m^{(n)}\right)}, \quad n, m = 0, 1, 2, 3, \ldots,$$

$$B_{nm} = \frac{2\omega B_n \int_0^{r_0} r J_n\left(\frac{\omega r}{a}\right) J_n\left(\frac{\mu_m^{(n)} r}{r_0}\right)}{a r_0 \mu_m^{(n)} J_n'^2\left(\mu_m^{(n)}\right)}, \quad n = 1, 2, 3, \ldots, \quad m = 0, 1, 2, \ldots$$

Method. First find the particular solution of the equation

$$\frac{\partial^2 u}{\partial t^2} = a^2 \left\{\frac{\partial^2 u}{\partial r^2} + \frac{1}{r}\frac{\partial u}{\partial r} + \frac{1}{r^2}\frac{\partial^2 u}{\partial \phi^2}\right\},$$

satisfying the boundary condition

$$u(r_0, \phi, t) = F(\phi) \sin \omega t.$$

This particular solution is naturally sought in the form

$$U(r, \phi, t) = V(r, \phi) \sin \omega t.$$

88. $u(r, \phi, z, t) = \left\{ \sum_{m=0}^{+\infty} A_m I_n \left(r \sqrt{\frac{m^2\pi^2}{l^2} - \frac{\omega^2}{a^2}} \right) \cos \frac{m\pi z}{l} \right\} \cos n\phi \cos \omega t +$

$$+ \sum_{m,k=0}^{+\infty} A_{mk} J_n \left(\frac{\mu_k^{(n)} r}{r_0} \right) \cos \frac{m\pi z}{l} \cos n\phi \cos at \sqrt{\frac{m^2\pi^2}{l^2} + \frac{\mu_k^{(n)2}}{r_0^2}},$$

$\mu_k^{(n)}$ are positive roots of the equation $J_n'(\mu) = 0$,

$$A_m = \frac{2 \int_0^l f(z) \cos \frac{m\pi z}{l} dz}{l \sqrt{\frac{m^2\pi^2}{l^2} - \frac{\omega^2}{a^2}} I_n' \left(r_0 \sqrt{\frac{m^2\pi^2}{l^2} - \frac{\omega^2}{a^2}} \right)}, \quad A_0 = \frac{\int_0^l f(z) dz}{\frac{l\omega}{a} J_n' \left(\frac{\omega r_0}{a} \right)},$$

$$A_{mk} = \frac{-2A_m \int_0^{r_0} r I_n \left(r \sqrt{\frac{m^2\pi^2}{l^2} - \frac{\omega^2}{a^2}} \right) J_n \left(\frac{\mu_k^{(n)} r}{r_0} \right) dr}{r_0^2 \left[1 - \frac{n^2}{\mu_k^{(n)2}} \right] J_n^2(\mu_k^{(n)})},$$

$$m = 1, 2, \ldots, \quad k = 0, 1, \ldots,$$

$$A_{0k} = \frac{-2A_0 \int_0^{r_0} r J_n \left(\frac{\omega r}{a} \right) J_n \left(\frac{\mu_k^{(n)} r}{r_0} \right) dr}{r_0^2 \left[1 - \frac{n^2}{\mu_k^{(n)2}} \right] J_n^2(\mu_k^{(n)})}, \quad k = 0, 1, 2, \ldots$$

89. $u(r, \phi, z, t)$

$$= \left\{ \sum_{k=0}^{+\infty} A_k J_n \left(\frac{\mu_k^{(n)} r}{r_0} \right) \cosh z \sqrt{\frac{\mu_k^{(n)2}}{r_0^2} - \frac{\omega^2}{a^2}} \right\} \cos n\phi \cos \omega t +$$

$$+ \sum_{m,k=0}^{+\infty} A_{mk} J_n \left(\frac{\mu_k^{(n)} r}{r_0} \right) \cos \frac{m\pi z}{l} \cos n\phi \cos at \sqrt{\frac{m^2\pi^2}{l^2} + \frac{\mu_k^{(n)2}}{r_0^2}},$$

$\mu_k^{(n)}$ are positive roots of the equation $J_n'(\mu) = 0$,

$$A_k = \frac{2 \int_0^{r_0} r f(r) J_n \left(\frac{\mu_k^{(n)} r}{r_0} \right) dr}{r_0^2 \left[1 - \frac{n^2}{\mu_k^{(n)2}} \right] J_n^2(\mu_k^{(n)}) \sqrt{\frac{\mu_k^{(n)2}}{r_0^2} - \frac{\omega^2}{a^2}} \sinh l \sqrt{\frac{\mu_k^{(n)2}}{r_0^2} - \frac{\omega^2}{a^2}}},$$

$$k = 0, 1, 2, \ldots,$$

$$A_{mk} = -\frac{2}{l} A_k \int_0^l \cosh z \sqrt{\frac{\mu_k^{(n)2}}{r_0^2} - \frac{\omega^2}{a^2}} \cos \frac{m\pi z}{l} \, dz, \quad m = 1, 2, \ldots,$$

$$k = 0, 1, 2, \ldots,$$

$$A_{0k} = -\frac{1}{l} A_k \int_0^l \cosh z \sqrt{\frac{\mu_k^{(n)}}{r_0^2} - \frac{\dot{\omega}^2}{a^2}} \, dz, \quad k = 0, 1, 2, \ldots$$

90. $u(r, \phi, t)$

$$= \frac{4K}{r_0 \phi_0 \rho} \sum_{k, n=1}^{+\infty} \frac{J_{\frac{n\pi}{\phi_0}}\left(\frac{\mu_k^{(n)} r}{r_0}\right) J_{\frac{n\pi}{\phi_0}}\left(\frac{\mu_k^{(n)} r_1}{r_0}\right)}{\mu_k^{(n)} J'^2_{\frac{n\pi}{\phi_0}}\left(\mu_k^{(n)}\right)} \sin \frac{n\pi\phi}{\phi_0} \sin \frac{n\pi\phi_1}{\phi_0} \sin \frac{\mu_k^{(n)} at}{r_0}, \quad (1)$$

where (r_1, ϕ_1) is the point at which the impulse K is imparted, $\mu_k^{(n)}$ are the roots of the equation $J_{n\pi/\phi_0}(\mu_k^{(n)}) = 0$.

91. $u(r, \phi, t)$

$$= \frac{\pi^2 K}{a\rho\phi_0} \sum_{n, k=1}^{+\infty} \frac{\lambda_k^{(n)} J^2_{\frac{n\pi}{\phi_0}}(\lambda_k^{(n)} r_1) R_{nk}(\tilde{r}) R_{nk}(r)}{J^2_{\frac{n\pi}{\phi_0}}(\lambda_k^{(n)} r_2) - J^2_{\frac{n\pi}{\phi_0}}(\lambda_k^{(n)} r_1)} \sin \frac{n\pi\tilde{\phi}}{\phi_0} \sin \frac{n\pi\phi}{\phi_0} \sin \lambda_k^{(n)} at,$$

where $R_{nk}(r) = Z_{n\pi/\phi_0}(\lambda_k^{(n)} r)$ and $\lambda_k^{(n)}$ have the same meaning as in problem 45 chapter V, and $(\tilde{r}, \tilde{\phi})$ is the point at which the impulse K is applied to the membrane, ρ is the surface density of the mass of the membrane (mass per unit area).

92. The velocity potential of particles of gas is a solution of the boundary-value problem

$$\frac{\partial^2 u}{\partial t^2} = a^2 \left\{ \frac{\partial^2 u}{\partial r^2} + \frac{1}{r} \frac{\partial u}{\partial r} + \frac{1}{r^2} \frac{\partial^2 u}{\partial \phi^2} \right\} \quad r_1 \leqslant r \leqslant r_2, \quad 0 \leqslant \phi \leqslant \phi_0,$$

$$0 \leqslant t < +\infty, \quad (1)$$

$$\left.\frac{\partial u}{\partial r}\right|_{r=r_1} = \left.\frac{\partial u}{\partial r}\right|_{r=r_2} = 0, \quad \left.\frac{\partial u}{\partial \phi}\right|_{\phi=0} = \left.\frac{\partial u}{\partial \phi}\right|_{\phi=\phi_0} = 0, \quad (2)$$

$$\left.\begin{aligned} u(r, \phi, 0) &= f(r, \phi), \\ u_t(r, \phi, 0) &= F(r, \phi), \end{aligned}\right\} \quad r_1 \leqslant r \leqslant r_2, \quad 0 \leqslant \phi \leqslant \phi_0. \quad (3)$$

The solution of the boundary-value problem (1), (2), (3) may be represented in the form

$$u(r, \phi, t) = \sum_{n, k=0}^{+\infty} \{A_{nk} \cos a\lambda_k^{(n)} t + B_{nk} \sin a\lambda_k^{(n)} t\} R_{nk}(r) \cos \frac{n\pi\phi}{\phi_0}, \quad (4)$$

where

$$R_{nk}(r) = N'_{\frac{n\pi}{\phi_0}}(\lambda_k^{(n)} r_1) J_{\frac{n\pi}{\phi_0}}(\lambda_k^{(n)} r) - J'_{\frac{n\pi}{\phi_0}}(\lambda_k^{(n)} r_1) N_{\frac{n\pi}{\phi_0}}(\lambda_k^{(n)} r), \tag{5}$$

$\lambda_k^{(n)}$ are positive roots of the equation

$$N'_{\frac{n\pi}{\phi_0}}(\lambda r_1) J'_{\frac{n\pi}{\phi_0}}(\lambda r_2) - J'_{\frac{n\pi}{\phi_0}}(\lambda r_1) N'_{\frac{n\pi}{\phi_0}}(\lambda r_2) = 0, \tag{6}$$

$$A_{nk} = \frac{2}{\phi_0 \int\limits_{r_1}^{r_2} r R_{nk}^2(r)\,\mathrm{d}r} \int\limits_{r_1}^{r_2} \int\limits_0^{\phi_0} f(r, \phi) R_{nk}(r) \cos \frac{n\pi\phi}{\phi_0} r \,\mathrm{d}r\,\mathrm{d}\phi, \quad n > 0, \tag{7}$$

$$A_{0k} = \frac{1}{\phi_0 \int\limits_{r_1}^{r_2} r R_{0k}^2(r)\,\mathrm{d}r} \int\limits_{r_1}^{r_2} \int\limits_0^{\phi_0} f(r, \phi) R_{0k}(r) r \,\mathrm{d}r\,\mathrm{d}\phi, \tag{8}$$

$$\int\limits_{r_1}^{r_2} r R_{nk}^2(r)\,\mathrm{d}r = \frac{2}{\pi^2 \lambda_k^{(n)2}} \cdot \frac{J_n'^2(\lambda_k^{(n)} r_2) - J_n'^2(\lambda_k^{(n)} r_1)}{J_n'^2(\lambda_k^{(n)} r_2)}. \tag{9}$$

93. *Solution.* Let us place the origin of a spherical system of coordinates at the centre of the vessel and let us direct the axis $\theta = 0$ parallel to the velocity of motion of the vessel for $t < 0$. Then the velocity potential u of gas particles will not depend on the angle ϕ and we obtain the boundary-value problem for u

$$\frac{\partial^2 u}{\partial t^2} = a^2 \left\{ \frac{1}{r^2} \frac{\partial}{\partial r} \left(r^2 \frac{\partial u}{\partial r} \right) + \frac{1}{r^2 \sin \theta} \frac{\partial}{\partial \theta} \left(\sin \theta \frac{\partial u}{\partial \theta} \right) \right\},$$

$$0 \leqslant r \leqslant r_0, \quad 0 \leqslant \theta \leqslant \pi, \quad 0 \leqslant t < +\infty, \tag{1}$$

$$u_r(r_0, \theta, t) = 0, \quad 0 \leqslant \theta \leqslant \pi, \quad 0 < t < +\infty, \tag{2}$$

$$u(r, \theta, 0) = vr \cos \theta, \quad u_t(r, \theta, 0) = 0, \quad 0 \leqslant r \leqslant r_0, \quad 0 \leqslant \theta \leqslant \pi. \tag{3}$$

Naturally one looks for the solution of the boundary-value problem (1), (2), (3) in the form

$$u(r, \theta, t) = w(r, t) \cos \theta. \tag{4}$$

This leads to the following boundary-value problem for w:

$$\frac{\partial^2 w}{\partial t^2} = a^2 \left\{ \frac{1}{r^2} \frac{\partial}{\partial r} \left(r^2 \frac{\partial w}{\partial r} \right) - \frac{2w}{r^2} \right\}, \quad 0 \leqslant r \leqslant r_0, \quad 0 < t < +\infty, \tag{5}$$

$$w_r(r_0, t) = 0, \quad 0 < t < +\infty, \tag{6}$$

$$w(r, 0) = vr, \quad w_t(r, 0), \quad 0 \leqslant r \leqslant r_0, \tag{7}$$

which is solved by the method of separation of variables. An expression is obtained for $w(r, t)$

$$w(r, t) = \sum_{k=1}^{+\infty} A_k \frac{J_{\frac{3}{2}}\left(\frac{\mu_n r}{r_0} \right)}{\sqrt{r}} \cos \frac{a\mu_k t}{r_0}, \tag{8}$$

where μ_k are positive roots of the equation

$$\mu J'_{\frac{3}{2}}(\mu) - \frac{1}{2} J_{\frac{3}{2}}(\mu) = 0, \tag{9}$$

$$A_k = \frac{v \int\limits_0^{r_0} r^{\frac{5}{2}} J_{\frac{3}{2}}\left(\frac{\mu_k r}{r_0}\right) dr}{\frac{r_0^2}{2} J^2_{\frac{3}{2}}(\mu_k)\left[1 - \frac{2}{\mu_k^2}\right]}, \quad k = 1, 2, 3, \ldots \tag{10}$$

94. The velocity potential u of the gas is a solution of the boundary-value problem

$$\frac{\partial^2 u}{\partial t^2} = a^2 \left\{ \frac{1}{r^2} \frac{\partial}{\partial r}\left(r^2 \frac{\partial u}{\partial r}\right) + \frac{1}{r^2 \sin\theta} \frac{\partial}{\partial \theta}\left(\sin\theta \frac{\partial u}{\partial \theta}\right)\right\},$$

$$0 \leqslant r \leqslant r_0, \quad 0 \leqslant \theta \leqslant \pi, \quad 0 < t < +\infty, \tag{1}$$

$$\left.\frac{\partial u}{\partial r}\right|_{r=r_0} = \omega A \cos\theta \cos\omega t, \tag{2}$$

$$u(r, \theta, 0) = 0, \quad u_t(r, \theta, 0) = 0. \tag{3}$$

The solution of the boundary-value problem (1), (2), (3) may be represented in the form

$$u(r, \theta, t) = \left\{ \omega A r + \sum_{n=1}^{+\infty} A_n \frac{J_{\frac{3}{2}}\left(\frac{\mu_n r}{r_0}\right)}{\sqrt{r}} \right\} \cos\theta \cos\omega t +$$

$$+ \sum_{k=1}^{+\infty} C_k \frac{J_{\frac{3}{2}}\left(\frac{\mu_k r}{r_0}\right)}{\sqrt{r}} \cos\theta \cos\frac{a\mu_k t}{r_0}, \tag{4}$$

where μ_n are positive roots of the equation

$$\mu J'_{\frac{3}{2}}(\mu) - \frac{1}{2} J_{\frac{3}{2}}(\mu) = 0, \tag{5}$$

$$A_k = \frac{\frac{\omega^3 A}{a^2} \int\limits_0^{r_0} r^{\frac{5}{2}} J_{\frac{5}{2}}\left(\frac{\mu_k r}{r_0}\right) dr}{\left(\frac{\mu_k^2}{r_0^2} - \frac{\omega^2}{a^2}\right) \frac{r_0^2}{2} J^2_{\frac{3}{2}}(\mu_k)\left[1 - \frac{2}{\mu_k^2}\right]}, \quad k = 1, 2, 3, \ldots, \tag{6}$$

$$C_k = \frac{- \int\limits_0^{r_0} \left\{ \omega A r + \sum\limits_{n=1}^{+\infty} A_n \frac{J_{\frac{3}{2}}\left(\frac{\mu_n r}{r_0}\right)}{\sqrt{r}} \right\} r^{\frac{3}{2}} J_{\frac{3}{2}}\left(\frac{\mu_k r}{r_0}\right) dr}{\frac{r_0^2}{2} J^2_{\frac{3}{2}}(\mu_k)\left[1 - \frac{2}{\mu_k^2}\right]} =$$

$$= -\omega A \frac{\int_0^{r_0} r^{\frac{5}{2}} J_{\frac{3}{2}}\left(\frac{\mu_k r}{r_0}\right) dr}{\frac{r_0^2}{2} J_{\frac{3}{2}}^2(\mu_k)\left[1-\frac{2}{\mu_k^2}\right]} - A_k, \quad k = 1, 2, 3, \ldots \tag{7}$$

Note. The term

$$\left\{\omega A r + \sum_{n=1}^{+\infty} A_n \frac{J_{\frac{3}{2}}\left(\frac{\mu_n r}{r_0}\right)}{\sqrt{r}}\right\} \cos\theta \cos\omega t, \tag{8}$$

appearing in (4) is a solution of equation (1), satisfying the boundary condition (2), but not satisfying the initial conditions (3).

The function $\omega A r \cos\theta \cos\omega t$ satisfies the boundary conditions (2), but does not satisfy equation (1).

95. $$u(r, \theta, t) = \left\{\frac{Ar^n}{nr_0^{n-1}} + \sum_{i=1}^{+\infty} A_i \frac{J_{n+\frac{1}{2}}\left(\frac{\mu_i^{(n)} r}{r_0}\right)}{\sqrt{r}}\right\} P_n(\cos\theta)\cos\omega t +$$

$$+ \sum_{k=1}^{+\infty} C_k \frac{J_{n+\frac{1}{2}}\left(\frac{\mu_k^{(n)} r}{r_0}\right)}{\sqrt{r}} P_n(\cos\theta)\cos\frac{a\mu_k^{(n)} t}{r_0}, \tag{1}$$

where $\mu_k^{(n)}$ are positive roots of the equation

$$\mu J'_{n+\frac{1}{2}}(\mu) - \frac{1}{2} J_{n+\frac{1}{2}}(\mu) = 0, \tag{2}$$

$$A_k = -\frac{\frac{\omega^2}{a^2}\frac{A}{nr_0^{n-1}} \int_0^{r_0} r^{n+\frac{3}{2}} J_{n+\frac{1}{2}}\left(\frac{\mu_k^{(n)} r}{r_0}\right) dr}{\frac{r_0^2}{2} J_{n+\frac{1}{2}}^2\left(\mu_k^{(n)}\right)\left[1-\frac{n(n+1)}{\mu_k^{(n)2}}\right]}, \tag{3}$$

$$C_k = -\frac{\int_0^{r_0}\left\{\frac{Ar^n}{nr_0^{n-1}} + \sum_{i=1}^{+\infty} A_i \frac{J_{n+\frac{1}{2}}\left(\frac{\mu_i^{(n)} r}{r_0}\right)}{\sqrt{r}}\right\} r^{\frac{3}{2}} J_{n+\frac{1}{2}}\left(\frac{\mu_k^{(n)} r}{r_0}\right) dr}{\frac{r_0^2}{2} J_{n+\frac{1}{2}}^2\left(\mu_k^{(n)}\right)\left[1-\frac{n(n+1)}{\mu_k^{(n)2}}\right]}$$

$$= -\frac{A}{nr_0^{n-1}} \frac{\int_0^{r_0} r^{n+\frac{3}{2}} J_{n+\frac{1}{2}}\left(\frac{\mu_k^{(n)} r}{r_0}\right) dr}{\frac{r_0^2}{2} J_{n+\frac{1}{2}}^2(\mu_k)\left[1-\frac{n(n+1)}{\mu_k^{(n)2}}\right]} - A_k. \tag{4}$$

Note. See the note to the answer to the preceding problem.

96. The velocity potential of the gas particles is a solution of the boundary-value problem

$$\frac{\partial^2 u}{\partial t^2} = a^2 \left\{ \frac{1}{r^2} \frac{\partial}{\partial r} \left(r^2 \frac{\partial u}{\partial r} \right) + \frac{1}{r^2 \sin\theta} \frac{\partial}{\partial \theta} \left(\sin\theta \frac{\partial u}{\partial \theta} \right) \right\},$$

$$0 \leqslant r \leqslant r_0, \quad 0 \leqslant \theta \leqslant \pi, \quad 0 < t < +\infty, \tag{1}$$

$$u_r(r_0, \theta, t) = P_n(\cos\theta) f(t), \quad 0 \leqslant \theta \leqslant \pi, \quad 0 < t < +\infty, \tag{2}$$

$$u(r, \theta, 0) = u_t(r, \theta, 0) = 0, \quad 0 \leqslant r \leqslant r_0, \quad 0 \leqslant \theta \leqslant \pi. \tag{3}$$

The solution of the boundary-value problem (1), (2), (3) may be represented in the form

$$u(r, \theta, t) = \left\{ \frac{r^n f(t)}{n r_0^{n-1}} + \sum_{k=1}^{+\infty} \psi_k(t) \frac{J_{n+\frac{1}{2}}\left(\frac{\mu_k^{(n)} r}{r_0} \right)}{\sqrt{r}} \right\} P_n(\cos\theta), \tag{4}$$

where $\mu_k^{(n)}$ are positive roots of the equation

$$\mu J'_{n+\frac{1}{2}}(\mu) - \tfrac{1}{2} J_{n+\frac{1}{2}}(\mu) = 0, \tag{5}$$

$$\psi_k(t) = \frac{r_0 A_k}{a \mu_k^{(n)}} \int_0^t f''(\tau) \sin \frac{a \mu_k^{(n)}}{r_0} (t - \tau) \, d\tau, \quad k = 1, 2, 3, \ldots, \tag{6}$$

$$A_k = -\frac{1}{n r_0^{n-1}} \frac{\int_0^{r_0} r^{n+\frac{3}{2}} J_{n+\frac{1}{2}}\left(\frac{\mu_k^{(n)} r}{r_0} \right) dr}{\frac{r_0^2}{2} J^2_{n+\frac{1}{2}}(\mu_k^{(n)}) \left[1 - \frac{n(n+1)}{\mu_k^{(n)2}} \right]}, \quad k = 1, 2, 3, \ldots \tag{7}$$

97. $$u(r, \theta, t) = \left\{ \sum_{n=1}^{+\infty} A_n \frac{J_{n+\frac{1}{2}}\left(\frac{\omega r}{a} \right)}{\sqrt{r}} P_n(\cos\theta) \right\} \cos \omega t +$$

$$+ \sum_{n, k=1}^{+\infty} A_{nk} \frac{J_{n+\frac{1}{2}}\left(\frac{\mu_k^{(n)} r}{r_0} \right)}{\sqrt{r}} P_n(\cos\theta) \cos \frac{a \mu_k^{(n)} t}{r_0}, \tag{1}$$

$$A_n = \frac{2n+1}{2 R_n'(r_0)} \int_0^{\pi} f(\theta) P_n(\cos\theta) \sin\theta \, d\theta,$$

$$R_n(r) = \frac{J_{n+\frac{1}{2}}\left(\frac{\omega r}{a} \right)}{\sqrt{r}}, \quad n = 0, 1, 2, \ldots, \tag{2}$$

$$A_{nk} = \frac{A_n \int_0^{r_0} r J_{n+\frac{1}{2}}\left(\frac{\omega r}{a}\right) J_{n+\frac{1}{2}}\left(\frac{\mu_k^{(n)} r}{r_0}\right) dr}{\frac{r_0^2}{2} J^2_{n+\frac{1}{2}}\left(\mu_k^{(n)}\right)\left[1 - \frac{n(n+1)}{\mu_k^{(n)2}}\right]},$$

$$k = 1, 2, 3, \ldots, \qquad n = 0, 1, 2, 3, \ldots, \tag{3}$$

where $\mu_k^{(n)}$ are positive roots of the equation

$$\mu J'_{n+\frac{1}{2}}(\mu) - \tfrac{1}{2} J_{n+\frac{1}{2}}(\mu) = 0. \tag{4}$$

98. The velocity potential of gas particles is a solution of the boundary-value problem

$$\frac{\partial^2 u}{\partial t^2} = a^2 \left\{\frac{1}{r^2}\frac{\partial}{\partial r}\left(r^2 \frac{\partial u}{\partial r}\right) + \frac{1}{r^2 \sin\theta}\frac{\partial}{\partial \theta}\left(\sin\theta \frac{\partial u}{\partial \theta}\right) + \frac{1}{r^2 \sin^2\theta}\frac{\partial^2 u}{\partial \phi^2}\right\}, \tag{1}$$

$$\left.\frac{\partial u}{\partial r}\right|_{r=r_0} = A P_{nm}(\cos\theta)\cos m\phi \cos\omega t, \tag{2}$$

$$u|_{t=0} = u_t|_{t=0} = 0. \tag{3}$$

The solution of the boundary-value problem (1), (2), (3) may be represented in the form

$$u(r, \theta, \phi, t) = \left\{\frac{A r_0^{\frac{3}{2}} J_{n+\frac{1}{2}}\left(\frac{\omega r}{a}\right)\cos\omega t}{\left[\frac{\omega r_0}{a} J'_{n+\frac{1}{2}}\left(\frac{\omega r_0}{a}\right) - J_{n+\frac{1}{2}}\left(\frac{\omega r_0}{a}\right)\right]\sqrt{r}} + \right.$$

$$\left. + \sum_{k=1}^{+\infty} A_k \frac{J_{n+\frac{1}{2}}\left(\frac{\mu_k r}{r_0}\right)}{\sqrt{r}} \cos\frac{a\mu_k t}{r_0}\right\} P_{nm}(\cos\theta)\cos m\phi, \tag{4}$$

where μ_k are positive roots of the equation

$$\mu J'_{n+\frac{1}{2}}(\mu) - \tfrac{1}{2} J_{n+\frac{1}{2}}(\mu) = 0, \tag{5}$$

$$A_k = -\frac{A r_0^{\frac{3}{2}} \int_0^{r_0} r J_{n+\frac{1}{2}}\left(\frac{\omega r}{a}\right) J_{n+\frac{1}{2}}\left(\frac{\mu_k r}{a}\right) dr}{\left[\frac{\omega r_0}{a} J'_{n+\frac{1}{2}}\left(\frac{\omega r_0}{a}\right) - J_{n+\frac{1}{2}}\left(\frac{\omega r_0}{a}\right)\right] \frac{r_0^2}{2} J^2_{n+\frac{1}{2}}(\mu_k)\left[1 - \frac{n(n+1)}{\mu_k^2}\right]},$$

$$k = 1, 2, \ldots \tag{6}$$

99. $$u(r,\theta,\phi,t) = \cos m\phi \cos \omega t \sum_{n=m}^{+\infty} A_n \frac{J_{n+\frac{1}{2}}\left(\frac{\omega r}{a}\right)}{\sqrt{r}} P_{nm} \cos\theta) +$$

$$+ \cos m\phi \sum_{n=m}^{+\infty}\sum_{k=1}^{+\infty} A_{nk} \frac{J_{n+\frac{1}{2}}\left(\frac{\mu_k^{(n)} r}{r_0}\right)}{\sqrt{r}} P_{nm}(\cos\theta) \cos \frac{a\mu_k^{(n)} t}{r_0}, \qquad (1)$$

where $\mu_k^{(n)}$ are positive roots of the equation

$$\mu J'_{n+\frac{1}{2}}(\mu) - \tfrac{1}{2} J_{n+\frac{1}{2}}(\mu) = 0. \qquad (2)$$

$$A_n = \frac{r_0^{\frac{3}{2}} \int\limits_0^{\pi} f(\theta) P_{nm}(\cos\theta) \sin\theta \, d\theta}{\frac{2}{2n+1} \frac{(n+m)!}{(n-m)!} \left[\frac{\omega r_0}{a} J'_{n+\frac{1}{2}}\left(\frac{\omega r_0}{a}\right) - \tfrac{1}{2} J_{n+\frac{1}{2}}\left(\frac{\omega r_0}{a}\right)\right]}, \quad n \geqslant m, \qquad (3)$$

$$A_{nk} = -\frac{A_n \int\limits_0^{r_0} r J_{n+\frac{1}{2}}\left(\frac{\omega r}{a}\right) J_{n+\frac{1}{2}}\left(\frac{\mu_k^{(n)} r}{r_0}\right) dr}{\frac{r_0^2}{2} J^2_{n+\frac{1}{2}}(\mu_k^{(n)}) \left[1 - \frac{n(n+1)}{\mu_k^{(n)2}}\right]}, \quad n \geqslant m, k = 1, 2, \ldots \qquad (4)$$

100. The velocity potential of gas particles is a solution of the boundary-value problem

$$\frac{\partial^2 u}{\partial t^2} = a^2 \left\{\frac{1}{r^2}\frac{\partial}{\partial r}\left(r^2 \frac{\partial u}{\partial r}\right) + \frac{1}{r^2 \sin\theta}\frac{\partial}{\partial\theta}\left(\sin\theta \frac{\partial u}{\partial\theta}\right) + \frac{1}{r^2\sin^2\theta}\frac{\partial^2 u}{\partial\phi^2}\right\}, \qquad (1)$$

$$\left.\frac{\partial u}{\partial r}\right|_{r=r_0} = f(t) P_{nm}(\cos\theta)\cos m\phi, \quad f(0) = f'(0) = 0, \qquad (2)$$

$$u|_{t=0} = u_t|_{t=0} = 0. \qquad (3)$$

The solution of the boundary-value problem (1), (2), (3) may be represented in the form

$$u(r,\theta,t) = \left\{\frac{r^n f(t)}{n r_0^{n-1}} + \sum_{k=1}^{+\infty} \psi_k(t) \frac{J_{n+\frac{1}{2}}\left(\frac{\mu_k^{(n)} r}{r_0}\right)}{\sqrt{r}}\right\} P_{nm}(\cos\theta)\cos m\phi, \qquad (4)$$

where $\mu_k^{(n)}$ are positive roots of the equation

$$\mu J'_{n+\frac{1}{2}}(\mu) - \tfrac{1}{2} J_{n+\frac{1}{2}}(\mu) = 0, \qquad (5)$$

$$\psi_k(t) = \frac{r_0 A_k}{a\mu_k^{(n)}} \int\limits_0^t f''(\tau) \sin \frac{a\mu_k^{(n)}}{r_0} (t-\tau)\, dt, \quad k = 1, 2, 3, \ldots, \qquad (6)$$

$$A_k = -\frac{1}{nr_0^{n-1}} \cdot \frac{\int\limits_0^{r_0} r^{n+\frac{3}{2}} J_{n+\frac{1}{2}}\left(\frac{\mu_k^{(n)} r}{r_0}\right) dr}{\frac{r_0^2}{2} J_{n+\frac{1}{2}}^2\left(\mu_k^{(n)}\right)\left[1 - \frac{n(n+1)}{\mu_k^{(n)2}}\right]}, \quad k = 1, 2, 3, \ldots \tag{7}$$

101. $$u(r, \theta, t) = \left\{\sum_{k=1}^{+\infty} A_k \frac{\alpha_k J_{\frac{3}{2}}(\lambda_k r) - \beta_k N_{\frac{3}{2}}(\lambda_k r)}{\sqrt{r}} \cos a\lambda_k t\right\} \cos\theta, \tag{1}$$

where λ_k are positive roots of the equation

$$[\lambda r_1 J'_{\frac{3}{2}}(\lambda r_1) - \tfrac{1}{2} J_{\frac{3}{2}}(\lambda r_1)][\lambda r_2 N'_{\frac{3}{2}}(\lambda r_2) - \tfrac{1}{2} N_{\frac{3}{2}}(\lambda r_2)] -$$

$$-[\lambda r_2 J'_{\frac{3}{2}}(\lambda r_2) - \tfrac{1}{2} J_{\frac{3}{2}}(\lambda r_2)][\lambda r_1 N'_{\frac{3}{2}}(\lambda r_1) - \tfrac{1}{2} N_{\frac{3}{2}}(\lambda r_1)] = 0, \tag{2}$$

$$\alpha_k = \lambda_k r_1 N'_{\frac{3}{2}}(\lambda_k r_1) - \tfrac{1}{2} N_{\frac{3}{2}}(\lambda_k r_1), \qquad \beta_k = \lambda_k r_2 J'_{\frac{3}{2}}(\lambda_k r_2) - \tfrac{1}{2} J_{\frac{3}{2}}(\lambda_k r_2), \tag{3}$$

$$A_k = \frac{v \int\limits_{r_1}^{r_2} r^{\frac{5}{2}} [\alpha_k J_{\frac{3}{2}}(\lambda_k r) - \beta_k N_{\frac{3}{2}}(\lambda_k r)]\, dr}{\int\limits_{r_1}^{r_2} r[\alpha_k J_{\frac{3}{2}}(\lambda_k r) - \beta_k N_{\frac{3}{2}}(\lambda_k r)]^2\, dr}, \quad k = 1, 2, 3, \ldots \tag{4}$$

102. $$u(r, \theta, t) = \left\{\frac{\alpha J_{\frac{3}{2}}\left(\frac{\omega r}{a}\right) - \beta N_{\frac{3}{2}}\left(\frac{\omega r}{a}\right)}{\sqrt{r}} \cos\omega t + \right.$$

$$\left. + \sum_{k=1}^{+\infty} A_k \frac{\alpha_k J_{\frac{3}{2}}(\lambda_k r) - \beta_k N_{\frac{3}{2}}(\lambda_k r)}{\sqrt{r}} \cos a\lambda_k t\right\} \cos\theta, \tag{1}$$

where $\lambda_k, \alpha_k, \beta_k$ have the same meaning as in the answer to the preceding problem, and

$$\alpha = \omega A \frac{r_1^{\frac{3}{2}}\left[\frac{\omega r_2}{a} N'_{\frac{3}{2}}\left(\frac{\omega r_2}{a}\right) - \frac{1}{2} J_{\frac{3}{2}}\left(\frac{\omega r_2}{a}\right)\right] - r_2^{\frac{3}{2}}\left[\frac{\omega r_1}{a} J'_{\frac{3}{2}}\left(\frac{\omega r_1}{a}\right) - \frac{1}{2} J_{\frac{3}{2}}\left(\frac{\omega r_1}{a}\right)\right]}{W(\omega, a, r_1, r_2)}, \tag{2}$$

$$\beta = \omega A \frac{r_1^{\frac{3}{2}}\left[\frac{\omega r_2}{a} J'_{\frac{3}{2}}\left(\frac{\omega r_2}{a}\right) - \frac{1}{2} N_{\frac{3}{2}}\left(\frac{\omega r_2}{a}\right)\right] - r_2^{\frac{3}{2}}\left[\frac{\omega r_1}{a} N'_{\frac{3}{2}}\left(\frac{\omega r_1}{a}\right) - \frac{1}{2} N_{\frac{3}{2}}\left(\frac{\omega r_1}{a}\right)\right]}{W(\omega, a, r_1, r_2)}, \tag{3}$$

$$W(\omega, a, r_1, r_2) = \begin{vmatrix} \left[\frac{\omega r_1}{a} J'_{\frac{3}{2}}\left(\frac{\omega r_1}{a}\right) - \frac{1}{2} J_{\frac{3}{2}}\left(\frac{\omega r_1}{a}\right)\right] & \left[\frac{\omega r_1}{a} N'_{\frac{3}{2}}\left(\frac{\omega r_1}{a}\right) - \frac{1}{2} N_{\frac{3}{2}}\left(\frac{\omega r_1}{a}\right)\right] \\ \left[\frac{\omega r_2}{a} J'_{\frac{3}{2}}\left(\frac{\omega r_2}{a}\right) - \frac{1}{2} J_{\frac{3}{2}}\left(\frac{\omega r_2}{a}\right)\right] & \left[\frac{\omega r_2}{a} N'_{\frac{3}{2}}\left(\frac{\omega r_2}{a}\right) - \frac{1}{2} N_{\frac{3}{2}}\left(\frac{\omega r_2}{a}\right)\right] \end{vmatrix}, \tag{4}$$

$$A_k = -\frac{\int\limits_{r_1}^{r_2} r\left[\alpha J_{\frac{3}{2}}\left(\frac{\omega r}{a}\right) - \beta N_{\frac{3}{2}}\left(\frac{\omega r}{a}\right)\right]\left[\alpha_k J_{\frac{3}{2}}(\lambda_k r) - \beta_k N_{\frac{3}{2}}(\lambda_k r)\right] dr}{\int\limits_{r_1}^{r_2} r\left[\alpha_k J_{\frac{3}{2}}(\lambda_k r) - \beta_k N_{\frac{3}{2}}(\lambda_k r)\right]^2 dr}. \tag{5}$$

(b) *Inhomogeneous media*

103. The solution of the boundary-value problem

$$\left.\begin{aligned} \rho_1 \frac{\partial^2 u_1}{\partial t^2} &= T_0 \left\{\frac{\partial^2 u_1}{\partial r^2} + \frac{1}{r}\frac{\partial u_1}{\partial r} + \frac{1}{r^2}\frac{\partial^2 u_1}{\partial \phi^2}\right\}, \quad 0 \leqslant r \leqslant r_1, \\ \rho_2 \frac{\partial^2 u_2}{\partial t^2} &= T_0 \left\{\frac{\partial^2 u_2}{\partial r^2} + \frac{1}{r}\frac{\partial u_2}{\partial r} + \frac{1}{r^2}\frac{\partial^2 u_2}{\partial \phi^2}\right\}, \quad r_1 \leqslant r \leqslant r_2, \end{aligned}\right\} \begin{aligned} &0 \leqslant \phi < 2\pi, &(1) \\ &0 < t < +\infty, &(1') \end{aligned}$$

$$\left.\begin{aligned} u(r_1-0, \phi, t) &= u(r_1+0, \phi, t), \\ u_r(r_1-0, \phi, t) &= u_r(r_1+0, \phi, t), \end{aligned}\right\} \begin{aligned} &0 \leqslant \phi \leqslant 2\pi, &(2) \\ &0 < t < +\infty, &(2') \end{aligned}$$

$$u(r_2, \phi, t) = 0, \tag{2''}$$

$$u(r, \phi, 0) = f(r, \phi), \quad u_t(r, \phi, 0) = F(r, \phi), \quad 0 \leqslant r \leqslant r_2, \quad 0 \leqslant \phi < 2\pi \tag{3}$$

is:

$$u(r, \phi, t) = \sum_{m, n=1}^{+\infty} R_{mn}(r)\{[\bar{a}_{mn} \cos n\phi + \bar{b}_{mn} \sin n\phi] \cos \lambda_{mn} t + \\ + [\bar{\bar{a}}_{mn} \cos n\phi + \bar{\bar{b}}_{mn} \sin n\phi] \sin \lambda_{mn} t\}, \tag{4}$$

where λ_{mn} are roots of the transcendental equation

$$\begin{vmatrix} J_n(\bar{\omega} r_1) & N_n(\bar{\bar{\omega}} r_1) & J_n(\bar{\bar{\omega}} r_1) \\ \bar{\omega} J'_n(\bar{\omega} r_1) & \bar{\bar{\omega}} N'_n(\bar{\bar{\omega}} r_1) & \bar{\bar{\omega}} J'_n(\bar{\bar{\omega}} r_1) \\ 0 & N_n(\bar{\bar{\omega}} r_2) & J_n(\bar{\bar{\omega}} r_2) \end{vmatrix} = 0, \tag{5}$$

$$\bar{\omega} = \frac{\rho_1 \lambda}{T_0}, \quad \bar{\bar{\omega}} = \frac{\rho_2 \lambda}{T_0}, \tag{6}$$

$$R_{mn}(r) = \begin{cases} [J_n(\bar{\bar{\omega}}_{mn} r_1) N_n(\bar{\bar{\omega}}_{mn} r_2) - N_n(\bar{\bar{\omega}}_{mn} r_1) J_n(\bar{\bar{\omega}}_{mn} r_2)] J_n(\bar{\omega}_{mn} r), \\ \qquad 0 \leqslant r \leqslant r_1, \\ [J_n(\bar{\bar{\omega}}_{mn} r) N_n(\bar{\bar{\omega}}_{mn} r_2) - N_n(\bar{\bar{\omega}}_{mn} r) J_n(\bar{\bar{\omega}}_{mn} r_2)] J_n(\bar{\omega}_{mn} r_1), \\ \qquad r_1 \leqslant r \leqslant r_2, \end{cases} \tag{7}$$

$$\bar{a}_{mn} = \frac{\int\limits_0^{2\pi} d\phi \int\limits_0^{r_2} \mu(r) f(r, \phi) R_{mn}(r) \cos n\phi \, dr}{\varepsilon_n \pi \int\limits_0^{r_2} \mu(r) R_{mn}^2(r) \, dr}, \quad \varepsilon_n = \begin{cases} 1 & \text{for } n \neq 0, \\ 2 & \text{for } n = 0, \end{cases} \tag{8}$$

$$\bar{b}_{mn} = \frac{\int\limits_0^{2\pi} d\phi \int\limits_0^{r_2} \mu(r) f(r, \phi) R_{mn}(r) \sin n\phi \, dr}{\pi \int\limits_0^{r_2} \mu(r) R_{mn}^2(r) dr}. \tag{9}$$

Formulae for $\bar{\bar{a}}_{mn}$ and $\bar{\bar{b}}_{mn}$ are obtained from formulae (8) and (9) by replacing the function $f(r, \phi)$ under the integral sign by $F(r, \phi)$ and by adding the factor λ_{mn} to the denominator.

§ 4. The Method of Integral Representations

1. The Application of the Fourier Integral

(a) *The Fourier transform*

We recall that the Fourier transform of the function $F(x, y)$ with kernel $e^{i(\lambda\xi+\mu\eta)}$ is the function

$$\bar{F}(\lambda, \mu) = \frac{1}{2\pi} \iint\limits_{-\infty}^{+\infty} e^{i(\lambda\xi+\mu\eta)} F(\xi, \eta) d\xi \, d\eta. \tag{I}$$

The original function is recovered by means of the inversion formula

$$F(x, y) = \frac{1}{2\pi} \iint\limits_{-\infty}^{+\infty} e^{-i(\lambda x+\mu y)} \bar{F}(\lambda, \mu) \, d\lambda \, d\mu. \tag{II}$$

The Fourier transform in space† is defined similarly.

104. *Solution.* Applying the Fourier transform of form (I) to equation (1) and the initial conditions (2) of the problem, we obtain an ordinary differential equation and initial conditions

$$\frac{d^2\bar{u}(\lambda, \mu, t)}{dt^2} + a^2(\lambda^2+\mu^2)\bar{u}(\lambda, \mu, t) = 0, \tag{1}$$

$$\bar{u}(\lambda, \mu, 0) = \bar{\Phi}(\lambda, \mu), \quad \frac{d\bar{u}(\lambda, \mu, 0)}{dt} = \bar{\Psi}(\lambda, \mu), \tag{2}$$

where $\bar{u}$, $\bar{\Phi}$, $\bar{\Psi}$ are Fourier transforms of the functions u, Φ, Ψ. The solution of equation (1) for the initial conditions (2) is written in the form

$$\bar{u} = \bar{\Phi}(\lambda, \mu) \cos a\rho t + \bar{\Psi}(\lambda, \mu) \frac{\sin a\rho t}{a\rho}, \quad \rho = \sqrt{\lambda^2+\mu^2}. \tag{3}$$

† For more detail see chapter V, § 3.

Applying the inverse Fourier transform, we find:

$$u(x, y, t) = \frac{1}{2\pi}\left\{\iint_{-\infty}^{+\infty} \Phi(\lambda, \mu) \cos a\rho t\, e^{-i(\lambda x+\mu y)}\, d\lambda\, d\mu + \right.$$

$$\left. + \iint_{-\infty}^{+\infty} \overline{\Psi}(\lambda, \mu) \frac{\sin a\rho t}{a\rho} e^{-i(\lambda x+\mu y)}\, d\lambda\, d\mu\right\}. \quad (4)$$

Substituting the values of $\overline{\Phi}(\lambda, \mu)$ and $\Psi(\lambda, \mu)$, we obtain the equality

$$u(x, y, t) = \frac{1}{(2\pi)^2}\iiiint_{-\infty}^{+\infty}\left\{\Phi(\xi, \eta) \cos a\rho t + \Psi(\xi, \eta)\frac{\sin a\rho t}{a\rho}\right\}\times$$

$$\times e^{i[\lambda(x-\xi)+\mu(y-\eta)]}\, d\xi\, d\eta\, d\lambda\, d\mu, \quad (5)$$

where $\rho = \sqrt{\lambda^2+\mu^2}$.

We introduce polar coordinates by means of the relations

$$\left.\begin{aligned} \xi - x &= r\cos\phi, & \lambda &= \rho\cos\theta, \\ \eta - y &= r\sin\phi, & \mu &= \rho\sin\theta, \end{aligned}\right\} \quad (6)$$

and obtain:

$$\lambda(\xi-x)+\mu(\eta-y) = \rho r\cos(\theta-\phi) = \rho r\cos\phi',$$

where ϕ' is the angle indicated in Fig. 53.

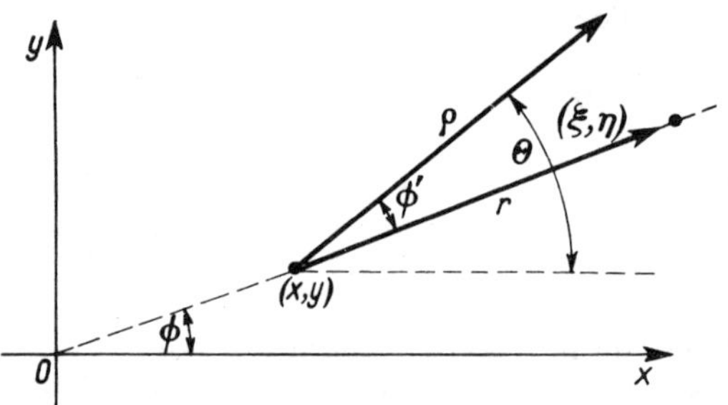

FIG. 53

We denote the first and second terms on the right-hand side of (5) by $u_1(x, y, t)$ and $u_2(x, y, t)$. By virtue of (6) we have:

$$u_2(x, y, t) = \frac{1}{(2\pi)^2}\int_0^{+\infty}\int_0^{+\infty}\int_0^{2\pi}\int_0^{2\pi} \Psi(\xi, \eta)\frac{\sin a\rho t}{a\rho} e^{i\rho r\cos\phi'} \rho r\, dr\, d\rho\, d\phi\, d\phi'. \quad (7)$$

Using†

$$\frac{1}{2\pi}\int_0^{2\pi} e^{i\rho r\cos\phi'}\, d\phi' = J_0(\rho r) \quad (8)$$

† See [7], page 665, (13′).

we obtain from (7):

$$u_2(x, y, t) = \frac{1}{2\pi a} \int_0^{+\infty} \int_0^{+\infty} \int_0^{2\pi} \Psi(\xi, \eta) \sin apt J_0(\rho r)\, r\, \mathrm{d}r\, \mathrm{d}\rho\, \mathrm{d}\phi. \tag{9}$$

But†

$$\int_0^{+\infty} J_0(\rho r) \sin a\rho t\, \mathrm{d}\rho = \begin{cases} 0 & \text{for} \quad at < r, \\ \dfrac{1}{\sqrt{a^2t^2-r^2}} & \text{for} \quad at > r, \end{cases} \tag{10}$$

therefore

$$u_2(x, y, t) = \frac{1}{2\pi a} \int_0^{at} \int_0^{2\pi} \frac{\Psi(\xi, \eta) r\, \mathrm{d}r\, \mathrm{d}\phi}{\sqrt{a^2t^2-r^2}}; \tag{11}$$

u_1 may be obtained from u_2 by differentiation with respect to t, if $\Psi(\xi, \eta)$ is replaced by $\Phi(\xi, \eta)$. Thus

$$u(x, y, t) = \frac{1}{2\pi a} \frac{\partial}{\partial t} \int_0^{at} \int_0^{2\pi} \frac{\Phi(\xi, \eta)\, r\, \mathrm{d}r\, \mathrm{d}\phi}{\sqrt{a^2t^2-r^2}} + \frac{1}{2\pi a} \int_0^{at} \int_0^{2\pi} \frac{\Psi(\xi, \eta) r\, \mathrm{d}r\, \mathrm{d}\phi}{\sqrt{a^2t^2-r^2}}, \tag{12}$$

$$r = \sqrt{(x-\xi)^2+(y-\eta)^2}.$$

105. *Solution.* Applying the Fourier transform in the same way as in the solution of the preceding problem, we obtain:

$$u(x, y, z, t) = u_1(x, y, z, t) + u_2(x, y, z, t), \tag{1}$$

where

$$u_1(x, y, z, t)$$
$$= \frac{1}{(2\pi)^3} \iiiiii_{-\infty}^{+\infty} \Phi(\xi, \eta, \zeta) \cos a\rho t\, \mathrm{e}^{i[\lambda(x-\xi)+\mu(y-\eta)+\nu(z-\zeta)]}\, \mathrm{d}\xi\, \mathrm{d}\eta\, \mathrm{d}\zeta\, \mathrm{d}\lambda\, \mathrm{d}\mu\, \mathrm{d}\nu, \tag{2}$$

$$u_2(x, y, z, t) = \frac{1}{(2\pi)^3} \iiiiii_{-\infty}^{+\infty} \Psi(\xi, \eta, \zeta) \frac{\sin a\rho t}{a\rho} \times$$
$$\times \mathrm{e}^{i[\lambda(x-\xi)+\mu(y-\eta)+\nu(z-\zeta)]}\, \mathrm{d}\xi\, \mathrm{d}\eta\, \mathrm{d}\zeta\, \mathrm{d}\lambda\, \mathrm{d}\mu\, \mathrm{d}\nu \; (\rho = \sqrt{\lambda^2+\mu^2+\nu^2}). \tag{3}$$

Changing to polar coordinates by the formulae

$$\left.\begin{aligned} \xi-x &= r \sin\theta \cos\phi, & \eta-y &= r\sin\theta \sin\phi, & \zeta-z &= r\cos\theta, \\ \lambda &= \rho \sin\theta' \cos\phi', & \mu &= \rho\sin\theta' \sin\phi', & \nu &= \rho\cos\theta', \end{aligned}\right\} \tag{4}$$

where θ is the angle between the positive direction of the z-axis and the vector $\boldsymbol{r} = (\xi-x)\,\boldsymbol{i}+(\eta-y)\,\boldsymbol{j}+(\zeta-z)\,\boldsymbol{k}$, and θ' is the angle between $\boldsymbol{r}$ and $\rho = \lambda \boldsymbol{i}+$

† See [7], page 673, (12) and (13).

$+\mu\boldsymbol{j}+\nu\boldsymbol{k}$ (Fig. 54), i.e. taking the direction of the z-axis as the positive direction of the polar axis in a spherical system of coordinates, we obtain:

$$u_2 = \frac{1}{(2\pi)^3}\int_0^{+\infty}\int_0^{+\infty}\int_0^{\pi}\int_0^{\pi}\int_0^{2\pi}\int_0^{2\pi}\Psi(x+r\sin\theta\cos\phi,\ y+r\sin\theta\sin\phi,\ z\cos\theta)\times$$

$$\times\frac{\sin a\rho t}{a\rho}\,e^{i\rho r\cos\theta'}\rho^2 r^2\sin\theta\ \sin\theta'\,d\rho\,dr\,d\theta\,d\theta'\,d\phi\,d\phi'.$$

Integration with respect to ϕ' and θ' gives:

$$u_2 = \frac{1}{2\pi^2 a}\int_0^{+\infty}\int_0^{+\infty}\int_0^{\pi}\int_0^{2\pi}\Psi(x+r\sin\theta\cos\phi,\ y+r\sin\theta\sin\phi,\ z+r\cos\theta)\times$$

$$\times\sin\rho r\sin a\rho t\ r\sin\theta\,d\rho\,dr\,d\theta\,d\phi$$

$$= \frac{1}{4\pi^2 a}\int_0^{+\infty}\int_0^{+\infty}\int_0^{\pi}\int_0^{2\pi}\Psi(x+r\sin\theta\cos\phi,\ y+r\sin\theta\sin\phi,\ z+r\cos\theta)\times$$

$$\times\{\cos\rho(r-at)-\cos\rho(r+at)\}\ r\sin\theta\,dr\,d\rho\,d\theta\,d\phi.$$

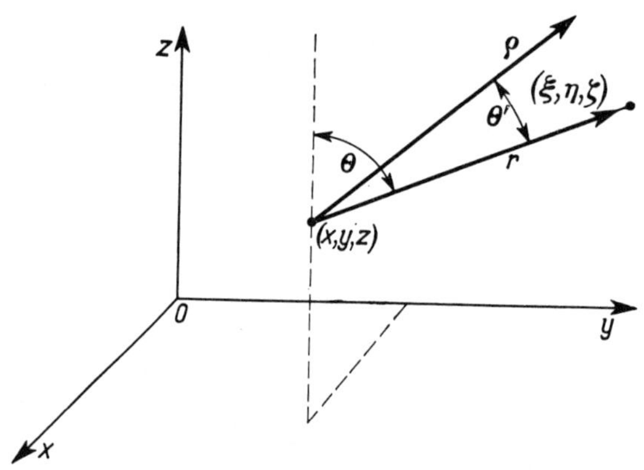

FIG. 54

The integral

$$\frac{1}{\pi}\int_0^{+\infty}d\rho\int_0^{+\infty}r\Psi(x+r\sin\theta\cos\phi,\ y+r\sin\theta\sin\phi,\ z+r\cos\theta)\cos\rho(r-at)\,dr \qquad (5)$$

may be evaluated by means of the Fourier integral

$$\frac{1}{\pi}\int_0^{+\infty}d\rho\int_0^{+\infty}f(r)\cos\rho(r-at)\,dr = f(at),$$

putting

$$f(r) = \begin{cases} r\Psi(x+r\sin\theta\cos\phi,\, y+r\sin\theta\sin\phi,\, z+r\cos\theta) & \text{for} \quad r \geqslant 0, \\ 0 & \text{for} \quad r \leqslant 0. \end{cases}$$

If $f(r)$ satisfies the expansion conditions in the Fourier integral and is continuous, then integral (5) equals

$$at\Psi(x+at\sin\theta\cos\phi,\, y+at\sin\theta\sin\phi,\, z+at\cos\theta) \quad \text{for} \quad t \geqslant 0$$

and zero for $t \leqslant 0$. Similarly

$$\frac{1}{\pi}\int\limits_0^{+\infty} \mathrm{d}\rho \int\limits_0^{+\infty} r\Psi(x+r\sin\theta\cos\phi,\, y+r\sin\theta\sin\phi,\, z+r\cos\theta)\cos\rho(r+at)\,\mathrm{d}r$$

$$= \begin{cases} -at\Psi(x-at\sin\theta\cos\phi,\, y-at\sin\theta\sin\phi,\, z-at\cos\theta) & \text{for} \quad t \leqslant 0, \\ 0 & \text{for} \quad t \geqslant 0. \end{cases}$$

Thus

$$u_2(x, y, z, t)$$

$$= \frac{t}{4\pi}\int\limits_0^{\pi}\int\limits_0^{2\pi} \Psi(x+at\sin\theta\cos\phi,\, y+at\sin\theta\sin\phi,\, z+at\cos\theta)\sin\theta\,\mathrm{d}\theta\,\mathrm{d}\phi;$$

$u_1(x, y, z, t)$ may be obtained from $u_2(x, y, z, t)$ by differentiation with respect to t after replacing Ψ first of all by Φ:

$$u_1(x, y, z, t)$$

$$= \frac{\partial}{\partial t}\left\{\frac{t}{4\pi}\int\limits_0^{\pi}\int\limits_0^{2\pi} \Phi(x+at\sin\theta\cos\phi,\, y+at\sin\theta\sin\phi,\, z+at\cos\theta)\sin\theta\,\mathrm{d}\theta\,\mathrm{d}\phi\right\}.$$

106. $$u(x,y,t) = \frac{1}{2\pi a}\int\limits_0^t \mathrm{d}\tau \iint\limits_{r\leqslant a(t-\tau)} \frac{f(\xi,\eta,\tau)}{\sqrt{a^2(t-\tau)-r^2}}\,\mathrm{d}\xi\,\mathrm{d}\eta,$$

$$r = \sqrt{(x-\xi)^2+(y-\eta)^2}. \tag{1}$$

107. $$u(x,y,z,t) = \frac{1}{4\pi a^2}\iiint\limits_{r\leqslant at} \frac{f\left(\xi,\eta,\zeta,t-\dfrac{r}{a}\right)}{r}\,\mathrm{d}\xi\,\mathrm{d}\eta\,\mathrm{d}\zeta,$$

$$r = \sqrt{(x-\xi)^2+(y-\eta)^2+(z-\zeta)^2}. \tag{1}$$

Changing to spherical coordinates, as in the preceding problem, integrate with respect to ϕ', θ', ρ and r.

108. $$u(x,y,t) = \frac{1}{\pi}\iint\limits_{-\infty}^{+\infty} \Phi(x+2\tilde{\xi}\sqrt{bt},\, y+2\tilde{\eta}\sqrt{bt})\sin t(\tilde{\xi}^2+\tilde{\eta}^2)\,\mathrm{d}\tilde{\xi}\,\mathrm{d}\tilde{\eta}+$$

$$+\frac{1}{\pi}\int\limits_0^t \mathrm{d}t \iint\limits_{-\infty}^{+\infty} \Psi(x+2\tilde{\xi}\sqrt{bt},\, y+2\tilde{\eta}\sqrt{bt})\sin t(\tilde{\xi}^2+\tilde{\eta}^2)\,\mathrm{d}\tilde{\xi}\,\mathrm{d}\tilde{\eta}. \tag{1}$$

Method. Applying the Fourier transform, the following expression for $u(x, y, t)$ is readily derived:

$$u(x,y,t) = \frac{1}{(2\pi)^3}\iiiint_{-\infty}^{+\infty} \Phi(\xi,\eta)\cos\lambda(x-\xi)\cos\mu(y-\eta)\cos bt(\lambda^2+\mu^2)\times$$

$$\times\, d\xi\, d\eta\, d\lambda\, d\mu + \frac{1}{(2\pi)^3}\int_0^t d\tau \iiiint_{-\infty}^{+\infty} \Psi(\xi,\eta)\cos\lambda(x-\xi)\cos\mu(y-\eta)\times$$

$$\times \cos b\tau(\lambda^2+\mu^2)\, d\xi\, d\eta\, d\lambda\, d\mu, \quad (2)$$

if it is taken into account that similar integrals in which in place of $\cos\lambda(x-\xi)$ or $\cos\mu(y-\eta)$ there is $\sin\lambda(x-\xi)$ or $\sin\mu(y-\eta)$, equal zero. Further transformations may be made by means of the equalities

$$\left.\begin{aligned} \int_{-\infty}^{+\infty} \cos p\sigma^2 \cos q\sigma\, d\sigma &= \sqrt{\frac{\pi}{p}}\cos\left(\frac{\pi}{4}-\frac{q^2}{4p}\right), \\ \int_{-\infty}^{+\infty} \sin p\sigma^2 \cos q\sigma\, d\sigma &= \sqrt{\frac{\pi}{p}}\sin\left(\frac{\pi}{4}-\frac{q^2}{4p}\right), \end{aligned}\right\} p > 0 \quad (3)$$

and changes of the variables

$$\frac{\xi - x}{2\sqrt{bt}} = \tilde{\xi}, \quad \frac{\eta - y}{2\sqrt{bt}} = \tilde{\eta}. \quad (4)$$

Equation (3) may be obtained, for instance, from Fresnel's integrals

$$\int_{-\infty}^{+\infty} \cos x^2\, dx = \sqrt{\frac{\pi}{2}} \quad \text{and} \quad \int_{-\infty}^{+\infty} \sin x^2\, dx = \sqrt{\frac{\pi}{2}} \quad (5)$$

by changing to the variable integration of σ in the formula

$$x = \sigma\sqrt{p} - \frac{q}{2\sqrt{p}}. \quad (6)$$

(b) *The Fourier–Bessel* (*Hankel*) *transform*

We recall that the Fourier–Bessel transform of the function $f(r)$, $0 \leqslant r < +\infty$, with kernel $J_\nu(\lambda\xi)$† is defined by

$$\overline{f}(\lambda) = \int_{-\infty}^{+\infty} \xi f(\xi) J_\nu(\lambda\xi)\, d\xi. \quad (1)$$

The original function is recovered by means of the inversion formula

$$f(r) = \int_0^{+\infty} \lambda \overline{f}(\lambda) J_\nu(\lambda r)\, d\lambda. \quad (2)$$

† Or, more briefly, the Fourier–Bessel transform of νth order.

109. $u(r, t)$

$$=\frac{A}{\sqrt{2}}\left\{\frac{1}{\left[\left(1+\frac{r^2-a^2t^2}{b^2}\right)+4\left(\frac{at}{b}\right)^2\right]^{1/2}}+\frac{1+\frac{r^2-a^2t^2}{b^2}}{\left[\left(1+\frac{r^2-a^2t^2}{b^2}\right)+4\left(\frac{at}{b}\right)^2\right]}\right\}^{1/2}.$$

Method. Applying the Fourier–Bessel transform of zero order to the boundary-value problem

$$\frac{\partial^2 u}{\partial t^2}=a^2\left(\frac{\partial^2 u}{\partial r^2}+\frac{1}{r}\frac{\partial u}{\partial r}\right), \quad 0<r, t<+\infty, \tag{1}$$

$$u(r, 0)=f(r), \quad u_t(r, 0)=g(r), \quad 0\leqslant r<+\infty, \tag{2}$$

the following expression for its solution is readily obtained:

$$u(r, t)=\int_0^{+\infty}\lambda\bar{f}(\lambda)\cos(a\lambda t)J_0(\lambda r)\,d\lambda+\frac{1}{a}\int_0^{+\infty}\bar{g}(\lambda)\sin(a\lambda t)J_0(\lambda r)\,d\lambda. \tag{3}$$

In our case $u_t(r, 0)=g(r)=0$, hence,

$$\bar{g}(\lambda)=0, \quad f(r)=\frac{A}{\sqrt{1+\frac{r^2}{b^2}}}.$$

For $\bar{f}(\lambda)$ the expression

$$\bar{f}(\lambda)=\frac{Ab}{\lambda}e^{-\lambda b} \tag{4}$$

is readily obtained. To do this we use the relation

$$\int_0^{+\infty}e^{-\omega x}J_0(\rho x)\,dx=\frac{1}{\sqrt{\rho^2+\omega^2}}\dagger$$

and the inversion formula. Substitution of (4) into (3) gives:

$$u(r, t)=Ab\int_0^{+\infty}e^{-\lambda b}\cos(a\lambda t)J_0(\lambda r)\,d\lambda=\operatorname{Re}\left\{Ab\int_0^{+\infty}e^{-\lambda(b+iat)}J_0(\lambda r)\,d\lambda\right\}, \tag{5}$$

where $\operatorname{Re}(p+qi)=p$ is the real part of the complex number $p+qi$.

110. The solution of the boundary-value probleme (1), (2) (see the condition of the problem) is:

$$u(r, t)=\frac{1}{2bt}\int_0^{+\infty}\rho f(\rho)J_0\left(\frac{\rho r}{2bt}\right)\sin\left(\frac{\rho^2+r^2}{4bt}\right)d\rho. \tag{1}$$

---

† See [7], page 672.

In particular, for $f(\rho) = Ae^{-\rho^2/a^2}$ we obtain:

$$w(r, t) = \frac{Ae^{-R^2/(1+\tau^2)}}{1+\tau^2}\left(\cos\frac{R^2\tau}{1+\tau^2} + \tau\sin\frac{R^2\tau}{1+\tau^2}\right), \tag{2}$$

where

$$\tau = \frac{4bt}{a^2} \quad \text{and} \quad R = \frac{r}{a}. \tag{3}$$

Method. We obtain an expression for the Fourier–Bessel transform of zero order of the solution of the boundary-value problem (1), (2) (see the condition)

$$\bar{u}(\lambda, t) = \cos(b\lambda^2 t)\int_0^{+\infty} \xi f(\xi)\, J_0(\lambda\xi)\, d\xi. \tag{4}$$

If, after applying the inversion formula, one uses Weber's integral

$$\int_0^{+\infty} \xi J_0(\lambda\xi) J_0(\lambda r)\, e^{-p\xi^2} d\xi = \frac{1}{2p} e^{-(\lambda^2+r^2)/4p} J_0\left(\frac{\lambda r}{2p}\right), \tag{5}$$

assuming $p = -ibt$, then expression (1) is obtained for $u(r, t)$. If one substitutes $f(\xi) = Ae^{-\xi^2/a^2}$ in (4), then the expression

$$\bar{u}(\lambda, t) = A\cos(b\lambda^2 t)\int_0^{+\infty} \xi e^{-\xi^2/a^2} J_0(\lambda\xi)\, d\xi = \frac{1}{2}Aa^2 e^{-\lambda^2 a^2/2}\cos(bt\lambda^2), \tag{6}$$

is obtained for $u(\lambda, t)$. This may be evaluated using Hankel's integral†

$$\int_0^{+\infty} J_\nu(at)\, e^{-pt^2} t^{\mu-1} dt = \frac{a^\nu \Gamma\left(\frac{1}{2}\mu + \frac{1}{2}\nu\right)}{2^{\nu+1} p^{\frac{1}{2}\mu+\frac{1}{2}\nu}\, \Gamma(1+\nu)} F_1\left(\frac{1}{2}\mu + \frac{1}{2}\nu; \nu+1; -\frac{a^2}{4p}\right)$$

with $\nu = 0$, $p = 1/a^2$, $\mu = 2$, $a = \lambda$.

Applying the inversion formula and using the same Hankel integral with $\mu = 2$, $\nu = 0$, $a = r$ and

$$p = \frac{a^2}{4}\left(1 + \frac{4ibt}{a^2}\right),$$

we obtain expression (2) for $u(r, t)$.

111. If the point $r = 0$ moves according to the law $u(0, t) = \phi(t)$, $0 < t < +\infty$, then an expression for $u(r, t)$ may be obtained from formula (1) of the answer to the preceding problem, if it is assumed

$$f(r) = \frac{2}{\pi}\int_0^{+\infty} \frac{\psi(t)}{t}\left(\frac{r^2}{4bt}\right) dt.$$

---

† See [42], item 393.

If $\psi(t)$ is given by relations (1) (see the condition), then

$$f(r) = \frac{2At_0}{\pi}\left\{\frac{\pi}{2} - \mathrm{Si}\left(\frac{r^2}{4bt_0}\right) + \frac{r}{4bt_0}\,\mathrm{Ci}\left(\frac{r^2}{4bt_0}\right)\right\},$$

where the integral sine and the integral cosine are given by the relations

$$\mathrm{Si}(x) = \int_0^x \frac{\sin\zeta}{\zeta}\,\mathrm{d}\zeta, \qquad \mathrm{Ci}(x) = -\int_0^{+\infty} \frac{\cos\zeta}{\zeta}\,\mathrm{d}\zeta.$$

Method. Assuming $r = 0$ in equation (1) of the answer to the preceding problem, we obtain an integral equation for determining $f(r)$. If it is assumed that $\rho = \sqrt{\xi}$ in it, then it is transformed into an integral equation, which is readily solved by means of the Fourier sine-transform†.

112. The equation of the forced transverse vibrations of the lamina‡ for $p(r, t) = 16\rho hbf(r)\psi'(t)$ takes the form

$$\frac{\partial^2 u}{\partial t^2} + b^2\left(\frac{\partial^2}{\partial r^2} + \frac{1}{r}\frac{\partial}{\partial r}\right)^2 u = 8bf(r)\psi'(t), \qquad -\infty < t < +\infty, \quad 0 \leqslant r < +\infty.$$

Its solution is:

$$u(r, t) = 4\int_{-\infty}^{t} \frac{\psi(\tau)\,\mathrm{d}\tau}{t-\tau}\int_0^{+\infty} \rho f(\rho)\sin\left[\frac{\rho^2+r^2}{4b(t-\tau)}\right] J_0\left(\frac{\rho r}{2b(t-\tau)}\right)\mathrm{d}\rho. \tag{1}$$

For the special cases we obtain the following expressions for $u(r, t)$:

(a) $$u(r, t) = \frac{2}{\pi}\int_{-\infty}^{t} \psi(\tau)\sin\left[\frac{r^2}{4b(t-\tau)}\right]\frac{\mathrm{d}\tau}{t-\tau};$$

(b) $$u(r, t) = \frac{8b}{\pi a}\int_{-\infty}^{t} \psi(\tau)\,\mathrm{d}\tau\int_0^{+\infty} J_0(\xi r)\,J_1(\xi a)\cos\left[b(t-\tau)\,\xi^2\right]\mathrm{d}\xi;$$

(c) $$u(r, t) = \frac{8\psi_0}{\pi ab}\int_0^{+\infty} J_0(\xi r)\,J_1(\xi a)\,\frac{1-\cos(bt\xi^2)}{\xi^4}\,\mathrm{d}\xi;$$

if $r \ll a$, then for $t < t_0$ we have

$$J_0\left(\frac{r\xi}{a}\right) = 1 - \frac{1}{4}\left(\frac{r}{a}\right)\xi^2 + O\left(\frac{r^4}{a^4}\right)$$

and

$$u(r, t) \approx \frac{2\psi_0 t}{\pi}\left\{\frac{2bt}{a^2}\left[1 - \cos\left(\frac{a^2}{4bt}\right)\right] + F\left(\frac{a^2}{4bt}\right) + \frac{a^2}{8bt}G\left(\frac{a^2}{4bt}\right) - \frac{r^2}{4bt}G\left(\frac{a^2}{4bt}\right)\right\},$$

---

† In connection with the definition of the Fourier sine-transform see chapter II, § 4.

‡ See problem 18.

where

$$F(x) = \frac{\pi}{2} - \mathrm{Si}(x) - x\,\mathrm{Ci}(x), \qquad G(x) = \frac{\sin x}{x} - \mathrm{Ci}(x);$$

a similar expression is obtained for $t > t_0$; if $r \gg a$, then

$$J_1\left(\frac{\xi a}{r}\right) \approx \frac{1}{2}\,\xi\,\frac{a}{r} - \frac{1}{16}\,\xi^2\left(\frac{a}{r}\right)^2$$

and

$$u(r,t) \approx \frac{2\psi_0 t}{\pi}\left[F\left(\frac{r^2}{4bt}\right) + \frac{a^2}{8bt}\mathrm{Ci}\left(\frac{r^2}{4bt}\right)\right], \qquad 0 \leqslant t \leqslant t_0,$$

$$u(r,t) \approx \frac{2\psi_0}{\pi}\left\{tF\left(\frac{r^2}{4bt}\right) - (t-t_0)F\left(\frac{r^2}{4b(t-t_0)}\right) + \right.$$

$$\left. + \frac{a^2}{8b}\mathrm{Ci}\left(\frac{r^2}{4bt}\right) - \frac{a^2}{8b}\mathrm{Ci}\left(\frac{r^2}{4b(t-t_0)}\right)\right\}, \qquad t_0 \leqslant t < +\infty;$$

(d) $$u(r,t) = A\int\limits_{-\infty}^{t} f(\tau)\,H(t-\tau)\,\mathrm{d}\tau,$$

where

$$H(x) = \frac{2p^2\,\mathrm{e}^{-\frac{pq}{p^2+x^2}}}{c^2(p^2+x^2)}\left[\cos\left(\frac{qx}{p^2+x^2}\right) + \frac{x}{p}\sin\left(\frac{qx}{p^2+x^2}\right)\right],$$

$$p = \frac{c}{4b}, \qquad q = \frac{r}{4b},$$

(e) $$\frac{\partial u(r,t)}{\partial t} = \frac{2Ap^2\,\mathrm{e}^{-\frac{pq}{p^2+t^2}}}{c^2(p^2+t^2)}\left[\cos\left(\frac{qt}{p^2+t^2}\right) + \frac{t}{p}\sin\left(\frac{qt}{p^2+t^2}\right)\right] = AH(t).$$

Method. In case (a) it is assumed that $f(r) = \delta(r)/2\pi r$, where $\delta(r)$ is the Dirac delta-function; in case (b) find first the Fourier–Bessel form of $u(r,t)$:

$$\overline{u}(\lambda,t) = \begin{cases} \dfrac{8\psi_0 bJ_1(a\lambda)}{\pi a\lambda}\cdot\dfrac{1-\cos(b\lambda^2 t)}{b^2\lambda^4}, & 0 \leqslant t \leqslant t_0, \\[2ex] \dfrac{8\psi_0 bJ_1(a\lambda)}{\pi a\lambda}\cdot\dfrac{\cos[b(t-t_0)\lambda^2] - \cos(bt\lambda^2)}{b^2\lambda^4}, & t_0 \leqslant t < +\infty, \end{cases}$$

and then apply the inversion formula. In deriving the asymptotic formulae use the integrals

$$\int\limits_0^{+\infty} J_1(x)\cos(bx^2)\,\mathrm{d}x = 1 - \cos\left(\frac{1}{4b}\right), \qquad \int\limits_0^{+\infty} J_1(x)\sin(bx^2)\,\mathrm{d}x = \sin\left(\frac{1}{4b}\right),$$

$$\int\limits_0^{+\infty} xJ_0(x)\cos(bx^2)\,\mathrm{d}x = \frac{1}{2b}\sin\left(\frac{1}{4b}\right), \qquad \int\limits_0^{+\infty} xJ_0(x)\sin(bx^2)\,\mathrm{d}x = \frac{1}{2b}\cos\left(\frac{1}{4b}\right).$$

2. Formation and Application of the Functions of the Effect of Concentrated Sources

(a) *Green's functions for an impulse*

113.
$$\varkappa(x, y, z, t) = \frac{1}{4\pi a}\frac{\delta(r-at)}{r} = \frac{1}{4\pi a^2}\cdot\frac{\delta\left(t-\dfrac{r}{a}\right)}{r}\dagger, \tag{1}$$

where $r = \sqrt{x^2+y^2+z^2}$.

In the case where the instantaneous point impulse occurs not at the origin of coordinates at time $t = 0$, but at the point ξ, η, ζ at time $t = \tau$, then the source function takes the form

$$\varkappa(x, y, z, \xi, \eta, \zeta, t-\tau) = \frac{1}{4\pi a^2}\,\frac{\delta\left(t-\tau-\dfrac{r}{a}\right)}{r}, \tag{2}$$

where
$$r = \sqrt{(x-\xi)^2+(y-\eta)^2+(z-\zeta)^2}.$$

Method. It is necessary to make use of the fact that‡

$$\frac{1}{\pi}\int_0^{+\infty} \cos k(x-x_0)\,dk = \delta(x-x_0),$$

and also the fact that $\delta(x+x_0) = 0$, if x and $x_0 > 0$ simultaneously. Expression (2) is obtained from (1) by replacing x by $x-\xi, \dots, t$ by $t-\tau$, which is valid since the equation $u_{tt} = a^2\Delta_3 u$ is invariant with respect to this substitution.

114. $\varkappa(x, y, z, \xi, \eta, \zeta, t-\tau)$

$$= \frac{1}{4\pi a^2}\left\{\frac{\delta\left(t-\tau-\dfrac{r}{a}\right)}{r} - \frac{c}{a}\,\frac{J_1\left(c\sqrt{(t-\tau)^2-\dfrac{r^2}{a^2}}\right)}{\sqrt{(t-\tau)^2-\dfrac{r^2}{a^2}}}\,\sigma_0\left(t-\tau-\frac{r}{a}\right)\right\}, \tag{1}$$

if in the original equation a plus sign is taken in front of c^2u, and

$\varkappa(x, y, z, \xi, \eta, \zeta, t-\tau)$

$$= \frac{1}{4\pi a^2}\left\{\frac{\delta\left(t-\tau-\dfrac{r}{a}\right)}{r} - \frac{c}{a}\,\frac{I_1\left(c\sqrt{(t-\tau)^2-\dfrac{r^2}{a^2}}\right)}{\sqrt{(t-\tau)^2-\dfrac{r^2}{a^2}}}\,\sigma_0\left(t-\tau-\frac{r}{a}\right)\right\}, \tag{2}$$

---

† One must take into account the fact that the δ-function is even and that $\delta(x) = \dfrac{1}{\alpha}\,\delta\left(\dfrac{x}{\alpha}\right)$, where α is an arbitrary positive constant. The latter statement is verified by integration with respect to x from $-\infty$ to $+\infty$.

‡ See [7], page 295.

if in the original equation a minus sign is taken before c^2u:

$$r = \sqrt{(x-\xi)^2+(y-\eta)^2+(z-\zeta)^2}.$$

Method. The problem is solved in the same way as the preceding one; firstly an expression is obtained for the function of the effect of an instantaneous concentrated impulse, occurring at the origin of coordinates $x = y = z = 0$ at time $t = 0$, and then, as in the solution of the preceding problem, a translation is made to the more general case of an instantaneous concentrated impulse at the point ξ, η, ζ at time τ. The function

$$\sigma_0(x) = \begin{cases} 0 & \text{for} \quad -\infty < x < 0, \\ 1 & \text{for} \quad 0 < x < +\infty \end{cases}$$

is connected to the function $\delta(x)$ by the relation

$$\sigma_0'(x) = \delta(x).$$

115†. $\varkappa(x, y, \xi, \eta, t-\tau)$

$$= \frac{1}{2\pi a} \frac{\left(\sigma_0 t-\tau-\dfrac{r}{a}\right)}{\sqrt{a^2(t-\tau)-r^2}}, \qquad r = \sqrt{(x-\xi)^2+(y-\eta)^2}.$$

116‡. $\varkappa(x, y, \xi, \eta, t-\tau)$

$$= \frac{1}{2\pi a} \frac{\sigma_0\left(t-\tau-\dfrac{r}{a}\right)}{\sqrt{a^2(t-\tau)^2-r^2}} \cosh\left\{c\sqrt{(t-\tau)^2-\frac{r^2}{a^2}}\right\}, \quad r = \sqrt{(x-\xi)^2+(y-\eta)^2},$$

if a plus sign is taken in front of c^2u in the equation. If a minus sign is taken in front of c^2u in the equation, then it is necessary to replace cosh by cos in the answer.

117. First let us consider a rectangular membrane

(1) for the first boundary-value problem, where $\varkappa = 0$ at the boundary of the rectangle,

$\varkappa(x, y, \xi, \eta, t-\tau)$

$$= \frac{4}{l_1 l_2} \sum_{m,n=1}^{+\infty} \sin\frac{m\pi x}{l_1} \sin\frac{n\pi y}{l_2} \sin\frac{m\pi\xi}{l_1} \sin\frac{n\pi\eta}{l_2} \cdot \frac{\sin\left\{\pi a(t-\tau)\sqrt{\dfrac{m^2}{l_1^2}+\dfrac{n^2}{l_2^2}}\right\}}{\pi a\sqrt{\dfrac{m^2}{l_1^2}+\dfrac{n^2}{l_2^2}}},$$

† As in the case of problems 113 and 114, the Green function is at first obtained for an impulse occurring at time $t = 0$ at the origin of coordinates and then transition is made to an impulse, occurring at an arbitrary point at an arbitrary time $t = \tau$.

‡ See the second footnote on the preceding page.

(2) for the second boundary-value problem where $\left.\frac{\partial \varkappa}{\partial x}\right|_{x=0} = \left.\frac{\partial \varkappa}{\partial x}\right|_{x=l_1} = 0$, $\left.\frac{\partial \varkappa}{\partial y}\right|_{y=0} = \left.\frac{\partial \varkappa}{\partial y}\right|_{y=l_2} = 0$,

$$\varkappa(x, y, \xi, \eta, t-\tau) = \frac{t-\tau}{l_1 l_2} +$$

$$+\frac{1}{l_1 l_2} \sum_{\substack{m,n=0\\ m+n\neq 0}}^{+\infty} \varepsilon_{m,n} \cos\frac{m\pi x}{l_1} \cos\frac{n\pi y}{l_2} \cos\frac{m\pi\xi}{l_1} \cos\frac{n\pi\eta}{l_2} \frac{\sin\left\{\pi a(t-\tau)\sqrt{\frac{m^2}{l_1}+\frac{n^2}{l_2^2}}\right\}}{\pi a\sqrt{\frac{m^2}{l_1^2}+\frac{n^2}{l_2^2}}},$$

where $\varepsilon_{m,n} = 2$ for $mn = 0$ and $\varepsilon_{m,n} = 4$ for $mn \neq 0$;

(3) for the third boundary-value problem when

$$\left.\frac{\partial \varkappa}{\partial x}\right|_{x=0} - \alpha_1 \varkappa|_{x=0} = 0, \quad \left.\frac{\partial \varkappa}{\partial x}\right|_{x=l_1} + \beta_1 \varkappa|_{x=l_1} = 0,$$

$$\left.\frac{\partial \varkappa}{\partial y}\right|_{y=0} - \alpha_2 \varkappa|_{y=0} = 0, \quad \left.\frac{\partial \varkappa}{\partial y}\right|_{y=l_2} + \beta_2 \varkappa|_{y=l_2} = 0,$$

$$\varkappa(x, y, \xi, \eta, t-\tau)$$

$$= 4\sum_{m,n=0}^{+\infty} \frac{\sin(\mu_m x+\phi_m)\sin(\nu_n y+\psi_n)\sin(\mu_n\xi+\phi_n)\sin(\nu_n\eta+\psi_n)}{\left[l_1+\frac{(\alpha_1\beta_1+\mu_m^2)(\alpha_1+\beta_1)}{(\alpha_1+\mu_m^2)(\beta_1+\mu_m^2)}\right]\left[l_2+\frac{(\alpha_2\beta_2+\nu_n^2)(\alpha_2^2+\beta_2^2)}{(\alpha_2^2+\nu_n^2)(\beta_2^2+\nu_n^2)}\right]} \times$$

$$\times \frac{\sin\left\{a(t-\tau)\sqrt{\mu_m^2+\nu_n^2}\right\}}{a\sqrt{\mu_m^2+\nu_n^2}},$$

where μ_m are positive roots of the equation

$$\frac{1}{\alpha_1+\beta_1}\left(\mu - \frac{\alpha_1\beta_1}{\mu}\right) = \cot l_1\mu,$$

ν_n positive roots of the equation

$$\frac{1}{\alpha_2+\beta_2}\left(\nu - \frac{\alpha_2\beta_2}{\nu}\right) = \cot l_2\nu,$$

$$\phi_m = \arctan\frac{\mu_m}{l_1}, \qquad \psi_m = \arctan\frac{\nu_n}{l_2}.$$

For a circular membrane

(1) in the case of the first boundary-value problem where $\varkappa|_{r=r_0} = 0$,

$$\varkappa(r, \phi, r', \phi', t-\tau) = \frac{2}{\pi a r_0}\sum_{n,\,k=0}^{+\infty} \frac{J_n\left(\frac{\mu_k^{(n)} r}{r_0}\right) J_n\left(\frac{\mu_k^{(n)} r'}{r_0}\right)}{\varepsilon_n \mu_k^{(n)} J_n'^2(\mu_k^{(n)})} \cos n(\phi-\phi') \sin\frac{a\mu_k^{(n)} t}{r_0},$$

where $\mu_k^{(u)}$ are positive roots of the equation $J_n(\mu) = 0$,

$$\varepsilon_n = \begin{cases} 2 & \text{and} \quad n = 0, \\ 1 & \text{and} \quad n \neq 0; \end{cases}$$

(2) in the case of the second boundary-value problem where $\left.\dfrac{\partial \varkappa}{\partial r}\right|_{r=r_0} = 0$

$\varkappa(r, \phi, r', \phi', t-\tau)$

$$= \frac{t-\tau}{\pi r_0^2} + \frac{2}{\pi a r_0} \sum_{n,k=0}^{+\infty} \frac{J_n\left(\dfrac{\mu_k^{(n)} r}{r_0}\right) J_n\left(\dfrac{\mu_k^{(n)} r'}{r_0}\right)}{\varepsilon_n \mu_k^{(n)} \left[1 - \dfrac{n^2}{\mu_k^{(n)2}}\right] J_n^2(\mu_k^{(n)})} \cos n(\phi-\phi') \sin \frac{a\mu_k^{(n)} t}{r_0},$$

where $\mu_k^{(n)}$ are positive roots of the equation $J_n(\mu) = 0$ and ε_n takes the same value as in case (1);

(3) in the case of the third boundary-value problem, where $\left.\dfrac{\partial \varkappa}{\partial r}\right|_{r=r_0} + \alpha \varkappa|_{r=r_0} = 0$,

$\varkappa(r, \phi, r', \phi', t-\tau)$

$$= \frac{2}{\pi a r_0} \sum_{n,k=0}^{+\infty} \frac{J_n\left(\dfrac{\mu_k^{(n)} r}{r_0}\right) J_n\left(\dfrac{\mu_k^{(n)} r'}{r_0}\right)}{\varepsilon_n \mu_k^{(n)} \left[1 + \dfrac{\alpha^2 - n^2}{\mu_k^{(n)2}}\right] J_n^2(\mu_k^{(n)})} \cos n(\phi-\phi') \sin \frac{a\mu_k^{(n)} t}{r_0},$$

where $\mu_k^{(n)}$ are positive roots of the equation $\mu J_n'(\mu) + \alpha J_n(\mu) = 0$, and ε_n takes the same value as in case (1).

118. Performing even reflections, we obtain, starting from the source function for an infinite plane:

$\varkappa(r, \phi, r_0, \phi_0, t-\tau)$

$$= \frac{1}{2\pi a} \sum_{k=0}^{n-1} \frac{\sigma_0\left(t-\tau-\dfrac{r_k^+}{a}\right)}{\sqrt{a^2(t-\tau)^2 - \left(r_k^+\right)^2}} \cosh\left[c\sqrt{(t-\tau)^2 - \frac{\left(r_k^+\right)^2}{a^2}}\right] +$$

$$+ \frac{\sigma_0\left(t-\tau-\dfrac{r_k^-}{a}\right)}{\sqrt{a^2(t-\tau)^2 - \left(r_k^-\right)^2}} \cosh\left[c\sqrt{(t-\tau)^2 - \left(\frac{r_k^-}{a}\right)^2}\right],$$

$$r_k^+ = \sqrt{r^2 + r_0^2 - 2rr_0 \cos\left(\phi - \phi_0 + 2k\frac{\pi}{m}\right)},$$

$$r_k^- = \sqrt{r^2 + r_0^2 - 2rr_0 \cos\left(\phi + \phi_0 + 2k\frac{\pi}{m}\right)}.$$

119. (a) The solution of the first boundary-value problem

$$u_{tt} = a^2\Delta_2 u + f(x, y, t) \quad \text{inside region } G \text{ for} \quad 0 < t < +\infty, \tag{1}$$

$$u|_{t=0} = \phi(x, y), \quad u_t|_{t=0} = \psi(x, y) \quad \text{inside region } G, \tag{2}$$

$$u|_\Gamma = \mu(x, y, t) \quad \text{on boundary } \Gamma \text{ for} \quad 0 < t < +\infty \tag{3}$$

is expressed by means of the Green's function $\varkappa = \varkappa(x, y, \xi, \eta, t-\tau)$, satisfying the boundary condition $\varkappa|_\Gamma = 0$, in the following way:

$$u(x, y, t) = \iint_G [\phi(\xi, \eta)\varkappa_t(x, y, \xi, \eta, t) + \psi(\xi, \eta)\varkappa(x, y, \xi, \eta, t)]\, d\xi\, d\eta +$$

$$+ \int_0^t d\tau \iint_G f(\xi, \eta, \tau)\varkappa(x, y, \xi, \eta, t-\tau)\, d\xi\, d\eta -$$

$$- a^2 \int_0^t d\tau \oint_\Gamma \mu(\xi, \eta, \tau) \frac{\partial \varkappa(x, y, \xi, \eta, t-\tau)}{\partial n}\, ds. \tag{4}$$

Here

$$\frac{\partial}{\partial n} = \cos(n, \xi)\frac{\partial}{\partial \xi} + \cos(n, \eta)\frac{\partial}{\partial \eta}$$

denotes the derivative with respect to the outer normal.

(b) The solution of the second boundary-value problem, differing from problem (a) only by the boundary condition

$$\left.\frac{\partial u}{\partial n}\right|_\Gamma = \mu(x, y, t), \tag{3'}$$

using the corresponding Green's function $\varkappa$, satisfying the boundary condition

$$\left.\frac{\partial \varkappa}{\partial n}\right|_\Gamma = 0,$$

is expressed in the following way:

$$u(x, y, t) = I_1 + I_2 + a^2 \int_0^t d\tau \oint_\Gamma \mu(\xi, \eta, \tau)\varkappa(x, y, \xi, \eta, t-\tau)\, d\xi\, d\eta, \tag{4'}$$

where I_1 and I_2 denote the first and second terms in (4).

(c) The solution of the third boundary-value problem, differing from problem (a) only by the boundary condition

$$\left.\left[\frac{\partial u}{\partial n} + \alpha u\right]\right|_\Gamma = \mu(x, y, t), \tag{3''}$$

using the corresponding effect function $\varkappa$, satisfying the boundary condition

$$\left.\left[\frac{\partial \varkappa}{\partial n} + \alpha\varkappa\right]\right|_\Gamma = 0,$$

is expressed in the following way:

$$u(x, y, t) = I_1 + I_2 + a^2 \int_0^t d\tau \oint_\Gamma \mu(\xi, \eta, \tau) \varkappa(x, y, \xi, \eta, t-\tau)\, ds, \qquad (4'')$$

where I_1 and I_2 have the same meaning as in (4′).

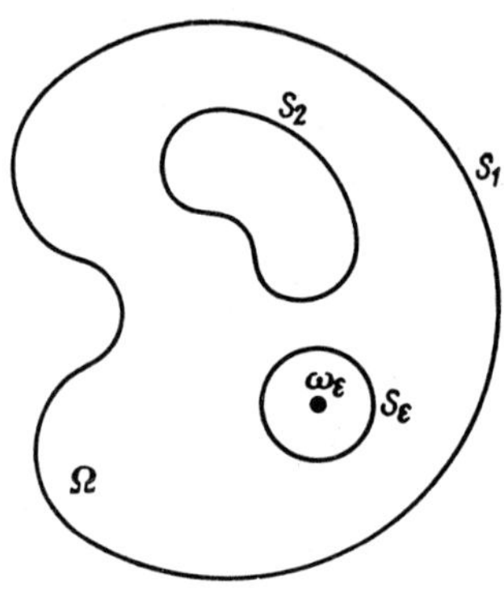

FIG. 55

Method. In the equations

$$\frac{\partial^2 \varkappa}{\partial t^2} = a^2 \underset{xy}{\Delta_2} \varkappa \pm c^2 \varkappa \qquad (5)$$

and

$$\frac{\partial^2 u}{\partial t^2} = a^2 \underset{xy}{\Delta_2} u \pm c^2 u + f(x, y, t) \qquad (6)$$

passing from x, y, t to ξ, η, τ and utilizing the initial conditions for $\varkappa(x, y, \xi, \eta, t-\tau)$:

$$\varkappa|_{t=\tau} = 0, \qquad \varkappa_t|_{t=\tau} = \delta(x-\xi)\delta(\eta-y),$$

it may be shown that

$$u(x, y, t) = \iint_G [u(\xi, \eta, 0) \varkappa_t(x, y, \xi, \eta, t) + u_t(\xi, \eta, 0) \varkappa(x, y, \xi, \eta, t)]\, d\xi\, d\eta +$$

$$+ \int_0^t d\tau \iint_G f(\xi, \eta, \tau) \varkappa(x, y, \xi, \eta, t-\tau)\, d\xi\, d\eta +$$

$$+ a^2 \int_0^t d\tau \oint_\Gamma \left[\varkappa(x, y, \xi, \eta, t-\tau) \frac{\partial u(\xi, \eta, \tau)}{\partial n} - \right.$$

$$\left. - u(\xi, \eta, \tau) \frac{\partial \varkappa(x, y, \xi, \eta, t-\tau)}{\partial n} \right] ds. \qquad (7)$$

To do this multiply equations (5) and (6) after transition to ξ, η, ζ by $u(\zeta, \eta, \tau)$ and $\varkappa(x, y, \xi, \eta, t-\tau)$ respectively, subtract one from the other and integrate the result with respect to ξ, η over the region G and with respect to τ from zero to t.

120. *Method.* Let the region Ω be bounded by the surfaces S_1 and S_2 (Fig. 55). We consider a sphere S_ε of radius ε and centre (x, y, z); and denote the volume bounded by it by ω_ε. Multiply the equation

$$\frac{\partial^2 u}{\partial \tau^2} = a^2 \underset{\xi\eta\zeta}{\Delta_3} u + f(\xi, \eta, \zeta, \tau)$$

by $\varkappa(x, y, z, \xi, \eta, \zeta, t-\tau) = \dfrac{1}{4\pi a^2} \dfrac{\delta\left(t-\tau-\dfrac{r}{a}\right)}{r}$, and the equation

$$\frac{\partial^2 \varkappa}{\partial \tau^2} = a^2 \underset{\xi\eta\zeta}{\Delta_3} \varkappa$$

by $u(\xi, \eta, \zeta, \tau)$, then subtract one from the other and integrate the result over the volume Ω excluding ω_ε and with respect to τ from $\tau_1 < 0$ to $\tau_2 > 0$.

(b) *Source functions of continuously acting concentrated sources*

121. $\omega(x, y, z, x_0, y_0, z_0, t) = \dfrac{1}{4\pi a^2} \dfrac{f\left(t-\dfrac{r}{a}\right)}{r}$

where

$$r = \sqrt{(x-x_0)^2+(y-y_0)^2+(z-z_0)^2}.$$

122. $\omega(x, y, x_0, y_0, t) = \begin{cases} \dfrac{1}{2\pi a} \displaystyle\int_0^{t-\frac{r}{a}} \dfrac{f(\tau)\, d\tau}{\sqrt{a^2(t-\tau)^2+r^2}} & \text{for} \quad t > \dfrac{r}{a}, \\ 0 & \text{for} \quad 0 < t < \dfrac{r}{a}, \end{cases}$

where $r = \sqrt{(x-x_0)^2+(y-y_0)^2}$.

123. $\tilde{\omega}(x, y, z, t) = \dfrac{1}{4\pi a^2} \displaystyle\sum_{r_k} \dfrac{f\left(t-\dfrac{r_k}{a}\right)}{\dfrac{1}{2}\left|\dfrac{dF(r)}{dr}\right|_{r=r_k}}$,

where

$$F(r) = \left\{X\left(t-\frac{r}{a}\right)-x\right\}^2 + \left\{Y\left(t-\frac{r}{a}\right)-y\right\}^2 + \left\{Z\left(t-\frac{r}{a}\right)-z\right\}^2 - r^2,$$

and r_k are positive roots of the equation $F(r) = 0$,

$$\frac{1}{2}\left|\frac{dF(r)}{dr}\right| = \left|r\left(1-\frac{w_r}{a}\right)\right|,$$

where w_r are the projections of the velocity of the source in the direction of the radius-vector r, drawn from the point of observation to the source; therefore $dF(r)/dr$ can reduce to zero only in that case where the velocity of motion of the source $w > a$; hence, only for this condition can the equation $F(r) = 0$ have more than one root.

If the source moves in a straight line with constant velocity v, then choosing the x-axis along the direction of motion of the source, we obtain:

(a) for $v < a$, i.e. for $M = va < 1$

$$\tilde{\omega}(x, y, z, t) = \frac{1}{4\pi a^2} \frac{f\left(t - \frac{M(x-vt)+\sqrt{(x-vt)^2+(1-M^2)(y^2+z^2)}}{a(1-M^2)}\right)}{\sqrt{(x-vt)^2+(1-M^2)(y^2+z^2)}};$$

(b) for $v > a$, i.e. for $M = va > 1$

$$\tilde{\omega}(x, y, z, t) = \frac{1}{4\pi a^2} \frac{f\left(t + \frac{M(x-vt)+\sqrt{(x-vt)^2-(M^2-1)(y^2+z^2)}}{a(M^2-1)}\right)}{\sqrt{(x-vt)^2-(M^2-1)(y^2+z^2)}} +$$

$$+ \frac{1}{4\pi a^2} \frac{f\left(t + \frac{M(x-vt)-\sqrt{(x-vt)^2-(M^2-1)(y^2+z^2)}}{a(M^2-1)}\right)}{\sqrt{(x-vt)^2-(M^2-1)(y^2+z^2)}}.$$

Note. From this equality the solution is determined inside a circular cone with apex at O', the axis of which is the negative semi-axis of x and the length OO' (see Fig. 56); $\cot \alpha = M^2-1$. In this cone the root $\sqrt{(x-vt)^2-(M^2-1)(y^2+z^2)}$

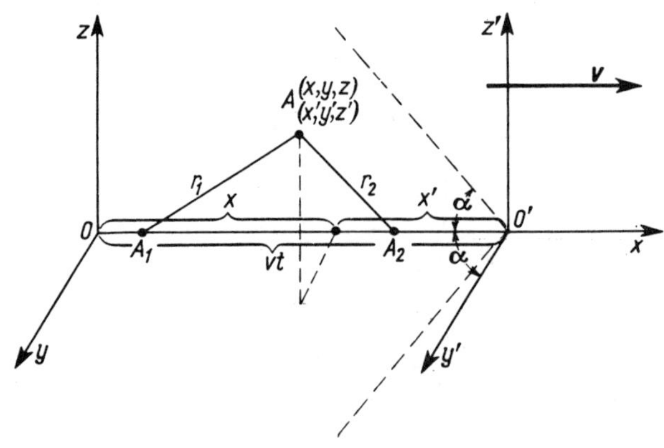

FIG. 56

is real. If the source began to act at time $t = 0$ when it was at point O, then the region, in which disturbances produced by it can be different from zero, is that part of space bounded by this cone and part of a sphere of radius with centre at the point O (the point O lying inside this region, and the cone touching the sphere). Disturbances which arrive at the point of "observation" $A(x, y, z)$ at time t for $v > a$ were sent out by the source from two positions: A_1 and A_2. The distances A_1A and A_2A equal

$$A_1A = r_1 = -\frac{M(x-vt)+\sqrt{(x-vt)^2-(M^2-1)(y^2+z^2)}}{M^2-1},$$

$$A_2A = r_2 = -\frac{M(x-vt)-\sqrt{(x-vt)^2-(M^2-1)(y^2+z^2)}}{M^2-1}.$$

The source is at the point A_1 at time $t_1 = t-r_1/a$, and it was at the point A_2 at time $t_2 = t-r_2/a$

If the magnitude of the source is constant and equal to q, then

(a) for $v < a$, i.e. $M = \dfrac{v}{a} < 1$,

$$\tilde{\omega}(x, y, z, t) = \frac{1}{4\pi a^2}\frac{q}{\sqrt{(x-vt)^2+(1-M^2)(y^2+z^2)}};$$

(b) for $v > a$, i.e. $M = \dfrac{v}{a} > 1$,

$$\tilde{\omega}(x, y, z, t) = \frac{1}{2\pi a^2}\frac{q}{\sqrt{(x-vt)^2-(M^2-1)(y^2+z^2)}}.$$

Method. In the formula giving the solution of equation (1) for the initial conditions (2)

$$\tilde{\omega}(x, y, z, t) = \frac{1}{4\pi a^2}\iiint\limits_{-\infty}^{+\infty}\frac{[f]}{r}\delta(\xi-[X])\delta(\eta-[Y])\delta(\zeta-[Z])\,d\xi\,d\eta\,d\zeta,$$

$$r = \sqrt{(x-\xi)^2+(y-\eta)^2+(z-\zeta)^2},$$

where $[\Phi]$ indicates that in the function Φ the argument t is replaced by $t-\dfrac{r}{a}$, one must change to new variables of integration α, β, γ:

$$\alpha = \xi-[X], \quad \beta = \eta-[Y], \quad \gamma = \zeta-[Z];$$

moreover in place of the Jacobian $\dfrac{D(\xi, \eta, \zeta)}{D(\alpha, \beta, \gamma)}$ it is best to use the Jacobian $\dfrac{D(\alpha, \beta, \gamma)}{D(\xi, \eta, \zeta)}$.

124. The source is located at the origin of a system of coordinates $O'x'y'z'$ moving with it, situated as indicated in Fig. 57, $x' = vt-x$, $y' = y$, $z' = z$.

(a) For $v < a$, i.e. $M = \dfrac{v}{a} < 1$, an equation of elliptic type

$$\frac{\partial^2 u}{\partial x'^2}+\frac{1}{1-M^2}\left(\frac{\partial^2 u}{\partial y'^2}+\frac{\partial^2 u}{\partial z'^2}\right) = -\frac{q}{a^2(1-M^2)}\delta(x')\delta(y')\delta(z')$$

is obtained; by substitution of the variables $x' = \xi, y' = \dfrac{\eta}{\sqrt{1-M^2}}, z = \dfrac{\zeta}{\sqrt{1-M^2}}$ it is transformed into Poisson's equation

$$\frac{\partial^2 u}{\partial \xi^2}+\frac{\partial^2 u}{\partial \eta^2}+\frac{\partial^2 u}{\partial \zeta^2} = -4\pi\left(\frac{q}{4\pi a^2}\right)\delta(\xi)\delta(\eta)\delta(\zeta)$$

the solution of which is the source function of a point source

$$U(\xi, \eta, \zeta) = \frac{q}{4\pi a^2\sqrt{\xi^2+\eta^2+\zeta^2}}$$

or in the original coordinates x', y', z':

$$u(x', y', z',) = \frac{q}{4\pi a^2\sqrt{x'^2+(1-M^2)(y'^2+z'^2)}}$$

(b) for $v > a$, i.e. for $M = \dfrac{v}{a} > 1$, an equation of hyperbolic type

$$\frac{\partial^2 u}{\partial x'^2} = \frac{1}{M^2-1}\left(\frac{\partial^2 u}{\partial y'^2}+\frac{\partial^2 u}{\partial z'^2}\right)+\frac{q}{a^2(M^2-1)}\delta(x')\delta(y')\delta(z')$$

is obtained. Solving it for the initial conditions

$$u|_{x'=0} = 0, \quad u_{x'}|_{x'=0} = 0$$

we obtain:

$$u(x', y', z') = \frac{1}{2\pi a^2}\frac{q}{\sqrt{x'^2-(M^2-1)(y'^2+z'^2)}}.$$

125. Let the electron move along the z-axis with velocity $v = \text{const.}$†, $a = c/\sqrt{\varepsilon} < v < c$, where c is the velocity of light in vacuum, and $a = c/\sqrt{\varepsilon}$ the velocity of light in the dielectric of dielectric constant ε. The scalar potential of the electromagnetic field, formed by the moving electron, equals

$$\phi = \begin{cases} \dfrac{2e}{\varepsilon\sqrt{(vt-z)^2-\gamma^2 r^2}} & \text{for} \quad vt-z > \gamma r, \\ 0 & \text{for} \quad vt-z < \gamma r. \end{cases} \tag{1}$$

Here e is the charge of the electron, $\gamma^2 = \dfrac{v^2}{a^2}-1$, $r = \sqrt{x^2+y^2}$, where it is assumed that at time $t = 0$ the electron existed at the point $x = y = z = 0$. The components of the vector potential equal

$$A_x = A_y = 0, \quad A_z = \varepsilon\frac{v}{c}\phi, \tag{2}$$

† In fact this velocity will vary due to radiation of energy by the electron. For more detail on this see [18].

where

$$H = \operatorname{rot} A, \quad E = -\operatorname{grad} \phi - \frac{1}{c}\frac{\partial A}{\partial t}. \tag{2'}$$

At each moment of time t the electromagnetic field, produced by the electron, differs from zero only in the lower cone with vertex at the electron (Fig. 57), and the equipotential surfaces inside the cone are hyperboloids of revolution

$$(vt-z)^2-\gamma^2 r^2 = \text{const.}$$

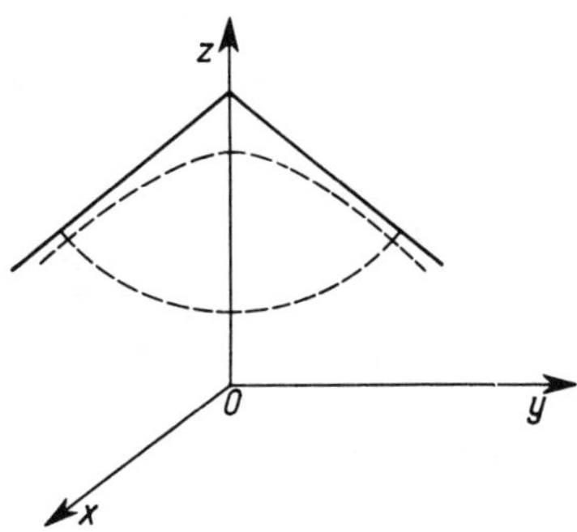

FIG. 57

Method. For the scalar and vector potentials the equations

$$\left.\begin{aligned} \Delta\phi - \frac{\varepsilon}{c^2}\frac{\partial^2\phi}{\partial t^2} &= -\frac{4\pi}{\varepsilon}\rho, \\ \Delta A - \frac{\varepsilon}{c^2}\frac{\partial^2 A}{\partial t^2} &= -\frac{4\pi}{c}j^{(e)}, \end{aligned}\right\} \tag{3}$$

hold, where†

$$\operatorname{div} A + \frac{\varepsilon}{c}\frac{\partial\phi}{\partial t} = 0. \tag{4}$$

In our case

$$j_x^{(e)} = j_y^{(e)} = 0, \quad j_z^{(e)} = v\rho = ve\delta(x)\delta(y)\delta(z-vt), \tag{5}$$

$$\rho = e\delta(x)\delta(y)\delta(z-vt). \tag{6}$$

Substitution of these values of ρ and j in equation (3) at once gives:

$$A_x = A_y = 0, \quad A_z = \varepsilon\frac{v}{c}\phi, \tag{7}$$

therefore equality (4) is transformed into the equality

$$\frac{\partial A_z}{\partial z} + \frac{\varepsilon}{c}\frac{\partial\phi}{\partial t} = 0, \tag{8}$$

† See [7], page 498.

which allows all the components of the electromagnetic field to be expressed in terms of the scalar potential ϕ.

126.

$$H_\phi = \begin{cases} -\dfrac{M_0\omega^2}{r^2a^2}\left[r\cos\omega\left(t-\dfrac{r}{a}\right)+\dfrac{a}{\omega}\sin\omega\left(t-\dfrac{r}{a}\right)\right]\sin\theta, & t>\dfrac{r}{a}, \\ 0 & t<\dfrac{r}{a}, \end{cases}$$

$$E_r = \begin{cases} -\dfrac{2M_0\omega^2}{r^3a}\left[r\sin\omega\left(t-\dfrac{r}{a}\right)-\dfrac{a}{\omega}\cos\omega\left(t-\dfrac{r}{a}\right)\right]\cos\theta, & t>\dfrac{r}{a}, \\ \dfrac{2M_0\cos\theta}{r^3} & t<\dfrac{r}{a}, \end{cases}$$

$$E_\theta = \begin{cases} -\dfrac{M_0\omega^2}{r^2a^2}\left[\left(r^2-\dfrac{a^2}{\omega^2}\right)\cos\omega\left(t-\dfrac{r}{a}\right)+\dfrac{ra}{\omega}\sin\omega\left(t-\dfrac{r}{a}\right)\right]\sin\theta, & t>\dfrac{r}{a}, \\ \dfrac{M_0\sin\theta}{r^3}, & t<\dfrac{r}{a}. \end{cases}$$

127. Assuming that the force $F(t)$ is applied at the origin of coordinates and is directed along the x-axis, we obtain for the displacements u, v, w along the x, y, z axes the expression

$$u = \frac{1}{4\pi\rho}\left[\frac{\partial^2}{\partial x^2}\left(\frac{1}{r}\right)\right]+\int_{r/a}^{r/b}\tau F(t-\tau)\,d\tau+$$
$$+\frac{x^2}{r^3}\left\{\frac{1}{a^2}F\left(t-\frac{r}{a}\right)-\frac{1}{b^2}F\left(t-\frac{r}{b}\right)+\frac{1}{b^2r}F\left(t-\frac{r}{b}\right)\right\},$$

$$v = \frac{1}{4\pi\rho}\left[\frac{\partial^2}{\partial x\partial y}\left(\frac{1}{r}\right)\right]\int_{r/a}^{r/b}\tau F(t-\tau)\,d\tau+\frac{xy}{r^3}\left\{\frac{1}{a^2}F\left(t-\frac{r}{a}\right)-\frac{1}{b^2}F\left(t-\frac{r}{b}\right)\right\},$$

$$w = \frac{1}{4\pi\rho}\left[\frac{\partial^2}{\partial x\partial y}\left(\frac{1}{r}\right)\right]\int_{r/a}^{r/b}\tau F(t-\tau)\,d\tau+\frac{xz}{r^3}\left\{\frac{1}{a^2}F\left(t-\frac{r}{a}\right)-\frac{1}{b^2}F\left(t-\frac{r}{b}\right)\right\},$$

where

$$a^2 = \frac{(2m-2)G}{(m-2)\rho} = \frac{\lambda+2\mu}{\rho}, \qquad b^2 = \frac{G}{\rho} = \frac{\mu}{\rho},$$

ρ is the mass density of the medium, a is the velocity of propagation of the longitudinal deformations, b is the velocity of propagation of transverse deformations.

Method. Elastic stresses, the resultant of which must be $F(t)$, are applied to the surface of a small sphere of radius r with centre at the origin of coordinates. Hence, for $r \to 0$ the stresses must be of order $1/r^2$ (provided $F(t) \not\equiv 0$ and is a finite quantity). Displacements, the derivatives of which are proportional to the stresses, must be of order $1/r$.

CHAPTER VII

EQUATIONS OF ELLIPTIC TYPE

$\Delta u + cu = -f$

§ 1. Problems for the Equation $\Delta_u - \varkappa^2 u = -f$

1. The concentration of the gas at a point $M(x, y, z)$, at a distance r from the source $P = P(\xi, \eta, \zeta)$, equals

$$u = \frac{Q_0}{4\pi D} \frac{e^{-\varkappa r}}{r}, \qquad r = \sqrt{(x-\xi)^2 + (y-\eta)^2 + (z-\zeta)^2},$$

where D is the diffusion coefficient, $\varkappa^2 = \beta/D.\beta$ is a constant of disintegration.

2. $u = (M, P) = \dfrac{Q_0}{4\pi D}\left(\dfrac{e^{-\varkappa r}}{r} + \dfrac{e^{-\varkappa r_1}}{r}\right),$

$$M = M(x, y, z), \qquad P = P(\xi, \eta, \zeta), \qquad r = \sqrt{(x-\xi)^2 + (y-\eta)^2 + (z-\zeta)^2},$$

where $r_1 = MP_1 = \sqrt{(x-\xi)^2 + (y-\eta)^2 + (r+\zeta)^2}, \quad P_1 = P_1(\xi, \eta, -\zeta).$

The source is situated at the point $P(\xi, \eta, \zeta)$.

Method. The condition of impermeability to gas of the wall $z = 0$

$$\left.\frac{\partial u}{\partial z}\right|_{z=0} = 0$$

shows that the solution must be even with respect to reflections in the plane.

3. The source function for the equation

$$\Delta_2 u - \varkappa^2 u = 0$$

in the plane (x, y) has the form

$$G(x, y; \xi, \eta) = \frac{1}{2\pi} K_0(\varkappa r),$$

where K_0 is a cylindrical function of imaginary argument of zero order of the second kind,

$$r = \sqrt{(x-\xi)^2 + (y-\eta)^2}.$$

A physical interpretation of the source function: A steady-state concentration, produced at the point x, y, z_0 by a source of unstable gas, uniformly distributed along an ifinnite straight line, parallel to the z-axis and passing through the point ξ, η, z_0; the magnitude of the source, referred to unit length, numerically equals D.

4. $G(x, y; \xi, \eta) = \frac{1}{2\pi}[K_4(\varkappa r)+K_0(\varkappa r_1)]$,

where

$$r = \sqrt{(x-\xi)^2+(y-\eta)^2}, \quad r_1 = \sqrt{(x-\xi)^2+(y+\eta)^2}.$$

5. If the source exists at the point (ξ, η, ζ), then

$$u = \frac{Q_0}{4\pi D}\sum_{n=-\infty}^{\infty}\left[\frac{e^{-\varkappa r_n}}{r_n}+\frac{e^{-\varkappa r_n'}}{r_n'}\right],$$

where

$$r_n = \sqrt{(x-\xi)^2+(y-\eta)^2+(z-\zeta_n)^2},$$

$$r_n' = \sqrt{(x-\xi)^2+(y-\eta)^2+(z-\zeta_n')^2},$$

$$\zeta_n = 2nl+\zeta, \quad \zeta_n' = 2nl-\zeta.$$

Method. The images in the planes $z = 0$ and $z = l$ are positive and situated at the points

$$(\xi, \eta, \zeta_n = 2nl+\zeta) \quad \text{and} \quad (\xi, \eta, \zeta_n' = 2nl-\zeta).$$

The convergence of the series is obvious because of the presence of the exponential factors $e^{-\varkappa r_n}$ and $e^{-\varkappa r_n'}$ under the sign of the sum.

6. If the source is located at the point (ξ, η), then

$$u = \frac{Q}{2\pi D}\sum_{n=-\infty}^{\infty}[K_0(\varkappa r_n)+K_0(\varkappa r_n')],$$

where

$$r_n = \sqrt{(x-\xi)^2+(y-\eta_n)^2},$$

$$r_n' = \sqrt{(x-\xi)^2+(y-\eta_n')^2},$$

$$\eta_n = 2nl+\eta, \quad \eta_n' = 2nl-\eta.$$

Method. See problem 5. The convergence of the series is obvious from the asymptotic formula

$$K_0(x) = \sqrt{\frac{1}{2\pi x}}e^{-x}+\dots$$

7.
$$u = \frac{Q_0}{2D}\sum_{n=1}^{\infty}\frac{\psi_n(M)\psi_n(P)}{\|\psi_n\|^2\sqrt{\lambda_n+\varkappa^2}}e^{-\sqrt{\lambda_n+\varkappa^2}\,|z-\zeta|},$$

where $(M(x, y);\ z)$ is the point of observation, $(P(\xi, \eta);\ \zeta)$ is the point at which the source is placed, λ_n and ψ_n are the eigenvalues and eigenfunctions of the plane problem

$$\Delta_2\psi_n+\lambda_n\psi_n = 0 \text{ in } S,$$

$$\frac{\partial\psi_n}{\partial\nu} = 0 \text{ on } C,$$

S is the cross-section of the tube, C is its boundary, and ν is the normal to C,

$$||\psi_n||^2 = \int_S \psi_n^2\, ds.$$

Method. The solution of the boundary-value problem

$$\left.\begin{aligned} D\Delta u-\beta u &= -f \\ \text{or}\quad \Delta u-\varkappa^2 u &= -f \end{aligned}\right\} \quad \text{inside} \quad \Sigma,$$

$$\frac{\partial u}{\partial\nu} = 0 \quad \text{on} \quad \Sigma,$$

where Σ is the surface of the tube, must be sought in the form

$$u = \sum_{n=1}^{\infty} f_n(z)\,\psi_n(z).$$

Knowing the solution of this problem, one readily passes to the limiting case of a point source.

8. If the source exists at the point (ρ, θ', ϕ'), then

$$u(r, \theta, \phi) = \frac{Q_0}{D}\, G(r, \theta, \phi;\ \rho, \theta', \phi'),$$

where $G(r, \theta, \phi;\ \rho, \theta', \phi')$ is the source function, defined by the formulae

$$G(r, \theta, \phi, \rho, \theta', \phi') = \sum_{n=0}^{\infty}\sum_{k=-n}^{n} \frac{Y_n^{(k)}(\theta, \phi)\, Y_n^{(k)}(\theta', \phi')}{||\, Y_n^{(k)}\, ||^2}\, G_n(r, \rho),$$

where

$$Y_n^{(k)} = P_n^{(k)}(\cos\theta)\begin{cases} \cos k\phi & \text{for} \quad k \geqslant 0, \\ \sin k\phi & \text{for} \quad k < 0 \end{cases}$$

are spherical functions, $P_n^{(k)}$ are the associated Legendre functions,

$$||\, Y_n^{(k)}\, ||^2 = \frac{2}{2n+1}\,\pi\varepsilon_k\,\frac{(n+k)!}{(n-k)!}, \quad \varepsilon_k = \begin{cases} 2 & \text{for} \quad k = 0, \\ 1 & \text{for} \quad k \neq 0, \end{cases}$$

$$G_n(r, \rho) = \begin{cases} \varkappa[\xi_n'(\varkappa a)\,\eta_n(\varkappa\rho)-\xi_n(\varkappa\rho)\,\eta_n'(\varkappa a)]\,\dfrac{\xi_n(\varkappa r)}{\xi_n'(\varkappa a)} & \text{for} \quad r < \rho, \\[2ex] \varkappa[\xi_n'(\varkappa a)\,\eta_n(\varkappa r)-\xi_n(\varkappa r)\,\eta_n'(\varkappa a)]\,\dfrac{\xi_n(\varkappa\rho)}{\xi_n'(\varkappa a)} & \text{for} \quad r > \rho; \end{cases}$$

$$\xi_n(x) = x^{-\frac{1}{2}} I_{n+\frac{1}{2}}(x), \qquad \eta_n(x) = x^{-\frac{1}{2}} K_{n+\frac{1}{2}}(x).$$

Method. The solution of the problem $\Delta u - \varkappa^2 u = -f$, $u_r|_{r-a} = 0$ is represented in the form

$$u(r, \theta, \phi) = \int_0^a \int_0^{2\pi} \int_0^{\pi} G(r, \theta, \phi, \rho, \theta', \phi') f(\rho, \theta', \phi')\, \rho^2 \mathrm{d}\rho \sin\theta' \mathrm{d}\theta' \mathrm{d}\phi'.$$

Assuming

$$u = \sum_{n=0}^{\infty} \sum_{k=-n}^{n} u_{nk}(r)\, Y_n^{(k)}(\theta, \phi), \qquad f = \sum_{n=0}^{\infty} \sum_{k=-n}^{n} f_{nk}(r)\, Y_n^{(k)}(\theta, \phi),$$

we obtain

$$Lu_{nk} = (r^2 u'_{nk})' - (\varkappa^2 r^2 + n(n+1))\, u_{nk} = -f_{nk}(r), \qquad u'_{nk}(a) = 0,$$

$$u_{nk}(r) = \int_0^a G_n(r, \rho) f_{nk}(\rho)\, \rho^2 \mathrm{d}\rho, \quad f_{nk} = \int_0^{2\pi} \int_0^{\pi} f(\rho, \theta', \phi')\, Y_n^{(k)}(\theta', \phi') \sin\theta' \mathrm{d}\theta' \mathrm{d}\phi',$$

$$LG_n = 0 \quad (r \neq \rho), \qquad G_n(\rho-0, \rho) = G_n(\rho+0, \rho),$$

$$G'_n(\rho+0, \rho) - G'_n(\rho-0, \rho) = -\frac{1}{\rho^2}, \qquad G'_n(a, \rho) = 0.$$

9. If the z-axis of a rectangular system of coordinates is directed along v_0, then

$$u(x, y, z) = \frac{Q_0}{D} \frac{\mathrm{e}^{-\varkappa r - \frac{v_0}{2D}(z-\zeta)}}{4\pi r},$$

where

$$\varkappa = \frac{\beta}{D}, \qquad r = \sqrt{(x-\xi)^2 + (y-\eta)^2 + (z-\zeta)^2}.$$

10. If $r = a$ is the radius of the cylinder, then

$$u(r) = u_0 \frac{I_0(\varkappa r)}{I_0(\varkappa a)}, \qquad r = \sqrt{x^2 + y^2}.$$

Method. It is required to find the bounded solution of the equation

$$\Delta_2 u - \varkappa^2 u = 0, \qquad r < a$$

for the condition

$$u|_{r=a} = u_0.$$

11. $u = u(r) = u_0 \dfrac{K_0(\varkappa r)}{K_0(\varkappa a)}$.

12. (a) $u = u_0 \dfrac{\sqrt{a}}{\sqrt{r}} \dfrac{I_{\frac{1}{2}}(\varkappa r)}{I_{\frac{1}{2}}(\varkappa a)} = u_0 \dfrac{a}{r} \dfrac{\sinh \varkappa r}{\sinh \varkappa a}$;

(b) $u = u_0 \dfrac{\sqrt{a}}{\sqrt{r}} \dfrac{I_{\frac{3}{2}}(\varkappa r)}{I_{\frac{3}{2}}(\varkappa a)} \cos\theta = u_0 \left(\dfrac{a}{r}\right)^2 \dfrac{\varkappa r \cosh \varkappa r - \sinh \varkappa r}{\varkappa a \cosh \varkappa a - \sinh \varkappa a} \cos\theta$.

13. (a) $u = u_0 \dfrac{\sqrt{a}}{\sqrt{r}} \dfrac{K_{\frac{1}{2}}(\varkappa r)}{K_{\frac{1}{2}}(\varkappa a)} = u_0 \dfrac{a e^{-\varkappa r}}{r e^{-\varkappa a}}$;

(b) $u = u_0 \dfrac{\sqrt{a}}{\sqrt{r}} \dfrac{K_{\frac{3}{2}}(\varkappa r)}{K_{\frac{3}{2}}(\varkappa a)} \cos\theta = u_0 \left(\dfrac{a}{r}\right)^2 \dfrac{\varkappa r+1 \cosh \varkappa r - \sinh \varkappa r}{\varkappa a+1 \cosh \varkappa a - \sinh \varkappa a} \cos\theta$.

14. (a) If the surface of the earth coincides with the plane $z = 0$, then the concentration of radioactive gas in the earth is given by the relation

$$u = \begin{cases} \dfrac{f_0}{\beta} e^{-\varkappa h} \sinh \varkappa z & \text{for} \quad 0 < z < h, \\[2ex] \dfrac{f_0}{\beta} (1 - e^{-\varkappa z} \cosh \varkappa h) & \text{for} \quad h < z < \infty, \end{cases}$$

where $\varkappa = \sqrt{\beta/D}$, β is the decay constant, D is the diffusion coefficient, f_0 is the density of the sources.

(b) The flow of gas through the surface of the earth is

$$q = D \left.\frac{\partial u}{\partial z}\right|_{z=0} = \frac{f_0}{\varkappa} e^{-\varkappa h}.$$

15. (a) If $z = 0$ is the surface of the earth, and the source exists at the point $(0, 0, h)$, then the concentration is

$$u = \frac{Q_0}{4\pi D} \left(\frac{e^{-\varkappa r_1}}{r_1} - \frac{e^{-\varkappa r_2}}{r_2} \right),$$

where

$$r_1 = \sqrt{\rho^2 + (z-h)^2}, \qquad r_2 = \sqrt{\rho^2 + (z+h)^2}, \qquad \rho = \sqrt{x^2+y^2}.$$

(b) The flow through the surface of the earth $z = 0$ equals

$$q(\rho) = D \left.\frac{\partial u}{\partial z}\right|_{z=0} = \frac{Q_0 h}{2\pi r_0^2} \left(\varkappa + \frac{1}{r_0}\right) e^{-\varkappa r_0},$$

where

$$r_0 = \sqrt{\rho^2 + h^2}.$$

16. It is required to determine Q_0 and h. In order to do this, using given observations, i.e. the value of $q(\rho)$, we find the total flow through the surface of the earth

$$Q = \int_0^\infty \int_0^{2\pi} q(\rho)\,\rho\,\mathrm{d}\rho\,\mathrm{d}\phi = Q_0 \mathrm{e}^{-\varkappa h}, \tag{1}$$

then the integral

$$I = \int_0^\infty \int_0^{2\pi} q\rho^2\,\mathrm{d}\rho\,\mathrm{d}\phi = -Q_0 h \left[\int_0^\infty \frac{\partial}{\partial \rho}\left(\frac{\mathrm{e}^{-\varkappa r}}{r}\right)\rho\,\mathrm{d}\rho\right]_{z=0}$$

$$= Q_0 h \int_0^\infty \frac{\mathrm{e}^{-\varkappa\sqrt{\rho^2+h^2}}}{\sqrt{\rho^2+h^2}}\,\mathrm{d}\rho = Q_0 h \int_0^\infty \mathrm{e}^{-\varkappa h \cosh \xi}\,\mathrm{d}\xi = Q_0 h K_0(\varkappa h),$$

or

$$I = Q_0 h K_0(\varkappa h). \tag{2}$$

From formulae (1) and (2) we have:

$$h K_0(\varkappa h)\,\mathrm{e}^{\varkappa h} = \frac{I}{Q}.$$

Hence, since I and Q are known, we can determine the value of h and then from formula (1) the magnitude of the source Q_0.

The position of the source in a horizontal plane is determined, obviously from the maximum observed flow $q(\rho)$.

§ 2. Some Problems on Natural Vibrations

1. Natural Vibrations of Strings and Rods

17. Let $v = v(x)$ denote the amplitude of the deflection of a point of the string with coordinate x. It is required to find the solution of the homogeneous equation

$$v'' + \lambda v = 0$$

for appropriate homogeneous boundary conditions.

(a) The boundary conditions

$$v(0) = 0, \quad v(l) = 0,$$

eigenvalues

$$\lambda_n = \left(\frac{\pi n}{l}\right)^2 \quad (n = 1, 2, \ldots),$$

eigenfunctions

$$v_n(x) = \sin \frac{\pi n}{l} x,$$

the square of the norm of the eigenfunctions

$$||v_n||^2 = \frac{l}{2}.$$

(b) The boundary conditions

$$v'(0) = 0, \quad v'(l) = 0,$$

eigenvalues

$$\lambda_n = \left(\frac{\pi n}{l}\right)^2 \quad (n = 0, 1, 2, \ldots),$$

eigenfunctions

$$v_n(x) = \cos \frac{\pi n}{l} x \quad (n = 0, 1, 2, \ldots).$$

the square of the norm

$$||v_n||^2 = \frac{l}{2} \varepsilon_n, \quad \varepsilon_n = \begin{cases} 2, & n = 0, \\ 1, & n \neq 0. \end{cases}$$

(c) The boundary conditions

$$v(0) = 0, \quad v'(l) = 0,$$

eigenvalues

$$\lambda_n = \left[\frac{\pi(2n+1)}{2l}\right]^2 \quad (n = 0, 1, 2, \ldots),$$

eigenfunctions

$$v_n(x) = \sin \frac{\pi(2n+1)}{2l} x \quad (n = 0, 1, 2, \ldots),$$

the square of the norm

$$||v_n||^2 = \frac{l}{2}.$$

(d) The boundary conditions

$$v'(0) - h_1 v(0) = 0, \quad v'(l) + h_2 v(l) = 0, \quad \text{where } h_1 > 0,\ h_2 > 0,$$

the eigenvalues λ_n are determined from the transcendental equation

$$\tan \sqrt{\lambda}\, l = \frac{(h_1 + h_2)\sqrt{\lambda}}{\lambda - h_1 h_2} \tag{1}$$

or

$$\tan \mu l = \frac{(h_1 + h_2)\mu}{\mu^2 - h_1 h_2}, \quad \lambda_n = \mu_n^2,$$

the eigenfunctions

$$v_n(x) = \frac{1}{\sqrt{\lambda_n + h_1^2}} [\sqrt{\lambda_n} \cos \sqrt{\lambda_n}\, x + h_1 \sin \sqrt{\lambda_n}\, x],$$

the square of the norm

$$||v_n||^2 = \frac{l}{2} + \frac{(h_1+h_2)(\mu_n^2 + hh_2)}{(\mu_n^2+h_1^2)(\mu_n^2+h_2^2)} \qquad (n = 1, 2, \ldots).$$

In the particular case for $h_1 = h_2$ equation (1) takes the form

$$\tan \mu = \frac{2h \dfrac{\mu}{l}}{\dfrac{\mu^2}{l^2} - h^2}, \qquad \lambda = \frac{\mu^2}{l^2}.$$

(e) The boundary conditions

$$v(0) = 0, \quad v'(l) + hv(l) = 0,$$

the eigenvalues λ_n are determined from the equation

$$\tan \mu = -\frac{\mu}{hl}, \quad \lambda_n = \frac{\mu_n^2}{l^2} \quad (n = 1, 2, \ldots),$$

the eigenfunctions

$$v_n(x) = \sin \sqrt{\lambda_n}\, x,$$

the square of the norm

$$||v_n||^2 = \frac{l}{2} + \frac{hl^2}{\mu_n^2 + h^2 l^2}.$$

(f) The boundary conditions

$$v'(0) = 0, \quad v'(l) + hv(l) = 0,$$

the eigenvalues λ_n are determined from the equation

$$\tan \mu = \frac{hl}{\mu}, \quad \lambda_n = \frac{\mu_n^2}{l^2},$$

the eigenfunctions

$$v_n(x) = \cos \sqrt{\lambda_n}\, x,$$

the square of the norm

$$||v_n||^2 = \frac{l}{2} + \frac{hl^2}{\mu_n^2 + h^2 l^2}.$$

18. The equation of the longitudinal natural vibrations of an inhomogeneous rod has the form

$$\frac{d}{dx}\left[E(x)\frac{dv}{dx}\right] + \lambda \rho v = 0, \quad E = \begin{cases} E_1, & (x < x_0) \\ E_2, & (x > x_0) \end{cases}, \quad \rho = \begin{cases} \rho_1, & (x < x_0) \\ \rho_2, & (x > x_0) \end{cases}.$$

(a) The boundary conditions

$$v(0) = 0, \quad v(l) = 0,$$

the eigenvalues are determined from the equation

$$a_1\rho_1 \cot \frac{\sqrt{\lambda}}{a_1} x_0 + a_2\rho_2 \cot \frac{\sqrt{\lambda}}{a_2}(l-x_0) = 0,$$

where

$$a_1 = \sqrt{\frac{E_1}{\rho_1}}, \quad a_2 = \sqrt{\frac{E_2}{\rho_2}},$$

the eigenfunctions

$$v_n(x) = \begin{cases} \dfrac{\sin \dfrac{\sqrt{\lambda_n}}{a_1} x}{\sin \dfrac{\sqrt{\lambda_n}}{a_1} x_0} & \text{for} \quad x < x_0, \\[2ex] \dfrac{\sin \dfrac{\sqrt{\lambda_n}}{a_2}(l-x)}{\sin \dfrac{\sqrt{\lambda_n}}{a_2}(l-x_0)} & \text{for} \quad x_0 < x < l, \end{cases}$$

the square of the norm

$$\|v_n\|^2 = \frac{\rho_1 x_0}{2\sin^2 \dfrac{\sqrt{\lambda}}{a_1} x_0} + \frac{\rho_2(l-x_0)}{2\sin^2 \dfrac{\sqrt{\lambda_n}}{a_2}(l-x_0)} \qquad (n = 1, 2, \dots).$$

(b) The boundary conditions

$$v'(0) = v'(l) = 0.$$

The eigenvalues are determined from the equation

$$a_1\rho_1 \tan \frac{\sqrt{\lambda}}{a_1} x_0 + a_2\rho_2 \tan \frac{\sqrt{\lambda}}{a_2}(l-x_0) = 0.$$

The eigenfunctions

$$v_n(x) = \begin{cases} \dfrac{\cos \dfrac{\sqrt{\lambda_n}}{a_1} x}{\cos \dfrac{\sqrt{\lambda_n}}{a_1} x_0} & \text{for} \quad 0 < x < x_0, \\[2ex] \dfrac{\cos \dfrac{\sqrt{\lambda_n}}{a_2}(l-x)}{\cos \dfrac{\sqrt{\lambda_n}}{a_2}(l-x_0)} & \text{for} \quad x_0 < x < l, \end{cases}$$

the square of the norm

$$||v_n||^2 = \frac{\rho_1 x_0}{2\cos^2 \frac{\sqrt{\lambda_n}}{a_1} x_0} + \frac{\rho_2(l-x_0)}{2\cos^2 \frac{\sqrt{\lambda_n}}{a_2}(l-x_0)} \qquad (n = 1, 2, \ldots).$$

(c) The boundary conditions

$$v'(0)-h_1 v(0) = 0, \qquad v'(l)+h_2 v(l) = 0.$$

The eigenvalues are determined from the equation

$$a_1\rho_1 \frac{h_1 - \frac{\sqrt{\lambda}}{a_1}\tan\frac{\sqrt{\lambda}}{a_1}x_0}{h_1\tan\frac{\sqrt{\lambda}}{a_1}x_0 + \frac{\sqrt{\lambda}}{a_1}} + a_2\rho_2 \frac{h_2 + \frac{\sqrt{\lambda}}{a_2}\tan\frac{\sqrt{\lambda}}{a_2}(l-x_0)}{-h_2\tan\frac{\sqrt{\lambda}}{a_2}(l-x_0) + \frac{\sqrt{\lambda}}{a_2}} = 0,$$

where

$$a_1 = \sqrt{\frac{E_1}{\rho_1}}, \qquad a_2 = \sqrt{\frac{E_2}{\rho_2}}.$$

The eigenfunctions

$$v_n(x) = \begin{cases} \dfrac{X_n(x)}{X_n(x_0)} & \text{for} \quad 0 < x < x_0, \\[2ex] \dfrac{Y_n(x)}{Y_n(x_0)} & \text{for} \quad x_0 < x < l, \end{cases} \qquad (n = 1, 2, \ldots)$$

$$X_n(x) = \sqrt{\lambda_n}\cos\frac{\sqrt{\lambda_n}}{a_1}x + a_1 h_1 \sin\frac{\sqrt{\lambda_n}}{a_1}x,$$

$$Y_n(x) = \sqrt{\lambda_n}\cos\frac{\sqrt{\lambda_n}}{a_2}(l-x) - a_2 h_2 \sin\frac{\sqrt{\lambda_n}}{a_2}(l-x).$$

The square of the norm

$$||v_n||^2 = \frac{\rho_1}{X_n^2(x_0)}\int_0^{x_0} X_n^2(x)dx + \frac{\rho_2}{Y_n^2(x_0)}\int_{x_0}^{l} Y_n^2(x)dx.$$

Method. It is required to find the non-trivial solutions

$$v(x) = \begin{cases} \bar{v}(x) & \text{for} \quad 0 < x < x_0, \\ \bar{\bar{v}}(x) & \text{for} \quad x_0 < x < l \end{cases}$$

of the homogeneous equations

$$\bar{v}'' + \frac{\lambda}{a_1^2}\bar{v} = 0, \qquad \bar{\bar{v}}'' + \frac{\lambda}{a_2^2}\bar{\bar{v}} = 0,$$

satisfying the boundary conditions (a) or (b) or (c) and the matching conditions at the point of discontinuity of the coefficients of the equation

$$\bar{v} = v, \quad E_1\bar{v}' = E_2\bar{\bar{v}}' \quad \text{for} \quad x = x_0.$$

The solution is conveniently sought in the form

$$v(x) = \begin{cases} \dfrac{X(x)}{X(x_0)} & \text{for} \quad 0 < x < x_0, \\[2ex] \dfrac{Y(x)}{Y(x_0)} & \text{for} \quad x_0 < x < l \end{cases}$$

where $X(x)$ satisfies the equation $X'' + \lambda X/a_1^2 = 0$ and the boundary condition for $x = 0$, and $Y(x)$ satisfies the equation $Y'' + \lambda Y/a_2^2 = 0$ and the boundary condition for $x = l$.

The eigenfunctions are orthogonal with weight $\rho(x)$:

$$\int_0^l v_n(x) v_m(x) \rho(x) \, dx = 0, \qquad m \neq n,$$

$$||v_n||^2 = \int_0^l v_n^2(x) \rho(x) dx = \frac{\rho_1}{X_n^2(x_0)} \int_0^{x_0} X_n^2(x) dx + \frac{\rho_2}{Y_n^2(x_0)} \int_{x_0}^l Y_n^2 \, dx.$$

19. The load is situated at the end $x = l$. The boundary condition at this end has the form

$$v'(l) = \frac{M}{\rho} \lambda v(l).$$

The eigenfunctions $\{v_n(x)\}$ satisfy the condition of orthogonality with the load

$$\int_0^l v_m(x) v_n(x) \rho(x) dx + M v_m(l) v_n(l) = 0 \quad \text{for} \quad m \neq n.$$

The square of the norm of the eigenfunction $v_n(x)$ is given by the formula

$$||v_n||^2 = \int_0^l v_n^2(x) \rho(x) dx + M v_n^2(l) \quad (n = 1, 2, 3, \dots).$$

(a) The end $x = 0$ is rigidly fixed, $v(0) = 0$.

The eigenvalues are determined from the equation $\cot \sqrt{\lambda_n} l = \dfrac{M}{\rho} \sqrt{\lambda_n}$, the eigenfunctions are

$$v_n(x) = \frac{\sin \sqrt{\lambda_n} x}{\sin \sqrt{\lambda_n} l} \qquad (n = 1, 2, 3, \dots),$$

the square of the norm

$$||v_n||^2 = \frac{l\rho}{2} + \frac{M^2}{2\rho}\lambda_n l + \frac{M}{2}.$$

Formulae for correction to the eigenvalues:

(1) if the load M is small, then

$$\lambda_n = \lambda_n^{(1)}\left(1 - \frac{2M}{\rho l} + \dots\right),$$

where $\lambda_n^{(1)} = \left[\frac{\pi(2n+1)}{2l}\right]^2$ are the eigenvalues of the unloaded rod with a free end;

(2) if the load M is large, then

$$\lambda_n = \lambda_n^{(2)} + \frac{2\rho}{Ml} + \dots,$$

where $\lambda_n^{(2)} = \left(\frac{\pi n}{l}\right)^2$ are the eigenvalues of the rod with a rigidly fixed end $x = l$.

(b) The end $x = 0$ is free, $v'(0) = 0$.

The eigenvalues are detemirned from the equation $\tan\sqrt{\lambda_n}l = -M\sqrt{\lambda_n}/\rho$. The eigenfunctions

$$v_n(x) = \frac{\cos\sqrt{\lambda_n}x}{\cos\sqrt{\lambda_n}l} \qquad (n = 1, 2, 3, \dots),$$

the square of the norm

$$||v_n||^2 = \frac{\rho l}{2}\left(1 + \frac{M^2}{\rho^2}\lambda_n\right) + \frac{M}{2}.$$

(c) The end $x = 0$ is fixed elastically, $v'(0) - hv(0) = 0$.

The eigenvalues λ_n are determined from the equation

$$\tan\sqrt{\lambda_n}l = \frac{h - \dfrac{M^2}{\rho}\lambda_n}{\left(1 + \dfrac{M}{\rho}h\right)\sqrt{\lambda_n}} \qquad (n = 1, 2, 3, \dots).$$

The eigenfunctions

$$v_n(x) = \frac{X_n(x)}{X_n(l)},$$

$$X_n(x) = \sqrt{\lambda_n}\cos\sqrt{\lambda_n}x + h\sin\sqrt{\lambda_n}x.$$

The square of the norm

$$||v_n||^2 = \frac{1}{X_n^2(l)}||X_n||^2.$$

Method. The dynamic condition of the load at the end $x = l$ has the form $Mu_{tt} = -Eu_x(l, t)$. Assuming $u(x, t) = v(x)T(t)$, we obtain for $v(x)$ after separation of the variables the equation

$$v''+\lambda v = 0, \quad v'(l) = \frac{M}{\rho}\lambda v(l).$$

The orthogonality condition follows from Green's theorem

$$\int_0^l [v_n L(v_m) - v_m L(v_n)]\, dx = [v_n v_m' - v_m v_n']_0^l,$$

where

$$L(v) = (Ev')'.$$

In order to calculate the norm one must use the characteritsic equation.

20. A point mass M exists at the point $x = x_0$.

(a) Both ends of the string are rigidly fixed,

$$v(0) = 0, \quad v(l) = 0.$$

The eigenvalues λ_n are determined from the equation

$$\cot\frac{\sqrt{\lambda_n}}{a}x_0 + \cot\frac{\sqrt{\lambda_n}}{a}(l-x_0) = \frac{M}{a\rho}\sqrt{\lambda_n}.$$

The eigenfunctions

$$v_n(x) = \begin{cases} \dfrac{\sin\dfrac{\sqrt{\lambda_n}}{a}x}{\sin\dfrac{\sqrt{\lambda_n}}{a}x_0} & \text{for} \quad 0 < x < x_0, \\[2ex] \dfrac{\sin\dfrac{\sqrt{\lambda_n}}{a}(l-x)}{\sin\dfrac{\sqrt{\lambda_n}}{a}(l-x_0)} & \text{for} \quad x_0 < x < l. \end{cases}$$

The square of the norm

$$\|v_n\|^2 = \frac{\rho x_0}{2\sin^2\dfrac{\sqrt{\lambda_n}}{a}x_0} + \frac{\rho(l-x_0)}{2\sin^2\dfrac{\sqrt{\lambda_n}}{a}(l-x_0)} + \frac{M}{2} \qquad (n = 1, 2, \ldots).$$

(b) Both ends of the string are free, $v'(0) = 0$, $v'(l) = 0$.

The eigenvalues λ_n are determined from the equation

$$\tan\frac{\sqrt{\lambda_n}}{a}x_0 + \tan\frac{\sqrt{\lambda_n}}{a}(l-x_0) = -\frac{M}{a\rho}\sqrt{\lambda_n}.$$

The eigenfunctions

$$v_n(x) = \begin{cases} \dfrac{\cos \dfrac{\sqrt{\lambda_n}}{a} x}{\cos \dfrac{\sqrt{\lambda_n}}{a} x_0} & \text{for} \quad 0 < x > x_0, \\[2ex] \dfrac{\cos \dfrac{\sqrt{\lambda_n}}{a} (l-x)}{\cos \dfrac{\sqrt{\lambda_n}}{a} (l-x_0)} & \text{for} \quad x_0 < x > l. \end{cases}$$

The square of the norm

$$||v_n||^2 = \frac{\rho x_0}{2\cos^2 \dfrac{\sqrt{\lambda_n}}{a} x_0} + \frac{\rho(l-x_0)}{2\cos^2 \dfrac{\sqrt{\lambda_n}}{a}(l-x_0)} + \frac{M}{2} \quad (n = 1, 2, \ldots).$$

(c) The ends of the string are fixed elastically,

$$v'(0) - h_1 v(0) = 0, \quad v'(l) + h_2 v(l) = 0.$$

The eigenvalues are determined from the equation

$$\frac{\sqrt{\lambda} \tan \sqrt{\dfrac{\lambda}{a}}(l-x_0) + ah_2}{\sqrt{\lambda} - ah_2 \tan \dfrac{\sqrt{\lambda}}{a}(l-x_0)} + \frac{-\sqrt{\lambda} \tan \dfrac{\sqrt{\lambda}}{a} x_0 + ah_1}{\sqrt{\lambda} + ah_1 \tan \dfrac{\sqrt{\lambda}}{a} x_0} = \frac{\sqrt{\lambda}}{k} M.$$

The eigenfunctions

$$v_n(x) = \begin{cases} \dfrac{X_n(x)}{X_n(x_0)} & \text{for} \quad 0 < x < x_0, \\[2ex] \dfrac{Y_n(x)}{Y_n(x_0)} & \text{for} \quad x_0 < x < l, \end{cases}$$

$$X_n(x) = \sqrt{\lambda_n} \cos \frac{\sqrt{\lambda_n}}{a} x + ah_1 \sin \frac{\sqrt{\lambda_n}}{a} x,$$

$$Y_n(x) = \sqrt{\lambda_n} \cos \frac{\sqrt{\lambda_n}}{a}(l-x) - ah_2 \sin \frac{\sqrt{\lambda_n}}{a}(l-x).$$

The square of the norm

$$||v_n||^2 = \int_0^l v_n^2(x) \rho \, dx + M v_n^2(x_0) \quad (n = 1, 2, \ldots).$$

21. The equation of the natural transverse vibrations of a homogeneous rod has the form

$$v^{(\mathrm{IV})} - \frac{\lambda}{a^2} v = 0,$$

where $a^2 = EJ/\rho S$, E is the modulus of elasticity, J is the moment of inertia of a cross-section about its horizontal axis, ρ is the density of the rod, S is its cross-sectional area.

(a) Both ends are rigidly fixed,

$$v = 0, \quad v' = 0 \quad \text{for} \quad x = 0, l,$$

$$\lambda_n = a^2 \frac{\mu_n^4}{l^4} \qquad (n = 1, 2, \ldots),$$

where μ_n is a root of the equation

$$\cosh \mu \cos \mu - \sin \mu \sinh \mu = 1.$$

The eigenfunction

$$v_n(x) = A_n \left\{ \left(\cosh \mu_n \frac{x}{l} - \cos \mu_n \frac{x}{l} \right) (\sinh \mu_n - \sin \mu_n) - \right.$$

$$\left. - (\cosh \mu_n - \cos \mu_n) \left(\sinh \mu_n \frac{x}{l} - \sin \mu_n \frac{x}{l} \right) \right\},$$

where A_n is an arbitrary constant.

(b) Both ends are free,

$$v'' = 0, \quad v''' = 0 \quad \text{for} \quad x = 0, \quad x = l,$$

$$\lambda_n = a^2 \frac{\mu_n}{l^4} \qquad (n = 1, 2, \ldots),$$

where μ_n is a root of the equation

$$\cosh \mu \cos \mu = 1,$$

$$v_n(x) = A_n \left\{ \left(\cosh \mu_n \frac{x}{l} + \cos \mu_n \frac{x}{l} \right) (\sinh \mu_n - \sin \mu_n) - \right.$$

$$\left. - \left(\sinh \mu_n \frac{x}{l} + \sin \mu_n \frac{x}{l} \right) (\cosh \mu_n - \cos \mu_n) \right\}.$$

(c) One end ($x = 0$) is fixed, the second end ($x = l$) free,

$$v = 0, \quad v' = 0' \quad \text{for} \quad x = 0; \quad v'' = 0; \quad v''' = 0 \quad \text{for} \quad x = l,$$

$$\lambda_n = a^2 \frac{\mu_n^4}{l^4}, \qquad (n = 1, 2, \ldots),$$

where μ_n is a root of the equation

$$\cosh \mu \cos \mu = -1,$$

$$v_n(x) = A_n \left\{ \left(\cosh \frac{\mu_n}{l} x - \cos \frac{\mu_n}{l} x \right) (\sinh \mu_n - \sin \mu_n) - \right.$$

$$\left. - (\cosh \mu_n - \cos \mu_n) \left(\sinh \frac{\mu_n}{l} x - \sin \frac{\mu_n}{l} x \right) \right\}.$$

2. Natural Vibrations of Volumes

22. Let $x = 0$, $x = a$, $y = 0$, $y = b$ be the sides of a rectangle.

(a) If the boundary of the membrane is rigidly fixed ($v = 0$ for $x = 0, a$; $y = 0, b$), then the eigenvalues

$$\lambda_{n,m} = \pi^2\left(\frac{m^2}{a^2}+\frac{n^2}{b^2}\right) \quad (m, n = 1, 2, \ldots),$$

the eigenfunctions

$$v_{m,n}(x, y) = \sin\frac{\pi m}{a}x \sin\frac{\pi n}{b}y$$

$$\|v_{m,n}\|^2 = \frac{ab}{4}.$$

(b) The boundary of the membrane is free ($v_x = 0$ for $x = 0, a$; $v_y = 0$ for $y = 0, b$),

$$\lambda_{m,n} = \pi^2\left(\frac{m^2}{a^2}+\frac{n^2}{b^2}\right) \quad (m, n, = 0, 1, 2, \ldots),$$

$$v_{m,n}(x, y) = \cos\frac{\pi m}{a}x \cos\frac{\pi n}{b}y,$$

$$\|v_{m,n}\|^2 = \frac{ab}{4}\varepsilon_m\varepsilon_n, \quad \varepsilon_0 = 2, \quad \varepsilon_k = 1, \quad k \neq 0.$$

(c) The two opposite sides $x = 0$, $x = a$ are rigidly fixed ($v = 0$ for $x = 0$, a) and the two other sides $y = 0$ and $y = b$ are free ($v_y = 0$ for $y = 0, b$),

$$\lambda_{m,n} = \pi^2\left(\frac{m^2}{a^2}+\frac{n^2}{b^2}\right) \quad (m = 1, 2, \ldots; n = 0, 1, 2, \ldots),$$

$$v_{m,n}(x, y) = \sin\frac{\pi m}{a}x \cos\frac{\pi n}{b}y,$$

$$\|v_{m,n}\|^2 = \frac{1}{4}ab\varepsilon_n.$$

(d) The two adjacent sides $x = 0$ and $y = 0$ are rigidly fixed, and the two other sides are free,

$$\lambda_{m,n} = \frac{\pi^2}{4}\left[\frac{(2m+1)^2}{a^2}+\frac{(2n+1)^2}{b^2}\right] \quad (m, n = 0, 1, 2, \ldots),$$

$$v_{m,n}(x, y) = \sin\frac{\pi(2m+1)}{2a}x \sin\frac{\pi(2n+1)}{2b}y,$$

$$\|v_{m,n}\|^2 = \frac{ab}{4}.$$

(e) All the sides of the rectangular membrane are fixed elastically,

$$v_x(0, y)-h_1v(0, y) = 0, \quad v_x(a, y)+h_2v(a, y) = 0,$$
$$v_y(x, 0)-h_3v(x, 0) = 0, \quad v_y(x, b)+h_4v(x, b) = 0.$$

The eigenvalues $\lambda_{m,n}$ are determined from the equation

$$\lambda_{m,n} = [\mu_m^{(1)}]^2+[\mu_n^{(2)}]^2 \qquad (n = 1, 2, \ldots),$$

where $\mu_m^{(1)}$ and $\mu_m^{(2)}$ are roots of the equations

$$\tan(\mu^{(1)}a) = \frac{(h_1+h_2)\mu^{(1)}}{(\mu^{(1)})^2-h_1h_2}, \quad \tan(\mu^{(2)}b) = \frac{(h_3+h_4)\mu^{(2)}}{(\mu^{(2)})^2+h_3h_4},$$

$$v_{m,n}(x, y) = [\mu_m^{(1)} \cos \mu_m^{(1)} x+h_1 \sin \mu_m^{(1)}x]\,[\mu_n^{(2)} \cos \mu_n^{(2)}y+h_3 \sin \mu_n^{(2)} y]\times$$
$$\times \frac{1}{\sqrt{(\mu_m^{(1)})^2+h_1^2}\,\sqrt{(\mu_n^{(2)})^2+h_4^2}},$$

$$||v_{m,n}||^2 = \left\{\frac{a}{2}+\frac{(h_1+h_2)\,[(\mu_m^{(1)})^2+h_1h_2]}{[(\mu_m^{(1)})^2+h_1^2]\,[(\mu_m^{(1)})^2+h_2^2]}\right\}\left\{\frac{b}{2}+\frac{(h_3+h_4)\,[(\mu_n^{(2)})^2+h_3h_4]}{[(\mu_n^{(2)})^2+h_3^2]\,[(\mu_n^{(2)})^2+h_4^2]}\right\}.$$

23. We choose the origin of a polar system of coordinates (ρ, ϕ) at the centre of the circle, the radius of the circle equals a.

(a) The edge of the membrane is rigidly fixed, $v|_{\rho=a} = 0$.

The eigenvalues $\lambda_{m,n} = \left(\mu_m^{(n)}/a\right)^2$ where $\mu_m^{(n)}$ are roots of the equation $J_n(\mu) = 0$,

$$v_{m,n} = J_n\left(\frac{\mu_m^{(n)}}{a}\rho\right)\begin{cases}\cos n\phi,\\ \sin n\phi\end{cases} \qquad (n = 0, 1, 2, \ldots;\ m = 1, 2, \ldots),$$

$$||v_{m,n}||^2 = \frac{a^2\pi\varepsilon_n}{2[\mu_m^{(n)}]^2}\,[J_n'(\mu_m^{(n)})]^2, \qquad \varepsilon_n = \begin{cases}2, & n = 0,\\ 1, & n \neq 0.\end{cases}$$

In particular, for $n = 0$

$$v = v_m = J_0\left(\frac{\mu_m^{(0)}}{a}\rho\right), \qquad ||v_m||^2 = \frac{a^2\pi}{[\mu_m^{(0)}]^2}J_1^2(\mu_m).$$

(b) The boundary $\rho = a$ of the membrane is free ($\partial v/\partial \rho = 0$ for $\rho = a$),

$$\lambda_{m,n} = \left(\frac{\mu_m^{(n)}}{a}\right)^2,$$

where $\mu_m^{(n)}$ is a root of the equation $J_n'(\mu) = 0$,

$$v_{m,n}(\rho, \phi) = J_n\left(\frac{\mu_m^{(n)}}{a}\rho\right)\begin{cases}\cos n\phi,\\ \sin n\phi,\end{cases}$$

$$||v_{m,n}||^2 = \frac{a^2\pi\varepsilon_n}{2[\mu_m^{(n)}]^2}\,[(\mu_m^{(n)})^2-n^2]J_n^2(\mu_m^{(n)}) \quad (m = 1, 2, \ldots;\ n = 0, 1, 2, \ldots).$$

(c) The boundary of the membrane is fixed elastically $(\partial v/\partial \rho + hv)_{\rho=a} = 0$,

$$\lambda_{m,n} = \left(\frac{\mu_m^{(n)}}{a}\right)^2,$$

where $\mu_m^{(n)}$ is the mth root of the equation

$$\mu J_n'(\mu) + ahJ_n(\mu) = 0,$$

$$v_{n,m}(\rho, \phi) = J_n\left(\frac{\mu_m^{(n)}}{a}\rho\right)\begin{cases}\cos n\phi, \\ \sin n\phi\end{cases} \quad (n = 0, 1, 2, \ldots;\ m = 1, 2, \ldots),$$

$$\|v_{n,m}\|^2 = \frac{\pi a^2 \varepsilon_n}{2[\mu_m^{(n)}]^2}[a^2h^2 + [\mu_m^{(n)}]^2 - n^2] J_n^2(\mu_m^{(n)}) \tag{1}$$

or

$$\|v_{n,m}\|^2 = \frac{\pi a^2 \varepsilon_n}{2}\left[1 + \frac{(\mu_m^{(n)})^2 - n^2}{a^2h^2}\right][J_n'(\mu_m^{(n)})]^2. \tag{2}$$

Formula (1) is suitable for small h, formula (2) for large h. A limiting transition in (1) for $h \to 0$ gives an expression for $\|v_{n,m}\|^2$ of problem (b); a limiting transition in formula (2) for $h \to \infty$ gives $\|v_{n,m}\|^2$ for the first boundary-value problem (a).

Method. It is required to solve the eigenvalue problem

$$\frac{1}{\rho}\frac{\partial}{\partial \rho}\left(\rho \frac{\partial v}{\partial \rho}\right) + \frac{1}{\rho^2}\frac{\partial^2 v}{\partial \phi^2} + \lambda v = 0 \quad (\rho < a),$$

$$|v(0, \phi)| < \infty,$$

(a) $v(a, \phi) = 0$, (b) $\dfrac{\partial v}{\partial \rho}(a, \phi) = 0$, (c) $\dfrac{\partial v}{\partial \rho}(a, \phi) + hv(a, \phi) = 0$.

The solution is sought in the form of a product $v(\rho, \phi) = R(\rho)\Phi(\phi)$. The method of calculating the norm is shown in [7] (page 650). Derive the general formula

$$\left\|J_n\left(\frac{\mu}{a}\rho\right)\right\|^2 = \frac{a^2}{2}\left[J_n'^2(\mu) + \left(1 - \frac{n^2}{\mu^2}\right)J_n^2(\mu)\right].$$

24. Let a, b, c denote the sides of the parallelepiped.

(a) If $v = 0$ for $x = 0, a$; $y = 0, b$; $z = 0, c$, then

$$\lambda_{m,n,k} = \pi^2\left(\frac{m^2}{a^2} + \frac{n^2}{b^2} + \frac{k^2}{c^2}\right) \quad (m, n, k = 1, 2, 3, \ldots),$$

$$v_{m,n,k} = \sin\frac{\pi m}{a}x \sin\frac{\pi n}{b}y \sin\frac{\pi k}{c}z, \quad \|v_{m,n,k}\|^2 = \frac{abc}{8}.$$

The normalized eigenfunctions

$$\bar{v}_{m,n,k} = \sqrt{\frac{8}{abc}}\, v_{m,n,k}.$$

(b) If boundary conditions of the second kind are given

$$v_x = 0 \quad \text{for} \quad x = 0, a; \; v_y = 0 \quad \text{for} \quad y = 0, b; \; v_z = 0 \quad \text{for} \quad z = 0, c,$$

then

$$\lambda_{m,n,k} = \pi^2 \left(\frac{m^2}{a^2} + \frac{n^2}{b^2} + \frac{k^2}{c^2} \right) \quad (m, n, k = 0, 1, 2, \ldots),$$

$$v_{m,n,k} = \cos \frac{\pi m}{a} x \cos \frac{\pi n}{b} y \cos \frac{\pi k}{c} z,$$

$$\|v_{m,n,k}\|^2 = \frac{abc}{8} \varepsilon_m \varepsilon_n \varepsilon_k, \qquad \varepsilon_j = \begin{cases} 2, & j = 0, \\ 1, & j \neq 0. \end{cases}$$

The normalized eigenfunctions

$$\bar{v}_{m,n,k} = \frac{1}{\|v_{m,n,k}\|} v_{m,n,k}.$$

(c) For the third boundary-value problem

$$(v_x - h_1 v)_{x=0} = 0, \quad (v_x + h_2 v)_{x=a} = 0, \quad (v_y - h_3 v)_{y=0} = 0, \quad (v_y + h_4 v)_{y=b} = 0,$$

$$(v_z - h_5 v)_{z=0} = 0, \quad (v_z + h_6 v)_{z=c} = 0$$

we have:

$$\lambda_{m,n,k} = [\mu_m^{(1)}]^2 + [\mu_n^{(2)}]^2 + [\mu_k^{(3)}]^2,$$

where $\mu_m^{(1)}$, $\mu_n^{(2)}$, $\mu_k^{(3)}$ are roots of the following equations:

$$\tan \mu^{(1)} a = \frac{(h_1 + h_2)\mu^{(1)}}{[\mu^{(1)}]^2 - h_1 h_2}, \quad \tan \mu^{(2)} b = \frac{(h_3 + h_4)\mu^{(2)}}{[\mu^{(2)}]^2 - h_3 h_4}, \quad \tan \mu^{(3)} c = \frac{(h_5 + h_6)\mu^{(3)}}{[\mu^{(3)}]^2 - h_5 h_6},$$

$$v_{m,n,k} = X_m(x) Y_n(y) Z_k(z),$$

$$X_m(x) = (\mu_m^{(1)} \cos \mu_m^{(1)} x + h_1 \sin \mu_m^{(1)} x) \frac{1}{\sqrt{(\mu_m^{(1)})^2 + h_1^2}},$$

$$Y_n(y) = (\mu_n^{(2)} \cos \mu_n^{(2)} y + h_3 \sin \mu_n^{(2)} y) \frac{1}{\sqrt{(\mu_m^{(2)})^2 + h_3^2}},$$

$$Z_k(z) = (\mu_k^{(3)} \cos \mu_k^{(3)} z + h_5 \sin \mu_k^{(3)} z) \frac{1}{\sqrt{(\mu_m^{(3)})^2 + h_5^2}},$$

$$\|v_{m,n,k}\|^2 = \left\{ \frac{a}{2} + \frac{(h_1 + h_2)[(\mu_m^{(1)})^2 + h_1 h_2]}{[(\mu_m^{(1)})^2 + h_1^2][(\mu_m^{(1)})^2 + h_2^2]} \right\} \times$$

$$\times \left\{ \frac{b}{2} + \frac{(h_3 + h_4)[(\mu_n^{(2)})^2 + h_3 h_4]}{[(\mu_n^{(2)})^2 + h_3^2][(\mu_n^{(2)})^2 + h_4^2]} \right\} \left\{ \frac{c}{2} + \frac{(h_5 + h_6)[(\mu_k^{(3)})^2 + h_5 h_6]}{[(\mu_k^{(3)})^2 + h_5^2][(\mu_k^{(3)})^2 + h_6^2]} \right\},$$

$$m, n, k = 1, 2, 3, \ldots$$

25. Let us choose a spherical system of coordinates (r, θ, ϕ) with origin at the centre of the sphere of radius a. The original equation is

$$\Delta v + \lambda v = 0$$

or

$$\frac{1}{r^2}\frac{\partial}{\partial r}\left(r^2\frac{\partial v}{\partial r}\right)+\frac{1}{r^2\sin\theta}\frac{\partial}{\partial\theta}\left(\sin\theta\frac{\partial v}{\partial\theta}\right)+\frac{1}{r^2\sin^2\theta}\frac{\partial^2 v}{\partial\phi^2}+\lambda v = 0.$$

(a) The first boundary-value problem: $v = 0$ for $r = a$,

$$\lambda_{n,m} = \left(\frac{\mu_m^{(n+\frac{1}{2})}}{a}\right)^2 \quad (n = 0, 1, 2, \ldots;\ m = 1, 2, \ldots), \tag{1}$$

where

$$\mu_m^{(n+\frac{1}{2})} \text{ is the } m\text{th root of the equation } J_{n+\frac{1}{2}}(\mu) = 0, \tag{2}$$

$$v_{n,m,l} = \psi_n\left(\frac{\mu_m^{(n+\frac{1}{2})}}{a}r\right)Y_n^{(l)}(\theta,\phi), \quad l = 0, \pm 1, \pm 2, \ldots \pm n, \tag{3}$$

where

$$Y_n^{(l)}(\theta,\phi) = P_n^{(l)}(\cos\theta)\begin{Bmatrix}\cos l\phi & \text{for} & l > 0,\\ \sin l\phi & \text{for} & l < 0,\end{Bmatrix} \tag{4}$$

$$P_n^{(l)}(x) = (1-x^2)^{l/2}\frac{\mathrm{d}^l}{\mathrm{d}x^l}P_n(x) \text{ is an associated function,}$$

$$P_n^{(0)} = P_n(x) \text{ is the Legendre polynomial,}$$

$$\psi_n(x) = \sqrt{\frac{\pi}{2x}}\,J_{n+\frac{1}{2}}(x), \tag{5}$$

$$\|v_{m,n,l}\|^2 = \left\|\psi_n\left(\frac{\mu_m^{(n+\frac{1}{2})}}{a}r\right)\right\|^2\|Y_n^{(l)}(\theta,\phi)\|^2, \tag{6}$$

$$\|Y_n^{(l)}(\theta,\phi)\|^2 = \frac{2\pi\varepsilon_l}{2n+1}\,\frac{(n+l)!}{(n-l)!}, \quad \varepsilon_l = \begin{Bmatrix}2, & l = 0,\\ 1, & l \neq 0,\end{Bmatrix} \tag{7}$$

$$\left\|\psi_n\left(\frac{\mu_m^{(n+\frac{1}{2})}}{a}r\right)\right\|^2 = \frac{\pi a}{2\mu_m^{(n+\frac{1}{2})}}\,\frac{a^2}{2}\left[J'_{n+\frac{1}{2}}\left(\mu_m^{(n+\frac{1}{2})}\right)\right]^2. \tag{8}$$

(b) The second boundary-value problem: $\partial v/\partial r = 0$ for $r = a$. Formulae (1), (3)–(7) still hold; only in this case $\mu_m^{(n+\frac{1}{2})}$ is a root of the equation

$$\psi_n'(\mu) = 0,$$

or

$$J'_{n+\frac{1}{2}}(\mu)-\frac{1}{2\mu}J_{n+\frac{1}{2}}(\mu)=0,$$

$$\left\|\psi_n\left(\frac{\mu_m^{(n+\frac{1}{2})}}{a}r\right)\right\|=\frac{\pi a^3}{4\left[\mu_m^{(n+\frac{1}{2})}\right]}\left(1-\frac{n(n+1)}{\left[\mu_m^{(n+\frac{1}{2})}\right]^2}\right)J^2_{n+\frac{1}{2}}\left(\mu_m^{(n+\frac{1}{2})}\right). \quad (9)$$

(c) The third boundary-value problem:

$$\frac{\partial v}{\partial r}+hv=0 \quad \text{for} \quad r=a.$$

All the formulae of problem 25(a), except formulae (2) and (8) still hold; $\mu_m^{(n+\frac{1}{2})}$ now denotes the root of the equation

$$\mu\psi_n'(\mu)+ah\psi_n(\mu)=0,$$

or

$$J'_{n+\frac{1}{2}}(\mu)-\frac{1-2ah}{2\mu}J_{n+\frac{1}{2}}(\mu)=0, \quad (10)$$

$$\left\|\psi_n\left(\frac{\mu_m^{(n+\frac{1}{2})}}{a}r\right)\right\|^2$$

$$=\frac{\pi a^3}{4\mu_m^{(n+\frac{1}{2})}}\left[1-\frac{n(n+1)+4ah(1-ah)}{\left[\mu_m^{(n+\frac{1}{2})}\right]^2}\right]J^2_{n+\frac{1}{2}}\left(\mu_m^{(n+\frac{1}{2})}\right). \quad (11)$$

26. Let us choose a cylindrical system of coordinates (ρ, ϕ, z) choosing the z-axis along the axis of the cylinder and placing the origin of coordinates on the lower base of the cylinder; a is the radius of the cylinder, l is the height of the cylinder. The original equation of the eigenvalue problem has the form

$$\Delta v+\lambda v=0$$

or

$$\frac{1}{\rho}\frac{\partial}{\partial\rho}\left(\rho\frac{\partial v}{\partial\rho}\right)+\frac{1}{\rho^2}\frac{\partial^2 v}{\partial\phi^2}+\frac{\partial^2 v}{\partial z^2}+\lambda v=0.$$

(a) The first boundary-value problem: $v=0$ for $\rho=a$, $z=0$, l,

$$\lambda_{m,n,k}=\left(\frac{\pi k}{l}\right)^2+\left(\frac{\mu_m^{(n)}}{a}\right)^2 \quad (n=0,1,2,\ldots;\ m,\ k=1,2,\ldots),$$

where $\mu_m^{(n)}$ is the mth root of the equation $J_n(\mu)=0$,

$$v_{n,m,k}=\sin\frac{\pi k}{l}zJ_n\left(\frac{\mu_m^{(n)}}{a}\rho\right)\begin{cases}\cos n\phi,\\ \sin n\phi,\end{cases}$$

$$\|v_{n,m,k}\|^2=\frac{l\pi a^2}{4}\varepsilon_n[J_n'(\mu_m^{(n)})]^2, \qquad \varepsilon_n=\begin{cases}2, & n=0,\\ 1, & n\neq 0.\end{cases}$$

(b) The second boundary-value problem: $\partial v/\partial r = 0$ for $r = a$, $\partial v/\partial z = 0$ for $z = 0, l$,

$$\lambda_{n,m,k} = \left(\frac{\mu_m^{(n)}}{a}\right)^2 + \left(\frac{\pi k}{l}\right)^2 \quad (n = 0, 1, 2, \ldots;\ k,\quad m = 1, 2, \ldots),$$

$$v_{n,m,k} = \cos \frac{\pi k}{l} z J_n\left(\frac{\mu_m^{(n)}}{a}\rho\right) \begin{cases} \cos n\phi, \\ \sin n\phi, \end{cases}$$

where $\mu_m^{(n)}$ is a root of the equation $J_n'(\mu) = 0$,

$$\|v_{n,m,k}\|^2 = \frac{l\pi a^2 \varepsilon_n}{4[\mu_m^{(n)}]^2}[(\mu_m^{(n)})^2 - n^2] J_n^2(\mu_m^{(n)}).$$

(c) The third boundary-value problem: $\partial v/\partial \rho + h_0 v = 0$ for $\rho = a$, $\partial v/\partial z - h_1 v = 0$ for $z = 0$, $\partial v/\partial z + h_2 v = 0$ for $z = 1$,

$$v_{n,m,k}(\rho, \phi, z) = Z_k(z) J_n\left(\frac{\mu_m^{(n)}}{a}\rho\right) \begin{cases} \cos n\phi, \\ \sin n\phi, \end{cases}$$

$$Z_k(z) = \frac{\nu_k \cos \nu_k z + h_1 \sin \nu_k z}{\sqrt{\nu_k^2 + h_1^2}},$$

ν_k is a root of

$$\tan \nu l = \frac{(h_1 + h_2)\nu}{\nu^2 - h_1 h_2},$$

$\mu_m^{(n)}$ is a root of the equation $\mu J_n'(\mu) + a h_0 J_n(\mu) = 0$,

$$\lambda_{n,m,k} = \nu_k^2 + \left(\frac{\mu_m^{(n)}}{a}\right)^2 \quad (m, k = 1, 2, \ldots;\quad n = 0, 1, 2, \ldots),$$

$$\|v_{n,m,k}\| = \pi \varepsilon_n \left\| J_n\left(\frac{\mu_m^{(n)}}{a}\rho\right)\right\|^2 \|Z_k(z)\|^2,$$

where

$$\left\| J_n\left(\frac{\mu_m^{(n)}}{a}\right)\right\|^2 = \frac{a^2}{2(\mu_m^{(n)})^2}[a^2 h_0^2 + (\mu_m^{(n)})^2 - n^2] J_n^2(\mu_m^{(n)}),$$

$$\|Z_k\|^2 = \frac{l}{2} + \frac{(h_1 + h_2)(\nu_k^2 + h_1 h_2)}{(\nu_k^2 + h_1^2)(\nu_k^2 + h_2^2)}.$$

Method. The solution is sought in the form of the product

$$v(\rho, \phi, z) = V(\rho, \phi) Z(z).$$

After separation of the variables we obtain problem 23 for $V(\rho, \phi)$, and problem 17 for $Z(z)$.

27. A cylindrical system of coordinates (ρ, ϕ) is chosen.

(a) The first boundary-value problem: $v = 0$ for $\rho = a$ and $\rho = b$,

$$v_{m,n}(\rho, \phi) = R_{m,n}(\rho) \begin{cases} \cos n\phi, \\ \sin n\phi, \end{cases} \quad (m = 1, 2, \ldots,\quad n = 0, 1, 2, \ldots),$$

$$R_{m,n}(\rho) = J_n(\mu_m^{(n)}\rho) N_n(\mu_m^{(n)} a) - J_n(\mu_m^{(n)} a) N_n(\mu_m^{(n)}\rho), \tag{1}$$

or

$$R_{m,n}(\rho) = \frac{J_n(\mu_m^{(n)}a)}{J_n(\mu_m^{(n)}b)}[J_n(\mu_m^{(n)}\rho)N_n(\mu_m^{(n)}b) - J_n(\mu_m^{(n)}b)N_n(\mu_m^{(n)}\rho)], \tag{2}$$

where $\mu_m^{(n)}$ is a root of the transcendental equation

$$J_n(a\mu)N_n(b\mu) - J_n(b\mu)N_n(a\mu) = 0, \tag{3}$$

which may also be written in the following form:

$$\frac{J_n(a\mu)}{J_n(b\mu)} = \frac{N_n(a\mu)}{N_n(b\mu)},$$

Here $N_n(x)$ is Neumann's function of nth order

$$\lambda_{m,n} = (\mu_m^{(n)})^2,$$

$$||v_{m,n}||^2 = \pi\varepsilon_n||R_{m,n}||^2 = \frac{2\varepsilon_n}{\pi[\mu_m^{(n)}]^2}\,\frac{J_n^2(\mu_m^{(n)}a) - J_n^2(\mu_m^{(n)}b)}{J_n^2(\mu_m^{(n)}b)}, \tag{4}$$

$$\varepsilon_n = \begin{cases} 2, & n = 0, \\ 1, & n \neq 0. \end{cases}$$

(b) The second boundary-value problem: $\partial v/\partial\rho = 0$ for $\rho = a$, $\rho = b$,

$$v_{m,n}(\rho, \phi) = R_{m,n}(\rho)\Phi_n(\phi),$$

$$\Phi_n(\phi) = \begin{cases} \cos n\phi, \\ \sin n\phi \end{cases} \quad (n = 0, 1, 2, \ldots),$$

$$R_{m,n}(\rho) = J_n(\mu_m^{(n)}\rho)N_n'(\mu_m^{(n)}a) - J_n'(\mu_m^{(n)}a)N_n(\mu_m^{(n)}\rho), \tag{5}$$

or

$$R_{m,n}(\rho) = \frac{J_n'(a\mu_m^{(n)})}{J_n'(b\mu_m^{(n)})}[J_n(\mu_m^{(n)}\rho)N_n'(\mu_m^{(n)}b) - J_n'(\mu_m^{(n)}b)N_n(\mu_m^{(n)}\rho)], \tag{5'}$$

where $\mu_m^{(n)}$ is the root of the equation

$$\frac{J_n'(a\mu)}{J_n'(b\mu)} = \frac{N_n'(a\mu)}{N_n'(b\mu)}. \tag{6}$$

The eigenvalue

$$\lambda_{m,n} = (\mu_m^{(n)})^2,$$

$$||v_{m,n}||^2 = \pi\varepsilon_n||R_{m,n}||^2$$

$$= \frac{2\varepsilon_n}{\pi[\mu_m^{(n)}]^2}\left\{\left(1 - \frac{n^2}{b^2[\mu_m^{(n)}]^2}\right)\frac{J_n'^2(\mu_m^{(n)}a)}{J_n'^2(\mu_m^{(n)}b)} - \left(1 - \frac{n^2}{a^2[\mu_m^{(n)}]^2}\right)\right\}. \tag{7}$$

In particular, for $n = 0$ we have:

$$||v_{m,n}||^2 = \pi\varepsilon_n||R_{m,n}||^2 = ||v_{m,n}||^2 = \frac{4}{\pi[\mu_m^{(0)}]^2}\,\frac{J_1^2(a\mu_m^{(0)}) - J_1^2(b\mu_m^{(0)})}{J_1^2(b\mu_m^{(0)})}.$$

(c) The third boundary-value problem: $\partial v/\partial \rho - hv = 0$ for $\rho = a$, $\partial v/\partial \rho + hv = 0$ for $\rho = b$,

$$v_{m,n}(\rho\phi) = R_{m,n}(\rho)\Phi_n(\phi), \qquad \Phi_n(\phi) = \begin{cases} \cos n\phi, \\ \sin n\phi, \end{cases}$$

$$R_{m,n}(\rho) = J_n(\mu_m^{(n)}\rho)\delta(a) - \Delta(a)N_n(\mu_m^{(n)}\rho), \tag{8}$$

or

$$R_{m,n}(\rho) = \frac{\Delta(a)}{\Delta(b)}[J_n(\mu_m^{(n)}\rho)\delta(b) - \Delta(b)N_n(\mu_m^{(n)}\rho)], \tag{9}$$

where

$$\left.\begin{aligned} \Delta(a) &= J_n'(\mu_m^{(n)}a) - \frac{h}{\mu_m^{(n)}} J_n(\mu_m^{(n)}a), \quad \Delta(b) = J_n'(\mu_m^{(n)}b) + \frac{h}{\mu_m^{(n)}} J_n(\mu_m^{(n)}b), \\ \delta(a) &= N_n'(\mu_m^{(n)}a) - \frac{h}{\mu_m^{(n)}} N_n(\mu_m^{(n)}a), \; \delta(b) = N_n'(\mu_m^{(n)}b) + \frac{h}{\mu_m^{(n)}} N_n(\mu_m^{(n)}b). \end{aligned}\right\} \tag{10}$$

The egienvalues are determined from the formula

$$\lambda_{m,n} = [\mu_m^{(n)}]^2, \qquad (n = 0, 1, 2, \ldots, \quad m = 1, 2, \ldots),$$

where $\mu_m^{(n)}$ is the mth root of the following equation:

$$\delta(a)\Delta(b) - \delta(b)\Delta(a) = 0, \qquad \text{or} \qquad \frac{\delta(a)}{\delta(b)} = \frac{\Delta(a)}{\Delta(b)},$$

$$||v_{m,n}||^2 = ||\Phi_n||^2||R_{m,n}||^2 = \pi\varepsilon_n||R_{m,n}||^2,$$

$$||R_{m,n}||^2 = \frac{2}{\pi^2[\mu_m^{(n)}]^2}\left\{\left[\frac{\Delta(a)}{\Delta(b)}\right]^2\left(1 - \frac{n^2 - b^2h^2}{b^2[\mu_m^{(n)}]^2}\right) - \left(1 - \frac{n^2 - a^2h^2}{a^2[\mu_m^{(n)}]^2}\right)\right\}.$$

For $h = 0$ we have:

$$\frac{\Delta(a)}{\Delta(b)} = \frac{J_n'(a\mu_m^{(n)})}{J_n'(b\mu_m^{(n)})},$$

and formulae (8) and (11) change into formulae (5) and (7).

Method. Assuming $v(\rho, \varphi) = R(\rho)\,\Phi(\phi)$, we obtain for the radial function $R(\rho)$ that

$$\frac{1}{\rho}\frac{d}{d\rho}\left(\rho\frac{dR}{d\rho}\right) + \left(\lambda - \frac{n^2}{\rho^2}\right)R = 0,$$

$$R(a) = R(b) = 0 \quad \text{in case (a)},$$

$$R'(a) = R'(b) = 0 \quad \text{in case (b)}.$$

The general solution of the equation has the form

$$R_n(\rho) = AJ_n(\mu\rho) + BN_n(\mu\rho), \qquad \mu = \sqrt{\lambda}.$$

Using the boundary conditions for $\rho = a$ and $\rho = b$, we obtain the expression, given in the answer for $R_n(\rho)$.

Calculation of the norm is carried out in the usual way.
For $\|R_{m,n}\|^2$ the general formula

$$\|R_{m,n}\|^2 = \frac{b^2}{2}\left[R'^2_{m,n}(\mu_m^{(n)}b)+\left(1-\frac{n^2}{b^2[\mu_m^{(n)}]^2}\right)R^2_{m,n}(\mu_m^{(n)}b)\right]-$$

$$-\frac{a^2}{2}\left[R'^2_{m,n}(\mu_m^{(n)}a)+\left(1-\frac{n^2}{a^2[\mu_m^{(n)}]^2}\right)R^2_{m,n}(\mu_m^{(n)}a)\right]$$

is obtained. In order to calculate $R_{m,n}(\mu_m^{(n)}\rho)$ and $R'_{m,n}(\mu_m^{(n)}\rho)$ for $\rho = a$ and $\rho = b$ one must make use of the expression for the Wronskian

$$J_n(x)N'_n(x)-N_n(x)J'_n(x) = \frac{2}{\pi x}.$$

28. It is required to solve the boundary-value problem

$$\frac{1}{\rho}\frac{\partial}{\partial\rho}\left(\rho\frac{\partial v}{\partial\rho}\right)+\frac{1}{\rho^2}\frac{\partial^2 v}{\partial\phi^2}+\lambda v = 0 \quad \text{for} \quad \rho > a, 0 < \phi < \phi_0$$

for the boundary conditions

(a) $v = 0$ for $\rho = a, \quad \phi = 0, \quad \phi = \phi_0;$

(b) $\dfrac{\partial v}{\partial\rho} = 0$ for $\rho = a, \quad \dfrac{\partial v}{\partial\phi} = 0$ for $\phi = 0, \quad \phi = \phi_0;$

(c) $\dfrac{\partial v}{\partial\rho}+h_0 v = 0$ for $\rho = a, \quad \dfrac{\partial v}{\partial\phi}-h_1 v = 0$ for $\phi = 0,$

$$\frac{\partial v}{\partial\phi}+h_2 v = 0 \quad \text{for} \quad \phi = \phi_0.$$

Answers:

(a) The first boundary-value problem for a sector gives

$$v_{m,n}(\rho,\phi) = J_{\frac{\pi n}{\phi_0}}\left(\frac{\mu_m^{(n)}}{a}\rho\right)\sin\frac{\pi n}{\phi_0}\phi \qquad (m, n = 1, 2, \ldots),$$

$$\lambda_{m,n} = \left(\frac{\mu_m^{(n)}}{a}\right)^2,$$

where $\mu_m^{(n)}$ is the mth root of the equation

$$J_{\frac{\pi n}{\phi_0}}(\mu) = 0,$$

$$\|v_{m,n}\|^2 = \frac{a^2\phi_0}{4}\left[J'_{\frac{\pi n}{\phi_0}}(\mu_m^{(n)})\right]^2.$$

(b) The second boundary-value problem for the sector gives

$$v_{m,n}(\rho,\phi) = J_{\frac{\pi n}{\phi_0}}\left(\frac{\mu_m^{(n)}}{a}\rho\right)\cos\frac{\pi n}{\varphi_0}\phi \; (n = 0, 1, 2, \ldots; m = 1, 2, \ldots),$$

$$\lambda_{m,n} = \left(\frac{\mu_m^{(n)}}{a}\right)^2, \quad \mu_m^{(n)} \text{ the } m\text{th root of the equation } J'_{\frac{\pi n}{\phi_0}}(\mu) = 0,$$

$$\|v_{m,n}\|^2 = \frac{\phi_0 \varepsilon_n}{2} \left\| J_{\frac{\pi n}{\phi_0}}\left(\frac{\mu_m^{(n)}}{a}\right) \right\|^2 = \frac{\phi_0 \varepsilon_n a^2}{4}\left(1 - \frac{n^2}{(\mu_m^{(n)})^2}\right) J^2_{\frac{\pi n}{\phi_0}}(\mu_m^{(n)}).$$

For the expression for $\left\| J_{\frac{\pi n}{\phi_0}}\left(\frac{\mu_m^{(n)}}{a}\rho\right) \right\|^2$ look in the answer to problem 23.

(c) The third boundary-value problem for the sector

$$v_{m,n}(\rho, \varphi) = J_{\nu_n}\left(\frac{\mu_m^{(n)}}{a}\rho\right)\Phi_n(\phi),$$

$$\Phi_n(\phi) = \frac{\nu_n \cos \nu_n \phi + h_1 \sin \nu_n \phi}{\sqrt{\nu_n^2 + h_1^2}},$$

$$\lambda_m^{(n)} = \left(\frac{\mu_m^{(n)}}{a}\right)^2 \qquad (m, n = 1, 2, \ldots),$$

where ν_n is a positive root of the equation $\tan \nu\phi_0 = (h_1 + h_2)\nu(\nu^2 - h_1 h_2)$, $\mu_m^{(n)}$ a root of the equation $\mu J'_{\nu_n}(\mu) + a h_0 J_{\nu_n}(\mu) = 0$,

$$\|v_{m,n}\|^2 = \|\Phi_n\|^2 \left\| J_{\nu_n}\left(\frac{\mu_m^{(n)}}{a}\rho\right) \right\|^2,$$

$$\|\Phi_n\|^2 = \frac{\phi_0}{2} + \frac{(h_1 + h_2)(\nu_n^2 + h_1 h_2)}{(\nu_n^2 + h_1^2)(\nu_n^2 + h_2^2)},$$

$$\left\| J_{\nu_n}\left(\frac{\mu_m^{(n)}}{a}\rho\right) \right\|^2 = \frac{a^2}{2}\left[1 + \frac{(\mu_m^{(n)})^2 - \nu_n^2}{a^2 h_0^2}\right][J'_n(\mu_m^{(n)})]^2$$
$$= \frac{a^2}{2}\left[1 + \frac{a^2 h_0^2 - \nu_n^2}{[\mu_m^{(n)}]^2}\right] J^2_{\nu_n}(\mu_m^{(n)})$$

29. We seek a solution of the equation

$$\frac{1}{\rho}\frac{\partial}{\partial \rho}\left(\rho \frac{\partial v}{\partial \rho}\right) + \frac{1}{\rho^2}\frac{\partial^2 v}{\partial \phi^2} + \lambda v = 0 \qquad (a < \rho < b, \quad 0 < \phi < \phi_0),$$

satisfying the homogeneous boundary conditions (a) of the first kind, (b) of the second kind, (c) of the third kind.

(a) The first boundary-value problem: $v = 0$ for $\rho = a, b$; $\phi = 0, \phi_0$,

$$v_{m,n}(\rho, \phi) = R_{m,n}(\rho)\Phi_n(\phi),$$

$$\Phi_n(\phi) = \sin \frac{\pi n}{\phi_0}\phi, \qquad (n = 1, 2, \ldots),$$

$$R_{m,n}(\rho) = J_{\frac{\pi n}{\phi_0}}(\mu_m^{(n)}\rho) N_{\frac{\pi n}{\phi_0}}(\mu_m^{(n)} a) - J_{\frac{\pi n}{\phi_0}}(\mu_m^{(n)} a) N_{\frac{\pi n}{\phi_0}}(\mu_m^{(n)}\rho),$$

or

$$R_{m,n}(\rho) = \frac{J_{\frac{\pi n}{\phi_0}}(\mu_m^{(n)}a)}{J_{\frac{\pi n}{\phi_0}}(\mu_m^{(n)}b)}\left[J_{\frac{\pi n}{\phi_0}}(\mu_m^{(n)}\rho)N_{\frac{\pi n}{\phi_0}}(\mu_m^{(n)}b) - J_{\frac{\pi n}{\phi_0}}(\mu_m^{(n)}b)N_{\frac{\pi n}{\phi_0}}(\mu_m^{(n)}\rho)\right],$$

$\lambda_{m,n} = [\mu_m^{(n)}]^2$ the mth root of the equation

$$\frac{J_{\frac{\pi n}{\phi_0}}(\mu a)}{J_{\frac{\pi n}{\phi_0}}(\mu b)} = \frac{N_{\frac{\pi n}{\phi_0}}(\mu a)}{N_{\frac{\pi n}{\phi_0}}(\mu b)},$$

$$||v_{m,n}||^2 = ||\Phi_n||^2||R_{m,n}||^2 = \frac{\phi_0}{2}||R_{m,n}||^2.$$

For the expression for $||R_{m,n}||^2$ see in problem 27 (formula (4)) with the substitution of J_n by $J_{\pi n/\phi}$.

Expressions for cases (b) and (c) are obtained in a similar manner.

30. (a) It is required to find the natural vibrations for the region $a \leqslant \rho \leqslant b$, $0 \leqslant z \leqslant l$, if $v = 0$ for $\rho = 0$, $\rho = a$ and $z = 0$, $z = l$. The eigenfunctions

$$v_{m,n,k}(\rho, \phi, z) = R_{m,n}(\rho)\Phi_n(\phi)Z_k(z),$$

where

$$\Phi_n(\phi) = \begin{cases}\cos n\phi, \\ \sin n\phi\end{cases} \quad (n = 0, 1, 2, \ldots),$$

$$Z_k(z) = \sin\frac{\pi k}{l}z \qquad (k = 1, 2, \ldots),$$

$$R_{m,n}(\rho) = J_n(\mu_m^{(n)}\rho)N_n(\mu_m^{(n)}a) - J_n(\mu_m^{(n)}a)N_n\mu_m^{(n)}\rho)$$

(see the answer to problem 27), $\mu_m^{(n)}$ a root of the equation

$$\frac{J_n(\mu a)}{J_n(\mu b)} = \frac{N_n(\mu a)}{N_n(\mu b)},$$

$$\lambda_{m,n} = (\mu_m^{(n)})^2 + \left(\frac{\pi k}{l}\right)^2,$$

$$||v_{m,n,k}||^2 = \frac{\pi l}{2}\varepsilon_n||R_{m,n}||^2.$$

For an expression for $||R_{m,n}||^2$ see in the answer to problem 27(a).

(b) In this case $\partial v/\partial\rho = 0$ for $\rho = a, b$, $\partial v/\partial z = 0$ for $z = 0$, $z = l$, so that

$$\Phi_n(\phi) = \begin{cases}\cos n\phi, \\ \sin n\phi\end{cases} \quad (n = 0, 1, 2, \ldots),$$

$$Z_k(z) = \cos\frac{\pi k}{l}z \quad (k = 0, 1, 2, \ldots).$$

$R_{m,n}(\rho)$ is given in the answer to problem 28(b),

$$\lambda_{m,n} = (\mu_m^{(n)})^2 + \left(\frac{\pi k}{l}\right)^2,$$

$$\|v_{m,n,k}\|^2 = \frac{\pi l}{2}\varepsilon_k \varepsilon_n \|R_{m,n}\|^2.$$

31. If λ_1 is the first eigenvalue of the ring-shaped membrane ($\varepsilon \leqslant \rho \leqslant a$) with fixed boundary, and λ_1^0 the first eigenvalue of the circular membrane $\rho \leqslant a$ with fixed boundary, then the correction

$$\Delta\lambda_1 = \lambda_1 - \lambda_1^0$$

is always positive, and for small ε equals

$$\Delta\lambda_1 = \frac{2}{a^2 J_1^2(\sqrt{\lambda_1^0}a)} \frac{1}{\ln \dfrac{1}{\sqrt{\lambda_1^0}\varepsilon}} + \ldots = 7{\cdot}41 \frac{1}{a^2 \ln \dfrac{a}{2{\cdot}4048\varepsilon}} + \ldots,$$

where terms of a higher order in ε are omitted.

From this formula we see that

$$\lim_{\varepsilon\to 0} \Delta\lambda_1 = 0.$$

Solution. The least eigenvalue λ_1^0 of a circular membrane rigidly fixed at the boundary $\rho = a$ is determined from the equation

$$J_0(\sqrt{\lambda_1^0}a) = 0,$$

and the first eigenvalue λ_1 of the ring-shaped membrane $\varepsilon < \rho < a$ with rigidly fixed boundary is determined from the equation

$$J_0(\sqrt{\lambda_1}\varepsilon)N_0(\sqrt{\lambda_1}a) - J_0(\sqrt{\lambda_1}a)N_0\sqrt{\lambda_1}\varepsilon) = 0. \tag{1}$$

Assuming

$$\sqrt{\lambda_1} = \sqrt{\lambda_1^0} + \alpha,$$

so that $\Delta\lambda_1 = 2\sqrt{\lambda_1^0}\,\alpha$, and taking into account the fact that

$$J_0(\sqrt{\lambda_1}\varepsilon) = 1 - \ldots, \qquad N_0\sqrt{\lambda_1}\varepsilon) = \frac{2}{\pi}\ln\sqrt{\lambda_1}\varepsilon + \ldots,$$

$$N_0(\sqrt{\lambda_1}a) = N_0(\sqrt{\lambda_1^0}a) - N_1(\sqrt{\lambda_1^0}a)\alpha a,$$

$$J_0(\sqrt{\lambda_1}a) = J_0(\sqrt{\lambda_1^0}a) - J_1(\sqrt{\lambda_1^0}a)\alpha a = -\alpha a J_1(\sqrt{\lambda_1^0}a).$$

From equation (1) we obtain:

$$\alpha = \frac{\pi}{2a}\frac{N_0(\sqrt{\lambda_1^0}a)}{J_1(\sqrt{\lambda_1^0}a)} \frac{1}{\ln \dfrac{1}{\sqrt{\lambda_1^0}\varepsilon}}.$$

Using the expression for the Wrónskian determinant

$$J_0(x)N_0'(x)-J_0'(x)N_0(x) = \frac{2}{\pi x};$$

since $J_0(\sqrt{\lambda_1^0}a) = 0$, hence we find:

$$N_0(\sqrt{\lambda_1^0}a) = \frac{2}{\pi\sqrt{\lambda_1^0}aJ_1(\sqrt{\lambda_1^0}a)}$$

and

$$\alpha = \frac{1}{\sqrt{\lambda_1^0}a^2J_1^2(\sqrt{\lambda_1^0}a)}\,\frac{1}{\ln\dfrac{1}{\sqrt{\lambda_1^0}\varepsilon}}.$$

Since the first root of the equation $J_0(\mu) = 0$ equals $\mu_1^0 = 2{\cdot}4048$ and $J_1(\mu_1^0) = 0{\cdot}5191$, then

$$\Delta\lambda_1 \approx 7{\cdot}41\frac{1}{a^2\ln\dfrac{a}{2{\cdot}4048\varepsilon}}.$$

32. If the load M is small, then

(a) $\lim\limits_{\mu\to 0}\lambda_1 = \lambda_1^0,$

(b) $\Delta\lambda_1 = \lambda_1-\lambda_1^0 \sim \dfrac{1}{\ln\dfrac{1}{\varepsilon}}.$

If the load M is large, then

(a) $\lim\limits_{M\to\infty}\lambda_1 = 0, \quad \lim\limits_{M\to\infty}\lambda_2 = \lambda_1^0,$

(b) $\lambda_1 = 2\pi C\dfrac{1}{M}+\dots,$

where C equals

$$C = -\frac{1}{\ln p} = \frac{1}{\ln\dfrac{1}{p}},$$

where

$$p = \frac{\varepsilon}{a},$$

so that

$$\lambda_1 = \frac{2\pi}{\ln\dfrac{1}{p}}\,\frac{1}{M}+\dots$$

Solution. The wave equation of the membrane has the form

$$\frac{d}{d\rho}\left(\rho\frac{du}{d\rho}\right)+\lambda\delta u = 0 \quad \text{for} \quad \varepsilon<\rho<a,$$

where δ is the mass density.

The boundary $\rho = a$ is fixed, so that

$$u\Big|_{\rho=a} = 0.$$

Let us denote $\sqrt{\lambda\delta}a = x, \quad \sqrt{\lambda\delta}\varepsilon = px, \quad p = \varepsilon/a$.

In order to obtain the second boundary condition, we modify the problem, replacing the circle of radius ε with centre at the point O by an absolutely rigid lamina of mass M, for which the equation of motion has the form

$$M\frac{\partial^2 u}{\partial t^2} = F,$$

where

$$F = \int\limits_{\rho=\varepsilon}^{2\pi} \frac{\partial u}{\partial \rho}\,\rho \mathrm{d}\phi = 2\pi\varepsilon\frac{\mathrm{d}u}{\mathrm{d}\rho}(\varepsilon),$$

or

$$-\lambda Mu\Big|_{\rho=\varepsilon} = 2\pi\varepsilon u'(\varepsilon),$$

since $\ddot{u} = -\lambda u$.

The general solution has the form

$$u = AJ_0(\sqrt{\lambda\delta}\rho)+BN_0(\sqrt{\lambda\delta}\rho)$$

or

$$u = N_0(x)J_0\left(\frac{x}{a}\rho\right)-J_0(x)N_0\left(\frac{x}{a}\rho\right).$$

The condition for $\rho = \varepsilon$ gives us an equation for determining λ:

$$\frac{x}{2p}\,\frac{J_0(px)N_0(x)-J_0(x)N_0(px)}{J_1(px)N_0(x)-N_1(px)J_0(x)} = S, \tag{1}$$

where

$$S = \frac{\delta\pi a^2}{M}, \qquad x = \sqrt{\lambda\delta}a.$$

The solution of the dispersion equation (1) may be found graphically. The graph of the function $S(x)$ has the form shown in Fig. 58.

Here k_0, k_1, k_2 are roots of the denominator of the function $S(x)$, and the dotted curve is the parabola

$$S = -\frac{1}{2}x^2\ln p.$$

Assigning a value of M (in Fig. 58 the horizontal AB), we find the corresponding roots of the dispersion equation.

For $M \to \infty$ the horizontal AB tends to the x-axis, the first root tends to zero, the remaining roots tend to the values $\lambda_1^0, \lambda_2^0, \lambda_3^0, \dots$, which are the roots of the numerator of the function $S(x)$. The quantities $\lambda_1^0, \lambda_2^0, \lambda_3^0, \ldots$ differ from $k_0, k_1, k_2, \ldots$ by amounts of the order of the quantity

$$C = \frac{1}{\ln \frac{\varepsilon}{a}}.$$

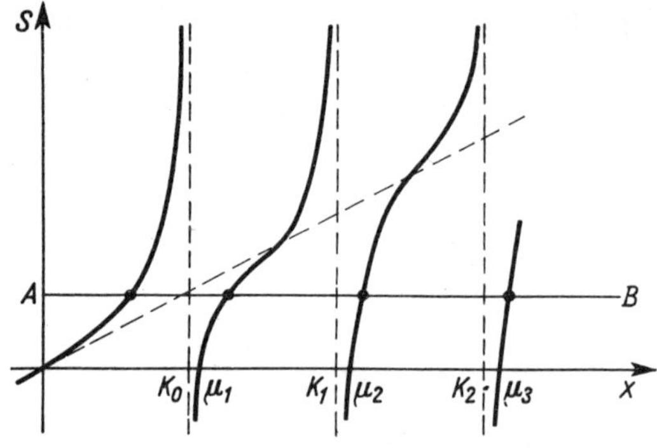

FIG. 58

For large masses M the first root will be small.
Let us expand the cylindrical functions near zero:

$$J_0(x) = 1 - \ldots, \qquad J_1(x) = \frac{1}{2}x + \ldots,$$

$$N_0(x) = \ln x + \ldots, \quad N_1(x) = -\frac{1}{x} + \ldots$$

Substituting these expressions in (1) we find:

$$\lambda_1 = \frac{2\pi}{M} \cdot \frac{1}{-\ln p} = 2\pi C \frac{1}{M} + \ldots,$$

where $C = -1/\ln p, \quad p = \varepsilon/a.$

33. Let the outer boundary of a ring-shaped membrane $\varepsilon \leqslant \rho \leqslant a$ be free, i.e.

$$\frac{\partial v}{\partial \rho}(a, \phi) = 0.$$

Then

$$\Delta\lambda_1 = \lambda_1 - \lambda_1^0 = \frac{2}{a^2 J_0^2(\sqrt{\lambda_1^0}\, a) \ln \frac{1}{\sqrt{\lambda_1^0}\, \varepsilon}} + \ldots \approx \frac{12 \cdot 3}{a^2 \ln \frac{a}{3 \cdot 83 \varepsilon}},$$

where $\lambda_1^0 = \left(\frac{\mu_1^0}{a}\right)^2$, μ_1^0 is the first root of the equation $J_1(\mu) = 0$, $\mu_1^0 = 3 \cdot 83$.

34. The problem on the natural vibrations of a circular membrane, drawn over the opening of a vessel of volume V_0 (drum) leads to the following equation:

$$\Delta_2 v + \lambda v = \frac{\delta_0 c_0^2}{V_0 T} \int_0^a \int_0^{2\pi} v\rho \, d\rho \, d\phi \qquad \left(\lambda = \frac{\omega^2}{c^2}\right), \tag{1}$$

if it is assumed that the velocity of the transverse waves is considerably less than the velocity of sound in air.

Here δ_0 is the density of the air in the vessel, c_0 the velocity of sound in air at the temperature and pressure in the vessel, corresponding to the stationary membrane $c = \sqrt{T/\rho}$, T the tension of the membrane, a its radius. The integral on the right indicates the additional pressure produced by the vibrations of the air in the vessel (see [38], page 217).

From (1) and the orthogonality of the trigonometric functions over $(0 \cdot 2\pi)$ it follows that for $n > 0$ the eigenfunctions of the membrane

$$v_{n,m} = J_n\left(\frac{\mu_m^{(n)}}{a}\rho\right)\begin{cases}\cos n\phi, \\ \sin n\phi\end{cases} \tag{2}$$

do not vary, in spite of the presence of an additional volume of air.

The eigenfunctions, possessing cylindrical symmetry ($n = 0$) have the form

$$v_m(\rho) = J_0\left(\frac{\mu_m}{a}\rho\right) - J_0(\mu_m) \qquad \left(\lambda = \frac{\mu_m^2}{a^2}\right), \tag{3}$$

where μ_m is a root of the equation

$$J_0(\mu) = -\frac{\chi}{\mu^2} J_2(\mu), \tag{4}$$

$$\chi = \frac{\pi \delta_0 c_0^2 a^4}{V_0 T},$$

which is obtained if (3) is substituted in equation (1).

The series for $J_2(\mu)$ gives:

$$\frac{J_2(\mu)}{\mu^2} = \frac{1}{8} - \frac{\mu^2}{96} + \frac{\mu^4}{3072}. \tag{5}$$

If $\nu_1 = 2 \cdot 4048$ is the first root of the equation $J_0(\nu) = 0$, then

$$J_0(\mu_1) = -J_1(\nu_1)\varepsilon = -0 \cdot 5191\varepsilon, \qquad (\varepsilon = \mu_1 - \nu_1).$$

From (4) and (5) we obtain:

$$\varepsilon = \frac{\left(\frac{1}{8} - \frac{\nu_1^2}{96} + \frac{\nu_1^4}{3072}\right)\chi}{J_1(\nu_1) + \left(\frac{\nu_1}{46} + \frac{\nu_1^3}{768}\right)\chi} = \frac{1}{6 \cdot 86186 + 0 \cdot 93047\chi}\chi.$$

This formula enables us to calculate the correction to the first eigenvalue due to the additional volume.

Thus, for instance,

$$\varepsilon = \begin{cases} 0\cdot 12833 & \text{for} \quad \chi = 1, \\ 0\cdot 08685 & \text{for} \quad \chi = 5 \end{cases}$$

and correspondingly

$$\mu_1 = 2\cdot 5331,$$

$$\mu_1 = 2\cdot 49165.$$

35. The fundamental frequency $\omega_1^0 \approx 1520 \text{ sec}^{-1}$. It can be increased $1\cdot 45$ times if the air volume is added

$$V_0 \approx 2918 \text{cm}^3 \quad (\chi = 10).$$

Method. See problem 34.

§ 3. Propagation and Radiation of Sound

The acoustic equations† have the form

$$\left.\begin{aligned} \boldsymbol{v}_t &= -c^2 \operatorname{grad} s, \\ s_t + \operatorname{div} \boldsymbol{v} &= 0, \end{aligned}\right\} \tag{1}$$

where $\boldsymbol{v}$ is the velocity vector of the gas, $s = (\rho - \rho_0)/\rho_0$ the condensation of the gas, $c = \sqrt{\gamma p_0/\rho_0}$ the velocity of sound, ρ_0 and p_0 the initial density and initial pressure, γ the adiabatic constant.

Assuming

$$\boldsymbol{v} = -\operatorname{grad} U \quad \text{or} \quad s = c^2 U_t,$$

where U is the velocity potential, we obtain the wave equation

$$U_{tt} = c^2 \Delta U.$$

The total pressure p is simply expressed in terms of the condensation s:

$$p = p_0(1 + \gamma s).$$

Denoting the excess pressure by $P = p - p_0$ we have:

$$P = \frac{\gamma p_0}{\rho_0} \rho_0 c^2 s = \rho_0 c^2 s.$$

Hence the excess pressure P also satisfies the wave equation

$$P_{tt} = c^2 \Delta P.$$

† See, for instance, [7], chapter II.

If the boundary Σ of the region, in which the solution is sought, is assumed absolutely rigid, then the normal component of the velocity equals zero:

$$v_n\Big|_{\Sigma} = 0 \quad \text{or} \quad \frac{\partial U}{\partial n}\Big|_{\Sigma} = 0, \quad \frac{\partial P}{\partial n}\Big|_{\Sigma} = 0,$$

where n is the normal to Σ.

The kinetic energy of the volume $\mathrm{d}x\,\mathrm{d}y\,\mathrm{d}z$ of the gas equals $\frac{1}{2}\rho_0 v^2\,\mathrm{d}x\,\mathrm{d}y\,\mathrm{d}z$. The potential energy is given by the expression

$$\frac{1}{2}|Ps| = \frac{1}{2\rho_0 c^2} P^2 .$$

The total energy per unit volume equals

$$W = \frac{1}{2}\rho_0 v^2 + \frac{1}{2\rho_0 c^2} P^2 .$$

Using Green's theorem, the law of conservation of energy is readily written down in the form

$$\frac{\partial}{\partial t}\int_T W\,\mathrm{d}\tau = -\int_S \boldsymbol{Y}\,\mathrm{d}\boldsymbol{S}, \qquad \boldsymbol{Y} = p\boldsymbol{v},$$

where T is some volume, bounded by the surface S. The vector $\boldsymbol{Y} = p\boldsymbol{v}$ is the energy flow per unit time across unit surface, called the Poynting vector.

The total energy, emitted by some source per unit time (the total output) equals

$$\Pi = \int_S \boldsymbol{Y}\,\mathrm{d}\boldsymbol{S},$$

where S is some closed surface surrounding the source. In the case of steady-state vibrations

$$P = p e^{i\omega t},$$

where p satisfies the wave equation

$$\Delta p + k^2 p = 0 .$$

In this section consider steady-state acoustic processes, i.e. we shall deal with wave equations, omitting the time factor $e^{i\omega t}$. If $s = s e^{i\omega t}$, $\boldsymbol{v} = \boldsymbol{v} e^{i\omega t}$, then s and $\boldsymbol{v}$ also satisfy the wave equation, where $p = ikc\rho_0 U$.

In the case of an harmonic dependence on time one usually uses quanitites which are values of the functions being considered averaged over a period. If the dependence on time is taken in the form $e^{i\omega t}$, then the amplitudes of p and v will be complex. Taking this into account, we obtain for the energy flow averaged with respect to time the expression

$$\overline{\boldsymbol{Y}} = \frac{1}{2}\operatorname{Re}(p\boldsymbol{v}^*),$$

called also the intensity or strength of the sound.

In acoustics the idea of impedance is widely used. As is known, the mechanical impedance of a system z is defined as the ratio of the pressure to the velocity. The quantity $\rho_0 c$ is called the acoustic radiation resistance. The dimensionless acoustic impedance is given by the ratio

$$\zeta = \frac{z}{\rho_0 c} \quad \text{or} \quad \zeta = \frac{p}{\rho_0 v c}.$$

1. Point Source

36. It is required to find the function

$$G(x, y, z, \xi, \eta, \zeta) = \frac{e^{-ikr}}{4\pi r} + v,$$

$$r = \sqrt{(x-\xi)^2+(y-\eta)^2+(z-\zeta)^2},$$

where v is the regular solution of the wave equation, which must be chosen so that for $z = 0$ one of the conditions: $G|_{z=0} = 0, \left.\frac{\partial G}{\partial z}\right|_{z=0} = 0$ is fulfilled.

(a) $$G(x, y, z, \xi, \eta, \zeta) = \frac{e^{-ikr}}{4\pi r} - \frac{e^{-ikr_1}}{4\pi r_1},$$

$$r_1 = \sqrt{(x-\xi)^2+(y-\eta)^2+(z+\zeta)^2}.$$

The solution of the first boundary-value problem will be:

$$u(x, y, z) = -\frac{z}{2\pi} \iint\limits_{-\infty}^{\infty} \left(ik + \frac{1}{R}\right) \frac{e^{-ikR}}{R} f(\xi, \eta)\, d\xi\, d\eta,$$

$$R = \sqrt{(x-\xi)^2+(y-\eta)^2+z^2};$$

(b) $$\hat{G}(x, y, z, \xi, \eta, \zeta) = \frac{e^{-ikr}}{4\pi r} + \frac{e^{-ikr_1}}{4\pi r_1}.$$

The solution of the second boundary-value problem will be:

$$u(x, y, z) = \iint\limits_{-\infty}^{\infty} \hat{G}|_{\zeta=0} f(\xi, \eta)\, d\xi\, d\eta = \frac{1}{2\pi} \iint\limits_{-\infty}^{\infty} \frac{e^{-ikR}}{R} f(\xi, \eta),\, d\xi\, d\eta.$$

Method. In order to find the source function G use the method of images.

37. (a) $$G(M, P) = -\frac{i}{4}[H_0^{(2)}(kr) - H_0^{(2)}(kr_1)],$$

(b) $$\hat{G}(M, P) = -\frac{i}{4}[H_0^{(2)}(kr) + H_0^{(2)}(kr_1)],$$

where

$$M = M(x, y), \qquad P = P(\xi, \eta),$$

$$r = \sqrt{(x-\xi)^2+(y-\eta)^2}, \qquad r_1 = \sqrt{(x-\xi)^2+(y+\eta)^2}.$$

Method. It is required to find the solution of the wave equation $\Delta_2 v+k^2 v = 0$, satisfying for $z=0$ the boundary condition $v=0$ or $\partial v/\partial z = 0$, at infinity the radiation condition

$$\lim_{r\to\infty} \sqrt{r}\left(\frac{\partial v}{\partial r}+ikv\right) = 0 \tag{1}$$

or

$$\lim_{r\to\infty} \sqrt{r}\left(\frac{\partial v}{\partial r}-ikv\right) = 0 \tag{2}$$

and having at $r\to 0$ a logarithmic singularity, i.e. representable in the form

$$G = -\frac{i}{4}H_0^{(2)}(kr)+v.$$

We use the radiation condition in the form (1) in connection with the choice of the time factor in the form $e^{i\omega t}$.

38. The velocity potential of a point source equals

$$U = Q_0\frac{e^{-ikr}}{4\pi r},$$

where Q_0 is the strength of the source.

The velocity $\boldsymbol{v} = -\text{grad } U$ has a radial component

$$v_r = \left(ik+\frac{1}{r}\right)U.$$

The pressure

$$p = ikc\rho_0 U.$$

The total energy radiated per unit time (the value averaged with respect to time)

$$\Pi = \frac{Q_0^2 k^2 c\rho_0}{8\pi}.$$

The dimensionless acoustic impedance

$$\zeta = \frac{ikc\rho_0}{\left(ik+\frac{1}{r}\right)c\rho_0} = \frac{1}{1+\frac{1}{ikr}}.$$

If r is sufficiently large, then

$$\zeta = 1+\frac{i}{kr}+\dots$$

Method. In order to calculate the velocity v and the excess pressure p use the formulae

$$v_r = -\frac{\partial U}{\partial r}, \quad p = ikc\rho_0 U.$$

The energy flow is calculated as the value averaged with respect to time of the product of the pressure and the velocity

$$\overline{Y} = \frac{1}{2}\mathrm{Re}(pv^*) = \frac{Q_0^2 c\rho_0 k^2}{32\pi^2 r^2}.$$

The total radiation energy

$$\Pi = \overline{Y} 4\pi r^2 = \frac{Q_0^2 k^2 c\rho_0}{8\pi}.$$

39. Let $P_0(0, 0, -a)$ be the rectangular coordinates of a point source of sound, $P_1(0, 0, a)$ its mirror image in the plane $z = 0$.

The velocity potential equals

$$U = \left(\frac{e^{-ikr}}{4\pi r} + \frac{e^{-ikr_1}}{4\pi r_1}\right) Q_0,$$

$$r = \sqrt{x^2+y^2+(z-a)^2}, \quad r_1 = \sqrt{x^2+y^2+(z+a)^2},$$

where

$$r_1^2 = \sqrt{r^2+4a^2+4ar\cos\theta},$$

where θ is the angle between P_0M and P_0P_1, $M(x, y, z)$ is the point of observation.

At large distances from the source (in the wave zone) we have:

$$r_1 = r + 2a\cos\theta,$$

so that

$$U = \frac{Q_0 e^{-ikr}}{4\pi r}(1+e^{-2iak\cos\theta}).$$

The radiation intensity

$$\overline{Y} = 2\overline{Y}_{38}[\cos(2ak\cos\theta)+1].$$

The total radiation energy

$$\Pi = 2\pi r^2 \int_0^{\frac{\pi}{2}} \overline{Y}\sin\theta\, d\theta = \frac{Q_0^2 k^2 c\rho_0}{8\pi}\left(1+\frac{\sin 2ak}{2ak}\right) = \Pi_{38}\left(1+\frac{\sin 2ak}{2ak}\right),$$

where $\overline{Y}_{38}$ and Π_{38} are the intensity and total energy of radiation of the point source considered in problem 38.

40. In this case for $z = 0$ the boundary condition $u = 0$ will hold for the velocity potential so that

$$U = \left(\frac{e^{-ikr}}{4\pi r} - \frac{e^{-ikr_1}}{4\pi r_1}\right) Q_0,$$

$$\Pi = \frac{Q_0^2 k^2 c\rho_0}{8\pi}\left(1-\frac{\sin 2ak}{2ak}\right) = \Pi_{38}\left(1-\frac{\sin 2ak}{2ak}\right).$$

41. *Method.* It is required to prove that

$$v_M(P) = v_P(M)$$

where $v_M(P)$ is the value at point P of the solution of the wave equation with the source at point M, $v_P(M)$ the solution at point M, the source at point P.

For the proof one must use Green's theorem.

42. It is required to find particular solutions of the equation

$$\Delta_2 u + \frac{\partial^2 u}{\partial z^2} + k^2 u = 0$$

for the condition

$$\left.\frac{\partial u}{\partial n}\right|_{\Sigma} = 0 \; (u \text{ is the velocity potential})$$

and the condition of the absence of waves arriving from infinity (radiation condition).

There exist particular solutions in the form of travelling waves

$$u_n(M, z) = A_n \psi_n(M) e^{i\gamma_n z} \quad \big(M = M(x, y)\big).$$

where

$$\gamma_n = \sqrt{k^2 - \lambda_n},$$

λ_n and $\psi_n(M)$ are the eigenvalues and eigenfunctions of a membrane having the shape of a perpendicular section S of the tube,

$$\Delta_2 \psi_n + \lambda_n \psi_n = 0 \text{ in } S, \quad \left.\frac{\partial \psi_n}{\partial n}\right|_C = 0 \quad (C \text{ is the boundary of } S).$$

If $\lambda_{n_0} < k^2$ and $\lambda_{n_0+1} > k^2$, then there exist n_0 travelling waves. For $n > n_0$ we have:

$$u = A_n \psi_n(M) e^{-p_n |z|}, \quad p_n = \sqrt{\lambda_n - k^2}$$

attenuated waves. We note that everywhere we shall assume the eigenfunctions to be normalized to unity. The maximum permissible wavelength, able to be propagated in the tube,

$$\Lambda_{\max} = \frac{2\pi}{\sqrt{\lambda_1}}$$

for a circular tube of radius a $\Lambda_{\max} \approx 2 \cdot 613a$.

The phase velocity

$$v_\phi = \frac{c}{\sqrt{1 - \frac{\lambda_n}{k^2}}} > C.$$

The excess pressure

$$P = -ikc\rho_0 u.$$

The velocity of particles along the z-axis equals

$$v_z = -i\gamma_n u.$$

The energy flow through a cross-section of the tube

$$\overline{Y}_n = \frac{1}{2} kc\rho_0\gamma_n |A_n|^2 \iint\limits_S \psi_n^2 \, dS \cdot = \frac{1}{2} kc\rho_0\gamma_n |A_n|^2 .$$

For a tube of circular section of radius a we have:

$$\lambda_n = \lambda_{m,n} = \left(\frac{\mu_m^{(n)}}{a}\right)^2 ,$$

$$\hat{\psi}_{m,n} = \sqrt{\frac{\varepsilon_n}{\pi a^2}} \frac{\hat{\mu}_m^{(n)}}{\sqrt{(\hat{\mu}_m^{(n)})^2 - n^2}} \frac{J_n\left(\frac{\hat{\mu}_m^{(n)} r}{a}\right)}{J_n(\hat{\mu}_m^{(n)})} \begin{matrix}\cos\\ \sin\end{matrix}\, n\phi, \qquad \varepsilon_n = \begin{cases} 1, & n = 0, \\ 2, & n \neq 0. \end{cases}$$

The energy flow

$$\overline{Y}_{m,n} = \frac{1}{2} |A_n|^2 kc\rho_0 \sqrt{k^2 - \left(\frac{\hat{\mu}_m^{(n)}}{a}\right)^2},$$

where $\mu_m^{(n)}$ is a root of the equation $J_n'(\mu) = 0$.

For a tube of rectangular section $0 \leqslant x \leqslant a$, $0 \leqslant y \leqslant b$ we have:

$$\lambda_{m,n} = \pi^2\left(\frac{m^2}{a^2} + \frac{n^2}{b^2}\right),$$

$$\psi_{m,n} = \sqrt{\frac{4\varepsilon_n \varepsilon_m}{ab}} \cos\frac{\pi m}{b} x \cos\frac{\pi n}{b} y \qquad (n, m = 0, 1, 2, \ldots).$$

The energy flow

$$\overline{Y}_{m,n} = \frac{1}{2} |A_{m,n}|^2 kc\rho_0 \sqrt{k^2 - \pi^2\left(\frac{m^2}{a^2} + \frac{n^2}{b^2}\right)} .$$

43. (a) $$G(M, P, z, \zeta) = \sum_{n=0}^{\infty} \frac{\psi_n(M)\psi_n(P)}{2\varkappa_n} e^{-\varkappa_n|z-\zeta|},$$

where $\varkappa_n = \sqrt{\lambda_n - k^2}$; λ_n and ψ_n are the eigenvalues and eigenfunctions of the first boundary-value problem: $\Delta_2\psi_n + \lambda_n\psi_n = 0$ in a cross-section S, $\psi_n = 0$ on the boundary C of the section S;

(b) $$\hat{G}(M, P, z, \zeta) = \sum_{n=0}^{\infty} \frac{\hat{\psi}_n(M)\hat{\psi}_n(P)}{2\hat{\varkappa}_n} e^{-\hat{\varkappa}|z-\zeta|}$$

where $\hat{\varkappa}_n = \sqrt{\hat{\lambda}_n - k^2}$; $\hat{\lambda}_n$ and $\hat{\psi}_n$ are the eigenvalues and eigenfunctions of the second boundary-value problem

$$\Delta_2\hat{\psi}_n + \hat{\lambda}_n\hat{\psi}_n = 0 \quad \text{in} \quad S, \qquad \left.\frac{\partial\hat{\psi}_n}{\partial\nu}\right|_C = 0 .$$

If S is a circle of radius a, then

$$\psi_n(M) = \psi_{m,n}(r, \phi) = \sqrt{\frac{\varepsilon_n}{\pi a^2}}\,\frac{J_n\left(\frac{\mu_m^{(n)}}{a}r\right)}{J_n'(\mu_m^{(n)})}\,{\cos \atop \sin} n\phi.$$

$$\hat{\psi}_n(M) = \hat{\psi}_{m,n}(r, \phi) = \sqrt{\frac{\varepsilon_n}{\pi a^2}}\,\frac{\hat{\mu}_m^{(n)}}{\sqrt{(\mu_m^{(n)})^2 - n^2}}\,\frac{J_n\left(\frac{\hat{\mu}_m^{(n)}}{a}r\right)}{J_n(\hat{\mu}_m^{(n)})}\,{\cos \atop \sin} n\phi,$$

where

$$\varepsilon_n = \begin{cases} 1 & \text{for} \quad n = 0, \\ 2 & \text{for} \quad n \neq 0, \end{cases}$$

$\mu_m^{(n)}$ is a root of the equation $J_n(\hat{\mu}) = 0$, and $\hat{\mu}_m^{(n)}$ is a root of the equation $J_n'(\mu) = 0$.

Method. One must apply the method of separation of variabies to the inhomogeneous equation

$$\Delta_2 v + \frac{\partial^2 v}{\partial z^2} + k^2 v = -f(M, z),$$

where f is an arbitrary function, and represents the solution in the form

$$v = \iint\limits_S \int\limits_{-\infty}^{\infty} G(M, P, z, \zeta) f(P, \zeta)\, d\sigma_P\, d\zeta.$$

If $v(M, z)$ is sought in the form

$$v(M, z) = \sum_{n=1}^{\infty} v_n(z) \psi_n(M),$$

then for $v_n(z)$ we obtain the equation

$$v_n'' - \varkappa_n^2 v_n(z) = -f_n(z), \quad f_n = \iint\limits_S f(P, z) \phi_n(P)\, d\sigma_P$$

solving which we find:

$$v_n = \frac{1}{2\varkappa_n} \int\limits_{-\infty}^{\infty} e^{-\varkappa_n |z-\zeta|} f_n(\zeta)\, d\zeta$$

or

$$v_n(z) = \int\limits_{\infty}^{\infty} \iint\limits_S \frac{e^{-\varkappa_n |z-\zeta|}}{2\varkappa_n} \psi_n(P) f(P, \zeta)\, d\sigma_P\, d\zeta.$$

44. The source function for the semi-infinite tube $z > 0$ of arbitrary section S

$$\text{(a)}\quad G(M, P, z, \zeta) = \sum_{n=1}^{\infty} \frac{\psi_n(M)\psi_n(P)}{2\varkappa_n}\, e^{-\varkappa_n \zeta} \sinh \varkappa_n z,$$

$$\text{(b)}\quad G(M, M', z, \zeta) = \sum_{n=1}^{\infty} \frac{\hat{\psi}_n(M)\hat{\psi}_n(M')}{2\hat{\varkappa}_n}\, e^{-\hat{\varkappa}_n \zeta} \cosh \hat{\varkappa}_n z.$$

Here ψ_n and $\hat{\psi}_n$ are eigenfunctions of the first and second boundary-value problems for the membrane S

$$\varkappa_n = \sqrt{\lambda_n - k^2}, \qquad \hat{\varkappa}_n = \sqrt{\hat{\lambda}_n - k^2}.$$

Method. Apply the method of images to the function

$$Z_n(z) = e^{-\varkappa_n |z-\zeta|},$$

so that

$$\text{(a)}\quad Z_n(z) = e^{-\varkappa_n(\zeta - z)} - e^{-\varkappa_n(\zeta + z)} = 2e^{-\varkappa_n \zeta} \sinh \varkappa_n z,$$

$$\text{(b)}\quad Z_n(z) = e^{-\hat{\varkappa}_n(\zeta - z)} + e^{-\hat{\varkappa}_n(\zeta + z)} = 2e^{-\hat{\varkappa}_n \zeta} \cosh \hat{\varkappa}_n z.$$

45. The source function, giving the spatial distribution for the velocity potential, equals

$$G(M, M', z, \zeta) = \sum_{n=1}^{\infty} \psi_n(M)\psi_n(M')K_n(z, \zeta),$$

where

$$K_n(z, \zeta) = \begin{cases} \dfrac{\cosh p_n z \cosh p_n(l-\zeta)}{p_n \sinh p_n l} & \text{for} \quad z < \zeta, \\[2ex] \dfrac{\cosh p_n(l-z) \cosh p_n \zeta}{p_n \sinh p_n l} & \text{for} \quad z > \zeta, \end{cases}$$

where $p_n = \sqrt{\lambda_n - k^2}$, $\psi_n(M)$ and λ_n are the eigenfunction and eigenvalue of the boundary-value problem

$$\Delta_2 \psi_n + \lambda_n \psi_n = 0 \text{ in } S, \qquad \frac{\partial \psi_n}{\partial \nu} = 0 \text{ on } C,$$

if S is the cross-section of the resonator, C the boundary of S.

Method. Considering the equation (see problem 42) for the velocity potential

$$\Delta^2 U + \frac{\partial^2 U}{\partial z^2} + k^2 U = -f(M, z)$$

with boundary conditions

$U|_\Sigma = 0$, $U_z|_{z=0,\, z=l} = 0$ (Σ the lateral face of the resonator) and assuming

$$U(M, z) = \sum_{n=1}^{\infty} v_n(z)\psi_n(M),$$

we obtain for $v_n(z)$ the equation

$$v_n'' - p_n^2 v_n = -f_n(z), \quad v_n'(0) = v_n'(l) = 0.$$

Its solution has the form

$$v_n = \int_0^l K_n(z, \zeta) f_n(\zeta)\, d\zeta,$$

where $K_n(z, \zeta)$ is the corresponding Green's function for the equation

$$v_n'' - p_n^2 v_n = 0.$$

For further detail see problem 43.

2. Radiation of Membranes, Cylinders and Spheres

46. The velocity

$$v = v_0\, e^{-ikz}, \quad z > 0.$$

The pressure

$$p = c\rho_0 v_0\, e^{-ikz}, \quad z > 0.$$

The energy flow (the average value with respect to time)

$$Y = 0{\cdot}5\, c\rho_0 v_0^2.$$

The specific acoustic impedance

$$\zeta = 1.$$

47. If at the boundary $z = 0$ the velocity is

$$v_z|_{z=0} = v_0(r) = \sum_{m=1}^{\infty} A_m J_0\left(\frac{\mu_m}{a} r\right),$$

where μ_m is a root of the equation $J_0(\mu) = 0$, then

$$v_z(r, z) = \sum_{m=1}^{\infty} A_m J_0\left(\frac{\mu_m}{a} r\right) e^{-i\gamma_m z} = \sum_{m=1}^{\infty} A_m J_0\left(\frac{\mu_m}{a} r\right) e^{-\varkappa_m z} \qquad (z > 0),$$

$$\gamma_m = \sqrt{k^2 - \frac{\mu_m^2}{a^2}}, \qquad \varkappa_m = \sqrt{\frac{\mu_m^2}{a^2} - k^2},$$

a is the radius of the tube.

The pressure

$$p = \sum_{m=0}^{\infty} B_m J_0\left(\frac{\mu_m}{a} r\right) e^{-i\gamma_m z} \qquad (z > 0),$$

where

$$B_m = \frac{\omega \rho_0}{\gamma_m} A_m .$$

If

$$v(r, 0) = A J_0\left(\frac{\mu_1}{a} r\right),$$

then

$$B_m = \begin{cases} 0 & \text{for} \quad m > 1, \\ \dfrac{\omega \rho_0 A}{\sqrt{k^2 - \dfrac{\mu_1^2}{a^2}}} & \text{for} \quad m = 1. \end{cases}$$

The average velocity of the piston

$$V = \frac{2J_1(\mu_1)}{\mu_1} A \approx 0{\cdot}428\, A .$$

The impedance

$$\zeta = \frac{p}{\rho_0 c v} = \frac{k}{\sqrt{k^2 - \dfrac{\mu_1^2}{a^2}}} .$$

The energy flow through a cross-section

$$\Pi = \frac{\pi \omega \rho_0 a^2}{2\mu_1^2 \sqrt{k^2 - \dfrac{\mu_1^2}{a^2}}} A^2 J_1^2(\mu_1) .$$

48. The radial velocity

$$v_r = v_0 \frac{H_1^{(2)}(kr)}{H_1^{(2)}(ka)} .$$

The pressure

$$p = i c \rho_0 v_0 \frac{H_0^{(2)}(kr)}{H_1^{(2)}(ka)} .$$

The impedance

$$\zeta = \frac{p}{\rho_0 c v_r} = i \frac{H_0^{(2)}(kr)}{H_1^{(2)}(kr)} = 1 + \ldots$$

At large distances, for $kr \gg 1$, we have:

$$v_r = \frac{v_0}{H_1^{(2)}(ka)}\sqrt{\frac{2}{\pi kr}}\,e^{-i\left(kr-\frac{3\pi}{4}\right)} = \frac{iv_0}{H_1^{(2)}(ka)}\sqrt{\frac{2}{\pi kr}}\,e^{-i\left(kr-\frac{\pi}{4}\right)}+\dots$$

$$p = \frac{ic\rho_0 v_0}{H_1^{(2)}(ka)}\sqrt{\frac{2}{\pi kr}}\,e^{-i\left(kr-\frac{\pi}{4}\right)}+\dots$$

The energy flow

$$\bar{Y} = 0{\cdot}5pv_r^* = \frac{c\rho_0 v_0^2}{\pi kr\left|H_1^{(2)}(ka)\right|^2}.$$

Method. It is required to solve the equation

$$\Delta_2 v + k^2 v = 0$$

in the region $r \geqslant a$ for the additional conditions

$$v|_{r=a} = v_0$$

and

$$\lim_{r\to\infty}\sqrt{r}\left(\frac{\partial v}{\partial r}+ikv\right) = 0 \quad \text{(radiation condition).}$$

49. The excess pressure equals

$$p \approx \pi^2 \nu a\rho_0 v_0 H_0^{(2)}(kr) \quad \left(\nu = \frac{\omega}{2\pi}\right).$$

In the wave zone

$$p = \sqrt{c\nu}\,\pi a\rho_0 v_0 \frac{e^{-i\left(kr-\frac{\pi}{4}\right)}}{\sqrt{r}}+\dots,$$

$$v_r = \pi a v_0\sqrt{\frac{\nu}{c}}\,\frac{e^{-i\left(kr-\frac{\pi}{4}\right)}}{\sqrt{r}}+\dots,$$

$$\zeta = 1.$$

The total radiated energy per unit length of the cylinder approximately equals

$$\Pi = \pi^3\rho_0\nu a^2 v_0^2.$$

Method. Use the expansion

$$H_1^{(2)}(x) = \frac{2i}{\pi x}+\dots \quad \text{for small } x,$$

$$H_0^{(2)}(x) = \sqrt{\frac{2}{\pi x}}\,e^{-i\left(x-\frac{\pi}{4}\right)} \quad \text{for large } x.$$

50. The pressure

$$p = A\cos\phi H_1^{(2)}(kr).$$

The radial velocity

$$v_r = \frac{iA}{c\rho_0}\cos\phi[H_1^{(2)}(kr)]', \qquad A = \frac{-ic\rho_0 v_0}{H_1^{(2)\prime}(ka)}.$$

The specific acoustic impedance

$$\zeta|_{r=a} = -i\frac{H_1^{(2)}(ka)}{H_1^{(2)\prime}(ka)}.$$

If $ka \ll 1$, then

$$A = \frac{2\pi^3 v^2 a^2 \rho_0 v_0}{c}, \qquad \zeta = \frac{\pi(ak)^2 k}{2} - ik = \zeta_0 - i\zeta_1,$$

where $\zeta_0 \ll \xi_1$.

The total radiation output per unit length

$$\Pi = \int_0^{2\pi} \overline{Y}r\,\mathrm{d}\phi = \frac{\pi^2}{4c}\rho_0\omega^3 a^4 v_0^2.$$

The reaction of the air per unit length of the cylinder in the direction of its motion

$$F = \int_0^{2\pi} ap(a, \phi)\cos\phi\,\mathrm{d}\phi = ika^2\pi c\rho_0 v_0\dagger.$$

Method. Consider that the boundary condition has the form

$$v_r|_{r=a} = v_0\cos\phi.$$

51. If

$$f(\phi) = \frac{a_0}{2} + \sum_{m=1}^{\infty}(a_m\cos m\phi + b_m\sin m\phi),$$

where

$$a_0 = \frac{1}{\pi}\int_0^{2\pi} f(\phi)\,\mathrm{d}\phi, \quad a_m = \frac{1}{2\pi}\int_0^{2\pi} f(\phi)\cos m\phi\,\mathrm{d}\rho \qquad (m = 1, 2, \ldots),$$

$$b_m = \frac{1}{2\pi}\int_0^{2\pi} f(\phi)\sin m\phi\,\mathrm{d}\phi,$$

then

$$p = \sum_{m=0}^{\infty}(A_m\cos m\phi + B_m\sin m\phi)H_m^{(2)}(kr),$$

† Here, as everywhere, the factor $e^{i\omega t}$ is omitted.

where

$$A_m = -\frac{ic\rho_0}{H_m^{(2)\prime}(ka)} a_m, \quad B_m = \frac{-ic\rho_0}{H_m^{(2)\prime}(ka)} b_m \quad (m = 1, 2, \ldots),$$

$$A_0 = -\frac{-ic\rho_0}{H_0^{(2)\prime}(ka)} \frac{a_0}{2}.$$

For

$$f(\phi) = v_0 = \frac{a_0}{2} \quad a_m = b_m = 0, \quad m > 0,$$

we obtain

$$p = \frac{ic\rho_0}{H_1^{(2)}(ka)} v_0 H_0^{(2)}(kr),$$

$$v = v_0 \frac{H_1^{(2)}(kr)}{H_1^{(2)}(ka)},$$

i.e. the solution of problem 48. Similarly the solutions of problems 49 and 50 are obtained.

52. The pressure

$$p = A\zeta_1^{(2)}(kr)P_1(\cos\theta).$$

The radial velocity

$$v_r = \frac{iA}{3c\rho_0}[\zeta_0^{(2)}(kr) - 2\zeta_2^{(2)}(kr)]P_1(\cos\theta),$$

where

$$\zeta_n^{(2)}(x) = \sqrt{\frac{\pi}{2x}} H_{n+\frac{1}{2}}^{(2)}(x),$$

$P_1(x) = x$ is a Legendre polynomial of first degree. If $ka \ll 1$, then

$$A = 0{\cdot}5\, c\rho_0(ak)^3 v_0.$$

The total force acting on the sphere in the direction of its vibrations

$$F = -i\omega \cdot \frac{2}{3}\pi\rho a^3 v_0.$$

The dimensionless specific impedance for $r = a$

$$\zeta = -\frac{i3\zeta_1^{(2)}(ka)}{\zeta_0^{(2)}(ka) - 2\zeta_2^{(2)}(ka)} = -\frac{2i}{3}ka + \ldots$$

In the wave zone

$$p = \frac{A}{kr} e^{-i(kr-\pi)} \cos\theta,$$

$$v_r = \frac{A}{c\rho_0 kr} e^{-i(kr-\pi)} \cos\theta.$$

The total energy radiated by the dipole per unit time

$$\overline{Y} = 0\cdot 5\frac{A^2\cos^2\theta}{c\rho_0(kr)^2} = \frac{1}{8}\frac{c\rho_0(ak)^6 v_0^2}{(kr)^2}\cos^2\theta.$$

The total output radiated by the acoustic dipole equals

$$\Pi = \frac{\pi}{6}c\rho_0 v_0^2 a^2(ak)^4,$$

i.e.

$$\Pi \sim k^4 \quad \text{or} \quad \Pi \sim \frac{1}{\lambda^4}.$$

53. If the expansion

$$f(\theta) = \sum_{m=0}^{\infty} A_m P_m(\cos\theta),$$

is possible, then the velocity

$$v = \sum_{m=0}^{\infty} \frac{A_m \zeta_m^{(2)\prime}(kr)}{\zeta_m^{(2)\prime}(ka)} P_m(\cos\theta),$$

the excess pressure

$$p = \sum_{m=0}^{\infty} \frac{B_m \zeta_m^{(2)}(kr)}{\zeta_m^{(2)\prime}(ka)} P_m(\cos\theta),$$

where

$$B_m = -ic\rho_0 A_m.$$

If $ka \ll 1$ then the total reaction of the medium on the sphere may be calculated from the formula

$$F = 2\pi a^2 \int_0^{\pi} (p)_{r=a}\cos\theta\sin\theta\, d\theta = \frac{4\pi a^2}{3} B_1 \frac{\zeta_1^{(2)}(ka)}{\zeta_2^{(2)\prime}(ka)} = \frac{ic\rho_0}{2}\left(\frac{4\pi a^3}{3}\right)A_1 k + \dots$$

If $f(\theta) = v_0$, then

$$A_m = \begin{cases} v_0 & \text{for} \quad m = 0, \\ 0 & \text{for} \quad m \neq 0 \end{cases}$$

and

$$F = 0.$$

Method. In order to calculate the total force acting on the sphere one must use the formulae

$$\zeta_1^{(2)}(\rho) \approx \frac{i}{\rho_2} + \dots; \quad \zeta_1^{(2)\prime}(\rho) \approx -\frac{2i}{\rho^3} + \dots \text{ for small } \rho.$$

54. If the velocity of the surface of the sphere equals zero everywhere except a small circular area of radius ε, around the point $\theta = 0$ (pole)

$$v(\theta) = \begin{cases} v_0 & \text{for} \quad 0 \leqslant \theta \leqslant \dfrac{\varepsilon}{a}, \\ 0 & \text{for} \quad \dfrac{\varepsilon}{a} < \theta < \pi, \end{cases}$$

where a is the radius of the sphere, then

$$p = -ic\rho_0 \sum_{m=0}^{\infty} A_m \frac{\zeta_m^{(2)}(kr)}{\zeta_m^{(2)\prime}(ka)} P_m(\cos\theta),$$

$$v_r = \sum_{m=0}^{\infty} A_m \frac{\zeta_m^{(2)\prime}(kr)}{\zeta_m^{(2)\prime}(ka)} P_m(\cos\theta),$$

$$A_m = \frac{2m+1}{2} v_0 \int_0^{\varepsilon/a} P_m(\cos\theta)\sin\theta\, d\theta \approx \frac{2m+1}{4}\left(\frac{\varepsilon}{a}\right)^2 v_0.$$

The total radiation energy equals

$$\Pi = \rho c v_0^2 \frac{\varepsilon^4}{32a^2} \frac{4\pi}{(ka)^2} \sum_{m=0}^{\infty} \frac{2m+1}{D_m^2}.$$

For very low frequencies

$$p \approx i\omega \frac{\rho_0}{4\pi r}(\pi\varepsilon^2 v_0)e^{ikr} = -i\omega \frac{\rho_0}{4\pi r} e^{ikr} Q_0,$$

where $Q_0 = \pi\varepsilon^2 v_0$ is the strength of the point source.

Method. An expression for D_m is obtained from formula (1) of problem 62.

55. The radial component of the velocity equals

$$v_r(r, \theta) = v_0 \frac{\zeta_2^{(2)\prime}(kr)}{\zeta_2^{(2)\prime}(ka)} P_2(\cos\theta).$$

The pressure

$$p = -ic\rho_0 v_0 \frac{\zeta_2^{(2)}(kr)}{\zeta_2^{(2)\prime}(ka)} P_2(\cos\theta).$$

For $ka \ll 1$ the intensity and strength of radiation of the quadrupole will equal:

$$\overline{Y} = \frac{c\rho_0 k^6 a^8}{162 r^2} v_0^2 P_2^2(\cos\theta),$$

$$\Pi = \frac{2\pi}{405} c\rho_0 k^6 a^8 v_0^2.$$

Method. Take into account the fact that $v|_{r=a}$ may be written in the form $v|_{r=a} = v_0 P_2(\cos\theta)$, and look for the pressure in the form

$$p = R(r)P_2(\cos\theta).$$

In order to calculate the energy flow and strength of radiation use the formulae

$$\zeta_2^{(2)}(x) = i\frac{3}{x^3} \qquad \text{for small } x,$$

$$\zeta_2^{(2)\prime}(x) = -i\frac{9}{x^4} \qquad \text{for small } x,$$

and also the asymptotic formulae for large x.

It is interesting to compare the formulae

$$\Pi = \frac{\pi}{6}c\rho_0 v_0^2 a^6 k^4 \sim \omega^4 \qquad \text{for dipole radiation,}$$

$$\Pi = \frac{2\pi}{405}c\rho_0 v_0^2 a^8 k^6 \sim \omega^6 \qquad \text{for quadruple radiation.}$$

56. If the velocity of the piston is v_0, then

$$p = \frac{ikc\rho_0 v_0}{2\pi}\int_0^a y\,\mathrm{d}y\int_0^{2\pi}\frac{\mathrm{e}^{-ikR}}{R}\,\mathrm{d}\psi,$$

where R is the distance of the point $M(y, \psi)$ from the point of observation M (Fig. 59).

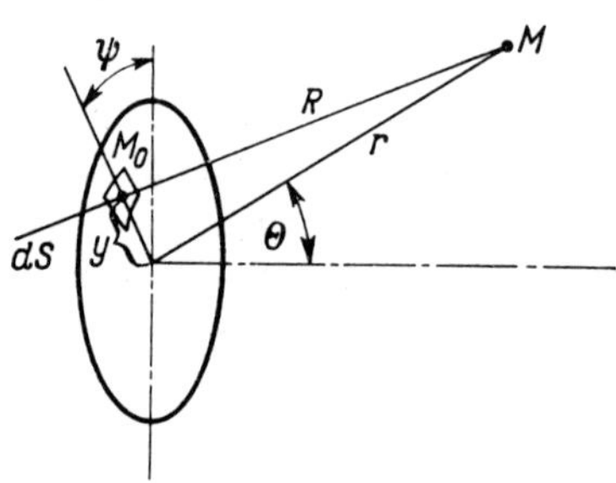

FIG. 59

If $a/R \ll 1$, then

$$p \approx \frac{ikc\rho_0 a^2}{2r}v_0\mathrm{e}^{-ikr}\left[\frac{2J_1(ka\sin\theta)}{ka\sin\theta}\right],$$

$$v_r \approx ika^2\frac{\mathrm{e}^{-ikr}}{2r}v_0\left[\frac{2J_1(ka\sin\theta)}{ka\sin\theta}\right].$$

The energy flow radiated by the piston

$$\overline{Y} = \frac{a^2 c\rho_0 v_0^2}{8r^2}\mu^2\left[\frac{2J_1(\mu\sin\theta)}{\mu\sin\theta}\right]^2,$$

$$\mu = ak.$$

If $\mu \ll 1$, then

$$\overline{Y} = \frac{a^2 c\rho_0 v_0^2}{8r^2}\mu^2\left(1-\frac{\mu^2\sin^2\theta}{8}\right).$$

In this case the total energy equals

$$\Pi = \frac{\pi a^2 c\rho_0 v_0^2}{4}\mu^2\left(1-\frac{\mu^2}{12}\right).$$

Method. The velocity potential, produced by the motion of the piston, is represented as the potential of a double layer

$$U = -\frac{v_0}{2\pi}\int_0^a\int_0^{2\pi}\frac{e^{-ikR}}{R}y\,dy\,d\psi.$$

The pressure

$$p = ikc\rho_0 U.$$

If $R \gg a$, then the function under the integral sign takes the form

$$\frac{e^{-ikR}}{R} = \frac{e^{ikr}\,e^{iky\sin\theta\cos\psi}}{r-y\cos\theta\cos\phi}\qquad (R = r-y\sin\theta\cos\psi),$$

so that

$$U = v_0\frac{e^{-ikr}}{2\pi r}\int_0^a y\,dy\int_0^{2\pi}e^{i(ky\sin\theta)\cos\psi}\,d\psi$$

$$= v_0\frac{e^{-ikr}}{2\pi r}\int_0^a y\,dy\,J_0(ky\sin\theta)2\pi = v_0\frac{e^{-ikr}}{2r}a^2\left[\frac{2J_1(ka\sin\theta)}{ka\sin\theta}\right].$$

Hence formulae for p and v_r are obtained.

57. The pressure at the surface of the lamina

$$p_{\,r=a} = \frac{c\rho_0}{2}v_0\{1-J_0(2\mu)+iM_0(2\mu)\},\quad \mu = ak, \tag{1}$$

where

$$M_0(x) = \frac{2}{\pi}\int_0^{\frac{\pi}{2}}\sin(x\sin\phi)\,d\phi$$

$$= \frac{2}{\pi}\left(x-\frac{x^3}{1^2\cdot 3^2}+\frac{x^5}{1^2\cdot 3^2\cdot 5^2}-\frac{x^7}{1^2\cdot 3^2\cdot 5^2\cdot 7^2}+\dots\right).$$

The force of reaction of the sound field on the lamina

$$F = 2\pi \int_0^a p|_{r=a'} a' \, da' = a^2 \pi c \rho_0 v_0 \left[1 - \frac{J_1(2\mu)}{\mu}\right] + i \frac{\pi c \rho_0 v_0}{2k^2} M_1(2\mu),$$

where

$$M_1(x) = \int_0^x M_0(\xi)\, \xi \, d\xi = \frac{2}{\pi}\left(\frac{x^3}{1^2 \cdot 3} - \frac{x^5}{1^2 . 3^2 . 5} + \frac{x^7}{1^2 . 3^2 . 5^2 . 7} + \dots\right).$$

If the area of the lamina is small, i.e. small a and therefore, small $\mu (\mu \ll 1)$, then

$$F \approx \frac{\pi c \rho_0 \mu^4}{2k^2} v_0 + i \frac{8 c \rho_0}{3k^2} \mu^3 v_0 .$$

The impedance

$$\zeta = i \frac{2\mu}{\pi} + \frac{\mu^2}{2}.$$

Solution. In order to calculate the potential of the double layer at the point $\rho = a'$ on the lamina it is convenient to choose polar coordinates ρ, ϕ with some point of the circumference $\rho = a'$ as pole.

Then

$$U = \frac{v_0}{2\pi} \int_{-\frac{\pi}{2}}^{\frac{\pi}{2}} \int_0^{2a' \cos\phi} e^{-ik\rho} \, d\rho \, d\phi = \frac{v_0}{ik\pi} \int_0^{\frac{\pi}{2}} \{1 - e^{-2ika' \cos\phi}\} \, d\phi .$$

We use the result that

$$\frac{2}{\pi} \int_0^{\frac{\pi}{2}} e^{-2ika' \cos\phi} \, d\phi = \frac{2}{\pi} \int_0^{\frac{\pi}{2}} e^{-2ika' \sin\phi} \, d\phi$$

$$= \frac{2}{\pi} \int_0^{\frac{\pi}{2}} \cos(2ka' \sin\phi) \, d\phi - \frac{2i}{\pi} \int_0^{\frac{\pi}{2}} \sin(2ka' \sin\phi) \, d\phi = J_0(2ka') - iM_0(2ka'),$$

where $M_0(x)$ is a function, given by the formula

$$M_0(x) = \frac{2}{\pi} \int_0^{\frac{\pi}{2}} \sin(x \sin\phi) \, d\phi = \frac{2}{\pi}\left(x - \frac{x^2}{1^2 . 3^2} + \frac{x^5}{1^2 . 3^2 . 5^2} - \dots\right).$$

At $x \to \infty$ the asymptotic formula for $M_0(x)$

$$M_0(x) = \sqrt{\frac{2}{\pi x}} \sin\left(x - \frac{\pi}{4}\right) + \frac{2}{\pi x} + O\left(\frac{1}{x^{3/2}}\right)$$

holds. Therefore, formula (1) is correct for $p|_{r=a}$. In order to find the force of reaction acting on the lamina it is necessary to evaluate the integral

$$F = 2\pi \int_0^a p|_{r=a'} a' \, da' = -i\pi c\rho v_0 \frac{1}{4k^2} \left\{ \int_0^{2\mu} (1 - J_0(\xi')) \xi' \, d\xi' - i \int_0^{2\mu} M_0(\xi) \xi \, d\xi \right\}$$

$$= \frac{1}{2} \pi a^2 \rho_0 c v_0 \left[1 - \frac{J_1(2\mu)}{\mu} \right] + \frac{i\rho_0 c v_0}{4k^2} M_1(2\mu),$$

where

$$M_1(x) = \int_0^x M_0(\xi) \xi \, d\xi .$$

58. In the wave zone (for $kr \gg 1$)

$$p = ikc\rho_0 a^2 \frac{e^{-ikr}}{r} \sum_{m=0}^{\infty} A_m \Phi_m(\theta),$$

$$v = ika^2 \frac{e^{-ikr}}{r} \sum_{m=0}^{\infty} A_m \Phi_m(\theta),$$

where A_m are expansion coefficients of $v|_{z=0} = f(\rho)$ in a series in $J_0\left(\frac{\mu_m}{a}\rho\right)$, equal to

$$A_m = \frac{2}{a^2 J_0^2(\mu_m)} \int_0^a f(\rho) J_0\left(\frac{\mu_m}{a}\rho\right) \rho \, d\rho,$$

$$\Phi_m(\theta) = \frac{2s J_1(s)}{s^2 - \mu_m^2}, \qquad s = ka \sin\theta,$$

μ_m is a root of the equation $J_1(\mu) = 0$.

Solution. Using the expansion of $f(\rho)$ in the series

$$v|_{z=0} = f(\rho) = \sum_{m=0}^{\infty} A_m J_0\left(\frac{\mu_m}{a}\rho\right),$$

we find:

$$p = -\frac{ikc\rho_0}{2\pi} \sum_{m=0}^{\infty} A_m \int_0^a J_0\left(\frac{\mu_m}{a}\rho\right) \rho \, d\rho \int_0^{2\pi} \frac{e^{-ikR}}{R} \, d\phi .$$

In the wave zone

$$\frac{e^{-ikR}}{R} \approx \frac{e^{-ikr}}{r} e^{ik\rho \sin\theta \cos\phi}$$

(see problem 56). Calculations give:

$$\int_0^a J_0\left(\frac{\mu_m}{a}\rho\right) \rho \, d\rho \int_0^{2\pi} e^{ik\rho \sin\theta \cos\phi} \, d\phi = 2\pi \int_0^a J_0\left(\frac{\mu_m}{a}\rho\right) J_0(k\rho \sin\theta) \rho \, d\rho .$$

To find this integral we use the formula

$$\int_0^a J_\nu(\alpha\rho)J_\nu(\beta\rho)\rho\,d\rho = \frac{a}{\beta^2-\alpha^2}[\beta J_\nu'(\alpha a)J_\nu(\beta a)-\alpha J_\nu'(\beta a)J_\nu(\alpha a)],$$

and put $\alpha = \mu_m/a$, $\beta = k\sin\theta$, $\nu = 0$, then we obtain:

$$\int_0^a J_0(\alpha\rho)J_0(\beta\rho)\rho\,d\rho = \frac{a^2 sJ_1(s)}{s^2-\mu_m^2}J_0(\mu_m) \qquad (J_1(\mu_m)=0),$$

so that

$$p = ikc\rho_0 a^2 \sum_{m=0}^{\infty} \frac{A_m J_0(\mu_m)sJ_1(s)}{s^2-\mu_m^2}\,\frac{e^{-ikr}}{r},$$

The first term ($m = 0$, $\mu_0 = 0$) of this series gives the solution of problem 56 on the vibrations of a rigid piston in an infinite screen

$$p = \frac{ikc\rho_0 a^2}{2}A_0\left[\frac{2J_1(s)}{s}\right]+\ldots,$$

$$s = ka\sin\theta,$$

where A_0, obviously, denotes the mean velocity of the piston.

3. Diffraction by a Cylinder and Sphere

59. If a plane wave is propagated along the x-axis perpendicularly to the axis of the cylinder (z-axis), then the pressure in it may be represented in the form

$$p_0 = Ae^{-ikx} = Ae^{-ikr\cos\phi} = A\left[J_0(kr)+2\sum_{n=1}^{\infty} i^m J_n(kr)\cos n\phi\right].$$

The pressure in the scattered wave

$$p_s = \sum_{m=0}^{\infty} B_m \cos m\phi H_m^{(2)}(kr),$$

where

$$B_0 = -\frac{J_1(ka)}{H_1^{(2)}(ka)}A, \quad B_m = -\frac{2i^m J_m'(ka)}{H_m^{(2)\prime}(ka)}A \qquad (m = 1, 2, \ldots).$$

The velocity of the gas in the scattered wave

$$vs_r = \frac{i}{c\rho_0}\sum_{m=0}^{\infty} B_m H_m^{(2)\prime}(kr)\cos m\phi.$$

In the wave zone (at large distances from the cylinder $kr \gg 1$),

$$p_s = \sqrt{\frac{2}{\pi kr}}\,e^{-i\left(kr-\frac{\pi}{4}\right)}\sum_{m=0}^{\infty} B_m i^m \cos m\phi,$$

$$v_{s_r} = \frac{1}{c\rho_0}\sqrt{\frac{2}{\pi kr}}\,e^{-i\left(kr-\frac{\pi}{4}\right)}\sum_{m=0}^{\infty} B_m i^m \cos m\phi = \frac{1}{c\rho_0}p_s.$$

60. The intensity of the scattered wave

$$Y_s = \frac{2Y_0}{\pi kr} |F_s|^2,$$

$$F_s = \sum_{m=0}^{\infty} \varepsilon_m \sin \gamma_m \, e^{-i\gamma_m} \cos m\phi, \quad \varepsilon_0 = 1, \; \varepsilon_m = 2, \; m > 0,$$

$$|F_s|^2 = \sum_{m=0}^{\infty} \sum_{n=0}^{\infty} \varepsilon_m \varepsilon_n \sin \gamma_m \sin \gamma_n \cos (\gamma_m - \gamma_n) \cos m\phi \cos n\phi,$$

γ_m is determined by the relations

$$\tan \gamma_0 = -\frac{J_1(\mu)}{N_1(\mu)},$$

$$\tan \gamma_m = \frac{J_{m-1}(\mu) - J_{m+1}(\mu)}{N_{m+1}(\mu) - N_{m-1}(\mu)}, \qquad \mu = ka,$$

N_m is Neumann's function. The total energy scattered per unit length of the cylinder,

$$\Pi_3 = \frac{4Y_0}{k} \sum_{m=0}^{\infty} \varepsilon_m \sin^2 \gamma_m, \qquad Y_0 = \frac{A^2}{2c\rho_0}.$$

Method. Using the relations

$$J'_m = 0.5(J_{m-1} - J_{m+1}), \qquad N'_m = 0.5(N_{m-1} - N_{m+1}),$$

the coefficients B_m are readily expressed in the form

$$B_m = -\varepsilon_m A i^{m+1} e^{-i\gamma_m} \sin \gamma_m,$$

where

$$\varepsilon_0 = 1, \quad \varepsilon_m = 2, \quad m \geqslant 1;$$

in the wave zone we have:

$$p_s \approx -\sqrt{\frac{4\rho_0 c Y_0 a}{\pi r}} F_s(\phi) e^{-ikr},$$

$$v_{s_r} = \frac{1}{\rho c} p_s.$$

61. If $ka \ll 1$, then in the wave zone we have:

$$p_s \approx \frac{i\sqrt{\pi} a^2 k^2 A}{2\sqrt{2kr}} e^{-i\left(kr - \frac{\pi}{4}\right)} (1 - 2\cos\phi) + \cdots,$$

$$v_{s_r} \approx \frac{1}{c\rho_0} p_s.$$

The intensity

$$\overline{Y}_s = \frac{\pi k^3 a^4}{8r}\overline{Y}_0(1-2\cos\phi)^2, \quad \overline{Y}_0 = \frac{A^2}{2c\rho_0}.$$

The total energy

$$\Pi_s = \frac{3}{4}\pi^2 k^3 a^4 Y_0 + \ldots = \frac{6\pi^5 a^4}{\lambda^3}Y_0 + \ldots,$$

where λ is the wavelength ($k = 2\pi/\lambda = \omega/c$). The total pressure on the surface of the cylinder

$$p_n = (p_0+p_s)|_{r=a} = \frac{4A}{\pi ka}\sum_{m=0}^{\infty}\frac{\cos m\phi}{\alpha_m}\,e^{-i\left(\gamma_m - \frac{\pi m}{2}\right)},$$

where

$$\alpha_m = \frac{\sqrt{[J_{m+1}(\mu)-J_{m-1}(\mu)]^2+[N_{m+1}(\mu)-N_{m-1}(\mu)]^2}}{2}.$$

The total force acting on unit length of the cylinder, directed along the path of propagation of the plane wave, equals

$$F = a\int_0^{2\pi} p_n \cos\phi \,d\phi = \frac{4A}{k\alpha_1}\,e^{-i\left(\gamma_1-\frac{\pi}{2}\right)}.$$

If $\mu = ka \ll 1$, then

$$p_n = A(1+2i\mu\cos\phi),$$
$$F = i4\pi^2 a^2 kA.$$

If $\mu = ka \gg 1$, then

$$F \approx \sqrt{4a\lambda}A\,e^{i\left(\mu-\frac{\pi}{4}\right)}.$$

Method. One must use the approximate formulae:
(a) for $\mu \gg m+\frac{1}{2}$

$$\alpha_0 \approx \sqrt{\frac{8}{\pi\mu}}, \quad \gamma_0 = \mu - \frac{\pi}{4}, \quad \alpha_m = \sqrt{\frac{2}{\pi\mu}}, \quad \gamma_m = \mu - \frac{1}{2}\pi\left(m+\frac{1}{2}\right),$$

(b) for $\mu = ka \ll m+\frac{1}{2}$

$$\alpha_0 \approx \frac{4}{\pi\mu}, \quad \gamma_0 \approx \frac{\pi\mu^2}{4}, \quad \alpha_m = \frac{m!}{2\pi}\left(\frac{2}{\mu}\right)^{m+1},$$

$$\gamma_m \approx -\frac{\pi m}{(m!)^2}\left(\frac{\mu}{2}\right)^{2m} \quad (m > 0).$$

62. Let a plane wave be propagated along the z-axis:

$$\overline{p}_0 = A\,e^{i(\omega t - kz)} = p_0\,e^{i\omega t}, \quad p_0 = A\,e^{-ikz} = A\,e^{-ikr\cos\theta}.$$

The pressure and radial velocity in the scattered wave are given by the formulae

$$p_s = \sum_{m=0}^{\infty} B_m \zeta_m^{(2)}(kr) P_m(\cos\theta),$$

$$v_{s_r} = -\frac{1}{ikc\rho_0}\frac{\partial p}{\partial r} = \frac{i}{c\rho_0}\sum_{m=0}^{\infty} B_m \zeta_m^{(2)\prime}(kr) P_m(\cos\theta),$$

where

$$\zeta_m^{(2)}(x) = \sqrt{\frac{\pi}{2x}}\, H^{(2)}_{m+\frac{1}{2}}(x),$$

$$B_m = -A(-i)^m(2m+1)\frac{\psi_m'(\mu)}{\zeta_m^{(2)\prime}(\mu)}, \qquad \mu = ka,$$

$$\psi_m(\mu) = \sqrt{\frac{\pi}{2\mu}}\, J_{\mu+\frac{1}{2}}(\mu).$$

The coefficients B_m are conveniently represented in the form

$$B_m = -A(-i)^{m+1}(2m+1)\sin\beta_m e^{i\beta_m},$$

where β_m are determined from the equations†

$$\left.\begin{aligned} mf_{m-1}(\mu)-(m+1)f_{m+1}(\mu) &= -(2m+1)D_m\cos\beta_m = -(2m+1)f_m'(\mu)\cos\beta_m,\\ (m+1)\psi_{m+1}(\mu)-m\psi_{m-1}(\mu) &= -(2m+1)D_m\sin\beta_m = -(2m+1)\psi_m'(\mu)\sin\beta_m. \end{aligned}\right\} \tag{1}$$

Here

$$f_m = \sqrt{\frac{\pi}{2\mu}}\, N_{m+\frac{1}{2}}(\mu).$$

In the wave zone $(r \to \infty)$

$$p_s = \frac{e^{-i\left(kr-\frac{\pi}{4}\right)}}{kr}\sum_{m=0}^{\infty} B_m i^m P_m(\cos\theta), \qquad v_s = \frac{1}{c\rho_0}p_s.$$

The intensity of the scattered wave

$$\overline{Y}_s = \overline{Y}_0\frac{1}{k^2r^2}\sum_{m,n=0}^{\infty}(2m+1)(2n+1)\sin\beta_m\sin\beta_n\cos(\beta_m-\beta_n)P_m(\cos\theta)P_n(\cos\theta).$$

The total energy radiated is

$$\Pi_s = \overline{Y}_0\frac{4\pi}{k^2}\sum_{n=0}^{\infty}(2n+1)\sin^2\beta_n, \qquad \overline{Y}_0 = \frac{A^2}{2c\rho_0}.$$

Method. Let us look for the solution in the form of a series

$$p_s = \sum_{m=0}^{\infty} B_m \zeta_m^{(2)}(kr) P_m(\cos\theta),$$

† See [38], page 351.

where B_m are coefficients, to be determined from the boundary condition $\left.\frac{\partial(p_0-p_s)}{\partial r}\right|_{r=a} = 0$. To do this it is necessary to obtain the expansion of the plane wave in a series. Let us prove that the formula

$$e^{-i\rho\cos\theta} = \sum_{m=0}^{\infty} C_m \psi_m(\rho) P_m(\cos\theta),$$

holds, where

$$C_m = (2m+1)(-i)^m.$$

In fact, we assume

$$C_m \psi_m(\rho) = \frac{2m+1}{2} \int_{-1}^{1} e^{-i\rho\xi} P_m(\xi) d\xi.$$

Integration by parts for large ρ gives:

$$\int_{-1}^{1} e^{-i\rho\xi} P_m(\xi) d\xi = 2(-i)^m \frac{\sin\left(\rho - \frac{m\pi}{2}\right)}{\rho} + O\left(\frac{1}{\rho}\right).$$

On the other hand,

$$\psi_m(\rho) = \frac{\sin\left(\rho - \frac{m\pi}{2}\right)}{\rho} + O\left(\frac{1}{\rho}\right),$$

where $O(1/\rho)$ are terms of higher order than $1/\rho$. From the equality of the asymptotic forms of the left and right-hand sides it follows that

$$C_m = (2m+1)(-i)^m.$$

63. If $\mu = ka \ll 1$, then

$$\left.\begin{aligned} p_s &\approx -\frac{Aa\mu^2}{3r}\left(1-\frac{3}{2}\cos\theta\right)e^{-ikr}, \\ v_s &= \frac{1}{c\rho}p_s \end{aligned}\right\} \quad \text{in the wave zone} \quad (kr \gg 1).$$

The intensity of the sound, scattered by the sphere,

$$\overline{Y}_s = \frac{\mu^4 a^2}{9r^2} Y_0 \left(1 - \frac{3}{2}\cos\theta\right)^2 + \ldots,$$

$$Y_0 = \frac{A^2}{2c\rho_0}.$$

The total energy of the sound scattered by the sphere,

$$\Pi_s = \frac{7\pi a^6 k^4}{9} Y_0 + \ldots = \frac{112}{9}\frac{\pi^5 a^6}{\lambda^4}\overline{Y}_0 + \ldots \quad (\lambda \text{ is the wavelength}).$$

For $ka \gg 1$ we have:

$$\overline{Y}_s = \frac{a^2}{4r^2}\left[1+\cot^2\frac{\theta}{2}\,J_1^2(ka\sin\theta)\right], \qquad \Pi_s \approx 2\pi a^2 Y_0.$$

The pressure on the surface of the sphere

$$p_\pi = (p_0+p_s)|_{r=a} = A\left(1-\frac{3}{2}\,i\mu\cos\theta\right)+\dots \quad \text{for} \quad \mu \ll 1.$$

Method. Expressions for p_s and $\overline{Y}_s$, Π_s may be obtained either by direct calculation, using the approximate formulae

$$\psi_0'(\mu) \approx -\frac{\mu}{3}, \quad \psi_1' \approx \frac{1}{3}, \quad \zeta_0^{2\prime} \approx -\frac{i}{\mu^2} \quad \zeta_1^{2\prime} \approx -\frac{2i}{\mu^3} \quad (\mu \ll 1),$$

or from the general formulae obtained in the solution of the preceding problem 62. In order to do this it is necsseary to consider the following approximate formulae for β_m and D_m.

If $\mu \gg m+\frac{1}{2}$, then

$$D_m \approx \frac{1}{\mu}, \quad \beta_m \approx \mu-\frac{1}{2}\pi(m+1).$$

If $\mu \ll m+\frac{1}{2}$, then $D_0 \approx 1/\mu^2$, $\beta_0 \approx \frac{1}{3}\mu^3$,

$$D_m \approx \frac{1.3.5\ldots(2m-1)(m+1)}{\mu^{m+2}}, \quad \beta_m \approx \frac{m\mu^{2m+1}}{1^2.3^2.5^2\ldots(2m-1)(2m+1)(m+1)}.$$

64. A plane wave is incident on the sphere

$$p_0 = A\mathrm{e}^{-ikr\cos\theta}.$$

The pressure in the scattered wave

$$p_s = 3i\,\frac{\rho_0\psi_1(\mu)-\rho_1\mu\psi_1'(\mu)}{\rho_0\zeta_1^{(2)}(\mu)-\rho_1\mu\zeta_1^{(2)\prime}(\mu)}\,\zeta_1^{(2)}(kr)\cos\theta = B_1\zeta_1^{(2)}(kr)\cos\theta,$$

where ρ_1 is the density of the sphere.

The radial component of gas velocity

$$v_s = -\frac{1}{ic\rho_0}\,B_1\zeta_1^{(2)\prime}(kr)\cos\theta$$

(in connection with the values ψ_1 and $\zeta_1^{(2)}$ see problem 62).

Solution. The equation of motion of the centre of gravity of the sphere under the action of air has the form

$$M\ddot{\xi} = -\iint(\overline{p}_0+\overline{p}_s)_{r=a}\cos\theta\, a^2\mathrm{d}\Omega,$$

where $M = 4\pi a^3\rho_1/3$ the mass of the sphere, or

$$M\omega^2\xi = a^2\iint(\overline{p}_0+\overline{p}_s)_{r=a}\cos\theta\,\mathrm{d}\Omega. \tag{1}$$

The boundary condition for $r = a$ may be written thus:

$$-\frac{i}{kc\rho_0}\frac{\partial}{\partial r}(\overline{p}_0+\overline{p}_s)\Big|_{r=a} = i\omega\xi\cos\theta. \qquad (2)$$

Multiplying (1) and (2), we eliminate ξ and obtain the boundary condition on the surface of the sphere

$$\frac{2a\rho_1}{3}\frac{\partial}{\partial r}(p_0+p_s)\Big|_{r=a} = \rho_0\cos\theta\int\limits_0^{\pi}(p_0+p_s)_{r=a}P_1(\cos\theta)\sin\theta\,d\theta, \qquad (3)$$

where

$$P_1(\cos\theta) = \cos\theta.$$

Using the expansion of the plane wave in spherical functions

$$p_0 = Ae^{-ikr\cos\theta} = \sum_{m=0}^{\infty} A_m\psi_m(kr)P_m(\cos\theta), \qquad A_m = (-1)^m(2m+1)$$

and assuming

$$p_s = \sum_{m=0}^{\infty} B_m\zeta_m^{(2)}(kr)P_m(\cos\theta),$$

we obtain from (3) by virtue of the orthogonality of the Legendre polynomials

$$B_m = \begin{cases} -\dfrac{\rho_0\psi_1(\mu)-\rho_1\psi_1'(\mu)\mu}{\rho_0\zeta_1^{(2)}(\mu)-\rho_1\mu\zeta_1^{(2)'}(\mu)}A_1 & \text{for} \quad m=1, \\ 0 & \text{for} \quad m\neq 1. \end{cases}$$

65. $p_s = B_1\zeta_1^{(2)}(kr)\cos\theta,$
where

$$B_1 = -A_1\frac{\rho_0\psi_1(\mu)-\rho_1\left(1-\dfrac{\mu_0^2}{\mu^2}\right)\mu\psi_1'(\mu)}{\rho_0\zeta_1^{(2)}(\mu)-\rho_1\left(1-\dfrac{\mu_0^2}{\mu^2}\right)\mu\zeta_1^{(2)}(\mu)}, \qquad \mu_0 = \frac{\omega_0}{c}a, \qquad \mu = ka.$$

At resonance, i.e. for $\mu = \mu_0$,

$$B_1 = A_1\frac{\psi_1(\mu)}{\zeta_1^{(2)}(\mu)}.$$

If a plane wave is incident on the sphere, then $A_1 = -3i$. If there is no external field, then we obtain the characteristic equation

$$\rho_0\zeta_1^{(2)}(\mu) = \rho_1\left(1-\frac{\mu_0^2}{\mu^2}\right)\mu\zeta_1^{(2)'}(\mu),$$

from which the frequency ω of the "free" vibrations of the sphere, produced by the external medium, is determined.

Method. See the preceding problem.

§ 4. Steady-state Electromagnetic Vibrations

1. Maxwell's Equations. Potentials. Green–Ostrogradskii Vector Formulae

66. Maxwell's equations in a non-conducting medium without sources are

$$\left.\begin{aligned} \operatorname{curl} \boldsymbol{H} &= \frac{1}{c}\frac{\partial \boldsymbol{D}}{\partial t}, & \operatorname{div} \boldsymbol{B} &= 0, & \boldsymbol{B} &= \mu \boldsymbol{H}, \\ \operatorname{curl} \boldsymbol{E} &= -\frac{1}{c}\frac{\partial \boldsymbol{B}}{\partial t}, & \operatorname{div} \boldsymbol{D} &= 0, & \boldsymbol{D} &= \varepsilon \boldsymbol{E}. \end{aligned}\right\} \tag{1}$$

In curvilinear orthogonal coordinates they have the form

$$\left.\begin{aligned} &\frac{\varepsilon}{c} h_2 h_3 \frac{\partial E_1}{\partial t} = \frac{\partial}{\partial x_2} h_3 H_3 - \frac{\partial}{\partial x_3} h_2 H_2, \qquad -\frac{\mu}{c} h_2 h_3 \frac{\partial H_1}{\partial t} = \frac{\partial}{\partial x_2} h_3 E_3 - \frac{\partial}{\partial x_3} h_2 E_2, \\ &\frac{\varepsilon}{c} h_3 h_1 \frac{\partial E_2}{\partial t} = \frac{\partial}{\partial x_3} h_1 H_1 - \frac{\partial}{\partial x_1} h_3 H_3, \qquad -\frac{\mu}{c} h_3 h_1 \frac{\partial H_2}{\partial t} = \frac{\partial}{\partial x_3} h_1 E_1 - \frac{\partial}{\partial x_1} h_3 E_3, \\ &\frac{\varepsilon}{c} h_1 h_2 \frac{\partial E_3}{\partial t} = \frac{\partial}{\partial x_1} h_2 H_2 - \frac{\partial}{\partial x_2} h_1 H_1, \qquad -\frac{\mu}{c} h_1 h_2 \frac{\partial H_3}{\partial t} = \frac{\partial}{\partial x_1} h_2 E_2 - \frac{\partial}{\partial x_2} h_1 E_1, \\ &\qquad \frac{\partial}{\partial x_1} h_2 h_3 H_1 + \frac{\partial}{\partial x_2} h_3 h_1 H_2 + \frac{\partial}{\partial x_3} h_1 h_2 H_3 = 0, \\ &\qquad \frac{\partial}{\partial x_1} h_2 h_3 E_1 + \frac{\partial}{\partial x_2} h_3 h_1 E_2 + \frac{\partial}{\partial x_3} h_1 h_2 E_3 = 0 \\ &\qquad (ds^2 = h_1^2 dx_1^2 + h_2^2 dx_2^2 + h_3 dx_3^2). \end{aligned}\right\} \tag{2}$$

If the dependence of the field on time is given by the factor $e^{-i\omega t}$, then in these equations one can make a subtitutions, using the relations

$$\frac{1}{c}\frac{\partial E_m}{\partial t} = -ik_0 E_m, \quad \frac{1}{c}\frac{\partial H_m}{\partial t} = -ik_0 H_m \quad \left(k_0 = \frac{\omega}{c}, \quad m = 1, 2, 3\right). \tag{3}$$

In a spherical system of coordinates $x_1 = r$, $x_2 = \theta$, $x_3 = \phi$ and $h_1 = 1$, $h_2 = r$, $h_3 = r \sin \theta$. In a cylindrical system of coordinates

$$x_1 = \rho, \quad x_2 = \phi, \quad x_3 = z \quad \text{and} \quad h_1 = 1, \quad h_2 = r, \quad h_3 = 1.$$

Method. Use expressions for the operators div and curl in a curvilinear system of coordinates (see the Supplement).

67. If the time dependence is given by the factor $e^{-i\omega t}$, then the vector potential and scalar potential satisfy the equations

$$\left.\begin{aligned} \Delta \boldsymbol{A} + k^2 \boldsymbol{A} &= -\frac{4\pi}{c}\mu \boldsymbol{j}, \\ & \qquad k^2 = \frac{\omega^2}{a^2}, \; a^2 = \frac{\varepsilon\mu}{c^2}, \\ \Delta \phi + k^2 \phi &= -\frac{4\pi}{\varepsilon}\rho, \end{aligned}\right\} \tag{1}$$

where

$$\phi = \frac{-ic}{\varepsilon\mu\omega} \operatorname{div} \boldsymbol{A}, \tag{2}$$

i.e. the scalar potential may be eliminated ($\boldsymbol{j}$ the vector of the current density).

An expression for $\Delta \boldsymbol{A}$ in an arbitrary orthogonal curvilinear system of coordinates has the form

$$\Delta \boldsymbol{A} = -\operatorname{curl}\operatorname{curl} \boldsymbol{A} + \operatorname{grad}\operatorname{div} \boldsymbol{A},$$

where

$$\operatorname{curl}\operatorname{curl} \boldsymbol{A} = \frac{1}{h_2 h_3}\left\{\frac{\partial}{\partial x_2}\frac{h}{h_1 h_2}\left[\frac{\partial}{\partial x_1}(h_2 A_2) - \frac{\partial}{\partial x_2}(h_1 A_1)\right] - \right.$$

$$\left. - \frac{\partial}{\partial x_3}\frac{h_2}{h_3 h_1}\left[\frac{\partial}{\partial x_3}(h_1 A_1) - \frac{\partial}{\partial x_1}(h_3 A_3)\right]\right\}\boldsymbol{i}_1 + \frac{1}{h_3 h_1}\left\{\frac{\partial}{\partial x_3}\frac{h}{h_2 h_3}\left[\frac{\partial}{\partial x_2}(h_3 A_3) - \right.\right.$$

$$\left.\left. - \frac{\partial}{\partial x_3}(h_2 A_2)\right] - \frac{\partial}{\partial x_1}\frac{h_3}{h_1 h_2}\left[\frac{\partial}{\partial x_1}(h_2 A_2) - \frac{\partial}{\partial x_2}(h_1 A_1)\right]\right\}\boldsymbol{i}_2 +$$

$$+ \frac{1}{h_1 h_2}\left\{\frac{\partial}{\partial x_1}\frac{h_2}{h_3 h_1}\left[\frac{\partial}{\partial x_3}(h_1 A_1) - \frac{\partial}{\partial x_1}(h_3 A_3)\right] - \right.$$

$$\left. - \frac{\partial}{\partial x_2}\frac{h_1}{h_2 h_3}\left[\frac{\partial}{\partial x_2}(h_3 A_3) - \frac{\partial}{\partial x_3}(h_2 A_2)\right]\right\}\boldsymbol{i}_3,$$

$$\operatorname{grad} \psi = \frac{1}{h_1}\frac{\partial \psi}{\partial x_1}\boldsymbol{i}_1 + \frac{1}{h_2}\frac{\partial \psi}{\partial x_2}\boldsymbol{i}_2 + \frac{1}{h_3}\frac{\partial \psi}{\partial x_3}\boldsymbol{i}_3,$$

$$\psi = \operatorname{div} \boldsymbol{A} = \frac{1}{h_1 h_2 h_3}\left[\frac{\partial}{\partial x_1}(h_2 h_3 A_1) + \frac{\partial}{\partial x_2}(h_1 h_3 A_2) + \frac{\partial}{\partial x_3}(h_1 h_2 A_3)\right],$$

where $\boldsymbol{i}_1$, $\boldsymbol{i}_2$, $\boldsymbol{i}_3$ are unit vectors of the coordinate system, A_1, A_2, A_3 are components of the vector $\boldsymbol{A}$.

68. In a homogeneous conducting medium Maxwell's equations have the form†

$$\boldsymbol{j}_{\text{str}} = 0, \qquad \rho = 0,$$

$$\left.\begin{aligned} \operatorname{curl} \boldsymbol{H} &= \frac{4\pi}{c}\sigma \boldsymbol{E} + \frac{\varepsilon}{c}\frac{\partial \boldsymbol{E}}{\partial t}, &\quad \operatorname{div} \boldsymbol{E} &= 0, \\ \operatorname{curl} \boldsymbol{E} &= -\frac{\mu}{c}\frac{\partial \boldsymbol{H}}{\partial t}, &\quad \operatorname{div} \boldsymbol{H} &= 0. \end{aligned}\right\} \tag{1}$$

Assuming

$$\boldsymbol{H} = \frac{1}{\mu}\operatorname{curl} \boldsymbol{A}, \quad \boldsymbol{E} = -\operatorname{grad} \phi - \frac{1}{c}\frac{\partial \boldsymbol{A}}{\partial t}, \tag{2}$$

† See [17], page 420.

we obtain for A and ϕ the equations

$$\Delta\phi = \frac{\varepsilon\mu}{c^2}\frac{\partial^2\phi}{\partial t^2} + \frac{4\pi\sigma\mu}{c^2}\frac{\partial\phi}{\partial t}, \tag{3}$$

$$\Delta A = \frac{\varepsilon\mu}{c^2}\frac{\partial^2 A}{\partial t^2} + \frac{4\pi\sigma\mu}{c^2}\frac{\partial A}{\partial t}, \tag{4}$$

where A and ϕ are connected by the Lorentz condition

$$\operatorname{div} A + \frac{\varepsilon\mu}{c}\frac{\partial\phi}{\partial t} + \frac{4\pi\sigma\mu}{c}\phi = 0. \tag{5}$$

If the time dependence is of the type $e^{-i\omega t}$, then

$$\Delta\phi + k^2\phi = 0, \quad k^2 = \frac{\varepsilon\mu}{c^2}\omega^2 + i\frac{4\pi\mu\sigma\omega}{c^2}, \tag{3'}$$

$$\Delta A + k^2 A = 0, \tag{4'}$$

$$\phi = \frac{-i\omega}{ck^2}\operatorname{div} A, \tag{5'}$$

i.e. for $\sigma \neq 0$ the wave number k is always complex.

69. If in vacuum ($\sigma = 0$, $\varepsilon = 1$, $\mu = 1$) there are no currents and no free charges, then, assuming $A = \frac{1}{c}\frac{\partial \boldsymbol{\Pi}}{\partial t}$, $\phi = -\operatorname{div}\boldsymbol{\Pi}$, we obtain:

$$H = \frac{1}{c}\operatorname{curl}\frac{\partial\boldsymbol{\Pi}}{\partial t}, \quad E = \operatorname{grad}\operatorname{div}\boldsymbol{\Pi} - \frac{1}{c^2}\frac{\partial^2\boldsymbol{\Pi}}{\partial t^2} \tag{1}$$

the polarization potential $\boldsymbol{\Pi}$ satisfies the equation

$$\Delta\boldsymbol{\Pi} - \frac{1}{c^2}\frac{\partial^2\boldsymbol{\Pi}}{\partial t^2} = 0. \tag{2}$$

For a time dependence of the type $e^{-i\omega t}$ we have:

$$H = -ik\operatorname{curl}\boldsymbol{\Pi}, \quad E = \operatorname{grad}\operatorname{div}\boldsymbol{\Pi} + k^2\boldsymbol{\Pi} \tag{1'}$$

and

$$\Delta\boldsymbol{\Pi} + k^2\boldsymbol{\Pi} = 0. \tag{2'}$$

The magnetic Hertzian vector is introduced by:

$$H' = \operatorname{grad}\operatorname{div}\boldsymbol{\Pi}' - \frac{1}{c^2}\frac{\partial^2\boldsymbol{\Pi}'}{\partial t^2}, \quad E' = -\frac{1}{c}\operatorname{curl}\frac{\partial\boldsymbol{\Pi}'}{\partial t}, \tag{3}$$

where

$$\Delta\boldsymbol{\Pi}' = \frac{1}{c^2}\frac{\boldsymbol{\Pi}\partial'}{\partial t^2}. \tag{4}$$

For a time dependence of the type $e^{-i\omega t}$ we have:

$$H' = \operatorname{grad}\operatorname{div}\boldsymbol{\Pi}' + k^2\boldsymbol{\Pi}', \quad E' = ik\operatorname{curl}\boldsymbol{\pi}', \tag{3'}$$

$$\Delta\boldsymbol{\Pi}' + k^2\boldsymbol{\pi}' = 0. \tag{4'}$$

Using equations (2′) and (4′) for $\boldsymbol{\Pi}$ and $\boldsymbol{\Pi}'$, it is possible to rewrite the formulae for $\boldsymbol{E}$ and $\boldsymbol{H}'$ differently:

$$\boldsymbol{E} = \text{curl curl}\, \boldsymbol{\Pi}, \quad \boldsymbol{H}' = \text{curl curl}\, \boldsymbol{\Pi}'.$$

In a conducting medium for steady-state fields ($\sim e^{-i\omega t}$) $\boldsymbol{\Pi}$ and $\boldsymbol{\Pi}'$ are formally introduced in the same way as for a vacuum; but in this case k^2 is the quantity

$$k^2 = \frac{\varepsilon\mu\omega^2}{c^2} + i\frac{4\pi\sigma\mu}{c^2}.$$

70. In a spherical system of coordinates we have:

(a) for a field of electric type ($H_r = 0$)

$$\left.\begin{aligned} &E_r = \frac{\partial^2 U}{\partial r^2} + k^2 U, \quad E_\theta = \frac{1}{r}\frac{\partial^2 U}{\partial r \partial\theta}, \quad E_\phi = \frac{1}{r\sin\theta}\frac{\partial^2 U}{\partial r\partial\phi}, \\ &H_r = 0, \quad H_\theta = \frac{-ik}{r\sin\theta}\frac{\partial U}{\partial\phi}, \quad H_\phi = \frac{ik}{r}\frac{\partial U}{\partial\theta}; \end{aligned}\right\} \tag{1}$$

(b) for a field of magnetic type ($E_r = 0$)

$$\left.\begin{aligned} &E_r' = 0, \quad E_\theta' = \frac{ik}{r\sin\theta}\frac{\partial U'}{\partial\phi} \quad E_\phi' = \frac{-ik}{r}\frac{\partial U'}{\partial\theta}, \\ &H_r' = \frac{\partial^2 U'}{\partial r^2} + k^2 U', \quad H_\theta' = \frac{1}{r}\frac{\partial^2 U'}{\partial r\partial\theta}, \quad H_\phi' = \frac{1}{r\sin\theta}\frac{\partial^2 U'}{\partial r\partial\phi}, \end{aligned}\right\} \tag{2}$$

where the potentials U and U' satisfy the equation

$$\frac{\partial^2 U}{\partial r^2} + \frac{1}{r^2\sin\theta}\frac{\partial}{\partial\theta}\left(\sin\theta\frac{\partial U}{\partial\theta}\right) + \frac{1}{r^2\sin^2\theta}\frac{\partial^2 U}{\partial\phi^2} k^2 U = 0, \tag{3}$$

or

$$\Delta U + k^2 U - \frac{2}{r}\frac{\partial U}{\partial r} = 0,$$

and the functions $u = Ur$, $u' = U'r$ satisfy the wave equation

$$\Delta u + k^2 u = 0.$$

In a cylindrical system of coordinates (r, ϕ, ρ)

(a) for a field of electric type ($H_z = 0$)

$$\left.\begin{aligned} &E_z = \frac{\partial^2 U}{\partial z^2} + k^2 U, \quad E_\phi = \frac{1}{\rho}\frac{\partial^2 U}{\partial\phi\partial z}, \quad E_\rho = \frac{\partial^2 U}{\partial\rho\partial z}, \\ &H_z = 0, \quad H_\phi = -ik\frac{\partial U}{\partial\rho}, \quad H_\rho = \frac{ik}{\rho}\frac{\partial U}{\partial\phi}: \end{aligned}\right\} \tag{4}$$

(b) for a field of magnetic type ($E_z = 0$) we have:

$$\left.\begin{aligned} &E_z' = 0, \quad E_\phi' = ik\frac{\partial U'}{\partial\rho}, \quad E_\rho' = \frac{-ik}{\rho}\frac{\partial U'}{\partial\phi}, \\ &H_z' = \frac{\partial^2 U'}{\partial z^2} + k^2 U', \quad H_\phi' = \frac{1}{\rho}\frac{\partial^2 U'}{\partial\phi\partial z}, \quad H_\rho' = \frac{\partial^2 U'}{\partial\rho\partial z}, \end{aligned}\right\} \tag{5}$$

where U and U' satisfy the equation

$$\frac{\partial^2 U}{\partial z^2}+\frac{1}{\rho}\frac{\partial}{\partial \rho}\left(\rho\frac{\partial U}{\partial \rho}\right)+\frac{1}{\rho^2}\frac{\partial^2 U}{\partial \phi^2}+k^2 U=0, \tag{6}$$

or

$$\Delta U+k^2 U=0.$$

Hence we see that

$$U=\Pi_z, \qquad U'=\Pi'_z.$$

In the spherical case

$$U\neq \Pi_r \quad \text{ɐup} \quad U'\neq \Pi'_r.$$

Method. For proof of the fundamental statement of the problem it is necessary to substitute the expressions for the components of the field by U (or U') in Maxwell's equations, written in an orthogonal curvilinear system of coordinates (see problem 66), and to require that they be satisfied; from this requirement the equation for U (or U') follows.

71.
$$E_1=k^2U+\frac{\partial^2 U}{\partial x_1^2}, \qquad E_2=\frac{1}{h_2}\frac{\partial^2 U}{\partial x_1 \partial x_2}+\frac{i\omega\mu}{c}\frac{\partial U'}{\partial x_3},$$

$$E_3=\frac{1}{h_3}\frac{\partial^2 U}{\partial x_1\partial x_3}-\frac{i\omega\mu}{c}\frac{\partial U'}{\partial x_2},$$

$$H_1=k^2U'+\frac{\partial^2 U'}{\partial x_1^2}, \qquad H_2=-\frac{ik^2c}{\omega\mu}\frac{1}{h_3}\frac{\partial U}{\partial x_3}+\frac{1}{h_2}\frac{\partial^2 U'}{\partial x_1\partial x_2},$$

$$H_3=\frac{ick^2}{\omega\mu}\frac{1}{h_2}\frac{\partial U}{\partial x_2}+\frac{1}{h_3}\frac{\partial^2 U'}{\partial x_1\partial x_3},$$

where U and U' are solutions of the equation

$$\frac{\partial^2 U}{\partial x_1^2}+\frac{1}{h_2h_3}\left[\frac{\partial}{\partial x_2}\frac{h_3}{h_2}\frac{\partial U}{\partial x_2}+\frac{\partial}{\partial x_3}\frac{h_2}{h_3}\frac{\partial U}{\partial x_3}\right]+k^2U=0,$$

$$k^2=\frac{\varepsilon\mu\omega^2}{c^2}+i\frac{4\pi\sigma\mu}{c^2}\,\omega.$$

Method. See problem 70.

72. At the boundary of separation of the two media for $r=a$ the conditions

$$\frac{k_1^2}{\mu_1}U_1=\frac{k_2^2}{\mu_2}U_2, \qquad \frac{k_1^2}{\mu_1}U_1'=\frac{k_2^2}{\mu_2}U_2',$$

$$\frac{\partial U_1}{\partial r}=\frac{\partial U_2}{\partial r}, \qquad \frac{\partial U_1'}{\partial r}=\frac{\partial U_2'}{\partial r},$$

must be fulfilled, where the sign 1 or 2 denotes the number of the medium (1 for $r<a$, 2 for $r>a$), k_1 and k_2 are determined from the formula

$$k_s^2=\frac{\varepsilon_s\mu_s\omega^2}{c^2}+i\frac{4\pi\sigma_s\mu_s\omega}{c^2}, \qquad s=1,2.$$

The functions U_s and U_s' satisfy the equation

$$\frac{\partial^2 U_s}{\partial r^2}+\frac{1}{r^2\sin\theta}\frac{\partial}{\partial\theta}\left(\sin\theta\frac{\partial U_s}{\partial\theta}\right)+\frac{1}{r^2\sin^2\theta}\frac{\partial^2 U}{\partial\psi^2}+k_s^2 U_s=0,\qquad s=1,2,$$

so that

$$\Delta u_s+k^2 u_s=0,\qquad u_s=\frac{U_s}{r}\quad(s=1,2).$$

Method. At the boundary of media 1 and 2 the tangential components of the electric field vector and the magnetic field vector, in the given case E_θ, E_ϕ, H_θ, H_ϕ, must be continuous.

73. *Method.* Use the vector formula

$$\operatorname{div}[\boldsymbol{ab}]=(\boldsymbol{b}\operatorname{curl}\boldsymbol{a})-(\boldsymbol{a}\operatorname{curl}\boldsymbol{b})\tag{1}$$

and Ostrogradskii's formula.

74. *Solution.* In the formula

$$\int_T\{\boldsymbol{W}\operatorname{curl}\operatorname{curl}\boldsymbol{U}-\boldsymbol{U}\operatorname{curl}\operatorname{curl}\boldsymbol{W}\}\mathrm{d}\tau=\int_\Sigma\{[\boldsymbol{U}\operatorname{curl}\boldsymbol{W}]-]\boldsymbol{W}\operatorname{curl}\boldsymbol{U}]\}\boldsymbol{n}\mathrm{d}\sigma\tag{1}$$

we assume:

$$\boldsymbol{W}=\boldsymbol{a}\phi,$$

where $\phi=\mathrm{e}^{ikr}/r$, $\boldsymbol{a}$ an arbitrary constant vector. Calculations give:

$$\operatorname{curl}\boldsymbol{W}=[\operatorname{grad}\phi,\boldsymbol{a}],\qquad \operatorname{curl}\operatorname{curl}\boldsymbol{W}=\boldsymbol{a}k^2\phi+\operatorname{grad}(\boldsymbol{a}\operatorname{grad}\phi),$$

$$\boldsymbol{U}\operatorname{curl}\operatorname{curl}\boldsymbol{W}=\boldsymbol{a}\{k^2\phi\boldsymbol{U}-\operatorname{grad}\phi\operatorname{div}\boldsymbol{U}\}+\operatorname{div}[(\boldsymbol{a}\operatorname{grad}\phi)\boldsymbol{U}],$$

$$[\boldsymbol{W}\operatorname{curl}\boldsymbol{U}]\boldsymbol{n}=[\operatorname{curl}\boldsymbol{U},\boldsymbol{n}]\boldsymbol{W},$$

$$[\boldsymbol{U}\operatorname{curl}\boldsymbol{W}]\boldsymbol{n}=[\boldsymbol{U}[\operatorname{grad}\phi,\boldsymbol{a}]]=(\boldsymbol{Aa})(\operatorname{grad}\phi,\boldsymbol{n})-(\boldsymbol{A}\operatorname{grad}\phi)(\boldsymbol{an}),$$

Because of Ostrogradskii's formula

$$\int_T\operatorname{div}[(\boldsymbol{a}\operatorname{grad}\phi)\boldsymbol{U}]\mathrm{d}\tau=\int_\Sigma(\boldsymbol{Un})(\operatorname{grad}\phi\boldsymbol{a})\mathrm{d}\sigma.\tag{2}$$

Under the sign of the surface integral in formula (1) there is the expression $\boldsymbol{Fa}$, where

$$\boldsymbol{F}=\boldsymbol{U}(\operatorname{grad}\phi\,\boldsymbol{n})-(\boldsymbol{U}\operatorname{grad}\phi)\boldsymbol{n}-[\operatorname{curl}\boldsymbol{U},\boldsymbol{n}]\phi+(\boldsymbol{Un})\operatorname{grad}\phi$$

$$=(\boldsymbol{Un})\operatorname{grad}\phi+[\operatorname{grad}\phi]\boldsymbol{Un}]]+[\boldsymbol{n}\operatorname{curl}\boldsymbol{U}]\phi.\tag{3}$$

The expression on the left-hand side under the integral sign has the form $\boldsymbol{\Phi a}$, where

$$\boldsymbol{\Phi}=(\operatorname{curl}\operatorname{curl}\boldsymbol{U}-k^2\boldsymbol{U})\phi+\operatorname{grad}\phi\operatorname{div}\boldsymbol{U}.$$

The vector $\boldsymbol{a}$ is, thus, a common factor for all terms of formula (1), and since it is arbitrary, it is possible to divide by it; as a result we obtain the formula

$$\int_T\boldsymbol{\Phi}\mathrm{d}\tau=\int_\Sigma\boldsymbol{F}\mathrm{d}\sigma,\tag{4}$$

if the point $M_0(r_{MM_0}=0)$ does not belong to the region T. If the point $M_0(r_{MM_0}=0)$ exists inside T, then we describe about this point a small sphere Σ_ε of radius ε and apply formula (4) to the region $T-T_\varepsilon$, bounded by the surfaces Σ and Σ_ε. We determine the value of $\boldsymbol{F}$ on Σ_ε.

We note that

$$\operatorname{grad}\phi|_{\Sigma_\varepsilon} = \left(\frac{1}{r} - ik\right)\phi_{\ \Sigma_\varepsilon}\boldsymbol{n} \approx \frac{1}{\varepsilon^2}\boldsymbol{n}, \qquad \phi|_{\Sigma_\varepsilon} \approx \frac{1}{\varepsilon}.$$

Therefore

$$\boldsymbol{F}|_{\Sigma_\varepsilon} = \left(\frac{1}{\varepsilon} - ik\right)\phi(\varepsilon)\{(\boldsymbol{Un})\boldsymbol{n} + [\boldsymbol{n}[\boldsymbol{Un}]]\} - [\operatorname{curl}\boldsymbol{Un}]\psi(\varepsilon) \approx \frac{\boldsymbol{U}}{\varepsilon^2}$$

and, hence,

$$\lim_{\varepsilon\to 0}\int_{\Sigma_\varepsilon} F\mathrm{d}\sigma = 4\pi\boldsymbol{U}(\mathrm{M}_0).$$

Since $\lim\limits_{\varepsilon\to 0}\int_{T_\varepsilon}\Phi\,\mathrm{d}\tau = 0$, we obtain in the limit

$$\boldsymbol{U}(M_0) = \frac{1}{4\pi}\int_T \boldsymbol{\Phi}\mathrm{d}\tau - \frac{1}{4\pi}\int_\Sigma \boldsymbol{F}\mathrm{d}\sigma$$

or

$$\boldsymbol{U}(M_0) = \frac{1}{4\pi}\int_T \{(\operatorname{curl}\operatorname{curl}\boldsymbol{U} - k^2\boldsymbol{U})\phi + \operatorname{grad}\phi\operatorname{div}\boldsymbol{U}\}\,\mathrm{d}\tau_M -$$

$$-\frac{1}{4\pi}\int_\Sigma \{(\boldsymbol{Un})\operatorname{grad}\phi + [[\boldsymbol{nU}]\operatorname{grad}\phi] + [\boldsymbol{n}\operatorname{curl}\boldsymbol{U}]\phi\}\,\mathrm{d}\sigma_M. \quad (5)$$

75. $$\boldsymbol{E}(M_0) = \frac{1}{c\omega\varepsilon}\int_T \{-k^2\phi\boldsymbol{j} + \operatorname{grad}\phi\operatorname{div}\boldsymbol{j}\}\,\mathrm{d}\tau -$$

$$-\frac{1}{4\pi}\int_\Sigma \{ik_0\mu\,[\boldsymbol{nH}]\phi + [[\boldsymbol{nE}]\operatorname{grad}\phi] + (\boldsymbol{nE})\operatorname{grad}\phi\}\,\mathrm{d}\sigma,$$

$$\boldsymbol{H}(M_0) = \frac{1}{c}\int_T [\boldsymbol{j}\operatorname{grad}\phi]\,\mathrm{d}\tau + \frac{1}{4\pi}\int_\Sigma \{ik_0\varepsilon\,[\boldsymbol{nE}] - [\boldsymbol{nH}]\operatorname{grad}\phi - (\boldsymbol{nH})\operatorname{grad}\phi\}\,\mathrm{d}\sigma,$$

where

$$\phi = \frac{\mathrm{e}^{ikr}}{r}, \quad k = \frac{\omega}{c}\sqrt{\varepsilon\mu}, \quad k_0 = \frac{\omega}{c}.$$

Method. In the general formula (5) in the answer to 74 one assumes respectively that $\boldsymbol{U} = \boldsymbol{E}$ and $\boldsymbol{U} = \boldsymbol{A}$. In the second case there is a term $\int_\Sigma [\boldsymbol{nj}]\phi\mathrm{d}\sigma$ on the right, which should be transformed into a volume integral by means of the relation

$$\int_\Sigma [\boldsymbol{nj}]\,\phi\,\mathrm{d}\sigma = \int_T \{-[\boldsymbol{j}\operatorname{grad}\phi] + \phi\operatorname{curl}\boldsymbol{j}\}\,\mathrm{d}\sigma. \quad (1)$$

To prove this it is necessary to multiply both side by the arbitrary vector $\boldsymbol{a}$ and use the relations

$$[\boldsymbol{nj}]\,\boldsymbol{a}\phi = \boldsymbol{n}\,[\boldsymbol{ja}]\,\phi,$$
$$\operatorname{div}[\boldsymbol{ja}\phi] = \boldsymbol{a}\phi \operatorname{curl}\boldsymbol{j} - (\boldsymbol{j}\operatorname{curl}\boldsymbol{a}\phi) = \boldsymbol{a}\,\{\phi\operatorname{curl}\boldsymbol{j} - [\boldsymbol{j}\operatorname{grad}\phi]\},$$

so that

$$\int_T \boldsymbol{a}\,[\boldsymbol{nj}]\,\phi\,\mathrm{d}\sigma = \int_T \operatorname{div}[\boldsymbol{j}, \boldsymbol{a}\phi]\,\mathrm{d}\tau = \boldsymbol{a}\int_T \{\phi\operatorname{curl}\boldsymbol{j} - [\boldsymbol{j}\operatorname{grad}\phi]\}\,\mathrm{d}\tau.$$

Hence by virtue of the arbitrary nature of $\boldsymbol{a}$ (1) follows.

2. Propagation of Electromagnetic Waves and Vibrations in Resonators

76. Let us choose the z-axis of a cylindrical system of coordinates ρ, ϕ, z along the axis of the cylinder.

Let ε, μ, σ be para eters of the surrounding medium. There exist waves of the form

$$\boldsymbol{E} = \boldsymbol{E}_0 \mathrm{e}^{-i\alpha z + i\omega t}\mathrm{e}^{-\beta|z|} \qquad (\alpha > 0),$$
$$\boldsymbol{H} = \boldsymbol{H}_0 \mathrm{e}^{-i\alpha z + i\omega t}\mathrm{e}^{-\beta|z|} \qquad (\beta > 0),$$

i.e. attenuation waves. Here the symbols

$$\alpha = \sqrt{\frac{\sqrt{\varepsilon^2\mu^2\omega^4 + 16\pi^2\sigma^2\mu^2\omega^2} + \varepsilon\mu\omega^2}{2c^2}}, \quad \beta = \sqrt{\frac{\sqrt{\varepsilon^2\mu^2\omega^4 + 16\pi^2\sigma^2\mu^2\omega^2} - \varepsilon\mu\omega^2}{2c^2}}$$

$$\boldsymbol{E}_0 = (E_{0\rho}, E_{0\phi}, 0), \qquad \boldsymbol{H}_0 = (H_{0\rho}, H_{0\phi}, 0),$$

are adopted, where

$$E_{0\rho} = \frac{Ak}{\rho}, \qquad H_{0\phi} = -\frac{ck}{\omega\mu} E_{0\rho},$$

$$E_{0\phi} = \frac{\mu\omega}{kc} H_{0\rho}, \qquad H_{0\rho} = \frac{Bk}{\rho},$$

where A and B are constant factors,

$$k^2 = \frac{\varepsilon\mu\omega^2}{c^2} - i\frac{4\pi\sigma\mu\omega}{c^2}, \qquad k = \alpha - i\beta.$$

If $\varepsilon = 1$, $\mu = 1$, $\sigma = 0$ (vacuum), then $k = \omega/c = k_0$—waves are propagated along such a conductor with the velocity of light:

$$\boldsymbol{E} = \boldsymbol{E}_0 \mathrm{e}^{i(\omega t - k_0 z)}, \qquad E_{0\rho} = \frac{Ak_0}{\rho}, \qquad H_{0\phi} = -E_{0\rho},$$

$$\boldsymbol{H} = \boldsymbol{H}_0\,\mathrm{e}^{i(\omega t - k_0 z)}, \qquad E_{0\phi} = H_{0\phi}, \qquad H_{0\rho} = \frac{Bk_0}{\rho}.$$

77. *Solution.* Let ε_1, μ_1, σ_1 be the characteristics of the conductor, ε_2, μ_2, σ_2 the characteristics of the surrounding medium.

We choose a cylindrical system of coordinates (ρ, ϕ, z), having directed the z-axis along the axis of the cylinder and having placed the origin of coordinates on the cylinder axis.

Denoting $\Pi_z = u$, $\Pi_z' = v$ and assuming that the dependence of u and v on z is given by the factor $e^{i\gamma z}$, i.e. $u = u^0 e^{i\gamma z}$, $v = v^0 e^{i\gamma z}$, etc., we obtain after division by this factor

$$\left.\begin{aligned} E_z^0 &= p^2 u^0, \quad E_\varphi^0 = \frac{iv}{\rho}\frac{\partial u^0}{\partial \phi} - \frac{i\omega\mu}{c}\frac{\partial v^0}{\partial \rho}, \quad E_\rho^0 = i\gamma\frac{\partial u^0}{\partial \rho} + \frac{i\omega\mu}{c\rho}\frac{\partial v^0}{\partial \phi}, \\ H_z^0 &= p^2 v^0, \quad H_\phi^0 = \frac{ick^2}{\omega\mu}\frac{\partial u^0}{\partial \rho} + \frac{i\gamma}{\rho}\frac{\partial v^0}{\partial \varphi}, \quad H_\rho = -\frac{ick^2}{\omega\mu}\frac{1}{\rho}\frac{\partial u^0}{\partial \phi} + i\gamma\frac{\partial v^0}{\partial \rho}, \end{aligned}\right\} \tag{1}$$

where $p^2 = k^2 - \gamma^2$, $k^2 = \varepsilon\mu\omega^2/c^2 - 4i\pi\sigma\mu\omega/c^2$, the functions

$$u^0 = \alpha\psi(\rho, \phi) \quad \text{and} \quad v^0 = \beta\psi(\rho, \phi), \tag{2}$$

where α and β are constants, and $\psi(\rho, \phi)$ is the solution of the equation

$$\frac{1}{\rho}\frac{\partial}{\partial \rho}\left(\rho\frac{\partial \psi}{\partial \rho}\right) + \frac{1}{\rho^2}\frac{\partial^2 \psi}{\partial \phi^2} + p^2\psi = 0.$$

Hence we find particular solutions of the form

$$\psi_n(\rho, \phi) = \begin{cases} J_n(p\rho)e^{in\phi} & \text{inside the cylinder}, \\ H_n^{(1)}(p\rho)e^{in\phi} & \text{outside the cylinder}. \end{cases} \tag{3}$$

Substituting the expression for ψ_n in formulae (1) and (2) we obtain: inside the cylinder

$$E_z = \alpha_1 p_1^2 J_n(p_1\rho)e^{in\phi},$$
$$H_z = \beta_1 p_1^2 J_n(p_1\rho)e^{in\phi},$$

$$E_\phi = -\left[\frac{\alpha_1 n\gamma}{\rho} J_n(p_1\rho) + \frac{i\omega\mu_1 p_1}{c}\beta_1 J_1'(p_1\rho)\right]e^{in\phi},$$

$$H_\phi = \left[\alpha_1\frac{ick_1^2 p_1}{\omega\mu_1} J_n'(p_1\rho) - \beta_1\frac{\gamma n}{\rho} J_n(p_1\rho)\right]e^{in\phi},$$

$$E_\rho = \left[\alpha_1 i\gamma p_1 J_n'(p_1\rho) - \beta\frac{\omega\mu_1 n}{c} J_n(p_1\rho)\right]e^{in\phi},$$

$$H_\rho = \left[\alpha_1\frac{ck_1^2}{\omega\mu_1}\frac{1}{\rho} J_n(p_1\rho) + i\beta p_1\gamma J_n'(p_1\rho)\right]e^{in\phi};$$

outside the cylinder

$$E_z = \alpha_2 p_2^2 H_n^{(1)}(p_2\rho)e^{in\phi},$$
$$H_z = \beta_2 p_2^2 H_n^{(1)}(p_2\rho)e^{in\phi},$$

$$E_\phi = -\left[\alpha\frac{n\gamma}{\rho} H_n^{(1)}(p_2\rho) + \frac{i\omega\mu_2 p_2}{c}\beta_2 H_n^{(1)\prime}(p_2\rho)\right]e^{in\phi},$$

$$H_\phi = \left[\alpha_2 \frac{ick^2 p_2}{\omega\mu_2} H_n^{(1)'}(p_2\rho) - \beta_2 \frac{\gamma n}{\rho} H_n^{(1)}(p_2\rho)\right] e^{in\phi},$$

$$E_\rho = \left[i\alpha_2 \gamma p_2 H_n^{(1)'}(p_2\rho) - \beta_2 \frac{\omega\mu_2 n}{c} H_n^{(1)}(p_2\rho)\right] e^{in\phi},$$

$$H_\rho = \left[\alpha_2 \frac{ck_2^2}{\omega\mu_2\rho} H_n^{(1)}(p_2\rho) + i\beta_2\gamma p_2 H_n^{(1)'}(p_2\rho)\right] e^{in\phi}.$$

On the boundary for $\rho = a$ the tangential components of $\boldsymbol{E}$ and $\boldsymbol{H}$ must be continuous. This gives four homogeneous equations with four unknowns α_1, α_2, β_1 and β_2. Letting the determinant of the system equal zero, we obtain the dispersion equation for γ

$$\left[\frac{k_1^2}{\mu_1\xi}\frac{J_n'(\xi)}{J_n(\xi)} - \frac{k_2^2}{\mu_2\eta}\frac{H_n^{(1)'}(\eta)}{H_n^{(1)}(\eta)}\right]\left[\frac{\mu_1}{\xi_1}\frac{J_n'(\xi)}{J_n(\xi)} - \frac{\mu_2}{\eta}\frac{H_n^{(1)'}(\eta)}{H_n^{(1)}(\eta)}\right] = n^2\gamma^2\left(\frac{1}{\eta^2} - \frac{1}{\xi^2}\right)^2, \quad (4)$$

where $\xi = p_1 a$, $\eta = p_2 a$, a is the radius of the cylinder. This equation has an innumerable number of roots γ_{nm} (see [35], page 460).

For the fundamental wave $n = 0$ the dispersion equation breaks down into two equations:

$$\frac{\eta H_0^{(1)}(\eta)}{H_1^{(1)}(\eta)} = \frac{k_2^2\mu_1}{k_1^2\mu_2}\frac{\xi J_0(\xi)}{J_1(\xi)} \quad (5)$$

and

$$\frac{\eta H_0^{(1)}(\eta)}{H_1^{(1)}(\eta)} = \frac{\mu_2}{\mu_1}\frac{\xi J_0(\xi)}{J_1(\xi)}. \quad (6)$$

The first of them corresponds to waves of magnetic type, and the second to waves of electric type.

78. Let Σ be the surface of the tube, S its perpendicular section, C the boundary of S. We choose the z-axis parallel to the axis of the tube. The dependence on time is $e^{-i\omega t}$.

Any field inside the waveguide may be represented as the sum of fields of electric type ($H_z = 0$) and magnetic type ($E_z = 0$), each of which is determined by the z-component of the corresponding Hertz vector (see problem 69).

If $H_z = 0$, then, assuming $\Pi_z = \Pi$, we obtain the problem for the scalar function

$$\Delta\Pi + k^2\Pi = 0 \text{ inside } \Sigma \left(k = \frac{\omega}{c}\right),$$

$$\Pi = 0 \text{ on } \Sigma.$$

If $E_z = 0$, then $\Pi_z' = \Pi'$ and

$$\Delta\Pi' + k^2\Pi' = 0 \text{ inside } \Sigma$$

$$\frac{\partial\Pi'}{\partial n} = 0 \text{ on } \Sigma.$$

There exist particular solutions of the form

$$\Pi(M, z) = \psi_n(M) e^{i\gamma_n z}, \quad \Pi'(M, z) = \hat{\psi}_n(M)\, e^{i\hat{\gamma}_n z},$$

where $\gamma_n = \sqrt{k^2-\lambda_n}$, $\hat{\gamma}_n = \sqrt{k^2-\hat{\lambda}_n}$, λ_n and $\hat{\lambda}_n$ = the eigenvalues of the boundary-value problems

$$\Delta_1 \psi_n + \lambda_n \psi_n = 0 \text{ in } S, \quad \psi_n = 0 \text{ on } C,$$

$$\Delta_2 \hat{\psi}_n + \hat{\lambda}_n \hat{\psi}_n = 0 \text{ in } S, \quad \frac{\partial \hat{\psi}_n}{\partial n} = 0 \text{ on } C.$$

If $\lambda_n \leqslant k^2$ for $n = 1, 2, \ldots, N$ and $\lambda_n > k^2$ for $n = N+1, N+2, \ldots$, then there exist N travelling waves, each of which is propagated with a phase velocity

$$v_n = \frac{kc}{\gamma_n} \frac{c}{\sqrt{1-\dfrac{\lambda_n}{k^2}}}.$$

If $\lambda_1 > k^2$, then there can be no travelling waves in the tube.

If $\Pi(M, z) = A_n \psi_n(M)\, e^{i\gamma_n z}$, then the energy flow through a cross-section equals

$$Y_z = |A_n|^2 \frac{ck}{8\pi} \gamma_n \lambda_n.$$

It is assumed that $\psi_n(M)$ are normalized to unity

$$\int_S \psi_n^2 \, dS = 1.$$

Method. If one introduces a rectangular system of coordinates, then

$$E_x = \frac{\partial^2 \Pi}{\partial x \partial z}, \quad E_y = \frac{\partial^2 \Pi}{\partial y \partial z}, \quad E_z = \frac{\partial^2 \Pi}{\partial z^2} + k^2 \Pi,$$

$$H_x = -ik \frac{\partial \Pi}{\partial y}, \quad H_y = ik \frac{\partial \Pi}{\partial x}, \quad H_z = 0.$$

The problem, obtained for Π', is similar to problem 42 on the propagation of acoustic waves in a cylindrical tube with rigid walls (see [7], page 595).

79. Travelling waves can exist if the following conditions are satisfied:

(a) If $\lambda_{m,n} = [\mu_m^{(n)}]^2 < k^2$, then there exist as many travelling waves as there are linearly independent solutions of the wave equation for $\lambda_{m,n}$, satisfying this inequality; here $\mu_m^{(n)}$ is a root of the equation

$$\frac{J_n(\mu a)}{J_n(\mu b)} = \frac{N_n(\mu a)}{N_n(\mu b)};$$

in this case there can be waves of electric type.

(b) For all eigenvalues $\hat{\lambda}_{m,n}$, for which the inequality

$$\hat{\lambda}_{m,n} = [\hat{\mu}_m^{(n)}]^2 < k^2,$$

is fulfilled, where $\hat{\mu}_m^{(n)}$ is a root of the equation

$$J_n'(\mu a)N_n'(\mu b) - J_n'(\mu b)N_n'(\mu a) = 0,$$

there exist travelling waves of magnetic type ($E_z = 0$).

For the fundamental wave of electric type ($n = 0$) we have:

$$\Pi_z = \Pi = A_m R_m(\rho)\,e^{i(\gamma_m z - \omega t)}, \qquad \gamma_m = k\sqrt{1 - \frac{\mu_m^2}{k^2}},$$

where A_m is a coefficient,

$$R_m(\rho) = J_0(\mu_m \rho)N_0(\mu_m a) - J_0(\mu_m a)N_0(\mu_m \rho),$$

μ_m the mth root of the equation

$$J_0(\mu a)N_0(\mu b) - J_0(\mu b)N_0(\mu a) = 0.$$

The energy flow across a cross-section equals

$$Y_z = \frac{c}{4\pi^3}|A_m|^2\sqrt{1 - \frac{\mu_m^2}{k^2}}\;\frac{J_0^2(\mu_m a) - J_0^2(\mu_m b)}{J_0^2(\mu_m b)},$$

The field components are given by the relations

$$E_z = \lambda_m \Pi, \qquad E_\phi = 0, \qquad E_\rho = i\gamma_m A_m R_m'(\rho)\,e^{(i\gamma_m z - \omega t)},$$

$$H_z = 0, \qquad H_\varphi = AikR_m'(\rho)e^{i(\gamma_m z - \omega t)}, \qquad H_\rho = 0,$$

so that

$$H_\phi = \frac{k}{\gamma_m}E_\rho.$$

Method. One must use the results of problem 78, after assuming that the region S has the shape of a ring with radii a and b. The eigenfunctions of the ring-shaped membrane with fixed and free boundaries are given respectively in the answer to problem 27.

80. Let the origin of a spherical system of coordinates (r, θ, ϕ) exist at the centre of the spherical resonator. The dependence on time is of the type $e^{-i\omega t}$.

Oscillations of electric type are determined by the formula

$$E_r = \frac{\partial^2}{\partial r^2}(ru) + k^2(ru), \qquad E_\theta = \frac{1}{r}\frac{\partial^2(ru)}{\partial r\partial\theta}, \qquad E_\phi = \frac{1}{r\sin\phi}\frac{\partial^2(ru)}{\partial r\partial\phi},$$

$$H_r = 0, \qquad H_\theta = \frac{-ik}{\sin\theta}\frac{\partial u}{\partial\phi}, \qquad H_\phi = ik\frac{\partial u}{\partial\theta},$$

where $u = u_{m,n}$ is an eigenfunction of the boundary-value problem

$$\Delta u + k^2 u\,0, \qquad u = 0 \quad \text{for} \quad r = a,$$

given by the formula

$$u_{m,n}(r, \theta, \phi) = \psi_n(k_{m,n}r)\,Y_n^{(m)}(\theta, \phi) \quad (n = 1, 2, \ldots;\ m = 0, \pm 1, \pm 2, \ldots, \pm n),$$

where $k_{m,n} = \omega_{m,n}/c$, the characteristic wave number, is a root of the equation

$$\frac{J_{n+\frac{1}{2}}(ka)}{J_{n-\frac{1}{2}}(ka)} = \frac{ka}{n}, \qquad \psi_n(\rho) = \sqrt{\frac{\pi}{2\rho}}\, J_{n+\frac{1}{2}}(\rho),$$

$Y_n^{(m)}(\theta, \phi) = P_n^{(m)}(\cos\theta) \cos \sin m\phi$ is a spherical function.

The lowest natural frequency corresponds to $n = 0$: $u_{m0}(r) = \psi_0(k_m r)$, where k_m is determined from the equation

$$J_{-\frac{1}{2}}(k_m a) = \sqrt{\frac{2}{\pi k_m a}} \cos k_m a = 0$$

and equals

$$k_m = \frac{(2m+1)\pi}{2a}, \qquad \omega_{1,0} = \frac{\pi}{2a} c.$$

For oscillations of magnetic type ($E_r = 0$) we have:

$$E_r = 0, \qquad E_\theta = \frac{ik}{\sin\theta} \frac{\partial v}{\partial \phi}, \qquad E_\phi = -ik \frac{\partial v}{\partial \theta},$$

$$H_r = \frac{\partial^2 (rv)}{\partial r^2} + k^2 (rv), \qquad H_\theta = \frac{1}{r} \frac{\partial^2 (rv)}{\partial r \partial \theta}, \qquad H_\phi = \frac{1}{r} \frac{\partial^2 (rv)}{\partial r \partial \phi},$$

where

$$v = v_{m,n} = \psi_n(k_{m,n} r)\, Y_n^{(m)}(\theta, \phi),$$

where $k_{m,n}$ is determined from the equation

$$J_{n+\frac{1}{2}}(ka) = 0.$$

For $n = 0$ we obtain:

$$v_{m,0} = \psi_0(k_m r),$$

where

$$k_m = \frac{\pi m}{a}, \qquad \omega_1 = c \frac{\pi}{a}.$$

Method. Compare with problem 25 on the characteristic acoustic vibrations of the sphere.

81. A section of cylindrical waveguide of arbitrary section, bounded by the two planes $z = \pm l$ is considered (the z-axis is parallel to the axis of the cylinder, see problem 78).

Oscillations of electric type ($H_z = 0$)

$$\Pi_z = \Pi_{m,n} = A_{m,n} \psi_n(M) \cos \frac{\pi m}{2l} (l - z),$$

where $\psi_n(M)$ is an eigenfunction of the boundary-value problem

$$\Delta_2 \psi + \lambda_n \psi_n = 0 \text{ in } S, \qquad \psi_n = 0 \text{ on } C.$$

The characteristic frequencies

$$\omega_{m,n} = c\sqrt{\lambda_n + \left(\frac{\pi m}{2l}\right)^2}.$$

Oscillations of magnetic type ($E_z = 0$)

$$\hat{\Pi}_z = \hat{\Pi}_{m,n}(M, z) = \hat{A}_{m,n}\hat{\psi}_n(M) \operatorname{sni} \frac{\pi m}{2l}(l-z),$$

where $\hat{\psi}_n(M)$ is an eigenfunction of the boundary-value problem

$$\Delta_2\hat{\psi}_n + \hat{\lambda}_n\hat{\psi}_n = 0 \text{ in } S, \qquad \frac{\partial\hat{\psi}_n}{\partial n} = 0 \text{ on } C.$$

The characteristic frequencies

$$\hat{\omega}_{m,n} = c\sqrt{\lambda_n + \left(\frac{\pi m}{2l}\right)^2}.$$

The electric energy averaged over a period in the standing wave equals the value of the magnetic energy averaged over a period

$$\bar{\mathcal{E}}_{\text{el}} = \bar{\mathcal{E}}_{\text{m}} = \frac{1}{16\pi} ck^2\lambda_n|A_n|^2.$$

The total energy of the standing wave is invariant with respect to time and equals

$$\mathcal{E} = \frac{1}{8\pi} ck^2\lambda_n|A_n|^2.$$

For a resonator with circular or rectangular sections the formulae for Π remain effective; there one must only substitute the explicit expression for the eigenfunction,

(a) for a rectangular section with sides a and b:

$$\psi_n(M) = \psi_{nm}(x, y) = \sqrt{\frac{4}{ab}} \sin\frac{\pi m}{a}x \operatorname{s\dot{}n}\frac{\pi n}{b}y,$$

$$\hat{\psi}_n(M) = \hat{\psi}_{nm}(x, y) = \sqrt{\frac{\varepsilon_m\varepsilon_n}{ab}} \cos\frac{\pi m}{a}x \sin\frac{\pi n}{b}y, \; \varepsilon_k = 2, \; k \neq 0, \; \varepsilon_0 = 1;$$

(b) for a circular section of radius a we have:

$$\psi_{nm}(r, \phi) = \sqrt{\frac{\varepsilon_n}{\pi a^2}} \frac{J_n\left(\frac{\mu_m^{(n)}}{a}r\right)}{J_n'(\mu_m^{(n)})} \begin{matrix}\cos\\ \sin\end{matrix} n\phi,$$

$$\hat{\psi}_{n,m}(r, \phi) = \sqrt{\frac{\varepsilon_n}{\pi a^2}} \frac{\hat{\mu}_m^{(n)}}{\sqrt{[\hat{\mu}_m^{(n)}]^2 - n^2}} \frac{J_n\left(\frac{\hat{\mu}_m^{(n)}}{a}r\right)}{J_n(\hat{\mu}_m^{(n)})} \begin{matrix}\cos\\ \sin\end{matrix} n\phi,$$

where $\mu_m^{(n)}$ is a root of the equation $J_n(\mu) = 0$, $\lambda_{mn} = [\mu_m^{(n)}]^2/a^2$, $\hat{\mu}_m^{(n)}$ a root of the equation $J_n'(\mu) = 0$.

The functions $\psi_{m,n}$ given above are normalized to unity.

Method. The functions Π and $\hat{\Pi}$ satisfy the wave equation $\Delta u + k^2 u = 0$ and the following boundary conditions:

$$\Pi = 0 \quad \text{on} \quad \Sigma; \quad \frac{\partial \Pi}{\partial z} = 0 \quad \text{for} \quad z = \pm l,$$

$$\frac{\partial \hat{\Pi}}{\partial n} = 0 \quad \text{on} \quad \Sigma; \quad \hat{\Pi} = 0 \quad \text{for} \quad z = \pm l.$$

In order to calculate the total energy one must use Green's theorem (see [7], 606–613).

82. Let the toroid be bounded by the surfaces $\rho = a$ and $\rho = b$ and the planes $z = -l$ and $z = l$. It is possible to treat it as a section of coaxial cable of length $2l$, considered in problem 79. For the polarization potentials $\boldsymbol{\Pi}$ and $\hat{\Pi}$ the formulae, obtained in the solution of problem 81, remain effective and for the eigenfunctions of a cross-section ψ_n and $\hat{\psi}_n$ one must take the expressions, deduced in the answer to problem 79,

$$\psi_{m,n}(\rho, \phi) = R_{m,n}(\rho) \begin{matrix} \cos n\phi \\ \sin n\phi \end{matrix}, \qquad \hat{\psi}_{m,n}(\rho, \phi) = \hat{R}_{m,n}(\rho) \begin{matrix} \cos n\phi \\ \sin n\phi \end{matrix},$$

where

$$R_{m,n}(\rho) = J_n(\mu_m^{(n)}\rho) N_n(\mu_m^{(n)}a) - J_n(\mu_m^{(n)}a) N_n(\mu_m^{(n)}\rho),$$

$$\hat{R}_{m,n}(\rho) = J_n(\hat{\mu}_m^{(n)}\rho) N_n'(\hat{\mu}_m^{(n)}a) - J_n'(\hat{\mu}_m^{(n)}a) N_n(\hat{\mu}_m^{(n)}\rho),$$

$\mu_m^{(n)}$ and $\hat{\mu}_m^{(u)}$ are determined respectively from the equations

$$R_{m,n}(b) = 0, \qquad \hat{R}'_{m,n}(b) = 0.$$

The characteristic frequencies of the oscillations equal

$$\omega_{m,n} = c\sqrt{[\mu_m^{(n)}]^2 + \left(\frac{\pi m}{2l}\right)^2}, \qquad \hat{\omega}_{m,n} = c\sqrt{[\hat{\mu}_m^{(n)}]^2 + \left(\frac{\pi m}{2l}\right)^2}.$$

Method. See problem 81.

83. *Diffraction by a cylinder.* The cylinder axis is directed along the z-axis; a plane wave is propagated along the x-axis, the intensity vector of the electric field in the incident wave is directed parallel to the axis of the conductor. Let ε_1, μ_1, σ_1 denote the parameters of the conductor, $\varepsilon_2 = 1$, $\mu_2 = 1$, $\sigma_2 = 0$ the parameters of the medium, and k_1 and k_2 the corresponding wave numbers, where

$$k^2 = \frac{\varepsilon\mu\omega^2 + i4\pi\sigma\mu\omega}{c^2}.$$

The dependence on time is of the type $e^{-i\omega t}$.

Only the z-component of the vector $\boldsymbol{E}$ differs from zero:

$$\boldsymbol{E} = (0, 0, E),$$

H_ρ and H_ϕ are expressed in terms of it:

$$H_\rho = -\frac{ic}{\mu\omega}\frac{1}{\rho}\frac{\partial E}{\partial \phi}, \qquad H_\phi = \frac{ic}{\mu\omega}\frac{\partial E}{\partial \rho}, \qquad H_z = 0.$$

For $E = E(\rho, \phi)$ we obtain:

$$E = \begin{cases} e^{ik_2 x} + \sum\limits_{m=-\infty}^{\infty} a_m H_m^{(1)}(k_2 r) e^{im\phi} & \text{for} \quad r > a, \\ \sum\limits_{m=-\infty}^{\infty} b_m J_m(k_1 r) e^{im\phi} & \text{for} \quad r < a, \end{cases} \tag{1}$$

where a is the radius of the conductor,

$$a_m = -i^m \frac{\frac{k_1}{\mu_1} J_m'(k_1 a) J_m(k_2 a) - k_2 J_m(k_1 a) J_m'(k_2 a)}{\frac{k_1}{\mu_1} J_m'(k_1 a) H_m^{(1)}(k_2 a) - k_2 H_m^{(1)\prime}(k_2 a) J_m(k_1 a)},$$

$$b_m = i^m \frac{J_m(k_2 a)}{J_m(k_1 a)} + \frac{H_m^{(1)}(k_2 a)}{J_m(k_1 a)} a_m.$$

If the conductor is ideally conducting, then

$$a_m = -i^m \frac{J_m(k_2 a)}{H_m^{(1)}(k_2 a)}, \qquad b_m = 0.$$

Method. It is required to find the solution of the equation

$$\Delta E^{(1)} + k_1^2 E^{(1)} = 0 \quad \text{for} \quad r < a,$$

$$\Delta E^{(2)} + k_2^2 E^{(2)} = 0 \quad \text{for} \quad r > a,$$

where

$$E^{(2)} = E_0 + u = e^{ik_2 x} + u,$$

satisfying on the surface of the conductor $r = a$ the conditions of continuity of E_z and H_ϕ, which gives:

$$E^{(1)} = E^{(2)}, \qquad \frac{1}{\mu_1}\frac{\partial E^{(1)}}{\partial \rho} = \frac{\partial E^{(2)}}{\partial \rho} \quad \text{for} \quad \rho = a.$$

Moreover, the function u must satisfy the radiation condition at infinity

$$\lim_{\rho \to \infty} \sqrt{\rho}\left(\frac{\partial u}{\partial \rho} - ik_2 u\right) = 0.$$

The solution is sought in the form (1). The coefficients a_m and b_m are calculated from the conditions for $\rho = a$, where the expansion of e^{ik_2x} in a series must be used:

$$e^{ik_2x} = e^{ik_2\rho \cos \phi} = \sum_{m=-\infty}^{\infty} i^m J_m(k_2 \rho) e^{im\phi}.$$

If the conductor is ideally conducting, then $k_1 = \infty$ and the boundary conditions reduce to the one:

$$E^{(2)} = e^{ik_2a \cos \phi} + u = 0 \quad \text{when} \quad \rho = a.$$

Therefore for a_m the expression is obtained

$$a_m = -i^m \frac{J_m(k_2 a)}{H_m^{(1)}(k_2 a)}.$$

84. *Diffraction by an ideally conducting sphere.* A plane wave is propagated in the direction of the polar axis of z of a spherical system of coordinates r, θ, ϕ, the electric field is polarized in the direction of the x-axis, and the magnetic field in the direction of the y-axis:

$$E_x^0 = H_y^0 = e^{ikz} = e^{ikr \cos \theta} = \sum_{n=0}^{\infty} (2n+1) i^n \psi_n(kr) P_n(\cos \theta),$$

$$E_r = \frac{\partial^2 U}{\partial r^2} + k^2 U, \qquad E_\theta = \frac{1}{r}\frac{\partial^2 U}{\partial r\, \partial \theta}, \qquad E_\phi = -\frac{ik}{r}\frac{\partial U'}{\partial \theta}, \tag{1}$$

$$H_r = \frac{\partial^2 U'}{\partial r^2} + k^2 U', \quad H_\theta = \frac{1}{r}\frac{\partial^2 U'}{\partial r\, \partial \theta}, \qquad H_\phi = \frac{ik}{r}\frac{\partial U}{\partial \theta}, \tag{2}$$

where

$$U = ru, \qquad U' = rv.$$

The functions $u = U/r$ and $v = U'/r$ are found from the wave equations $\Delta u + k^2 u = 0$ and $\Delta v + k^2 v = 0$ and the boundary conditions

$$\frac{\partial}{\partial r}(ru) = 0, \qquad v = 0 \qquad \text{for} \qquad r = a,$$

which are a consequence of the equalities

$$E_\theta = \frac{1}{r}\frac{\partial^2 (ru)}{\partial r\, \partial \theta} = 0, \qquad E_\phi = -\frac{ik}{r}\frac{\partial}{\partial \theta}(rv) = 0 \qquad \text{for} \qquad r = a.$$

In order to solve the problem, it is necessary, first of all, to find the potentials u^0 and v^0 for the incident wave. Since the electromagnetic field is completely determined by the values E_r and H_r, then we calculate:

$$E_r^0 = \frac{\partial x}{\partial r} E_x^0 = \sin \theta \cos \phi\, e^{ikr \cos \theta} = \frac{-\cos \phi}{ikr} \frac{\partial}{\partial \theta} e^{ikr \cos \theta}$$

$$= \sum_{n=0}^{\infty} (2n+1) i^n \frac{\psi_n(kr)}{ikr} P_n^{(1)}(\cos \theta) \cos \phi,$$

$$H_r^0 = \frac{\partial y}{\partial r} H_y^0 = \sin\theta \sin\phi \, e^{ikr\cos\theta} = \frac{-\sin\phi}{ikr} \frac{\partial}{\partial\theta} e^{ikr\cos\theta}$$

$$= \sum_{n=0}^{\infty} (2n+1) i^n \frac{\psi_n(kr)}{ikr} P_n^{(1)}(\cos\theta) \sin\phi,$$

where

$$P_n^{(1)}(\cos\theta) = -\frac{d}{d\theta} P_n(\cos\theta).$$

On the other hand

$$E_r^0 = \frac{\partial^2(ru^0)}{\partial r^2} + k^2 ru^0, \qquad H_r^0 = \frac{\partial^2(rv^0)}{\partial r^2} + k^2(rv^0).$$

Assuming

$$u^0 = \sum_{n=0}^{\infty} a_n \psi_n(kr) P_n^{(1)}(\cos\theta)\cos\phi,$$

$$v^0 = \sum_{n=0}^{\infty} b_n \psi_n(kr) P_n^{(1)}(\cos\theta)\sin\phi,$$

comparing both expressions for E_r^0 and H_r^0 and taking into account the equation

$$\frac{d^2}{dr^2}(r\psi_n) + k^2 r\psi_n = \frac{n(n+1)\psi_n}{r},$$

we obtain:

$$a_n = b_n = \frac{2n+1}{n(n+1)} \frac{i^{n-1}}{k}.$$

We shall now look for the solution of the problem in the form

$$u(r, \theta, \phi) = \sum_{n=0}^{\infty} a_n [\psi_n(kr) + \alpha_n \zeta_n^{(1)}(kr)] P_n^{(1)}(\cos\theta)\cos\phi,$$

$$v(r, \theta, \phi) = \sum_{n=0}^{\infty} a_n [\psi_n(kr) + \beta_n \zeta_n^{(1)}(kr)] P_n^{(1)}(\cos\theta)\sin\phi.$$

The boundary conditions for $r = a$ enable us to determine α_n and β_n:

$$\alpha_n = -\frac{\Psi_n(ka)}{Z_n^{(1)}(ka)}, \qquad \Psi_n(x) = \frac{d}{dx}[x\psi_n(x)], \qquad \psi_n(x) = \sqrt{\frac{\pi}{2x}} J_{n+\frac{1}{2}}(x),$$

$$\beta_n = -\frac{\psi_n(ka)}{\zeta_n^{(1)}(ka)}, \quad Z_n^{(1)}(x) = \frac{d}{dx}[x\zeta_n^{(1)}(x)], \quad \zeta_n^{(1)}(x) = \sqrt{\frac{\pi}{2x}} H_{n+\frac{1}{2}}^{(1)}(x).$$

85. *Diffraction by a conducting sphere.* If the system of coordinates and the incident wave are chosen as in the preceding problem, then the unknown Borgnitz' potentials† $U = ru$ and $U' = rv$ will be determined by the expressions

$$u = \begin{cases} \sum_{n=0}^{\infty} a_n[\psi_n(k_1 r) + \alpha_n \zeta_n^{(1)}(k_1 r)] P_n^{(1)}(\cos\theta)\cos\phi & \text{for} \quad r > a \text{ (air)} \\ \sum_{n=0}^{\infty} A_n \psi_n(k_2 r) P_n^{(1)}(\cos\theta)\cos\phi & \text{for} \quad r < a, \end{cases}$$

$$v = \begin{cases} \sum_{n=0}^{\infty} a_n[\psi_n(k_1 r) + \beta_n \zeta_n^{(1)}(k_1 r)] P_n^{(1)}(\cos\theta)\sin\phi & \text{for} \quad r > a, \\ \sum_{n=0}^{\infty} B_n \psi_n(k_2 r) P_n^{(1)}(\cos\theta)\sin\phi & \text{for} \quad r < a, \end{cases}$$

where

$$\alpha_n = \frac{\dfrac{k_1^2}{\mu_1}\psi_n(k_1 a)\Psi_n(k_2 a) - \dfrac{k_2^2}{\mu_2}\psi_n(k_2 a)\Psi_n(k_1 a)}{\Delta},$$

$$A_n = \frac{k_1^2[\Psi_n(k_1 a)\zeta_n^{(1)}(ak_1) - \psi_n(ak_1)Z_n^{(1)}(ak_1)]}{\mu_1 \Delta} a_n,$$

$$\Delta = \frac{k_2^2}{\mu_2} Z_n^{(1)}(k_1 a)\psi_n(k_2 a) - \frac{k_1^2}{\mu_1}\Psi_n(k_2 a)\zeta_n^{(1)}(k_1 a);$$

similarly expressions for B_n and β_n are written down; k_2 is the wave number of the sphere, k_1 the wave number of the medium.

The components of the electric and magnetic fields are calculated from formulae (1) and (2) of problem 84.

Method. One must use the expressions obtained in the solution of the preceding problem for the potentials u^0 and v^0 of the incident wave. The boundary conditions on the surface of the sphere have the form

$$\left.\begin{aligned} \frac{\partial}{\partial r}(ru_2) &= \frac{\partial}{\partial r}(ru_1), & \frac{k_1^2 u_1}{\mu_1} &= \frac{k_2^2 u_2}{\mu_2}, \\ \frac{\partial}{\partial r}(rv_2) &= \frac{\partial}{\partial r}(rv_1), & \frac{k_1^2 v_1}{\mu_1} &= \frac{k_2^2 v_2}{\mu_2} \end{aligned}\right\} \quad \text{for} \quad r = a.$$

3. Radiation of Electromagnetic Waves

86. *Electric dipole in infinite space.* Let $p = p_0 e^{-i\omega t}$ be the moment of the dipole. We choose a spherical system of coordinates r, θ, ϕ, place the dipole

† See problem 70.

at the origin of coordinates and direct the z-axis along the vector $\boldsymbol{p}_0$; then it is possible to write:

$$E_r = 2\cos\theta\left(\frac{1}{r^2}-\frac{ik}{r}\right)\Pi_0,$$

$$E_\phi = \sin\theta\left(\frac{1}{r^2}-\frac{ik}{r}-k^2\right)\Pi_0,$$

$$H_\phi = ik\sin\theta\left(ik-\frac{1}{r}\right)\Pi_0,$$

$$E_\phi = H_r = H_\theta = 0.$$

Here Π_0 is the component of the Hertz vector, directed along the z-axis,

$$\Pi_0 = p_0\frac{e^{ikr}}{r}e^{-i\omega t}.$$

In the wave zone ($kr \gg 1$) correct to terms of order $1/r^2$

$$E_r = 0, \quad E_\theta = H_\phi = -k^2\sin\theta\,\Pi_0.$$

The energy flow averaged over a period

$$\overline{Y} = 2\pi r^2\int_0^\pi \frac{c}{4\pi}\frac{1}{2}E_\theta H_\phi\sin\theta\,d\theta = \frac{p_0^2k^4c}{3}.$$

Method. See [7], page 502.

87. *Method.* Let the dipole be situated at the origin of a spherical system of coordinates r, θ, ϕ and its moment $\boldsymbol{p}_0$ directed along the z-axis ($\theta = 0$). Then

$$\left.\begin{aligned} &E_r = \frac{\partial^2}{\partial r^2}(ru)+k^2(ru), && E_\theta = \frac{1}{r}\frac{\partial^2}{\partial r\,\partial\theta}(ru), && E_\phi = \frac{1}{r}\frac{\partial^2}{\partial r\,\partial\phi}(ru) = 0,\\ &H_r = 0, && H_\theta = 0, && H_\phi = ik\frac{\partial u}{\partial\theta},\end{aligned}\right\} \quad (1)$$

where $u = U/r$ is a solution of the equation

$$\Delta u + k^2u = 0,$$

where

$$\lim_{r\to\infty} r\left(\frac{\partial u}{\partial r}-iku\right) = 0 \quad \text{(radiation condition).}$$

The excitation condition may be taken in the form

$$H_\phi \approx -p_0\frac{ik\sin\theta}{r^2} \quad \text{for small } r$$

or

$$H_\phi = -p_0k^2\frac{\sin\theta}{r}e^{ikr} \text{ for large } r.$$

This gives:

$$u = A\zeta_1^{(1)}(kr)\cos\theta, \qquad \zeta_1^{(1)}(kr) = \frac{i}{k^2}\frac{e^{ikr}}{r}\left(ik - \frac{1}{r}\right),$$

where

$$A = ik^2 p_0 .$$

Hence the formulae of problem 86 for the components of the field E_r, E_θ, H_ϕ follow.

88. Let the dipole with moment $\boldsymbol{p} = \boldsymbol{p}_0 e^{-i\omega t}$ be directed along the z-axis of the coordinate system r, θ, ϕ, the origin of which is situated at the centre of a sphere of radius a.

The function $u = u(r, \theta)$ is given by the formula

$$u(r,\theta) = A[\zeta_1(kr) + B\psi_1(kr)]P_1(\cos\theta),$$

where

$$A = ik^2 p_0, \qquad \psi_n(x) = \sqrt{\frac{\pi}{2x}}\, J_{n+\frac{1}{2}}(x)$$

$$B = -\frac{Z_1^{(1)}(ka)}{\Psi_1(ka)} = \frac{ka\zeta_1^{(1)\prime}(ka) + \zeta_1^{(1)}(ka)}{ka\psi_1'(ka) + \psi_1(ka)}, \qquad \zeta_n^{(1)}(x) = \sqrt{\frac{\pi}{2x}}\, H^{(1)}_{n+\frac{1}{2}}(x).$$

The field components are calculated from formulae (1) of problem 87.

Method. The problem differs from the preceding one in the fact that in place of the radiation condition at infinity the boundary condition $E_\theta = 0$ or $\partial/\partial r(ru) = 0$ on the surface of the sphere for $r = a$ appears. Therefore in the solution two linearly independent cylindrical functions must be included, for instance $H^{(1)}_{n+\frac{1}{2}}$ and $H^{(2)}_{n+\frac{1}{2}}$, $N_{n+\frac{1}{2}}$ and $J_{n+\frac{1}{2}}$, $H^{(1)}_{n+\frac{1}{2}}$ and $J_{n+\frac{1}{2}}$, etc. We choose the functions $J_{n+\frac{1}{2}}$ and $H^{(1)}_{n+\frac{1}{2}}$. The constant A is the same as in the preceding problem, constant B is chosen from the condition for $r = a$.

89. If a spherical system of coordinates r, θ, ϕ is chosen with origin at the centre of the sphere and the polar axis $\theta = 0$ directed along the dipole, then it is possible to write:

$$E_r = \frac{\partial^2}{\partial r^2}(ru) + k^2(ru), \qquad E_\theta = \frac{1}{r}\frac{\partial^2(ru)}{\partial r\,\partial\theta}, \qquad E_\phi = 0,$$

$$H_r = 0, \qquad H_\theta = 0, \qquad H_\phi = \frac{ick^2}{\mu\omega}\frac{\partial u}{\partial\theta},$$

where

$$k^2 = \begin{cases} k_1^2 = \dfrac{\varepsilon_1\mu_1\omega^2 + i4\pi\mu_1\sigma_1\omega}{c^2} & \text{for} \quad r > a, \\[2ex] k_2^2 = \dfrac{\varepsilon_2\mu_2\omega^2 + i4\pi\mu_2\sigma_2\omega}{c^2} & \text{for} \quad r < a. \end{cases}$$

The function

$$u = \begin{cases} u_1 & \text{for} \quad r > a, \\ u_2 & \text{for} \quad r < a \end{cases}$$

is given by the formulae

$$u_1 = C\zeta_1(k_1 r)\cos\theta,$$

$$u_2 = ip_0 k_2^2[\zeta_1(k_2 r)+B\psi_1(k_2 r)]\cos\theta,$$

where

$$B = \frac{\zeta_1^{(1)}(ak_2)\,Z_1^{(1)}(ak_1) - \dfrac{k_1^2\mu_2^2}{k_2^2\mu_1}\,\zeta_1^{(1)}(ak_1)\,Z_1^{(1)}(ak_2)}{\dfrac{k_1^2\mu_2}{k_2^2\mu_1}\,\zeta_1^{(1)}(ak_1)\,\Psi_1(ak_2) - \psi_1(ak_2)\,Z_1^{(1)}(ak_1)},$$

$$C = -\frac{\zeta_1^{(1)}(ak_2)\,\Psi_1(ak_2) - \psi_1(ak_2)\,Z_1^{(1)}(ak_2)}{\dfrac{k_1^2\mu_2}{k_2^2\mu_1}\,\zeta_1^{(1)}(ak_1)\,\Psi_1(ak_2) - \psi_1(ak_2)\,Z_1^{(1)}(ak_1)}\,p_0 k_2^2,$$

$$\Psi_1(x) = \frac{\mathrm{d}}{\mathrm{d}x}[x\psi_1(x)], \qquad Z_1^{(1)}(x) = \frac{\mathrm{d}}{\mathrm{d}x}[x\zeta_1^{(1)}(x)].$$

For $\sigma_1 \to \infty$ we have $C \to 0$, $B \to -Z_1^{(1)}(ak_2)/\psi_1(ak_2)$, i.e. we find the solution of problem 88.

For $a \to \infty$ we have $C \to 0$, $B \to 0$ and we obtain the solution of problem 86 on the dipole in infinite space.

90. Let us introduce a spherical system of coordinates r, θ, ϕ with origin at the centre of the sphere and the polar axis directed along the dipole. As in the preceding problem

$$E_\phi = H_r = H_\theta = 0, \qquad E_r = \frac{\partial^2}{\partial r^2}(ru)+k^2(ru), \qquad E_\theta = \frac{1}{r}\frac{\partial^2(ru)}{\partial r\,\partial\theta},$$

$$H_\phi = \frac{ick^2}{\mu\omega}\frac{\partial u}{\partial\theta},$$

where

$$u = \begin{cases} u_1 & \text{for} \quad r < a, \\ u_2 & \text{for} \quad a < r < b, \\ u_3 & \text{for} \quad r > b \end{cases}$$

is given by the expressions

$$u_1 = ip_0 k_0^2[\zeta_1^{(1)}(k_0 r)+A\psi_1(k_0 r)]\cos\theta,$$

$$u_2 = [B\psi_1(kr)+C\zeta_1^{(1)}(kr)]\cos\theta,$$

$$u_3 = D\zeta_1^{(1)}(k_0 r)\cos\theta.$$

The coefficients A, B, C, D are found from the solution of the set of the following four equations:

$$ip_0 k_0^2[\zeta_1^{(1)}(ak_0)+A\psi_1(ak_0)] = \frac{k^2}{k_0^2\mu}[B\psi_1(ak)+C\zeta_1^{(1)}(ak)],$$

$$D\zeta_1^{(1)}(k_0 b) = \frac{k^3}{k_0^2\mu}[B\psi_1(kb)+C\zeta_1^{(1)}(kb)],$$

$$ip_0k^2[Z_1^{(1)}(k_0a)+A\Psi_1(k_0a)] = B\Psi_1(ka)+CZ_1^{(1)}(ka),$$

$$DZ_1^{(1)}(k_0b) = B\Psi_1(kb)+CZ_1^{(1)}(kb).$$

Here the symbols

$$\Psi_1(x) = [x\psi_1(x)]', \quad Z_1^{(1)}(x) = [x\zeta_1^{(1)}(x)]'$$

are adopted.

Method. The potentials u_1, u_2, u_3 satisfy the equations

$$\Delta u_s+k_s^2u_s = 0 \quad (s = 1, 2, 3), \quad k_1=k_3=k_0, \quad k_2 = k$$

and the boundary conditions

$$\left.\begin{aligned} k_0^2u_1 &= \frac{k^2}{\mu}u_2, \\ \frac{\partial}{\partial r}(u_1r) &= \frac{\partial}{\partial r}(u_2r) \end{aligned}\right\} \text{for } r = a; \quad \left.\begin{aligned} \frac{k_2}{\mu}u_2 &= k_0^2u_3, \\ \frac{\partial r}{\partial}(ru_2) &= \frac{\partial}{\partial r}(ru_3) \end{aligned}\right\} \text{for } r = b.$$

On the choice of expressions for u_1, u_2, u_3 see the preceding problem.

91. *Solution.* We introduce a spherical system of coordinates r, θ, ϕ with origin at the centre of the sphere, the dipole exists at the point $r = r'$, $\theta = 0$. The field is independent of the angle ϕ and is defined in terms of the scalar potential $u(r, \theta)$:

$$E_r = \frac{\partial^2}{\partial r^2}(ru)+k^2(ru), \quad E_\theta = \frac{1}{r}\frac{\partial^2(ru)}{\partial r\partial\theta}, \quad E_\phi = 0,$$

$$H_r = 0, \quad H_\theta = 0, \quad H_\phi = \frac{ick^2}{\mu\omega}\frac{\partial u}{\partial\theta}.$$

The function

$$u = \begin{cases} u_1 & \text{for} \quad r < a, \\ u_2 & \text{for} \quad r > a \end{cases}$$

satisfies the wave equation $\Delta u+k^2u = 0$ where

$$k^2 = \begin{cases} k_0^2 = \dfrac{\omega^2}{c^2} & \text{for} \quad r < a. \\ \dfrac{\varepsilon\mu\omega^2+i4\pi\sigma\mu\omega}{c^2} & \text{for} \quad r > a. \end{cases}$$

On the surface of the sphere $r = a$ the tangential components of the vector E and vector H, i.e. E_θ and H_ϕ must be continuous:

$$\left.\begin{aligned} \frac{\partial^2}{\partial r\partial\theta}(ru_1) &= \frac{\partial^2}{\partial r\partial\theta}(ru_2), \\ k_0^2\frac{\partial u_1}{\partial\theta} &= \frac{k_2}{\mu}\frac{\partial u_2}{\partial\theta} \end{aligned}\right\} \quad \text{for} \quad r = a.$$

These conditions will be fulfilled, if it is required that $\partial(ru)/\partial r$ and k^2/μ be continuous:

$$\left.\begin{aligned}\frac{\partial}{\partial r}(ru_1) &= \frac{\partial}{\partial r}(ru_2),\\ k_0^2 u_1 &= \frac{k^2}{\mu}u_2\end{aligned}\right\}\quad \text{for} \quad r = a.$$

The function ru_1, obviously, has a singularity of the type $\dfrac{e^{ik_0R}}{R}$ at the source, where $R = \sqrt{r'^2+r^2-2rr'\cos\theta}$ ($(r, 0, \phi)$ the point of observation), i.e. $u_1 \sim \dfrac{1}{r'}\dfrac{e^{ik_0R}}{R}$.

Assuming $u_1 = \bar{u}_0+v_1$, where $\bar{u}_0 = \dfrac{\alpha}{r'}u_0 = \dfrac{\alpha}{r'}\dfrac{e^{ik_0R}}{ik_0R}$ (α is a normalizing factor which will be determined below), we obtain for v_1 and u_2:

$$\Delta v_1+k_0^2 v_1 = 0 \quad \text{for} \quad r < a, \qquad \Delta u_2+k^2u_2 = 0 \quad \text{for} \quad r > a,$$

$$\left.\begin{aligned}\frac{\partial}{\partial r}(rv_1)-\frac{\partial}{\partial r}(ru_2) &= -\frac{\partial}{\partial r}(r\bar{u}_0),\\ k_0^2(v_1+\bar{u}_0) &= \frac{k^2}{\mu}u_2\end{aligned}\right\}\quad \text{for} \quad r = a, \tag{1}$$

$$\lim_{r\to\infty} r\left(\frac{\partial u_2}{\partial r}-iku_2\right) = 0.$$

Particular solutions have the form

$$v_{1n} = [A_n\psi_n(k_0r)+A_n'\zeta_n^{(1)}(k_0r)]P_n(\cos\theta),$$

$$u_{2n} = [B_n\zeta_n^{(1)}(kr)+B_n'\psi_n(kr)]P_n(\cos\theta).$$

Because of the bounded nature of u_1 at $r = 0$ the coefficient $A_n' = 0$; from the radiation condition for $r\to\infty$ it follows that $B_n' = 0$. Therefore

$$\left.\begin{aligned}v_1(r,\theta) &= \sum_{n=0}^{\infty} A_n\psi_n(k_0r)P_n(\cos\theta),\\ u_2(r,\theta) &= \sum_{n=0}^{\infty} B_n\zeta_n^{(1)}(kr)P_n(\cos\theta).\end{aligned}\right\} \tag{2}$$

In order to determine the coefficients A_n and B_n from the boundary conditions for $r = a$, we make use of the expansion of the fundamental solution u_0 in a series in Legendre polynomials:

$$u_0 = \frac{e^{ik_0R}}{ik_0R} = \left\{\begin{aligned}&\sum_{n=0}^{\infty} a_n\zeta_n^{(1)}(k_0r)P_n(\cos\theta) \quad \text{for} \quad r < r',\\ &\sum_{n=0}^{\infty} b_n\psi_n(k_0r)P_n(\cos\theta) \quad \text{for} \quad r > r',\end{aligned}\right\} \tag{3}$$

$$a_n = (2n+1)\psi_n(k_0r'), \qquad b_n = (2n+1)\zeta_n^{(1)}(k_0r').$$

For $r' \to 0$ the condition

$$\bar{u}_0 \to u_{87} = ip_0 k_0^2 \zeta_1^{(1)}(k_0 r) P_1(\cos\theta) \qquad (p_0 \text{—is the dipole moment})$$

must be fulfilled. Taking into acount the fact that the first term in (3) for $n = 0$ must be rejected, since for it $H_\phi = E_r = E_\theta = 0$, and noting that

$$\lim_{r'\to 0} \frac{\alpha a_n}{r'} = \begin{cases} 0 & \text{for} \quad n > 1, \\ -0{\cdot}5k_0^2 & \text{for} \quad n = 1 \end{cases}$$

we find $\alpha = -2ip_0$. Substituting expressions (2) and (3) in condition (1) with $r = a$ (with $r = a > r'$), we obtain

$$\beta a_n Z_n^{(1)}(k_0 a) + A_n \Psi_n(k_0 a) = B_n Z_n^{(1)}(ka),$$

$$k_0^2[a_n \beta \zeta_n^{(1)}(k_0 a) + A_n \psi_n(k_0 a)] = \frac{k^2}{\mu} B_n \zeta_n^{(1)}(ka),$$

$$Z_n^{(1)}(\rho) = [\rho\zeta_n^{(1)}(\rho)]', \quad \Psi_n(\rho) = [\rho\psi_n(\rho)]', \quad \beta = \frac{\alpha}{r'} = -\frac{2ip_0}{r'}.$$

Hence we find

$$A_n = \left[\frac{k^2}{k_0^2\mu} Z_n^{(1)}(k_0 a)\zeta_n^{(1)}(ka) - \zeta_n^{(1)}(k_0 a) Z_n^{(1)}(ka)\right] \frac{\beta a_n}{\Delta},$$

$$B_n = [\psi_n(k_0 a) Z_n^{(1)}(k_0 a) - \zeta_n^{(1)}(k_0 a)\Psi_n(k_0 a)] \frac{\rho a_n}{\Delta},$$

$$\Delta = \psi_n(k_0 a) Z_n^{(1)}(ka) - \frac{k^2}{k_0^2\mu} \zeta_n^{(1)}(ka)\Psi_n(k_0 a).$$

If $\sigma \to \infty$ $(k \to \infty)$, then $B_n = 0$,

$$A_n = -\frac{Z_n^{(1)}(k_0 a)}{\Psi_n(k_0 a)} \beta a_n,$$

and we arrive at the solution of the problem on the dipole, situated at the point $(r', 0, \phi)$ inside an ideally conducting sphere.

92. *Vertical electric antenna above the spherical earth.* The antenna (point dipole) is situated at the point $r' = a + h$ $(h > 0)$, $\theta = 0$ and is oriented along the axis $\theta = 0$. The moment of the dipole equals $p = p_0 e^{-i\omega t}$. The time factor $e^{-i\omega t}$ is omitted.

For the potential $u = U/r$ we have:
inside the earth $(r < a)$

$$u_1 = \sum_{n=0}^{\infty} A_n \psi_n(kr) P_n(\cos\theta),$$

outside the earth ($r > a$)

$$u_2 = \beta \frac{e^{ik_0R}}{ik_0R} + \sum_{n=0}^{\infty} B_n \zeta_n^{(1)}(k_0 r) P_n(\cos\theta)$$

$$= \begin{cases} \sum_{n=0}^{\infty} (\beta a_n + B_n)\zeta_n^{(1)}(k_0 r) P_n(\cos\theta) & (r > r'), \\ \sum_{n=0}^{\infty} [\beta b_n \psi_n(k_0 r) + B_n \zeta_n^{(1)}(k_0 r)] P_n(\cos\theta) & (r < r'), \end{cases}$$

where

$$A_n = \frac{\zeta_n^{(1)}(k_0 a)\Psi_n(k_0 a) - \psi_n(k_0 a) Z_n^{(1)}(k_0 a)}{\frac{k_0^2}{k^2}\zeta_n^{(1)}(k_0 a)\Psi_n(ka) - Z_n^{(1)}(k_0 a)\psi_n(ka)} \frac{k_0^2}{k^2}\beta b_n, \qquad k_0 = \frac{\omega}{c},$$

$$B_n = \frac{\psi_n(ka)\Psi_n(k_0 a) - \frac{k_0^2}{k^2}\psi_n(k_0 a)\Psi_n(ka)}{\frac{k_0^2}{k^2}\zeta_n^{(1)}(k_0 a)\Psi_n(ka) - Z_n^{(1)}(k_0 a)\psi_n(ka)}\beta b_n, \qquad k^2 = \frac{\varepsilon\omega^2 + 4\pi\sigma\omega i}{c^2}, \quad \mu = 1,$$

$$a_n = (2n+1)\psi_n(k_0 r'), \qquad b_n = (2n+1)\zeta_n^{(1)}(k_0 r'), \qquad \beta = \frac{-2ip_0}{a+h}.$$

If the earth is ideally conducting, then

$$A_n = 0, \qquad B_n = -\frac{\Psi_n(k_0 a)}{Z_n^{(1)}(k_0 a)}\beta b_n.$$

As a result

$$u_1 = 0,$$

$$u_2 = \bar{u}_0 - \sum_{n=0}^{\infty} \beta b_n \frac{\psi_n(k_0 a)\zeta_n^{(1)}(k_0 r)}{\zeta_n^{(1)}(k_0 a)} P_n(\cos\theta).$$

See problem 91.

93. *Vertical electric antenna on the spherical earth.* The antenna is situated at the point $r' = a$, $\theta = 0$ on the surface of the earth.

Inside the earth ($r < a$)

$$u_1 = \frac{2p_0 k_0}{a^2 k^2} \sum_{n=0}^{\infty} \frac{(2n+1)\zeta_n^{(1)}(k_0 a)}{\psi_n(ka)[Z_n^{(1)}(k_0 a) - C_n]} \psi_n(kr) P_n(\cos\theta),$$

outside the earth ($r > a$)

$$u_2 = \frac{2p_0}{a_2 k} \sum_{n=0}^{\infty} \frac{(2n+1)\zeta_n^{(1)}(k_0 r)}{Z_n^{(1)}(k_0 a) - C_n} P_n(\cos\theta).$$

Here C_n denotes the expression

$$C_n = \frac{k_0^2}{k^2}\frac{\Psi_n(ka)}{\psi_n(ka)}\zeta_n^{(1)}(k_0 a).$$

Method. It is necessary to perform a limiting transition for $h \to 0$ in the solution of the preceding problem. In the process of the calculations make use of the Wronskian expression

$$\psi_n(x)\zeta_n^{(1)\prime}(x) - \zeta_n^{(1)}(x)\psi_n'(x) = \frac{i}{x^2}.$$

A limiting transition for $h \to 0$ gives:

$$\lim_{h\to 0}(a_n\beta + B_n) = \frac{2p_0}{a^2k_0}\frac{2n+1}{Z_n^{(1)}(k_0 a) - C_n}.$$

4. Antenna on the Plane Earth

94. The electric Hertz vector $\boldsymbol{\Pi}$ is introduced, directed along the antenna. In a cylindrical system of coordinates ρ, ϕ, z we have:

$$\Pi_\rho = \Pi_\phi = 0, \qquad \Pi_z = \Pi.$$

Since the problem possesses axial symmetry,

$$E_\rho = \frac{\partial^2 \Pi}{\partial\rho\partial z}, \qquad E_\phi = 0, \qquad E_z = \frac{\partial^2\Pi}{\partial z^2} + k^2\Pi = -\frac{1}{\rho}\frac{\partial}{\partial\rho}\left(\rho\frac{\partial\Pi}{\partial\rho}\right),$$

$$\Delta\Pi + k^2\Pi = 0,$$

$$H_\rho = H_z = 0, \qquad H_\phi = -\frac{ick^2}{\omega}\frac{\partial\Pi}{\partial\rho}.$$

On the surface of the earth with $z = 0$

$$k_0^2\Pi_0 = k^2\Pi, \qquad \frac{\partial\Pi_0}{\partial z} = \frac{\partial\Pi}{\partial z},$$

where

Π_0, $k_0^2 = \dfrac{\omega^2}{c^2}$ are quantities in the atmosphere,

Π, $k^2 = \dfrac{\varepsilon\omega^2 + i4\pi\sigma\omega}{c^2}$ are quantities in the earth

$$(\mu = 1).$$

The dipole moment $p = p_0 e^{-i\omega t}$, $p_0 = 1$; the factor $e^{-i\omega t}$ is everywhere omitted.

95. The electromagnetic field is expressed in terms of the magnetic Hertz vector, for which only the component along the axis of the antenna $\Pi_z = \Pi$ differs from zero, therefore $E_z = 0$. Because of the axial symmetry

$$E_\rho = 0, \qquad E_\phi = -i\frac{\omega}{c}\frac{\partial\Pi}{\partial\rho},$$

$$H_\rho = \frac{\partial^2\Pi}{\partial\rho\,\partial z}, \qquad H_\phi = 0, \qquad H_z = k^2\Pi + \frac{\partial^2\Pi}{\partial z^2}.$$

The potential Π satisfies the equation

$$\Delta\Pi+k^2\Pi = 0, \quad \text{where} \quad k^2 = \begin{cases} k_0^2 = \dfrac{\omega^2}{c^2} & \text{for} \quad z > 0, \\ \dfrac{\varepsilon\omega^2+i4\pi\sigma\omega}{c^2} & \text{for} \quad z < 0, \end{cases}$$

and the matching conditions on the surface of the earth

$$\Pi_0 = \Pi, \quad \frac{\partial \Pi_0}{\partial z} = \frac{\partial \Pi}{\partial z} \quad \text{for} \quad z = 0,$$

where

$$\Pi_0 = \frac{e^{ik_0R}}{R}+\Pi_{0\,\text{sec}},$$

$$\Pi = \frac{e^{ikR}}{R}+\Pi_{\text{sec}},$$

where $R = \sqrt{r^2+z^2}$. The first terms in our expressions give the Hertz potential for the dipole in an infinite medium with a corresponding wave number (k or k_0), $\Pi_{0\,\text{sec}}$ and Π_{sec} is the secondary radiation.

96. We introduce the coordinate system x, y, z, choosing the z-axis perpendicular to the earth's surface, and the x-axis along the antenna.

$$\mathbf{E} = \text{grad div}\, \boldsymbol{\Pi}+k^2\boldsymbol{\Pi}, \quad \mathbf{H} = -ik\, \text{curl}\, \boldsymbol{\Pi},$$

$$\boldsymbol{\Pi} = (\Pi_x, 0, \Pi_z),$$

where Π_x and Π_z satisfy the wave equation

$$E_x = k^2\Pi_x+\frac{\partial}{\partial x}\left(\frac{\partial \Pi_x}{\partial x}+\frac{\partial \Pi_z}{\partial z}\right), \quad E_y = \frac{\partial}{\partial y}\left(\frac{\partial \Pi_x}{\partial x}+\frac{\partial \Pi_z}{\partial z}\right),$$

$$H_x = -\frac{ick^2}{\omega}\frac{\partial \Pi_z}{\partial y}, \quad H_y = -\frac{ick^2}{\omega}\left(\frac{\partial \Pi_x}{\partial z}-\frac{\partial \Pi_z}{\partial x}\right), \quad H_z = \frac{ick^2}{\omega}\frac{\partial \Pi_x}{\partial y}.$$

The boundary conditions at $z = 0$ (on the earth's surface)

$$k_0^2\Pi_{0z} = k^2\Pi_z, \quad k_0^2\frac{\partial \Pi_{0x}}{\partial z} = k^2\frac{\partial \Pi_x}{\partial z},$$

$$k_0^2\Pi_{0x} = k^2\Pi_x, \quad \frac{\partial \Pi_{0x}}{\partial x}+\frac{\partial \Pi_{0z}}{\partial z} = \frac{\partial \Pi_x}{\partial x}+\frac{\partial \Pi_z}{\partial z}.$$

Usually in place of Π_z the function F is introduced:

$$\Pi_{0z} = \frac{\partial F_0}{\partial x}, \quad \Pi_z = \frac{k_0^2}{k^2}\frac{\partial F}{\partial x}.$$

The first and last boundary conditions give:

$$F_0 = F, \quad \Pi_{0x}+\frac{\partial F_0}{\partial z} = \Pi_x+\frac{k_0^2}{k^2}\frac{\partial F}{\partial z}.$$

97. Let the current coil be situated in the plane x, z, so that the normal to the coil is directed along the y-axis. The field vectors are expressed in terms of the magnetic Hertz vector

$$E = -i\frac{\omega}{c}\,\text{curl}\,\boldsymbol{\Pi}, \qquad H = k^2\boldsymbol{\Pi}+\text{grad div}\,\boldsymbol{\Pi},$$

the components Π_y and Π_z of the vector $\boldsymbol{\Pi}$ differ from zero, so that

$$E_x = \frac{i\omega}{c}\left(\frac{\partial \Pi_z}{\partial y}-\frac{\partial \Pi_y}{\partial z}\right), \qquad E_y = -\frac{i\omega}{c}\frac{\partial \Pi_z}{\partial x},$$

$$H_x = \frac{\partial}{\partial x}\left(\frac{\partial \Pi_y}{\partial y}+\frac{\partial \Pi_z}{\partial z}\right), \qquad H_y = k^2\Pi_y+\frac{\partial}{\partial y}\left(\frac{\partial \Pi_y}{\partial y}+\frac{\partial \Pi_z}{\partial z}\right),$$

$$H_z = k^2\Pi_z+\frac{\partial}{\partial z}\left(\frac{\partial \Pi_y}{\partial y}+\frac{\partial \Pi_z}{\partial z}\right).$$

The boundary conditions at $z = 0$

$$\Pi_{z0} = \Pi_z, \qquad k_0^2\Pi_{0y} = k^2\Pi_y,$$

$$\frac{\partial \Pi_{0y}}{\partial z} = \frac{\partial \Pi_y}{\partial z}, \qquad \frac{\partial \Pi_{0y}}{\partial y}+\frac{\partial \Pi_{0z}}{\partial z} = \frac{\partial \Pi_y}{\partial y}+\frac{\partial \Pi_z}{\partial z}.$$

If it is assumed

$$\Pi_{0z} = \frac{\partial F_0}{\partial y}, \qquad \Pi_z = \frac{\partial F}{\partial y},$$

then in place of the first and fourth conditions there is obtained:

$$F_0 = F, \qquad \Pi_{0y}+\frac{\partial F_0}{\partial z} = \Pi_y+\frac{\partial F}{\partial z}.$$

98. Let us place the origin of coordinates in the antenna. Then above the earth

$$\Pi_0 = \frac{2k^2}{k_0^2+k^2}\frac{e^{ik_0R}}{R}+\int_0^\infty f_0(\lambda)J_0(\lambda r)e^{-\sqrt{\lambda^2-k_0^2}z}\,d\lambda \quad (z>0),$$

in the earth

$$\Pi = \frac{2k_0^2}{k_0^2+k^2}\frac{e^{ikR}}{R}+\int_0^\infty f(\lambda)J_0(\lambda r)e^{+\sqrt{\lambda^2-k^2}z}\,d\lambda \quad (z<0),$$

where

$$f_0(\lambda) = \frac{2k_0^2k^2}{k_0^2+k^2}\frac{\lambda}{\sqrt{\lambda^2-k_0^2}}\frac{\sqrt{\lambda^2-k_0^2}-\sqrt{\lambda^2-k^2}}{k^2\sqrt{\lambda^2-k_0^2}+k_0^2\sqrt{\lambda^2-k^2}},$$

$$f(\lambda) = -\frac{2k_0^2k^2}{k_0^2+k^2}\frac{\lambda}{\sqrt{\lambda^2-k^2}}\frac{\sqrt{\lambda^2-k_0^2}-\sqrt{\lambda^2-k^2}}{k^2\sqrt{\lambda^2-k_0^2}+k_0^2\sqrt{\lambda^2-k^2}},$$

$$R = \sqrt{r^2+z^2}.$$

Solution. As in problem 94 we introduce the electric Hertz vector $\boldsymbol{\Pi} = (0, 0, \Pi_z = \Pi)$, where

$$\Pi_0 = \frac{2k_0^2}{k_0^2+k^2}\frac{e^{ik_0R}}{R}+\Pi_{0\,\mathrm{sec}}, \qquad \Pi = \frac{2k^2}{k_0^2+k^2}\frac{e^{ikR}}{R}+\Pi_{\mathrm{sec}}.$$

Let us use the integral expansion of the primary potential

$$\frac{e^{ikR}}{R} = \int_0^\infty J_0(\lambda r)\frac{e^{-\sqrt{\lambda^2-k^2}|z|}\lambda\, d\lambda}{\sqrt{\lambda^2-k^2}},$$

and let us look for the secondary cxcitation in the form

$$\Pi_{0\,\mathrm{sec}} = \int_0^\infty f_0(\lambda)J_0(\lambda r)e^{\sqrt{\lambda^2-k_0^2}z}\, d\lambda \qquad (z>0),$$

$$\Pi_{\mathrm{sec}} = \int_0^\infty f(\lambda)J_0(\lambda r)e^{\sqrt{\lambda^2-k^2}z}\, d\lambda \qquad (z<0).$$

$\Pi_{0\,\mathrm{sec}}$ and Π_{sec}, represented by these integrals, obviously, satisfy the equations

$$\Delta\Pi_{0\,\mathrm{sec}}+k_0^2\Pi_{0\,\mathrm{sec}} = 0, \qquad \Delta\Pi_{\mathrm{sec}}+k^2\Pi_{\mathrm{see}} = 0.$$

Requiring the boundary conditions to be satisfied

$$k_0^2\Pi_0 = k^2\Pi, \qquad \frac{\partial\Pi_0}{\partial z} = \frac{\partial\Pi}{\partial z} \quad \text{for} \quad z=0,$$

we obtain:

$$\int_0^\infty J_0(\lambda r)\left[\frac{2k_0^2k^2}{k_0^2+k^2}\frac{\lambda}{\mu_0}+f_0(\lambda)\right]d\lambda = \int_0^\infty \left[\frac{2k_0^2k^2}{k_0^2+k^2}\frac{\lambda}{\mu}+f(\lambda)\right]J_0(\lambda r)\, d\lambda$$

and

$$\int_0^\infty [\mu_0 f_0(\lambda)+\mu f(\lambda)]J_0(\lambda r)\, d\lambda = 0 \qquad (\mu = \sqrt{\lambda^2-k^2}),$$

where $\mu^2 = \lambda^2-k^2$, ${\mu^2}_0 = \lambda^2-k_0^2$.

Hence we find

$$f_0(\lambda) = \frac{2k_0^2k^2}{k_0^2+k^2}\frac{\lambda}{\mu_0}\frac{\mu_0-\mu}{k^2\mu_0+k_0^2\mu},$$

$$f(\lambda) = -\frac{2k_0^2k^2}{k_0^2+k^2}\frac{\lambda}{\mu}\frac{\mu_0-\mu}{k^2\mu_0+k_0^2\mu}.$$

Special cases:

(1) $k=\infty$, the earth is ideally conducting,

$$f(\lambda) = 0, \qquad f_0(\lambda) = \frac{2\lambda}{\mu_0},$$

$$\Pi_0 = 2\int_0^\infty J_0(\lambda r)e^{-\mu_0|z|}\frac{\lambda\,d\lambda}{\mu_0} = 2\frac{e^{ik_0R}}{R},$$

$\Pi = 0$ (in the earth).

The primary excitation of the antenna is reflected from the surface of the earth.

(2) $k = k_0$, the antenna in a homogeneous medium (in air). In this case

$$f_0(\lambda) = 0, \quad f(\lambda) = 0,$$

$\Pi = \dfrac{e^{ik_0R}}{R}$ over all space.

99. The magnetic Hertz vector $\boldsymbol{\Pi} = (0, 0, \Pi)$ is defined in the following way: above eth earth

$$\Pi_0 = \frac{e^{ik_0R}}{R} + \int_0^\infty f_0(\lambda)J_0\lambda r)e^{-\mu_0 z}\,d\lambda \quad (z > 0),$$

in the earth

$$\Pi = \frac{e^{ikR}}{R} + \int_0^\infty f(\lambda)J_0(\lambda r)e^{\mu z}\,d\lambda \quad (z < 0),$$

where

$$f_0(\lambda) = \frac{\lambda}{\mu_0}\frac{\mu_0-\mu}{\mu_0+\mu}, \quad f(\lambda) = \frac{\lambda}{\mu}\frac{\mu-\mu_0}{\mu+\mu_0}, \quad \mu_0 = \sqrt{\lambda^2-k_0^2}, \quad \mu = \sqrt{\lambda^2-k^2}.$$

Expressions for Π_0 and Π may be written differently:

$$\Pi_0 = \int_0^\infty \frac{2J_0(\lambda r)}{\mu+\mu_0}e^{-\mu_0 z}\lambda\,d\lambda \quad \text{for} \quad z > 0,$$

$$\Pi = \int_0^\infty \frac{2J_0(\lambda r)}{\mu+\mu_0}e^{\mu z}\lambda\,d\lambda \quad \text{for} \quad z < 0.$$

In the case of ideally conducting earth $k = \infty$, $\mu = \infty$ and $\Pi = \Pi_0 = 0$. The action of the magnetic antenna is compensated by eddy currents, originating in the earth.

Method. See problems 95 and 98.

100. If the antenna is directed along the x-axis, then as in problem 96 the Hertz vector $\boldsymbol{\Pi} = (\Pi_x 0, \Pi_z)$, where

$$\Pi_{0x} = \int_0^\infty \frac{2J_0(\lambda r)}{N'}e^{-\mu_0 z}\lambda\,d\lambda \quad \text{for} \quad z > 0,$$

$$\Pi_x = \frac{k_0^2}{k^2}\int_0^\infty \frac{2J_0(\lambda r)}{N'}e^{\mu z}\lambda\,d\lambda \quad \text{for} \quad z < 0,$$

$$\Pi_{0z} = 2(k^2-k_0^2)\cos\phi \int_0^\infty \frac{J_0'(\lambda r)}{NN'} e^{-\mu_0 z}\lambda\, d\lambda, \qquad z>0,$$

$$\Pi_z = \frac{2k_0^2}{k^2}(k^2-k_0^2)\cos\phi \int_0^\infty \frac{J_0'(\lambda r)}{NN'} e^{\mu z}\lambda\, d\lambda, \qquad z<0,$$

where

$$N' = \mu+\mu_0, \quad N = k^2\mu_0+k_0^2\mu, \quad \mu = \sqrt{\lambda^2-k^2}, \quad \mu_0 = \sqrt{\lambda^2-k_0^2}.$$

Method. The function Π_x is determined by the equation $\Delta u+k^2u=0$ and the boundary conditions

$$k_0^2\Pi_{0x} = k^2\Pi_x, \quad k_0^2\frac{\partial \Pi_{0x}}{\partial z} = k^2\frac{\partial \Pi_x}{\partial z} \quad \text{for} \quad z=0.$$

Hence we see that the functions $\Pi_0 = k_0^2\Pi_{0x}$ and $\Pi = k^2\Pi_x$ agree with the expression in the solution of the preceding problem 99.

For $\Pi_z = \partial F/\partial x$ we have:

$$F_0 = F, \quad \Pi_{0x}+\frac{\partial F_0}{\partial z} = \Pi_x+\frac{k_0^2}{k^2}\frac{\partial F}{\partial z} \quad \text{for} \quad z=0.$$

Assuming

$$F_0 = \int_0^\infty f_0(\lambda)J_0(\lambda r)e^{-\mu_0 z}\, d\lambda \quad (z>0),$$

$$F = \int_0^\infty f(\lambda)J_0(\lambda r)e^{\mu z}\, d\lambda \quad (z<0)$$

and using the expressions already found for $\boldsymbol{\Pi}_{0x}$ and $\boldsymbol{\Pi}_x$, we obtain:

$$f_0(\lambda) = f(\lambda) = \frac{2(k^2-k_0^2)\lambda}{NN'}.$$

The function $\boldsymbol{\Pi}_z$ is calculated from the formula

$$\Pi_{0z} = \frac{\partial F_0}{\partial x} = \cos\phi\,\frac{\partial F_0}{\partial r},$$

$$\Pi_z = \frac{k_0^2}{k^2}\cos\phi\,\frac{\partial F}{\partial r}.$$

101. We use all the symbols of problem 97. In this case the fields E and H are expressed in terms of the magnetic Hertz vector $\boldsymbol{\Pi} = (0, \Pi_y, \Pi_z)$, where

$$\Pi_{0y} = \int_0^\infty \frac{2k^2}{N} J_0(\lambda r)e^{-\mu_0 z}\lambda\, d\lambda \quad \text{for} \quad z>0,$$

$$\Pi_y = \int_0^\infty \frac{2k_0^2}{N} J_0(\lambda r) e^{\mu z} \lambda \, d\lambda \quad \text{for} \quad z < 0,$$

$$\Pi_{0z} = 2(k^2 - k_0^2) \sin\phi \int_0^\infty \frac{J_0'(\lambda r)}{NN'} e^{-\mu_0 z} \lambda^2 \, d\lambda \quad \text{for} \quad z > 0,$$

$$\Pi_z = 2(k^2 - k_0^2) \sin\phi \int_0^\infty \frac{J_0'(\lambda r)}{NN'} e^{\mu z} \lambda^2 \, d\lambda \quad \text{for} \quad z < 0.$$

The values of N and N' are given in problem 100.

102. The polarization potential $\boldsymbol{\Pi} = (0, 0, \Pi_z = \Pi)$ determines the components of the electromagnetic field by the formulae

$$\boldsymbol{E} = \text{grad div}\, \boldsymbol{\Pi} + k^2 \boldsymbol{\Pi}, \quad \boldsymbol{H} = -ik \,\text{curl}\, \boldsymbol{\Pi}.$$

For the potential

$$\Pi = \begin{cases} \Pi_1 & \text{for} \quad z > a, \\ \Pi_2 & \text{for} \quad 0 < z < a \end{cases}$$

we obtain:

$$\Pi = \Pi_{1\,\text{prim}} + \Pi_{1\,\text{sec}},$$

where

$$\Pi_{1\,\text{prim}} = \int_0^\infty J_0(\lambda r) e^{-\mu_1 |z - z_0|} \frac{\lambda \, d\lambda}{\mu_1},$$

$$\Pi_{1\,\text{sec}} \doteq \int_0^\infty f_1(\lambda) J_0(\lambda r) e^{-\mu_1 (z + z_0 - 2a)} \frac{\lambda \, d\lambda}{\mu_1},$$

$$f_1(\lambda) = \frac{k_2^2 \mu_1 - k_1^2 \mu_2 \tan \mu_2 a}{k_2^2 \mu_1 + k_1^2 \mu_2 \tan \mu_2 a}, \quad \mu_1 = \sqrt{\lambda^2 - k_1^2}, \quad \mu_2 = \sqrt{\lambda^2 - k_2^2}.$$

We find an expression for Π_2 similarly.

103. Let $I = I_0 f(s)\, e^{-i\omega t}$ ($f(s) \leqslant 1$) be the current intensity in a linear conductor $-l \leqslant s \leqslant l$ of length $2l$. A cylindrical system of coordinates is chosen so that the linear current is directed along the z-axis and is symmetrical with respect to the origin of coordinates. The Hertz vector $\boldsymbol{\Pi} = (0, 0, \Pi)$ is given by the relation

$$\Pi(\rho, \phi, z) = \frac{I_0}{-ikc} \int_{-l}^{l} \Pi^0[\rho, z, \phi; \xi, \psi, \zeta] f(\zeta) \, d\zeta,$$

where $\Pi^0 = e^{ikR}/R$, R is the distance between the points $[M(\rho, \phi), z]$ and $[M_0(\xi, \eta), \zeta]$,

$$\boldsymbol{E} = \text{grad div}\, \boldsymbol{\Pi} + k^2 \boldsymbol{\Pi}, \quad \boldsymbol{H} = -ik \,\text{curl}\, \boldsymbol{\Pi}.$$

The radiation resistance equals

$$R = -\frac{1}{I_0}\left\{\int_{-l}^{l} \Pi[f''+k^2 f]\,dz - \Pi f'\Big|_{-l}^{l}\right\},$$

if

$$f(-l) = f(l) = 0.$$

Method. Normalization of Π is obtained from the condition $H_\phi \approx 2I/c\rho$ in the neighbourhood of the current.

The input impedance for a linear current is determined by the following formula using the method of induced e.m.f.'s:

$$R = -\frac{1}{I_0}\int_{-l}^{l} E_z(M_0, M_0; z) f(z)\,dz.$$

Substituting here in place of E_z the expression

$$E_z = \frac{\partial^2 \Pi}{\partial z^2} + k^2 \Pi$$

and integrating by parts, we obtain the expression for R deduced above.

104. If the dipole is half-wave, then $I = I_0 f(z)$ for $-l \leqslant z \leqslant l$, where

$$f(z) = \cos kz, \quad k = \frac{\omega}{c},$$

$$\Pi = \frac{I_0}{-ikc}\int_{-l}^{l} \Pi^0(M, M_0, z-\zeta)\cos k\zeta\,d\zeta,$$

the input impedance of the half-wave dipole

$$R = \frac{1}{c}\left\{\int_0^{2\pi} \frac{1-\cos \alpha}{\alpha}\,d\alpha - i\int_0^{2\pi} \frac{\sin \alpha}{\alpha}\,d\alpha\right\}.$$

The active component of the input impedance or radiation resistance

$$R_a = \frac{1}{c}\int_0^{2\pi} \frac{1-\cos \alpha}{\alpha}\,d\alpha.$$

The reactive component or reactance

$$R_r = -\frac{1}{c}\int_0^{2\pi} \frac{\sin \alpha}{\alpha}\,d\alpha.$$

Solution. In order to calculate R one uses $E_z = \partial^2\Pi/\partial z^2 + k^2\Pi$, where

$$E_z = \frac{I_0}{-ick}\int_{-l}^{l}\left[\frac{\partial^2 \Pi^0}{\partial z^2} + k^2 \Pi^0(M, M_0; z-\zeta)\right] f(\zeta)\,d\zeta.$$

Taking into account the fact that $\partial^2\Pi^0/\partial z^2 = \partial^2\Pi^0/\partial\zeta^2$, and integrating further by parts, we obtain:

$$E_z(M, M_0, z) = \frac{I_0}{-ick}\left\{\int_{-l}^{l} \Pi^0(M, M_0;\, z-\zeta)\,[f''(\zeta)+k^2 f(\zeta)]\,\mathrm{d}\zeta + \right.$$

$$\left. +\Pi^0(M, M_0;\, l+z)f'(-l)-\Pi^0(M, M_0, l-z)f'(l)\right\}.$$

This is true if $f''(z)$ is piecewise continuous.

For the half-wave dipole

$$f''(z)+k^2 f = 0, \quad f(\pm l) = 0, \quad f'(-l) = -f'(l) = k.$$

Therefore

$$E_z(M, M_0;\, z) = \frac{I_0}{-ic}\{\Pi^0(M, M_0, l+z)+\Pi^0(M, M_0;\, l-z)\}.$$

Substitution of this value of E_z in the equation

$$R = -\frac{1}{I_0}\int_{-l}^{l} E_z(M_0, M_0;\, z)f(z)\,\mathrm{d}z$$

gives:

$$R = -\frac{2i}{c}\int_{-l}^{l} \Pi^0(M_0, M_0;\, l+z)f(z)\,\mathrm{d}z,$$

where

$$\Pi^0(M_0, M_0;\, l+z) = \frac{\mathrm{e}^{ik(l+z)}}{l+z}.$$

Assuming $l+z = a$, after simple transformations we obtain the formula for R deduced above. In particular, in a practical system of units

$$R = 30\left\{\int_0^{2\pi} \frac{1-\cos\alpha}{\alpha}\,\mathrm{d}\alpha - i\int_0^{2\pi}\frac{\sin\alpha}{\alpha}\,\mathrm{d}\alpha\right\}\,\Omega.$$

105. Let the z-axis coincide with the guide axis, and let the dipole exist in the plane $z = \zeta$ at the point M_0 and is directed parallel to the z-axis. The field is defined by the one only z-component of the electric Hertz vector

$$\boldsymbol{\Pi} = (0, 0, \Pi),$$

where

$$\Pi = \frac{4\pi p_0}{-ikc}\Pi^0(M, M_0;\, z-\zeta), \quad p_0 = I_0 l\text{—dipole moment},$$

M and M_0 are points in the plane of a perpendicular section,

$$\Pi^0(M, M_0;\, z-\zeta) = \sum_{n=1}^{\infty} \frac{\psi(M)\psi_n(M_0)}{2p_n}\,\mathrm{e}^{-p_n|z-\zeta|}, \tag{1}$$

$p_n = \sqrt{\lambda_n - k^2}$, $k = \omega/c$, λ_n is an eigenvalue, and ψ_n are the normalized eigenfunctions of the boundary-value problem

$$\Delta_2\psi_n + \lambda_n\psi_n = 0 \text{ in } S,$$
$$\psi_n = 0 \text{ on } C,$$

S is the cross-section of the guide, C is the boundary of the region S.

The radiation resistance

$$R^{(a)} = \lim_{z\to\infty} \frac{1}{I_0^2} \int\limits_{S_z} \int\limits_{+S_{-z}} \frac{c}{4\pi}[EH^*]\,d\sigma$$

equals

$$R^{(a)} = \frac{4\pi}{c}\frac{l^2}{k}\sum_{n=1}^{N}\frac{\lambda_n\psi_n^2(M_0)}{2\sqrt{k^2-\lambda_n}},$$

where N is the maximum number of travelling waves in the waveguide

$$(\lambda_N < k^2, \quad \lambda_{N+1} > k^2).$$

If the dipole exists on the axis of a circular guide of radius a, then

$$R^{(a)} = \frac{2}{ck^2a_n}\left(\frac{l}{a}\right)^2\sum_{m=1}^{N}\frac{\mu_m^2}{J_1^2(\mu_m)\sqrt{1-\dfrac{\mu_m^2}{a^2k^2}}},$$

where μ_m is a root of the equation $J_0(\mu) = 0$.

Method. The formula $\Pi = \dfrac{4\Pi p_0}{-ikc}\Pi_0$ follows from the general formula for Π, deduced in the answer to problem 103. The source function Π^0 for the wave equation

$$\Delta u + k^2 u = 0$$

in an arbitrary cylindrical region with zero boundary conditions was found in problem 45.

To calculate $R^{(a)}$ the formula

$$R^{(a)} = \lim_{z\to\infty} \frac{c}{4\pi I_0^2} \int\limits_{S_z} \int\limits_{+S_{-z}} \{E_x H_y^* - E_y H_x^*\}\,dx\,dy†$$

is used and Green's first theorem.

106. For an arbitrary linear current $I = I_0 f(z)$ for $-l \leqslant z \leqslant l$ the Hertz function

$$\Pi = \frac{4\pi I_0}{-ikc}\int_{-l}^{l}\Pi^0(M, M_0; z-\zeta)f(\zeta)\,d\zeta,$$

where $\Pi^0(M, M_0; z-\zeta)$, is given by formula (1) of the answer of problem 105.

† See [7], page 542.

The magnitude of the radiation energy $W_r = I_0^2 R^{(a)}$, where

$$R^{(a)} = \frac{4\pi}{ck}\sum_{n=1}^{N}\frac{\lambda_n^2\psi_n^2(M_0)}{2\varkappa_n}\left\{\left[\int_{-l}^{l} f(\zeta)\cos\varkappa_n\zeta\,d\zeta\right]^2+\left[\int_{-l}^{l} f(\zeta)\sin\varkappa_n\zeta\,d\zeta\right]^2\right\},$$

$$\varkappa_n = \sqrt{k^2-\lambda_n}.$$

In the general case for a half-wave dipole in a guide of arbitrary section S the formulae

$$R^{(a)} = \frac{4\pi}{c}\sum_{n=1}^{N}\frac{\psi_n^2(M_0)\left(1+\cos\pi\sqrt{1-\gamma_n^2}\right)}{\lambda_n\sqrt{1-\gamma_n^2}},$$

$$R^{(r)} = -\frac{4\pi}{c}\sum_{n=1}^{N}\frac{\psi_n^2(M_0)\sin\pi\sqrt{1-\gamma_n^2}}{\lambda_n\sqrt{1-\gamma_n^2}}+\frac{4\pi}{c}\sum_{n=N+1}^{\infty}\frac{\psi_n^2(M_0)\left(1+e^{-\pi\sqrt{\gamma_n^2-1}}\right)}{\lambda_n\sqrt{\gamma_n^2-1}},$$

are obtained, $\Delta\psi_n+\lambda_n\psi_n = 0$ on S, $\psi_n = 0$ on C, $\int_S \psi_n^2 dS = 1$, $\gamma_n^2 = \lambda_n/k^2$, C is the boundary of S, $\gamma_N < 1$, $\gamma_{N+1} > 1$. See problems 45 and 103.

107. For a half-wave dipole, lying on the axis of a circular waveguide we have:

the resistive part of the input impedance

$$R^{(a)} = \frac{4}{c}\sum_{m=1}^{N}\frac{1+\cos\pi\sqrt{1-\gamma_m^2}}{J_1^2(\mu_m)\mu_m^2\sqrt{1-\gamma_m^2}} \qquad (\gamma_N < 1,\ \gamma_{N+1} > 1),$$

the reactance

$$R^{(r)} = -\frac{4}{c}\sum_{m=1}^{N}\frac{\sin\pi\sqrt{1-\gamma_m^2}}{J_1^2(\mu_m)\mu_m^2\sqrt{1-\gamma_m^2}}+\frac{4}{c}\sum_{m=N+1}^{\infty}\frac{1-e^{-\pi\sqrt{\gamma_m^2-1}}}{J_1^2(\mu_m)\mu_m^2\sqrt{\gamma_m^2-1}},$$

where $\gamma_m^2 = \mu_m^2/a^2k^2$, μ_m is a root of the equation $J_0(\mu) = 0$, a is the radius of the waveguide.

108. Let $S(0 \leqslant x \leqslant a,\ 0 \leqslant y \leqslant b)$ be the waveguide section.

(a) An infinitely small dipole is oriented along the y-axis at the point $M_0(d, y_0)$. The radiation resistance of this dipole is given by the formula

$$R^{(a)} = \frac{l^2}{\omega\varepsilon}\sum_{n=1}^{N}\sum_{n=1}^{N'}\frac{\varkappa_{mn}\left[\frac{\partial\psi_{mn}}{\partial y_0}(M_0)\right]^2}{2\lambda_{mn}}+\omega\mu l^2\sum_{m=0}^{N_1}\sum_{n=0}^{N_1'}\frac{\left[\frac{\partial\hat{\psi}_{mn}}{\partial x_0}(M_0)\right]^2}{2\hat{\varkappa}_{mn}\hat{\lambda}_{mn}}, \qquad (1)$$

where

$$\psi_n(M) = \psi_{mn}(x, y) = \sqrt{\frac{4}{ab}}\sin\frac{\pi m}{a}x\sin\frac{\pi n}{b}y,$$

$$\hat{\psi}_n(M) = \hat{\psi}_{mn}(x, y) = \sqrt{\frac{\varepsilon_m \varepsilon_n}{ab}} \cos \frac{\pi m}{a} x \cos \frac{\pi n}{b} y \quad \left(\varepsilon_j = \begin{cases} 1, & j = 0, \\ 2, & j \neq 0 \end{cases} \right),$$

$$\lambda_{mn} = \pi^2 \left(\frac{m^2}{a^2} + \frac{n^2}{b^2} \right) \quad (m, n = 1, 2, \ldots),$$

$$\hat{\lambda}_{mn} = \pi^2 \left(\frac{m^2}{a^2} + \frac{n^2}{b^2} \right) \quad (m, n = 0, 1, 2, \ldots),$$

$$\varkappa_{mn} = \sqrt{k^2 - \lambda_{mn}}, \quad \hat{\varkappa}_{mn} = \sqrt{k^2 - \hat{\lambda}_{mn}}.$$

The limits N and N', N_1 and N_1' are such that λ_{NN}, $\hat{\lambda}_{N_1 N_1'}$ are the maximum eigenvalues for which $\varkappa_{mn}$ and $\hat{\varkappa}_{mn}$ are real.

Of most interest in practice is the case of the H_{10} wave. We have:

$$\psi_{10} = 0, \quad \hat{\psi}_{10}(x, y) = \sqrt{\frac{2}{ab}} \cos \frac{\pi}{a} x, \quad \hat{\lambda}_{10} = \left(\frac{\pi}{a} \right)^2$$

and for $R^{(a)}$ we obtain Slater's formula

$$R^{(a)} = \frac{l^2}{ab} \sqrt{\frac{\mu}{\varepsilon}} \frac{\sin^2 \frac{\pi}{a} d}{\sqrt{1 - \left(\frac{\pi}{ak} \right)^2}} \tag{2}$$

(formulae (1) and (2) are given in a practical system of units).

(b) Let the half-wave dipole be oriented along the y-axis, with its ends at the points $M_1(d, y_1)$ and $M_2(d, y_2)$, where $y_2 - y_1 = \lambda/2 = \pi/k$. The current distribution in the dipole is given by the relation

$$I = I_0 \sin k(y - y_1).$$

The radiation resistance equals

$$R^{(a)} = \frac{8\pi}{c} k \sum_{\substack{m=1 \\ (\lambda_{mn} < k^2)}} \sum_{n=1} \frac{2\varepsilon_n}{ab} \frac{\sin^2 \frac{\pi m}{a} d \cos^2 \frac{\pi n}{b} \left(y_1 + \frac{\pi}{2k} \right) \cos^2 \frac{\pi n}{2bk}}{p_{mn} \left[k^2 - \left(\frac{\pi n}{b} \right)^2 \right]},$$

where

$$p_{mn} = \sqrt{\lambda_{mn} - k^2}, \quad \lambda_{mn} = \pi^2 \left(\frac{m^2}{a^2} + \frac{n^2}{b^2} \right), \quad \varepsilon_n = \begin{cases} 1, & n = 0, \\ 2, & n \neq 0. \end{cases}$$

The upper limits of summation are found from the condition $\lambda_{mn} < k^2$.

SUPPLEMENT

I. Different Orthogonal Systems of Coordinates

Let x, y, z be the cartesian coordinates of some point, and x_1, x_2, x_3 the curvilinear orthogonal coordinates of this point. The square of an element of length is expressed by the relation

$$ds^2 = dx^2+dy^2+dz^2 = h_1^2 dx_1^2+h_2^2 dx_2^2+h_3^2 dx_3^2,$$

where

$$h_i = \sqrt{\left(\frac{\partial x}{\partial x_i}\right)^2+\left(\frac{\partial y}{\partial x_i}\right)^2+\left(\frac{\partial z}{\partial x_i}\right)^2} \qquad (i = 1, 2, 3)$$

are metric coefficients, or Lamé coefficients. The orthogonal coordinate system is completely described by the three metric coefficients h_1, h_2, h_3.

We give a general expression for the operators grad, div, curl, Δ in an orthogonal curvilinear system of coordinates:

$$\operatorname{grad} u = \sum_{j=1}^{3} \frac{1}{h_j}\frac{\partial u}{\partial x_j}\boldsymbol{i}_j,$$

$$\operatorname{div} \boldsymbol{A} = \frac{1}{h_1 h_2 h_3}\left[\frac{\partial}{\partial x_1}(h_2 h_3 A_1)+\frac{\partial}{\partial x_2}(h_3 h_1 A_2)+\frac{\partial}{\partial x_3}(h_1 h_2 A_3)\right],$$

$$\operatorname{rot} \boldsymbol{A} = \frac{1}{h_1 h_2 h_3}\begin{vmatrix} h_1\boldsymbol{i}_1 & h_2\boldsymbol{i}_2 & h_3\boldsymbol{i}_3 \\ \dfrac{\partial}{\partial x_1} & \dfrac{\partial}{\partial x_2} & \dfrac{\partial}{\partial x_3} \\ h_1 A_1 & h_2 A_2 & h_3 A_3 \end{vmatrix},$$

$$\Delta u = \frac{1}{h_1 h_2 h_3}\left[\frac{\partial}{\partial x_1}\left(\frac{h_2 h_3}{h_1}\frac{\partial u}{\partial x_1}\right)+\frac{\partial}{\partial x_2}\left(\frac{h_3 h_1}{h_2}\frac{\partial u}{\partial x_2}\right)+\frac{\partial}{\partial x_3}\left(\frac{h_1 h_2}{h_3}\frac{\partial u}{\partial x_3}\right)\right],$$

where $\boldsymbol{i}_1$, $\boldsymbol{i}_2$, $\boldsymbol{i}_3$ are unit basic vectors, $\boldsymbol{A} = (A_1, A_2, A_3)$ an arbitrary vector, u a scalar.

1. Rectangular Coordinates

$$x_1 = x, \quad x_2 = y, \quad x_3 = z, \quad h_1 = 1, \quad h_2 = 1, \quad h_3 = 1,$$

$$\operatorname{grad} u = \frac{\partial u}{\partial x}\boldsymbol{i}+\frac{\partial u}{\partial y}\boldsymbol{j}+\frac{\partial u}{\partial z}\boldsymbol{k}, \qquad \operatorname{div} \boldsymbol{A} = \frac{\partial A_x}{\partial x}+\frac{\partial A_y}{\partial y}+\frac{\partial A_z}{\partial z},$$

$$\operatorname{curl} A = \begin{vmatrix} i & j & k \\ \dfrac{\partial}{\partial x} & \dfrac{\partial}{\partial y} & \dfrac{\partial}{\partial z} \\ A_x & A_y & A_z \end{vmatrix} = \left(\frac{\partial A_z}{\partial y} - \frac{\partial A_y}{\partial z}\right) i + \ldots,$$

$$\Delta u = u_{xx} + u_{yy} + u_{zz},$$

where $\boldsymbol{i}$, $\boldsymbol{j}$ and $\boldsymbol{k}$ are the directional unit vectors of the x, y, z axes.

2. Cylindrical Coordinates

$$x_1 = r, \quad x_2 = \phi, \quad x_3 = z$$

are connected to the rectangular coordinates by the equations

$$x = r\cos\phi, \quad y = r\sin\phi, \quad z = z.$$

Coordinate surfaces: $r =$ const. cylinders, $\phi =$ const. planes, $z =$ const. planes.

The metric coefficients are

$$h_1 = 1, \quad h_2 = r, \quad h_3 = 1,$$

so that

$$\operatorname{grad} u = \frac{\partial u}{\partial r} i_1 + \frac{1}{r}\frac{\partial u}{\partial \phi} i_2 + \frac{\partial u}{\partial z} i_3,$$

$$\operatorname{div} A = \frac{1}{r}\frac{\partial}{\partial r}(rA_1) + \frac{1}{r}\frac{\partial A_2}{\partial \theta} + \frac{\partial A_3}{\partial z},$$

$$\operatorname{curl} A = \left(\frac{1}{r}\frac{\partial A_3}{\partial \phi} - \frac{\partial A_2}{\partial z}\right) i_1 + \left(\frac{\partial A_1}{\partial z} - \frac{\partial A_3}{\partial r}\right) i_2 + \left[\frac{1}{r}\frac{\partial}{\partial r}(rA_2) - \frac{1}{r}\frac{\partial A_1}{\partial \phi}\right] i_3,$$

$$\Delta u = \frac{1}{r}\frac{\partial}{\partial r}\left(r\frac{\partial u}{\partial r}\right) + \frac{1}{r^2}\frac{\partial^2 u}{\partial \phi^2} + \frac{\partial^2 u}{\partial z^2}.$$

3. Spherical Coordinates

$$x_1 = r, \quad x_2 = \theta, \quad x_3 = \phi$$

are connected to the rectangular coordinates by the relations

$$x = r\sin\theta\cos\phi, \quad y = r\sin\theta\sin\phi, \quad z = r\cos\theta.$$

Coordinate surfaces: concentric spheres $r =$ const., planes $\phi =$ const., cones $\theta =$ const.

The metric coefficients equal

$$h_1 = 1, \quad h_2 = r, \quad h_3 = r\sin\theta,$$

so that

$$\operatorname{grad} u = \frac{\partial u}{\partial r} i_1 + \frac{1}{r}\frac{\partial u}{\partial \theta} i_2 + \frac{1}{r\sin\theta}\frac{\partial u}{\partial \phi} i_3,$$

$$\operatorname{div} \boldsymbol{A} = \frac{1}{r^2}\frac{\partial}{\partial r}(r^2 A_1) + \frac{1}{r \sin\theta}\frac{\partial}{\partial \theta}(\sin\theta A_2) + \frac{1}{r\sin\theta}\frac{\partial A_3}{\partial \phi},$$

$$\operatorname{curl} \boldsymbol{A} = \frac{1}{r\sin\theta}\left[\frac{\partial}{\partial A}(\sin\theta A_3) - \frac{\partial A_2}{\partial \phi}\right]\boldsymbol{i}_1 + \frac{1}{r}\left[\frac{1}{\sin\theta}\frac{\partial A_1}{\partial\phi} - \frac{\partial}{\partial r}(rA_3)\right]\boldsymbol{i}_2 +$$

$$+\frac{1}{r}\left[\frac{\partial}{\partial r}(rA_2) - \frac{\partial A_1}{\partial\theta}\right]\boldsymbol{i}_3,$$

$$\Delta u = \frac{1}{r^2}\frac{\partial}{\partial r}\left(r^2\frac{\partial u}{\partial r}\right) + \frac{1}{r^2\sin\theta}\frac{\partial}{\partial\theta}\left(\sin\theta\frac{\partial u}{\partial\theta}\right) + \frac{1}{r^2\sin^2\theta}\frac{\partial^2 u}{\partial\phi^2}.$$

4. Elliptic Coordinates

$$x_1 = \lambda, \quad x_2 = \mu, \quad x_3 = z$$

are determined by means of the conversion formulae

$$x = c\lambda\mu, \quad y = c\sqrt{(\lambda^2-1)(1-\mu^2)}, \quad z = z,$$

where c is a scaling factor.

The metric coefficients are

$$h_1 = c\sqrt{\frac{\lambda^2-\mu^2}{\lambda^2-1}}, \quad h_2 = c\sqrt{\frac{\lambda^2-\mu^2}{1-\mu^2}}, \quad h_3 = 1.$$

Coordinate surfaces: $\lambda =$ const. cylinders of elliptic section with foci at the points $x = \pm c$, $y = 0$, $u =$ const. a family of confocal hyperbolic cylinders, $z =$ const. planes.

5. Parabolic Coordinates

If r, θ are polar coordinates of a point in a plane, then the parabolic coordinates can be introduced by means of the formulae

$$x_1 = \lambda = \sqrt{2r}\sin\frac{\theta}{2}, \quad x_2 = \mu = \sqrt{2r}\cos\frac{\theta}{2}, \quad x_3 = z.$$

The coordinate surfaces $\lambda =$ const. and $\mu =$ const. represent intersecting parabolic cylinders with axes parallel to the z-axis.

The formulae

$$x = \frac{1}{2}(\mu^2-\lambda^2), \quad y = \lambda\mu, \quad z = z$$

give a connection with cartesian coordinates.

The metric coefficients

$$h_1 = h_2 = \sqrt{\lambda^2+\mu^2}, \quad h_3 = 1.$$

6. Ellipsoidal Coordinates

They are introduced by means of the equations $(a > b > c)$:

$$\frac{x^2}{a^2+\lambda}+\frac{y^2}{b^2+\lambda}+\frac{z^2}{c^2+\lambda}=1 \quad (\lambda > -c^2) \text{ (equation of the ellipsoid)},$$

$$\frac{x^2}{a^2+\mu}+\frac{y^2}{b^2+\mu}+\frac{z^2}{c^2+\mu}=1 \quad (-c^2 > \mu > -b^2) \text{ (equation of the one-sheet hyperboloid)},$$

$$\frac{x^2}{a^2+\nu}+\frac{y^2}{b^2+\nu}+\frac{z^2}{c^2+\nu}=1 \quad (-b^2 > \nu > -a^2) \text{ (equation of the two-sheet hyperboloid)}.$$

To each point (x, y, z) there corresponds only one set of values λ, μ, ν. The parameters

$$x_1 = \lambda, \quad x_2 = \mu, \quad x_3 = \nu$$

are called ellipsoidal coordinates. The coordinates x, y, z are expressed clearly in terms of λ, μ, ν:

$$x = \pm\sqrt{\frac{(\lambda+a^2)(\mu+a^2)(\nu+a^2)}{(b^2-a^2)(c^2-a^2)}},$$

$$y = \pm\sqrt{\frac{(\lambda+b^2)(\mu+b^2)(\nu+b^2)}{(c^2-b^2)(a^2-b^2)}},$$

$$z = \pm\sqrt{\frac{(\lambda+c^2)(\mu+c^2)(\nu+c^2)}{(a^2-c^2)(b^2-c^2)}}.$$

The Lamé coefficients equal

$$h_1 = \frac{1}{2}\sqrt{\frac{(\lambda-\mu)(\lambda-\nu)}{R^2(\lambda)}}, \quad h_2 = \frac{1}{2}\sqrt{\frac{(\mu-\nu)(\mu-\lambda)}{R^2(\mu)}},$$

$$h_3 = \frac{1}{2}\sqrt{\frac{(\nu-\lambda)(\nu-\mu)}{R^2(\nu)}},$$

where

$$R(s) = \sqrt{(s+a^2)(s+b^2)(s+c^2)} \quad (s = \lambda, \mu, \nu).$$

The Laplacian operator may be represented in the form

$$\Delta u = \frac{4}{(\lambda-\mu)(\lambda-\nu)(\mu-\nu)}\left[(\mu-\nu)R(\lambda)\frac{\partial}{\partial\lambda}\left(R(\lambda)\frac{\partial u}{\partial\nu}\right)+\right.$$

$$\left.+(\nu-\lambda)R(\mu)\frac{\partial}{\partial\mu}\left(R(\mu)\frac{\partial u}{\partial\mu}\right)+(\lambda-\mu)R(\nu)\frac{\partial}{\partial\nu}\left(R(\nu)\frac{\partial u}{\partial\nu}\right)\right].$$

A particular solution of Laplace's equation, dependent only on λ, $U = U(\lambda)$ is given by the formula

$$U = A\int\frac{d\lambda}{R(\lambda)}+B,$$

where A and B are arbitrary constants.

7. Degenerate Ellipsoidal Coordinates

(a) The degenerate ellipsoidal coordinates (α, β, ϕ) for an elongated ellipsoid of revolution are defined by means of the formulae

$$x = c\sin\beta\cos\phi, \quad y = c\sin\alpha\sin\beta\sin\phi, \quad z = c\cosh\alpha\cos\beta,$$

where c is a scaling factor, $0 \leqslant \alpha < \infty$, $0 \leqslant \beta \leqslant \pi$, $-\pi < \phi \leqslant \pi$. Coordinate surfaces: elongated ellipsoids of revolution $\alpha =$ const., two-sheet hyperboloids of revolution $\beta =$ const. and the planes $\phi =$ const.

The square of a linear element is given by the expression

$$ds^2 = c^2(\sinh^2\alpha+\sin^2\beta)(d\alpha^2+d\beta^2)+c^2\sinh^2\alpha\sin^2\beta\, d\phi^2,$$

from which the values

$$h_1 = h_2 = c\sqrt{\sinh^2\alpha+\sin^2\beta}, \quad h_3 = h_\phi = c\sinh\alpha\sin\beta.$$

are obtained for the metric coefficients.

Laplace's equation has the form

$$\Delta u = \frac{1}{c^2(\sinh^2\alpha+\sin^2\beta} \left[\frac{1}{\sinh\alpha}\frac{\partial}{\partial\alpha}\left(\sinh\alpha\frac{\partial u}{\partial\alpha}\right)+\frac{1}{\sin\beta}\frac{\partial}{\partial\beta}\left(\sin\beta\frac{\partial u}{\partial\beta}\right)+\right.$$
$$\left.+\left(\frac{1}{\sinh^2\alpha}+\frac{1}{\sin^2\beta}\right)\frac{\partial^2 u}{\partial\phi^2}\right] = 0.$$

(b) The set of degenerate ellipsoidal coordinates (α, β, ϕ) for an oblate ellipsoid of revolution is defined by means of the equalities

$$x = c\cosh\alpha\sin\beta\cos\phi, \quad y = c\cosh\alpha\sin\beta\sin\phi, \quad z = c\sinh\alpha\cos\phi,$$
$$0 \leqslant \alpha < \infty, \quad 0 \leqslant \beta \leqslant \pi, \quad -\pi < \phi \leqslant \pi.$$

Coordinate surfaces: oblate ellipsoids of revolution $\alpha =$ const., one-sheet hyperboloids of revolution $\beta =$ const. and the planes $\phi =$ const., passing through the z-axis.

The square of a linear element and the Laplacian operator in the coordinate system being considered have the form

$$ds^2 = c^2(\cosh^2\alpha-\sin^2\beta)(d\alpha^2+d\beta^2)+c^2\cosh^2\alpha\sin^2\beta\, d\phi^2,$$

$$\Delta u = \frac{1}{c^2(\cosh^2\alpha-\sin^2\beta)} \left[\frac{1}{\cosh\alpha}\frac{\partial}{\partial\alpha}\left(\cosh\alpha\frac{\partial u}{\partial\alpha}\right)+\frac{1}{\sin\beta}\frac{\partial}{\partial\beta}\left(\sin\beta\frac{\partial u}{\partial\beta}\right)+\right.$$
$$\left.+\left(\frac{1}{\sin^2\alpha}-\frac{1}{\cosh^2\alpha}\right)\frac{\partial^2 u}{\partial\phi^2}\right].$$

8. Toroidal Coordinates

The system of toroidal coordinates (α, β, ϕ) are defined by means of the formulae

$$x = \frac{c\sinh\alpha\cos\phi}{\cosh\alpha-\cos\beta}, \quad y = \frac{c\sinh\alpha\sin\phi}{\cosh\alpha-\cos\beta}, \quad z = \frac{c\sin\beta}{\cosh\alpha-\cos\beta},$$

where c is a scaling factor $0 \leqslant \alpha < \infty$, $-\pi < \beta \leqslant \pi$, $-\pi < \phi \leqslant \pi$.

Coordinate surfaces are the tores $\alpha = \text{const.}$

$$(\rho - c\coth\alpha)^2 + z^2 = \left(\frac{c}{\sinh\alpha}\right)^2 \qquad (\rho = \sqrt{x^2+y^2}),$$

spheres $\beta = \text{const.}$

$$(z - c\cot\beta)^2 + \rho^2 = \left(\frac{c}{\sin\beta}\right)^2,$$

planes $\phi = \text{const.}$

The square of a linear element in a toroidal system of coordinates has the form

$$ds^2 = \frac{c^2}{(\cosh\alpha - \cos\beta)^2}[d\alpha^2 + d\beta^2 + \sinh^2\alpha\, d\phi^2],$$

the metric coefficients equal

$$h_\alpha = h_\beta = \frac{c}{\cosh\alpha - \cos\beta}, \qquad h_\phi = \frac{c\sinh\alpha}{\cosh\alpha - \cos\beta},$$

and the Laplacian operator is given by the following expression:

$$\Delta u = \frac{\partial}{\partial\alpha}\left(\frac{\sinh\alpha}{\cosh\alpha - \cos\beta}\frac{\partial u}{\partial\alpha}\right) + \frac{\partial}{\partial\beta}\left(\frac{\sinh\alpha}{\cosh\alpha - \cos\beta}\frac{\partial u}{\partial\beta}\right) + \\ + \frac{1}{(\cosh\alpha - \cos\beta)\sinh\alpha}\frac{\partial^2 u}{\partial\phi^2}.$$

It is convenient to introduce in place of u a new function v by means of the relation

$$u = \sqrt{2\cosh\alpha - 2\cos\beta}\, v.$$

By this the equation $\Delta u = 0$ reduces to the equation

$$v_{\alpha\alpha} + v_{\beta\beta} + v_\alpha \coth\alpha + \frac{1}{4}v + \frac{1}{\sinh^2\alpha}v_{\phi\phi} = 0.$$

9. Bipolar Coordinates

(a) Bipolar coordinates in a plane.

The variables

$$x_1 = \alpha, \qquad x_2 = \beta, \qquad x_3 = z$$

are called bipolar coordinates if the equalities

$$x = \frac{a\sinh\alpha}{\cosh\alpha - \cos\beta}, \qquad y = \frac{a\sin\beta}{\cosh\alpha - \cos\beta}, \qquad z = z$$

hold. The metric coefficients equal

$$h_1 = h_2 = \frac{a}{\cosh\alpha - \cos\beta}, \qquad h_3 = 1.$$

(b) The bispherical coordinates

$$x_1 = \alpha, \quad x_2 = \beta, \quad x_3 = \phi$$

are defined by means of the formulae

$$x = \frac{c \sin \alpha \cos \phi}{\cosh \beta - \cos \alpha}, \quad y = \frac{c \sin \alpha \sin \phi}{\cosh \beta - \cos \alpha}, \quad z = \frac{c \sinh \beta}{\cosh \beta - \cos \alpha},$$

where c is a constant factor $0 \leqslant \alpha \leqslant \beta$, $-\infty < \beta < \infty$, $-\pi < \phi \leqslant \pi$.

These formulae may be represented in concise form

$$z + i\rho = ci \cot \frac{\alpha + i\beta}{2} \quad (\rho = \sqrt{x^2+y^2}).$$

The coordinate surfaces are: tapered surfaces of revolution $\alpha = \text{const.}$:

$$(\rho - c \cot \alpha)^2 + z^2 = \left(\frac{c}{\sin \alpha}\right)^2,$$

spheres $\beta = \text{const.}$:

$$\rho^2 + (z - c \cot \beta)^2 = \left(\frac{c}{\sinh \beta}\right)^2,$$

planes $\rho = \text{const.}$

An expression for the square of a linear element in three-dimensional bipolar coordinates has the form

$$ds^2 = \frac{c^2}{(\cosh \beta - \cos \alpha)^2} [d\alpha^2 + d\beta^2 + \sin^2 \alpha \, d\phi^2],$$

from which it follows

$$h_1 = h_2 = \frac{c}{\cosh \beta - \cos \alpha}, \quad h_3 = \frac{c \sin \alpha}{\cosh \beta - \cos \alpha}$$

and Laplace's equation takes the form

$$\frac{\partial}{\partial \alpha}\left(\frac{\sin \alpha}{\cosh \beta - \cos \alpha} \frac{\partial u}{\partial \alpha}\right) + \frac{\partial}{\partial \beta}\left(\frac{\sin \alpha}{\cosh \beta - \cos \alpha} \frac{\partial u}{\partial \beta}\right) + $$

$$+ \frac{1}{\sin \alpha (\cosh \beta - \cos \alpha)} \frac{\partial^2 u}{\partial \phi^2} = 0.$$

In the solution of Laplace's equation the substitution

$$u = \sqrt{2 \cosh \beta - 2 \cos \alpha}\, v$$

is suitable. Then for the function v the equation

$$v_{\alpha\alpha} + v_{\beta\beta} + v_\alpha \cot \alpha - \frac{1}{4} v + \frac{1}{\sin^2 \alpha} v_{\phi\phi} = 0$$

is obtained.

10. Spheroidal Coordinates

(a) The elongated spheroidal coordinates

$$x_1 = \lambda, \quad x_2 = \mu, \quad x_3 = \phi,$$

$$x = c\lambda\mu, \quad y = c\sqrt{(\lambda^2-1)(1-\mu^2)}\cos\phi, \quad z = c\sqrt{(\lambda^2-1)(1-\mu^2)}\sin\phi,$$

$$\lambda \geqslant 1, \quad -1 \leqslant \mu \leqslant 1, \quad 0 \leqslant \phi \leqslant 2\pi,$$

$$h_1 = c\sqrt{\frac{\lambda^2-\mu^2}{\lambda^2-1}}, \quad h_2 = c\sqrt{\frac{\lambda^2-\mu^2}{1-\mu^2}}, \quad h_3 = c\sqrt{(\lambda^2-1)(1-\mu^2)}.$$

(b) The oblate spheroidal coordinates

$$x_1 = \lambda, \quad x_2 = \mu, \quad x_3 = \phi,$$

$$x = c\lambda\mu\sin\phi, \quad y = c\sqrt{(\lambda^2-1)(1-\mu^2)}, \quad z = c\lambda\mu\cos\phi.$$

The surfaces $\lambda =$ const. are oblate spheroids, $\mu =$ const. one-sheet hyperboloids.

The metric coefficients

$$h_1 = c\sqrt{\frac{\lambda^2-\mu^2}{\lambda^2-1}}, \quad h_2 = c\sqrt{\frac{\lambda^2-\mu^2}{1-\mu^2}}, \quad h_3 = c\lambda\mu.$$

11. Paraboloidal Coordinates

The variables

$$x_1 = \lambda, \quad x_2 = \mu, \quad x_3 = \phi,$$

defined by the relations

$$x = \lambda\mu\cos\phi, \quad y = \lambda\mu\sin\phi, \quad z = \frac{1}{2}(\lambda^2-\mu^2),$$

are called paraboloidal coordinates. The metric coefficients equal

$$h_1 = h_2 = \sqrt{\lambda^2+\mu^2}, \quad h_3 = \lambda\mu.$$

The coordinate surfaces $\lambda =$ const., $\mu =$ const. are parabolae of revolution about the axis of symmetry Oz.

II. Some Formulae of Vector Analysis

Symbols: $\boldsymbol{a}$ is a vector function, u is a scalar function.

$$[[\boldsymbol{ab}]\,\boldsymbol{c}] = (\boldsymbol{ac})\boldsymbol{b}-(\boldsymbol{bc})\boldsymbol{a},$$

$$[\boldsymbol{a}\,[\boldsymbol{bc}]] = \boldsymbol{b}(\boldsymbol{ac})-\boldsymbol{c}(\boldsymbol{ab}),$$

$$\operatorname{grad}(uv) = u\operatorname{grad}v+v\operatorname{grad}u,$$

$$\operatorname{div}(u\boldsymbol{a}) = \boldsymbol{a}\operatorname{grad}u+u\operatorname{div}\boldsymbol{v},$$

$$\operatorname{curl}(u\boldsymbol{a}) = [\operatorname{grad}u\boldsymbol{a}]+u\operatorname{curl}\boldsymbol{a},$$

$$\begin{aligned}
\operatorname{div}[\boldsymbol{ab}] &= \boldsymbol{b}\operatorname{curl}\boldsymbol{a}-\boldsymbol{a}\operatorname{curl}\boldsymbol{b},\\
\operatorname{curl}\operatorname{curl}\boldsymbol{a} &= \operatorname{grad}\operatorname{div}\boldsymbol{a}-\Delta\boldsymbol{a},\\
\operatorname{grad}(\boldsymbol{a},\boldsymbol{b}) &= \boldsymbol{a}\operatorname{div}\boldsymbol{b}+\boldsymbol{b}\operatorname{div}\boldsymbol{a}+[\boldsymbol{a}\operatorname{curl}\boldsymbol{b}]+[\boldsymbol{b}\operatorname{curl}\boldsymbol{a}],\\
\operatorname{curl}[\boldsymbol{ab}] &= \boldsymbol{a}\operatorname{div}\boldsymbol{b}-\boldsymbol{b}\operatorname{div}\boldsymbol{a}+(\boldsymbol{b}\nabla)\boldsymbol{a}-(\boldsymbol{a}\nabla)\boldsymbol{b},
\end{aligned}$$

where

$$(\boldsymbol{b}\nabla)\boldsymbol{a} = b_x\frac{\partial\boldsymbol{a}}{\partial x}+b_y\frac{\partial\boldsymbol{a}}{\partial y}+b_z\frac{\partial\boldsymbol{a}}{\partial z}.$$

III. Special Functions

1. Trigonometric Functions

$$\cos z = 1-\frac{1}{2!}z^2+\frac{1}{4!}z^4-\ldots = \frac{1}{2}(e^{iz}+e^{-iz}) = \cosh(iz),$$

$$\sin z = z-\frac{1}{3!}z^3+\frac{1}{5!}z^5-\ldots = \frac{1}{2i}(e^{iz}-e^{-iz}) = -i\sinh(iz),$$

$$\begin{aligned}
\cos(x+y) &= \cos x\cos y-\sin x\sin y,\\
\sin(x+y) &= \sin x\cos y+\cos x\sin y,\\
\cos x\cos y &= \frac{1}{2}\cos(x+y)+\frac{1}{2}\cos(x-y),\\
\sin x\sin y &= -\frac{1}{2}\cos(x+y)+\frac{1}{2}\cos(x-y).
\end{aligned}$$

2. Hyperbolic Functions

$$\cosh z = 1+\frac{1}{2!}z^2+\frac{1}{4!}z^4+\ldots = \frac{1}{2}(e^{z}+e^{-z}) = \cos(iz),$$

$$\sinh z = z+\frac{1}{3!}z^3+\frac{1}{5!}z^5+\ldots = \frac{1}{2}(e^{z}-e^{-z}) = -i\sin(iz),$$

$$\begin{aligned}
\cosh^2 z-\sinh^2 z &= 1,\\
\cosh(x+y) &= \cosh x\cos y+\sinh x\sinh y,\\
\sinh(x+y) &= \sinh x\cosh y+\cosh x\sinh y.
\end{aligned}$$

3. The Error Integral

$$\Phi(z) = \frac{2}{\sqrt{\pi}}\int_0^z e^{-\alpha^2}\,d\alpha.$$

The expansion in a series for small z

$$\Phi(z) = \frac{2}{\sqrt{\pi}}\left(z-\frac{z^3}{1!\,3}+\frac{z^5}{2!\,5}-\ldots\right).$$

The asymptotic expansion for large z

$$\Phi(z) = 1 - \frac{1}{\sqrt{\pi}} \frac{e^{-z^2}}{z} \left(1 - \frac{1}{2z^2} + \frac{3.4}{(2z^2)} - \frac{4.5.6}{(2z)^6} + \dots\right).$$

In Table 1 values are given of $\Phi(z)$ for $0 \leqslant z \leqslant 2.3$.

4. Gamma-functions

$$\Gamma(z) = \int_0^\infty e^{-t} t^{z-1} dt \qquad (\operatorname{Re} z > 0), \qquad \Gamma(z+1) = z\Gamma(z),$$

$$\Gamma(n+1) = n!, \quad \Gamma(1) = 1, \quad \Gamma\left(\frac{1}{2}\right) = \sqrt{\pi},$$

$$\Gamma(z)\Gamma(1-z) = \frac{\pi}{\sin \pi z}, \qquad \Gamma(2z) = \frac{2^{2z-1}}{\sqrt{\pi}} \Gamma(z)\Gamma\left(z+\frac{1}{2}\right),$$

$$\Gamma(z) \to \sqrt{2\pi}\, z^{z-\frac{1}{2}} e^{-z} \quad \text{for} \quad z \gg 1,$$

The Beta-function

$$B(x, y) = \int_0^1 t^{x-1}(1-t)^{y-1} dt = \frac{\Gamma(x)\Gamma(y)}{\Gamma(x+y)}$$

$$= 2\int_0^{\frac{\pi}{2}} \sin^{2x-1}\phi \cos^{2y-1}\phi \, d\phi \qquad (\operatorname{Re} x > 0, \quad \operatorname{Re} y > 0).$$

5. Elliptic Functions

$$\int_0^x \frac{dx}{\sqrt{(1-x^2)(1-k^2x^2)}} = \operatorname{sn}^{-1} x, k),$$

$$\int_x^1 \frac{dx}{\sqrt{(1-x^2)(x^2+k^2-1)}} = \operatorname{dn}^{-1}(x, k),$$

$$\int_x^1 \frac{dx}{\sqrt{(1-x^2)(1-k^2+k^2x^2)}} = \operatorname{cn}^{-1}(x, k),$$

$$\int_0^x \frac{dx}{\sqrt{(1+x^2)(1+k'^2x^2)}} = \operatorname{tn}^{-1}(x, k),$$

$$K = \int_0^1 \frac{dx}{\sqrt{(1-x^2)(1-k^2x^2)}},$$

$$K' = \int_0^1 \frac{dx}{\sqrt{(1-x^2)(1-(k')^2x^2)}}, \qquad k = \sin\alpha, \qquad k' = \cos\alpha.$$

6. Bessel Functions

$$J_n(z) = \frac{1}{n!}\left(\frac{z}{2}\right)^n \left[1 - \frac{\left(\frac{z}{2}\right)^2}{1!\,(n+1)} + \frac{\left(\frac{z}{2}\right)^4}{2!(n+1)(n+2)} - \dots\right]$$

$$= \sum_{k=0}^{\infty} \frac{(-1)^k \left(\frac{z}{2}\right)^{2k+n}}{\Gamma(k+1)\Gamma(k+n+1)} \underset{z\to\infty}{\longrightarrow} \sqrt{\frac{2}{\pi z}} \cos\left[z - \frac{1}{2}\pi\left(n+\frac{1}{2}\right)\right],$$

(Re $z > 0$),

$$N_n(z) = \frac{1}{\pi}\left[2\ln\left(\frac{z}{2}\right) + \gamma - \psi(n+1)\right] J_n(z) - \frac{1}{\pi}\sum_{m=0}^{n-1} \frac{(n-m-1)!}{m!\left(\frac{z}{2}\right)^{n-2m}} -$$

$$-\frac{1}{\pi}\sum_{m=1}^{\infty}(-1)^m \frac{\left(\frac{z}{2}\right)^{n+2m}}{m!(n+m)!}\left[\sum_{k=1}^{m}\left(\frac{1}{k}+\frac{1}{k+n}\right)\right],$$

where $\psi(n) = \dfrac{\Gamma'(n)}{\Gamma(n)}$

$$N_0(z) \underset{z\to 0}{\longrightarrow} \frac{2}{\pi}(\ln z - 0.11593), \qquad n = 0,$$

$$N_n(z) \underset{z\to 0}{\longrightarrow} \frac{(n-1)!}{\pi}\left(\frac{2}{z}\right)^n, \qquad n = 1, 2, 3, \dots,$$

$$N_n(z) \underset{z\to\infty}{\longrightarrow} \sqrt{\frac{2}{\pi z}} \sin\left[z - \frac{1}{2}\pi\left(n+\frac{1}{2}\right)\right] \qquad (\mathrm{Re}\, z > 0),$$

$$J_{-n}(z) = (-1)^n J_n(z), \qquad N_{-n}(z) = (-1)^n N_n(z),$$

$$\left.\begin{aligned} J_n(z)\,N_n'(z) - J_n' N_n(z) &= \Delta(J_n, N_n) = \frac{2}{\pi z}, \\ N_{n-1}(z) J_n(z) - N_n(z) J_{n-1}(z) &= \Delta(J_n, N_n) = \frac{2}{\pi z}, \end{aligned}\right\} \text{(Wronskian)},$$

$$e^{-iz\sin\phi} = \sum_{n=-\infty}^{\infty} J_n(z)\, e^{-in\phi}, \; J_n(z) = \frac{1}{2\pi}\int_{-\pi}^{\pi} e^{-iz\sin\phi + in\phi}\, d\phi \text{ (n is an integer)},$$

$$H_n^{(1)}(z) = J_n(z)+iN_n(z)\xrightarrow[z\to\infty]{}\sqrt{\frac{2}{\pi z}}\,e^{i\left[z-\frac{\pi}{2}n-\frac{\pi}{4}\right]},$$

$$H_n^{(2)}(z) = J_n(z)-iN_n(z)\xrightarrow[z\to\infty]{}\sqrt{\frac{2}{\pi z}}\,e^{-i\left[z-\frac{\pi}{2}n-\frac{\pi}{4}\right]}.$$

Recurrence formulae

$$\frac{2n}{x}Z_n(x) = Z_{n-1}+Z_{n+1}, \qquad 2Z_n'(x) = Z_{n-1}-Z_{n+1}$$

or

$$\frac{d}{dx}[x^n Z_n(x)] = x^n Z_{n-1}, \qquad \frac{d}{dx}\left[\frac{Z_n(x)}{x^n}\right] = -\frac{Z_{n+1}}{x^n},$$

$$Z_0' = -Z_1, \qquad [xZ_1(x)]' = xZ_0(x),$$

$$\int Z_0^2(\alpha x)x\,dx = \frac{1}{2}x^2[Z_0^2(\alpha x)+Z_1^2(\alpha x)],$$

$$\int Z_m^2(\alpha x)x\,dx = \frac{1}{2}x^2[Z_m^2(\alpha x)-Z_{m-1}(\alpha x)Z_{m+1}(\alpha x)].$$

Here $Z_n(x) = AJ_n(x)+BN_n(x)$ is any solution of Bessel's equation

$$\frac{1}{x}\frac{d}{dx}\left(x\frac{dZ_n}{dx}\right)+\left(1-\frac{n^2}{x^2}\right)Z_n = 0.$$

Functions of imaginary argument

$$I_n(x) = \left(\frac{1}{i}\right)^n J_n(ix)\xrightarrow[x\to\infty]{}\frac{1}{\sqrt{2\pi x}}e^x, \qquad I_n(x) = \sum_{k=0}^{\infty}\frac{\left(\frac{x}{2}\right)^{2k+n}}{\Gamma(k+1)\Gamma(k+n+1)},$$

$$K_n(x) = \frac{1}{2}xi^{n+1}H_n^{(1)}(ix)\xrightarrow[x\to\infty]{}\sqrt{\frac{\pi}{2x}}\,e^{-x}, \qquad K_n(x)\xrightarrow[x\to 0]{}\frac{(n-1)!}{2}\left(\frac{2}{x}\right)^n+\dots,$$

$$I_{-n}(x) = I_n(x), \qquad K_n(x) = K_{-n}(x),$$

$$I_n(x)K_n'(x)-I_n'(x)K_n(x) = \Delta(I_n, K_n) = -\frac{1}{x},$$

$$I_n(x)K_{n-1}(x)+I_{n-1}(x)K_n(x) = -\Delta(I_n, K_n) = \frac{1}{x},$$

$$I_{n-1}(x)-I_{n+1}(x) = \frac{2n}{x}I_n(x), \qquad I_{n-1}(x)+I_{n+1}(x) = 2I_n'(x),$$

$$Z_n''+\frac{1}{x}Z_n'-\left(1+\frac{n^2}{x^2}\right)Z_n = 0.$$

Some integrals

$$\int_0^\infty e^{-at} J_n(bt)\frac{dt}{t} = \frac{1}{nb^n}[\sqrt{a^2+b^2}-a]^n,$$

$$\int_0^\infty e^{-at} J_n(bt)t^n\,dt = \frac{(2b)^n\,\Gamma\left(n+\frac{1}{2}\right)}{\Gamma\left(\frac{1}{2}\right)(a^2+b^2)^{n+\frac{1}{2}}},$$

$$\int_0^\infty e^{-at} N_0(bt)\,dt = \frac{-\frac{2}{\pi}}{\sqrt{a^2+b^2}}\ln\frac{a+\sqrt{a^2+b^2}}{a} \quad (a, b \text{ are real and positive}).$$

7. Legendre Polynomials

The equation $[(1-x^2)\,y']'+n(n+1)\,y = 0 \quad (-1 < x < 1)$,

$$\frac{1}{R} = \frac{1}{\sqrt{r_0^2+r^2-2rr_0\cos\theta}} = \begin{cases} \sum\limits_{n=0}^{\infty}\frac{1}{r_0}\left(\frac{r}{r_0}\right)^n P_n(\cos\theta) & \text{for} \quad r < r_0, \\ \sum\limits_{n=0}^{\infty}\frac{1}{r}\left(\frac{r_0}{r}\right)^n P_n(\cos\theta) & \text{for} \quad r > r_0, \end{cases}$$

$$(n+1)P_{n+1}(x)-x(2n+1)P_n(x)+nP_{n-1}(x) = 0,$$

$$P_n(x) = \frac{1}{2^n n!}\frac{d^n}{dx^n}[(x^2-1)^n].$$

The equation of the associated functions

$$[(1-x^2)y']'+\left(n(n+1)-\frac{m^2}{1-x^2}\right)y = 0,$$

$$\int_{-1}^{1} P_n^{(m)}(x)P_k^{(m)}(x)\,dx = \frac{2}{2n+1}\frac{(n+m)!}{(n-m)!}\delta_{kn}, \qquad \delta_{kn} = \begin{cases}1, & k = n,\\ 0, & k \neq n,\end{cases}$$

$$P_n^{(m)}(x) = (1-x^2)^{\frac{m}{2}}\frac{d^m}{dx^m}P_n(x),$$

$$e^{i\rho\cos\vartheta} = \sum_{n=0}^{\infty}(2n+1)\,i^n\psi_n(\rho)P_n(\cos\theta),$$

$$\psi_n(\rho) = \sqrt{\frac{\pi}{2\rho}}J_{n+\frac{1}{2}}(\rho),$$

$$\frac{e^{ikR}}{ikR} = \begin{cases} \sum\limits_{n=0}^{\infty} (2n+1)\zeta_n^{(1)}(\rho_0)\psi_n(\rho)P_n(\cos\theta), & r < r_0, \\ \sum\limits_{n=0}^{\infty} (2n+1)\psi_n(\rho_0)\zeta_n^{(1)}(\rho)P_n(\cos\theta), & r > r_0, \end{cases}$$

$$\zeta_n^{(1)}(\rho) = \sqrt{\frac{\pi}{2\rho}}\, H^{(1)}_{n+\frac{1}{2}}(\rho),$$

$$R = \sqrt{r^2+r_0^2-2rr_0\cos\theta,}$$

$$\rho = kr, \qquad \rho_0 = kr_0.$$

$$\psi_n(\rho)\zeta_n^{(1)\prime}(\rho)-\zeta_n^{(1)}\psi_n'(\rho) = \frac{i}{\rho^2},$$

$$\psi_0(\rho) = \frac{\sin\rho}{\rho}, \qquad \psi_1(\rho) = \frac{1}{\rho}\left(\frac{\sin\rho}{\rho}-\cos\rho\right),$$

$$\zeta_0^{(1)}(\rho) = \frac{e^{i\left(\rho-\frac{\pi}{2}\right)}}{\rho}, \qquad \zeta_1^{(1)}(\rho) = \frac{e^{i\left(\rho-\frac{\pi}{2}\right)}}{\rho}\left(\frac{1}{\rho}-i\right).$$

8. The Hypergeometric Function $F(\alpha, \beta, \gamma)$

The equation

$$z(1-z)y''+[\gamma-(\alpha+\beta+1)z]y'-\alpha\beta y = 0,$$

$$y = y_1 = F(\alpha, \beta, \gamma, z) = 1+\frac{\alpha\beta}{1\cdot\gamma}z+\frac{\alpha(\alpha+1)\beta(\beta+1)}{1\cdot2\cdot\gamma(\gamma+1)}z^2+\dots,$$

$$y = y_2 = z^{1-\gamma}F(\alpha-\gamma+1, \beta-\gamma+1, 2-\gamma, z),$$

$$P_n(z) = F\left(-n, n+1, 1, \frac{1-z}{2}\right) = (-1)^n F\left(-n, n+1, 1, \frac{1+z}{2}\right),$$

$$P_n^{(m)}(z) = \frac{(n+m)!}{2^m m!(n-m)!}(1-z^2)^{\frac{m}{2}} F\left(m-n, m+n+1, m+1, \frac{1-z}{2}\right).$$

The confluent (degenerate) hypergeometric function

$$F = F(\alpha, \gamma, \rho) = 1+\frac{\alpha}{\gamma}\frac{\rho}{1}+\frac{\alpha(\alpha+1)\rho^2}{\gamma(\gamma+1)z!}+\dots,$$

$$\rho\frac{d^2F}{d\rho^2}+(\gamma-\rho)\frac{dF}{d\rho}-\alpha F = 0.$$

IV. Tables

TABLE 1

THE ERROR INTEGRAL $\Phi(z) = \frac{2}{\sqrt{\pi}} \int_0^z e^{-\alpha^2} d\alpha, \quad 0 \leqslant z \leqslant 2{\cdot}3$

z	$\Phi(z)$	z	$\Phi(z)$	z	$\Phi(z)$	z	$\Phi(z)$
0·00	0·0000	0·40	0·4284	0·80	0·7421	1·20	0·9103
0·01	0·0113	0·41	0·4380	0·81	0·7480	1·21	0·9130
0·02	0·0226	0·42	0·4475	0·82	0·7538	1·22	0·9155
0·03	0·0338	0·43	0·4569	0·83	0·7595	1·23	0·9181
0·04	0·0451	0·44	0·4662	0·84	0·7651	1·24	0·9205
0·05	0·0564	0·45	0·4755	0·85	0·7707	1·25	0·9229
0·06	0·0676	0·46	0·4847	0·86	0·7761	1·26	0·9252
0·07	0·0789	0·47	0·4937	0·87	0·7814	1·27	0·9275
0·08	0·0901	0·48	0·5027	0·88	0·7867	1·28	0·9297
0·09	0·1013	0·49	0·5117	0·89	0·7918	1·29	0·9319
0·10	0·1125	0·50	0·5205	0·90	0·7969	1·30	0·9340
0·11	0·1236	0·51	0·5292	0·91	0·8019	1·31	0·9361
0·12	0·1348	0·52	0·5379	0·92	0·8086	1·32	0·9381
0·13	0·1459	0·53	0·5465	0·93	0·8116	1·33	0·9400
0·14	0·1569	0·54	0·5549	0·94	0·8163	1·34	0·9419
0·15	0·1680	0·55	0·5633	0·95	0·8209	1·35	0·9438
0·16	0·1790	0·56	0·5716	0·96	0·8254	1·36	0·9456
0·17	0·1900	0·57	0·5798	0·97	0·8299	1·37	0·9473
0·18	0·2009	0·58	0·5879	0·98	0·8342	1·38	0·9490
0·19	0·2118	0·59	0·5959	0·99	0·8385	1·39	0·9507
0·20	0·2227	0·60	0·6039	1·00	0·8427	1·40	0·9523
0·21	0·2335	0·61	0·6117	1·01	0·8468	1·41	0·9539
0·22	0·2443	0·62	0·6194	1·02	0·8508	1·42	0·9554
0·23	0·2550	0·63	0·6270	1·03	0·8548	1·43	0·9569
0·24	0·2657	0·64	0·6346	1·04	0·8586	1·44	0·9583
0·25	0·2763	0·65	0·6420	1·05	0·8624	1·45	0·9597
0·26	0·2869	0·66	0·6494	1·06	0·8661	1·46	0·9611
0·27	0·2974	0·67	0·6566	1·07	0·8698	1·47	0·9624
0·28	0·3079	0·68	0·6633	1·08	0·8733	1·48	0·9637
0·29	0·3183	0·69	0·6708	1·09	0·8768	1·49	0·9649
0·30	0·3286	0·70	0·6778	1·10	0·8802	1·50	0·9661
0·31	0·3389	0·71	0·6847	1·11	0·8835	1·51	0·9661
0·32	0·3491	0·72	0·6914	1·12	0·8868	1·6	0·9763
0·33	0·3593	0·73	0·6981	1·13	0·8900	1·7	0·9838
0·34	0·3694	0·74	0·7047	1·14	0·8931	1·8	0·9891
0·35	0·3794	0·75	0·7112	1·15	0·8961	1·9	0·9928
0·36	0·3893	0·76	0·7175	1·16	0·8991	2·0	0·9953
0·37	0·3992	0·77	0·7238	1·17	0·9020	2·1	0·9970
0·38	0·4090	0·78	0·7300	1·18	0·9048	2·2	0·9981
0·39	0·4187	0·79	0·7361	1·19	0·9076	2·3	0·9989

TABLE 2

ROOTS OF THE CHARACTERISTIC EQUATIONS $J_0(\mu) = 0$ AND $J_1(\mu) = 0$

n	Roots μ_n of the equation $J_0(\mu) = 0$	Roots μ_n of the equation $J_1(\mu) = 0$	n	Roots μ_n of the equation $J_0(\mu) = 0$	Roots μ_n of the equation $J_1(\mu) = 0$
1	2·4048	3·8317	6	18·0711	19·6159
2	5·5201	7·0156	7	21·2116	22·7601
3	8·6537	10·1735	8	24·3525	25·9037
4	11·7915	13·3237	9	27·4935	29·0468
5	14·9309	16·4706	10	30·6346	32·1897

TABLE 3

ROOTS OF THE CHARACTERISTIC EQUATION $\frac{J_0(\mu)}{J_1(\mu)} = \frac{1}{lh}\mu$

$lh=h$	μ_1	μ_2	μ_3	μ_4	μ_5	μ_6
0·0	0·0000	3·8317	7·0156	10·1735	13·3237	16·4706
0·01	0·1412	3·8343	7·0170	10·1745	13·3244	16·4712
0·02	0·1995	3·8369	7·0184	10·1754	13·3252	16·4718
0·03	0·2814	3·8421	7·0213	10·1774	13·3267	16·4731
0·06	0·3438	3·8473	7·0241	10·1794	13·3282	16·4743
0·08	0·3960	3·8525	7·0270	10·1813	13·3297	16·4755
0·10	0·4417	3·8577	7·0298	10·1833	13·3312	16·4767
0·15	0·5376	3·8706	7·0369	10·1882	13·3349	16·4797
0·20	0·6170	3·8835	7·0440	10·1931	13·3387	16·4828
0·30	0·7465	3·9091	7·0582	10·2029	13·3462	16·4888
0·40	0·8516	3·9344	7·0723	10·2127	13·3537	16·4949
0·50	0·9408	3·9594	7·0864	10·2225	13·3611	16·5010
0·60	1·0184	3·9841	7·1004	10·2322	13·3686	16·5070
0·70	1·0873	4·0085	7·1143	10·2419	13·3761	16·5131
0·80	1·1490	4·0325	7·1282	10·2519	13·3835	16·5191
0·90	1·2048	4·0562	7·1421	10·2613	13·3910	16·5251
1·0	1·2558	4·0795	7·1558	10·2710	13·3984	16·5312
1·5	1·4569	4·1902	7·2223	10·3188	13·4353	16·5612
2·0	1·5994	4·2910	7·2884	10·3658	13·4719	16·5910
3·0	1·7887	4·4634	7·4103	10·4566	13·5434	16·6499
4·0	1·9081	4·6018	7·5201	10·5423	13·6125	16·7073
5·0	1·9898	4·7131	7·6177	10·6223	13·6786	16·7630
6·0	2·0490	4·8033	7·7039	10·6964	13·7414	16·8168
7·0	2·0937	4·8772	7·7797	10·7646	13·8008	16·8684
8·0	2·1286	4·9384	7·8464	10·8271	13·8566	16·9179
9·0	2·1566	4·9897	7·9051	10·8842	13·9090	16·9650
10·0	2·1795	5·0332	7·9569	10·9363	13·9580	17·0099

TABLE 3 (continued)

$lh=h$	μ_1	μ_2	μ_3	μ_4	μ_5	μ_6
15·0	2·2509	5·1773	8·1422	11·1367	14·1576	17·2008
20·0	2·2880	5·2568	8·2534	11·2677	14·2983	17·3442
30·0	2·3261	5·3410	8·3771	11·4221	14·4748	17·5348
40·0	2·3455	5·3846	8·4432	11·5081	14·5774	17·6508
50·0	2·3572	5·4112	8·4840	11·5621	14·6433	17·7272
60·0	2·3651	5·4291	8·5116	11·5990	14·6889	18·7807
80·0	2·3750	5·4516	8·5466	11·6461	14·7475	17·8502
100·0	2·3809	5·4652	8·5678	11·6747	14·7834	17·8931
∞	2·4048	5·5201	8·6537	11·7915	14·9309	17·0711

TABLE 4

ROOTS μ_n OF THE CHARACTERISTIC EQUATION $J_0(\mu)\,N_0(k\mu)-N_0(\mu)\,J_0(k\mu) = 0$

k	μ_1	μ_2	μ_3	μ_4	μ_5
1·2	15·7014	31·4126	47·1217	62·8304	78·5385
1·5	6·2702	12·5598	18·8451	25·1294	31·4133
2·0	3·1230	6·2734	9·4182	12·5614	15·7040
2·5	2·0732	4·1773	6·2754	8·3717	10·4672
3·0	1·5485	3·1291	4·7038	6·2767	7·8487
3·5	1·2339	2·5002	3·7608	5·0196	6·2776
4·0	1·0244	2·0809	3·1322	4·1816	5·2301

TABLE 5

ROOTS OF THE CHARACTERISTIC EQUATION $\tan\mu = -\mu/h$

h	μ_1	μ_2	μ_3	μ_4	μ_5	μ_6
0	1·5708	4·7124	7·8540	10·9956	14·1372	17·2788
0·1	1·6320	4·7335	7·8667	11·0047	14·1443	17·2845
0·2	1·6887	4·7544	7·8794	11·0137	14·1513	17·2903
0·3	1·7414	4·7751	7·8920	11·0228	14·1584	17·2961
0·4	1·7906	4·7956	7·9046	11·0318	14·1654	17·3019
0·5	1·8366	4·8158	7·9171	11·0409	14·1724	17·3076
0·6	1·8798	4·8358	7·9295	11·0498	14·1795	17·3134
0·7	1·9203	4·8556	7·9419	11·0588	14·1865	17·3192
0·8	1·9586	4·8751	7·9542	11·0677	14·1935	17·3249
0·9	1·9947	4·8943	7·9665	11·0767	14·2005	17·3306
1·0	2·0288	4·9132	7·9787	11·0856	14·2075	17·3364
1·5	2·1746	5·0037	8·0382	11·1296	14·2421	17·3649
2·0	2·2889	5·0870	8·0965	11·1727	14·2764	17·3932

TABLE 5 (continued)

h	μ_1	μ_2	μ_3	μ_4	μ_5	μ_6
3·0	2·4557	5·2329	8·2045	11·2560	14·3434	17·4490
4·0	2·5704	5·3540	8·3029	11·3349	14·4080	17·5034
5·0	2·6537	5·4544	8·3914	11·4086	14·4699	17·5562
6·0	2·7165	5·5378	8·4703	11·4773	14·5288	17·6072
7·0	2·7654	5·6078	8·5406	11·5408	14·5847	17·6562
8·0	2·8044	5·6669	8·6031	11·5994	14·6374	17·7032
9·0	2·8363	5·7172	8·6587	11·6532	14·6860	17·7481
10·0	2·8628	5·7606	8·7083	11·7027	14·7335	17·7908
15·0	2·9476	5·9080	8·8898	11·8959	14·9251	17·9742
20·0	2·9930	5·9921	9·0019	12·0250	15·0625	18·1136
30·0	3·0406	6·0831	9·1294	12·1807	15·2380	18·3018
40·0	3·0651	6·1311	9·1986	12·2688	15·3417	18·4180
50·0	3·0801	6·1606	9·2420	12·3247	15·4090	18·4953
60·0	3·0901	6·1805	9·2715	12·3632	15·4559	18·5497
80·0	3·1028	6·2058	9·3089	12·4124	15·5164	18·6209
100·0	3·1105	6·2211	9·3317	12·4426	15·5537	18·6650
∞	3·1416	6·2832	9·4248	12·5664	15·7080	18·8496

REFERENCES

I. General Courses and Collections of Problems on the Equations of Mathematical Physics

1. A. N. Krylov, *On some differential equations in mathematical physics, having an application in technical problems* (*O nekotorykh differentsial'nykh uravneniyakh matematicheskoi fiziki, imeyushchikh prilozheniye v tekhnicheskikh voprosakh*) Gostekhizdat (1950).

2. V. I. Smirnov, *A course of higher mathematics* (*Kurs vysshei matematiki*) vols. II, III, IV, Gostekhizdat (1949–1951). English translation of complete course (Vols. I to V), by Pergamon Press (1964).

3. S. L. Sobolev, *Partial differential equations of mathematical physics* (*Uravneniya matematicheskoi fiziki*) Gostekhizdat (1954). English translation, Pergamon Press (1964).

4. I. G. Petrovskii, *Lectures on partial differential equations* (*Lektsii ob uravneniakh s chastnymi proizvodnymi*) Gostekhizdat (1953).

5. V. I. Levin and Yu. N. Grosberg, *Differential equations of mathematical physics* (*Differentsial'nyye uravneniya matematicheskoi fiziki*) Gostekhizdat (1951).

6. N. S. Koshlyakov, *Fundamental differential equations of mathematical physics* (*Osnovnyye differentsial'nyye uravneniya matematicheskoi fiziki*) United Scientific and Technical Press (1936).

7. A. N. Tikhonov and A. A. Samarskii, *Equations of mathematical physics*, (*Uravneniya matematicheskoi fiziki*) Gostekhizdat (1953). Eglish translation, Pergamon Press (1963).

8. R. Courant and D. Hilbert, *Methods of mathematical physics*, vols. I and II, Gostekhizdat (1951). Original publication in German: Methoden der mathematischen Physik, Springer, Berlin (Vol. I, 1931; Vol. II, 1937). Ist English ed., Interscience Publishers, New York (1953).

9. A. Webster and G. Sege, *Partial differential equations of mathematical physics*, State Scientific and Technical Press (1934).

10. N. M. Gyunter and R. O. Kuz'min, *Collection of problems on higher mathematics*, part III, Gostekhizdat (1951).

11. M. M. Smirnov, *Problems on the equations of mathematical physics* (*Zadachi po uravneniyam matematicheskoi fiziki*) Gostekhizdat (1954).

12. B. M. Budak, *Collection of problems on mathematical physics* (*Sbornik zadach po matematicheskoi fizike*) Moscow State University—Moscow Math. Institute (1952).

II. Monographs on Special Problems

13. N. M. Gyunter, *Potential theory and its application to the fundamental problems of mathematical physics* (*Teoriya potentsiala i yeye primeneniye k osnovnym zadacham matematicheskoi fiziki*) Gostekhizdat (1953)

14. N. Ye. Kochin, *Vector analysis and the fundamentals of tensor analysis* (*Vektornoye ischisleniye i nachala tenzornogo ischisleniya*) Academy of Sciences U.S.S.R. (1951).

15. N. Ye. Kochin, I. A. Kibel' and N. V. Roze, *Theoretical hydromechanics* (*Teoreticheskaya gidromekhanika*) Gostekhizdat (1948).

16. L. D. Landau and E. M. Lifshits, *Mechanics of continuous media* (*Mekhanika sploshnykh sred*) Gostekhizdat (1953). English translation of this book (and the Complete *Course of theoretical physics*) published by Pergamon Press.

17. I. Ye. Tamm, *Fundamentals of the theory of electricity* (*Osnovy teorii elektrichestva*) Gostekhizdat (1954).

18. D. D. Ivanenko and A. A. Sokolov, *Classical field theory* (*Klassicheskaya teoriya polya*) Gostekhizdat (1944).

19. L. I. Sedov, *Methods of similarity and dimensions in mechanics* (*Metody podobiya i razmernosti v mekhanike*) Gostekhizdat (1954).

20. L. N. Sretenskii, *Theory of the Newtonian potential* (*Teoriya n'yutonovskogo potentsiala*) Gostekhizdat (1946).

21. G. A. Grinberg, *Selected problems on the mathematical theory of electric and magnetic phenomena* (*Izbrannyye voprosy matematicheskoi teorii elektricheskikh i magnitnykh yavlenii*) Academy of Sciences, U.S.S.R. (1948).

22. A. I. Lur'e, *Operational analysis* (*Operatsionnoye ischisleniye*) Gostekhizdat (1950).

23. P. Ya. Polubarinova-Kochin, *Theory of the motion of subterranean waters* (*Teoriya dvizheniya gruntovykh vod*) Gostekhizdat (1952).

24. S. P. Timoshenko, *Wave theory in engineering* (*Teoriya kolebanii v inzhenernom dele*) State Scientific and Technical Press (1932).

25. A. V. Lykov, *Theory of heat conduction* (*Teoriya teploprovodnosti*) Gostekhizdat (1952).

26. N. I. Muskhelishvili, *Some fundamental problems on the mathematical theory of elasticity* (*Nekotoryye osnovnyye zadachi matematicheskoi teorii uprugosti*) Academy of Sciences, U.S.S.R. (1949).

27. N. N. Lebedev, *Special functions and their applications* (*Spetsial'nyye funktsii i ikh prilozheniya*) Gostekhizdat (1953).

28. R. O. Kuz'min, *Bessel functions* (*Funktsii Besselya*) United Scientific and Technical Press (1935).

29. B. A. Vvedenskii and A. G. Arenberg, *Waveguides* (*Radiovolnovody*) Gostekhizdat (1946).

30. D. I. Blokhintsev, *Fundamentals of quantum mechanics* (*Osnovy kvantovoi mekhaniki*) Gostekhizdat (1949).

31. F. Frank and R. von Mises, *Differential and integral equations of mathematical physics*, United Scientific and Technical Press (1937). Original German edition: *Die Differentialgleichungen der Mechanik und Physik*, Braunschweig (1935).

32. A. Lyav, *Mathematical theory of elasticity* (*Matematicheskaya teoriya uprugosti*) United Scientific and Technical Press (1935).

33. Treffts, *Mathematical theory of elasticity* (*Matematicheskaya teoriya uprugosti*) State Scientific and Technical Press (1932).

34. H. Lamb, *Hydrodynamics*, 6th ed., Cambridge University Press (1932).

35. J. A. Stratton, *Electromagnetic theory*, McGraw-Hill (1941).

36. Lord Rayleigh, *The theory of sound*, Macmillan, London (1896).

37. W. R. Smythe, *Static and dynamic electricity*, McGraw-Hill (1939).

38. Louis de Broglie, *Electromagnetic waves in waveguides and hollow resonators*, Foreign Literature Publishing House (1948).

39. P. M. Morse, *Vibrations and sound.*

40. B. Van der Pol and H. Bremmer, *Operational calculus as the basis of the two-sided Laplace transform.*

41. H. S. Carslaw and J. C. Jaeger, *Operational methods in applied mathematics*, Oxford University Press (2nd ed., (1948).

42. H. S. Carslaw, and J. C. Jaeger, *Conduction of heat in solids*, Oxford University Press (1st ed., 1946).

43. G. N. Watson, *The theory of Bessel functions*, Cambridge University Press.

44. E. V. Hobson, *Theory of spherical and ellipsoidal harmonics*, Cambridge University Press (1931).

45. I. A. Charnyi, *Irregular motion of a real fluid in pipes* (*Neustanovivsheyesya dvizheniye real'noi zhidkosti v trubakh*) Gostekhizdat (1951).

46. N. N. Lebedev, Ya. S. Uflyand and I. P. Skal'skaya, *Collection of problems on mathematical physics* (*Sbornik zadach po matematicheskoi fizike*) Gostekhizdat (1955). English translation in preparation by Pergamon Press.

47. I. N. Sneddon, *Fourier transforms*, McGraw-Hill (1951).

INDEX

A CATALOG OF SELECTED DOVER BOOKS IN ALL FIELDS OF INTEREST

THE BOOK OF BEASTS: Being a Translation from a Latin Bestiary of the Twelfth Century, T. H. White. Wonderful catalog real and fanciful beasts: manticore, griffin, phoenix, amphivius, jaculus, many more. White's witty erudite commentary on scientific, historical aspects. Fascinating glimpse of medieval mind. Illustrated. 296pp. 5⅜ × 8¼. (Available in U.S. only) 24609-4 Pa. $5.95

FRANK LLOYD WRIGHT: ARCHITECTURE AND NATURE With 160 Illustrations, Donald Hoffmann. Profusely illustrated study of influence of nature—especially prairie—on Wright's designs for Fallingwater, Robie House, Guggenheim Museum, other masterpieces. 96pp. 9¼ × 10¾. 25098-9 Pa. $7.95

FRANK LLOYD WRIGHT'S FALLINGWATER, Donald Hoffmann. Wright's famous waterfall house: planning and construction of organic idea. History of site, owners, Wright's personal involvement. Photographs of various stages of building. Preface by Edgar Kaufmann, Jr. 100 illustrations. 112pp. 9¼ × 10. 23671-4 Pa. $7.95

YEARS WITH FRANK LLOYD WRIGHT: Apprentice to Genius, Edgar Tafel. Insightful memoir by a former apprentice presents a revealing portrait of Wright the man, the inspired teacher, the greatest American architect. 372 black-and-white illustrations. Preface. Index. vi + 228pp. 8¼ × 11. 24801-1 Pa. $9.95

THE STORY OF KING ARTHUR AND HIS KNIGHTS, Howard Pyle. Enchanting version of King Arthur fable has delighted generations with imaginative narratives of exciting adventures and unforgettable illustrations by the author. 41 illustrations. xviii + 313pp. 6⅛ × 9¼. 21445-1 Pa. $5.95

THE GODS OF THE EGYPTIANS, E. A. Wallis Budge. Thorough coverage of numerous gods of ancient Egypt by foremost Egyptologist. Information on evolution of cults, rites and gods; the cult of Osiris; the Book of the Dead and its rites; the sacred animals and birds; Heaven and Hell; and more. 956pp. 6⅛ × 9¼. 22055-9, 22056-7 Pa., Two-vol. set $21.90

A THEOLOGICO-POLITICAL TREATISE, Benedict Spinoza. Also contains unfinished *Political Treatise*. Great classic on religious liberty, theory of government on common consent. R. Elwes translation. Total of 421pp. 5⅜ × 8½. 20249-6 Pa. $6.95

INCIDENTS OF TRAVEL IN CENTRAL AMERICA, CHIAPAS, AND YUCATAN, John L. Stephens. Almost single-handed discovery of Maya culture; exploration of ruined cities, monuments, temples; customs of Indians. 115 drawings. 892pp. 5⅜ × 8½. 22404-X, 22405-8 Pa., Two-vol. set $15.90

LOS CAPRICHOS, Francisco Goya. 80 plates of wild, grotesque monsters and caricatures. Prado manuscript included. 183pp. 6⅝ × 9⅝. 22384-1 Pa. $4.95

AUTOBIOGRAPHY: The Story of My Experiments with Truth, Mohandas K. Gandhi. Not hagiography, but Gandhi in his own words. Boyhood, legal studies, purification, the growth of the Satyagraha (nonviolent protest) movement. Critical, inspiring work of the man who freed India. 480pp. 5⅜ × 8½. (Available in U.S. only) 24593-4 Pa. $6.95

ILLUSTRATED DICTIONARY OF HISTORIC ARCHITECTURE, edited by Cyril M. Harris. Extraordinary compendium of clear, concise definitions for over 5,000 important architectural terms complemented by over 2,000 line drawings. Covers full spectrum of architecture from ancient ruins to 20th-century Modernism. Preface. 592pp. 7½ × 9⅜. 24444-X Pa. $14.95

THE NIGHT BEFORE CHRISTMAS, Clement Moore. Full text, and woodcuts from original 1848 book. Also critical, historical material. 19 illustrations. 40pp. 4⅝ × 6. 22797-9 Pa. $2.50

THE LESSON OF JAPANESE ARCHITECTURE: 165 Photographs, Jiro Harada. Memorable gallery of 165 photographs taken in the 1930's of exquisite Japanese homes of the well-to-do and historic buildings. 13 line diagrams. 192pp. 8⅜ × 11¼. 24778-3 Pa. $8.95

THE AUTOBIOGRAPHY OF CHARLES DARWIN AND SELECTED LETTERS, edited by Francis Darwin. The fascinating life of eccentric genius composed of an intimate memoir by Darwin (intended for his children); commentary by his son, Francis; hundreds of fragments from notebooks, journals, papers; and letters to and from Lyell, Hooker, Huxley, Wallace and Henslow. xi + 365pp. 5⅜ × 8. 20479-0 Pa. $5.95

WONDERS OF THE SKY: Observing Rainbows, Comets, Eclipses, the Stars and Other Phenomena, Fred Schaaf. Charming, easy-to-read poetic guide to all manner of celestial events visible to the naked eye. Mock suns, glories, Belt of Venus, more. Illustrated. 299pp. 5¼ × 8¼. 24402-4 Pa. $7.95

BURNHAM'S CELESTIAL HANDBOOK, Robert Burnham, Jr. Thorough guide to the stars beyond our solar system. Exhaustive treatment. Alphabetical by constellation: Andromeda to Cetus in Vol. 1; Chamaeleon to Orion in Vol. 2; and Pavo to Vulpecula in Vol. 3. Hundreds of illustrations. Index in Vol. 3. 2,000pp. 6⅛ × 9¼. 23567-X, 23568-8, 23673-0 Pa., Three-vol. set $37.85

STAR NAMES: Their Lore and Meaning, Richard Hinckley Allen. Fascinating history of names various cultures have given to constellations and literary and folkloristic uses that have been made of stars. Indexes to subjects. Arabic and Greek names. Biblical references. Bibliography. 563pp. 5⅜ × 8½. 21079-0 Pa. $7.95

THIRTY YEARS THAT SHOOK PHYSICS: The Story of Quantum Theory, George Gamow. Lucid, accessible introduction to influential theory of energy and matter. Careful explanations of Dirac's anti-particles, Bohr's model of the atom, much more. 12 plates. Numerous drawings. 240pp. 5⅜ × 8½. 24895-X Pa. $4.95

CHINESE DOMESTIC FURNITURE IN PHOTOGRAPHS AND MEASURED DRAWINGS, Gustav Ecke. A rare volume, now affordably priced for antique collectors, furniture buffs and art historians. Detailed review of styles ranging from early Shang to late Ming. Unabridged republication. 161 black-and-white drawings, photos. Total of 224pp. 8⅜ × 11¼. (Available in U.S. only) 25171-3 Pa. $12.95

VINCENT VAN GOGH: A Biography, Julius Meier-Graefe. Dynamic, penetrating study of artist's life, relationship with brother, Theo, painting techniques, travels, more. Readable, engrossing. 160pp. 5⅜ × 8½. (Available in U.S. only) 25253-1 Pa. $3.95

HOW TO WRITE, Gertrude Stein. Gertrude Stein claimed anyone could understand her unconventional writing—here are clues to help. Fascinating improvisations, language experiments, explanations illuminate Stein's craft and the art of writing. Total of 414pp. 4⅜ × 6⅜. 23144-5 Pa. $5.95

ADVENTURES AT SEA IN THE GREAT AGE OF SAIL: Five Firsthand Narratives, edited by Elliot Snow. Rare true accounts of exploration, whaling, shipwreck, fierce natives, trade, shipboard life, more. 33 illustrations. Introduction. 353pp. 5⅜ × 8½. 25177-2 Pa. $7.95

THE HERBAL OR GENERAL HISTORY OF PLANTS, John Gerard. Classic descriptions of about 2,850 plants—with over 2,700 illustrations—includes Latin and English names, physical descriptions, varieties, time and place of growth, more. 2,706 illustrations. xlv + 1,678pp. 8½ × 12¼. 23147-X Cloth. $75.00

DOROTHY AND THE WIZARD IN OZ, L. Frank Baum. Dorothy and the Wizard visit the center of the Earth, where people are vegetables, glass houses grow and Oz characters reappear. Classic sequel to *Wizard of Oz*. 256pp. 5⅜ × 8. 24714-7 Pa. $4.95

SONGS OF EXPERIENCE: Facsimile Reproduction with 26 Plates in Full Color, William Blake. This facsimile of Blake's original "Illuminated Book" reproduces 26 full-color plates from a rare 1826 edition. Includes "The Tyger," "London," "Holy Thursday," and other immortal poems. 26 color plates. Printed text of poems. 48pp. 5¼ × 7. 24636-1 Pa. $3.50

SONGS OF INNOCENCE, William Blake. The first and most popular of Blake's famous "Illuminated Books," in a facsimile edition reproducing all 31 brightly colored plates. Additional printed text of each poem. 64pp. 5¼ × 7. 22764-2 Pa. $3.50

PRECIOUS STONES, Max Bauer. Classic, thorough study of diamonds, rubies, emeralds, garnets, etc.: physical character, occurrence, properties, use, similar topics. 20 plates, 8 in color. 94 figures. 659pp. 6⅛ × 9¼. 21910-0, 21911-9 Pa., Two-vol. set $15.90

ENCYCLOPEDIA OF VICTORIAN NEEDLEWORK, S. F. A. Caulfeild and Blanche Saward. Full, precise descriptions of stitches, techniques for dozens of needlecrafts—most exhaustive reference of its kind. Over 800 figures. Total of 679pp. 8⅛ × 11. Two volumes. Vol. 1 22800-2 Pa. $11.95
Vol. 2 22801-0 Pa. $11.95

THE MARVELOUS LAND OF OZ, L. Frank Baum. Second Oz book, the Scarecrow and Tin Woodman are back with hero named Tip, Oz magic. 136 illustrations. 287pp. 5⅜ × 8½. 20692-0 Pa. $5.95

WILD FOWL DECOYS, Joel Barber. Basic book on the subject, by foremost authority and collector. Reveals history of decoy making and rigging, place in American culture, different kinds of decoys, how to make them, and how to use them. 140 plates. 156pp. 7⅞ × 10¾. 20011-6 Pa. $8.95

HISTORY OF LACE, Mrs. Bury Palliser. Definitive, profusely illustrated chronicle of lace from earliest times to late 19th century. Laces of Italy, Greece, England, France, Belgium, etc. Landmark of needlework scholarship. 266 illustrations. 672pp. 6⅛ × 9¼. 24742-2 Pa. $14.95

ILLUSTRATED GUIDE TO SHAKER FURNITURE, Robert Meader. All furniture and appurtenances, with much on unknown local styles. 235 photos. 146pp. 9 × 12. 22819-3 Pa. $7.95

WHALE SHIPS AND WHALING: A Pictorial Survey, George Francis Dow. Over 200 vintage engravings, drawings, photographs of barks, brigs, cutters, other vessels. Also harpoons, lances, whaling guns, many other artifacts. Comprehensive text by foremost authority. 207 black-and-white illustrations. 288pp. 6 × 9. 24808-9 Pa. $8.95

THE BERTRAMS, Anthony Trollope. Powerful portrayal of blind self-will and thwarted ambition includes one of Trollope's most heartrending love stories. 497pp. 5⅜ × 8½. 25119-5 Pa. $8.95

ADVENTURES WITH A HAND LENS, Richard Headstrom. Clearly written guide to observing and studying flowers and grasses, fish scales, moth and insect wings, egg cases, buds, feathers, seeds, leaf scars, moss, molds, ferns, common crystals, etc.—all with an ordinary, inexpensive magnifying glass. 209 exact line drawings aid in your discoveries. 220pp. 5⅜ × 8½. 23330-8 Pa. $4.50

RODIN ON ART AND ARTISTS, Auguste Rodin. Great sculptor's candid, wide-ranging comments on meaning of art; great artists; relation of sculpture to poetry, painting, music; philosophy of life, more. 76 superb black-and-white illustrations of Rodin's sculpture, drawings and prints. 119pp. 8⅜ × 11¼. 24487-3 Pa. $6.95

FIFTY CLASSIC FRENCH FILMS, 1912–1982: A Pictorial Record, Anthony Slide. Memorable stills from Grand Illusion, Beauty and the Beast, Hiroshima, Mon Amour, many more. Credits, plot synopses, reviews, etc. 160pp. 8¼ × 11. 25256-6 Pa. $11.95

THE PRINCIPLES OF PSYCHOLOGY, William James. Famous long course complete, unabridged. Stream of thought, time perception, memory, experimental methods; great work decades ahead of its time. 94 figures. 1,391pp. 5⅜ × 8½. 20381-6, 20382-4 Pa., Two-vol. set $19.90

BODIES IN A BOOKSHOP, R. T. Campbell. Challenging mystery of blackmail and murder with ingenious plot and superbly drawn characters. In the best tradition of British suspense fiction. 192pp. 5⅜ × 8½. 24720-1 Pa. $3.95

CALLAS: PORTRAIT OF A PRIMA DONNA, George Jellinek. Renowned commentator on the musical scene chronicles incredible career and life of the most controversial, fascinating, influential operatic personality of our time. 64 black-and-white photographs. 416pp. 5⅜ × 8¼. 25047-4 Pa. $7.95

GEOMETRY, RELATIVITY AND THE FOURTH DIMENSION, Rudolph Rucker. Exposition of fourth dimension, concepts of relativity as Flatland characters continue adventures. Popular, easily followed yet accurate, profound. 141 illustrations. 133pp. 5⅜ × 8½. 23400-2 Pa. $3.50

HOUSEHOLD STORIES BY THE BROTHERS GRIMM, with pictures by Walter Crane. 53 classic stories—Rumpelstiltskin, Rapunzel, Hansel and Gretel, the Fisherman and his Wife, Snow White, Tom Thumb, Sleeping Beauty, Cinderella, and so much more—lavishly illustrated with original 19th century drawings. 114 illustrations. x + 269pp. 5⅜ × 8½. 21080-4 Pa. $4.50

SUNDIALS, Albert Waugh. Far and away the best, most thorough coverage of ideas, mathematics concerned, types, construction, adjusting anywhere. Over 100 illustrations. 230pp. 5⅜ × 8½. 22947-5 Pa. $4.50

PICTURE HISTORY OF THE NORMANDIE: With 190 Illustrations, Frank O. Braynard. Full story of legendary French ocean liner: Art Deco interiors, design innovations, furnishings, celebrities, maiden voyage, tragic fire, much more. Extensive text. 144pp. 8⅜ × 11¾. 25257-4 Pa. $9.95

THE FIRST AMERICAN COOKBOOK: A Facsimile of "American Cookery," 1796, Amelia Simmons. Facsimile of the first American-written cookbook published in the United States contains authentic recipes for colonial favorites—pumpkin pudding, winter squash pudding, spruce beer, Indian slapjacks, and more. Introductory Essay and Glossary of colonial cooking terms. 80pp. 5⅜ × 8½. 24710-4 Pa. $3.50

101 PUZZLES IN THOUGHT AND LOGIC, C. R. Wylie, Jr. Solve murders and robberies, find out which fishermen are liars, how a blind man could possibly identify a color—purely by your own reasoning! 107pp. 5⅜ × 8½. 20367-0 Pa. $2.50

THE BOOK OF WORLD-FAMOUS MUSIC—CLASSICAL, POPULAR AND FOLK, James J. Fuld. Revised and enlarged republication of landmark work in musico-bibliography. Full information about nearly 1,000 songs and compositions including first lines of music and lyrics. New supplement. Index. 800pp. 5⅜ × 8¼. 24857-7 Pa. $14.95

ANTHROPOLOGY AND MODERN LIFE, Franz Boas. Great anthropologist's classic treatise on race and culture. Introduction by Ruth Bunzel. Only inexpensive paperback edition. 255pp. 5⅜ × 8½. 25245-0 Pa. $5.95

THE TALE OF PETER RABBIT, Beatrix Potter. The inimitable Peter's terrifying adventure in Mr. McGregor's garden, with all 27 wonderful, full-color Potter illustrations. 55pp. 4¼ × 5½. (Available in U.S. only) 22827-4 Pa. $1.75

THREE PROPHETIC SCIENCE FICTION NOVELS, H. G. Wells. *When the Sleeper Wakes, A Story of the Days to Come* and *The Time Machine* (full version). 335pp. 5⅜ × 8½. (Available in U.S. only) 20605-X Pa. $5.95

APICIUS COOKERY AND DINING IN IMPERIAL ROME, edited and translated by Joseph Dommers Vehling. Oldest known cookbook in existence offers readers a clear picture of what foods Romans ate, how they prepared them, etc. 49 illustrations. 301pp. 6⅛ × 9¼. 23563-7 Pa. $6.50

SHAKESPEARE LEXICON AND QUOTATION DICTIONARY, Alexander Schmidt. Full definitions, locations, shades of meaning of every word in plays and poems. More than 50,000 exact quotations. 1,485pp. 6½ × 9¼. 22726-X, 22727-8 Pa., Two-vol. set $27.90

THE WORLD'S GREAT SPEECHES, edited by Lewis Copeland and Lawrence W. Lamm. Vast collection of 278 speeches from Greeks to 1970. Powerful and effective models; unique look at history. 842pp. 5⅜ × 8½. 20468-5 Pa. $11.95

THE BLUE FAIRY BOOK, Andrew Lang. The first, most famous collection, with many familiar tales: Little Red Riding Hood, Aladdin and the Wonderful Lamp, Puss in Boots, Sleeping Beauty, Hansel and Gretel, Rumpelstiltskin; 37 in all. 138 illustrations. 390pp. 5⅜ × 8½. 21437-0 Pa. $5.95

THE STORY OF THE CHAMPIONS OF THE ROUND TABLE, Howard Pyle. Sir Launcelot, Sir Tristram and Sir Percival in spirited adventures of love and triumph retold in Pyle's inimitable style. 50 drawings, 31 full-page. xviii + 329pp. 6½ × 9¼. 21883-X Pa. $6.95

AUDUBON AND HIS JOURNALS, Maria Audubon. Unmatched two-volume portrait of the great artist, naturalist and author contains his journals, an excellent biography by his granddaughter, expert annotations by the noted ornithologist, Dr. Elliott Coues, and 37 superb illustrations. Total of 1,200pp. 5⅜ × 8.
Vol. I 25143-8 Pa. $8.95
Vol. II 25144-6 Pa. $8.95

GREAT DINOSAUR HUNTERS AND THEIR DISCOVERIES, Edwin H. Colbert. Fascinating, lavishly illustrated chronicle of dinosaur research, 1820's to 1960. Achievements of Cope, Marsh, Brown, Buckland, Mantell, Huxley, many others. 384pp. 5¼ × 8¼. 24701-5 Pa. $6.95

THE TASTEMAKERS, Russell Lynes. Informal, illustrated social history of American taste 1850's–1950's. First popularized categories Highbrow, Lowbrow, Middlebrow. 129 illustrations. New (1979) afterword. 384pp. 6 × 9.
23993-4 Pa. $6.95

DOUBLE CROSS PURPOSES, Ronald A. Knox. A treasure hunt in the Scottish Highlands, an old map, unidentified corpse, surprise discoveries keep reader guessing in this cleverly intricate tale of financial skullduggery. 2 black-and-white maps. 320pp. 5⅜ × 8½. (Available in U.S. only) 25032-6 Pa. $5.95

AUTHENTIC VICTORIAN DECORATION AND ORNAMENTATION IN FULL COLOR: 46 Plates from "Studies in Design," Christopher Dresser. Superb full-color lithographs reproduced from rare original portfolio of a major Victorian designer. 48pp. 9¼ × 12¼. 25083-0 Pa. $7.95

PRIMITIVE ART, Franz Boas. Remains the best text ever prepared on subject, thoroughly discussing Indian, African, Asian, Australian, and, especially, Northern American primitive art. Over 950 illustrations show ceramics, masks, totem poles, weapons, textiles, paintings, much more. 376pp. 5⅜ × 8. 20025-6 Pa. $6.95

SIDELIGHTS ON RELATIVITY, Albert Einstein. Unabridged republication of two lectures delivered by the great physicist in 1920–21. *Ether and Relativity* and *Geometry and Experience*. Elegant ideas in non-mathematical form, accessible to intelligent layman. vi + 56pp. 5⅜ × 8½. 24511-X Pa. $2.95

THE WIT AND HUMOR OF OSCAR WILDE, edited by Alvin Redman. More than 1,000 ripostes, paradoxes, wisecracks: Work is the curse of the drinking classes, I can resist everything except temptation, etc. 258pp. 5⅜ × 8½. 20602-5 Pa. $4.50

ADVENTURES WITH A MICROSCOPE, Richard Headstrom. 59 adventures with clothing fibers, protozoa, ferns and lichens, roots and leaves, much more. 142 illustrations. 232pp. 5⅜ × 8½. 23471-1 Pa. $3.95

PLANTS OF THE BIBLE, Harold N. Moldenke and Alma L. Moldenke. Standard reference to all 230 plants mentioned in Scriptures. Latin name, biblical reference, uses, modern identity, much more. Unsurpassed encyclopedic resource for scholars, botanists, nature lovers, students of Bible. Bibliography. Indexes. 123 black-and-white illustrations. 384pp. 6 × 9. 25069-5 Pa. $8.95

FAMOUS AMERICAN WOMEN: A Biographical Dictionary from Colonial Times to the Present, Robert McHenry, ed. From Pocahontas to Rosa Parks, 1,035 distinguished American women documented in separate biographical entries. Accurate, up-to-date data, numerous categories, spans 400 years. Indices. 493pp. 6½ × 9¼. 24523-3 Pa. $9.95

THE FABULOUS INTERIORS OF THE GREAT OCEAN LINERS IN HISTORIC PHOTOGRAPHS, William H. Miller, Jr. Some 200 superb photographs capture exquisite interiors of world's great "floating palaces"—1890's to 1980's: *Titanic, Ile de France, Queen Elizabeth, United States, Europa,* more. Approx. 200 black-and-white photographs. Captions. Text. Introduction. 160pp. 8⅞ × 11¾. 24756-2 Pa. $9.95

THE GREAT LUXURY LINERS, 1927–1954: A Photographic Record, William H. Miller, Jr. Nostalgic tribute to heyday of ocean liners. 186 photos of Ile de France, Normandie, Leviathan, Queen Elizabeth, United States, many others. Interior and exterior views. Introduction. Captions. 160pp. 9 × 12. 24056-8 Pa. $9.95

A NATURAL HISTORY OF THE DUCKS, John Charles Phillips. Great landmark of ornithology offers complete detailed coverage of nearly 200 species and subspecies of ducks: gadwall, sheldrake, merganser, pintail, many more. 74 full-color plates, 102 black-and-white. Bibliography. Total of 1,920pp. 8⅜ × 11¼. 25141-1, 25142-X Cloth. Two-vol. set $100.00

THE SEAWEED HANDBOOK: An Illustrated Guide to Seaweeds from North Carolina to Canada, Thomas F. Lee. Concise reference covers 78 species. Scientific and common names, habitat, distribution, more. Finding keys for easy identification. 224pp. 5⅜ × 8½. 25215-9 Pa. $5.95

THE TEN BOOKS OF ARCHITECTURE: The 1755 Leoni Edition, Leon Battista Alberti. Rare classic helped introduce the glories of ancient architecture to the Renaissance. 68 black-and-white plates. 336pp. 8⅜ × 11¼. 25239-6 Pa. $14.95

MISS MACKENZIE, Anthony Trollope. Minor masterpieces by Victorian master unmasks many truths about life in 19th-century England. First inexpensive edition in years. 392pp. 5⅜ × 8½. 25201-9 Pa. $7.95

THE RIME OF THE ANCIENT MARINER, Gustave Doré, Samuel Taylor Coleridge. Dramatic engravings considered by many to be his greatest work. The terrifying space of the open sea, the storms and whirlpools of an unknown ocean, the ice of Antarctica, more—all rendered in a powerful, chilling manner. Full text. 38 plates. 77pp. 9¼ × 12. 22305-1 Pa. $4.95

THE EXPEDITIONS OF ZEBULON MONTGOMERY PIKE, Zebulon Montgomery Pike. Fascinating first-hand accounts (1805–6) of exploration of Mississippi River, Indian wars, capture by Spanish dragoons, much more. 1,088pp. 5⅜ × 8½. 25254-X, 25255-8 Pa. Two-vol. set $23.90

A CONCISE HISTORY OF PHOTOGRAPHY: Third Revised Edition, Helmut Gernsheim. Best one-volume history—camera obscura, photochemistry, daguerreotypes, evolution of cameras, film, more. Also artistic aspects—landscape, portraits, fine art, etc. 281 black-and-white photographs. 26 in color. 176pp. 8⅜ × 11¼. 25128-4 Pa. $12.95

THE DORÉ BIBLE ILLUSTRATIONS, Gustave Doré. 241 detailed plates from the Bible: the Creation scenes, Adam and Eve, Flood, Babylon, battle sequences, life of Jesus, etc. Each plate is accompanied by the verses from the King James version of the Bible. 241pp. 9 × 12. 23004-X Pa. $8.95

HUGGER-MUGGER IN THE LOUVRE, Elliot Paul. Second Homer Evans mystery-comedy. Theft at the Louvre involves sleuth in hilarious, madcap caper. "A knockout."—Books. 336pp. 5⅜ × 8½. 25185-3 Pa. $5.95

FLATLAND, E. A. Abbott. Intriguing and enormously popular science-fiction classic explores the complexities of trying to survive as a two-dimensional being in a three-dimensional world. Amusingly illustrated by the author. 16 illustrations. 103pp. 5⅜ × 8½. 20001-9 Pa. $2.25

THE HISTORY OF THE LEWIS AND CLARK EXPEDITION, Meriwether Lewis and William Clark, edited by Elliott Coues. Classic edition of Lewis and Clark's day-by-day journals that later became the basis for U.S. claims to Oregon and the West. Accurate and invaluable geographical, botanical, biological, meteorological and anthropological material. Total of 1,508pp. 5⅜ × 8½. 21268-8, 21269-6, 21270-X Pa. Three-vol. set $25.50

LANGUAGE, TRUTH AND LOGIC, Alfred J. Ayer. Famous, clear introduction to Vienna, Cambridge schools of Logical Positivism. Role of philosophy, elimination of metaphysics, nature of analysis, etc. 160pp. 5⅜ × 8½. (Available in U.S. and Canada only) 20010-8 Pa. $2.95

MATHEMATICS FOR THE NONMATHEMATICIAN, Morris Kline. Detailed, college-level treatment of mathematics in cultural and historical context, with numerous exercises. For liberal arts students. Preface. Recommended Reading Lists. Tables. Index. Numerous black-and-white figures. xvi + 641pp. 5⅜ × 8½. 24823-2 Pa. $11.95

28 SCIENCE FICTION STORIES, H. G. Wells. Novels, *Star Begotten* and *Men Like Gods,* plus 26 short stories: "Empire of the Ants," "A Story of the Stone Age," "The Stolen Bacillus," "In the Abyss," etc. 915pp. 5⅜ × 8½. (Available in U.S. only) 20265-8 Cloth. $10.95

HANDBOOK OF PICTORIAL SYMBOLS, Rudolph Modley. 3,250 signs and symbols, many systems in full; official or heavy commercial use. Arranged by subject. Most in Pictorial Archive series. 143pp. 8⅛ × 11. 23357-X Pa. $5.95

INCIDENTS OF TRAVEL IN YUCATAN, John L. Stephens. Classic (1843) exploration of jungles of Yucatan, looking for evidences of Maya civilization. Travel adventures, Mexican and Indian culture, etc. Total of 669pp. 5⅜ × 8½. 20926-1, 20927-X Pa., Two-vol. set $9.90

DEGAS: An Intimate Portrait, Ambroise Vollard. Charming, anecdotal memoir by famous art dealer of one of the greatest 19th-century French painters. 14 black-and-white illustrations. Introduction by Harold L. Van Doren. 96pp. 5⅜ × 8½.
25131-4 Pa. $3.95

PERSONAL NARRATIVE OF A PILGRIMAGE TO ALMANDINAH AND MECCAH, Richard Burton. Great travel classic by remarkably colorful personality. Burton, disguised as a Moroccan, visited sacred shrines of Islam, narrowly escaping death. 47 illustrations. 959pp. 5⅜ × 8½. 21217-3, 21218-1 Pa., Two-vol. set $17.90

PHRASE AND WORD ORIGINS, A. H. Holt. Entertaining, reliable, modern study of more than 1,200 colorful words, phrases, origins and histories. Much unexpected information. 254pp. 5⅜ × 8½. 20758-7 Pa. $5.95

THE RED THUMB MARK, R. Austin Freeman. In this first Dr. Thorndyke case, the great scientific detective draws fascinating conclusions from the nature of a single fingerprint. Exciting story, authentic science. 320pp. 5⅜ × 8½. (Available in U.S. only) 25210-8 Pa. $5.95

AN EGYPTIAN HIEROGLYPHIC DICTIONARY, E. A. Wallis Budge. Monumental work containing about 25,000 words or terms that occur in texts ranging from 3000 B.C. to 600 A.D. Each entry consists of a transliteration of the word, the word in hieroglyphs, and the meaning in English. 1,314pp. 6⅛ × 10.
23615-3, 23616-1 Pa., Two-vol. set $27.90

THE COMPLEAT STRATEGYST: Being a Primer on the Theory of Games of Strategy, J. D. Williams. Highly entertaining classic describes, with many illustrated examples, how to select best strategies in conflict situations. Prefaces. Appendices. xvi + 268pp. 5⅜ × 8½. 25101-2 Pa. $5.95

THE ROAD TO OZ, L. Frank Baum. Dorothy meets the Shaggy Man, little Button-Bright and the Rainbow's beautiful daughter in this delightful trip to the magical Land of Oz. 272pp. 5⅜ × 8. 25208-6 Pa. $4.95

POINT AND LINE TO PLANE, Wassily Kandinsky. Seminal exposition of role of point, line, other elements in non-objective painting. Essential to understanding 20th-century art. 127 illustrations. 192pp. 6½ × 9¼. 23808-3 Pa. $4.50

LADY ANNA, Anthony Trollope. Moving chronicle of Countess Lovel's bitter struggle to win for herself and daughter Anna their rightful rank and fortune—perhaps at cost of sanity itself. 384pp. 5⅜ × 8½. 24669-8 Pa. $6.95

EGYPTIAN MAGIC, E. A. Wallis Budge. Sums up all that is known about magic in Ancient Egypt: the role of magic in controlling the gods, powerful amulets that warded off evil spirits, scarabs of immortality, use of wax images, formulas and spells, the secret name, much more. 253pp. 5⅜ × 8½. 22681-6 Pa. $4.50

THE DANCE OF SIVA, Ananda Coomaraswamy. Preeminent authority unfolds the vast metaphysic of India: the revelation of her art, conception of the universe, social organization, etc. 27 reproductions of art masterpieces. 192pp. 5⅜ × 8½.
24817-8 Pa. $5.95

CHRISTMAS CUSTOMS AND TRADITIONS, Clement A. Miles. Origin, evolution, significance of religious, secular practices. Caroling, gifts, yule logs, much more. Full, scholarly yet fascinating; non-sectarian. 400pp. 5⅜ × 8½.
23354-5 Pa. $6.50

THE HUMAN FIGURE IN MOTION, Eadweard Muybridge. More than 4,500 stopped-action photos, in action series, showing undraped men, women, children jumping, lying down, throwing, sitting, wrestling, carrying, etc. 390pp. 7⅞ × 10⅝.
20204-6 Cloth. $19.95

THE MAN WHO WAS THURSDAY, Gilbert Keith Chesterton. Witty, fast-paced novel about a club of anarchists in turn-of-the-century London. Brilliant social, religious, philosophical speculations. 128pp. 5⅜ × 8½. 25121-7 Pa. $3.95

A CEZANNE SKETCHBOOK: Figures, Portraits, Landscapes and Still Lifes, Paul Cezanne. Great artist experiments with tonal effects, light, mass, other qualities in over 100 drawings. A revealing view of developing master painter, precursor of Cubism. 102 black-and-white illustrations. 144pp. 8¾ × 6⅜. 24790-2 Pa. $5.95

AN ENCYCLOPEDIA OF BATTLES: Accounts of Over 1,560 Battles from 1479 B.C. to the Present, David Eggenberger. Presents essential details of every major battle in recorded history, from the first battle of Megiddo in 1479 B.C. to Grenada in 1984. List of Battle Maps. New Appendix covering the years 1967–1984. Index. 99 illustrations. 544pp. 6½ × 9¼. 24913-1 Pa. $14.95

AN ETYMOLOGICAL DICTIONARY OF MODERN ENGLISH, Ernest Weekley. Richest, fullest work, by foremost British lexicographer. Detailed word histories. Inexhaustible. Total of 856pp. 6½ × 9¼.
21873-2, 21874-0 Pa., Two-vol. set $17.00

WEBSTER'S AMERICAN MILITARY BIOGRAPHIES, edited by Robert McHenry. Over 1,000 figures who shaped 3 centuries of American military history. Detailed biographies of Nathan Hale, Douglas MacArthur, Mary Hallaren, others. Chronologies of engagements, more. Introduction. Addenda. 1,033 entries in alphabetical order. xi + 548pp. 6½ × 9¼. (Available in U.S. only)
24758-9 Pa. $11.95

LIFE IN ANCIENT EGYPT, Adolf Erman. Detailed older account, with much not in more recent books: domestic life, religion, magic, medicine, commerce, and whatever else needed for complete picture. Many illustrations. 597pp. 5⅜ × 8½.
22632-8 Pa. $8.95

HISTORIC COSTUME IN PICTURES, Braun & Schneider. Over 1,450 costumed figures shown, covering a wide variety of peoples: kings, emperors, nobles, priests, servants, soldiers, scholars, townsfolk, peasants, merchants, courtiers, cavaliers, and more. 256pp. 8⅜ × 11¼. 23150-X Pa. $7.95

THE NOTEBOOKS OF LEONARDO DA VINCI, edited by J. P. Richter. Extracts from manuscripts reveal great genius; on painting, sculpture, anatomy, sciences, geography, etc. Both Italian and English. 186 ms. pages reproduced, plus 500 additional drawings, including studies for *Last Supper, Sforza* monument, etc. 860pp. 7⅞ × 10¾. (Available in U.S. only) 22572-0, 22573-9 Pa., Two-vol. set $25.90